Quantum Mechanics
for Engineers and
Material Scientists

An Introduction

Quantum Mechanics for Engineers and Material Scientists

An Introduction

M. P. Anantram
University of Washington, Seattle, USA

Daryoush Shiri
Chalmers University of Technology, Sweden

World Scientific

NEW JERSEY · LONDON · SINGAPORE · BEIJING · SHANGHAI · HONG KONG · TAIPEI · CHENNAI · TOKYO

Published by

World Scientific Publishing Co. Pte. Ltd.

5 Toh Tuck Link, Singapore 596224

USA office: 27 Warren Street, Suite 401-402, Hackensack, NJ 07601

UK office: 57 Shelton Street, Covent Garden, London WC2H 9HE

British Library Cataloguing-in-Publication Data
A catalogue record for this book is available from the British Library.

QUANTUM MECHANICS FOR ENGINEERS AND MATERIAL SCIENTISTS
An Introduction

Copyright © 2024 by World Scientific Publishing Co. Pte. Ltd.

For photocopying of material in this volume, please pay a copying fee through the Copyright Clearance Center, Inc., 222 Rosewood Drive, Danvers, MA 01923, USA. In this case permission to photocopy is not required from the publisher.

ISBN 978-981-12-7438-1 (hardcover)
ISBN 978-981-12-7532-6 (paperback)
ISBN 978-981-12-7439-8 (ebook for institutions)
ISBN 978-981-12-7440-4 (ebook for individuals)

For any available supplementary material, please visit
https://www.worldscientific.com/worldscibooks/10.1142/13356#t=suppl

Desk Editor: Rhaimie Wahap

Typeset by Stallion Press
Email: enquiries@stallionpress.com

To Vk.

To Anita and my great teachers,
Seyyed Mohammad Ramin Azad, Mansour Ajudani, and
Mohammad Jafar Tahvildari.

PREFACE

Our introductory textbook is tailored for students as well as practicing engineers providing them with the necessary foundation in practical quantum mechanics. We have tried to provide a comprehensive resource that aims to bridge the gap between the conceptual challenges of quantum mechanics and the required mathematical fundamentals. It combines mathematical rigor with real-world examples and analogies to facilitate learning and comprehension.

Here are some key features and highlights of the book:

1. **Target Audience:** The book is intended for students in engineering, applied physics and material science, specifically those in Bachelor's, Master's, and PhD programs. It caters to students with varying levels of prior experience in quantum mechanics. We have taught almost all chapters from the book in various courses at the University of Washington for over a decade. The practicing engineers who want to get quickly to the crux of the matter, can use this book. If you wonder what quantum computer is, after a fast-paced review of chapters 1-6, you can read chapters 7 and 8. Calculation of band structure of materials are explained in lucid manner. Learning this helps you to understand the concept of bandgap and be ready for advanced solid-state books, semiconductor device physics, and VLSI circuits. The Stern-Gerlach experiment makes the concept of spin as well as qubit clear and lays the foundation for advanced topics like spin transistors and quantum gates.

2. **Conceptual and Mathematical Challenges:** Quantum mechanics is conceptually challenging, and so examples are provided to help reinforce the concepts. The book acknowledges that readers need reminders about the required mathematical fundamentals. We only assume that readers have introductory knowledge of differential equations and linear algebra, at the sophomore level.

3. **Detailed Derivations:** The book provides step-by-step mathematical derivations leading to important results. We find that this can be helpful for readers who may need a more thorough understanding of the underlying mathematics.

4. **Concept Boxes:** Concept boxes are used to help readers reflect on key concepts and results, making the material more accessible and aiding in comprehension.

5. **Modern Examples:** The book incorporates well-designed, contemporary examples to help readers grasp and internalize the concepts discussed. This is particularly valuable for students in engineering and material science who may be looking to apply quantum mechanics to real-world problems.

   The book covers the basics of quantum computing, quantum gates, two-level systems, orbitals, spin, periodic solids, tunneling, and the Fermi golden rule.

6. **Applications to Nanomaterials:** The book explains how quantum mechanics is applied to understand electronic and optical properties of nanomaterials. It avoids referring readers to research articles with different notations and units, making the content more accessible. The book covers a wide range of nanomaterials and their properties e.g., band structures and density of electronic states. It covers topics like carbon nanotubes, graphene, superconducting qubits, scanning tunneling microscopy, and more.

7. **Analogies and Engineering Concepts:** We use analogies based on familiar engineering concepts e.g., electric circuits, to make the material more relatable and understandable.

8. **End-of-Chapter Problems:** Each chapter includes carefully designed problems to test and reinforce the understanding of

the material. These problems are valuable for self-assessment and classroom instruction.

9. **Instructor Resources:** We provide solutions to the end-of-chapter problems and slides for instructors, which can be useful for teaching and course development.

In summary, we hope that our textbook serves as a well-structured resource for students and curious readers looking to build a foundation in quantum mechanics with a focus on applications.

M. P. Anantram
University of Washington

Daryoush Shiri
Chalmers University of Technology

October 2023

AUTHOR BIOGRAPHY

M. P. Anantram is a Professor of Electrical and Computer Engineering at the University of Washington, Seattle. Prior to this, he was a Professor at the University of Waterloo, Canada, and worked at the Center for Nanotechnology at the NASA Ames Research Center. His group works on quantum mechanical methods to model nanoscale materials and devices. He has been teaching quantum mechanics for over fifteen years to seniors and first-year graduate students.

Daryoush Shiri is a scientist at Chalmers University of Technology and works on both quantum computing and superconducting microwave circuits. Before this he was a postdoctoral fellow at the Institute for Quantum Computing, Waterloo and at Chalmers University. He received his Ph.D. degree in electrical and computer engineering in 2013 from the University of Waterloo, working on nano-device engineering. Daryoush worked at start-up companies as a radio-frequency/analog CMOS design engineer and team leader. He has taught various courses in nano-engineering, circuits, and VLSI. He has co-supervised several Ph.D. and master's degree students. Daryoush is unique with his deep knowledge of VLSI, Analog CMOS, circuit theory and quantum computing.

ACKNOWLEDGEMENTS

We thank Aseem Gauri, Ashwin Kaliyaperumal, Austin Hong, Diya Kumar, Jessica Kim, Manjari Anant, Matthew Chen, Mrunal Shenoy, Oguz Tanatar, Rose Liu, and Sophia Gershaft for proof reading, Hrishikesh Seshadri for creating the glossary, Mrunal Shenoy and Sophia Gershaft for drawing diagrams, and Manjari Anant for suggesting concept boxes and providing a few examples. We thank Austin Hong for also checking several equations in the book.

We thank Anita Fadavi Roudsari (Chalmers University of Technology) for her work on figures and feedback, Jean-Francois Van Huele (Brigham Young University) for discussion of uncertainty principle, and Xiaodong Xu (University of Washington) and Debashish Ghoshal (Delhi University) for discussions on the derivation of spin operators.

We would like to thank Rhaimie Wahap of World Scientific Publishing Company for help with editing and Zvi Ruder for help with the planning of the book.

CONSTANTS TABLE

Symbol	Name	Value
ϵ_0	Permittivity of free space	$8.854 \times 10^{-12}\,(F/m)$
μ_0	Permeability of free space	$1.257 \times 10^{-6}\,(N/A^2)$
		$4\pi \times 10^{-7} \mathrm{H/m}$
π	Pi	$3.1415\ldots$
a_0	Bohr radius	$5.292 \times 10^{-11}\,m$
c	Speed of light	$2.99792458 \times 10^8\,ms^{-1}$
e	Euler's constant	$2.718\ldots$
h	Planck's constant	$6.626 \times 10^{-34}\,Js$
		$4.1357 \times 10^{-15} eVs$
$\hbar$	Reduced Planck's constant	$1.0546 \times 10^{-34} Js$
		$6.5821 \times 10^{-16} eVs$
i	Imaginary unit	$\sqrt{-1}$
k_B	Boltzmann constant	$1.38 \times 10^{-23}\,(J/K)$
		$8.617 \times 10^{-5}\,\frac{eV}{K}$
m_o	Rest mass of an electron	$9.109 \times 10^{-31}\,kg$
q	Magnitude of charge of an electron	$1.602 \times 10^{-19}\,C$

MATHEMATICAL GLOSSARY

- $\otimes$ — Tensor or Kronecker product
- $\dagger$ — Hermitian conjugate
- $|-\rangle$ — Qubit state $\frac{|0\rangle - |1\rangle}{\sqrt{\{2\}}}$ (corresponds to $-45°$ polarized photons)
- $|+\rangle$ — Qubit state $\frac{|0\rangle + |1\rangle}{\sqrt{\{2\}}}$ (corresponds to $+45°$ polarized photons)
- $|\uparrow\rangle$ — Up component of spin
- $|\downarrow\rangle$ — Down component of spin
- α_{so} — spin-orbit coupling constant
- $\alpha(\omega)$ — Absorption coefficient
- Δ — Detuning frequency
- $\Delta\widehat{A}$ — Uncertainty or standard deviation in measurement of $\widehat{A}$
- ΔE — Standard deviation in the measurement of Energy / Change in energy / Difference in eigenenergies (contextual)
- Δp — Standard deviation or uncertainty in the measurement of momentum
- Δt — Time interval / lifetime
- Δx — Standard deviation or uncertainty in the measurement of position / change in position (contextual)
- $\delta(E - E_o)$ — Dirac delta function with argument $E - E_o$
- $\widehat{\mu}$ — Magnetic moment operator
- $\widehat{\mu}_j$ — Magnetic moment operator due to spin along any direction
- μ_B — Bohr magneton
- μ_r — Relative magnetic permeability of the medium
- $\vec{\eta}$ — Unit polarization vector
- $\vec{\sigma}$ — Total Pauli operator (matrix)
- $\sigma_x, \sigma_y, \sigma_z$ — x, y and z Pauli matrices or Pauli operators

- σ — Standard deviation
- $\widehat{\Phi}$ — Magnetic flux operator
- $\psi^*(r)$ or $\psi(r)^*$ — Complex conjugate of $\psi(r)$
- Ω — Rabi frequency
- ω — Resonant frequency
- ω_o — Larmor frequency
- ω_o — Resonant frequency
- ω_o — Angular frequency of the perturbing potential
- ω_{21} — Natural frequency of a two-level system
- ω_{pn} — Resonant angular frequency $(= (E_p - E_n)/\hbar)$
- ϵ_r — Relative electric permittivity of the medium
- $\langle \widehat{A} \rangle$ — Expectation or average value of an operator $\widehat{A}$
- $\widehat{A}^\dagger$ — Hermitian conjugate of operator $\widehat{A}$
- $\widehat{A}$ — Arbitrary operator
- $\bar{A}$ — Magnetic vector potential
- $|B_1\rangle, |B_2\rangle, |B_3\rangle, |B_4\rangle$ — Bell states
- $\bar{B}(\bar{r})$ — Magnetic field
- B_z — z-component of the magnetic field, or longitudinal field or pulse
- $\widehat{CCN}$ — Controlled-Controlled NOT gate
- $\widehat{CH}$ — Controlled-H gate
- $\widehat{CS}$ — Controlled-S gate
- $\widehat{CX}$ — Controlled-NOT gate
- $\widehat{CY}$ — Controlled-Y gate
- $\widehat{CZ}$ — Controlled-Z gate
- DOS — Density of states
- E_F — Fermi energy
- E_{FL} — Fermi energies in the left contact
- E_{FR} — Fermi energies in the right contact
- E_{cmin} — Conduction band minimum
- E_{vmax} — Valence band maximum
- E_c — Conduction band minimum
- E_n — Energy eigenvalue. The corresponding eigen functions are usually ϕ_n, χ_n or $|n>$
- E_v — Valence band maximum

- $E(\bar{k})$ — Energy dispersion relation, bandstructure, energy as a function of wave vector $\bar{k}$
- $f(E)$ — Fermi-Dirac distribution function (probability of finding an electron at energy E)
- $f_L(E)$ — Fermi function in the left contact
- $f_R(E)$ — Fermi function in the right contact
- f_o — Frequency of the perturbing potential
- f_{ps} — Resonant frequency $(= (E_p - E_s)/h)$
- $|GHZ\rangle = \frac{1}{\sqrt{2}}(|000\rangle + |111\rangle)$ — Greenberger-Horne-Zeilinger state
- G_Q — Quantum of conductance
- G — Small bias conductance (also used for Generation rate)
- $\langle \widehat{H} \rangle$ — Expectation value of energy, average energy
- $\widetilde{H}_{mn}$ — Hamiltonian matrix element on m^{th} row and n^{th} column
- $\widetilde{H}_{nm}$ — Hamiltonian element on n^{th} row and m^{th} column
- $\widehat{H}_o$ — Time-independent or unperturbed Hamiltonian
- $\widehat{H}$ — Hadamard gate
- $\widehat{H}$ — Hamiltonian operator
- $H(i,j)$ — Hopping integral between single orbitals centered on atoms i and j
- $H_{n,m}(i,j)$ — Hopping integral between orbitals n and m centered on atoms i and j.
- $H_n(y)$ — Hermite polynomial with argument y
- $\widehat{I}$ — Identity operator
- I — Identity matrix (sometimes refers to current)
- $Im(A)$ — Imaginary component of complex number A
- $\widehat{\imath}$ — Unit vector along the x-axis
- $\bar{J}$ — Generalized angular momentum
- $\widehat{J}$ — Generalized angular momentum operator (sometimes refers to current operator)
- $J_P(x,y,z,E)$ — Probability current density (sometimes refers to electron flux)
- $\widehat{J}_-$ — Generalized angular momentum lowering operator
- $\widehat{J}_+$ — Generalized angular momentum raising operator
- $\widehat{J}_x, \widehat{J}_y, \widehat{J}_z$ — x, y and z components of generalized angular momentum operator

- j — Integer or half-integer (Generalized angular momentum)
- j — Current density
- $\hat{j}$ — Unit vector along the y-axis
- $\widehat{KE}$ — Kinetic energy operator
- KE — Kinetic Energy
- $\hat{k}$ — Unit vector along the z-axis
- k or $\bar{k}$ — Wave vector
- $\bar{L}$ — Orbital angular momentum
- $\hat{L}$ — Orbital angular momentum operator
- $\hat{L}_-$ — Orbital angular momentum lowering operator
- $\hat{L}_+$ — Orbital angular momentum raising operator
- $\hat{L}_p, \hat{L}_q$ — Arbitrary component of orbital angular momentum operator
- $\hat{L}_x, \hat{L}_y, \hat{L}_z$ — x, y and z components of orbital angular momentum operator
- l — Angular/Orbital quantum number
- **M** — Total scattering matrix
- m^* — Effective mass
- m_c — Effective mass (of an electron in the conduction band)
- m_v — Effective mass of a hole (in the valence band)
- $m^*_{xx}, m^*_{yy}, m^*_{zz}$ — Effective mass in x, y and z directions
- n_r — Refractive index of the medium
- n — (Principal) Quantum Number / Integer (contextual)
- $\wp(r)$ — Radial probability density
- $\widehat{PE}$ — Potential energy operator
- PE — Potential Energy
- P_l^m — Legendre polynomials
- P_n — Projection operator
- $P_{ps}(t)$ — Probability of finding an electron in state ϕ_{po} when it started in state ϕ_{so}
- $\hat{p}$ — Momentum operator
- $\bar{p}$ — Momentum vector
- $\bar{p}_{\bar{k}}$ — Momentum eigenvalues
- $\langle \hat{p} \rangle$ — Expectation or average value of momentum
- $\hat{p}_a$ and $\hat{p}_b$ — Arbitrary component of momentum operator
- $\hat{p}_x, \hat{p}_y, \hat{p}_z$ — x, y and z components of the momentum operator

- p_x, p_y, p_z — x, y and z types of the p-orbital
- p_x, p_y, p_z — x, y and z components of the momentum
- $\widehat{Q}$ — Charge operator
- Q — Charge
- R^3 — 3D vector space of coordinate system
- $R_{\{p\}s}$ — Transition probability of going to any one of the many final states ϕ_{po} from initial state s
- R_{ps} — Transition rate to go from initial state s to final state p
- R — Reflection probability / recombination rate / radius
- R — Transition probability
- $Re(A)$ — Real part of complex number A
- $\widehat{r}$ — Position operator
- $\langle r \rangle$ — Average value of the radial location of an electron
- $\bar{r}$ — Position vector of a point in space
- S — Spin quantum number / Surface (contextual)
- $S(i, j)$ — Overlap integral between the single orbital on atoms i and j
- $\widehat{S}_j$ — Spin angular momentum operator due to spin along any direction
- $\widehat{S}_x, \widehat{S}_y, \widehat{S}_z$ — x, y and z component of the spin angular momentum operator
- $\widehat{S}$ — Phase gate (also, total spin operator)
- $\bar{S}$ — Spin angular momentum
- $S_{n,m}(i, j)$ — Overlap integral between n and m orbital wavefunctions centered on atoms i and j
- s — Orbital index of an atom
- s — Spin
- $\widehat{T}$ — T-gate (also translation operator)
- T — Transmission / Transmission probability / Absolute temperature in Kelvin / Rotation period (contextual)
- $T(E)$ — Transmission versus energy
- t, t_x, t_y, t_z — Hopping energy
- t — Time / transmission amplitude
- $\widehat{U}$ — Unitary transform
- U_{nk} — Matrix element of perturbation U between states $|n\rangle$ and $|k\rangle$

- $U(\bar{r}, t)$ — Potential energy at the point $(\bar{r}, t)$ / time-dependent perturbing Hamiltonian (contextual)
- $\hat{u}$ — Normalized vector
- $u_{ij}, \tilde{u}_{ij}$ — Matrix element
- $u(E)$ — Step function with argument E
- $u(\bar{r})$ — Amplitude of the potential created by a perturbing AC potential
- $u_{\bar{k}}(\bar{r})$ — Wavefunction component that is periodic with the underlying lattice (Bloch's theorem)
- $\hat{V}$ — Voltage operator for an LC oscillator
- V_B — Bias voltage
- V — Voltage / volume (contextual)
- $\bar{v}_g$ — Group velocity
- $\hat{X}$ — $\hat{X}$-gate
- $\hat{x}$ — Position operator or unit vector along the x-direction
- $\hat{x}^\dagger$ — Hermitian conjugate of position operator $\hat{x}$
- $\langle \hat{x} \rangle$ — Expected or average value of position after measurement
- $|\uparrow x\rangle$ — $+x$ directed spin ket vector
- $|\downarrow x\rangle$ — $-x$ directed spin ket vector
- $\hat{Y}$ — $\hat{Y}$-gate
- $Y_l^m(\theta, \phi)$ — Spherical harmonic
- $\hat{y}$ — Unit vector along the y-axis
- $|\uparrow y\rangle$ — $+y$ directed spin ket vector
- $|\downarrow y\rangle$ — $-y$ directed spin ket vector
- $\hat{Z}$ — $\hat{Z}$-gate
- $|\uparrow z\rangle$ — $+z$ directed spin ket vector
- $|\downarrow z\rangle$ — $-z$ directed spin ket vector

CONTENTS

NOTE TO INSTRUCTOR

This course often marks a students' first comprehensive introduction to quantum mechanics. The primary challenges students encounter are largely conceptual, as this subject profoundly reshapes their approach to understanding physics. Most of the mathematical content in this textbook is not overly demanding for students who have already completed coursework in differential equations, matrices, and vectors. The utilization of operators may be unfamiliar to students at first, but with practice problems, they swiftly adapt and become proficient.

In our book, we have meticulously outlined the mathematical steps to assist instructors in focusing on core concepts and guiding students through derivations efficiently. Encouraging back-and-forth discussions with students to address questions and clarify concepts is integral to the learning process.

Our book encompasses enough material to support either a two-quarter or two-semester course. While all the chapters have undergone testing with students, we offer below an outline tailored for the initial course.

Slides: Instructors have access to presentation slides for all chapters.

Homework: Solutions are made available to instructors. We typically give homework to emphasize the learning of important concepts and problem solving every week except for the week when a midterm exam is held.

Python notebook: Python notebooks that help students learn basic quantum mechanics by solving simple problems numerically (especially for Chapters 1 and 2) and qiskit problems for quantum information (Chapters 7 and 8) are made available along with solutions to the instructors.

Glossary: The glossary helps students keep track of mathematical symbols.

What do we cover in the first course on quantum mechanics? This is discussed below.

Chapter 1, Schrödinger Equation: Students appreciate the historical introduction, which includes Planck's hypothesis, the wave nature of particles, and the hydrogen atom. It is essential to introduce both the time-dependent and time-independent Schrödinger equations, with a focus on properties 1 and 2 of the wavefunction (derivation of these properties can be omitted, along with property 3). Working through examples of the free particle and plane wave is crucial. We can skip the simple harmonic oscillator.

Students find the discretization of the Schrödinger equation valuable, as it aids in understanding and reviewing matrices. In our teaching approach, we provide Python code to students, enabling them to solve the Schrödinger equation on a one-dimensional grid. This code is available to all instructors. Sections 1.1.1, 1.1.3, 1.1.4, 1.2, 1.3 (without derivations), 1.4, 1.5.1, 1.5.3, 1.6 are covered.

Chapter 2, Wave Function Properties: The deliberate overlap with Chapter 1 serves to underscore key concepts and the fundamental principles of superposition and quantum measurement. Mathematically, we emphasize the orthonormality of wavefunctions. While the entire chapter is covered, worked out examples are not presented in the lecture.

Chapter 3, Operators and Expectation Values: This chapter covers operators, important observables, and expectation values. We skip the derivation of the LC oscillator and commutator brackets. We cover sections 3.1, 3.2, 3.3. In section 3.5, only the central results of the examples are emphasized.

Chapter 4, Hilbert Space: Students are introduced to Hilbert spaces, building upon their existing knowledge of 3D vector spaces and Fourier transforms. We place strong emphasis on both bracket notation and matrix notation, highlighting the relationship between them. Additionally, we discuss Hermitian operators and their eigenvalues. Sections 4.1, 4.2, and 4.3 are covered in lecture. While example 3 from section 4.4 is covered in lecture, some of the other examples are given as homework.

Chapter 5, Uncertainty Principle: We do not emphasize the uncertainty principle in our teaching, as it has limited relevance in problem-solving. When we introduce the concept of spin, we discuss the generalized uncertainty principle.

Chapter 6, Projection Operators: We do not cover projection operators as a standalone topic. Instead, we introduce them within the context of Chapter 7, which focuses on quantum information.

Chapters 7 and 8, Quantum Information and Quantum Gates: Students are enthusiastic about learning quantum information and computing, and we cover these chapters in their entirety. We also assign homework problems based on qiskit for these chapters.

Chapters 9, Two-level System: The chapter on two-level systems instructs students in solving a practical yet straightforward problem using the time-dependent Schrödinger equation. It includes examples such as the evolution of qubit states and introduces students to concepts like Rabi oscillations. The entire chapter is covered.

Chapter 10, Tunneling: We prioritize the practical application of tunneling formulae instead of focusing on derivations. We provide a qualitative and quantitative discussion of tunneling probability as a function of energy for single and two-barrier systems. Our curriculum includes the applications of scanning tunneling microscopy and resonant tunneling diodes. Although we do not cover tunneling through multiple barriers in class, Master's and PhD students are assigned a homework problem in which they write code based on the theory presented in section 10.7. The end-of-chapter problem on this topic offers valuable instruction. In this first course, we do not include topics such as tunneling time and the Breit-Wigner Formula. Sections 10.1, 10.2, 10.4, and 10.5 are covered in the lecture.

Chapter 11, Quantum dots, wells, and nanowires: We find it important to provide a concise introduction to effective mass and energy bands, particularly for students who have not taken a course on semiconductors. While we do present all derivations based on the separation of variables, we keep it accessible to students and guide them through one such derivation. Our primary focus is on discussing the impact of size quantization on energy levels, which constitutes the majority of our class time. Additionally, we discuss the application of quantum wells in Terahertz sources. The entire chapter is covered by providing a material science perspective.

Chapter 12, Density of States: The concept of density of states (DOS) is essential in various applications, such as density and current calculations, as well as optical absorption coefficient determination. During our lectures, we instruct students on deriving the 1D DOS for both conduction and valence bands. Additionally, we assign homework problems that cover DOS for 1D, 2D, and 3D systems. In class, we provide a qualitative discussion of DOS plots for all cases as they are important to material science students.

Chapter 13, Electron Density and Current: We present the framework of Büttiker's scattering theory, elucidating how wavefunctions (Density of State) and the Fermi distribution function in the contacts lead to expressions for density and current in a nanodevice. We discuss the derived expression for the quantum of conductance, which depends solely on fundamental physical constants. We cover sections 13.1 (excluding the derivation) and 13.2.

Chapter 14, Periodic system: We provide a brief description of orbitals and solve the Schrödinger equation to find the band structure of 1D, 2D and 3D periodic solids using Bloch's theorem, focusing on a material science perspective. Our coverage includes the case of a 1D chain of atoms, resulting in both metal and semiconductor properties. We also derive the band structure of graphene as an example of a 2D periodic system. Sections 14.1, 14.2, 14.3.1, 14.3.2, 14.3.3, and 14.4.1 are included in our teaching.

Chapter 15, Spin: We cover the Stern-Gerlach experiment, properties of spin, the Schrödinger equation with spin, and the characteristics of fermions and bosons. We opt not to present the derivation for spin operators due to its time-consuming nature. In this chapter, we introduce the generalized uncertainty principle from Chapter 5 and explain to students that components of spin cannot be simultaneously measured. As an optional topic, we may discuss the spin-orbit and Rashba interactions, which is important to both material science and device technologies. Our class covers Sections 15.1, 15.2, 15.3, and 15.6, with Sections 15.4 and 15.5 considered depending on available time.

Chapter 16, Angular Momentum, Chapter 17, Time-independent Perturbation theory, Chapter 18, Scattering Rate and Fermi Golden Rule: We previously included this material in our classes, but we haven't done so recently due to the focus on quantum computing and two-level systems. These topics would be more suitable for a second course on quantum mechanics. Chapter 18 contains discussions on scattering rates and light absorption coefficients, which are particularly relevant to engineering and material science students.

Chapter 19, Hydrogen atom: This information is provided as a reference to students. Instead of delving into rigors of solving Schrödinger equation in spherical coordinates, we introduce orbitals by using detailed examples. We currently introduce orbitals to students in Chapter 14 and the discussion in this chapter is more detailed both qualitatively and quantitatively, with a derivation included.

Chapter 20, Dirac and Kronecker Delta functions: This is a reference provided to students. They are encouraged to become familiar with the properties and usage of Kronecker and Dirac delta functions.

Chapter 1

THE SCHRÖDINGER EQUATION

Contents

1.1 Planck's Hypothesis and History

Several experiments in the early 20th century fundamentally changed scientists' understanding of waves and particles. Classical interpretation of physics left gaps in certain results from the experiments and could not explain all physical observations and assertions. This chapter will describe crucial discoveries that led to the birth of quantum mechanics and the development of the Schrödinger equation, one of the governing equations of quantum physics. The Schrödinger equation is central in many branches of engineering and science that deal with nanoscale materials and devices.

This section begins by discussing a revolutionary hypothesis made by Max Planck on the quantized (discrete) nature of photon energy. Then, this section will describe the use of Planck's hypothesis by other scientists to make new predictions about the quantum nature of matter.

We hope that this section will help the reader build intuition for the consequences of Planck's hypothesis and provide a brief historical development that predates Schrödinger equation. The main topics discussed in this section are as follows:

(1) photoelectric effect;
(2) de Broglie Principle, which proposed the duality of particles and waves;
(3) Bohr model for the energy spectrum of the Hydrogen atom;
(4) quantum tunneling.

1.1.1 *Planck's hypothesis*

Prior to the 20th century, scientists could not explain the spectrum of radiation (light) emitted by a blackbody using classical physics. However, in year the 1900, Max Planck successfully explained the phenomenon by making a revolutionary assumption about the discrete nature of light energy. This assumption is now called *Planck's hypothesis*.

Planck's hypothesis states that light of a wavelength λ in free space is composed of only *discrete* or *quantized* bursts of energy called photons. The energy of each photon is equal to

$$E = nh\frac{c}{\lambda} = nhf \tag{1.1}$$

where $h = 6.625 \times 10^{-34}$ Js (Joule-seconds) is a fundamental constant called Planck's constant, $c = 3 \times 10^8$ m/s is the speed of light in a vacuum, λ is the wavelength of light, $f = \frac{c}{\lambda}$ is the frequency of light, and n is any positive integer $(n = 0, 1, 2, 3, \ldots)$. The above equation suggests that light energy can occur only in an integer number of packets with energy equal to hf. Light energy cannot occur as a fraction of hf. It must only occur in an integer number of single "unit" chunks. $n = 0$ corresponds to the absence of a photon.

Another commonly used form of expressing equation (1.1) is

$$E = n\hbar\omega \tag{1.2}$$

where $\hbar = \frac{h}{2\pi}$ (pronounced "h bar") and $\omega = 2\pi f$ is the angular frequency of the light wave. To illustrate the importance of Planck's hypothesis, let us use a simple example below.

Example 1. A laser pointer with a red color light beam has a wavelength of 700 nm. What is the minimum energy that can be emitted in a light pulse from the laser pointer? Express your answer in electron volts, where $1 \text{ eV} = 1.602 \times 10^{-19}$ J.

Solution: The minimum nonzero energy corresponds to a single photon, i.e., $n = 1$. So, the minimum energy of emitted light is

$$E = nh\frac{c}{\lambda} = 1 \, (6.625 \times 10^{-34} \text{ Js}) \, \frac{3 \times 10^8 \text{ m/s}}{700 \times 10^{-9} \text{ m}}$$

$$= 2.84 \times 10^{-19} \text{ J} = 1.77 \text{ eV}$$

From a classical perspective, this result is surprising because it means that we cannot emit a burst of photon energy with a wavelength of 700 nm (red color) lower than 1.77 eV. Until Planck, we believed that a red light beam's energy could be made arbitrarily small (say, even smaller than 0.0001 eV!).

Concept: Planck's hypothesis states that light (photon) in free space exists only in discrete or quantized bursts of energy. The central formula is $E = nh\frac{c}{\lambda} = nhf$.

1.1.2   *The Photoelectric effect*

Scientists observed that shining light on a metal ejects electrons from the metal, a phenomenon called the photoelectric effect (Figure 1.1(a)). Classically, this should be possible using light of any frequency. However, experiments revealed otherwise. Experiments reveal that electrons are ejected into the vacuum only if the frequency of light is larger than a minimum value. The ejected electrons could be measured as current in a vacuum lamp (Figure 1.1(b)). Measuring this current under different voltages, light frequencies, and intensities shows the following features:

(i) Electrons are emitted (or current is induced) only if the frequency of incident light exceeds a threshold frequency f_{Th} (Figure 1.1(a) and (c)).

(ii) For frequencies smaller than f_{Th}, electrons are not emitted even if the light source's intensity (photons per second) increases. This is contrary to classical expectations that the intensity of the light does not affect the ejection of electrons if the frequency of light is smaller than f_{Th}.

(iii) For frequencies higher than the threshold frequency, i.e., $f > f_{Th}$, increasing the intensity of light (sending more photons per second) results in an increase in the number of emitted electrons. The increase follows a linear relationship, as shown in Figure 1.1(d).

The contradiction between the behavior of the photoelectric effect and classical approaches was explained using an extension of Planck's hypothesis by Albert Einstein in 1905. Einstein hypothesized that electrons could only absorb a quantum of light equal to $E = hf$. The probability for a single electron to absorb multiple photons simultaneously is miniscule. So, the probability for electron emission is almost zero if incident photons have energy smaller than hf_{Th}.

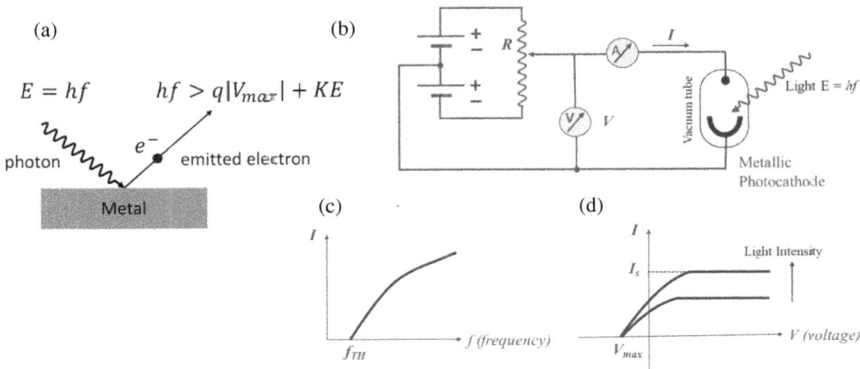

Figure 1.1. (a) Photons incident on metal can eject electrons only if the incident photon frequency is larger than a threshold value. (b) Apparatus to measure ejected electrons from the metallic photocathode in the vacuum tube as a current. (c) Electric current is measured in the circuit only when the photon frequency is above a threshold, $f > f_{Th}$. (d) Applying a negative voltage using the adjustable potentiometer in (b) can shut down the current. Current and voltage are recorded by the meters in series and parallel with the vacuum tube, respectively. As all current–voltage plots converge to a single negative voltage (V_{max}), this voltage is the measure of the minimum binding energy (work function) which is $q|V_{max}|$. The charge of an electron $-q$ is equal to -1.602×10^{-19} C.

If the incident photon energy hf is at least equal to the binding energy, electron emission occurs. Experimentally, one photon results in the ejection of either no electron or precisely one electron. Mathematically, this is expressed as follows:

$$hf \geq \text{Binding Energy} + \text{KE} \tag{1.3}$$

Here, KE is the kinetic energy of the emitted electron, which is a positive number. The binding energy is the minimum energy needed to free an electron from the material and move it infinitely far away. So, the threshold energy (hf) for electron emission from the metal is $\geq$ Binding Energy. This corresponds to a photon frequency of $f_{Th} \geq \frac{\text{Binding Energy}}{h}$.

Figure 1.1(d) illustrates the main points of increasing the photon intensity. If the condition in equation (1.3) holds so that the incident photon energy is sufficient, an increase in photon intensity will cause an increase in the rate of electron emission from the metal and thus current. We reiterate that it is unlikely that a single photon will result in the ejection of more than one electron.

Concept: The photoelectric effect supports Planck's hypothesis by proving the quantum nature of photons. It shows that a minimum frequency (and hence energy) of a single photon is required to eject an electron from a metal.

1.1.3   *de Broglie principle*

As Planck and Einstein successfully proved that light (a wave) also behaves like particles, called photons, scientists asked if material particles can also behave like waves. Louis de Broglie proposed that there is also a wave nature associated with every particle with a mass. The wavelength (λ) is related to the momentum of the particle (p) via the relationship,

$$\lambda = \frac{h}{p} \tag{1.4}$$

where h is Planck's constant. The momentum p is related to mass (m) and speed (v) by $p = mv$. The wavelength λ of the particle is called the de Broglie wavelength. From equation (1.4), we can see that a larger momentum corresponds to a smaller particle wavelength.

In light and sound waves, a state variable will always change with time. For example, electromagnetic waves stem from electric and magnetic fields changing in space and time; sound waves involve particle density changing in space and time. We understand these intuitively now. However, when it comes to the quantum mechanical wave associated with a particle, we will see later what is changing with time (a quantity called the wave function that is governed by the Schrödinger equation).

Example 2. An electron and a proton are traveling at a velocity of 1% of light, and a tennis ball is traveling at $57\,\text{m/s}$. You are given their masses. Express the wavelength in the units of Å. An Ångström (Å) is a length scale defined by $1\,\text{Å} = 1 \times 10^{-10}\,\text{m}$.

Table 1.1. de Broglie wavelength of three different particles.

	Mass (kg)	Speed (m/s)	Momentum (kg m/s) $(p = mv)$	Wavelength (Å) $\lambda = h/p$
Electron	9.1×10^{-31}	3×10^6	2.7×10^{-24}	2.42
Proton	1.67×10^{-27}	3×10^6	5.01×10^{-21}	0.0013
Tennis ball	57×10^{-3}	57	3.25	2.04×10^{-24}

Solution: The de Broglie wavelengths associated with matter is inversely proportional to its mass. A higher particle mass results in a shorter wavelength, as seen in Table 1.1.

Concept: de Broglie's hypothesis says that a particle of mass (m) shows a wave-like property with wavelength $\lambda = \frac{h}{p} = \frac{h}{mv}$.

A classic experiment conducted in the early 1800s called Young's double-slit experiment clearly showed that coherent light behaved as a wave rather than as particles. Coherent light from a source impinges on a double slit. Classically, one would expect the pattern of the intensity of particles on the screen to be as shown in Figure 1.2(a) if light were particles. However, Young's double-slit experiment showed an intensity pattern expected for waves, as shown in Figure 1.2(b). Light from the source travels to the double slit where it is diffracted. The resulting interference pattern on the screen creates a bright fringe in the middle if the two slits are equidistant from the source. Experiments inspired by Young's double-slit experiment have now been performed using particles (atoms, molecules). These experiments show that for a small particle number, a few random dots appear on the screen as shown in Figure 1.2(c). However, as the number of particles increases, a clear wave-like interference pattern emerges even with particles, as shown in Figure 1.2(d).

If particles behave as waves, a central question asked by the founders of quantum mechanics was whether particles such as electrons, atoms and molecules would also show an interference

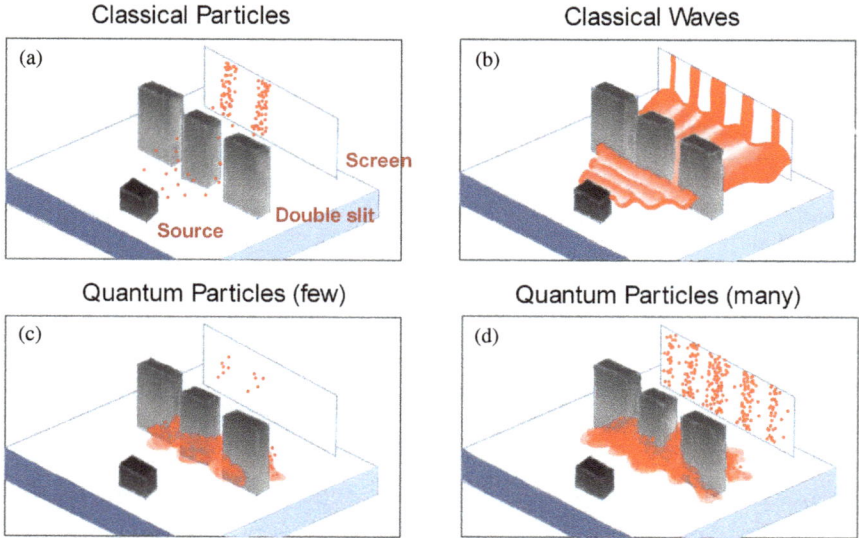

Figure 1.2. *Classical particles and quantum particles* correspond to our interpretation of particles before and after the birth of wave–particle duality. Classical waves correspond to coherent light waves, acoustic waves or even water waves in a pond. Before the birth of quantum mechanics, if atoms were incident on a double slit, the expectation would have been the interference pattern shown in (a). After the birth of quantum mechanics and wave–particle duality, very carefully performed experiments show that the interference pattern in (d) is obtained with particles. Art by Sophia Gershaft and Mrunal Shenoy. Reference: www.toutestquantique.fr (Jubobroff).

pattern in a double-slit-like experiment. The answer is a definite yes though the experiments are significantly more difficult than experiments with light.

Figure 1.3 shows the results of an exciting investigation proving the wave nature of a large molecule phthalocyanine (PcH_2). In this experiment, excited PcH_2 molecules are shot toward the double-slit screen. Each molecule detected on the screen (CCD camera) appears as a dot and signifies its particle nature. If one looks at the pattern after a long time, an interference pattern that unmistakably resembles the intensity pattern in Young's double-slit experiment with light emerges. See (Juffmann, 2005) and (Tonomura, 1989).

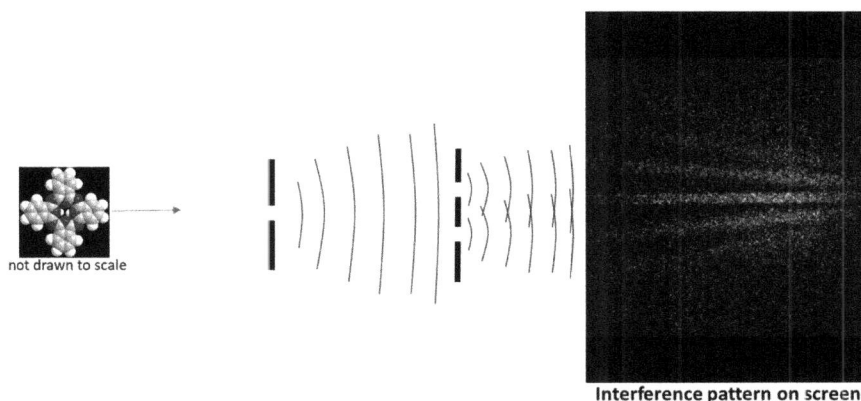

Interference pattern on screen

Figure 1.3. A double-slit experiment with Phthalocyanine molecules reveals their wave nature. Molecules are incident on the apparatus one at a time. When a few molecules are sent through the double slit, individual points of impact can be detected on the screen. However, after 90 minutes, when many molecules have hit the screen, an interference pattern as shown on the right is observed. The Phthalocyanine molecule ($C_{32}H_{18}N_8$) is heavy and has a mass of 514 atomic mass units. Courtesy of Thomas Juffmann, University of Vienna (Juffmann, 2012).

Concept: An interference pattern can be seen in a double-slit experiment with atoms and molecules, which proves the wave nature of particles.

The wave nature of particles embodied in the de Broglie principle (equation (1.4)) is called the wave particle duality.

1.1.4   *Energy levels in a hydrogen atom*

In discussing the hydrogen atom, we do not take the historical perspective. Instead, we will use Newton's laws and de Broglie principle to derive the energy levels.

Hydrogen is the simplest atom — it consists of only one proton and one electron. Due to its simplicity, hydrogen received much attention. Experiments showed the following key features:

- In the classical model for a hydrogen atom, a single proton is at the center of the atom (the nucleus). Electrons revolve around the nucleus in a circular orbit akin to a planet revolving around the sun

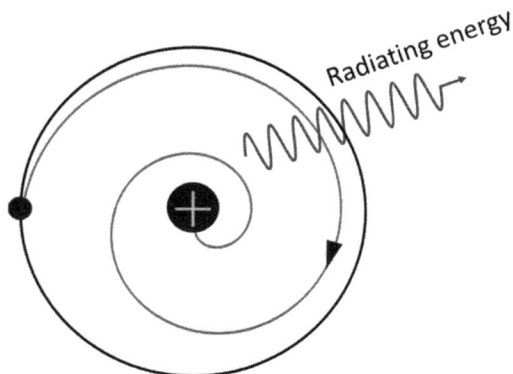

Figure 1.4. An electron moving around the positively charged nucleus is an accelerating particle. According to classical electrodynamics, it should lose energy and spiral into the nucleus.

(see blue circle in Figure 1.4). By the definition of acceleration, an electron rotating around the nucleus is accelerating at any moment because the direction of its velocity constantly changes. Predictions from the field of electrodynamics showed that an accelerating electron should lose energy and should eventually spiral into the nucleus (Figure 1.4).

• Hydrogen atoms absorb and emit only discrete energies of photons.

To prevent the electron from spiraling into the nucleus, Bohr hypothesized that: Electrons move around the nucleus of a hydrogen atom in stable discrete orbits where they will not lose energy, except when making a transition to another orbit. We will use de Broglie's principle to treat electrons as waves. If an electron revolves around the nucleus in a circular path, the circumference should be equal to an integer number of (de Broglie) wavelengths,

Circumference of orbit = Integer × Wavelength of electron

$$2\pi r = n\lambda \tag{1.5}$$

Here, r is the radius of orbit and λ is the electron's wavelength, which is related to its momentum (mass times velocity) by de Broglie's formula.

We can use equation (1.5) to derive an expression for the energy levels in a hydrogen atom through the following process:

Total energy = Kinetic energy (KE) + Potential energy (PE)
Newton's law, $F = ma$
de Broglie's equation, $\lambda = \frac{h}{p}$
The derivation starts by writing the kinetic energy (KE) of the electron which is

$$KE = \frac{1}{2}mv^2 \tag{1.6}$$

The electrostatic potential energy of an electron at a distance r from the nucleus is

$$PE = -\frac{q^2}{4\pi\varepsilon_0 r} \tag{1.7}$$

where ε_0 is the vacuum dielectric constant.

$$\text{The total energy, } E = KE + PE = \frac{1}{2}mv^2 - \frac{q^2}{4\pi\varepsilon_0 r} \tag{1.8}$$

Newton's law implies that for a stable orbit of the electron, the centripetal force is equal to the Coulombic force between the nucleus and electron. The centripetal force for a particle of mass m, speed v, and radius of orbit r is $\frac{mv^2}{r}$ and the Coulombic attractive force between the nucleus and electron is $\frac{q^2}{4\pi\varepsilon_0 r^2}$. So,

$$F_{\text{centripetal}} = F_{\text{Coulombic}} \tag{1.9}$$

$$\frac{mv^2}{r} = \frac{q^2}{4\pi\varepsilon_0 r^2} \tag{1.10}$$

From equation (1.10), we obtain the speed of the electron (v) to be

$$v = \sqrt{\frac{q^2}{4\pi\varepsilon_0 m r}} \tag{1.11}$$

Substituting equation (1.11) into equation (1.8), gives us the total energy,

$$E = -\frac{q^2}{8\pi\varepsilon_0 r} \tag{1.12}$$

The above equation still does not give us the values for the energy levels as the value of r is not known. To find the possible values of r, we use equation (1.5):

$$2\pi r = n\lambda \tag{1.5}$$

and de Broglie's principle

$$\lambda = \frac{h}{p} = \frac{h}{mv} \tag{1.4}$$

Substituting equations (1.4) and (1.5) into equation (1.10), we must find that the radius of electron orbit depends on the integer n and is equal to

$$r_n = n^2 \left(\frac{4\pi\varepsilon_0 \hbar^2}{q^2 m} \right) = n^2 a_o \tag{1.13}$$

A subscript n is added to the radius to denote there are many values of the radius of orbit depending on the value of n. Note that an electron cannot have *any* radius while rotating around the nucleus; the radii of the energy levels are calculable from fundamental constants and a discrete set of integer numbers (n). The Bohr radius (a_o) is defined to be

$$a_o = \frac{4\pi\varepsilon_0 \hbar^2}{q^2 m} = 0.53 \text{ Å} \tag{1.14}$$

the radius for the smallest orbit corresponding to $n = 1$ in the hydrogen atom. Substituting equation (1.14) in equation (1.12), we find that the energy levels of an electron in the hydrogen atom are given by

$$E_n = -\frac{E_o}{n^2} \quad \text{where } E_o = \frac{q^2}{8\pi\varepsilon_0 a_o} \tag{1.15}$$

$$\vdots$$

$$E_3 = \frac{-13.6\ eV}{9} \qquad \qquad \qquad n = 3$$

$$E_2 = \frac{-13.6\ eV}{4} \qquad \qquad \qquad n = 2$$

$$E_1 = -13.6\ eV \qquad \qquad \qquad n = 1$$

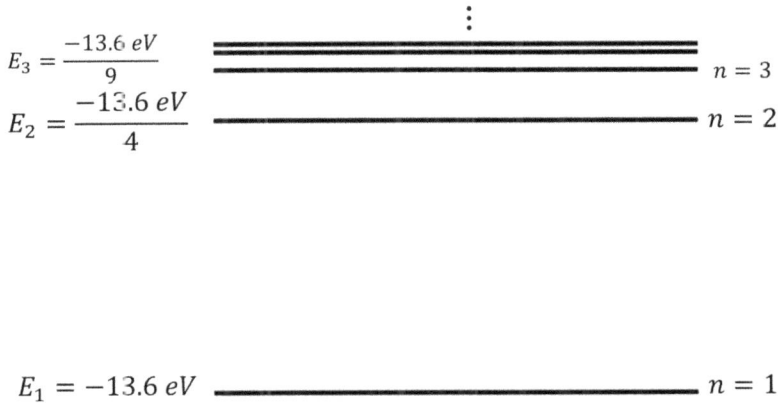

Figure 1.5. Discrete energy levels in a hydrogen atom.

The subscript n shows that the energy levels' value depends on n, which can take only integer values $1, 2, 3, \ldots$

The allowed energy levels and corresponding radii of orbit correspond to (Figure 1.5)

$$n = 1, \quad E_1 = -\frac{E_c}{1} = -13.6\,\text{eV}, \quad r_1 = a_o = 0.53\,\text{Å} = \text{Bohr radius} \tag{1.16}$$

$$n = 2, \quad E_2 = -\frac{E_c}{4} = -3.4\,\text{eV}, \quad r_2 = n^2 a_o = 4a_o \approx 2\,\text{Å} \tag{1.17}$$

and so on.

Now consider an electron in a hydrogen atom in the lowest energy state $n = 1$ If photons of energy $h\upsilon = E_m - E_1$ shine on the hydrogen atom, the electron in state $n = 1$ can absorb the photon and go to quantum state m with energy E_m. Additionally, the electron in an excited state n with energy E_m can emit a photon with energy $hf = E_m - E_n$ and transition from quantum state m to quantum state n which has a lower energy E_n. From equation (1.15), we can see that as the quantum number n increases, the energy level spacing becomes smaller and the energy spectrum becomes a continuum.

Concept: Bohr's model of the hydrogen atom correctly predicted how photons are emitted/absorbed by hydrogen. The main equation is $E_n = -\frac{E_o}{n^2}$ where $E_o = \frac{q^2}{8\pi\varepsilon_0 a_o}$.

1.1.5 *Frequency and wave vector of a free particle*

A free particle (FP) is defined to be a particle whose potential energy is the same all over space. So, the velocity of a FP is independent of its location.

We discuss the relationship between wave vector, momentum, and energy of a FP (free electron, free atom, etc.) with a nonzero mass. The de Broglie principle relates momentum (p) of a FP to its wave vector (k) as discussed earlier is

$$p = \frac{h}{\lambda} = \hbar\frac{2\pi}{\lambda} = \hbar k \tag{1.18}$$

Here, λ is the wavelength and $k = \frac{2\pi}{\lambda}$ is its wave vector or wave number.

The energy (E) as per Planck's hypothesis is

$$E = \hbar\omega$$

Recall that the angular frequency $\omega = 2\pi f$, where f is the frequency. We also know that the energy of a FP is

$$E = \frac{p^2}{2m} = \frac{\hbar^2 k^2}{2m}$$

From the previous two equations, we find the relationship between frequency f and wave vector k,

$$\hbar\omega = \frac{\hbar^2 k^2}{2m}$$

$$f = \frac{hk^2}{8m\pi^2} \tag{1.19}$$

$$f = \frac{h}{2m\lambda^2}$$

So, for a FP of mass m, the frequency varies quadratically with wave vector or inversely as the wavelength squared. In contrast, for light, we know that the frequency,

$$f = \frac{c}{\lambda} \tag{1.20}$$

varies inversely with the wavelength or is directly proportional to the wave vector. Photons do not follow equation (1.17) because they do not have a mass.

Concept: For a free particle of mass m, the frequency varies directly with the square of the wave vector or inversely with the square of the wavelength, as given by

$$f = \frac{hk^2}{8m\pi^2} \quad \text{and} \quad f = \frac{h}{2m\lambda^2}$$

1.2 Schrödinger Wave Equation and the Wave Function

In the earlier section we saw that particles display wave-like behavior as embodied in de Broglie's principle. Wave equations for sound, light, and vibrations on a string were proven and tested before 1900. To describe the wave nature of particles, Erwin Schrödinger proposed a wave equation that is now fundamental to quantum mechanics. Schrödinger primarily researched acoustic waves before he engaged with quantum physics. His in-depth knowledge of acoustic waves made him very well prepared to propose the central wave equation in quantum mechanics, which now bears his name (Gross, 2014).

The Schrödinger wave equation, which defines the wave nature of particles (including electrons), is given by

$$i\hbar \frac{\partial \Psi\left(\bar{r}, t\right)}{\partial t} = -\frac{\hbar^2}{2m} \nabla^2 \Psi\left(\bar{r}, t\right) + U\left(\bar{r}, t\right) \Psi\left(\bar{r}, t\right) \tag{1.21}$$

In the above equation,

- m is the mass of particle.
- $i = \sqrt{-1}$ (a pure imaginary number).

- $\hbar = \frac{h}{2\pi}$ is the reduced Planck's constant.
- $\bar{r} = (x, y, z)$ is a vector that is the spatial location; the bar $(\bar{})$ on top of r is to remind us that r is a vector.
- t is time.
- $\Psi(\bar{r}, t)$ represents the wave assigned to the particle. It is called the wave function.

The left-hand side (LHS) of equation (1.19) is the time derivative of the function $\Psi(\bar{r}, t)$ that represents the wave. Ψ is called the wave function, and will be discussed shortly. The first term on the right-hand side (RHS) is proportional to gradient-squared of $\Psi(\bar{r}, t)$. The second term on the RHS is the product of the potential energy $U(\bar{r}, t)$ felt by the particle at $(\bar{r}, t)$ and $\Psi(\bar{r}, t)$.

Let us analyze the units in the above equation. Planck's constant is in Js (Joule-second), and time t has the units of seconds. As a result, the differential operator on the LHS $i\hbar\frac{\partial}{\partial t}$ has units of Joules, which is the unit of energy. On the RHS, you can perform dimensional analysis to convince yourself that $\frac{\hbar^2}{2m}\nabla^2$ also has the units of Joules. The potential energy $U(\bar{r}, t)$ also has the units of Joules.

It is important to note that on the RHS of the above equation, the order of the variables is essential for the first term because the *differential operator* is acting on $\Psi(\bar{r}, t)$. We know from calculus

$$\nabla^2 \Psi(\bar{r}, t) \neq \Psi(\bar{r}, t) \nabla^2 \qquad (1.22)$$

The order of the variables is not essential for the second term on the RHS of the Schrödinger wave equation because both $U(\bar{r}, t)$ and $\Psi(\bar{r}, t)$ are scalar functions that depend on both space and time. As a result of the commutative property,

$$U(\bar{r}, t)\Psi(\bar{r}, t) = \Psi(\bar{r}, t)U(\bar{r}, t) \qquad (1.23)$$

We often write Schrödinger wave equation by factoring out $\Psi(\bar{r}, t)$ on the RHS,

$$i\hbar\frac{\partial \Psi(\bar{r}, t)}{\partial t} = \left[-\frac{\hbar^2}{2m}\nabla^2 + U(\bar{r}, t)\right]\Psi(\bar{r}, t) \qquad (1.24)$$

The symbol $\widehat{H}$ represents the sum in the square bracket and is called the Hamiltonian operator,

$$\widehat{H} = -\frac{\hbar^2}{2m}\nabla^2 + U(\bar{r}, t) \tag{1.25}$$

The hat $(\,\hat{}\,)$ on top of H is used to remind us that the Hamiltonian is an operator (here, a differential operator). For an operator acting on a scalar function, we cannot interchange the *order of operations* (unless the operator is also a scalar function). This is because an operator acts on all terms or functions directly to the right of it, so if the operator were moved one space to the left, it would act on more terms than before. For example, $g(x)\frac{d}{dx}f(x) \neq \frac{d}{dx}f(x)g(x)$, because while $g(x)$ and $f(x)$ are each scalar functions, $\frac{d}{dx}$ is an operator and affects all terms to the right of it. For the Hamiltonian operator specifically, $\widehat{H}\Psi(\bar{r}, t) \neq \Psi(\bar{r}, t)\widehat{H}$ because of the gradient-squared term. $\widehat{H}$ is also referred to as the energy operator since it is the sum of the kinetic and potential energies. While the potential energy term $U(\bar{r}, t)$ is not new, we will see in Chapter 2 that $-\frac{\hbar^2}{2m}\nabla^2$ stands for the kinetic energy operator.

Using the symbol $\widehat{H}$, Schrödinger wave equation is compactly expressed as

$$i\hbar\frac{\partial \Psi(\bar{r}, t)}{\partial t} = \widehat{H}\Psi(\bar{r}, t) \tag{1.26}$$

The function representing the wave Ψ is a complex quantity (it has amplitude and phase) and is called the wave function. It describes the entire state of the particle and is central to calculating the values of all observable quantities of relevance to the particle, such as the following:

(i) probability of finding the particle at any $(\bar{r}, t)$;
(ii) energy of the particle;
(iii) momentum of the particle;
(iv) spatial location of the particle.

The probability of finding the particle in a small volume dv at location $\bar{r}$ and time t is equal to $|\Psi(\bar{r}, t)|^2 \, dv$. Note that $|\Psi(\bar{r}, t)|^2 = \Psi(\bar{r}, t)\Psi^*(\bar{r}, t)$, where $\Psi^*(\bar{r}, t)$ is the complex conjugate of $\Psi(\bar{r}, t)$.

$|\Psi(\bar{r}, t)|^2$ is called the probability density and it has dimensions of $1/\text{volume}$ $(1/\text{m}^3)$. However, in one-dimensional problems, $|\Psi(x, t)|^2$ has a dimension of $1/\text{m}$ and $|\Psi(x, t)|^2 \, dx$ is the probability of finding the particle in interval dx.

$$i\hbar \frac{\partial \Psi(\bar{r}, t)}{\partial t} = -\frac{\hbar^2}{2m} \nabla^2 \Psi(\bar{r}, t) + U(\bar{r}, t) \Psi(\bar{r}, t)$$

- m is the mass of particle.
- $i = \sqrt{-1}$ (a pure imaginary number).
- $\hbar = \frac{h}{2\pi}$ is reduced Planck's constant.
- $\bar{r} = (x, y, z)$ is a vector that is the spatial location; the bar $(\bar{})$ on top of r is to remind us that $\bar{r}$ is a vector.
- t is time.
- $\Psi(\bar{r}, t)$ represents the particle-wave. It is called the wave function.

Concept: Solving Schrödinger wave equation gives us the wave function $\Psi(\bar{r}, t)$ of the quantum particle. The wave function holds all information about the quantum system.

The function representing the wave Ψ is a complex quantity (it has amplitude and phase) and is called the wave function.

The probability of finding the particle in an infinitesimally small volume dv (where $dv = dx \, dy \, dz$) at location $\bar{r}$ and time t is equal to $|\Psi(\bar{r}, t)|^2 dv$.

1.3   Properties of the Wave Function (Normalization and Continuity)

Given the Schrödinger equation and interpretation of the wave function, we can derive the mathematical properties that the wave function obeys. We will summarize these properties and then derive them. A student learning quantum mechanics for the first time may choose to read "Property 1," and then skip the remainder of this section. You can come back to Properties 2 and 3 at a later time.

The first property does not have a derivation, but simply follows the meaning of probability density. Equation (1.27) says that since

$|\Psi(\bar{r},t)|^2 \, dv$ is the probability of finding the particle in an infinitesimally small volume dv, the sum of the probability of finding it somewhere in space must be 1.

Property 1. Integral of probability density (**Normalization**). A particle in the state $\Psi(\bar{r},t)$ must exist somewhere in space. So, the integral of probability density over all space must be unity (1). That is,

$$\int dv \, |\Psi(\bar{r},t)|^2 = 1 \tag{1.27}$$

If a given wave function satisfies the above criteria it is called a normalized wave function.

Property 2. The wave function is a **single-valued** function at any location in space. That is,

$$\Psi(\bar{r} + d\bar{r}_1) = \Psi(\bar{r} + d\bar{r}_2) \text{ when } d\bar{r}_1 \ \& \ d\bar{r}_2 \to 0 \tag{1.28}$$

Property 3. The derivative of the wave function is **continuous** unless the potential energy is infinite:

$$\nabla\Psi(\bar{r},t)|_{\bar{r}+d\bar{r}_1} = \nabla\Psi(\bar{r},t)|_{\bar{r}+d\bar{r}_2} \text{ when } d\bar{r}_1 \ \& \ d\bar{r}_2 \to 0 \tag{1.29}$$

What does this third condition say physically? Later on, we will learn that the derivative of the wave function with respect to position is proportional to the particle's momentum. If the derivative has a jump at a given point in space, it means that the momentum of the particle has a sudden jump. A jump (or discontinuity) in momentum is possible if the particle hits a barrier with infinite potential energy. Mathematically speaking, a hard barrier is modeled by an infinite potential facing the particle, one which the particle finds impenetrable.

Figure 1.6 shows some typical cases of allowed and unallowed wave functions. $\Psi(\bar{r},t)$ will always be continuous while $\nabla\Psi(\bar{r},t)$ may or may not be continuous.

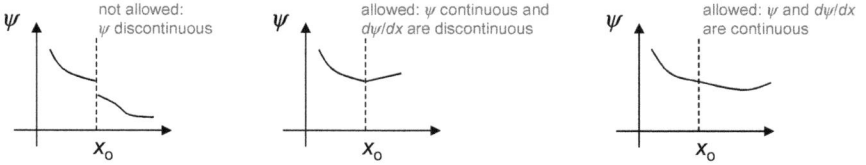

Figure 1.6. Allowed and unallowed shapes of wave functions in dimension. These properties hold in 3D space as discussed in the text. The middle wave function is allowed only if there is a delta function in the potential energy.

Proof of Property of 3[1]: The derivative of Ψ is continuous unless the potential energy felt by the particle in Schrödinger equation is infinite.

Let us start with Schrödinger equation and consider only the 1D case:

$$i\hbar\frac{\partial\Psi(x,t)}{\partial t} = -\frac{\hbar^2}{2m}\frac{d^2\Psi(x,t)}{dx^2}+U(x)\Psi(x,t) \qquad \text{(same as (1.24))}$$

We first integrate equation (1.24) around a small region $2\Delta x$ centered at $x = x_0$, which gives

$$\int_{x_0-\Delta x}^{x_0+\Delta x} dx\, i\hbar\frac{\partial\Psi(x,t)}{\partial t} = \int_{x_0-\Delta x}^{x_0+\Delta x} dx\frac{d^2\Psi(x,t)}{dx^2}$$
$$+ \int_{x_0-\Delta x}^{x_0+\Delta x} dxU(x)\Psi(x,t) \qquad (1.30)$$

In this small region $2\Delta x$, unless $U(x)$ becomes infinite, we can assume both $U(x)$ and $\Psi(x,t)$ do not change in the interval from $x_0 - \Delta x$ to $x_0 + \Delta x$. As a result, we can simplify the terms in equation (1.30).

The LHS is rewritten as

$$i\hbar\int_{x_0-\Delta x}^{x_0+\Delta x} dx\frac{\partial\Psi(x,t)}{\partial t} = i\hbar\frac{\partial}{\partial t}\Psi(x_0,t)\cdot 2\Delta x \qquad (1.31)$$

---

[1]You may skip this theorem in the first reading.

The second term on the RHS is approximated by

$$\int_{x_0-\Delta x}^{x_0+\Delta x} dx\, U(x)\Psi(x,t) = U(x_0)\Psi(x_0,t) \cdot 2\Delta x \tag{1.32}$$

The first term on the RHS of equation (1.30) depends on the second derivative of the wave function with respect to position. If the first derivative is discontinuous, the second derivative can become infinite. We are deriving the continuity property of the first derivative. So, we cannot assume that $\int_{x_0-\Delta x}^{x_0+\Delta x} \frac{\partial^2 \Psi}{\partial x^2} dx$ is equal to $\left.\frac{\partial^2 \Psi}{\partial x^2}\right|_{x_0} 2\Delta x$. Instead, we will have to first perform the integral and then set the upper and lower limits. We know that

$$\int \frac{\partial^2 \Psi}{\partial x^2} dx = \frac{\partial \Psi}{\partial x} \quad \text{and}$$

$$\int_{x_0-\Delta x}^{x_0+\Delta x} \frac{\partial^2 \Psi}{\partial x^2} dx = \left.\frac{\partial \Psi}{\partial x}\right|_{x_0+\Delta x} - \left.\frac{\partial \Psi}{\partial x}\right|_{x_0-\Delta x} \tag{1.33}$$

Substituting equations (1.31), (1.32), and (1.33) in equation (1.30), we have

$$i\hbar \frac{\partial}{\partial t} \Psi(x_0,t) \cdot 2\Delta x = \left.\frac{\partial \Psi}{\partial x}\right|_{x_0+\Delta x} - \left.\frac{\partial \Psi}{\partial x}\right|_{x_0-\Delta x} + U(x_0)\Psi(x_0 t) \cdot 2\Delta x \tag{1.34}$$

If $U(x_0)$ is finite, we set the limit $\Delta x = 0$, to get,

$$\lim_{\Delta x \to 0} \left.\frac{\partial \Psi}{\partial x}\right|_{x_0+\Delta x} = \left.\frac{\partial \Psi}{\partial x}\right|_{x_0-\Delta x} \tag{1.35}$$

That is, the derivative of the wave function is continuous unless the particle has infinite potential energy, which is what we set out to prove. Note that when $U(x_0) = \infty$, then $U(x_0)\Psi(x_0 t) \cdot 2\Delta x$ can be a finite number and the derivative of the wave function can be discontinuous.

Concept: The wave function should be normalized to unity, single-valued at all points in space, and its derivative should be continuous unless the potential energy of the particle is infinite.

1.4   Schrödinger Equation: Time-Independent Version

The time-independent Schrödinger equation is useful as there are many problems where the potential energy, $U(\bar{r}, t)$, does not vary with time. For such potential energies, which are also extensively used to study quantum systems, solving the Schrödinger equation is much simpler. So, it is useful to discuss this in some detail.

When potential energy does not vary with time, we can simplify the potential energy term to

$$U(\bar{r}, t) = U(\bar{r}) \tag{1.36}$$

Substituting this in equation (1.24) gives

$$i\hbar \frac{\partial}{\partial t} \Psi(\bar{r}, t) = \left[-\frac{\hbar^2}{2m} \nabla^2 + U(\bar{r}) \right] \Psi(\bar{r}, t) \tag{1.37}$$

We see that the operator $i\hbar\frac{\partial}{\partial t}$ on the LHS of equation (1.37) is only time dependent while the operator $-\frac{\hbar^2}{2m}\nabla^2 + U(\bar{r})$ on the RHS is only space dependent. However, Ψ on both sides of the equation is still a function of $\bar{r}$ and t. Thus, our method of solving Schrödinger equation starts by predicting solutions of the form,

$$\Psi(\bar{r}, t) = R(\bar{r})T(t) \tag{1.38}$$

Equation (1.38) is a product of two functions; the first function $R(\bar{r})$ depends only on space and the second function $T(t)$ depends only on time. Substituting equation (1.38) in equation (1.37), we have

1. $\dfrac{\partial}{\partial t} \Psi(\bar{r}, t) = \dfrac{\partial}{\partial t} R(\bar{r}) T(t)$ where $R(\bar{r})$ can be treated as a constant

$$= R(\bar{r}) \frac{\partial}{\partial t} T(t)$$

2. $\nabla^2 \Psi(\bar{r}, t) = \nabla^2 R(\bar{r}) T(t)$ where $T(t)$ can be treated as a constant

$$= T(t) \nabla^2 R(\bar{r}).$$

In order to fully separate the variables onto two sides of a single equation, we can substitute equation (1.38) and the above two equations into equation (1.37):

$$i\hbar R\left(\bar{r}\right)\frac{\partial}{\partial t}T(t) = -\frac{\hbar^2}{2m}T(t)\nabla^2 R\left(\bar{r}\right) + U\left(\bar{r}\right)R\left(\bar{r}\right)T(t)$$

Left-multiplying both the LHS and RHS of the resulting equation by $1/R\left(\bar{r}\right)T(t)$ gives us

$$i\hbar\frac{1}{T(t)}\frac{\partial}{\partial t}T(t) = \frac{1}{R(\bar{r})}\left[-\frac{\hbar^2}{2m}\nabla^2 + U(\bar{r})\right]R(\bar{r}) \qquad (1.39)$$

The LHS of equation (1.39) is only a function of time and the RHS is only a function of position, or space. Changing time t can change $T(t)$ and its derivative on the LHS since they're time dependent. Since the RHS does not depend on time, changing time t cannot change the terms on the RHS. The same variable independence occurs in the equation if we try to change the position. Only the function $U(\bar{r})$ and $R(\bar{r})$ on the RHS will be affected with changes in position because the LHS does not depend on position. This characteristic of the time-independent version is useful because it allows us to realize that both sides of the equation must stay constant to any changes in time or position. Thus, we will set both the LHS and RHS of the above equation to be equal to same constant that we will call E. The procedure to solve this equation is called the *separation of variables* in partial differential equations. While the procedure should be clear below, you may consult (Brown, 2011) to review the method. That is,

$$i\hbar\frac{1}{T(t)}\frac{\partial}{\partial t}T(t) = \frac{1}{R(\bar{r})}\left[-\frac{\hbar^2}{2m}\nabla^2 + U(\bar{r})\right]R(\bar{r}) = E \qquad (1.40)$$

The LHS and RHS of equation (1.40) can be notated separately as

$$i\hbar\frac{\partial}{\partial t}T(t) = ET(t) \qquad (1.41)$$

and

$$\left[-\frac{\hbar^2}{2m}\nabla^2 + U(\bar{r})\right]R(\bar{r}) = ER(\bar{r}) \qquad (1.42)$$

Note that we do not know the values of E or the functions $T(t)$ and $R(\bar{r})$. The solution of equation (1.41) is an exponential function of time,

$$T(t) = Ae^{-\frac{iE}{\hbar}t} \qquad (1.43)$$

where A is constant and is unknown at present. Solving equation (1.42) for a given form of $U(\bar{r})$ gives us the allowed values of E in equation (1.43). The spatial part of the wave function $R(\bar{r})$ will also be obtained by solving equation (1.42). Substituting equation (1.43) in equation (1.38), the time-dependent wave function is

$$\Psi(\bar{r}, t) = e^{-\frac{iE}{\hbar}t} R(\bar{r}) \qquad (1.44)$$

Note that the constant A in the above equation is hidden in the solution $R(\bar{r})$.

Equation (1.42) has the general form of an eigenvalue problem.

Eigenvalue problem
The general mathematical form of an eigenvalue problem is

Operator [Eigenfunction] = (Eigenvalue) [Eigenfunction].

The operator can be differential operator or a $N \times N$ matrix.
In the case of a matrix the eigenfunctions are $N \times 1$ vectors.
The eigenvalues are scalars.
There is one eigenfunction corresponding to each eigenvalue.
Degenerate eigenvalues have different eigenfunctions.
The eigenfunctions can be chosen to be orthogonal to each other.

In our case, the operator is the Hamiltonian,

$$\hat{H} R(\bar{r}) = ER(\bar{r}) \qquad (1.45)$$

$$\left[-\frac{\hbar^2}{2m} \nabla^2 + U(\bar{r}) \right] R(\bar{r}) = ER(\bar{r}) \qquad (1.46)$$

We define the eigenvalue and the eigenfunction pairs by an index n, called the quantum number. This index n is different from the same symbol used in Planck's hypothesis. Equations (1.42) and (1.43) can be written as

$$\left[-\frac{\hbar^2}{2m}\nabla^2 + U(\bar{r})\right] R_n(\bar{r}) = E_n R_n(\bar{r}) \tag{1.47}$$

$$T_n(t) = Ae^{-\frac{iE_n t}{\hbar}} \tag{1.48}$$

Equation (1.44) for an eigenfunction with quantum number n can be rewritten as

$$\Psi(\bar{r}, t) = e^{-\frac{iE_n t}{\hbar}} R_n(\bar{r}) \tag{1.49}$$

It is easy to verify that equation (1.49) satisfies the Schrödinger wave equation for a system with a time-independent potential energy (depends only on space). The probability density of the eigenfunctions given by equation (1.49) are known as *stationary states* because their probability density $|\Psi(\bar{r}, t)|^2 = |R_n(\bar{r})|^2$ is stationary (does not depend on time).

Meaning of eigenvalues E_n: In equations (1.40)–(1.44), E has units of energy. E_n for all n represent the *energy eigenvalues*. Note that energy eigenvalues are often referred to as *eigenenergies* or *allowed energy levels*.

Meaning of eigenfunctions (also called eigenstates) $R_n(\bar{r})$ is the *eigenfunction* corresponding to *eigenvalue E_n*. They represent only the *spatial part*. The full wave function of a particle with an energy eigenvalue E_n is $e^{-\frac{iE_n t}{\hbar}} R_n(\bar{r})$. The probability density of finding this particle at $(\bar{r}, t)$ is

$$\left|e^{-\frac{iE_n t}{\hbar}} R_n(\bar{r})\right|^2 = |R_n(\bar{r})|^2 \tag{1.50}$$

The above equation is valid only for an energy eigenstate of a time-independent Hamiltonian. To see the above equality, we have used $\left|e^{-\frac{iE_n t}{\hbar}} R_n(\bar{r})\right|^2 = \left|e^{-\frac{iE_n t}{\hbar}}\right|^2 |R_n(\bar{r})|^2$. Using Euler's formula $e^{i\theta} = \cos(\theta) + i\sin(\theta)$, the previous equation becomes

$$\left|e^{-\frac{iE_n t}{\hbar}} R_n(\bar{r})\right|^2 = \left|\cos\left(\frac{E_n t}{\hbar}\right) - i\sin\left(\frac{E_n t}{\hbar}\right)\right|^2 |R_n(\bar{r})|^2 = |R_n(\bar{r})|^2$$

The probability density in an eigenstate of a time-independent Hamiltonian is also time-independent. That is,

$$\frac{d}{dt}|\Psi(\bar{r}, t)|^2 = 0$$

In the eigenvalue problem given by equation (1.42), the eigenfunctions can be chosen to be orthogonal to each other. The *orthogonality* of eigenfunctions reads

$$\int dv\, R_n^*(\bar{r}) R_m(\bar{r}) = \delta_{mn} = \begin{cases} 1 & \text{if } n = m \\ 0 & \text{if } n \neq m \end{cases}$$

δ_{mn} is called the Kronecker delta and is discussed further in Chapter 20. Finally, we remark that when the Hamiltonian ($\widehat{H}$) is time independent, an elegant expression for the wave function's time evolution is

$$\Psi(\bar{r}, t) = e^{-i\frac{\widehat{H}(t-t_o)}{\hbar}} \Psi(\bar{r}, t_o) \tag{1.51}$$

If we know the wave function at time $t = t_o$, the above equation gives us the wave function at any later time at the same point in space. The reader should verify that equation (1.51) is the solution to $i\hbar\frac{\partial \Psi(\bar{r}, t)}{\partial t} = \widehat{H}\Psi(\bar{r}, t)$, when $\widehat{H}$ is time independent.

Concept: When the potential energy of a particle is independent of time, solving Schrödinger wave equation gives the allowed energy levels (*energy eigenvalues*) E_n and the corresponding wave functions

$$\psi(\bar{r}, t) = e^{-\frac{iE_n t}{\hbar}} R_n(\bar{r}).$$

$R_n(\bar{r})$ is the eigenfunction corresponding to energy eigenvalue E_n.

Terminology:

(a) E_n: Energy eigenvalue (also known as eigenenergies or allowed energy levels)
(b) $\psi_n(\bar{r})$: **Eigenfunction** corresponding to energy eigenvalue E_n. The eigenfunction is also called the **eigenstate**.
(c) The subscript n is called the **quantum number**.

1.4.1 *Difference between wave function and eigenfunction*

You will encounter the terms *wave function* and *eigenfunction* in studying quantum mechanics. As this can be confusing, we would like to differentiate between them here. To do this, we consider the case of a guitar string. The equation for waves traveling on a guitar string is an eigenvalue problem. The solutions result in eigenfunctions, which describe the guitar string's highly symmetric vibrations. Note that eigenfunctions are also called eigenmodes or normal modes. For example, the first ($n = 1$) eigen or normal mode is half a sine wave; the second ($n = 2$) eigen or normal mode is a full sine wave. See Figure 1.7. The displacement of the string at $x = 0$ and $x = L$ is zero for all n because it is clamped at the two ends. The eigenfunction or eigenmodes are given by (A is the amplitude),

$$A \sin\left(\frac{n\pi}{L}x\right)$$

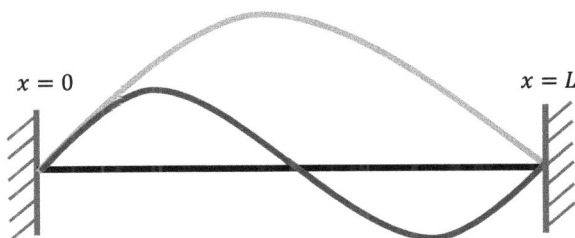

Figure 1.7. The eigenfunctions/eigenmodes/normal modes (all equivalent terms) of a guitar string. The black line shows the string at rest. The green line shows the lowest ($n = 1$), and the red line shows the second lowest ($n = 2$) eigenfunctions.

During an exciting piece of music, the guitar string is in a superposition (or linear combination) of many eigenmodes. An example of this would be a guitar string in a state that has both $n = 1$ and $n = 2$ eigenmodes; the wave amplitude at any instant of time would be given by

$$A_1 \sin\left(\frac{\pi}{L}x\right) + A_2 \sin\left(\frac{2\pi}{L}x\right)$$

where A_1 and A_2 are the amplitudes of the two eigenmodes.

Similarly, *eigenfunctions* are the solutions of the eigenvalue equation $\widehat{H}R(\bar{r}) = ER(\bar{r})$. These solutions, which are represented by $R_n(\bar{r})$ earlier in this section, correspond to the different eigenfunctions. A quantum system's general wave function is also a superposition (or linear combination) of eigenfunctions, akin to a guitar string. The terminology *wave function* stands for a superposition of eigenfunctions (we show the explicit time dependence next),

$$\Psi(\bar{r}, t) = \sum_n A_n e^{-\frac{iE_n t}{\hbar}} R_n(\bar{r})$$

A specific example involving only the two lowest eigenmodes $n = 1$ and 2 ($A_n = 0$ for $n \geq 3$) is

$$\Psi(\bar{r}, t) = A_1 e^{-\frac{iE_1 t}{\hbar}} R_1(\bar{r}) + A_2 e^{-\frac{iE_2 t}{\hbar}} R_2(\bar{r})$$

Note that $|A_1|^2 + |A_2|^2 = 1$ because the wave function $\Psi(\bar{r}, t)$ should be normalised because of Property 1 (this will be discussed further in Chapter 2). At $t = 0$, we have

$$\Psi(\bar{r}, t = 0) = A_1 R_1(\bar{r}) + A_2 R_2(\bar{r})$$

If one of the amplitudes is zero, say $A_1 = 0$ and $A_2 = 1$, we have

$$\Psi(\bar{r}, t = 0) = R_2(\bar{r})$$

The above equation says that the eigenfunction is a particular case of the wave function.

You will learn more about the properties of the wave function in Chapter 2.

Notation: We use the symbol (capital Psi), $\Psi(\bar{r}, t)$ to denote the time-dependent wave function when the Hamiltonian is either time-dependent or time-independent.

We use the symbols R, ψ, χ, and ϕ to denote the spatial part of the eigenfunction in the time-independent Schrödinger equation.

Summary of The Time-independent Schrödinger equation

To obtain the eigenfunction of a particle in a time-independent potential energy $U(\bar{r})$, we should solve the eigenvalue problem,

$$\left[-\frac{\hbar^2}{2m}\nabla^2 + U(\bar{r})\right] R(\bar{r}) = ER(\bar{r}).$$

(same as equations (1.42) and (1.101))

Each $U(\bar{r})$ results in a distinct set of eigenfunctions $R_n(r)$ and corresponding energy eigenvalues E_n. These eigenfunction and eigenenergy pairs can be written as:

$(R_1(\bar{r}), E_1), (R_2(\bar{r}), E_2), (R_3(\bar{r}), E_3), \ldots$ ad infinitum.

The above eigenvalue problem is often written as

$$\left[\frac{\hbar^2}{2m}\nabla^2 + U(\bar{r})\right] \psi_n(\bar{r}) = E_n \psi_n(\bar{r}).$$

(we have replaced $R_n(\bar{r})$ by $\psi_n(\bar{r})$)

The wave function of a particle in an energy eigenfunction is $e^{-\frac{iE_n t}{\hbar}} \psi_n(\bar{r})$.

If a particle is in energy level E_n, the **probability** of finding the particle in a small volume dv at location $\bar{r}$ is $|R_n(\bar{r})|^2 dv$, which is independent of time.

Two eigenfunctions $\psi_n(\bar{r})$ and $\psi_m(\bar{r})$, are *orthonormal* (orthogonal and normal):

$$\int dv \psi_n^*(\bar{r})\psi_m(\bar{r}) = \delta_{mn} = \begin{cases} 1 & if \ n=m \\ 0 & if \ n \neq m \end{cases}$$

(Continued)

(Continued)

Normal implies that the eigenfunctions are normalized: $\int dv |\psi_n(\bar{r})|^2 = 1$. That is the sum of the probabilities over all space is unity.

Orthogonal implies $\int dv \psi_n^*(\bar{r}) \psi_{m \neq n}(\bar{r}) = 0$, which means that eigenfunctions corresponding to two different quantum numbers $(n \neq m)$ are orthogonal.

More than one eigenfunction can correspond to the same energy eigenvalue. That is, orthogonal eigenfunctions $\psi_n(\bar{r})$ and $\psi_m(\bar{r})$ can have energy eigenvalues which obey $E_n = E_m$. That is, the energy eigenvalues are *degenerate*.

1.5   Solving Schrödinger Equation for Some Simple Potentials

We will discuss the solution of the time-independent Schrödinger equation for some simple examples. For the cases of the particle in a box (PiB) and the FP, we will derive the values for the energy eigenvalues (energy levels) and eigenfunctions. For the case of a simple harmonic oscillator (SHO), we will only discuss the solution.

1.5.1   *Particle in a box*

The one-dimensional PiB is a simple example that helps clarify the steps involved in solving Schrödinger equation:

- setting up the differential equation;
- using properties of the wave function and boundary conditions;
- quantum numbers;
- eigenfunctions; and
- energy levels allowed by the Hamiltonian.

One can think of the PiB as a particle of mass m bound in a finite one-dimensional box of length L. The particle can only move along

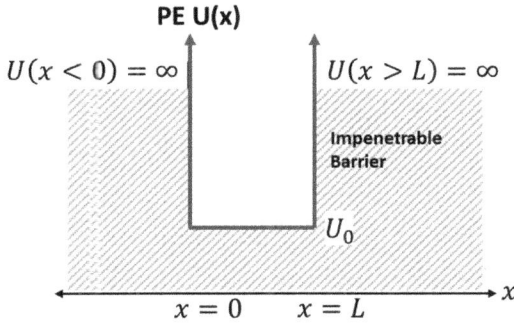

PE U(x)

$U(x < 0) = \infty$

$U(x > L) = \infty$

Impenetrable
Barrier

U_0

$x = 0$ $x = L$

x

Figure 1.8. Potential energy (PE) versus position for a PiB with length L. The particle can only move along the x-axis and cannot exist outside the range of the box.

the x-axis. The PE of the particle is defined by (see Figure 1.8)

$$\text{PE} = \begin{cases} \infty & x < 0 \\ U_o & 0 < x < L \\ \infty & x > L \end{cases} \tag{1.52}$$

The potential energy U_o is a constant.

The Schrödinger equation for Figure 1.7 is

$$\left[-\frac{\hbar^2}{2m}\frac{\partial^2}{\partial x^2} + U_o \right]\psi(x) = E\psi(x) \quad \text{for } 0 < x < L \tag{1.53}$$

$$\left[-\frac{\hbar^2}{2m}\frac{\partial^2}{\partial x^2} + \infty \right]\psi(x) = E\psi(x) \quad \text{for } x < 0 \text{ and } x > L \tag{1.54}$$

Solving equation (1.54) is straightforward because the PE is infinite for $x < 0$ and $x > L$. The probability of finding an electron in these regions must be zero

$$\psi(x) = 0 \quad \text{for } x < 0 \text{ and } x > L \tag{1.55}$$

We will now solve for the wave function in the region $0 < x < L$. Equation (1.53) can be rewritten as

$$-\frac{\hbar^2}{2m}\frac{\partial^2}{\partial x^2}\psi(x) = (E - U_o)\psi(x) \tag{1.56}$$

Multiplying the above equation by $\frac{2m}{\hbar^2}$, we get

$$-\frac{\partial^2}{\partial x^2}\psi(x) = k^2\psi(x) \tag{1.57}$$

where

$$k^2 = \frac{2m(E - U_o)}{\hbar^2} \tag{1.58}$$

The solution of equation (1.57) is either $\{e^{ikx} \text{ and } e^{-ikx}\}$ or $\{sin\,(kx)$ and $cos\,(kx)\}$, where k is the wave vector. The general solution is of the form

$$\psi(x) = Ce^{ikx} + De^{-ikx} \quad \text{for } 0 < x < L$$

Since $\sin\,(kx) = \frac{1}{2i}(e^{ikx} - e^{-ikx})$ and $\cos\,(kx) = \frac{1}{2}(e^{ikx} + e^{-ikx})$, we can rewrite the above equation as

$$\psi(x) = A\sin\,(kx) + B\cos\,(kx) \quad \text{for } 0 < x < L \tag{1.59}$$

We will now determine the three quantities A, B, and k in equation (1.59). To find these quantities, we use the given information and properties of the wave function discussed earlier. From equation (1.55), we can use the fact that the value of the wave function is zero at $x = 0$ and $x = L$ as boundary conditions. Again, this is because the probability of finding an electron outside the boundaries is zero. Furthermore, we can also use Property 2 of the eigenfunction (equation (1.28)), which says that the eigenfunction should be single valued at any point. From these two equations, we can conclude that the limiting value of the eigenfunction ψ should be zero when approaching the point $x = 0$ either from the right or left side. This condition of continuity can be mathematically written as

$$\lim_{x \to 0^+} \psi(x) = \lim_{x \to 0^-} \psi(x) \tag{1.60}$$

Using equations (1.28) and (1.55), we can rewrite equation (1.59) as

$$A\sin(0) + B\cos(0) = 0 \tag{1.61}$$

As $\sin(0) = 0$, the RHS of equation (1.61) can be zero only if $B = 0$. So, equation (1.59) becomes

$$\psi(x) = A\sin(kx) \quad \text{for } 0 < x < L \tag{1.62}$$

Now that we know the value of B, we should find the values of A and k to determine the eigenfunction. The value of k can be found from the continuity of the eigenfunction at $x = L$, which is given by

$$\lim_{x \to L^+} \psi(x) = \lim_{x \to L^-} \psi(x) \tag{1.63}$$

As the eigenfunction at $x = L^+$ is zero (equation (1.55)), we find from equation (1.63) that

$$A\sin(kL) = 0 \tag{1.64}$$

For equation (1.64) to be true, either $A = 0$ or $\sin(kL) = 0$. If we took $A = 0$, we get $\psi(x) = 0$ for $0 < x < L$. This combined with equation (1.55), which says that $\psi(x) = 0$ for $x <$ and $x > L$, would imply that the eigenfunction is zero everywhere. The solution $\psi(x) = 0$ over all space is a trivial solution, which is of no use. So, we conclude that

$$\sin(kL) = 0 \tag{1.65}$$

For $\sin(kL) = 0$ to be true, the following condition must be true:

$$kL = n\pi \quad \text{where } n = \text{integer} \tag{1.66}$$

That is, the only allowed values of wave vector k for a PiB of length L are

$$k = \frac{n\pi}{L} \tag{1.67}$$

Equation (1.67) implies that the allowed wave vector values are quantized. Note that the entire continuum of possible wave vector values is not allowed, like we will see in the case of a FP (Section 1.5.3). Now that we've defined the value of k in equation (1.67), substituting it into equation (1.62), we get

$$\psi_n(x) = A\sin\left(\frac{n\pi}{L}x\right) \quad \text{for } 0 < x < L \tag{1.68}$$

$$\psi_n(x) = 0 \quad \text{for } x < 0 \text{ and } x > L \tag{1.69}$$

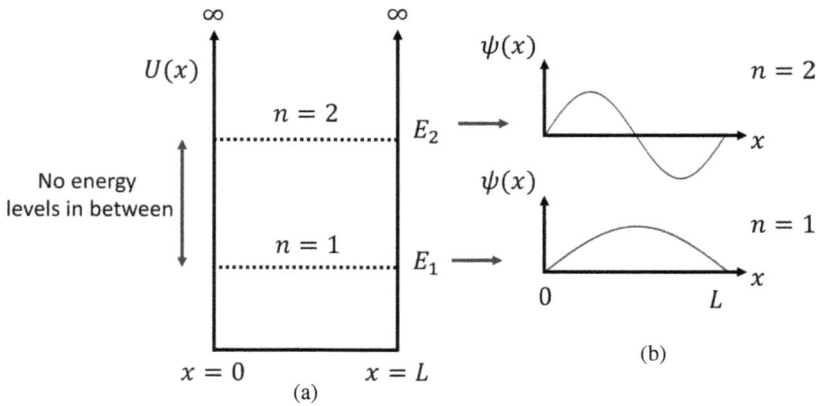

Figure 1.9. (a) The discrete energy levels corresponding to $n = 1$ and $n = 2$ are shown. We have assumed that $U_o = 0$. (b) The lowest two energy eigenfunctions, ψ_1 and ψ_2 corresponding to $n = 1$ and $n = 2$ are shown. ψ_1 has nodes at $x = 0$ and L and ψ_2 has nodes at $x = 0$, $L/2$ and L.

where subscript n is added to ψ to make clear that many values of the quantum number are possible. $n = 1, 2, 3, \ldots$ **are referred to as the quantum numbers of the PiB**. The solution of the eigenfunction given by equation (1.68) corresponds to a standing wave inside the PiB. See Figure 1.9. Note that $n = 0$ does not make sense because the eigenfunction will be zero as per equation (1.68).

The electron/particle in the box behaves like an electromagnetic wave in a waveguide or cavity. When the boundary is fully reflective, the electromagnetic wave's amplitude goes to zero at the boundary. Another analogy is the waves corresponding to allowed fundamental modes in a guitar string, where the amplitude of vibration is zero at the two clamped ends of the string.

To find the value of A, we use the normalization condition (Property 1, of eigenfunctions discussed earlier in equation (1.27):

$$\int_{-\infty}^{+\infty} dx \, |\psi_n(x)|^2 = 1 \tag{1.70}$$

Using the wave function from equations (1.68) and (1.69) in equation (1.70) gives

$$\int_0^L dx\, |A|^2 \sin^2\left(\frac{n\pi}{L}x\right) = 1 \tag{1.71}$$

Noting that the value of A is independent of position x and using $\int_0^L dx \sin^2\left(\frac{n\pi}{L}x\right) = \frac{L}{2}$ (independent of the value of n) in equation (1.71), we find that

$$A = \sqrt{\frac{2}{L}} \tag{1.72}$$

Substituting the value of A we found in equation (1.72) into equation (1.68),

$$\psi_n(x) = \sqrt{\frac{2}{L}}\sin\left(\frac{n\pi}{L}x\right) \quad \text{for } 0 < x < L \tag{1.73}$$

$$\psi_n(x) = 0 \quad \text{for } x < 0 \text{ and } x > L \tag{1.74}$$

Substituting the value of k in equation (1.67) into equation (1.58) gives us the energy level corresponding to quantum number n:

$$E_n = \frac{\hbar^2}{2m}\left(\frac{n\pi}{L}\right)^2 + U_o \tag{1.75}$$

A comparison between a quantum and classical PiB is discussed in Table 1.2.

Connection to Property 3 of the wave function: Equations (1.73) and (1.74) represent an example where the derivative of the wave function is discontinuous at $x = 0$ because the potential is infinity for $x < 0$ and finite for $x > 0$:

$$\frac{d\psi(x = 0^-)}{dx} = 0$$

$$\frac{d\psi(x = 0^+)}{dx} = \frac{n\pi}{L}\sqrt{\frac{2}{L}}$$

Table 1.2. Comparison of a quantum and classical PiB. For the quantum case, we have only considered eigenfunctions and not arbitrary superpositions.

Property	PiB: Quantum	PiB: Classical
Lowest energy	$\dfrac{\hbar^2}{2m}\left(\dfrac{\pi}{L}\right)^2 + U_o,$	U_o
Energy levels	Quantized, $$E_n = \frac{\hbar^2}{2m}\left(\frac{n\pi}{L}\right)^2 + U_o$$	Continuum, $U_o < E < \infty$
Energy level spacing between consecutive energy levels	$E_{n+1} - E_n = \dfrac{\hbar^2}{2m}\left(\dfrac{\pi}{L}\right)^2 (2n+1)$	Tends to zero
The probability of finding the particle at location x	Nonuniform, $$\|\psi_n\|^2\,dx = \frac{2}{L}\sin^2\left(\frac{n\pi}{L}x\right) dx$$ Zero when $\sin\left(\dfrac{n\pi}{L}x\right) = 0$	Uniform, $\dfrac{dx}{L}$

Concept: A quantum PiB behaves differently from a classical PiB: (a) The probability of finding a particle in space is non-uniform. (b) The lowest energy level is nonzero. (c) The energy eigenvalues form a discrete spectrum.

1.5.2   *Simple harmonic oscillator*

The one-dimensional SHO is a classic problem in quantum mechanics that has a large number of applications including phonons or atomic vibrations in a solid. The potential energy corresponds to a mass m attached to a spring with a spring-constant $\mathcal{K}$. The PE of the system is given by

$$\text{PE} = \frac{1}{2}\mathcal{K}x^2 \quad -\infty < x < +\infty \qquad (1.76)$$

Schrödinger equation can then be written as

$$\left[-\frac{\hbar^2}{2m}\frac{\partial^2}{\partial x^2} + \frac{1}{2}\mathcal{K}x^2\right]\psi(x) = E\psi(x) \quad -\infty < x < +\infty \qquad (1.77)$$

Here, we will only summarize the solution for the SHO. The eigenenergy levels are

$$E_n = \left(n + \frac{1}{2}\right)\hbar\omega \tag{1.78}$$

where the quantum numbers are

$$n = 0, 1, 2, \ldots \tag{1.79}$$

The angular frequency ω is related to the spring constant $\mathcal{K}$ and the mass m via the usual classical expression,

$$\omega = \sqrt{\frac{\mathcal{K}}{m}} \tag{1.80}$$

For a more detailed solution of the differential equation and derivation of the eigenfunction of the SHO, the reader may consult (Shankar, 2008). The eigenfunction is

$$\psi_n(x) = \frac{1}{\sqrt{2^n n!}}\left(\frac{m\omega}{\pi\hbar}\right)^{1/4} e^{-\frac{m\omega x^2}{2\hbar}} H_n\left(\sqrt{\frac{m\omega}{\hbar}}x\right) \tag{1.81}$$

where n has been defined in equation (1.79). The function $H_n\left(\sqrt{\frac{m\omega}{\hbar}}x\right)$ are called Hermite polynomials and are given by $\left(y = \sqrt{\frac{m\omega}{\hbar}}x\right)$:

$$H_n(y) = (-1)^n e^{y^2}\frac{d^n}{dy^n}(e^{-y^2}) \tag{1.82}$$

Figure 1.10 shows the energy levels and eigenfunctions. The horizontal lines correspond to different energy eigenvalues, which are equally spaced by $\hbar\omega$. The eigenfunctions are vertically shifted to make each one visible. The eigenfunctions (equation (1.81)) corresponding to quantum number n have n nodes, excluding the exponentially approaching nodes at $x \to -\infty$ and $x \to +\infty$. In contrast to the infinite wall in the PiB example, the quadratic potential is smooth. Therefore, the first derivative of eigenfunctions is continuous at all points in space, and property 3 of the wave function holds at all points. Note that the wave function is not zero outside the boundary of the quadratic potential. Look at Table 1.3 for a comparison of a quantum and classical SHO.

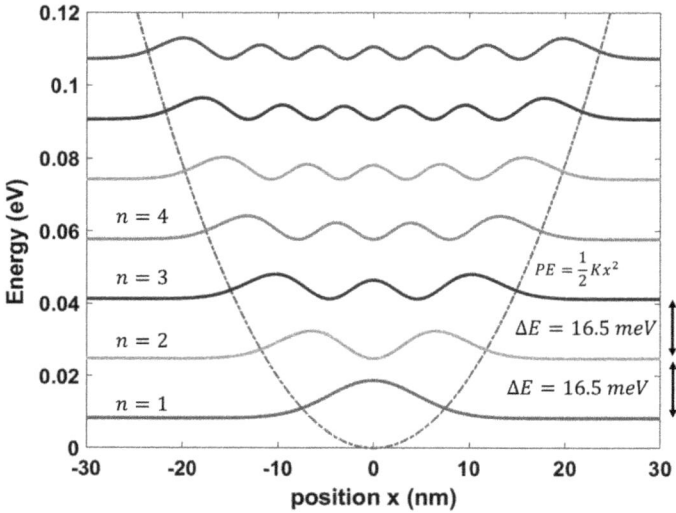

Figure 1.10. Eigenfunction and energy eigenvalues of a quantum simple harmonic oscillator. The spring constant and mass are $\mathcal{K} = 2\pi \times 10^{-4}$ N/m and $m = 10 \times 10^{-31}$ kg.

Table 1.3. Comparison of quantum and classical Simple Harmonic Oscillator.

Property	Quantum	Classical
Lowest Energy possible (called zero-point energy)	$\frac{1}{2}\hbar\omega$	Zero
Allowed energy levels	Discrete, $E_n = \left(n + \frac{1}{2}\right)\hbar\omega$	Continuum
Energy level spacing between consecutive energy levels	$E_{n+1} - E_n = \hbar\omega$ $\omega = \sqrt{\mathcal{K}/m}$	Tends to zero
Probability of finding the particle outside $-\sqrt{\dfrac{2E_n}{k}} < x < +\sqrt{\dfrac{2E_n}{k}}$	Nonzero	Zero

Example 3. Show that the eigenfunctions of SHO are normalized.

Solution: We have to show that $\int_{-\infty}^{+\infty} dx \, |\psi_n(x)|^2 = 1$

$$\int_{-\infty}^{+\infty} dx \left| \frac{1}{\sqrt{2^n n!}} \left(\frac{m\omega}{\pi\hbar} \right)^{1/4} e^{-\frac{m\omega x^2}{2\hbar}} H_n \left(\sqrt{\frac{m\omega}{\hbar}} x \right) \right|^2$$

$$= \frac{1}{2^n n!} \left(\frac{m\omega}{\pi\hbar} \right)^{1/2} \int_{-\infty}^{+\infty} dx e^{-\frac{m\omega x^2}{\hbar}} \left| H_n \left(\sqrt{\frac{m\omega}{\hbar}} x \right) \right|^2$$

We assume $z = \sqrt{\frac{m\omega}{\hbar}} x$, and use the following property of Hermite polynomials:

$$\int_{-\infty}^{+\infty} dz e^{-z^2} |H_n(z)|^2 = 2^n n!$$

Then by changing the variable of x to z, the earlier integral becomes

$$\int_{-\infty}^{+\infty} dx \, |\psi_n(x)|^2 = \frac{1}{2^n n!} \int_{-\infty}^{+\infty} dz e^{-z^2} |H_n(z)|^2 = \frac{2^n n!}{2^n n!} = 1$$

Concept: A quantum SHO behaves differently from a classical SHO: (a) The probability of finding a particle in space is non uniform. (b) The lowest energy level is nonzero. (c) The energy eigenvalues form a discrete spectrum.

1.5.3 *Free particle*

We now consider the case of a FP which moves in a one-dimensional space with a constant potential,

$$PE = U(r) = U_o = \text{constant} \quad \text{for} \ -\infty < x < +\infty \qquad (1.83)$$

Schrödinger equation for a FP is

$$\left[-\frac{\hbar^2}{2m} \frac{\partial^2}{\partial x^2} + U_o \right] \psi(x) = E\psi(x) \quad \text{for all } x \qquad (1.84)$$

Equation (1.84) can be rewritten as

$$-\frac{\hbar^2}{2m}\frac{\partial^2}{\partial x^2}\psi\,(x) = (E - U_o)\psi\,(x) \tag{1.85}$$

Multiplying by $\frac{2m}{\hbar^2}$, the above equation can be written in a more condensed form

$$-\frac{\partial^2}{\partial x^2}\psi(x) = k^2\psi(x) \tag{1.86}$$

where

$$k^2 = \frac{2m(E - U_o)}{\hbar^2} \tag{1.87}$$

Like in the case of a PiB, we note that the solution of equation (1.86) can be written either as $\{e^{ikx}$ and $e^{-ikx}\}$ or $\{\sin(kx)$ and $\cos(kx)\}$. For the FP, we choose the set of solutions $\{e^{ikx}$ and $e^{-ikx}\}$ because they are traveling waves,

$$\psi_k(x) = A_k\,e^{ikx} \tag{1.88}$$

For energy $E \geq U_o$, equation (1.88) satisfies equation (1.86) for all real values of wave vector k

$$-\infty < k < +\infty.$$

k, the quantum number, is the FP's wave vector. In contrast to the PiB, where the quantized value of the wave vector is $\frac{n\pi}{L}$, a FP has a continuous distribution of wave vectors. The energy eigenvalues of the FP from equation (1.87) are given by

$$E_k = \frac{\hbar^2 k^2}{2m} + U_o \quad \text{where } -\infty < k < +\infty \tag{1.89}$$

The values of $k < 0$ correspond to traveling waves going from right to left and the values of $k > 0$ correspond to traveling waves going from left to right.

At this point, let us pause and write down the time-dependent wave functions discussed in equation (1.49). For $k > 0$, we have, $\psi_k = e^{i(kx-\omega t)}$ and for $k < 0$, we have $e^{-i(kx+\omega t)}$, where $\omega = E_k/\hbar$. The reader should note that the arguments of these wave $kx - \omega t$ and $kx + \omega t$ correspond to write and left propagating waves. A common point of confusion is why we did not write down the solutions in equation (1.88) simply as $\sin(kx)$ and $\cos(kx)$. Well, we could have done this too and it is equally valid. Note that a sum of the left and right propagating waves,

$$\psi_k(x) = A_k\, e^{ikx} + B_k\, e^{-ikx} \quad (k \geq 0) \tag{1.90}$$

satisfies Schrödinger equation. The above equation can also be written as a sum of sine and cosine functions using Euler's identity as

$$\psi_k(x) = C_k \sin(kx) + D_k \cos(kx)$$

where $C_k = i(A_k - B_k)$ and $D_k = A_k + B_k$

Normalization of the eigenfunction: The eigenfunction in equation (1.88) is not normalized yet. We know from Property 1 of the wave function (equation (1.27)) that it must be normalized. Normalizing it over all space (we have dropped the subscript k), we get

$$\int_{-\infty}^{+\infty} dx\, |\psi(x)|^2 = |A|^2 \int_{-\infty}^{+\infty} dx \left|e^{ikx}\right|^2 = 1 \tag{1.91}$$

The quantity in the integrand $\left|e^{ikx}\right|^2$ is equal to one. So, the above equation is

$$|A|^2\, \infty = 1 \rightarrow |A|^2 \rightarrow \frac{1}{\infty} \tag{1.92}$$

Substituting this into equation (1.88), we have

$$\psi(x) \rightarrow 0 \tag{1.93}$$

This means that the eigenfunction of a free particle is spread all over all space. So, the probability of finding the particle in a small volume at any location is vanishingly small.

In calculations using the eigenfunction (say to find the energy, momentum, and so on), it does not make sense to use $\psi(x) \to 0$. To circumvent this problem, assume an imaginary box of length L, in which the eigenfunction is normalized. Then, set the length L equal to infinity at the end of the calculation (after performing all the integrals). The quantities calculated should not depend on the length L at the end of the analysis. For simplicity, we will take the box to lie from $-L/2$ to $+L/2$. Then, equation (1.91) gives

$$\int_{-L/2}^{+L/2} dx\, |\psi(x)|^2 = |A|^2 \int_{-L/2}^{+L/2} dx\, \left| e^{ikx} \right|^2$$

$$= |A|^2 \int_{-L/2}^{+L/2} dx = |A|^2 L = 1 \qquad (1.94)$$

or

$$A \to \frac{1}{\sqrt{L}} e^{i\phi} \qquad (1.95)$$

where ϕ is an arbitrary phase factor. Substituting equation (1.95) into equation (1.88)

$$\psi(x) = \frac{e^{i\phi}}{\sqrt{L}} e^{ikx} \qquad (1.96)$$

Do not forget to set $L = \infty$ at the end of the calculation.

The probability of finding a particle at a given location and the values of the allowed energy eigenvalues are the same as in the classical case as follows:

- The spacing between two consecutive energy levels tends to zero as k is a continuous variable, and the energy eigenvalues can take any value $E \geq U_o$.
- The probability of finding the particle is the same at all locations (independent of x).

What is the momentum carried by an FP? We now show that the momentum operator's definition is consistent with equation (1.88), which corresponds to an FP with momentum $\hbar k$. The momentum operator (defined in Chapter 3) is $\hat{p} = \frac{\hbar}{i}\nabla$. Applying the momentum operator on equation (1.88) gives us

$$\hat{p}\psi = \frac{\hbar}{i}\nabla\psi = \frac{\hbar}{i}\frac{\partial}{\partial x}\left(A\,e^{ikx}\right) = \hbar k\left(A\,e^{ikx}\right) = \hbar k\psi \qquad (1.97)$$

So $\psi = A\,e^{ikx}$ is indeed the eigenfunction of the momentum, and its eigenvalue is $\hbar k$ for any real k. We also know that the Hamiltonian operating on ψ gives the energy of the FP (E_k) multiplied by ψ. So, $A\,e^{ikx}$ is an eigenfunction of both energy and momentum operators in the case of an FP.

Discrete and continuous quantum numbers: It is important to remember that the quantum number is a general term used to define the solution of the Schrödinger equation and can represent both discrete and continuous numbers. For a PiB and SHO, the quantum number n forms a discrete spectrum while for a FP, the quantum number k forms a continuous spectrum. Note that a single Hamiltonian can also have both discrete and continuous eigenvalues (see Problem 11).

Concept: A free particle: (a) is equally likely to exist at all points in space. (b) the lowest energy level is zero like in the classical case. (c) the energy eigenvalues form a continuous spectrum like in the classical case.

1.6 Solving The 1D Schrödinger Equation on a Computer: Discretized Version

So far, we have considered only two simple cases, the PiB and FP, which have analytical solutions for the eigenfunctions and energy eigenvalues. A typical engineering problem has a more complicated potential energy function for which an analytical solution may not exist. Here, we will describe the process of solving the Schrödinger

$$a$$

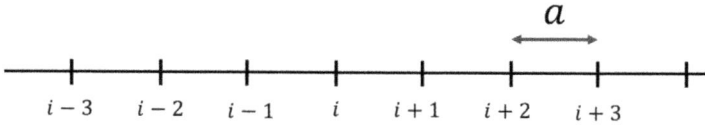

Figure 1.11. The x-axis is discretized into grid points numbered i, $i+1$, and so on. The distance between neighboring grid points is a. Note that i is the site index (grid point).

equation for finite systems with arbitrary potentials on the computer. As a result, we will be able to handle complicated potential profiles encountered in problems involving devices, molecules, etc. Our procedure starts with the Schrödinger equation and discretizes it using the finite difference (FD) method. After discretization, the resulting set of linear equations gives the corresponding energy eigenvalues and eigenfunctions.

The first step is to discretize the x-axis, as shown in Figure 1.11. We will choose a discretization scheme where the distance between any two neighboring grid points is a.

Schrödinger equation is

$$\left[-\frac{\hbar^2}{2m} \frac{\partial^2}{\partial x^2} + U(x) \right] \psi(x) = E\psi(x) \tag{1.98}$$

The first derivative at grid point i is

$$\left. \frac{\partial \psi}{\partial x} \right|_{\text{grid point } i} = \frac{\psi(i+1) - \psi(i-1)}{2a} \tag{1.99}$$

where $2a$ is the distance between grid points $i - 1$ and $i + 1$. The second derivative at grid point i can be written as

$$\left. \frac{\partial^2 \psi}{\partial x^2} \right|_{\text{grid point } i} = \left. \frac{\partial}{\partial x} \frac{\partial \psi}{\partial x} \right|_{\text{grid point } i} = \frac{\left. \frac{\partial \psi}{\partial x} \right|_{\text{grid point } i+\frac{1}{2}} - \left. \frac{\partial \psi}{\partial x} \right|_{\text{grid point } i-\frac{1}{2}}}{a}$$

$$\tag{1.100}$$

But are there values for grid points at $i - \frac{1}{2}$ and $+\frac{1}{2}$? There are not, but these are only imaginary points to help us calculate the second derivative! We are not going to store any values at $i - \frac{1}{2}$ and $i + \frac{1}{2}$.

The first derivatives at these imaginary points are

$$\left.\frac{\partial\psi}{\partial x}\right|_{\text{grid point } i+\frac{1}{2}} = \frac{\psi(i+1) - \psi(i)}{a} \tag{1.101}$$

and

$$\left.\frac{\partial\psi}{\partial x}\right|_{\text{grid point } i-\frac{1}{2}} = \frac{\psi(i) - \psi(i-1)}{a} \tag{1.102}$$

Substituting equations (1.101) and (1.102) into equation (1.100), the second derivative at grid point i can be written as

$$\left.\frac{\partial^2\psi}{\partial x^2}\right|_{\text{grid point } i} = \left.\frac{\partial}{\partial x}\frac{\partial\psi}{\partial x}\right|_{\text{grid point } i} = \frac{\psi(i+1) - 2\psi(i) + \psi(i-1)}{a^2} \tag{1.103}$$

To discretize Schrödinger equation, we first note that the second derivative $\frac{\partial^2\psi}{\partial x^2}$ is

$$-\frac{\hbar^2}{2m}\frac{\partial^2}{\partial x^2}\psi(x) = -\frac{\hbar^2}{2ma^2}[\psi(i+1) + \psi(i-1) - 2\psi(i)] \tag{1.104}$$

The second term on the LHS of Schrödinger equation is

$$U(x)\psi(x) = U(i)\psi(i) \tag{1.105}$$

where $U(i)$ is the potential evaluated at grid point i. The RHS of Schrödinger equation,

$$E\psi(x) = E\psi(i) \tag{1.106}$$

Substituting equations (1.104), (1.105), and (1.106) into equation (1.98) gives us the linear algebraic equation

$$t\psi(i-1) - \varepsilon(i)\psi(i) + t\psi(i+1) = E\psi(i) \tag{1.107}$$

which is valid for all grid points i. In equation (1.107),

$$t = -\frac{\hbar^2}{2ma^2} \tag{1.108}$$

and

$$\varepsilon(i) = -2t + U(i) \tag{1.109}$$

If our system uses N grid points for discretization, the values of i vary from $i = 1$ to $i = N$, in equation (1.107). One is then left wondering what the form of equation (1.107) is for $i = 1$ which has a term $\psi(0)$, and $i = N$ which has a term $\psi(N+1)$. Note that grid points $i = 0$ and $i = N+1$ are not present in the above discretization scheme. This is physically equivalent to terminating the system with infinite potential energy to the left of $i = 1$ and right of $i = N$ by setting $\psi(0) = \psi(N+1) = 0$ in equation (1.107). The system of linear algebraic equations with all quantities defined can now be written as

$$\varepsilon(1)\psi(1) + t\psi(2) = E\psi(1) \tag{1.110}$$

$$t\psi(i-1) + \varepsilon(i)\psi(i) + t\psi(i+1) = E\psi(i) \text{ for } 2 \leq i \leq N-1 \tag{1.111}$$

$$t\psi(N-1) + \varepsilon(N)\psi(N) = E\psi(N)$$

We can write equation (1.111) in matrix form as

$$H\psi = E\psi \tag{1.112}$$

where

$$H = \begin{pmatrix} \varepsilon(1) & t & 0 & . & . & . & . & . & 0 \\ t & \varepsilon(2) & t & 0 & & & & & 0 \\ 0 & t & \varepsilon(3) & t & 0 & & & & 0 \\ . & & & & & . & & & . \\ . & & & & & . & & & . \\ . & & & & & . & & & . \\ . & & & & & . & & & . \\ 0 & . & . & . & . & 0 & t & \varepsilon(N-1) & t \\ 0 & . & . & . & . & . & & t & \varepsilon(N) \end{pmatrix}$$

$$\text{and} \quad \psi = \begin{bmatrix} \psi(1) \\ \psi(2) \\ . \\ . \\ . \\ \psi(N) \end{bmatrix} \tag{1.113}$$

The eigenfunctions ψ and energy values E result from solving equation (1.112), which is an eigenvalue problem, for which there are standard methods available in most numerical packages. The summary of the above procedure is as follows:

- Solving $H\psi = E\psi$ over N grid points corresponds to solving N linear equations where the matrices H and ψ are shown in equation (1.113).
- There are N eigenvalues and eigenfunctions represented by (E_p, ψ_p), where $p = \{1, 2, 3, \ldots, N\}$ is the quantum number corresponding to the discrete energy values.
- Notation wise, $\psi_p(i)$ is the value of the eigenfunction represented by quantum number p at grid point i.
- The eigenfunctions obey $\psi_p^\dagger \psi_m = \delta_{pm}$, which is the condition for normalization. Note that $\psi_p^\dagger$ is the conjugate and transpose of ψ_p. p and m are quantum numbers.
- Plotting each ψ_p vector versus $i = 1, 2, 3, \ldots, N$ shows the shape of the eigenfunction.
- The probability density is found by plotting $|\psi_p|^2$ versus $i = 1, 2, 3, \ldots, N$.

Summary of computational procedure is as follows:

(1) Discretize Schrödinger equation:
   a. Choose the grid spacing a
   b. Find the values of $U(i)$ for all grid points i
   c. Evaluate $\varepsilon(i)$ for all grid points and t
   d. Construct the matrix H in equation (1.112) according to equation (1.113).

(2) Solve the eigenvalue problem in equation (1.112) to obtain energy eigenvalues and eigenfunctions.
(3) Ensure that the value of grid spacing a is small enough to avoid large numerical errors due to FD approximations. One way of doing this is to make the grid spacing half of the earlier value to ensure that no significant changes in energy levels occur.

Example 4. A charged particle exists in a one-dimensional potential with a constant electric field $E_0 = 0.5 \, \text{V}/\mu\text{m}$ along the x-axis, between $x = 0$ and $x = 1$ nm. Outside of this region, the potential is infinite. Assume that the particle is 10,000 times heavier than a free electron and has a charge equal to an electron $(-1.602 \times 10^{-19}$ C$)$. Computationally, find the eigenenergies and eigenfunctions of an electron in this tilted potential well.

Solution: Following the above computational procedure, we build a Hamiltonian matrix after discretizing the box $(0 < x < L)$ with 400 grid points. The first contribution to Hamiltonian is the discretized version of $-\frac{\hbar^2}{2m}\frac{\partial^2}{\partial x^2}$. The potential energy $V(x) = -qEx$ is the second contribution to the Hamiltonian. The Hamiltonian is a 400×400 matrix and is easily diagonalizable with a linear algebra package.

The probability densities for the five eigenfunctions in this box with a nonuniform potential is shown in Figure 1.12. The titled potential energy profile due to $U(x) = -qEx$ increases from 0 meV

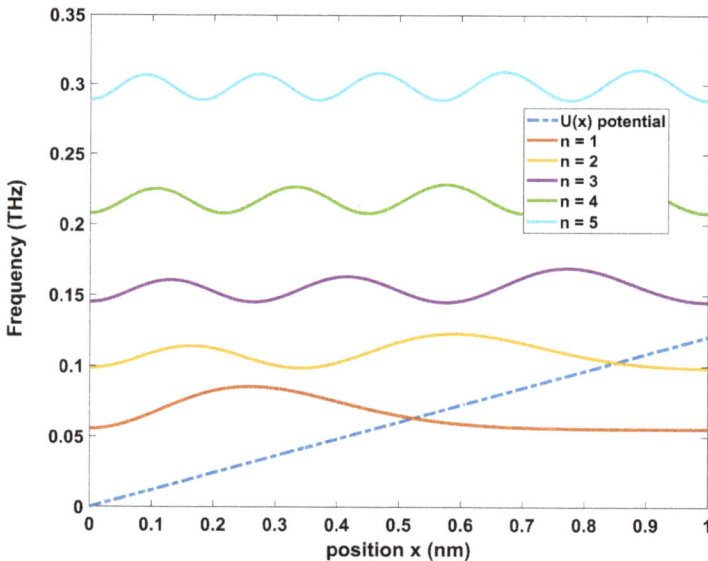

Figure 1.12. This composite figure shows both the first five energy eigenvalues and the corresponding wave functions. The plots are shifted by their eigenenergy for visibility. The dashed blue line is the potential energy. To convert the y-axis units to meV, multiply the frequency by $\frac{h}{e} = 4.136 \times 10^{-15}$. Wave functions are multiplied by 100 for visibility.

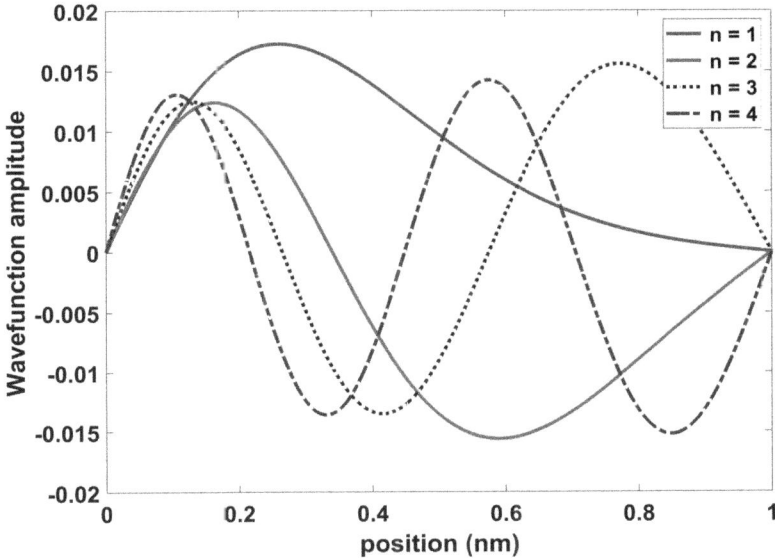

Figure 1.13. A close-up view of the lowest four eigenfunctions. The $n = 1$, $n = 2$, $n = 3$, and $n = 4$ eigenfunctions have $(n + 1)$ nodes. However, the wave functions are no longer symmetric or antisymmetric about the box's center as in the PiB of Section 1.5.1. Note that the $n = 4$ eigenfunction hardly feels the titled potential and resembles that of a PiB.

$(f = 0\,\text{THz})$ at $x = 0$ to $0.496\,\text{meV}$ $(f = 0.12\,\text{THz})$ at $x = 1\,\text{nm}$. Note that the eigenfunctions have $(n + 1)$ nodes as in the PiB case, but the eigenfunctions are no longer perfect even or odd functions (about the center of the box), as in the case of the symmetric PiB in Section 1.5.1.

As can be seen for $n = 1, 2$, and 3, the particle exists primarily inside the tilted potential well around $x = 0$. For higher quantum numbers $(n = 4$ and $n = 5)$, the particle's energy is much higher than $0.12\,\text{THz}$, and the particle almost behaves as if it is inside a PiB discussed earlier. Figure 1.13 shows the wave function amplitudes of the first four eigenfunctions.

The frequency corresponding to the eigenenergy values are $E_1 = 0.055\,\text{THz}$, $E_2 = 0.098\,\text{THz}$, $E_3 = 0.144\,\text{THz}$, $E_4 = 0.206\,\text{THz}$, $E_5 = 0.287\,\text{THz}$.

1.7 Problems

Section 1.1 and Some Fundamentals

(1) How many electron volts (eV) make one Joule?

(2) If an electron moves from the positive to the negative terminal of a 1.5 V battery, how much does the potential energy of the electron change?

(3) Consider a photon with a wavelength of 500 nm. What is its energy in Joules and eV?

(4) You have a source (say a light emitter) that emits photons of wavelength 500 nm. What is the minimum energy that can be received by us in a time interval of 1 s? (Exclude the trivial answer of zero.)

(5) How fast should a tennis ball travel so that its wavelength is 1 Å? (Assume Newtonian mechanics. That is, neglect that particles cannot travel with a speed larger than the speed of light.) What quantity is this wave in?

(6) This problem introduces the student to the concept of *excitons*, a bound electron–hole pair. Consider a semiconductor where an electron has been excited from the valence to the conduction band. The excitation leaves behind a hole in the valence band. As the electron and the hole have opposite charges, they form a bound system like the hydrogen atom. That is, the electron goes around the hole due to their mutual electrical attraction. This bound system is called an exciton. For simplicity, assume that the hole is infinitely heavy, and the electron has a mass of 9.1×10^{-31} kg. Note that in reality holes have a finite mass just like the electron and cannot be assumed to be at rest. But we will make this approximation anyway.

Using an analysis like Bohr's model for an atom, estimate the lowest energy level of the electron by assuming excitons in technologically important semiconductors: (a) silicon, (b) GaAs, and (c) germanium. Assume the correct dielectric constant of the material but for simplicity, assume the rest mass of the electron.

Section 1.2

(7) $\psi(x)$ is the eigenfunction of the one-dimensional Schrödinger equation. What are the (a) units of the eigenfunction, (b) units of $|\psi(x)|^2$, and (c) meaning of $|\psi(x)|^2 dx$?

Section 1.4

(8) Consider a time-independent Hamiltonian. If $\Psi(\bar{r}, t_o)$ is the wave function at time $t = t_o$, show that $\Psi(\bar{r}, t) = e^{-i\frac{\hat{H}(t-t_o)}{\hbar}}\Psi(\bar{r}, t_o)$ is a valid solution of the Schrödinger equation, $i\hbar\frac{\partial \Psi(\bar{r}, t)}{\partial t} = \hat{H}\Psi(\bar{r}, t)$.

Section 1.5

(9) The effective mass of an electron in GaAs (Gallium Arsenide) is $0.067 \times 9.1 \times 10^{-31}$ kg. Design a one-dimensional GaAs PiB such that when you put an electron in the $n = 2$ energy level, it will emit green light of wavelength of 550 nm while transitioning to the $n = 1$ energy level.

(10) Consider a potential: $U(x) = \begin{cases} \infty & \text{for } x < 0 \\ 0 & \text{for } x > 0 \end{cases}$

(a) Derive expressions for the eigenfunctions and energy eigenvalues. You do not have to normalize the eigenfunctions. (Hint: Determine the boundary condition that you will use at $x = 0$.)

(b) Plot one example of the eigenfunction.

(11) Consider a potential (take $L = 8$ nm):

$$U(x) = \begin{cases} U_o = 400 \text{ meV} & \text{for } x < 0 \text{ and } x > L \\ 0 & \text{for } 0 < x < L \end{cases}$$

(a) By solving Schrödinger equation, find the lowest eigenvalue (energy level) and the corresponding eigenfunction.

(b) Plot the eigenfunction using a plotting package. Ensure that you present a zoomed-in plot of the eigenfunction for $x < 0$ and $x > L$.

(c) Show that the eigenvalues form a discrete spectrum for energies less than U_o and a continuous spectrum for higher energies. That is, for this Hamiltonian, there are both discrete quantum numbers with eigenenergies smaller than U_o and a continuum of quantum numbers with eigenenergies larger than U_o.

(12) Consider an electron constrained to lie on a purely one-dimensional ring with a circumference of L. The potential energy on the one-dimensional ring is zero. (This is similar to the PiB, but the electron is in a ring.)

(a) Derive an expression for the energy eigenvalues and eigenfunctions. Discuss the quantum numbers by including both a discussion of their sign and magnitude. Do these quantum numbers mean anything physical?

(b) What are the energy levels and eigenfunctions for $L = 8\,\text{nm}$?

(c) Compare the energy levels and eigenfunctions of this problem with a PiB, where the box length (L) is $8\,\text{nm}$.

(Hint: Pay attention to finding the correct boundary conditions. Using circular coordinates might make it easier. A particle in a ring classically will travel either clockwise or anticlockwise, so the eigenfunctions could potentially be clockwise or anticlockwise propagating waves in a ring.)

(13) Consider a three-dimensional problem with the potential energy:

$$U(x, y, z) = \begin{cases} \infty & \text{for } x < 0 \\ \infty & \text{for } x > L \\ 0 & \text{for } 0 < x < L \end{cases}$$

For the three-dimensional problem, the Hamiltonian is

$$H = -\frac{\hbar^2}{2m}\left(\frac{\partial^2}{\partial x^2} + \frac{\partial^2}{\partial y^2} + \frac{\partial^2}{\partial z^2}\right) + U(x, y, z)$$

Show that the eigenfunction

$$\psi_{n,ky,kz} = \begin{cases} A\sin\left(\dfrac{n\pi x}{L}\right) e^{ik_y y} e^{ik_z z} & \text{for } 0 < x < L \\ 0 & \text{outside of } 0 < x < L \end{cases}$$

satisfies the Schrödinger equation. What are the energy eigenvalues?

(14) Given a potential barrier $U = U_o\,\delta(x)$ (assume that U_o is a number and $\delta(x)$ is the Dirac delta function),

(a) Is the eigenfunction continuous across $x = 0$? What is the physical reason for your answer?
(b) Is the derivative of the eigenfunction continuous across $x = 0$? If not, derive a relationship between the derivative of the eigenfunction at $x = 0^+$ and $x = 0^-$.

Section 1.6

(15) Solve this problem numerically by discretizing the Schrödinger equation. Consider a potential:

$$U(x) = \begin{cases} 400\,\text{meV} & \text{for } x < 0 \text{ and } x > L \\ 0 & \text{for } 0 < x < L \end{cases}$$

Take $L = 8\,\text{nm}$. In solving this problem make sure to include at least $10\,\text{nm}$ to the left of $x = 0$ and to the right of $x = L$.

Find the lowest two eigenvalues by discretizing Schrödinger equation. Plot the corresponding eigenfunctions. What happens if you make the grid spacing two times smaller? How does your answer compare to Problem 11?

(16) Consider a **double quantum well**, where the wells are made of GaAs, and the barriers are made of AlGaAs, as shown in the following figure. (In solving the problem, you will only need the potential profile and mass of an electron in GaAs and AlGaAs. These values are supplied below.)

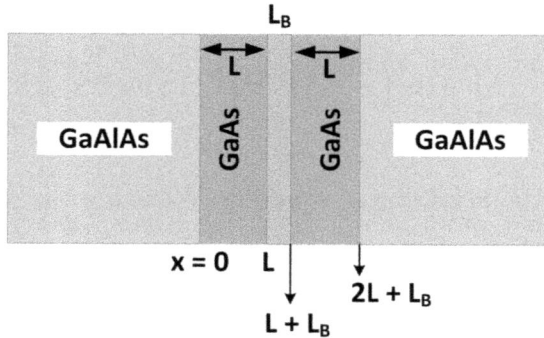

Figure 1.14.

The potential energy is defined by (units meV):

$$U(x) = \begin{cases} 300 & x < 0 \\ 300 & L < x < L + L_B \\ 300 & x > 2L + L_B \\ 0 & 0 < x < L \\ 0 & L + L_B < x < 2L + L_B \end{cases}$$

Take $L = 120\,\text{Å}$ and $L_B = 20\,\text{Å}$. Assume that the mass of an electron in both GaAs and AlGaAs is $0.067 \times 9.1 \times 10^{-31}\,\text{kg}$.

(a) What are the lowest two eigenvalues? Plot the eigenfunctions corresponding to these eigenvalues. Are these eigenfunctions symmetric or antisymmetric about the center of the structure?

(b) What happens to the energy level separation in (a) if L_B is $10\,\text{Å}$?

(c) Why the energy level difference between the lowest two eigenvalues is larger for $L_B = 10\,\text{Å}$ compared to $L_B = 20\,\text{Å}$?

References

Brown, J. and Churchill, R. (2011). *Fourier Series, and Boundary Value Problems*, 8th edn. McGraw-Hill Education.

Gross, R. and Marx, A. (2014). *Festkörperphysik*, 1st edn. De Gruyter Verlag.

Juffmann, T., Milic, A., Müllneritsch, M., Asenbaum, P., Tsukernik, A., Tüxen, J., Mayor, M., Cheshnovsky, O., and Arndt, M. (2012). Real-time single-molecule imaging of quantum interference. *Nature Nanotechnology* **7**, 297–300.

Shankar, R. (2008). *Principles of Quantum Mechanics*, 2nd edn. Springer.

Tonomura, A., Endo, J., Matsuda, T., and Kawasaki, T. (1989). Demonstration of single-electron buildup of an interference pattern. *American Journal of Physics* **57**, 117.

Chapter 2

WAVE FUNCTION PROPERTIES

Contents

In this chapter, we will discuss the fundamental properties of the wave function. These properties can be categorized into those that can be derived from Schrödinger's equation and those that follow from physical interpretations of the wave function and can only be verified experimentally.

Let us summarize some of the results from Chapter 1 first. Solving the time-independent Schrödinger's equation

$$\widehat{H}\psi(\bar{r}) = E\psi(\bar{r}) \quad \text{or}$$

$$\left[-\frac{\hbar^2}{2m}\nabla^2 + U(\bar{r})\right]\psi(\bar{r}) = E\psi(\bar{r})$$

gives us:

(a) energy eigenvalue E_n and
(b) eigenfunction $\psi_n(\bar{r})$ corresponding to energy eigenvalue E_n. The eigenfunction is also called the eigenstate;
(c) The subscript n is called the quantum number.

$\psi_n(\bar{r})$ and E_n satisfy the eigenvalue equation, $\widehat{H}\psi_n(\bar{r}) = E_n\psi_n(\bar{r})$. The quantum number n can span discrete or continuous values as follows:

- *Bound eigenfunctions* of a Hamiltonian have discrete-valued quantum numbers. Consequently, the energy eigenvalues form a discrete spectrum. The PiB and SHO discussed in Chapter 1 are examples of eigenfunctions with integer-valued quantum numbers n.
- *Unbound/traveling wave eigenfunctions* of a Hamiltonian have a continuous distribution of quantum numbers. The free particle discussed in Chapter 1 is an example, where the wave vector k and the corresponding energy eigenvalues form a continuum.

Terminology:

(a) E_n: Energy **eigenvalue**.
(b) $\psi_n(\bar{r})$: **Eigenfunction** corresponding to energy eigenvalue E_n. The eigenfunction is also called the **eigenstate**.
(c) Subscript n: **quantum number**.

2.1   Probability

The probability density to find a particle at a location $\bar{r}$ at time t is,

$$\text{Probability density} = |\Psi(\bar{r}, t)|^2 \tag{2.1}$$

The unit of probability density is the probability per unit volume in 3D (or $1/\text{m}^3$). If the particle is in an energy eigenfunction n, then the corresponding time-dependent wave function is $\Psi(\bar{r}, t) = e^{-\frac{iE_n t}{\hbar}} \psi_n(\bar{r})$ (Section 1.4). The above equation then results in a time-independent probability density for a particle in an eigenfunction.

$$\text{Probability density} = |\Psi(\bar{r}, t)|^2 = |\psi_n(\bar{r})|^2 \tag{2.2}$$

The **probability** of finding the electron in a small volume dv at location $\bar{r}$ is the probability density multiplied by the volume element dv. That is,

$$\text{Probability} = |\Psi(\bar{r}, t)|^2 dv \tag{2.3}$$

Note that for 1D and 2D problems, the probability density has units of $1/m$ and $1/m^2$, respectively. In these cases, the probability of finding a particle in an element of length (dx) or an element of area (dA) is given by

$$\text{Probability} = |\psi_n(x)|^2 dx \qquad (1D)$$

$$\text{Probability} = |\psi_n(x, y)|^2 dA \qquad (2D)$$

Concept: For the time-independent Schrödinger equation, the probability density is only a function of spatial location (not time) if the particle is in an energy eigenfunction.

Example 1. Find and plot the probability density function of a particle in a box at (x, t) in eigenfunctions $(a)\, n = 1$, $(b)\, n = 2$, and $(c)\, n = 3$. Assume a 1D PiB of length $L = 5\,\text{nm}$.

Solution: From Chapter 1, we know that for PiB, the wave function corresponding to an arbitrary quantum number n is,

$$\psi_n(x) = \sqrt{\frac{2}{L}} \sin\left(\frac{n\pi}{L}x\right) \text{ for } 0 < x < L$$

Using equation (2.2), the probability density is $|\psi_n(x)|^2 = \frac{2}{L}\sin^2\left(\frac{n\pi}{L}x\right)$. The probability density for quantum numbers $n = 1, 2$, and 3 is plotted in Figure 2.1. One interesting feature to note in the figure is that the probability density for even quantum numbers $(n = 2)$ is zero at the center of the box, while odd quantum numbers have a maxima $(n = 1$ and $n = 3)$.

Example 2. Consider a free particle with a wave vector (quantum number) k. What is the probability of finding the particle at (x, t)?

Solution: From Chapter 1, we know that the wave function of the 1D free particle with wave vector k is

$$\phi_k(x) = \frac{e^{i\phi}}{\sqrt{L}} e^{ikx} \text{ (where } L \text{ tends to infinity)}$$

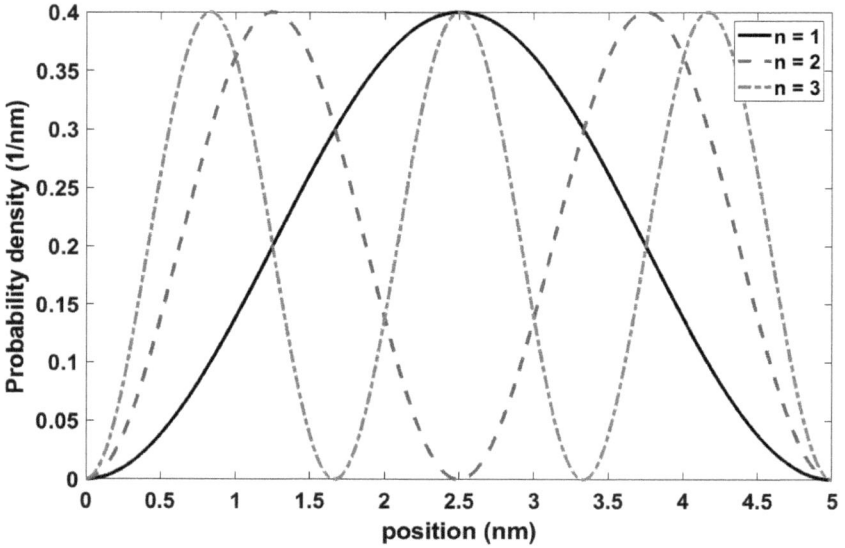

Figure 2.1. The probability density for a PiB in an eigenfunction corresponding to $n = 1$, $n = 2$, or $n = 3$. The box length is $L = 5\,\mathrm{nm}$. $n = 2$ has a minimum in the middle of the box.

Using equation (2.2), the probability density $|\phi_k(x)|^2 = \frac{1}{L}$. Since the probability density is independent of position, the free particle, we are equally likely to find the particle anywhere.

Concept: The probability density of a free particle does not depend on the spatial location x.

2.2   Orthonormality of Wave Functions

The eigenfunctions of the time-independent Schrödinger equation form an orthonormal set. That is,

$$\int dv\, \psi_n^*(\bar{r})\, \psi_m(\bar{r}) = \delta_{mn} \tag{2.4}$$

Orthonormal functions have two properties which are as follows:

$$\text{For } n = m,\ \int dv\, |\psi_n(\bar{r})|^2 = 1 \tag{2.5}$$

That is, the eigenfunction is normlized

$$\text{For } n \neq m, \quad \int dv\, \psi_n^*(\bar{r})\psi_m(\bar{r}) = 0 \qquad (2.6)$$

That is, the eigenfunctions are orthogonal to each other.

Concept: The eigenfunctions of the time-independent Schrödinger equation form an orthonormal set.

Example 3. Verify that the eigenfunctions of a PiB form an orthonormal set.

Solution: For a PiB, the eigenfunction for quantum number n is

$$\psi_n(x) = \sqrt{\frac{2}{L}} \sin\left(\frac{n\pi}{L}x\right) \text{ for } 0 < x < L$$

The LHS of the orthonormality relationship (equation (2.4)) is

$$\int dx\, \psi_n^*(x)\psi_m(x) = \frac{2}{L} \int dx\, \sin\left(\frac{n\pi}{L}x\right)\sin\left(\frac{m\pi}{L}x\right)$$

If $m \neq n$, then the above integral reduces to

$$\int dx\, \psi_n^*(x)\psi_m(x)$$

$$= \frac{2}{L} \cdot \frac{1}{2} \int dx \left[\cos\left(\frac{(m-n)\pi}{L}x\right) - \cos\left(\frac{(m+n)\pi}{L}x\right)\right]$$

$$= \frac{2}{L} \cdot \frac{1}{2} \left[\frac{L}{(m-n)\pi}\sin\left(\frac{(m-n)\pi}{L}x\right)\right.$$

$$\left. - \frac{L}{(m+n)\pi}\sin\left(\frac{(m+n)\pi}{L}x\right)\right]_{x=0}^{x=L} = 0$$

because $\sin(0) = \sin(m-n)\pi = \sin(m+n)\pi = 0$.
  If $m = n$,

$$\int dx\, \psi_n^*(x)\psi_m(x) = \frac{2}{L} \int dx \left\{\sin\left(\frac{n\pi}{L}x\right)\right\}^2$$

$$= \frac{1}{L} \int_{x=0}^{x=L} dx \left(1 - \cos\left(\frac{2n\pi}{L}x\right)\right)$$

$$= 1 - \left[\frac{1}{2n\pi}\sin\left(\frac{2n\pi}{L}x\right)\right]_{x=0}^{x=L} = 1 - 0 = 1$$

Therefore, we have shown that

$$\int dx \psi_n^*(x)\psi_m(x) = \begin{cases} 1 & \text{if } m = n \\ 0 & \text{if } m \neq n \end{cases}$$

and the verification of orthonormality for the PiB is complete.

Example 4. Verify that the wave functions of the 1D free particle form an orthonormal set.

Solution: For a free particle, the wave function is $\psi_k(x) = \frac{e^{i\phi}}{\sqrt{L}}e^{ikx}$. The LHS of the orthonormality relationship (equation (2.4)) is

$$\int dx \, \psi_k^*(x)\psi_{k'}(x) = \lim_{L\to\infty} \frac{1}{L} \int_{-L/2}^{+L/2} dx e^{i(k-k')x} = \begin{cases} 1 & \text{if } k = k' \\ 0 & \text{if } k \neq k' \end{cases}$$

This is because if $k = k'$, then

$$\int dx \, \psi_k^*(x)\psi_{k'}(x) = \lim_{L\to\infty} \frac{1}{L} \int_{-\frac{L}{2}}^{+\frac{L}{2}} dx = \lim_{L\to\infty} \frac{1}{L} \cdot L = 1$$

and if $k \neq k'$,

$$\int dx \psi_k^*(x)\psi_{k'}(x) = \lim_{L\to\infty} \frac{1}{L} \frac{2\sin((k-k')L/2)}{(k-k')} = 0$$

as the numerator is finite.

2.3 The Superposition Principle

Although we have previously studied the eigenfunction $\psi_n(\bar{r})$ in some detail, it only corresponds to one energy level of the system with quantum number n. To characterize an arbitrary allowed state of a particle in quantum mechanics, we must study an arbitrary wave function, which is a linear combination (or superposition) of the eigenfunctions,

$$\Psi(\bar{r}, t) = \sum_n a_n(t)\psi_n(\bar{r}) \tag{2.7}$$

One thing to note is that for each eigenfunction $\psi_n(\bar{r})$, the coefficient $a_n(t)$ changes with time. One reason for their time dependence is the

$e^{-\frac{iE_n t}{\hbar}}$ component of the wave function, as discussed in Chapter 1. A second reason for their time dependence could be due to an externally applied field, such as a light pulse or a time-dependent potential. For example, in the presence of a laser field, the wave function for an electron in a PiB evolves with time (e.g., the electron leaves one eigenfunction $\psi_1(\bar{r})$ and moves to another $\psi_2(\bar{r})$ as time passes). We will learn more about calculating the coefficients $a_n(t)$ in the presence of an external time-dependent field in Chapter 9.

An example of a superposition state is shown in Figure 2.2. At time $t = 0$, it consists of three different eigenfunctions ψ_1, ψ_2, and ψ_3, which are weighted by coefficients $a_{1o} = 1/\sqrt{7}$, $a_{2o} = -2/\sqrt{7}$, and $a_{3o} = \sqrt{2}/\sqrt{7}$, respectively,

$$\Psi(x, t = 0) = a_{1o}\psi_1 + a_{2o}\psi_2 + a_{3o}\psi_3$$

$$\Psi(x, t = 0) = \frac{1}{\sqrt{7}}\psi_1 - \frac{2}{\sqrt{7}}\psi_2 + \frac{\sqrt{2}}{\sqrt{7}}\psi_3$$

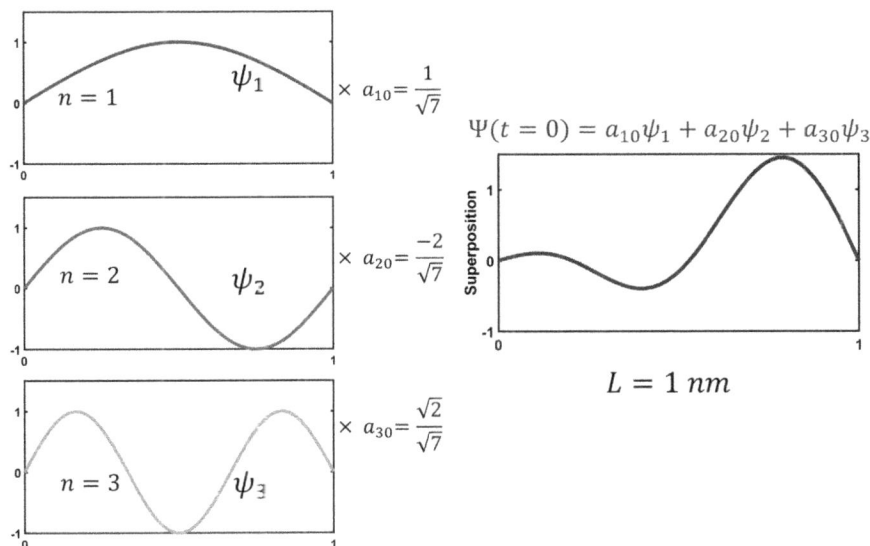

Figure 2.2. Illustration of superposition. The eigenfunctions corresponding to $n = 1$, $n = 2$, and $n = 3$ (left). Coefficients $a_{10} = 1/\sqrt{7}$, $a_{20} = (-2)/\sqrt{7}$, and $a_{30} = \sqrt{2}/\sqrt{7}$ multiply these eigenfunctions to yield the wave function on the right, which is a superposition state.

The reader should verify that this is a properly normalized wave function, $|a_{1o}|^2 + |a_{2o}|^2 + |a_{3o}|^2 = 1$. Figure 2.2 shows that the particle in this superposition state has a higher probability to be found in the right half of the box.

Using the fact that any wave function $\Psi(\bar{r}, t) = \sum_n a_n(t)\psi_n(\bar{r})$ should be normlized, we can show that

$$\sum_n |a_n(t)|^2 = 1 \tag{2.8}$$

To show that equation (2.8) holds, we first note that the wave function normalization requires

$$\int dv \Psi(\bar{r}, t)^* \Psi(\bar{r}, t) = 1 \tag{2.9}$$

Substituting equation (2.7) in equation (2.9) yields

$$\int dv \left[\sum_n a_n(t)\psi_n(\bar{r}) \right]^* \left[\sum_m a_m(t)\psi_m(\bar{r}) \right] = 1 \tag{2.10}$$

Note that if you do not use different indices m and n in the above equation, many terms in the expansion of $\Psi(\bar{r}, t)^* \Psi(\bar{r}, t)$ will be missing (write it out and try it). Because the coefficients $a_n(t)$ do not depend on the spatial location of the wave functions, they can be moved out of the integral sign to get

$$\sum_{n,m} a_n^*(t)a_m(t) \int dv\, \psi_n^*(\bar{r})\psi_m(\bar{r}) = 1 \tag{2.11}$$

Note that $\sum_{n,m} A_{mn} = \sum_n \sum_m A_{mn}$. Using the orthonormality property, $\int dv\, \psi_n^*(\bar{r})\psi_m(\bar{r}) = \delta_{mn}$ in equation (2.11), we have

$$\sum_{n,m} a_n^*(t)a_m(t)\delta_{mn} = 1 \tag{2.12}$$

Using the sifting property of the Kronecker delta function (see Chapter 20)

$$\sum_m a_m \delta_{mn} = a_n \tag{2.13}$$

we can turn equation (2.12) into

$$\sum_n a_n^*(t)a_n(t) = \sum_n |a_n(t)|^2 = 1 \tag{2.14}$$

This ends the proof.

We will see in Section 3.3 (and again in Chapter 6 when we learn about projection operators) that $|a_n(t)|^2$ is the probability of finding an arbitrary wave function $\Psi(\bar{r}, t)$ in the nth eigenfunction $\psi_n(\bar{r})$ upon measurement. Equation (2.14) says that the sum of these probabilities is one as expected.

Concept: A general wave function of a quantum system can be in a *superposition* (*linear combination*) of the eigenfunctions of the Hamiltonian.

Example 5. Consider a time-independent Hamiltonian. The wave function has been engineered at $t = 0$ to be the superposition $\Psi(\bar{r}, t = 0) = \sum_n a_{no}\psi_n(\bar{r})$, where a_{no} is time independent. Use equation (1.51) to find the time-dependent wave function for $t > 0$.

Solution: We know from Section 1.4 of Chapter 1 that when the Hamiltonian is time independent, the wave function evolves in time as

$$\Psi(\bar{r}, t) = e^{-i\frac{\hat{H}(t-t_o)}{\hbar}}\Psi(\bar{r}, t_o)$$

Choosing $t_o = 0$, we have

$$\Psi(\bar{r}, t) = e^{-\frac{i\hat{H}t}{\hbar}}\Psi(\bar{r}, t = 0) = \sum_n a_{no}e^{-\frac{i\hat{H}t}{\hbar}}\psi_n(\bar{r}) \tag{2.15}$$

Using the Taylor expansion $e^{-\frac{i\hat{H}t}{\hbar}} = 1 - \frac{1}{1!}\frac{i\hat{H}t}{\hbar} + \frac{1}{2!}\frac{i^2\hat{H}^2t^2}{\hbar^2} + \cdots$ and because $\hat{H}\psi_n(\bar{r}) = E_n\psi_n(\bar{r})$, the above equation becomes

$$\Psi(\bar{r}, t) = \sum_n a_{no}e^{-\frac{iE_nt}{\hbar}}\psi_n(\bar{r}) \tag{2.16}$$

Equation (2.16) means that at any later time, the wave function has changed in a nontrivial manner, where each eigenfunction $\psi_n(\bar{r})$ has been multiplied by a phase factor $e^{-\frac{iE_nt}{\hbar}}$. Note that the probability density to find the particle at a given location $\bar{r}$ will vary with time if the wave function is in a superposition state $\Psi(\bar{r}, t)$. We will see this in Example 7.

Example 6. You are given the following two superposition wave functions for a PiB. Which of these wave functions is valid, and what is the reason?

$$\Psi_a(x, t = 0) = 1.0\,\psi_1(x) + 1.0\,\psi_2(x) \tag{2.17}$$

$$\Psi_b(x, t = 0) = 0.8\,\psi_1(x) + 0.6\,\psi_2(x) \tag{2.18}$$

Solution: The normalized eigenfunctions for a PiB are $\psi_n(x) = \sqrt{\frac{2}{L}}\sin(\frac{n\pi}{L}x)$. For a wave function to be valid, it must be normalized. That is, equation (2.14) should hold for the wave function.

For Ψ_a, $\displaystyle\sum_n |a_n(t)|^2 = |1|^2 + |1|^2 = 2$

For Ψ_b, $\displaystyle\sum_n |a_n(t)|^2 = |0.8|^2 + |0.6|^2 = 0.64 + 0.36 = 1$

So, Ψ_a is not a valid wave function because it is not normalized. However, Ψ_a can be normalized by dividing it by the norm of the wave function, $\sqrt{\int_{x=0}^{x=L} |\Psi_a(x,0)|^2\,dx}$. The normalized version is given by

$$\Psi_{a,\text{normalized}}(x,0) = \frac{\Psi_a(x,0)}{\sqrt{\int_{x=0}^{x=L} |\Psi_a(x,0)|^2\,dx}} = \frac{\Psi_a(x,0)}{\sqrt{2}}$$

where

$$\int_{x=0}^{x=L} |\Psi_a(x,0)|^2\,dx = \int_{x=0}^{x=L} \Psi_a^*(x,0).\Psi_a(x,0)dx$$

$$= \sum_{n=1,2} |a_n(t=0)|^2 = 1 + 1 = 2$$

Example 7. What is the probability density to find a particle in a PiB at (x, t) if the wave function at time $t = 0$ is

$$\Psi(\bar{r}, t = 0) = a_{2o}\psi_2(\bar{r}) + a_{3o}\psi_3(\bar{r})?$$

Assume that a_{2o} and a_{3o} are real. Also, assume the box length is 1 nm, and the mass of the particle is six times larger than the free electron mass. Plot the probability density at various times.

Solution: We know that the wave function evolves with time as per $\Psi(x,t) = e^{-i\frac{\hat{H}(t-t_o)}{\hbar}}\Psi(x,t_o)$. Substituting the value of $\Psi(x,t_o)$ where $t_o = 0$ and using $e^{-i\frac{\hat{H}(t-t_o)}{\hbar}}\psi_n(x) = e^{-i\frac{E_n(t-t_o)}{\hbar}}\psi_n(x)$, we find that

$$\Psi(x,t) = a_{2o}e^{-\frac{iE_2t}{\hbar}}\psi_2(x) + a_{3o}e^{-\frac{iE_3t}{\hbar}}\psi_3(x)$$

The probability density to find the particle at (x,t) using equation (2.1) is

Probability density at (x,t)
$$= |\Psi(x,t)|^2 = |a_{2o}|^2 |\psi_2(x)|^2 + |a_{3o}|^2 |\psi_3(x)|^2$$
$$+ 2\text{Re}[a_{2o}{}^* a_{3o}e^{-\frac{i(E_2-E_3)t}{\hbar}}\psi_2(x)^*\psi_3(x)]$$

where $\text{Re}(A)$ stands for the real part of A. Using $\psi_n(x) = \sqrt{\frac{2}{L}}\sin\left(\frac{n\pi}{L}x\right)$ for a PiB, and noting that a_{2o} and a_{3o} are real, the above equation becomes

Probability density at (x,t)

$$= \frac{2}{L}\left[a_{2o}^2\sin^2\left(\frac{2\pi}{L}x\right) + a_{3o}^2\sin^2\left(\frac{3\pi}{L}x\right)\right.$$
$$\left. + 2a_{2o}.a_{3o}\sin\left(\frac{2\pi}{L}x\right)\sin\left(\frac{3\pi}{L}x\right)\cos\left(\frac{(E_2-E_3)t}{\hbar}\right)\right] \quad (2.19)$$

In a superposition state, the probability density at a specific location varies periodically with time, even for a time-independent Hamiltonian. Although the first and second terms ($n = 2$ and $n = 3$) are time independent, the third term is time dependent, which is due to the *interference* between $n = 2$ and $n = 3$ eigenfunctions.

Classically, if an experiment has two different outcomes, A and B, the total probability is the sum of the probability of event A happening or the probability of event B happening. Equation (2.19) shows that there is an additional term in quantum mechanics that comes from the interference of A and B (the third term on the RHS of equation (2.19)). Furthermore, opposite to the classical notion, this interference term can cancel the other terms and make the total probability zero at certain (x,t), even if $P(A)$ and $P(B)$ are nonzero!

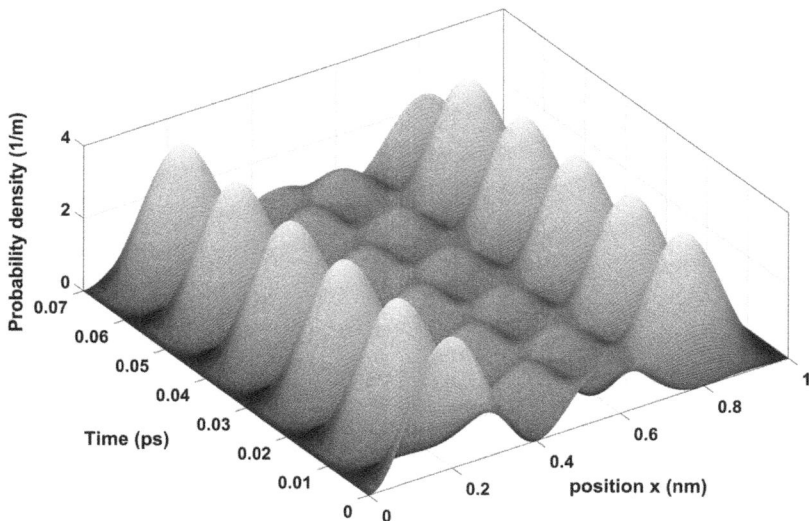

Figure 2.3. Probability density of an electron in a superposition of $n = 2$ and $n = 3$ eigenfunctions of a PiB versus time. The wavefunction is $\Psi(x,t) = \frac{1}{\sqrt{2}} e^{\frac{-iE_2 t}{\hbar}} \psi_2(x) + \frac{1}{\sqrt{2}} e^{\frac{-iE_3 t}{\hbar}} \psi_3(x)$. As time proceeds, the electron oscillates back and forth between the two ends of the box ($L = 1\,\text{nm}$).

Figure 2.3 shows the probability density. It is assumed that the coefficients a_{2o} and a_{3o} are both equal to $1/\sqrt{2}$. As it can be seen, the maximum probability density oscillates between the two ends of the box as time proceeds. The period of oscillation (T) is determined by the energy difference between the energy eigenvalues of the $n = 2$ and $n = 3$ quantum states,

$$\frac{2\pi}{T} = 2\pi f = \frac{E_3 - E_2}{\hbar}$$

where f is the frequency. For the PiB, the energy eigenvalues are given by $E_n = \frac{\hbar^2}{2m}\left(\frac{n\pi}{L}\right)^2$. Using $L = 1\,\text{nm}$ and $m = 6 \times 9.1 \times 10^{-31}\,\text{kg}$, we have

$$E_2 = 250.4\,\text{meV} \quad \text{and} \quad E_3 = 563.5\,\text{meV}$$

which yields a time period (T),

$$T = 13.2\,\text{fs}$$

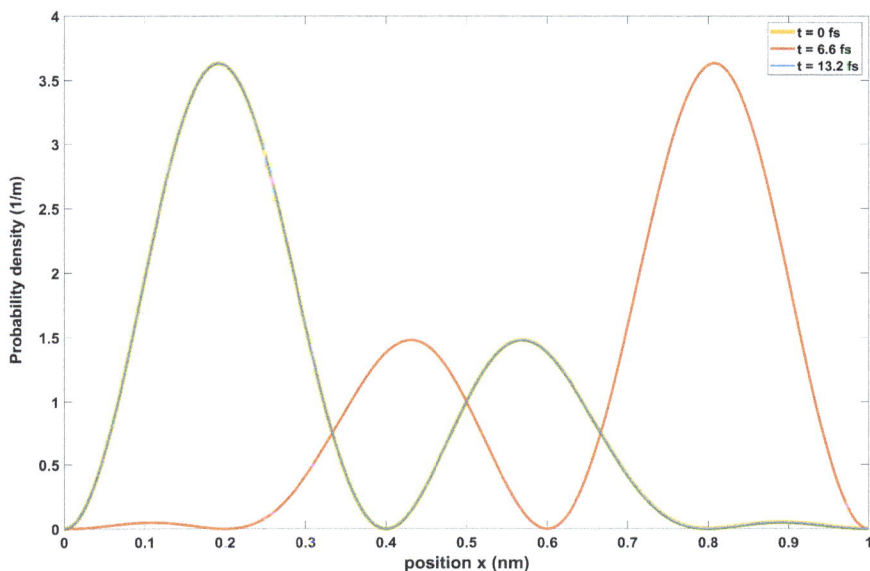

Figure 2.4. The probability density versus position at three different times, corresponding to cross-sections of Figure 2.3. Compare the curves and note that the time period of oscillation is $T = 13.2$ fs. Note that the graphs of $t = 0$ fs (yellow) and $t = 13.2$ fs (blue) lie on top of each other, since they are one time period apart.

In this superposition of the $n = 2$ and $n = 3$ eigenfunctions, the particle changes its position from the left to right ends of the box every $\frac{T}{2} = 6.6$ fs $= 6.6 \times 10^{-15}$ s. The oscillation in probability density is seen more clearly in the line plot of Figure 2.4, which shows snapshots of the probability density at three different times $(t = 0, T/2, \text{ and } T)$.

Concept: In Chapter 1, we saw that the probability density in an eigenfunction is time independent. Here, we see that the probability density of a wave function in a superposition of eigenfunctions oscillates with time even if the Hamiltonian is time independent.

Example 8. Consider a time-independent Hamiltonian. The wave function has been engineered at $t = 0$ to be the superposition $\Psi(\bar{r}, t = 0) = \sum_n a_{no} \psi_n(\bar{r})$, where a_{no} is time independent. Derive a general expression for how the probability density changes with time.

Solution: We know from Section 1.4 of Chapter 1 that when the Hamiltonian is time independent, the wave function evolves with time as

$$\Psi(\bar{r}, t) = e^{-i\frac{\hat{H}(t-t_o)}{\hbar}} \Psi(\bar{r}, t_o)$$

Using the superposition state at $t_o = 0$, we have

$$\Psi(\bar{r}, t) = e^{-\frac{i\hat{H}t}{\hbar}} \Psi(\bar{r}, t = 0) = \sum_n a_{no} e^{-\frac{i\hat{H}t}{\hbar}} \psi_n(\bar{r})$$

Since $\hat{H}\psi_n(\bar{r}) = E_n \psi_n(\bar{r})$, the above equation becomes

$$\Psi(\bar{r}, t) = \sum_n a_{no} e^{-\frac{iE_n t}{\hbar}} \psi_n(\bar{r})$$

The probability density to find the particle at a given location $\bar{r}$ varies with time due to the interference terms as given below:

$$|\Psi(\bar{r}, t)|^2 = \sum_n |a_{no}|^2 |\psi_n(\bar{r})|^2 + \sum_{m,n} \text{Re}\left[a_{no}^* a_{mo} \exp\left(\frac{i(E_m - E_n)t}{\hbar}\right)\right]$$

2.4   Quantum Measurement

Consider a superposition state for a quantum system governed by a time-independent Hamiltonian,

$$\Psi(\bar{r}, t) = \sum_n a_n(t)\psi_n(\bar{r}) \qquad \text{(same as equation (2.7))}$$

where ψ_n are the eigenfunctions of the Hamiltonian. We saw in Chapter 1 that the only time dependence is from the phase factor $e^{-\frac{iE_n t}{\hbar}}$. That is,

$$a_n(t) = e^{-\frac{iE_n t}{\hbar}} a_{no}$$

where a_{no} is time independent. **We will now discuss the measurement of energy.**

Recall that each eigenfunction $\psi_n(\bar{r})$ is associated with an energy eigenvalue E_n. The probability of finding a particle at an energy eigenvalue of E_n is $|a_n(t)|^2 = |a_{no}|^2$. Further, if the measurement yielded energy E_n, the wave function collapses to

ψ_n after measurement. For example, if we measure the energy of the particle to be E_2, then the particle will "collapse" into the energy eigenfunction $\psi_2(\bar{r})$. After measurement, the particle ceases to exist in the superposition state described in equation (2.7). Instead, the new collapsed wave function involves only the eigenfunction corresponding to energy eigenvalue E_2,

$$\Psi(\bar{r}, t) = e^{-\frac{iE_2 t}{\hbar}} \psi_2(\bar{r})$$

We say that *the wave function of the system has collapsed* from $\psi(\bar{r}, t)$ to $\psi_2(\bar{r})$. Subsequent measurements of energy of the particle will show that the eigenfunction $\psi_2(\bar{r})$ occurs with 100% probability and the particle always has energy E_2.

Various classical apparatuses can measure the observables of a quantum system. The current belief is that the interaction of the quantum system with the classical measuring apparatus forces the quantum system to collapse into the measured energy eigenvalue (e.g., E_2 in the earlier example). The probability that the energy E_m is measured is $|a_{mo}|^2$, where a_{mo} is the coefficient of $\psi_m(\bar{r})$ in the original wave function $\Psi(\bar{r}, t)$. If E_m is measured, then the system collapses into the wave function $\psi_m(\bar{r})$. After this collapse, the particle continues to stay in the eigenfunction,

$$\Psi(\bar{r}, t) = e^{-\frac{iE_m t}{\hbar}} \psi_m(\bar{r}) \text{ (up to a constant multiplicative}$$
$$\text{phase factor } e^{i\phi})$$

Concept: A single measurement of a wave function gives only one of the many possible eigenvalues. After the measurement, the system collapses to the eigenfunction corresponding to the eigenvalue measured.

The energy measured in each measurement can be the eigenenergies $E_1, E_2, E_3, \ldots$. The probability of measuring energy E_n is $|a_n(t)|^2$. As a result, in a time-independent potential, the *average energy* (represented by $\langle H \rangle$ or $\langle E \rangle$) is given by

$$\langle H \rangle \text{ or } \langle E \rangle = \sum_n |a_n(t)|^2 E_n = \sum_n |a_{no}|^2 E_n \qquad (2.20)$$

This *average energy* is the weighted average of the measured values E_n, where $|a_{no}|^2$, the weighting factor, is the probability of finding the system to be in quantum number n. The *average energy* is also referred to as the *expectation value* or expected value of energy. It is important to notice that experimentally, the average is calculated from the results of measurements on many identically prepared quantum systems, all of which have the same initial wave function.

Concept: The average energy of a quantum system is the weighted average of the energy eigenvalues, which is $\langle E \rangle = \Sigma_n |a_{no}|^2 E_n$.

Next, consider two experimenters in different labs within the same department who receive identical PiB superposition states, $\Psi(\bar{r}, t) = e^{-\frac{iE_1 t}{\hbar}} a_{1o} \psi_1(\bar{r}) + e^{-\frac{iE_2 t}{\hbar}} a_{2o} \psi_2(\bar{r})$. The two PiB systems and other conditions in their lab are exactly the same. Upon measurement of their individual PiB, Experimenter A may find the system to have eigenfunction ψ_1 with energy E_1 and Experimenter B may find the system to have eigenfunction ψ_2 with energy E_2.

Now, if each experimenter keeps receiving the same state $\Psi(\bar{r}, t)$ in their labs, they would both measure the eigenenergy E_1 a fraction of $|a_{1o}|^2$ times and the eigenfunction E_2 a fraction of $|a_{2o}|^2$ times. This means that they would both measure the same *average energy* $|a_{1o}|^2 E_1 + |a_{2o}|^2 E_2$.

That is, individual measurements of identical systems can give the two experimenters different outcomes, but repeatedly **preparing** *the quantum system with the same wave function and measuring an observable (here, energy) gives both experimenters the same average value.*

Classically, the measurements above would be impossible. For example, if two coins are flipped at the exact same time in the same precise manner with precisely the same surrounding conditions (gravity and air molecules are the same in these experiments), then Newton's law predicts that the coins would always land on the same face, independent of the number of repetitions and the location of the labs.

Example 9. For the PiB in the superposition state $\Psi(x, t = 0) = \frac{1}{\sqrt{7}}\psi_1 - \frac{2}{\sqrt{7}}\psi_2 + \frac{\sqrt{2}}{\sqrt{7}}\psi_3$ shown in Figure 2.2:

(a) Write down the wave function at any later time.
(b) At time $t = 0$, if you make a single measurement of energy, what are the possible energy values that you can measure?
(c) If you make a large number of measurements, write down the probability of measuring various energies.
(d) What is the average energy that will be measured in the state $\Psi(x, t = 0)$?

Find how the answer to (a)–(d) change at any later time.

Solution:

(a) We know from Section 1.4 of Chapter 1 that when the Hamiltonian is time independent, the wave function evolves in time as $\Psi(\bar{r}, t) = e^{-i\frac{\hat{H}(t-t_o)}{\hbar}}\Psi(\bar{r}, t_o)$. Using this, we immediately find that

$$\Psi(x, t) = \frac{1}{\sqrt{7}}e^{-\frac{iE_1 t}{\hbar}}\psi_1 - \frac{2}{\sqrt{7}}e^{-\frac{iE_2 t}{\hbar}}\psi_2 + e^{-\frac{iE_3 t}{\hbar}}\frac{\sqrt{2}}{\sqrt{7}}\psi_3$$

(b) We can rewrite the original state given to us as

$$\Psi(x, t = 0) = \frac{1}{\sqrt{7}}\psi_1 - \frac{2}{\sqrt{7}}\psi_2 + \frac{\sqrt{2}}{\sqrt{7}}\psi_3 + 0\psi_4 + 0\psi_5 + 0\psi_6 + \cdots$$

That is, only the coefficients of the $n = 1, 2$, and 3 wave functions are nonzero, and the coefficients of all other quantum numbers or eigenfunctions are zero. So, we will only be able to measure the energies corresponding to $n = 1, 2$, and 3, which are E_1, E_2, and E_3. Energy levels E_4 and higher energy levels will not be measured.

(c) If you make a large number of measurements, the probabilities of measuring E_1, E_2, and E_3 will be the coefficient squared of the various eigenfunctions. So, the probability of measuring the various states will be as shown in the following table:

Energy	Quantum number	Coefficient	Probability $= \|\text{Coefficient}\|^2$
E_1	$n = 1$	$\frac{1}{\sqrt{7}}$	$\frac{1}{7}$
E_2	$n = 2$	$-\frac{2}{\sqrt{7}}$	$\frac{4}{7}$
E_3	$n = 3$	$\frac{\sqrt{2}}{\sqrt{7}}$	$\frac{2}{7}$
E_4 and higher	$n = 4, 5, \ldots$	0	0

(d) If you make a large number of measurements, the average energy given by equation (2.20) will be

$$\langle E \rangle = \frac{1}{7}E_1 + \frac{4}{7}E_2 + \frac{2}{7}E_3$$

which is the sum of the eigenenergies weighted by the probability of their occurrence. Energy levels E_4 and higher energies are never measured and they do not contribute to the average energy in the given state.

2.5 Problems

Section 2.1

(1) You have a PiB with a box length $L = 4\,\text{nm}$ and the mass of the particle is six times larger than the free electron mass. Plot the probability densities $(|\psi(x)|^2)$ for $n = 1$ and $n = 1000$. Assume that we have a measurement apparatus which can only give us the location of a particle averaged over 0.5 nm, i.e., at each given point x_o, the apparatus measures the following quantity:

$$P_{x_o} = \int_{x_o - 0.25\,\text{nm}}^{x_o + 0.25\,\text{nm}} |\psi(x)|^2 \, dx$$

Now, for $n = 1$, plot the probability to measure the particle at x_o (P_{x_o}) when the particle is at points $x_o = 0.25, 0.5, 0.75, 1, 1.25,$ 1.5, 1.75, 2, 2.25, 2.5, 2.75, 3, 3.25, 3.5, and 3.75 (all in nm). Repeat the same for $n = 1000$. What is the difference between these two cases? In this problem, you expect to see that when $n = 1$, the probability of finding the particle at each point changes significantly depending on the locations of the particle. However, for very large quantum numbers (n), you see that the value of P_{x_o} oscillates less and almost constant, as if the particle is equally probable to be found everywhere in the box (with uniform probability distribution) as one would expect without quantum mechanics. This is called the *correspondence principle*. The principle says that in many situations, quantum mechanical calculations predict the same results as classical ones when you move to very large quantum numbers. Check the above exercise for $n = 10,000$. In other words, to see the quantum mechanical effects, the quantum numbers must be very small, i.e., around $n = 1, 2, \ldots$. That is why many quantum experiments must be performed at very low temperatures; the thermal noise can excite the system to very high energy levels (with very high n), and as a result, the quantum effects will be hidden.

Section 2.2

(2) Show that the wave functions of the $n = 1$ and $n = 2$ states of the PiB, $\psi_1(x) = \sqrt{\frac{2}{L}} sin\left(\frac{\pi}{L}x\right)$ and $\psi_2(x) = \sqrt{\frac{2}{L}} sin\left(\frac{2\pi}{L}x\right)$ valid for $0 < x < L$, are orthogonal to each other.

(3) Show that the wave function of a 1D free particle with quantum numbers $k = 10^9 \, \mathrm{m}^{-1}$ and $k' = 10^8 \, \mathrm{m}^{-1}$ are orthogonal to each other.

(4) Consider a free particle in one dimension with the Hamiltonian $\hat{H} = -\frac{\hbar^2}{2m}\frac{\partial^2}{\partial x^2}$. The eigenfunctions of the Hamiltonian are $\phi_k = \frac{1}{\sqrt{L}}e^{ikx}$. The wave function of this system when the particle is localized at the origin $x = 0$ is $\delta(x - 0)$. This wave function can be expanded in terms of the eigenfunctions of the free particle. Determine the coefficients in the expansion.

Section 2.3

(5) What is the probability density to find a particle in a PiB at (x, t) if the wave function at time $t = 0$ is

$$\Psi(\bar{r}, t = 0) = \frac{1}{\sqrt{2}}\psi_1(\bar{r}) + \frac{1}{\sqrt{2}}\psi_2(\bar{r})?$$

Assume that the box length is 1 nm, and the mass of the particle is six times larger than the free electron mass. Plot the probability density at various times.

Section 2.4

(6) Understanding quantum measurement:

(i) You have a million copies of a PiB in the state given by eigenfunction ψ_5 at time $t = 0$. So, the wave function is $\Psi(x, t = 0) = \psi_5(x)$ (same as equation (2.7)). If you measure the energy of the particle, what values will you find? Does the answer depend on if you make the measurement at $t = 0$ or at a later time?

(ii) You have a million copies of a PiB in the state given by $\Psi(x, t = 0) = \psi_5(x)$. If you measure (or calculate) the average energy of the million copies, what value will you find?

(iii) You have a million copies of a PiB in the state given by $\Psi(x, t = 0) = (3\psi_5 + 4\psi_7)/5$. If you measure the energy of the particle, what values will you find? What is the average energy? Does the answer depend on if you make the measurement at $t = 0$ or at a later time?

Chapter 3

OPERATORS AND
EXPECTATION VALUES

Contents

3.1 Operators in Differential Form

In Chapter 1, we discussed that the energy eigenvalues of a quantum system can be found by solving an eigenvalue problem $\widehat{H}\psi_n = E_n\psi_n$ corresponding to the Hamiltonian operator $\widehat{H}$. To emphasize again, if the energy of a quantum system is measured, only values corresponding to the energy eigenvalues E_n can be obtained.

In quantum mechanics, there is an operator $(\hat{O})$ corresponding to every observable (O), not just the energy operator (Hamiltonian) as discussed in Chapter 1. Examples of other observables are position, momentum, angular momentum, and so on. The hat on top of O is used to represent *operator*, though the hat is dropped often for simplicity. The values obtained in a measurement are real numbers.

The eigenvalue problem corresponding to an observable O is

$$\hat{O}\phi_n = \alpha_n\phi_n \tag{3.1}$$

where the eigenfunction ϕ_n has eigenvalue α_n. As per quantum mechanics, a single measurement of observable $\hat{O}$ on a superposition state

$$\Phi = \sum_n a_n\phi_n$$

can only yield one of the many eigenvalues α_m. Further, after measurement, it will collapse into the eigenfunction ϕ_m corresponding to the measured eigenvalue α_m.

The differential operator for *position* and *momentum* followed from classical mechanics, which predates quantum mechanics. Here, we will state the expressions for various operators in quantum mechanics and encourage the reader to study alternate sources for a historical perspective.

Position operator

We will look at simple 1D problems where the particle is present on a single coordinate. Examples are a particle that lives along a straight line or on a circle's circumference. We will call this single coordinate x. In these 1D problems, the position operator $(\hat{x})$ is defined by

$$\hat{x}\phi(x) = x\phi(x) \tag{3.2}$$

The position operator simply returns the scalar coordinate x, which is the eigenvalue.

In 3D space, however, the position operator is a vector operator because the position has $x, y,$ and z components. It is defined by

$$\hat{r}\phi(\overline{r}) = \overline{r}\phi(\overline{r}) \tag{3.3}$$

where $\hat{r}$ is a vector operator and the eigenvalue $\overline{r} = x\hat{i} + y\hat{j} + z\hat{k}$ or (x, y, z) is a vector. $\hat{i}, \hat{j},$ and $\hat{k}$ are unit vectors along the $x, y,$ and z

axes. The position operator's eigenvalue is the vector $\bar{r}$. The above equation consists of three separate equations,

$$\hat{x}\phi(\bar{r}) = x\phi(\bar{r})$$

$$\hat{y}\phi(\bar{r}) = y\phi(\bar{r})$$

$$\hat{z}\phi(\bar{r}) = z\phi(\bar{r})$$

Momentum operator

The momentum operator in 1D is defined by

$$\hat{p}_x = \frac{\hbar}{i}\frac{\partial}{\partial x}, (i = \sqrt{-1}) \tag{3.4}$$

The momentum operator is simply the differential operator scaled by $\frac{\hbar}{i}$. The eigenvalue equation for the momentum operator is

$$\hat{p}_x\phi_k(x) = p_k\phi_k(x) \tag{3.5}$$

Substituting for the momentum operator, we have the following differential equation:

$$\frac{\hbar}{i}\frac{\partial}{\partial x}\phi_k(x) = p_k\phi_k(x) \tag{3.6}$$

To find ϕ_k, we move coefficients to the RHS and then separate variables,

$$\frac{1}{\phi_k}\partial\phi_k = \frac{ip_k}{\hbar}\partial x$$

Integrating this gives

$$\ln(\phi_k) = \frac{ip_k}{\hbar}x + C$$

$$\phi_k = A\,e^{\frac{ip_k x}{\hbar}}$$

where C and A are constants, or we can write this as,

$$\phi_k(x) = A\,e^{ikx} \tag{3.7}$$

The scalar k is a quantum number whose allowed values are real numbers ranging from $-\infty$ to $+\infty$. Note that k is the wave vector because e^{ikx} is a periodic function (wave) with a wavelength equal to $2\pi/k$. The eigenvalues of the momentum operator are

$$p_k = \hbar k \tag{3.8}$$

In 3D space, the $x, y,$ and z components of the momentum operator are

$$\hat{p}_x = \frac{\hbar}{i}\frac{\partial}{\partial x}, \hat{p}_y = \frac{\hbar}{i}\frac{\partial}{\partial y}, \text{ and } \hat{p}_z = \frac{\hbar}{i}\frac{\partial}{\partial z}$$

So, the differential operator for momentum $(\hat{p} = \hat{i}\hat{p}_x + \hat{j}\hat{p}_y + \hat{k}\hat{p}_z)$ can be written as

$$\hat{p} = \frac{\hbar}{i}\left(\hat{i}\frac{\partial}{\partial x} + \hat{j}\frac{\partial}{\partial y} + \hat{k}\frac{\partial}{\partial z}\right)$$

Using the definition, $\nabla \equiv \hat{i}\frac{\partial}{\partial x} + \hat{j}\frac{\partial}{\partial y} + \hat{k}\frac{\partial}{\partial z}$, the momentum operator is written as

$$\hat{p} = \frac{\hbar}{i}\nabla \tag{3.9}$$

The eigenvalue equation for the momentum operator is

$$\hat{p}\phi_{\bar{k}}(\bar{r}) = \bar{p}_{\bar{k}}\phi_{\bar{k}}(\bar{r}) \tag{3.10}$$

where the vectors $\bar{p}_{\bar{k}}$ are the momentum eigenvalues and $\phi_{\bar{k}}(\bar{r})$ are the momentum eigenfunctions. Substituting for the momentum operator $\hat{p} = \frac{\hbar}{i}\nabla$, we have

$$\frac{\hbar}{i}\nabla\phi_{\bar{k}}(\bar{r}) = \bar{p}_{\bar{k}}\phi_{\bar{k}}(\bar{r}) \tag{3.11}$$

Using the trial solution $\phi_{\bar{k}}(\bar{r}) = \phi_1(x)\phi_2(y)\phi_3(z)$, the above equation can be split into three equations:

$$\frac{\hbar}{i}\frac{\partial}{\partial x}\phi_1(\bar{r}) = p_{k_x}\phi_1(\bar{r}), \quad \frac{\hbar}{i}\frac{\partial}{\partial y}\phi_2(\bar{r}) = p_{k_y}\phi_2(\bar{r}),$$

$$\text{and} \quad \frac{\hbar}{i}\frac{\partial}{\partial z}\phi_3(\bar{r}) = p_{k_z}\phi_3(\bar{r})$$

where $p_{k_x} = \hbar k_x$, $p_{k_y} = \hbar k_y$, and $p_{k_z} = \hbar k_z$ are the components of the momentum eigenvalue. Solving these three equations, we find that the eigenfunctions are

$$\phi_{\bar{k}}(\bar{r}) = A e^{ik_x x} e^{ik_y y} e^{ik_z z} = A e^{i\bar{k}\cdot\bar{r}} \qquad (3.12)$$

where $\bar{k} = k_x\hat{i} + k_y\hat{j} + k_z\hat{k}$ is the wave vector. (k_x, k_y, k_z) stands for the three quantum numbers corresponding to the three spatial coordinates. The allowed values of k_x, k_y, and k_z are real numbers ranging from $-\infty$ to $+\infty$. The eigenvalues in vector form are

$$\bar{p}_{\bar{k}} = \hbar\bar{k} \qquad (3.13)$$

Concept: In quantum mechanics, there is an operator corresponding to every observable. Eigenfunctions of the position operator are the scalar spatial locations. Eigenfunctions of the momentum operator are plane waves.

Hamiltonian (energy operator)

The Hamiltonian operator $(\widehat{H})$ which corresponds to the total energy is obtained by adding the kinetic $(\widehat{KE})$ and potential $(\widehat{PE})$ energy operators

$$\widehat{H} = \widehat{KE} + \widehat{PE} \qquad (3.14)$$

From the position and momentum operators, we can find the kinetic and potential energy operators. The classical expression for kinetic energy is

$$KE = \frac{1}{2}mv^2 = \frac{(mv)^2}{2m} = \frac{p^2}{2m} \qquad (3.15)$$

where $p = mv$ is the momentum. In quantum mechanics, the kinetic energy operator follows by replacing p in the above equation by the momentum operator $\hat{p}$,

$$\text{Kinetic energy: } \widehat{KE} = \frac{\hat{p}^2}{2m} = \frac{\hat{p}\cdot\hat{p}}{2m} \qquad (3.16)$$

Substituting the expression for the 3D momentum operator from equation (3.9) into equation (3.16), we have

$$\widehat{KE} = \frac{1}{2m}\left(\frac{\hbar}{i}\nabla\right)\cdot\left(\frac{\hbar}{i}\nabla\right) = -\frac{\hbar^2}{2m}\left(\frac{\partial}{\partial x}\frac{\partial}{\partial x} + \frac{\partial}{\partial y}\frac{\partial}{\partial y} + \frac{\partial}{\partial z}\frac{\partial}{\partial z}\right)$$

$$= -\frac{\hbar^2}{2m}\left(\frac{\partial^2}{\partial x^2} + \frac{\partial^2}{\partial y^2} + \frac{\partial^2}{\partial z^2}\right)$$

$$\widehat{KE} = -\frac{\hbar^2}{2m}\nabla^2 \tag{3.17}$$

The operator for potential energy (PE) is a function of position and the classical form is $PE = U(\bar{r}, t)$. In quantum mechanics, we turn $\bar{r}$ to the operator form $\hat{r}$ to obtain the PE operator,

$$\widehat{PE} = U(\hat{r}, t) \tag{3.18}$$

Using equation (3.3), we know that $\hat{r}f(\bar{r}) = \bar{r}f(\bar{r})$, where $f(\bar{r})$ is an arbitrary wave function. So,

$$\widehat{PE}f(\bar{r}) = U(\hat{r}, t)f(\bar{r}) = U(\bar{r}, t)f(\bar{r})$$

So, if the potential energy operator is acting on a wavefunction expressed in terms of position $\bar{r}$, the potential energy operator can be written as

$$\widehat{PE} = U(\bar{r}, t) \tag{3.19}$$

From the expressions for $\widehat{KE}$ and $\widehat{PE}$ above, we have

$$\widehat{H} = -\frac{\hbar^2}{2m}\nabla^2 + U(\bar{r}, t) \tag{3.20}$$

Solving the following differential equation gives us the eigenvalues and eigenfunctions of the Hamiltonian,

$$\left[-\frac{\hbar^2}{2m}\nabla^2 + U(\bar{r})\right]\psi_n(\bar{r}) = E_n\psi_n(\bar{r}) \tag{3.21}$$

which is the time-independent Schrödinger equation. The eigenvalues and mathematical form of the eigenfunctions of the Hamiltonian depend on the precise form of the potential energy $U(\bar{r}, t)$. Analogous to the discussion of position and momentum operators in 1D, we can write the Hamiltonian in 1D by replacing the position vectors with scalars and the gradient with a partial differential,

$$\widehat{H} = -\frac{\hbar^2}{2m}\frac{\partial^2}{\partial x^2} + U(x) \tag{3.22}$$

The corresponding eigenvalue equation in 1D is

$$\left[-\frac{\hbar^2}{2m}\frac{\partial^2}{\partial x^2} + U(x) \right] \psi_n(x) = E_n \psi_n(x) \tag{3.23}$$

Concept: The Hamiltonian (energy) operator is the sum of the kinetic energy and the potential energy operators. For different situations, the eigenfunction will depend on the expression for the potential energy because the kinetic energy operator will be the same.

We will discuss operators for other quantities such as current density, spin angular momentum, and orbital angular momentum in later chapters. Note that charge, flux, voltage, and current in a circuit are also quantum mechanical operators. In Section 3.3, we discuss the Hamiltonian of the LC resonator using electric charge and flux operators.

3.2 Operators Corresponding to Observables in Quantum Mechanics

Physical quantities observed in nature are quantified by real numbers. Examples include position, momentum, energy, and so on. So, in quantum mechanics, operators, corresponding to observables should have real eigenvalues. The class of operators that yield real eigenvalues in equation (3.1) are called Hermitian operators.

An operator $\hat{O}$ is Hermitian if it satisfies the following relation:

$$\int dv \psi(\overline{r})^* \left[\hat{O}\phi(\overline{r}) \right] = \int dv \left[\hat{O}\psi(\overline{r}) \right]^* \phi(\overline{r}) \tag{3.24}$$

where the asterisk signifies the complex conjugate. Let us demonstrate that Hermitian operators have real eigenvalues. Consider an eigenfunction $\psi_n(\bar{r})$ of a Hermitian operator $\hat{O}$ with eigenvalue α_n,

$$\hat{O}\psi_n(\bar{r}) = \alpha_n \psi_n(\bar{r}) \tag{3.25}$$

We can use equation (3.25) to verify that

$$\int dv \psi_n(\bar{r})^* \left[\hat{O}\psi_n(\bar{r})\right] - \int dv \left[\hat{O}\psi_n(\bar{r})\right]^* \psi_n(\bar{r})$$

$$= (\alpha_n - \alpha_n^*) \int dv \psi_n(\bar{r})^* \psi_n(\bar{r})$$

Assuming that the eigenfunctions ψ_n are normalized, $\int dv \psi_n(\bar{r})^* \psi_n(\bar{r}) = 1$. So, the above equation becomes

$$\int dv \psi_n(\bar{r})^* \left[\hat{O}\psi_n(\bar{r})\right] - \int dv \left[\hat{O}\psi_n(\bar{r})\right]^* \psi_n(\bar{r}) = (\alpha_n - \alpha_n^*)$$

If an operator obeys the definition of being Herminitian (equation (3.24)), then the LHS of the above equation is zero, which means that $\alpha_n = \alpha_n^*$. Therefore, the eigenvalues of a Hermitian operator $\hat{O}$ are real. We will leave it to the reader to show that the eigenfunctions of a Hermitian operator are orthonormal to each other.

In quantum mechanics, operators that correspond to observables are linear. An operator $\hat{O}$ is linear if it satisfies the following properties:

Distributive property $\hat{O}\left[\psi(\bar{r}) + \phi(\bar{r})\right] = \hat{O}\psi(\bar{r}) + \hat{O}\phi(\bar{r})$

$$\tag{3.26}$$

Scalar multiplication $\hat{O}\left[c\psi(\bar{r})\right] = c\hat{O}\psi(\bar{r})$ $\qquad$ (3.27)

Equation (3.26) reflects the distributive property of a linear operator; that is, operator $\hat{O}$ acting on the sum of two functions ψ and ϕ is equal to the sum of $\hat{O}$ acting on ψ and $\hat{O}$ acting on ϕ individually. In equation (3.27), c is a complex scalar; if $\hat{O}$ is a linear operator, then c can be factored out.

Example 1. Show that the 1D momentum operator $\hat{p}_x = \frac{\hbar}{i}\frac{\partial}{\partial x}$ is Hermitian.

Solution: To show this, we substitute the 1D momentum operator $\hat{p}_x$ for $\hat{O}$ in the LHS of equation (3.24) and show that it is equal to the RHS. Using integration by parts, the LHS is equal to

$$\int dx \psi(x)^* \left[\frac{\hbar}{i} \frac{\partial}{\partial x} \phi(x) \right]$$

$$= \int dx \left[\frac{\hbar}{i} \frac{\partial}{\partial x} \left(\psi(x)^* \phi(x) \right) \right] - \int dx \phi(x) \left[\frac{\hbar}{i} \frac{\partial}{\partial x} \psi(x)^* \right]$$

$$= \left[\frac{\hbar}{i} \psi(x)^* \phi(x) \right]_{-\infty}^{+\infty} - \int dx \phi(x) \left[\frac{\hbar}{i} \frac{\partial}{\partial x} \psi(x)^* \right]$$

Since the wave functions $\psi(x)$ and $\phi(x)$ are both bounded (that is, the wave functions at $x = -\infty$ and $+\infty$ are zero), the first term above is zero. Then, the RHS can be rewritten as

$$= \int dx \phi(x) \left[\frac{\hbar}{i} \frac{\partial}{\partial x} \psi(x) \right]^*$$

So, the momentum operator in 1D satisfies equation (3.24), and is Hermitian. The reader should verify by themselves that while x is Hermitian, $\frac{\partial}{\partial x}$ is not Hermitian.

Orthogonal eigenfunctions: Consider two eigenfunctions of a Hermitian operator $\hat{O}$, $\psi_n(\bar{r})$, and $\psi_m(\bar{r})$ with two different eigenvalues a_n and a_m, respectively. We will now show that these eigenfunctions are **orthogonal** to each other. That is,

$$\textbf{Orthogonality} \quad \int dv \psi_n(\bar{r})^* \psi_m(\bar{r}) = 0 \qquad (3.28)$$

To prove this, we first note from equation (3.24) that

$$\int dv \psi_n(\bar{r})^* \left[\hat{O} \psi_m(\bar{r}) \right] = \int dv \left[\hat{O} \psi_n(\bar{r}) \right]^* \psi_m(\bar{r})$$

Using $\hat{O} \psi_m(\bar{r}) = a_m \psi_m(\bar{r})$ and $\hat{O} \psi_n(\bar{r}) = a_n \psi_n(\bar{r})$ in the above equation, we get

$$(a_n - a_m) \int dv \psi_n(\bar{r})^* \psi_m(\bar{r}) = 0$$

As eigenvalue $a_n \neq a_m$, we see that $\int dv \psi_n(\bar{r})^* \psi_m(\bar{r}) = 0$.

3.3   Expectation Value

Consider an experiment involving tossing a perfect dice. Then, the probability of getting numbers $n = 1, 2, \ldots, 6$ are all equal to $p(n) = 1/6$. If we are interested in finding the average (or mean value) of the numbers we get in many tossing trials, i.e., $\bar{n}$ or $\langle n \rangle$, then according to probability theory, we can write

$$\langle n \rangle \text{ or } \bar{n} = \sum_{n=1}^{6} p(n)n \qquad (3.29)$$

The above quantity is called *average value, mean value* or *expectation value*. Simply put, the average is summing all outcomes of the experiment (here n) weighted by the probability of each one occuring (here $1/6$). How can we generalize the above concept and define *expectation value* in a quantum mechanical sense?

If we have an electron described by the wave function $\Psi(x, t)$, then the probability of finding it in the position interval $[x, x + dx]$ is given by $p(x) = |\Psi(x, t)|^2 dx$. This is because $|\Psi(x, t)|^2$ is the probability density (probability per unit length), which after being multiplied by the length of the position interval, i.e., dx, returns the probability. Now if the electron traverses through different positions like $x_1, x_2, \ldots, x_N$, then the expectation (average) value of position is found from summing up the positions weighted by the probability as given in the following (similar to the dice example):

$$\langle x \rangle \text{ or } \bar{x} = \sum_{n=1}^{N} p(x_n)x_n = \sum_{n=1}^{N} |\Psi(x_n, t)|^2 x_n dx = \int_{-\infty}^{+\infty} |\Psi(x, t)|^2 x dx$$

$$(3.30)$$

In the case of continuous x (infinite possible positions), the summation is turned into integral. Note that since the probability density or norm of the wave function is $|\Psi(x, t)|^2 = \Psi(x, t)^* \Psi(x, t)$, we can rewrite equation (3.30) as follows:

$$\langle x \rangle = \int_{-\infty}^{+\infty} \Psi(x,t)^* \Psi(x,t) x dx = \int_{-\infty}^{+\infty} \Psi(x,t)^* x \Psi(x,t) dx$$

$$= \int_{-\infty}^{+\infty} \Psi(x,t)^* \hat{x} \Psi(x,t) dx \tag{3.31}$$

If we generalize the above and assert that the expectation value of a quantity described by quantum mechanical operator, $\hat{O}$, in a system with wave function $\Psi(\bar{r}, t)$ is

$$\langle O \rangle = \int dv \Psi(\bar{r}, t)^* \hat{O} \Psi(\bar{r}, t) \tag{3.32}$$

If $\hat{O} = \hat{r}$, which is the position operator, the **expectation value of the position** of the particle in state $\Psi(\bar{r}, t)$ is

$$\langle r \rangle = \int dv \Psi(\bar{r}, t)^* \hat{r} \Psi(\bar{r}, t) \tag{3.33}$$

Because the position operator is simply multiplicative as given by equation (3.3), the above equation becomes

$$\langle r \rangle = \int dv \Psi(\bar{r}, t)^* \bar{r} \Psi(\bar{r}, t) \tag{3.34}$$

$$\langle r \rangle = \int dv \bar{r} |\Psi(\bar{r}, t)|^2 \tag{3.35}$$

In the above equation, $|\Psi(\bar{r}, t)|^2$ is the probability density to find the particle at location $\bar{r}$.

If $\hat{O} = \hat{p}$, which is the momentum operator, the **expectation value of momentum** of the particle in state $\Psi(\bar{r}, t)$ is

$$\langle p \rangle = \int dv \Psi(\bar{r}, t)^* \hat{p} \Psi(\bar{r}, t) \tag{3.36}$$

Using equation (3.9), the above equation can be written as

$$\langle p \rangle = \frac{\hbar}{i} \int dv \Psi(\bar{r}, t)^* \nabla \Psi(\bar{r}, t) \tag{3.37}$$

As ∇ is a differential operator, the order of three terms inside the integral in the above equation is important. If a polynomial of operators $\hat{p}$ or $\hat{r}$ is given, then the expectation values of these polynomials are:

$$\langle F(\hat{r}) \rangle = \int dv \Psi(\bar{r}, t)^* F(\bar{r}) \Psi(\bar{r}, t) \tag{3.38}$$

$$\langle F(\hat{p}) \rangle = \int dv \Psi(\bar{r}, t)^* F\left(\frac{\hbar}{i} \nabla\right) \Psi(\bar{r}, t) \tag{3.39}$$

Proving equations (3.38) and (3.39) is quite straightforward.

If $\hat{O} = \hat{H}$, which is the Hamiltonian, the **expectation value of energy** is

$$\langle H \rangle = \int dv \Psi(\bar{r}, t)^* \hat{H} \Psi(\bar{r}, t) \tag{3.40}$$

For a time-independent Hamiltonian, we have

$$\langle H \rangle = \int dv \Psi(\bar{r}, t)^* [-\frac{\hbar^2}{2m} \nabla^2 + U(\bar{r})] \Psi(\bar{r}, t) \tag{3.41}$$

The kinetic and potential energy terms of the Hamiltonian are polynomials of the $\hat{p}$ and $\hat{r}$, operators i.e., $\hat{H} = \hat{p}^2/2m + U(\hat{r})$. Therefore we can use equations (3.38) and (3.39) which lead to the above equation. We again take the superposition state to be given by

$$\Psi(\bar{r}, t) = \sum_n a_n(t) \psi_n(\bar{r}) \tag{3.42}$$

where $a_n(t) = e^{-\frac{iE_n t}{\hbar}} a_{n0}$, ψ_n is the eigenfunction of the Hamiltonian with eigenvalue E_n, and a_{n0} is a complex number. Substituting equation (3.42) in equation (3.40), we have

$$\langle H \rangle = \int dv \left[\sum_m a_m(t) \psi_m(\bar{r}) \right]^* \hat{H} \sum_n a_n(t) \psi_n(\bar{r})$$

$$= \sum_{m,n} a_m(t)^* a_n(t) \int dv \psi_m(\bar{r})^* \hat{H} \psi_n(\bar{r})$$

Using $\hat{H} \psi_n = E_n \psi_n$ in the above equation, we get

$$\langle H \rangle = \sum_{m,n} a_m(t)^* a_n(t) \int dv \psi_m(\bar{r})^* E_n \psi_n(\bar{r})$$

$$= \sum_{m,n} a_m(t)^* a_n(t) E_n \int dv \psi_m(\bar{r})^* \psi_n(\bar{r})$$

Using the orthonormality of the eigenfunctions in the above equation, we have

$$\langle H \rangle = \sum_{m,n} a_m(t)^* a_n(t) E_n \delta_{m,n}$$

$$= \sum_n |a_n(t)|^2 E_n$$

$$= \sum_n |a_{n0}|^2 E_n \qquad (3.43)$$

which agrees with our discussion of the expectation value in Chapter 2. The above equation means that the average energy is the weighted sum of all the eigenenergies. The weight factors $|a_{n0}|^2$ are the probabilities of finding the system in those eigenenergies.

Concept: The *average* or *expectation value* $\langle O \rangle$ of an operator $\hat{O}$ is calculated using the formula $\langle O \rangle = \int dv \Psi(\bar{r}, t)^* \hat{O} \Psi(\bar{r}, t)$.

3.4   Commutator Bracket

The commutator bracket of two operators $\hat{B}$ and $\hat{C}$ is defined by

$$[\hat{B}, \hat{C}] = \hat{B}\hat{C} - \hat{C}\hat{B} \text{ or}$$

$$[\hat{B}, \hat{C}]f = \hat{B}\hat{C}f - \hat{C}\hat{B}f$$

where f is a function on which the operators act. The operator immediately to the left of f acts before the operators further to the left. So,

$$\hat{B}\hat{C}f = \hat{B}\lfloor \hat{C}f \rfloor$$

Two operators $\hat{B}$ and $\hat{C}$ are said to *commute* if $\hat{B}\hat{C} = \hat{C}\hat{B}$.

As an example, using the definition of momentum and position operators, we will show that $\hat{y}$ and $\hat{p}_x$ commute. That is, $\hat{y}\hat{p}_x = \hat{p}_x\hat{y}$. This can be verified by evaluating $[\hat{y}\hat{p}_x - \hat{p}_x\hat{y}]f(x, y)$. Because x and y are independent variables, $y\left(-i\hbar \frac{d}{dx}\right) f(x, y) = \left(-i\hbar \frac{d}{dx}\right) yf(x, y)$, which proves that $\hat{y}$ and $\hat{p}_x$ commute.

Next, we will show that $\hat{x}$ and $\hat{p}_x$ do not commute and we will find that $\hat{x}\hat{p}_x - \hat{p}_x\hat{x} = i\hbar \hat{I}$, where $\hat{I}$ is the identity operator. Identity operator leaves a state unchanged, $\hat{I}\psi = \psi$.

First, we apply $\hat{x}\hat{p}_x$ to the wave function $\psi(x, t)$ to find

$$\hat{x}\hat{p}_x\psi(x, t) = \hat{x}\left(-i\hbar \frac{d}{dx}\psi(x, t)\right) = -i\hbar\hat{x}\left(\frac{d}{dx}\psi(x, t)\right)$$

$$= -i\hbar x \frac{d}{dx}\psi(x, t)$$

In the above, the operator $\hat{x}$ is allowed to extract its eigenvalue x from the derivative of the wave function. On the other hand, when $\hat{p}_x\hat{x}$ operates on the wave function, we have

$$\hat{p}_x\hat{x}\psi(x, t) = \hat{p}_x(x\psi(x, t)) = -i\hbar \frac{d}{dx}(x\psi(x, t))$$

$$= -i\hbar\psi(x, t) - i\hbar x \frac{d}{dx}\psi(x, t)$$

By subtracting the latter from the former expression, we have

$$\hat{x}\hat{p}_x\psi(x, t) - \hat{p}_x\hat{x}\psi(x, t) = -i\hbar\psi(x, t)$$

which shows

$$[\hat{x}, \hat{p}_x] = \hat{x}\hat{p}_x - \hat{p}_x\hat{x} = i\hbar\hat{I}$$

meaning that two operators do not commute.

3.5 Examples

In this section, we will work out examples of the expectation value to further illustrate this concept.

Example 2 (Expectation value of position (PiB)). Find the expectation value of a particle's position in a box of length L. The particle is in an eigenfunction of the Hamiltonian.

Solution: From the definition of the expectation value of position in equation (3.33), we will evaluate

$$\langle \bar{r} \rangle = \int dv \Psi(\bar{r}, t)^* \hat{r} \Psi(\bar{r}, t)$$

Note that in our case $\hat{r} = \hat{x}$ because the box is 1D.

$$\langle x \rangle = \int dx \Psi(x, t)^* \hat{x} \Psi(x, t)$$

The eigenfunction of the PiB as a function of (x, t) is

$$\Psi(x, t) = e^{-\frac{iE_n t}{\hbar}} \sqrt{\frac{2}{L}} \sin\left(\frac{n\pi x}{L}\right)$$

By plugging this into the definition for the expectation value in equation (3.32) and noting that $(e^{-ia})^* \cdot (e^{-ia}) = e^{ia}e^{-ia} = e^0 = 1$, we have

$$\langle x \rangle = \int_{x=0}^{x=L} \left\{ e^{-\frac{iE_n t}{\hbar}} \sqrt{\frac{2}{L}} \sin\left(\frac{n\pi x}{L}\right) \right\}^* \hat{x} e^{-\frac{iE_n t}{\hbar}} \sqrt{\frac{2}{L}} \sin\left(\frac{n\pi x}{L}\right) dx$$

Using the definition of the position operator in equation (3.2), we have $\hat{x} \sin\left(\frac{n\pi x}{L}\right) = x \sin\left(\frac{n\pi x}{L}\right)$. So, the above equation for $\langle x \rangle$

becomes

$$\langle x \rangle = \frac{2}{L} \int_{x=0}^{x=L} x \sin^2 \left(\frac{n\pi x}{L} \right) dx$$

Using the trigonometric identity $\sin^2(\theta) = \frac{1-\cos(2\theta)}{2}$, we have

$$\langle x \rangle = \frac{2}{L} \int_{x=0}^{x=L} \frac{x}{2} dx + \frac{2}{L} \int_{x=0}^{x=L} \frac{-x}{2} \cos \left(\frac{2n\pi x}{L} \right) dx$$

Using integration by parts, you can show that the second term is zero due to the periodicity of the cosine function. The first term is nonzero:

$$\langle x \rangle = \frac{2}{L} \int_{x=0}^{x=L} \frac{x}{2} dx = \frac{1}{L} \left[\frac{x^2}{2} \right]_{x=0}^{x=L} = \frac{L}{2} \qquad (3.44)$$

The above equation means the expectation (or average) value of position in any eigenstate (n) of the Hamiltonian is $\frac{L}{2}$. If we consider the particle with quantum number $n = 1$, we know that the corresponding eigenfunction $\psi_1(x) = \sqrt{\frac{2}{L}} \sin \left(\frac{\pi x}{L} \right)$ is peaked at and symmetric about the position $x = \frac{L}{2}$. So, it makes intuitive sense that the probability density is maximum at $\frac{L}{2}$. However, for a particle with quantum number $n = 2$, we know that the eigenfunction $\psi_2(x) = \sqrt{\frac{2}{L}} \sin \left(\frac{2\pi x}{L} \right)$ is zero at $x = \frac{L}{2}$ while the probability density is maximum at $x = \frac{L}{4}$ and $\frac{3L}{4}$. However, if we use equation (3.33), to solve for the expectation value in eigenfunction $\psi_2(x)$, we also get $\langle x \rangle = \frac{L}{2}$ even though the probability density is zero there. This is because the probability density $\frac{2}{L} \sin^2(\frac{n\pi x}{L})$ is symmetric about the point $x = \frac{L}{2}$.

Example 3 (Expectation value of position (free particle)).
Find the expectation value of the *position* of a free particle in an eigenstate of the Hamiltonian.

Solution: From the definition of the expectation value of position in equation (3.33), we evaluate,

$$\langle x \rangle = \int dx \Psi(x,t)^* \hat{x} \Psi(x,t)$$

The wave function of a free particle as a function of (x,t) is

$$\Psi(x,t) = e^{-\frac{iE_k t}{\hbar}}\sqrt{\frac{1}{L}}e^{ikx}$$

By plugging this into the definition for $\langle x \rangle$, we have

$$\langle x \rangle = \lim_{L\to\infty} \int_{x=-L/2}^{x=L/2} \left\{ e^{-\frac{iE_k t}{\hbar}}\sqrt{\frac{1}{L}}e^{ikx} \right\}^* \hat{x} e^{-\frac{iE_k t}{\hbar}}\sqrt{\frac{1}{L}}e^{ikx} dx$$

$$= \lim_{L\to\infty} \frac{1}{L} \int_{x=-L/2}^{x=L/2} x dx = 0$$

This result is expected because the probability density $|\Psi(x,t)|^2 = \frac{1}{L}$ is constant over all space. It will be equally likely to find the particle at $-x$ and $+x$, thus the expectation (average) value will be at $x = 0$.

Example 4 (Expectation value of momentum (PiB)). Find the expectation value of *momentum* of a particle in a box of length L. The particle is in an eigenfunction of the Hamiltonian.

Solution: From the definition of the expectation value of the momentum operator in equation (3.36), we must calculate the following quantity:

$$\langle p_x \rangle = \int dx\, \Psi(x,t)^* \hat{p}_x \Psi(x,t)$$

The wave function is $\Psi(x,t) = e^{-i\frac{E_n t}{\hbar}}\sqrt{\frac{2}{L}}\sin\left(\frac{n\pi x}{L}\right)$.

As the box is 1D, we have the momentum operator only along the x direction. If we substitute the wave function in the integral and note that the momentum operator $-i\hbar\frac{d}{dx}$ (equation (3.4)) acts on the wave function $\Psi(x,t)$ to its right-hand side:

$$\hat{p}_x\Psi(x,t) = -i\hbar\frac{d}{dx}\Psi(x,t) = -i\hbar\left(\frac{n\pi}{L}\right)e^{-i\frac{E_n t}{\hbar}}\sqrt{\frac{2}{L}}\cos\left(\frac{n\pi x}{L}\right)$$

$$\langle p_x \rangle = -i\hbar\left(\frac{n\pi}{L}\right)\frac{2}{L}\int_{x=0}^{x=L}\sin\left(\frac{n\pi x}{L}\right)\cos\left(\frac{n\pi x}{L}\right)dx$$

Using $\sin(2\theta) = 2\sin(\theta)\cos(\theta)$, we have

$$\langle p_x \rangle = -i\hbar \left(\frac{n\pi}{L}\right) \frac{1}{L} \int_{x=0}^{x=L} \sin\left(\frac{2n\pi x}{L}\right) dx$$

$$= \frac{-i\hbar}{2L} \left[\cos\left(\frac{2n\pi x}{L}\right)\right]_{x=0}^{x=L} = 0 \tag{3.45}$$

That is, the expectation (or average) value of momentum along x direction is zero.

Example 5 (Expectation value of momentum (free particle)). Find the expectation value of the *momentum* of a free particle. The particle is in an eigenstate of the Hamiltonian, which is a traveling wave.

Solution: From the definition of the expectation value of momentum in equation (3.36), we evaluate

$$\langle p_x \rangle = \int dx \Psi(x,t)^* \hat{p}_x \Psi(x,t)$$

The eigenfunction of the free particle as a function of (x,t) is

$$\Psi(x,t) = e^{-\frac{iE_k t}{\hbar}} \sqrt{\frac{1}{L}} e^{ikx}$$

By plugging this into the definition for $\langle p_x \rangle$ and using equations (3.5) and (3.8), we have

$$\langle p_x \rangle = \lim_{L \to \infty} \int_{x=-L/2}^{x=L/2} \left\{ e^{-\frac{iE_k t}{\hbar}} \sqrt{\frac{1}{L}} e^{ikx} \right\}^* \hat{p}_x e^{-\frac{iE_k t}{\hbar}} \sqrt{\frac{1}{L}} e^{ikx} dx$$

$$= \lim_{L \to \infty} \int_{x=-L/2}^{x=L/2} \left\{ e^{-\frac{iE_k t}{\hbar}} \sqrt{\frac{1}{L}} e^{ikx} \right\}^* \left(-i\hbar \frac{d}{dx}\right) e^{-\frac{iE_k t}{\hbar}} \sqrt{\frac{1}{L}} e^{ikx} dx$$

$$= \lim_{L \to \infty} \int_{x=-L/2}^{x=L/2} \left\{ e^{-\frac{iE_k t}{\hbar}} \sqrt{\frac{1}{L}} e^{ikx} \right\}^* (\hbar k) e^{-\frac{iE_k t}{\hbar}} \sqrt{\frac{1}{L}} e^{ikx} dx$$

$$= \lim_{L \to \infty} \frac{\hbar k}{L} \int_{x=-L/2}^{x=L/2} dx$$

$$= \lim_{L \to \infty} \frac{\hbar k}{L} L$$

$$= \hbar k$$

That is, the free particle has a well-defined expectation value momentum of $\hbar k$.

Example 6 (Expectation value of energy (PiB)). A particle in a box is in the superposition of the first and the third eigenfunctions as follows:

$$\Psi(x, t = 0) = \frac{1}{\sqrt{2}} \psi_1 + \frac{1}{\sqrt{2}} \psi_3$$

Find the wave function at a later time t.

Calculate the expectation (average) value of energy in terms of meV and its corresponding frequency in units of THz. The particle's mass is six times the rest mass of an electron, and the length of the box (L) is 10 Å.

Solution: The time evolution of the wave function given in Chapter 1 is

$$\Psi(x, t) = e^{-i\frac{\hat{H}t}{\hbar}} \Psi(x, t = 0) = e^{-i\frac{\hat{H}t}{\hbar}} \left(\frac{1}{\sqrt{2}} \psi_1 + \frac{1}{\sqrt{2}} \psi_3 \right)$$

$$= \frac{e^{-iE_1 t/\hbar}}{\sqrt{2}} \psi_1 + \frac{e^{-iE_3 t/\hbar}}{\sqrt{2}} \psi_3$$

The expectation value of energy is the sum of eigenenergies (E_1 and E_3) weighted by the probability of finding the corresponding eigenstates (ψ_1 and ψ_3) as given in equation (3.43),

$$\langle E \rangle = \sum_n |a_n(t)|^2 E_n = \left| \frac{e^{-iE_1 t/\hbar}}{\sqrt{2}} \right|^2 E_1 + \left| \frac{e^{-iE_3 t/\hbar}}{\sqrt{2}} \right|^2 E_3 = \frac{E_1 + E_3}{2}$$

Using the energy $E_n = \frac{n^2\pi^2\hbar^2}{mL^2}$ of the PiB,

$$\langle E \rangle = \frac{E_1 + E_3}{2} = \frac{\pi^2\hbar^2}{mL^2}\left(\frac{1+9}{2}\right)$$

$$= 5 \times \frac{\pi^2 \times (1.0545 \times 10^{-34}\,\text{J}\cdot\text{s})^2}{6 \times (9.11 \times 10^{-31}\,\text{kg}) \times (10 \times 10^{-10}\,\text{m})^2}$$

Converting with Joule $= \frac{\text{kg·m}^2}{\text{s}^2}$,

$$\langle E \rangle = 10^{-19}\frac{(\text{J}\cdot\text{s})^2}{\text{kg·m}^2} = 10^{-19}\frac{\text{kg·m}^2}{\text{s}^2} = 10^{-19}\,\text{J}$$

With the conversion $1\,\text{eV} = 1.602 \times 10^{-19}\,\text{J}$, we have

$$\langle E \rangle = \frac{10^{-19}\,\text{J}}{1.602 \times 10^{-19}\,\text{J/eV}} = 624\,\text{meV}$$

By dividing $\langle E \rangle$ expressed in units of Joule by Planck's constant, we can find which frequency this energy corresponds to

$$\langle E \rangle = hf \rightarrow f = \frac{10^{-19}\,\text{J}}{6.6267 \times 10^{-34}\,\text{J}\cdot\text{s}} = 150.9\,\text{THz}$$

Example 7 (Expectation value of energy (PiB)). We are given the wave function at time $t = 0$ of a PiB in a superposition of three eigenfunctions of the Hamiltonian:

$$\Psi(x, t = 0) = a_{10}\psi_1(x) + a_{20}\psi_2(x) + a_{30}\psi_3(x)$$

where $a_{10} = \frac{1}{\sqrt{7}}$, $a_{20} = \frac{2}{\sqrt{7}}$, and $a_{30} = \frac{\sqrt{2}}{\sqrt{7}}$.

(a) Show that the wave function is normalized.
(b) What is the expectation value of the energy of the state $\Psi(x, t)$ at any time?

(c) Find the probability density of the wave function as it evolves in time.

Solution:

(a) First, note that the initial superposition state is normalized because

$$\int_{x=0}^{x=L} |\Psi(x, t = 0)|^2 dx = |a_{10}|^2 + |a_{20}|^2 + |a_{30}|^2 = \frac{1}{7} + \frac{4}{7} + \frac{2}{7} = 1$$

where we have used the orthonormality of eigenfunctions.

In order to obtain the time evolution of the superposition state, we operate the wave function at time $t = 0$ by $e^{-\frac{i\hat{H}t}{\hbar}}$.

$$\Psi(x, t) = e^{-i\frac{\hat{H}t}{\hbar}} (a_{10}\psi_1(x) + a_{20}\psi_2(x) + a_{30}\psi_3(x))$$

$$= e^{-i\frac{E_1 t}{\hbar}} a_{10}\psi_1(x) + e^{-i\frac{E_2 t}{\hbar}} a_{20}\psi_2(x) + e^{-i\frac{E_3 t}{\hbar}} a_{30}\psi_3(x)$$

$$(3.46)$$

The expectation value of energy is found simply by adding energies weighted by the probabilities of each eigenenergy (see equation (3.43)):

$$\langle E \rangle = \sum_n |a_n(t)|^2 E_n = \left| \frac{e^{-\frac{iE_1 t}{\hbar}}}{\sqrt{7}} \right|^2 E_1 + \left| 2\frac{e^{-\frac{iE_2 t}{\hbar}}}{\sqrt{7}} \right|^2 E_2 + \left| \sqrt{2}\frac{e^{-\frac{iE_3 t}{\hbar}}}{\sqrt{7}} \right|^2 E_3$$

$$= \frac{E_1 + 4E_2 + 2E_3}{7} = \frac{\pi^2\hbar^2}{mL^2} \left(\frac{1 + 4 \times 4 + 2 \times 9}{7} \right) = 5\frac{\pi^2\hbar^2}{mL^2}$$

If $L = 10\,\text{Å}$ and the particle's mass is six times the mass of a free electron, similar to the previous example, you get $\langle E \rangle = 10^{-19}\,\text{J} = 624\,\text{meV} = h(150.9\,\text{THz})$.

(b) The probability density at (x, t) is given by

$$|\Psi(x, t)|^2 = \Psi^*(x, t)\Psi(x, t)$$

$$= |a_{10}|^2|\psi_1(x)|^2 + |a_{20}|^2|\psi_2(x)|^2 + |a_{30}|^2|\psi_3(x)|^2$$

$$+ 2\text{Re}\left[a_{10}^*a_{20}e^{-\frac{i(E_1-E_2)t}{\hbar}}\psi_1(x)^*\psi_2(x)\right]$$

$$+ 2\text{Re}\left[a_{10}^*a_{30}e^{-\frac{i(E_1-E_3)t}{\hbar}}\psi_1(x)^*\psi_3(x)\right]$$

$$+ 2\text{Re}\left[a_{20}^*a_{30}e^{-\frac{i(E_2-E_3)t}{\hbar}}\psi_2(x)^*\psi_3(x)\right]$$

where we have used the following identities:

$$(a + b + c)^*(a + b + c) = a^*a + b^*b + c^*c + a^*b + ab^*$$

$$+ a^*c + ac^* + c^*b + cb^* \text{ and}$$

$$a^*b + ab^* = 2\text{Re}(ab)$$

$\text{Re}(z)$ stands for the real part of z. Given that $\psi_n(x) = \sqrt{\frac{2}{L}}\sin\left(\frac{n\pi}{L}x\right)$ for a PiB, noting that a_{1o}, a_{2o}, and a_{3o} are all real values, the expression for the probability density becomes

Probability density at (x, t)

$$= \frac{2}{L}\left[a_{10}^2\sin^2\left(\frac{1\pi}{L}x\right) + a_{20}^2\sin^2\left(\frac{2\pi}{L}x\right) + a_{30}^2\sin^2\left(\frac{3\pi}{L}x\right)\right.$$

$$+ 2a_{10}\cdot a_{20}\sin\left(\frac{1\pi}{L}x\right)\sin\left(\frac{2\pi}{L}x\right)\cos\left(\frac{(E_1 - E_2)t}{\hbar}\right)$$

$$+ 2a_{10}\cdot a_{30}\sin\left(\frac{1\pi}{L}x\right)\sin\left(\frac{3\pi}{L}x\right)\cos\left(\frac{(E_1-E_3)t}{\hbar}\right)$$

$$+ 2a_{20}\cdot a_{30}\sin\left(\frac{2\pi}{L}x\right)\sin\left(\frac{3\pi}{L}x\right)\cos\left.\left(\frac{(E_2 - E_3)t}{\hbar}\right)\right]$$

Remember that the probability is not merely the sum of three independent probabilities $|a_{10}|^2\sin^2\left(\frac{1\pi}{L}x\right) + |a_{20}|^2\sin^2\left(\frac{2\pi}{L}x\right) + |a_{30}|^2\sin^2\left(\frac{3\pi}{L}x\right)$; instead, it has an additional cosine term for each interference between the three eigenfunctions. Interference is introduced from the phase difference between the time-dependent eigenstates comprising the superposition $\Psi(x,t)$ in equation (3.46). In the example corresponding to Figures 2.3 and 2.4, we demonstrated the oscillation of probability density with time.

Example 8 (Expectation value of energy (SHO)). Find the expectation value of the energy of a SHO at time $t = 0$ if the wave function that we are given is in a superposition of two eigenfunctions as in

$$\Psi(x, t = 0) = a_o\psi_0(x) + a_2\psi_2(x)$$

where $a_o = \frac{1}{\sqrt{2}}$ and $a_2 = \frac{1}{\sqrt{2}}$.

Solution: The time-evolved superposition state at any time is

$$\Psi(x,t) = e^{-i\frac{\hat{H}t}{\hbar}}(a_o\psi_0(x) + a_2\psi_2(x))$$

$$= e^{-i\frac{E_0 t}{\hbar}}a_o\psi_0(x) + e^{-i\frac{E_2 t}{\hbar}}a_2\psi_2(x)$$

Since eigenenergies of the SHO are

$$E_n = \left(n + \frac{1}{2}\right)\hbar\omega \rightarrow E_0 = \frac{1}{2}\hbar\omega, \quad E_2 = \frac{5}{2}\hbar\omega$$

the expectation value of energy is (equation (3.43))

$$\langle E \rangle = \sum_n |a_n(t)|^2 E_n = \left|\frac{e^{-i0.5\omega t}}{\sqrt{2}}\right|^2 \left(\frac{1}{2}\hbar\omega\right)$$

$$+ \left|\frac{e^{-i2.5\omega t}}{\sqrt{2}}\right|^2 \left(\frac{5}{2}\hbar\omega\right) = \frac{3}{2}\hbar\omega$$

This state has an average energy corresponding to the state $n = 1$ even though a single energy measurement will never yield this value (see Section 2.4).

Example 9 (Translation operator). Show that applying the operator,

$$\hat{T} = e^{-\frac{ib\hat{p}_x}{\hbar}} \tag{3.47}$$

shifts the position of the particle by b. That is,

$$\hat{T}\psi(x) = \psi(x - b) \tag{3.48}$$

$\hat{T}$ is called the translation operator because it shifts the position of the particle by b.

Solution: Let us expand the exponential function using Taylor series and then apply every term of the expansion to the wave function:

$$\hat{T}\psi(x) = \left[\hat{I} - i\frac{b\hat{p}_x}{\hbar} + \frac{1}{2!}\left(-i\frac{b\hat{p}_x}{\hbar}\right)^2 + \frac{1}{3!}\left(-i\frac{b\hat{p}_x}{\hbar}\right)^3 + \cdots\right]\psi(x)$$

$$= \left[\hat{I} - i\frac{b\hat{p}_x}{\hbar} - \frac{b^2}{\hbar^2 2!}(\hat{p}_x)^2 + \frac{ib^3}{\hbar^3 3!}(\hat{p}_x)^3 + \cdots\right]\psi(x)$$

Now we use the fact that the momentum operator acts as $\hat{p}_x = -i\hbar\frac{\partial}{\partial x}$

$$\hat{T}\psi(x) = \left[\hat{I} - b\frac{\partial}{\partial x} + \frac{b^2}{2!}\left(\frac{\partial}{\partial x}\right)^2 - \frac{b^3}{3!}\left(\frac{\partial}{\partial x}\right)^3 + \cdots\right]\psi(x)$$

$$= \psi(x) - b\frac{\partial\psi(x)}{\partial x} + \frac{b^2}{2!}\frac{\partial^2\psi(x)}{\partial x^2} - \frac{b^3}{3!}\frac{\partial^3\psi(x)}{\partial x^3} + \cdots$$

The terms on the right-hand side of the above equation are simply the Taylor series expansion of $\psi(x - b)$. That is,

$$\hat{T}\psi(x) = \psi(x - b)$$

Example 10 (Time evolution of a wavefunction with an imaginary energy). This example is an advanced topic which the reader is encouraged to skip on the first reading. We know that a Hamiltonian, which is a Hermitian operator, can have only real eigenvalues. However, we will make an exception and consider an imaginary eigenvalue for energy and illustrate what happens to the system's wave function. Consider a system with two energy eigenvalues which are $E_1 = 3\hbar$ and $E_2 = \left(5 - \frac{i}{2}\right)\hbar$.

Find the eigenfunctions $\varphi_1(x, t)$ and $\varphi_2(x, t)$ and their probability densities as a function of time t.

What is the physical meaning of the negative imaginary part of the particle's energy?

Find the expectation value of energy for the superposition state $\Psi(x, t) = \frac{1}{\sqrt{2}}\varphi_1(x, t) + \frac{1}{\sqrt{2}}\varphi_2(x, t)$.

Solution: We know that the wave function evolves in time according to the time-dependent Schrödinger equation:

$$i\hbar \frac{d}{dt}|\Psi(t)\rangle = \hat{H}|\Psi(t)\rangle \quad \rightarrow \quad |\Psi(t)\rangle = e^{-i\frac{\hat{H}t}{\hbar}}|\Psi(0)\rangle$$

The time-evolved eigenfunctions at a later time are

$$\varphi_1(x, t) = e^{-i\frac{\hat{H}t}{\hbar}}|\varphi_1(x, 0)\rangle = e^{-i\frac{E_1 t}{\hbar}}|\varphi_1(x, 0)\rangle = e^{-i3t}|\varphi_1(x, 0)\rangle \quad \text{and}$$
$$(3.49)$$

$$\varphi_2(x, t) = e^{-i\frac{\hat{H}t}{\hbar}}|\varphi_2(x, 0)\rangle = e^{-i\frac{E_2 t}{\hbar}}|\varphi_2(x, 0)\rangle = e^{-i5t}e^{-0.5t}|\varphi_2(x, 0)\rangle$$
$$(3.50)$$

Since the amplitudes of exponential functions with purely imaginary exponents are 1, the probability density changes with time as

$$|\varphi_1(x, t)|^2 = |e^{-i3t}|^2|\varphi_1(x, 0)|^2 = 1 \times |\varphi_1(x, 0)|^2 = |\varphi_1(x, 0)|^2 \text{ and}$$
$$(3.51)$$

$$|\varphi_2(x, t)|^2 = |e^{-i3t}|^2|e^{-0.5t}|^2|\varphi_2(x, 0)|^2 = e^{-t}|\varphi_2(x, 0)|^2 \qquad (3.52)$$

For a particle in $\varphi_1(x, 0)$, the probability amplitude stays constant. However, for the particle in $\varphi_2(x, 0)$, the probability amplitude decreases as time increases. This decay of probability means that the particle leaves the state φ_2 and goes somewhere else. A negative imaginary part added to the eigenenergy models the decay of probability amplitude due to the interaction of the quantum system with the outside world.

For the expectation value of energy, we can use the following method, which adds the eigenenergies weighted by their corresponding probability amplitudes:

$$\Psi(x,t) = \frac{1}{\sqrt{2}}\varphi_1(x,t) + \frac{1}{\sqrt{2}}\varphi_2(x,t)$$

$$\langle H \rangle = \langle E \rangle = \sum_n |a_n(t)|^2 E_n$$

$$= \left| \frac{1}{\sqrt{2}} e^{-i3t} \right|^2 E_1 + \left| \frac{1}{\sqrt{2}} e^{-i5t} e^{-0.5t} \right|^2 E_2$$

The above equation yields

$$\langle H \rangle = \langle E \rangle = \frac{1}{2} E_1 + \frac{1}{2} e^{-t} E_2 = \frac{\hbar}{2}(3 + 5e^{-t}) \tag{3.53}$$

Note that we only added real energy eigenvalues to obtain the average or expectation value of energy. The imaginary part of the eigenenergy results in a decaying term, which means that the energy relaxes (dissipates) as time increases. The above equation shows that the average energy of the system relaxes from $4\hbar$ to $\frac{3\hbar}{2}$ as $e^{-\frac{t}{1s}}$, meaning that the *relaxation time constant* is $1\,s$.

3.6   LC Oscillator

In an LC oscillator (Figure 3.1) without dissipation, we know that the energy oscillates between magnetic and electric forms. We will now describe an LC oscillator using quantum mechanics and discuss the essential consequences. The various symbols used are

Voltage: V,   Charge: Q,   Capacitance: $\mathbb{C}$
Current: $\mathbb{I}$,   Magnetic flux: Φ,   Inductance: $\mathcal{L}$

Recalling that the electric energy stored in the capacitor is

$$\text{Electric energy} = \frac{1}{2}\mathbb{C}V^2 \tag{3.54}$$

and the magnetic energy stored in the inductor is

$$\text{Magnetic energy} = \frac{1}{2}\mathcal{L}\mathbb{I}^2 \tag{3.55}$$

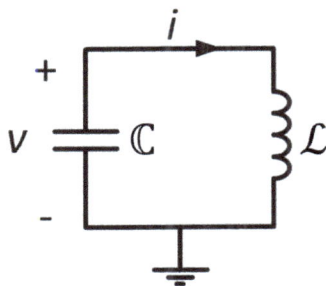

Figure 3.1. An LC resonator for which $H = \frac{\Phi^2}{2\mathcal{L}} + \frac{Q^2}{2\mathbb{C}}$.

The total energy, which is the Hamiltonian, is the sum of the electric and magnetic components,

$$H = \frac{1}{2}\mathbb{C}V^2 + \frac{1}{2}\mathcal{L}\mathbb{I}^2 \tag{3.56}$$

Using expressions from classical electricity and magnetism,

$$Q = \mathbb{C}V \text{ and} \tag{3.57}$$

$$\Phi = \mathcal{L}\mathbb{I} \tag{3.58}$$

the total energy in equation (3.56) can be rewritten as

$$H = \frac{Q^2}{2\mathbb{C}} + \frac{\Phi^2}{2\mathcal{L}} \tag{3.59}$$

We observe that the derivative of the Hamiltonian with respect to charge on the capacitor Q gives us

$$\frac{\partial H}{\partial Q} = \frac{Q}{\mathbb{C}} \tag{3.60}$$

Again from electricity and magnetism, we know that $\frac{Q}{\mathbb{C}}$ is the voltage across the capacitor (V) and the voltage across the inductor is $V = \frac{\partial \Phi}{\partial t}$

$$V = \frac{Q}{\mathbb{C}} \quad \text{and} \quad \text{(capacitor)} \tag{3.61}$$

$$V = \frac{\partial \Phi}{\partial t} \quad \text{(inductor)} \tag{3.62}$$

As the capacitor and inductor in the LC circuit have the same voltage across them. Therefore,

$$\frac{\partial \Phi}{\partial t} = \frac{Q}{\mathbb{C}} \tag{3.63}$$

Substituting equation (3.63) in equation (3.60),

$$\frac{\partial H}{\partial Q} = \frac{\partial \Phi}{\partial t} \tag{3.64}$$

Similarly, we observe that the derivative of the Hamiltonian with respect to the flux Φ gives us

$$\frac{\partial H}{\partial \Phi} = \frac{\Phi}{\mathcal{L}} \tag{3.65}$$

We know that for an inductor,

$$\Phi = \mathcal{L}\mathbb{I} = \mathcal{L}\frac{\partial Q}{\partial t} \tag{3.66}$$

where we have used that current in the circuit is $\mathbb{I} = \frac{\partial Q}{\partial t}$. Substituting equation (3.66) in equation (3.65), we have

$$\frac{\partial H}{\partial \Phi} = \frac{\partial Q}{\partial t} \tag{3.67}$$

In physics, variables (Q, Φ) that satisfy equations (3.64) and (3.67) are called *canonical variables*. A set of canonical variables that you have used many times are (x, p_x), position and momentum. You should verify that for the Hamiltonian of a SHO,

$$H = \frac{p_x^2}{2m} + \frac{1}{2}\mathcal{K}x^2 \quad \text{(Hamiltonian of SHO)} \tag{3.68}$$

$$\frac{\partial H}{\partial x} = \frac{\partial p_x}{\partial t} \tag{3.69}$$

$$\frac{\partial H}{\partial p_x} = \frac{\partial x}{\partial t} \tag{3.70}$$

where we have used $p_x = m \times \text{velocity} = m\frac{\partial x}{\partial t}$ and force $F = \mathcal{K}x = \frac{\partial p_x}{\partial t}$.

In quantum mechanics, the charge (Q) and flux (Φ) are operators. So, we add an extra hat on top of each quantity, and the total energy in equation (3.59) becomes the Hamiltonian operator,

$$\widehat{H} = \frac{\hat{Q}^2}{2\mathbb{C}} + \frac{\hat{\Phi}^2}{2\mathcal{L}} \tag{3.71}$$

Just as canonical variables (x, p_x) are related by

$$\hat{p}_x = \frac{\hbar}{i} \frac{d}{dx}$$

the canonical variables (Φ, Q) are related by

$$\hat{Q} = \frac{\hbar}{i} \frac{d}{d\Phi} \tag{3.72}$$

Substituting equation (3.72) in equation (3.71), the Hamiltonian can be rewritten as

$$\widehat{H} = -\frac{\hbar^2}{2\mathbb{C}} \frac{\partial^2}{\partial\Phi^2} + \frac{1}{2\mathcal{L}} \hat{\Phi}^2 \tag{3.73}$$

Concept: Flux ($\hat{\Phi}$) and charge ($\hat{Q}$) are canonical variables, just as position ($\hat{x}$) and momentum ($\hat{p}_x$) are.

The Hamiltonian of a quantum mechanical LC oscillator can be defined in terms of charge and flux operators.

3.6.1 *Solution of Schrödinger equation for LC oscillator*

The Hamiltonian of the LC oscillator in equation (3.73) has an identical form to the SHO discussed in Chapter 1. Schrödinger equation for the LC oscillator then reads

$$\left[-\frac{\hbar^2}{2\mathbb{C}} \frac{\partial^2}{\partial\Phi^2} + \frac{1}{2\mathcal{L}} \Phi^2 \right] u(\Phi) = E u(\Phi) \tag{3.74}$$

which is identical in form to equation (1.77). Table 3.1 shows the close correspondence between the two problems. The operators for position $\hat{x}$ and momentum $\hat{p} = \frac{\hbar}{i} \frac{d}{dx}$ in the SHO have been replaced by the operators for flux $\hat{\Phi}$ and charge $\hat{Q} = \frac{\hbar}{i} \frac{d}{d\Phi}$ in the LC oscillator.

Table 3.1. Correspondence between various terms for simple harmonic oscillator and LC oscillator.

Correspondence between simple harmonic and LC oscillators	Simple harmonic oscillator	LC oscillator
	Mass: m Spring constant: $\mathcal{K}$ Position: x	Capacitance: $\mathbb{C}$ 1/inductance: $1/\mathcal{L}$ Flux: Φ
	Momentum: $\hat{p}_x = \dfrac{\hbar}{i}\dfrac{d}{dx}$	Charge: $\hat{Q} = \dfrac{\hbar}{i}\dfrac{d}{d\Phi}$
	Kinetic energy: $-\dfrac{\hbar^2}{2m}\dfrac{\partial^2}{\partial x^2}$	Capacitive energy: $-\dfrac{\hbar^2}{2\mathbb{C}}\dfrac{d^2}{d\Phi^2}$
	Potential energy: $\dfrac{1}{2}\mathcal{K}$	Inductive energy: $\dfrac{1}{2\mathcal{L}}\Phi^2$
Hamiltonian	$-\dfrac{\hbar^2}{2m}\dfrac{\partial^2}{\partial x^2} + \dfrac{1}{2}\mathcal{K}x^2$	$-\dfrac{\hbar^2}{2\mathbb{C}}\dfrac{d^2}{d\Phi^2} + \dfrac{1}{2\mathcal{L}}\Phi^2$
Resonant frequency	$\omega = \sqrt{\dfrac{\mathcal{K}}{m}}$	$\omega = \dfrac{1}{\sqrt{\mathcal{L}\mathbb{C}}}$
Eigenenergies	$E_n = \left(n+\dfrac{1}{2}\right)\hbar\omega$	$E_n = \left(n+\dfrac{1}{2}\right)\hbar\omega$
Eigenfunctions	$\dfrac{1}{\sqrt{2^n n!}}\left(\dfrac{m\omega}{\hbar}\right)^{1/4} e^{-\frac{m\omega x^2}{2\hbar}} H_n\left(\sqrt{\dfrac{m\omega}{\hbar}}\,x\right)$	$\dfrac{1}{\sqrt{2^n n!}}\left(\dfrac{\omega\mathbb{C}}{\hbar}\right)^{1/4} e^{-\frac{\omega\mathbb{C}\Phi^2}{2\hbar}} H_n\left(\sqrt{\dfrac{\omega\mathbb{C}}{\hbar}}\,\Phi\right)$

Also, note the potential energy term in the LC oscillator, $\frac{1}{2\mathcal{L}}\Phi^2$, is similar to the PE energy term for an SHO $\frac{1}{2}\mathcal{K}x^2$, where the spring constant $\mathcal{K}$ in SHO corresponds to $\frac{1}{\mathcal{L}}$. The KE energy-like term in the LC oscillator $-\frac{\hbar^2}{2\mathbb{C}}\frac{\partial^2}{\partial\Phi^2}$ is similar to the KE energy term for an SHO $-\frac{\hbar^2}{2m}\frac{\partial^2}{\partial x^2}$, where m in the SHO corresponds to the capacitance $\mathbb{C}$.

The eigenfunctions $u(\Phi)$ and eigenenergies E_n, obtained by solving equation (3.74), have an identical mathematical form to equations (1.78)–(1.82) of the SHO in Chapter 1. The eigenenergies can be written down in analogy to be

$$E_n = \left(n + \frac{1}{2}\right)\hbar\omega \tag{3.75}$$

where the quantum numbers are

$$n = 0, 1, 2, 3, \cdots \tag{3.76}$$

The angular frequency ω is related to the inductance $\mathcal{L}$ and capacitance $\mathbb{C}$ via the usual classical expression,

$$\omega = \frac{1}{\sqrt{\mathcal{L}\mathbb{C}}} \tag{3.77}$$

The eigenfunction again by analogy to equation (1.81) is

$$u_n(\Phi) = \frac{1}{\sqrt{2^n n!}}\left(\frac{\omega\mathbb{C}}{\hbar}\right)^{1/4} e^{-\frac{\omega\mathbb{C}\Phi^2}{2\hbar}} H_n\left(\sqrt{\frac{\omega\mathbb{C}}{\hbar}}\Phi\right) \tag{3.78}$$

where H_n are the Hermite polynomials defined in Chapter 1.

Concept: There is a one-to-one mapping between the Hamiltonian, eighenfunctions, and eigenvalues of a simple harmonic oscillator and a LC oscillator.

Example 11 (Quantized energies in LC oscillator). If a resonator's capacitance is 600 fF (femto Farads) and the inductance is 1200 pH (pico Henri), find the LC oscillator's resonance frequency in GHz. What is the (a) minimum energy of the resonator and (b) energy difference between two consecutive eigenstates in eV units?

Solution: Using the given values of capacitance and inductance, we have

$$f = \frac{1}{2\pi\sqrt{\mathcal{L}\mathbb{C}}} = \frac{1}{2\pi\sqrt{(1200 \times 10^{-12} \times 600 \times 10^{-15})}} \cong 5.9\,\text{GHz}$$

The difference between two consecutive energy levels is

$$E_{n+1} - E_n = \hbar\omega = hf = 24.4\,\mu\text{eV}$$

Example 12 (Numerical solution of the wave function and energy levels of an *LC* oscillator). Use the methods you learned in Section 1.6 and solve the *LC* oscillator's Schrödinger equation using the finite difference method. Take the capacitance and inductance to be 96.82 fF and 4.088 nH, respectively. Find the eigenenergies and eigenfunctions for $n = 1, 2, 3$, and 4.

Solution: We assume that the magnetic flux (Φ) varies from $-0.25\Phi_0$ to $0.25\Phi_0$ and we discretize Φ using one thousand points, e.g., $N = 1000$. Φ_0 is the quantum of magnetic flux, and it is equal to $\Phi_0 = 2.7 \times 10^{-15}$ Weber. The first term of the Schrödinger equation (equation (3.73)) contributes to both the diagonal and off-diagonal terms of the Hamiltonian matrix. The second term is an inductive energy $\frac{1}{2L}\Phi^2$ and it is added only to the main diagonal of the Hamiltonian.

Using the given values of L and C, we calculate the resonance frequency, which is

$$f = \frac{1}{2\pi\sqrt{\mathcal{L}\mathbb{C}}} \cong 8\text{GHz}$$

Figure 3.2 shows the spacing between consecutive eigenfrequencies is 8 GHz. To convert the frequency to energy, you can multiply by Planck's constant ($E(\text{J}) = hf$).

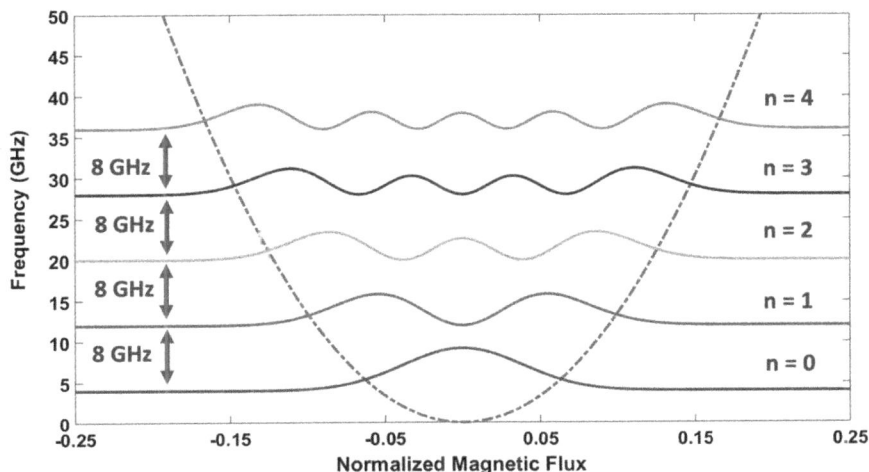

Figure 3.2. The five solutions of the LC resonator in our numerical example. The potential energy (divided by Planck's constant) versus normalized flux (Φ/Φ_0) is the red-dashed parabola. The probability density corresponding to $n = 0, 1, 2, 3$, and 4 is labeled. The probability density has been shifted vertically for clarity. The flat lines on the right and left are eigenfrequencies, which are 8 GHz apart. They are 4, 12, 20, and 28 GHz. Multiplying them by Planck's constant converts them into energy values in terms of Joule.

3.7 Problems

Section 3.1

(1) (Problem to help the reader understand quantum measurement) You have a million copies of a PiB in the state given by $(3\psi_1 + 4\psi_2)/5$. If you measure the momentum of the particle, what values will you find?

(2) What are the eigenfunctions and eigenvalues of the momentum operator? How do they relate to the eigenfunctions of the Hamiltonian of a free particle? You can address the 1D free particle case.

Section 3.2

(3) Show that x is a Hermitian operator.

(4) Show that $\frac{\partial}{\partial x}$ is not a Hermitian operator.

(5) Show that the 3D momentum operator $\hat{p} = \frac{\hbar}{i}\nabla$ is Hermitian.

(6) Show that the Hamiltonian of a SHO in one dimension is Hermitian. For an SHO, $\hat{H} = -\frac{\hbar^2}{2m}\frac{\partial^2}{\partial x^2} + \frac{1}{2}kx^2$.

Section 3.3

(7) Find the expectation value of the momentum operator for a state with eigenfunction Ae^{ikx}, where A is the normalization constant.

(8) Find $\langle\hat{p}\rangle$ (expectation value of momentum) for the following eigenfunctions of a 1D particle in a box:

  (i) $n = 1$
  (ii) $n = 2$

(9) Consider a 1D problem with potential energy:

$$U(x) = \begin{cases} \infty & \text{for } x < 0 \\ \infty & \text{for } x > L \\ 0 & \text{for } 0 < x < L \end{cases}$$

The eigenfunction corresponding to quantum number n is $\psi_n = A_n \sin\left(\frac{n\pi}{L}x\right)$, where A_n is the normalization constant. Find the expectation value of the position operator for an electron in the state:

  (i) ψ_2. Express your answer in terms of the box length L.
  (ii) $\left(\frac{3}{\sqrt{2}}\psi_1 - \frac{3}{\sqrt{2}}\psi_2 + 4\psi_3\right)/5$ at $t = 0$. Express your answer in terms of the box length L.
  (iii) Recalculate (ii) by including the time dependent part of the wave function. Express your answer in terms of the box length L, and the eigenenergies E_1, E_2 and E_3, and mass of the particle m.
  (iv) By using that the mass of the particle $m = 9.1 \times 10^{-31} kg$ and the length of the box is $3\,nm$, plot the expectation value of position as a function of time.

(10) Prove equations (3.38) and (3.39) in the main text.

Section 3.4

(11) The operators $\hat{O}_1$ and $\hat{O}_2$ are defined as follows:

$$\hat{O}_1 \psi(x) = x^3 \psi(x)$$

$$\hat{O}_2 \psi(x) = x \frac{d}{dx} \psi(x)$$

Show that $[\hat{O}_1, \hat{O}_2] = \hat{O}_1 \hat{O}_2 - \hat{O}_2 \hat{O}_1 = -3x^3$

Section 3.6.1

(12) Show that $\hat{U} x^n = (x + \lambda)^n$ if the operator is $\hat{U} = \exp\left(\lambda \frac{d}{dx}\right)$ and λ is a real number.

Chapter 4

HILBERT SPACE AND BRAKET NOTATION

Contents

4.1 An Introduction to Hilbert Space

Before reading about Hilbert space, it would be helpful to review 3D coordinate systems and Fourier expansion. This will help the reader understand Hilbert spaces, which is a vector space used in quantum mechanics. A more indepth treatment can be found in (Reza, 1971), (Steeb, 1998) and (Zeidler, 1991).

4.1.1   *A brief review of the 3D coordinate system, a simple vector space*

Any arbitrary **vector** in three-dimensional space can be defined as

$$\bar{r} = a\hat{i} + b\hat{j} + c\hat{k} \tag{4.1}$$

where a, b, and c are real numbers.

Unit vectors $\{\hat{i}, \hat{j}, \hat{k}\}$ are called the **basis vectors**. They are the minimum set of orthogonal vectors required to span the **vector space**, meaning that any vector can be written as a superposition of these basis vectors. The basis vectors have a unit **norm** (norm is the length of the vector). So, the dot product (or inner product) of the vector with itself is one:

$$\hat{i} \cdot \hat{i} = \hat{j} \cdot \hat{j} = \hat{k} \cdot \hat{k} = 1 \tag{4.2}$$

The basis vectors are orthogonal to each other, meaning, the pairwise dot product of two different basis vectors is 0:

$$\hat{i} \cdot \hat{j} = \hat{i} \cdot \hat{k} = \hat{j} \cdot \hat{k} = 0 \tag{4.3}$$

The components along the basis vectors are obtained by using:

$$a = \hat{i} \cdot \bar{r}, \quad b = \hat{j} \cdot \bar{r}, \quad \text{and} \quad c = \hat{k} \cdot \bar{r}$$

We can also say that $\bar{r}$ is a superposition or combination of the unit (basis) vectors along the x, y, and z axes with projections a, b, and c, respectively. Note that the set $\{\hat{i}, \hat{j}, \hat{k}\}$ which satisfies the above equations is called a *complete set*.

If $\bar{r}_1$ and $\bar{r}_2$ are vectors in the vector space represented by $\mathbb{R}$, then their linear combination also belongs to the vector space:

$$\alpha\bar{r}_1 \in \mathbb{R} \quad \text{and} \quad \beta\bar{r}_2 \in \mathbb{R} \tag{4.4}$$

$$\alpha\bar{r}_1 + \beta\bar{r}_2 \in \mathbb{R} \tag{4.5}$$

where α and β are real numbers.

The **Inner product** of $\bar{r}_1 = a\hat{i} + b\hat{j} + c\hat{k}$ and $\bar{r}_2 = d\hat{i} + e\hat{j} + f\hat{k}$ is their dot product, which is defined by

$$\bar{r}_1 \cdot \bar{r}_2 = ad + be + cf$$

All you need to do is multiply the corresponding unit vector's coefficients and sum up the product terms together. The result should be a real number.

Example 1. Normalize the vector $\bar{r}_1 = a\hat{i} + b\hat{j} + c\hat{k}$. The term *normalize* here refers to finding a vector in the same direction as $\bar{r}_1$ but having a unit norm.

Solution: Recall that the dot product of a vector with itself is its norm squared (or absolute value squared or length squared). The norm or length of $\bar{r}_1$ is found by first taking the length squared,

$$|\bar{r}_1|^2 = \bar{r}_1 \cdot \bar{r}_1 = a \cdot a + b \cdot b + c \cdot c = a^2 + b^2 + c^2$$

and then taking its square root,

$$|\bar{r}_1| = \sqrt{a^2 + b^2 + c^2}$$

To normalize a vector, we divide the vector $\bar{r}_1$ by its norm or length:

$$\hat{u} = \frac{a\hat{i} + b\hat{j} + c\hat{k}}{|r_1|} = \frac{a\hat{i} + b\hat{j} + c\hat{k}}{\sqrt{a^2 + b^2 + c^2}}$$

Convince yourself that $\hat{u}$ is now normalized by showing that $\hat{u} \cdot \hat{u} = 1$. The above process of dividing a given vector by its length is called *normalization*. Here, $\hat{u}$ is a unit vector and not an operator.

Concept: An arbitrary vector in the 3D Cartesian coordinate system is expressed as a linear combination (superposition) of a vector space generated by unit vectors $\{\hat{i}, \hat{j}, \hat{k}\}$ weighted by real numbers a, b, and c,

$$\bar{r} = a\hat{i} + b\hat{j} + c\hat{k}$$

4.1.2 *A brief review of Fourier expansion, a vector space of functions*

The Fourier expansion of a function $g(x)$ in terms of harmonics e^{ikx} is

$$g(x) = \int dk \, G(k)e^{ikx} \tag{4.6}$$

provided the following condition holds

$$\int_{-\infty}^{+\infty} |g(x)|^2 dx < \infty \tag{4.7}$$

$G(k)$ is called the Fourier transform of $g(x)$. Equation (4.6) is a superposition of the basis functions $\{e^{ikx}\}$ with coefficients $G(k)$ to yield the desired function $g(x)$. The wave vector k consists of real numbers, $-\infty < k < \infty$. The various e^{ikx} form a *complete set*[1] of **basis functions** for an infinite-dimensional vector space; they are orthonormal to each other, as expressed in equation (4.9). Infinite dimensional refers to the infinite values of k each of which generates a unique basis function. The condition in equation (4.7) is called the boundedness of energy.[2] For a signal with bounded energy, the total energy in the signal in x domain can be found by integrating the Fourier transform over k space, hence the quantity $|G(k)|^2$ is called the spectral energy density (see Problem 1),

$$\int_{-\infty}^{+\infty} dx\, g^*(x)g(x) = \int_{-\infty}^{+\infty} dx\, |g(x)|^2 = \int_{-\infty}^{+\infty} dk\, G^*(k)G(k)$$

$$= \int_{-\infty}^{+\infty} dk\, |G(k)|^2 \tag{4.8}$$

The above equation is known as **Parseval's theorem**. The symbol $*$ stands for complex conjugate.

The orthonormality condition of the basis functions in position (x) and wave vector (k) space are as follows:

$$(\text{orthonormality}) \int dk\, e^{ikx} e^{-ikx'} = \delta(x - x') \tag{4.9}$$

$$(\text{orthonormality}) \int dx\, e^{ikx} e^{-ik'x} = \delta(k - k') \tag{4.10}$$

x' and k' also refer to position and wave vector. If $x = x'$, the first integral above is infinite, and similarly if $k = k'$, the second integral above is infinite. Conversely, if $x \neq x'$, the first integral is zero, and the second integral is zero when $k \neq k'$. To find out the coefficient

---

[1]A *complete set* of *basis functions* is the minimum set of functions required to represent a function $g(x)$ with a finite spectral density.

[2]A reader familiar with signal theory should note that a time-dependent voltage signal $V(t)$ where $\int_{-\infty}^{+\infty} \frac{1}{R}|V(t)|^2 dt < \infty$ (similar to equation (4.7)) represents a signal with a bounded energy.

$G(k)$ in the expansion of equation (4.6), we multiply both sides by $e^{-ik'x}$ and integrate over all x,

$$\int dx\, g(x)e^{-ik'x} = \int dx \int dk\, G(k)e^{i(k-k')x}$$

Using equation (4.10) and the sifting property of the delta function, the above equation becomes

$$\int dx\, g(x)e^{-ik'x} = \int dk \left[G(k) \int dx\, e^{i(k-k')x} \right]$$

$$= \int dk\, G(k)\delta(k-k') = G(k') \qquad (4.11)$$

Hence to extract or "project out" the Fourier coefficient or component, we can use:

$$G(k') = \int dx\, g(x)e^{-ik'x}$$

Similarly, a function g in discrete space is the discrete sum of harmonics called the Fourier series, which is defined by

$$g(x_j) = \sum_{\alpha=1}^{N} G(k_\alpha)e^{ik_\alpha x_j} \qquad (4.12)$$

Again, in the discrete version, the **basis vectors** $e^{ik_\alpha x_j}$ form a complete set and are orthonormal to each other as expressed in equation (4.13). The size of the vector space here is finite and equal to N, which is the number of x_i points. The orthonormality conditions of the basis vectors in real (x_i) and wave vector (k_α) space are expressed as follows:

$$\text{(orthonormality)} \sum_{\alpha=1}^{N} e^{ik_\alpha x_i} e^{-ik_\alpha x_j} = \delta_{ij} \qquad (4.13)$$

$$\text{(orthonormality)} \sum_{i=1}^{N} e^{ik_\alpha x_i} e^{-ik_\beta x_i} = \delta_{\alpha\beta} \qquad (4.14)$$

Like the above discussion, we can "project out" the component of $e^{ik_\alpha x_i}$ in equation (4.12) by taking the inner product of $g(x_i)$ with

$e^{ik_\alpha x_i}$ (Table 4.1),

$$G(k_\alpha) = \sum_{i=1}^{N} e^{-ik_\alpha x_i} g(x_i)$$

by using equation (4.13).

Concept: Underlying the Fourier expansion of a function $g(x)$ is a vector space of functions formed by the set of exponentially oscillatory functions $\{e^{ikx}\}$.

The superposition (linear combination) of the set of $\{e^{ikx}\}$ gives us

$$g(x) = \int dk \, G(k) e^{ikx}$$

where $G(k)$ are complex coefficients of e^{ikx} and are extracted using equation (4.11).

4.1.3   *Review of Hilbert space, a vector space of functions*

The wave functions obtained by solving **the time-independent Schrödinger equation** form a vector space referred to as the Hilbert space, which is named after the mathematician David Hilbert. Unlike real space (x, y, z), which is a three-dimensional vector space, the Hilbert space can be of any dimension depending on the Hamiltonian. Further, the Hilbert space is a functional space, meaning that it depends on variables; in our discussion, it will be a function of position, though other variables such as the wave vector are also valid. They can be infinite or finite dimensional, just like Fourier expansions in continuous or discrete space. The main properties of Hilbert space closely resemble the Fourier expansion and are summarized in the following. The reader can verify the properties by using the PiB as an example to consolidate their understanding of the underlying concepts.

The basis vectors of the Hilbert space are the eigenfunctions $\{\psi_1(\bar{r}), \psi_2(\bar{r}), \psi_3(\bar{r}), \ldots\}$ of the Hamiltonian operator. The Hilbert

Table 4.1. A comparison of 3D vector spaces, Fourier analysis, and Hilbert spaces.

	Basis vectors	Orthonormality	Expansion (superposition of eigenvectors)	Components (coefficients of expansion)
3D vector space	$\{\hat{i}, \hat{j}, \hat{k}\}^1$	$\hat{i}\cdot\hat{i} = \hat{j}\cdot\hat{j} = \hat{k}\cdot\hat{k} = 1$ $\hat{i}\cdot\hat{j} = \hat{i}\cdot\hat{k} = \hat{j}\cdot\hat{k} = 0$	$\bar{r} = a\hat{i} + b\hat{j} + c\hat{k}$	$a = \hat{i}\cdot\bar{r},\, b = \hat{j}\cdot\bar{r},\, c = \hat{k}\cdot\bar{r}$
Fourier analysis	e^{ikx}, where $-\infty < k < \infty^2$	$\int dk e^{ikx} e^{-ikx'} = \delta(x - x')$ $\int dx e^{ikx} e^{-ik'x} = \delta(k - k')$	$g(x) = \int dk G(k) e^{ikx}$	$G(k) = \int dx g(x) e^{-ikx}$
Hilbert space	$\{\psi_n(r)\}$, where $n = 1, 2, 3, \ldots N^3$	$\int dv \psi_n^*(r)\psi_m(r) = \delta_{mn}$	$\psi(r) = \sum_n a_n \psi_n(r)$	$a_n = \int dv \psi_n^*(r)\psi(r)$
Braket notation	**bra** $\langle n|$ or $\langle\psi_n|$ **ket** $|n\rangle$ or $|\psi_n\rangle$	$\langle n|m\rangle = \delta_{mn}$	$|\psi\rangle = \sum_n a_n|n\rangle$ $\langle\psi| = \sum_m a_m^*\langle m|$	$a_n = \langle n|\psi\rangle$

Note: The braket notation is shown in the last row of the table.

[1] The basis vectors are not a function of position. There are only three basis vectors.

[2] The basis vectors are a function of position x. There are infinite basis vectors corresponding to the infinite k-values.

[3] The basis vectors are functions of position $\bar{r}$. There are N basis vectors. N can be either finite or infinite.

space can be either finite or infinite dimensional:

- For the PiB, the eigenfunctions of the Hilbert space are $\psi_n(x) = \sqrt{\frac{2}{L}}\sin(\frac{n\pi x}{L})$, where quantum numbers $n = 1, 2, 3, \ldots$ form a discrete spectrum.
- For a free particle, the eigenfunctions of the Hilbert space are $\phi_k(x) = A_k e^{ikx}$, where the quantum numbers $-\infty < k < \infty$ form a continuous spectrum.

The Hilbert space defined by basis functions (eigen functions) $\{\psi_n(\bar{r}), n = 1, 2, 3, \ldots\}$ is called a *functional vector space* because they are functions of position.

First, recall the inner product's definition (same as dot product) of two vectors in 3D space. If $\bar{r}_1 = a\hat{i} + b\hat{j} + c\hat{k}$ and $\bar{r}_2 = d\hat{i} + e\hat{j} + f\hat{k}$, then their inner product is found by adding the products of the corresponding components:

$$\text{Inner product } (\bar{r}_1, \bar{r}_2) = \bar{r}_1 \cdot \bar{r}_2 = ad + be + cf \qquad (4.15)$$

To define the inner product of two functions, we will first replace r with a single coordinate x for brevity (without loss of generality). We will then discretize the x-coordinate into equally spaced grid points $x_1, x_2, x_3, \ldots, x_N$ so that we represent each function $\psi_n(x)$ as an N-component vector,

$$\psi_n(x) = \begin{bmatrix} \psi_n(x_1) \\ \psi_n(x_2) \\ \vdots \\ \psi_n(x_N) \end{bmatrix}$$

where the grid spacing $x_{i+1} - x_i = \Delta x$ is equal for all $i = 1, 2, 3, \ldots, N$. To find the inner product of two real functions $\psi_n(x)$ and $\psi_m(x)$, let us first see what happens when we multiply the two functions in a manner like the way we multiplied the two vectors in equation (4.15),

$$(\psi_n(x), \psi_m(x)) = \psi_n(x_1)\psi_m(x_1) + \psi_n(x_2)\psi_m(x_2) + \cdots$$

$$+ \psi_n(x_N)\psi_m(x_N) \qquad (4.16)$$

The above summation depends on the number of grid points and if N increases, the sum diverges. If the above summation is multiplied by the grid spacing Δx, the summation converges and becomes independent of the number of grid points. Therefore, the correct definition of the inner product is

$$(\psi_n(x), \psi_m(x)) = [\psi_n(x_1)\psi_m(x_1) + \psi_n(x_2)\psi_m(x_2) + \cdots$$
$$+ \psi_n(x_N)\psi_m(x_N)]\Delta x \qquad (4.17)$$

In the limit that the spacing between grid points becomes very small, the above sum for the inner product becomes an integral,

$$(\psi_n(x), \psi_m(x)) = \int dx\, \psi_n(x)\psi_m(x) \qquad (4.18)$$

In quantum mechanics, the wave functions $\{\psi_n(x), n = 1, 2, 3, \ldots\}$ have real and imaginary parts. The definition of an inner product in the previous equation is then modified to include a complex conjugate of the first term. By this, we have enforced that $\psi_n^*(x)\psi_n(x)$ is always a real number.

Inner product: $(\psi_n(x), \psi_m(x)) = \int dx\, \psi_n^*(x)\psi_m(x) \qquad (4.19)$

From the above, we can define the norm of a function $\psi(x)$ as the inner product of the function with itself:

$$\text{norm}^2 = \text{length}^2 = \int dx\, \psi^*(x)\psi(x) = \int dx |\psi(x)|^2 \qquad (4.20)$$

The basis vectors have unit **norm** and are **orthogonal** to each other, as expressed by

$$(\text{orthonormality}) \int dv\, \psi_n^*(\bar{r})\psi_m(\bar{r}) = \delta_{mn} \qquad (4.21)$$

The above property is called *orthonormality* (normalized and orthogonal), and was also discussed in Chapter 2.

An arbitrary vector (wave function) in the Hilbert space is a superposition of basis functions (or sum of the basis vectors),

$$\Psi(\bar{r}) = \sum_n a_n \psi_n(\bar{r}) = a_1 \psi_1(\bar{r}) + a_2 \psi_2(\bar{r}) + a_3 \psi_3(\bar{r}) + \cdots \qquad (4.22)$$

where a_n are complex numbers. Note that a_n can be time dependent as discussed in Chapters 1, 2, and 3 (even if the Hamiltonian is time independent).

The complex conjugate of $\Psi(\bar{r})$ is

$$\Psi(\bar{r})^* = \sum_n a_n^* \psi_n(\bar{r})^* = a_1^* \psi_1(\bar{r})^* + a_2^* \psi_2(\bar{r})^* + a_3^* \psi_3(\bar{r})^* + \cdots \quad (4.23)$$

To normalize a vector $\Psi(r)$, we divide it by its norm (see equation (4.20)):

$$\tilde{\Psi}(\bar{r}) = \frac{\Psi(\bar{r})}{\left[\int dv\, |\Psi(\bar{r})|^2\right]^{\frac{1}{2}}} \quad (4.24)$$

The denominator is just the norm (also called magnitude) of $\Psi(\bar{r})$ which is a positive (scalar) number.

Consider two wave functions (vectors) $\phi(r)$ and $\chi(r)$ in the Hilbert space,

$$\phi(\bar{r}) = \sum_n b_n \psi_n(\bar{r}) = b_1 \psi_1(\bar{r}) + b_2 \psi_2(\bar{r}) + b_3 \psi_3(\bar{r}) + \cdots \quad \text{and}$$

$$\chi(\bar{r}) = \sum_n c_n \psi_n(\bar{r}) = c_1 \psi_1(\bar{r}) + c_2 \psi_2(\bar{r}) + c_3 \psi_3(\bar{r}) + \cdots$$

Using the definition of inner product (equation (4.17)), we can show that

$$\int dv\, \phi^*(\bar{r})\chi(\bar{r}) = \sum_i b_i^* c_i \quad (4.25)$$

where we used the orthonormality of $\{\psi_n\}$ from equation (4.21).

We can project out the component of $\psi_n(\bar{r})$ in equation (4.22) by taking the inner product of $\Psi(\bar{r})$ with $\psi_n(\bar{r})$,

$$a_n = \int dv\, \psi_n^*(\bar{r})\Psi(\bar{r}) \quad (4.26)$$

Using equation (4.21), a_n is the projection or component of Ψ along basis (or eigen) function ψ_n.

Example 2. Write down detailed steps to show how a_3 is projected out starting from equation (4.19).

Solution: To project out an expansion coefficient a_3, both sides of equation (4 22) are multiplied by $\psi_3(\bar{r})^*$ from left,

$$\psi_3^*(\bar{r})\mathbf{\Psi}(\bar{r}) = \sum_n a_n \psi_3^*(\bar{r})\psi_n(\bar{r}) = a_1 \psi_3^*(\bar{r})\psi_1(\bar{r}) + a_2 \psi_3^*(\bar{r})\psi_2(\bar{r})$$

$$+ a_3 \psi_3^*(\bar{r})\psi_3(\bar{r}) + a_4 \psi_3^*(\bar{r})\psi_4(\bar{r}) + \cdots$$

Then, integrating over all space:

$$\int dv \psi_3^*(\bar{r})\mathbf{\Psi}(\bar{r}) = a_1 \int dv\, \psi_3^*(\bar{r})\psi_1(\bar{r}) + a_2 \int dv\, \psi_3^*(\bar{r})\psi_2(\bar{r})$$

$$- a_3 \int dv\, \psi_3^*(\bar{r})\psi_3(\bar{r}) + a_4 \int dv\, \psi_3^*(\bar{r})\psi_4(\bar{r}) + \cdots$$

Due to orthonormality all terms with different subscripts on the RHS are zero except for the third term, which is equal to 1. Hence,

$$a_3 \int dv\, \psi_3^*(\bar{r})\mathbf{\Psi}(\bar{r}) = a_3$$

Finally, let us discuss the **Closure theorem**, which states that if functions $\psi_n(\bar{r}')$ are a complete set of orthonormal wave functions, then

$$\sum_n \psi_n^*(\bar{r})\psi_n(\bar{r}') = \delta(\bar{r} - \bar{r}') \tag{4.27}$$

We prove this in the following. Decompose a given function $f(\bar{r})$ in terms of the basis function (vectors) $\{\psi_n(\bar{r}), n = 1, 2, 3, \ldots\}$

$$f(\bar{r}) = \sum_n a_n \psi_n(\bar{r})$$

where the coefficients are found by

$$a_n = \int_{\text{all space}} \psi_n^*(\bar{r}')f(\bar{r}')\, dv'$$

Substituting the above equation into the expression for $f(\bar{r})$, we have

$$f(\bar{r}) = \sum_n \psi_n(\bar{r}) \left[\int_{\text{all space}} \psi_n^*(\bar{r}')f(\bar{r}')\, dv' \right]$$

Since the summation over n and the integral over space are independent of each other, we can move the summation inside the integral sign and rearrange to obtain

$$f(\bar{r}) = \int_{\text{all space}} f(\bar{r}') \sum_n \psi_n^*(\bar{r}') \psi_n(\bar{r}) \, dv'$$

For the LHS to be equal to the RHS, the term $\sum_n \psi_n(\bar{r}')^* \psi_n(\bar{r})$ must be zero except when $\bar{r} = \bar{r}'$. That is,

$$\sum_n \psi_n^*(\bar{r}) \psi_n(\bar{r}') = \delta(\bar{r} - \bar{r}')$$

which completes the proof of equation (4.27).

Concept: Underlying an arbitrary wave function $\boldsymbol{\Psi}(\bar{r})$ of a quantum system represented by a Hamiltonian $\widehat{H}$ is a Hilbert space (vector space) formed by the set of energy eigenfunctions $\{\psi_n(\bar{r})\}$.

An arbitrary wave function $\boldsymbol{\Psi}(\bar{r})$ of the quantum system can be written as

$$\boldsymbol{\Psi}(\bar{r}) \sum_n a_n \psi_n(\bar{r})$$

where a_n are complex number coefficients.

Note: The arbitrary wave function $\boldsymbol{\Psi}(\bar{r})$ can be expanded in terms of eigenfunctions corresponding to other operators too and not just the energy eigenfunctions.

4.2   Braket Notation (Dirac's Notation)

The bra-ket (braket) notation is a convenient mathematical notation invented by Paul A. M. Dirac to make lengthy notation shorter and niftier (Dirac, 1958). They are used extensively in quantum mechanics. Interestingly, Dirac was an electrical engineer who became a mathematician. In an interview in the 1970s, he mentioned that his engineering education changed his outlook to a large extent (Schweber, 1994). In the next section, we show that the eigenfunctions and operators can be represented in matrix form. The braket notation is handy in calculating the operator's expectation values and matrix form.

We saw in Chapter 1 that to find the allowed energies of an electron in PiB, we had to solve a differential equation (Schrödinger equation) written in the form of $\hat{H}\psi_n = E_n\psi_n$. The allowed wave functions (solutions) of this Hamiltonian operator build a Hilbert space or a **spectrum** of 1D functions of x, i.e., $\{\psi_1, \psi_2, \psi_3, \ldots\}$. Recall that for a box of length L, we had

$$\psi_n = \sqrt{\frac{2}{L}} \sin\left(\frac{n\pi x}{L}\right)$$

There are infinite basis functions that can be counted by $n = 1, 2, 3$, *ad infinitum*.

Let's pause to see how operator $\hat{H}$ acts on an eigenfunction and then see why braket and matrix notations are helpful. Assume that the potential energy in the box is zero, then we apply the differential form of the Hamiltonian to one of the allowed eigenfunctions indexed by n, i.e., ψ_n, as follows:

$$\hat{H}\psi_n = -\frac{\bar{h}^2}{2m}\frac{d^2}{dx^2}\psi_n = -\frac{\hbar^2}{2m}\frac{d^2}{dx^2}\sqrt{\frac{2}{L}}\sin\left(\frac{n\pi x}{L}\right)$$

$$= \frac{n^2\hbar^2\pi^2}{2mL^2}\sqrt{\frac{2}{L}}\sin\left(\frac{n\pi x}{L}\right) = E_n\psi_n$$

Recall that the RHS of the above is $E_n\psi_n$ because we found that the allowed eigen energies of the PiB are $E_n = \frac{n^2\hbar^2\pi^2}{2mL^2}$ (in Chapter 1). The above says that the Hamiltonian acting on an eigenfunction returns the same function multiplied by a factor E_n.

As you remember from linear algebra, when a matrix is multiplied by a vector, the length and direction of the vector generally changes. However, if you manage to find special vectors that only stretch or compress along their original direction (without being rotated), these vectors are called eigenvectors of the matrix and the proportionality factors are called eigenvalues. In this case, we write $Ou = cu$ (O is a matrix, u is an eigenvector, and c is an eigenvalue).

It is quite natural if after seeing this similarity with linear algebra, we stop and ask (as the pioneers of quantum mechanics did) the following question: Is there a way to use the matrix notation for the Hamiltonian and wave functions instead of differential operators and differential equations? The answer is yes. Let's first assign a new name and symbol to each eigenfunction, the *ket* vector indexed by n, i.e., $\psi_n = |n\rangle$. So, we have the following:

$$\{\psi_1, \psi_2, \psi_3, \ldots\} = \{|1\rangle, |2\rangle, |3\rangle, \ldots\} \tag{4.28}$$

Now, instead of using the differential form of $\hat{H}$ every time we apply it on ψ_n, we simply use the ket notation:

$$\hat{H}|n\rangle = E_n|n\rangle \tag{4.29}$$

Next, we rename the complex conjugate of the above eigenfunctions (to get rid of carrying the conjugation superscript * around) using the *bra vector* notation, i.e., we say $\psi_n^* = \langle n|$. Thus, we have

$$\{\psi_1^*, \psi_2^*, \psi_3^*, \ldots\} = \{\langle 1|, \langle 2|, \langle 3|, \ldots\} \tag{4.30}$$

What is the use of the new notation introduced above? Let's say we want to find the inner product of two eigenfunctions, ψ_m and ψ_n. Recall that this inner product was

$$\int dv \, \psi_m^*(\bar{r})\psi_n(\bar{r}) = \begin{cases} 1 & \text{if } n = m \\ 0 & \text{if } n \neq m \end{cases} = \delta_{mn}$$

Instead of writing the integral in the above equation, using the bra and ket notation, we can compactly write it as

$$\int dv \, \psi_m^*(\bar{r})\psi_n(\bar{r}) = \langle m| \cdot |n\rangle = \langle m|n\rangle = \delta_{mn} \tag{4.31}$$

where the dot ($\cdot$) means the dot or inner product of a bra with ket vector. But for simplicity, we remove the dot to squeeze the notation for inner product to $\langle m|n\rangle$. From the Kronecker delta, it also follows that

$$\langle n|n\rangle = 1 \tag{4.32}$$

In summary, whenever you see a bra vector (e.g., $\langle m|$) on the left side of an expression, you must multiply the complex conjugate of a wave function corresponding to that index (e.g., ψ_m^*) and integrate over space.

Note that conjugating an inner product in braket form merely changes the order of indices. We prove it here and use it forever without worrying about where it comes from:

$$(\langle m|n\rangle)^* = \left(\int dv \, \psi_m^*(\bar{r})\psi_n(\bar{r})\right)^* = \int dv \, \psi_m(\bar{r})\psi_n^*(\bar{r})$$

$$= \int dv \, \psi_n^*(\bar{r})\psi_m(\bar{r}) = \langle n|m\rangle$$

In summary,

$$\langle m|n\rangle^* = \langle n|m\rangle \tag{4.33}$$

Note: Sometimes $|n\rangle$ is written as $|\psi_n\rangle$ but this is only a matter of taste.

Let's now assume we are given a wave function $\phi(x)$ which is a superposition of a few eigenfunctions of the PiB Hamiltonian. If we measure the energy of the particle when it is in this state, the expectation value (average) of energy is

$$\langle \hat{H} \rangle = \int_0^L dx \, \phi^*(x)\hat{H}\phi(x) \tag{4.34}$$

Using the braket notation we just learned, we can rewrite the expectation value of an operator as follows:

$$\langle \hat{H} \rangle = \langle \phi| \cdot \hat{H}|\phi\rangle = \langle \phi|\hat{H}|\phi\rangle \tag{4.35}$$

It is as if H operates on the right-most ket and then whatever the result is must be multiplied (inner product) to the bra vector from the left. Again, we squeezed things by removing the "dot" in the inner product and write the above equation as $\langle \hat{H} \rangle = \langle \phi|\hat{H}| \cdot \phi\rangle = \langle \phi|\hat{H}|\phi\rangle$.

Now assume that instead of superposition, $|\phi\rangle$ is one of the eigenfunctions of the PiB Hamiltonian, e.g., the one with $n = 5$. Then $|\phi\rangle = |5\rangle$ and the expectation value of the measured energy is expected to be E_5 as we see:

$$\langle \hat{H} \rangle = \langle 5|\hat{H}|5 \rangle = \langle 5|E_5|5 \rangle = E_5 \langle 5|5 \rangle = E_5 \times 1 = E_5$$

We used the fact that a scalar (E_5) can be moved to the left of the bra $\langle 5|$. Let's see if we want to use the differential form of the Hamiltonian and the explicit form of wave functions, how much extra work must be done:

$$\langle \hat{H} \rangle = \int_0^L dx \, \psi_5^*(x) \hat{H} \psi_5(x)$$

$$= \int_0^L dx \sqrt{\frac{2}{L}} \sin\left(\frac{5\pi x}{L}\right) \left(-\frac{\hbar^2}{2m}\frac{d^2}{dx^2}\right) \sqrt{\frac{2}{L}} \sin\left(\frac{5\pi x}{L}\right)$$

$$= \int_0^L dx \sqrt{\frac{2}{L}} \sin\left(\frac{5\pi x}{L}\right) \left(\frac{25\hbar^2\pi^2}{2mL^2}\right) \sqrt{\frac{2}{L}} \sin\left(\frac{5\pi x}{L}\right)$$

$$= \frac{25\hbar^2\pi^2}{2mL^2} \int_0^L dx \sqrt{\frac{2}{L}} \sin\left(\frac{5\pi x}{L}\right) \sqrt{\frac{2}{L}} \sin\left(\frac{5\pi x}{L}\right)$$

$$= \frac{25\hbar^2\pi^2}{2mL^2} = E_5$$

The things we said above hold true also for other quantum mechanical observables like momentum, position, angular momentum, etc.

Momentum: Recall that for a 1D free particle, the eigenfunction is $\phi_k(x) = e^{ikx}$. Here, k is the quantum number but it is not discrete like the quantum number ($n = 1, 2, 3, \ldots$) of a PiB. We build a set composed of an infinite number of orthonormal wave functions corresponding to the infinite possible values of k,

$$\{\ldots, \phi_k, \ldots\}$$

Let's apply momentum operator on one of these eigenfunctions,

$$\hat{p}\phi_k = \frac{\hbar}{i}\frac{d}{dx}e^{ikx} = \hbar k e^{ikx} = \hbar k \phi_k \tag{4.36}$$

It can be seen that ϕ_k is the eigenfunction of the momentum operator and the eigenvalue is $\hbar k$. Just like before, we can replace ϕ_k by $|k\rangle$.

Then, the above equation reduces to

$$\hat{p}|k\rangle = \hbar k|k\rangle = p|k\rangle \tag{4.37}$$

Position: The operator $\hat{x}$ acts on the wave functions of the Hamiltonian as follows:

$$\hat{x}\psi_n(x) = x\psi_n(x) \tag{4.38}$$

It multiplies x to the wave function. Using the braket notation, we can write

$$\hat{x}|n\rangle = x|n\rangle \tag{4.39}$$

The above equation is not precise and is not mathematically rigorous notation but for the sake of this book, we will adopt this casual notation. Any function of $\hat{x}$, e.g., the potential energy also acts on the Hamiltonian eigenfunction as follows:

$$U(\hat{x})|n\rangle = U(x)|n\rangle \tag{4.40}$$

With the above introductory remarks, we are prepared to rewrite everything we learned in Chapter 3 using braket formalism:

- The wave function ψ_n is represented by **ket vector** $|n\rangle$ or $|\psi_n\rangle$.
- The complex conjugate of ψ_n, ψ_n^* is represented by **bra vector** $\langle n|$ or $\langle \psi_n|$.
- A linear combination of ket or bra vectors results in a ket or bra vector, respectively:

$$|\psi\rangle = a_n|\psi_n\rangle + a_m|\psi_m\rangle \quad \text{or} \quad |\psi\rangle = a_n|n\rangle + a_m|m\rangle \tag{4.41}$$

- The inner product of two wave functions in the braket notation is defined to be

$$\langle \psi|\chi\rangle = \int dv\, \psi^*(\bar{r})\chi(\bar{r}) = \text{complex number} \tag{4.42}$$

Remember that the bra should be the complex conjugate.
- The inner product of eigenvectors $|m\rangle$ and $|n\rangle$ is defined by,

$$\langle \psi_m|\psi_n\rangle = \delta_{mn} \quad \text{or} \quad \langle m|n\rangle = \delta_{mn} \tag{4.43}$$

- The complex conjugate of an inner product is written as,

$$\langle \psi_m|\psi_n\rangle^* = \langle \psi_n|\psi_m\rangle \quad \text{or} \quad \langle m|n\rangle^* = \langle n|m\rangle \tag{4.44}$$

- A linear operator $\hat{O}$ acts on a sum of ket vectors as follows,

$$\hat{O}[|\phi\rangle + |\psi\rangle] = \hat{O}|\phi\rangle + \hat{O}|\psi\rangle \tag{4.45}$$

The identity operator $(\hat{I})$ can be written as,

$$\hat{I} = \sum_n |\psi_n\rangle\langle\psi_n| \quad \text{or} \quad \hat{I} = \sum_n |n\rangle\langle n| \tag{4.46}$$

The above equation is referred to as the **Closure theorem** and is verified below.

Any ket $|\phi\rangle$ can be written as a superposition of basis kets, i.e., $|\phi\rangle = \sum_k a_k|k\rangle$. If we apply the operator $\sum_n |n\rangle\langle n|$ from the left to $|\phi\rangle$,

$$\sum_n |n\rangle\langle n|\phi\rangle = \sum_n |n\rangle \left(\sum_k \langle n|a_k|k\rangle\right) = \sum_n |n\rangle \left(\sum_k a_k\langle n|k\rangle\right)$$

$$= \sum_n |n\rangle \left(\sum_k a_k\delta_{kn}\right) = \sum_n a_n|n\rangle = |\phi\rangle = \hat{I}|\phi\rangle$$

Thus, we have $\sum_n |n\rangle\langle n| = \hat{I}$, which proves equation (4.46). It is also said that the eigen (kets) vectors build a *complete* set.

Let's find the projection of a superposition state $|\phi\rangle = \sum_n a_n|n\rangle$ on another state like $|q\rangle$. This is the inner product of two states which can be written as

$$\langle q|\phi\rangle = \langle q|\hat{I}|\phi\rangle = \langle q|\sum_n |n\rangle\langle n||\phi\rangle = \sum_n \langle q|n\rangle\langle n|\phi\rangle$$

We have used the Closure theorem in the above. Now if the state q is the same as ϕ, we have

$$1 = \langle\phi|\phi\rangle = \sum_n \langle\phi|n\rangle\langle n|\phi\rangle = \sum_n (\langle n|\phi\rangle)^*\langle n|\phi\rangle = \sum_n (a_n)^*a_n$$

$$= \sum_n |a_n|^2$$

So, we showed that the sum of probabilities is 1. This is another way of showing that the norm of the state $|\phi\rangle$ is one.

- The integral of an operator $\hat{O}$ between states $|\psi_n\rangle$ and $|\psi_m\rangle$, $\int dv\, \psi_m^*(\bar{r},t)\hat{O}\psi_n(\bar{r},t)$, is represented in the braket notation by

$$O_{mn} = \langle\psi_m|\hat{O}|\psi_n\rangle \quad \text{or} \quad \langle m|\hat{O}|n\rangle \tag{4.47}$$

- The expectation or average value of an operator $\hat{O}$ that was studied in Chapter 3 is written in braket notation as

$$\langle\hat{O}\rangle = \int dv\, \psi^*(r,t)\hat{O}\psi(r,t) = \langle\psi|\hat{O}|\psi\rangle \tag{4.48}$$

We will now express the expectation value in commonly encountered forms. The wave function $|\psi\rangle$ can be written as

$$|\psi\rangle = \sum_n a_n|\psi_n\rangle \quad \text{or} \quad |\psi\rangle = \sum_n a_n|n\rangle \tag{4.49}$$

Note that complex conjugation yields

$$\langle\psi| = \sum_m a_m^*\langle\psi_m| \quad \text{or} \quad \langle\psi| = \sum_m a_m^*\langle m| \tag{4.50}$$

Substituting equations (4.49) and (4.50) in equation (4.48), the expectation value of $\hat{O}$ can be rewritten as

$$\langle\hat{O}\rangle = \langle\psi|\hat{O}|\psi\rangle = \sum_m \sum_n a_m^* a_n \langle m|\hat{O}|n\rangle = \sum_m \sum_n a_m^* a_n O_{mn} \tag{4.51}$$

where

$$O_{mn} = \langle m|\hat{O}|n\rangle \quad \text{(same as equation (4.47))}$$

In the following section, we show that an operator can be represented as a matrix, and O_{mn} corresponds to the matrix element corresponding to row m and column n of the operator O in matrix form.

Concept: The braket notation is a convenient mathematical notation to describe properties and mathematical operations in the Hilbert space.

4.3   Operators in Braket and Matrix Forms

Now let's get back to the Hamiltonian of PiB and assume for some reason we are only interested in the lowest four eigenenergies, eigenfunctions, and their linear combinations. Here, the Hilbert space is then composed of four eigenfunctions for which we use the ket notation:

$$\{\psi_1, \psi_2, \psi_3, \psi_4\} = \{|1\rangle, |2\rangle, |3\rangle, |4\rangle\}$$

Let's proceed further and assume we encode or assign each ket by a normalized column vector as follows:

$$\left\{ |1\rangle = \begin{bmatrix} 1 \\ 0 \\ 0 \\ 0 \end{bmatrix}, |2\rangle = \begin{bmatrix} 0 \\ 1 \\ 0 \\ 0 \end{bmatrix}, |3\rangle = \begin{bmatrix} 0 \\ 0 \\ 1 \\ 0 \end{bmatrix}, |4\rangle = \begin{bmatrix} 0 \\ 0 \\ 0 \\ 1 \end{bmatrix} \right\}$$

To satisfy relations like $\langle n|n\rangle = 1$, we define the corresponding bra of the above set as row vectors simply by transposing and complex conjugating them:

$$\{\langle 1|, \langle 2|, \langle 3|, \langle 4|\}$$

$$= \{[1 \quad 0 \quad 0 \quad 0], [0 \quad 1 \quad 0 \quad 0], [0 \quad 0 \quad 1 \quad 0], [0 \quad 0 \quad 0 \quad 1]\}$$

It is easy to show that the ket and bra vectors satisfy relations like $\langle m|n\rangle = \delta_{mn}$. For example,

$$\int dx\, \psi_2^*(x)\psi_4(x) = \langle 2|4\rangle = [0 \quad 1 \quad 0 \quad 0] \begin{bmatrix} 0 \\ 0 \\ 0 \\ 1 \end{bmatrix}$$

$$= 0 \cdot 0 + 1 \cdot 0 + 0 \cdot 0 + 1 \cdot 0 = 0$$

$$\int dx\, \psi_3^*(x)\psi_3(x) = \langle 3|3\rangle = [0 \quad 0 \quad 1 \quad 0] \begin{bmatrix} 0 \\ 0 \\ 1 \\ 0 \end{bmatrix}$$

$$= 0 \cdot 0 + 0 \cdot 0 + 1 \cdot 1 + 0 \cdot 0 = 1$$

With the above notation involving column and row vectors to represent the eigenfunctions, how is it possible to write something like

$\hat{H}|1\rangle = E_1|1\rangle, \hat{H}|2\rangle = E_2|2\rangle$, etc? We must find a way to represent the Hamiltonian operator with a 4×4 matrix.

We show that the following matrix can do the job. A matrix whose diagonal elements are the eigenenergies represents the matrix form of the Hamiltonian.

$$\hat{H} = \begin{bmatrix} E_1 & 0 & 0 & 0 \\ 0 & E_2 & 0 & 0 \\ 0 & 0 & E_3 & 0 \\ 0 & 0 & 0 & E_4 \end{bmatrix}$$

With this, we can write

$$\hat{H}|1\rangle = \begin{bmatrix} E_1 & 0 & 0 & 0 \\ 0 & E_2 & 0 & 0 \\ 0 & 0 & E_3 & 0 \\ 0 & 0 & 0 & E_4 \end{bmatrix} \begin{bmatrix} 1 \\ 0 \\ 0 \\ 0 \end{bmatrix} = \begin{bmatrix} E_1 \\ 0 \\ 0 \\ 0 \end{bmatrix} = E_1 \begin{bmatrix} 1 \\ 0 \\ 0 \\ 0 \end{bmatrix} = E_1|1\rangle$$

It is also easy to show $\hat{H}|2\rangle = E_2|2\rangle$ and so on. We multiply the above from the left side by a bra vector $\langle 1|$ to verify:

$$\langle 1|\hat{H}|1\rangle = \begin{bmatrix} 1 & 0 & 0 & 0 \end{bmatrix} \begin{bmatrix} E_1 & 0 & 0 & 0 \\ 0 & E_2 & 0 & 0 \\ 0 & 0 & E_3 & 0 \\ 0 & 0 & 0 & E_4 \end{bmatrix} \begin{bmatrix} 1 \\ 0 \\ 0 \\ 0 \end{bmatrix}$$

$$= \begin{bmatrix} 1 & 0 & 0 & 0 \end{bmatrix} \begin{bmatrix} E_1 \\ 0 \\ 0 \\ 0 \end{bmatrix} = E_1$$

And if we multiply the above from the left side by a bra vector $\langle 3|$,

$$\langle 3|\hat{H}|1\rangle = \begin{bmatrix} 0 & 0 & 1 & 0 \end{bmatrix} \begin{bmatrix} E_1 & 0 & 0 & 0 \\ 0 & E_3 & 0 & 0 \\ 0 & 0 & E_3 & 0 \\ 0 & 0 & 0 & E_4 \end{bmatrix} \begin{bmatrix} 1 \\ 0 \\ 0 \\ 0 \end{bmatrix}$$

$$= \begin{bmatrix} 0 & 0 & 1 & 0 \end{bmatrix} \begin{bmatrix} E_1 \\ 0 \\ 0 \\ 0 \end{bmatrix} = 0$$

Therefore, we say the element of Hamiltonian matrix on the m-th row and n-th column is returned by the following product:

$$\hat{H}_{mn} = \langle m|\hat{H}|n \rangle$$

As we saw before, the RHS is shorthand for,

$$\langle m|\hat{H}|n \rangle = \int dv\, \psi_m^*(r,t)\hat{H}\psi_n(r,t)$$

We now show how to write down the momentum matrix elements of a PiB using the four energy eigenfunctions discussed above. The momentum matrix would have the following form in a basis involving only the lowest four eigenfunctions:

$$\hat{p} = \begin{bmatrix} p_{11} & p_{12} & p_{13} & p_{14} \\ p_{21} & p_{22} & p_{23} & p_{24} \\ p_{31} & p_{32} & p_{33} & p_{34} \\ p_{41} & p_{42} & p_{43} & p_{44} \end{bmatrix}$$

Recall that the eigenfunction of the energy operator (Hamiltonian) of the PiB has the following general form for a given quantum number n:

$$|\psi_n\rangle = \psi_n(x) = \sqrt{\frac{2}{L}} \sin\left(\frac{n\pi x}{L}\right)$$

The off diagonal elements of this matrix p_{mn} represents the efficiency with which an electron in quantum state (n) can absorb/emit a photon and transition to state (m). Since the electron is already in quantum state n, its efficiency in transitioning to the same state is zero. This can also be shown mathematically:

$$p_{nn} = \int dx\, \psi_n(x)^* \hat{p}\psi_n(x) = \frac{\hbar}{i}\int dx\, \psi_n(x)^* \frac{\partial}{\partial x}\psi_n(x)$$

$$= \frac{\hbar}{i}\frac{2}{L}\int dx \sin\left(\frac{n\pi x}{L}\right)\frac{\partial}{\partial x}\sin\left(\frac{n\pi x}{L}\right)$$

$$= \frac{\hbar}{i}\frac{2n\pi}{L^2}\int_0^L \sin\left(\frac{n\pi x}{L}\right)\cos\left(\frac{n\pi x}{L}\right) dx$$

$$= \frac{\hbar}{i}\frac{n\pi}{L^2}\int_0^L \sin\left(\frac{2n\pi x}{L}\right) dx = 0$$

The above integral is zero because within the interval $(0, L)$, $\sin(\frac{2n\pi x}{L})$ is an odd function about $x = L/2$, which makes the area under the function zero.

For off-diagonal terms $(n \neq m)$, we have

$$p_{mn} = \frac{\hbar}{i} \int dx \, \psi_m(x)^* \frac{\partial}{\partial x} \psi_n(x) = \frac{\hbar}{i} \frac{2n\pi}{L^2} \int_0^L \sin\left(\frac{m\pi x}{L}\right)$$

$$\times \cos\left(\frac{n\pi x}{L}\right) dx = \frac{2\hbar}{iL} \frac{nm[1 - (-1)^{m+n}]}{(m^2 - n^2)}$$

$$= \begin{cases} 0 & \text{if } m + n \text{ is an even number} \\ \dfrac{4\hbar nm}{iL(m^2 - n^2)} & \text{if } m + n \text{ is an odd number} \end{cases}$$

As a result, the form of the momentum matrix is:

$$\hat{p} = \begin{bmatrix} 0 & \neq 0 & 0 & \neq 0 \\ \neq 0 & 0 & \neq 0 & 0 \\ 0 & \neq 0 & 0 & \neq 0 \\ \neq 0 & 0 & \neq 0 & 0 \end{bmatrix}$$

Calculating the nonzero matrix element is straightforward and is left to the reader. The momentum matrix element is a measure of the overlap between one wave function and the derivative of the second one. It is of practical importance in calculating photon absorption and emission in materials. The precise reason for this will become apparent in Chapter 18. It is used to calculate photon absorption/emission in solar cells and light-emitting diodes. p_{mn} decides the efficiency with which an electron in quantum state (m) can absorb/emit a photon and go to state (n).

General operator $(\hat{O})$: Now we generalize what we learned in previous sections and write the operators in matrix form. The form of matrix and the values of its elements depends on what eigenfunctions (eigenkets) we use to build the matrix. For example, in the previous example, we built the matrix form of the momentum operator using the energy eigenfunctions.

Consider an operator $\hat{O}$. We can pre- and post-multiply it by the identity operator $\hat{I}$ and leave $\hat{O}$ unchanged,

$$\hat{O} = \hat{I}\hat{O}\hat{I} \tag{4.52}$$

Think of the above as pre- and post-multiplying a square matrix A by the identity matrix $\hat{I}$. Now using equation (4.46) (closure theorem), we can rewrite the above equation as

$$\hat{O} = \sum_m |m\rangle\langle m|\hat{O} \sum_n |n\rangle\langle n|$$

$$= \sum_{m,n} |m\rangle\langle m|\hat{O}|n\rangle\langle n| \tag{4.53}$$

The only requirement for the closure theorem was that $\{|n\rangle\}$ is a complete orthonormal set of eigenfunctions. The term $\langle m|\hat{O}|n\rangle$ is represented by O_{mn} (as we saw in equation (4.47)), which is a complex number. Operator $\hat{O}$ can be written in braket notation as

$$\hat{O} = \sum_{m,n} O_{mn}|m\rangle\langle n| \tag{4.54}$$

Any operator $\hat{O}$ can be written in matrix form as

$$\hat{O} = \begin{bmatrix} \langle 1|\hat{O}|1\rangle & \langle 1|\hat{O}|2\rangle & \langle 1|\hat{O}|3\rangle & \cdot & \cdot & \cdot \\ \langle 2|\hat{O}|1\rangle & \langle 2|\hat{O}|2\rangle & \langle 2|\hat{O}|3\rangle & \cdot & \cdot & \cdot \\ \langle 3|\hat{O}|1\rangle & \langle 3|\hat{O}|2\rangle & \langle 3|\hat{O}|3\rangle & \cdot & \cdot & \cdot \\ \cdot & \cdot & \cdot & \cdot & \cdot & \cdot \\ \cdot & \cdot & \cdot & \cdot & \cdot & \cdot \\ \cdot & \cdot & \cdot & \cdot & \cdot & \cdot \end{bmatrix}$$

which is

$$\hat{O} = \begin{bmatrix} O_{11} & O_{12} & O_{13} & \cdot & \cdot & \cdot \\ O_{21} & O_{22} & O_{23} & \cdot & \cdot & \cdot \\ O_{31} & O_{32} & O_{33} & \cdot & \cdot & \cdot \\ \cdot & \cdot & \cdot & \cdot & \cdot & \cdot \\ \cdot & \cdot & \cdot & \cdot & \cdot & \cdot \\ \cdot & \cdot & \cdot & \cdot & \cdot & \cdot \end{bmatrix} \tag{4.55}$$

where $O_{mn} = \langle m|\hat{O}|n\rangle$ are defined in equation (4.47).

Equations (4.54) and (4.55) are two equivalent ways of writing an operator. The former is useful in deriving formulae while the latter is useful when performing computations.

Operators in the eigenfunction basis: If $\{|n\rangle\}$ is the eigenfunction basis of operator $\hat{O}$, then we show that the matrix is of diagonal form. We have

$$\hat{O}|n\rangle = \alpha_n|n\rangle \tag{4.56}$$

where α_n and $|n\rangle$ are the eigenvalues and eigenfunctions of the operator $\hat{O}$, respectively. In a basis consisting of the eigenfunction of operator $\hat{O}$,

$$O_{mn} = \langle m|\hat{O}|n\rangle = \alpha_n\langle m|n\rangle = \alpha_n\delta_{mn} \tag{4.57}$$

That is the off-diagonal components of $O_{mn}(m \neq n)$ are zero. Substituting the above equation in equation (4.54) gives us that in the eigenfunction basis, operator $\hat{O}$ can be written as

$$\hat{O} = \sum_n \alpha_n|n\rangle\langle n| \quad \text{(in an eigenfunction basis)} \tag{4.58}$$

While the matrix in equation (4.55) is valid in any basis state, the reader should note that in the eigenfunction basis of operator $\hat{O}$, equation (4.55) becomes a diagonal matrix. On its main diagonal, we have $\alpha_1, \alpha_2, \alpha_3 \ldots$, and the rest of the matrix elements are zero as follows:

$$\hat{O} = \begin{bmatrix} \alpha_1 & 0 & 0 & \cdot & \cdot & \cdot \\ 0 & \alpha_2 & 0 & \cdot & \cdot & \cdot \\ 0 & 0 & \alpha_3 & \cdot & \cdot & \cdot \\ \cdot & \cdot & \cdot & \cdot & \cdot & \cdot \\ \cdot & \cdot & \cdot & \cdot & \cdot & \cdot \\ \cdot & \cdot & \cdot & \cdot & \cdot & \cdot \end{bmatrix} \quad \text{(in an eigenfunction basis)}$$

Concept: An operator $(\hat{O})$ can be expressed as a matrix. The matrix elements are given by

$$O_{mn} = \langle m|\hat{O}|n\rangle$$

$\{|n\rangle\}$ is a complete basis of eigenfunctions for the physical problem at hand.

The matrix representation of an operator $(\hat{O})$ is extremely useful in performing calculations and creating models with finite-dimensional matrices.

If $\hat{O} = \widehat{H}$, then using equation (4.54), the Hamiltonian operator in braket form is

$$\widehat{H} = \sum_{m,n} H_{mn}|m\rangle\langle n|$$

Substituting $\widehat{H}$ for $\hat{O}$ in equation (4.55), the Hamiltonian operator in matrix form is

$$\widehat{H} = \begin{bmatrix} H_{11} & H_{12} & H_{13} & \cdot & \cdot & \cdot \\ H_{21} & H_{22} & H_{23} & \cdot & \cdot & \cdot \\ H_{31} & H_{32} & H_{33} & \cdot & \cdot & \cdot \\ \cdot & \cdot & \cdot & \cdot & \cdot & \cdot \\ \cdot & \cdot & \cdot & \cdot & \cdot & \cdot \\ \cdot & \cdot & \cdot & \cdot & \cdot & \cdot \end{bmatrix} \tag{4.59}$$

where

$$H_{mn} = \langle m|\widehat{H}|n\rangle = \int dx\, \psi_m(x)^* \widehat{H} \psi_n(x) \tag{4.60}$$

If $\{\psi_n\}$ are eigenfunctions of the Hamiltonian,

$$\widehat{H}|\psi_n\rangle = E_n|\psi_n\rangle \tag{4.61}$$

Substituting equation (4.61) in equation (4.60),

$$H_{mn} = \langle m|\widehat{H}|n\rangle = E_n\langle m|n\rangle = E_n\delta_{mn} \tag{4.62}$$

where we have used equation (4.43). Substituting equation (4.62) in equation (4.54), the Hamiltonian in operator form is

$$\widehat{H} = \sum_n E_n |n\rangle\langle n| \qquad \text{(in eigenfunction basis)}$$

Substituting equation (4.62) in equations (4.59), the Hamiltonian matrix is

$$\widehat{H} = \begin{bmatrix} E_1 & 0 & 0 & . & . & . \\ 0 & E_2 & 0 & . & . & . \\ 0 & 0 & E_3 & . & . & . \\ . & . & . & . & . & . \\ . & . & . & . & . & . \\ . & . & . & . & . & . \end{bmatrix} \qquad \text{(in eigenfunction basis)} \quad (4.63)$$

a diagonal matrix with the eigenvalues along the diagonal. If the basis functions used to write down the Hamiltonian matrix are not the eigenfunctions of the Hamiltonians, then the matrix representation will not be diagonal. We will see a few examples of this case later.

Concept: An operator $(\widehat{O})$ expressed as a matrix using its eigenfunctions is a diagonal matrix.

When $\widehat{O}$ is the Hamiltonian operator, its matrix representation in a basis consisting of the eigenfunctions of the Hamiltonian is a diagonal matrix containing the eigenenergies on the diagonals.

Now we express the matrix form of position operator in the basis set of energy eigenfunctions $\{\psi_n\}$. By substituting $\hat{x}$ for $\widehat{O}$ in equation (4.55), the matrix form of the position operator $\hat{x}$ is

$$\hat{x} = \begin{bmatrix} x_{11} & x_{12} & x_{13} & . & . & . \\ x_{21} & x_{22} & x_{23} & . & . & . \\ x_{31} & x_{32} & x_{33} & . & . & . \\ . & . & . & . & . & . \\ . & . & . & . & . & . \\ . & . & . & . & . & . \end{bmatrix},$$

where

$$x_{mn} = \langle \psi_m | \hat{x} | \psi_n \rangle = \int dx\, \psi_m(x)^* x\, \psi_n(x)$$

In writing the above equation, we have used the definition of the position operator, $\hat{x}\psi_n(x) = x\psi_n(x)$ from Chapter 3. Note that the eigenvalues of the position operator $\hat{x}$ are simply the position coordinate x.

Note that in the PiB problem, we have an infinite number of discrete eigenstates indexed by $n = 1, 2, 3, \ldots$. That means the matrix representation of Hamiltonian and other operators have infinite dimension. However, in special cases where we are interested only in a few first eigenstates, the size of the matrix is finite, for example in the 4×4 case discussed above.

4.3.1 *Hermitian conjugate operators in matrix form*

The Hermitian conjugate of operator $\hat{O}$ is found by conjugating and then transposing the operator. The act of conjugation and transposing is represented by a dagger in the superscript

$$\hat{O}^{\dagger} = (\hat{O}^T)^* \tag{4.64}$$

Also, the following holds:

$$(\hat{O}^{\dagger})^{\dagger} = \hat{O} \tag{4.65}$$

Since a vector is a special form of matrix, we can say the bra vector is conjugate and transpose of the column (ket) vector, i.e.,

$$[|n\rangle]^{\dagger} = \langle n| \tag{4.66}$$

With the above, following properties also hold:

$$[\hat{O}|\psi_n\rangle]^{\dagger} = \langle \psi_n|\hat{O}^{\dagger} \quad \text{or} \quad [\hat{O}|n\rangle]^{\dagger} = \langle n|\hat{O}^{\dagger} \tag{4.67}$$

$$[c|n\rangle]^{\dagger} = \langle n|c^* \quad \text{if } c \text{ is a scalar}$$

Note that Hermitian operators were defined in Chapter 3 using the integro-differential form of operators. Now we re-express equation

(3.24) in matrix formalism. Recall that we said an operator $\widehat{O}$ is Hermitian if the following identity holds:

$$\int dv\, \psi(\bar{r})^*[\widehat{O}\,\phi(\bar{r})] = \int dv[\widehat{O}\,\psi(\bar{r})]^*\, \phi(\bar{r}) \tag{4.68}$$

We start from the LHS to write

$$\text{LHS} = \langle\psi|\widehat{O}|\phi\rangle = [\widehat{O}^\dagger|\psi\rangle]^\dagger|\phi\rangle$$

The LHS of the above is equal to the RHS of equation (4.68) only if we have $\widehat{O}^\dagger = \widehat{O}$. Meaning that an operator is **Hermitian** only if it satisfies the following identity:

$$\widehat{O}^\dagger = \widehat{O} \tag{4.69}$$

An operator is called *unitary* only if its Hermitian conjugate is equal to its inverse, i.e.,

$$\widehat{O}^\dagger = \widehat{O}^{-1} \rightarrow \widehat{O}\widehat{O}^\dagger = \hat{I} \tag{4.70}$$

4.4   Examples Using Matrix Form

Let us consider two identical atoms interacting with each other. Each atom consists of a single s-orbital and a single electron. The energy of the s-orbital of the isolated atom is ε. When the two atoms are placed close by, the term in the Hamiltonian that represents the ability of an electron to *tunnel* or *hop* between the atoms is represented by t. The Hamiltonian of this two-atom system is nondiagonal and can be written approximately as

$$H = \begin{pmatrix} \varepsilon & t \\ t & \varepsilon \end{pmatrix}$$

Note that t in the above equation is not time but has units of energy. The eigenvalue equation $H\psi = E\psi$ is then

$$\begin{pmatrix} \varepsilon & t \\ t & \varepsilon \end{pmatrix}\begin{pmatrix} c_1 \\ c_2 \end{pmatrix} = E\begin{pmatrix} c_1 \\ c_2 \end{pmatrix} \tag{4.71}$$

Single Atom   Atom 1   Atom 2

$$E_2 = \varepsilon + t$$

ε ————————

2t

$$E_1 = \varepsilon - t$$

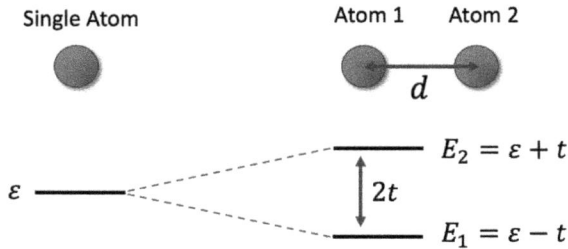

Figure 4.1. The energy levels of two far apart atoms that are degenerate (equal) split as they come close to each other. The amount of splitting depends on the off-diagonal term of the Hamiltonian, often called the hopping or tunneling energy, which is represented by t.

It is easy to verify that the solutions to the above eigenvalue equation are

Solution 1:

$$E_1 = \varepsilon - t \tag{4.72}$$

$$\begin{pmatrix} c_1 \\ c_2 \end{pmatrix} = \frac{1}{\sqrt{2}} \begin{pmatrix} 1 \\ -1 \end{pmatrix} \tag{4.73}$$

Solution 2:

$$E_2 = \varepsilon + t \tag{4.74}$$

$$\begin{pmatrix} c_1 \\ c_2 \end{pmatrix} = \frac{1}{\sqrt{2}} \begin{pmatrix} 1 \\ 1 \end{pmatrix} \tag{4.75}$$

This shows that when two identical atoms are brought together, their equal energy ε is split into $\varepsilon \pm t$. In the diagonal basis, we can write the Hamiltonian as (Figure 4.1)

$$\begin{pmatrix} \varepsilon - t & 0 \\ 0 & \varepsilon + t \end{pmatrix} \begin{pmatrix} a_1 \\ a_2 \end{pmatrix} = E \begin{pmatrix} a_1 \\ a_2 \end{pmatrix}$$

Example 4. Consider a quantum system with only two energy eigenvalues, g and e. Such a system is referred to as a two-level atom. The diagonalized Hamiltonian has the following form:

$$H = \begin{bmatrix} g & 0 \\ 0 & e \end{bmatrix}$$

(a) Find the energy eigenvectors[3] of the Hamiltonian, $\{u, v\}$, and show that they are orthonormal.

(b) Change the basis to $\left\{ \alpha = \frac{u+v}{\sqrt{2}}, \beta = \frac{u-v}{\sqrt{2}} \right\}$ and show that the new basis is still orthonormal.

(c) Write the Hamiltonian matrix in this new basis. Is the Hamiltonian diagonal?

Solution: (a) For energy eigenvalue (g), we calculate the eigenvectors using the definition of the eigenvalue problem,

$$H = \begin{bmatrix} g & 0 \\ 0 & e \end{bmatrix} \begin{bmatrix} a \\ b \end{bmatrix} = g \begin{bmatrix} a \\ b \end{bmatrix} \implies \begin{cases} ga = ga \\ eb = gb \end{cases}$$

Since g and e are different, b must be zero to satisfy the second equation $eb = gb$. On the other hand, a can take any value. The choice $a = 1$ and $b = 0$ results in an eigenvector of unity norm,

$$u = \begin{bmatrix} a \\ b \end{bmatrix} = \begin{bmatrix} 1 \\ 0 \end{bmatrix}$$

is the eigenvector corresponding to energy eigenvalue g. With the same method, we can show that the eigenvector corresponding to the energy eigenvalue of e is

$$v = \begin{bmatrix} a \\ b \end{bmatrix} = \begin{bmatrix} 0 \\ 1 \end{bmatrix}$$

You can show the orthonormality of the basis by evaluating the inner product,

$$\langle u|v \rangle = \begin{bmatrix} 1 & 0 \end{bmatrix} \begin{bmatrix} 0 \\ 1 \end{bmatrix} = 1 \times 0 + 0 \times 1 = 0$$

$$\langle u|u \rangle = \begin{bmatrix} 1 & 0 \end{bmatrix} \begin{bmatrix} 1 \\ 0 \end{bmatrix} = 1 \times 1 + 0 \times 0 = 1$$

$$\langle v|v \rangle = \begin{bmatrix} 0 & 1 \end{bmatrix} \begin{bmatrix} 0 \\ 1 \end{bmatrix} = 0 \times 0 + 1 \times 1 = 1$$

---

[3]When we deal with matrices, eigenfunctions are referred to as eigenvectors.

(b) The new basis is

$$\alpha = \frac{u+v}{\sqrt{2}} = \frac{1}{\sqrt{2}}\left(\begin{bmatrix}1\\0\end{bmatrix}+\begin{bmatrix}0\\1\end{bmatrix}\right) = \begin{bmatrix}\frac{1}{\sqrt{2}}\\\frac{1}{\sqrt{2}}\end{bmatrix} \text{ and}$$

$$\beta = \frac{u-v}{\sqrt{2}} = \begin{bmatrix}\frac{1}{\sqrt{2}}\\\frac{-1}{\sqrt{2}}\end{bmatrix}$$

It is easy to show that $\langle\alpha|\alpha\rangle = \langle\beta|\beta\rangle = 1$ and $\langle\alpha|\beta\rangle = 0$.

(c) Now we calculate the matrix elements of the Hamiltonian in the new basis:

$$\tilde{H}_{11} = \langle\alpha|H|\alpha\rangle = \begin{bmatrix}\frac{1}{\sqrt{2}} & \frac{1}{\sqrt{2}}\end{bmatrix}\begin{bmatrix}g & 0\\0 & e\end{bmatrix}\begin{bmatrix}\frac{1}{\sqrt{2}}\\\frac{1}{\sqrt{2}}\end{bmatrix} = \begin{bmatrix}\frac{1}{\sqrt{2}} & \frac{1}{\sqrt{2}}\end{bmatrix}\begin{bmatrix}\frac{g}{\sqrt{2}}\\\frac{e}{\sqrt{2}}\end{bmatrix} = \frac{g+e}{2}$$

$$\tilde{H}_{12} = \langle\alpha|H|\beta\rangle = \begin{bmatrix}\frac{1}{\sqrt{2}} & \frac{1}{\sqrt{2}}\end{bmatrix}\begin{bmatrix}g & 0\\0 & e\end{bmatrix}\begin{bmatrix}\frac{1}{\sqrt{2}}\\\frac{-1}{\sqrt{2}}\end{bmatrix} = \begin{bmatrix}\frac{1}{\sqrt{2}} & \frac{1}{\sqrt{2}}\end{bmatrix}\begin{bmatrix}\frac{g}{\sqrt{2}}\\\frac{-e}{\sqrt{2}}\end{bmatrix} = \frac{g-e}{2}$$

$$\tilde{H}_{21} = \langle\beta|H|\alpha\rangle = \begin{bmatrix}\frac{1}{\sqrt{2}} & \frac{-1}{\sqrt{2}}\end{bmatrix}\begin{bmatrix}g & 0\\0 & e\end{bmatrix}\begin{bmatrix}\frac{1}{\sqrt{2}}\\\frac{1}{\sqrt{2}}\end{bmatrix} = \begin{bmatrix}\frac{1}{\sqrt{2}} & \frac{-1}{\sqrt{2}}\end{bmatrix}\begin{bmatrix}\frac{g}{\sqrt{2}}\\\frac{e}{\sqrt{2}}\end{bmatrix} = \frac{g-e}{2}$$

$$\tilde{H}_{22} = \langle\beta|H|\beta\rangle = \begin{bmatrix}\frac{1}{\sqrt{2}} & \frac{-1}{\sqrt{2}}\end{bmatrix}\begin{bmatrix}g & 0\\0 & e\end{bmatrix}\begin{bmatrix}\frac{1}{\sqrt{2}}\\\frac{-1}{\sqrt{2}}\end{bmatrix} = \begin{bmatrix}\frac{1}{\sqrt{2}} & \frac{-1}{\sqrt{2}}\end{bmatrix}\begin{bmatrix}\frac{g}{\sqrt{2}}\\\frac{-e}{\sqrt{2}}\end{bmatrix} = \frac{g+e}{2}$$

Having all the matrix elements, we can build the matrix in the new basis:

$$\tilde{H} = \begin{bmatrix}\tilde{H}_{11} & \tilde{H}_{12}\\\tilde{H}_{21} & \tilde{H}_{22}\end{bmatrix} = \begin{bmatrix}\frac{g+e}{2} & \frac{g-e}{2}\\\frac{g-e}{2} & \frac{g+e}{2}\end{bmatrix}$$

The Hamiltonian in the new basis is not diagonal anymore because the new basis vectors are not its eigenvectors.

Example 5. The Hamiltonian matrix of a system in a 3D Hilbert space is $H = \begin{bmatrix}4 & 1 & 0\\1 & 4 & 0\\0 & 0 & 5\end{bmatrix}$ (eV) (a) Find the energy eigenvalues and eigenvectors of the systems.

(b) If the system is in a superposition state $|\psi\rangle = \frac{1}{\sqrt{6}}\begin{bmatrix}1\\-1\\2\end{bmatrix}$, express the state as a linear combination of the energy eigenvectors of the

Hamiltonian. Then, find the expectation value or average energy of the system.

Solution: (a) To find the energy eigenvalues of the Hamiltonian, we solve the eigenvalue equation, $Hu = Eu$ or equivalently $(H - EI)u = 0$. u is a 3×1 vector. To have nontrivial solutions, the determinant of the $H - EI$ matrix must be zero. Hence,

$$\det(H - EI) = \det \begin{bmatrix} 4 - E & 1 & 0 \\ 1 & 4 - E & 0 \\ 0 & 0 & 5 - E \end{bmatrix} = 0$$

$$(5 - E)((4 - E)^2 - 1) = 0 \implies (E - 5)(E - 5)(E - 3) = 0$$

Therefore, there are three energy eigenvalues, two of which are degenerate and are $E = 5\,\mathrm{eV}$, and the third one is $E = 3\,\mathrm{eV}$. To solve part (b), we have to find the eigenvectors corresponding to the energy eigenvalues we found above. For $E = 3$, we write $(H - EI)u = 0$ and solve the three equations to find the three components of u which we call a, b, c:

$$(H - 3I)u = \begin{bmatrix} 4 - 3 & 1 & 0 \\ 1 & 4 - 3 & 0 \\ 0 & 0 & 5 - 3 \end{bmatrix} \begin{bmatrix} a \\ b \\ c \end{bmatrix} = \begin{bmatrix} 1 & 1 & 0 \\ 1 & 1 & 0 \\ 0 & 0 & 2 \end{bmatrix} \begin{bmatrix} a \\ b \\ c \end{bmatrix} = \begin{bmatrix} 0 \\ 0 \\ 0 \end{bmatrix}$$

Multiplying out the matrix product above, we find that the equations which $a, b,$ and c must satisfy are:

$$\begin{cases} a + b = 0 \\ a + b = 0 \\ 2c = 0 \end{cases} \tag{4.76}$$

The solutions $a = 1, b = -1$, and $c = 0$ satisfy the above equation, so the normalized eigenvector for $E = 3$ is

$$u_3 = \frac{1}{\sqrt{2}} \begin{bmatrix} 1 \\ -1 \\ 0 \end{bmatrix}$$

We find one of the eigenvectors corresponding to $E = 5\,\text{eV}$ with the same method as above

$$(H - 5I)u = \begin{bmatrix} 4-5 & 1 & 0 \\ 1 & 4-5 & 0 \\ 0 & 0 & 5-5 \end{bmatrix} \begin{bmatrix} a \\ b \\ c \end{bmatrix}$$

$$= \begin{bmatrix} -1 & 1 & 0 \\ 1 & -1 & 0 \\ 0 & 0 & 0 \end{bmatrix} \begin{bmatrix} a \\ b \\ c \end{bmatrix} = \begin{bmatrix} 0 \\ 0 \\ 0 \end{bmatrix} \tag{4.77}$$

The eigenvector must satisfy equation (4.77). One plausible solution for this is $a = 0$, $b = 0$, and $c = 1$. So, the eigenvector which is normalized and orthogonal to u_3 (check this yourself) is

$$u_{51} = \begin{bmatrix} 0 \\ 0 \\ 1 \end{bmatrix}$$

As the energy eigenvalue of $E = 5\,\text{eV}$ is degenerate (occurs twice), we must find its eigenvectors which are orthogonal to u_{51} and u_3. We have equation (4.77) for the eigenvector again. The solution $a = b = 1$ and $c = 0$ satisfies (4.77) and is also orthogonal to u_{51}. So, the orthonormal eigenvector for the second eigenvalue of $5\,\text{eV}$ is

$$u_{52} = \frac{1}{\sqrt{2}} \begin{bmatrix} 1 \\ 1 \\ 0 \end{bmatrix}$$

Now the superposition state can be decomposed as follows:

$$|\psi\rangle = \frac{1}{\sqrt{6}} \begin{bmatrix} 1 \\ -1 \\ 2 \end{bmatrix} = \frac{\sqrt{2}}{\sqrt{6}} \begin{bmatrix} \frac{1}{\sqrt{2}} \\ \frac{-1}{\sqrt{2}} \\ 0 \end{bmatrix} + \frac{2}{\sqrt{6}} \begin{bmatrix} 0 \\ 0 \\ 1 \end{bmatrix} = \frac{\sqrt{2}}{\sqrt{6}} u_3 + \frac{2}{\sqrt{6}} u_{51}$$

Therefore, the average energy is equal to

$$\langle H \rangle = \sum_n |a_{no}|^2 E_n = \left(\frac{\sqrt{2}}{\sqrt{6}} \right)^2 E_3 + \left(\frac{2}{\sqrt{6}} \right)^2 E_5$$

$$= \frac{2}{6} \times 3 + \frac{4}{6} \times 5 = 4.33\,\text{eV}$$

Example 6. The Hamiltonian matrix of a particle is $H =$
$\hbar\omega \begin{bmatrix} 1 & 0 & 0 \\ 0 & 0 & 0 \\ 0 & 0 & -1 \end{bmatrix} \equiv \hbar\omega S_z$ and the initial state (at $t = 0$) of the particle
is given by $|\psi(0)\rangle = \frac{1}{\sqrt{2}} \begin{bmatrix} 1 \\ 0 \\ 1 \end{bmatrix}$.

Find the (a) state of the particle at later time t, i.e., $|\psi(t)\rangle$, and
(b) probability amplitude of finding the projection of the initial state
$|\psi(0)\rangle$ in the current state $|\psi(t)\rangle$.

Solution: (a) The evolution of the state is governed by the time-
dependent Schrödinger equation in Chapter 1.

$$\frac{d}{dt}|\psi(t)\rangle = -\frac{i}{\hbar}H|\psi(t)\rangle$$

The solution is

$$|\psi(t)\rangle = e^{\frac{-iHt}{\hbar}}|\psi(t = 0)\rangle = e^{\frac{-i\hbar\omega S_z t}{\hbar}}|\psi(t = 0)\rangle$$

$$= \frac{1}{\sqrt{2}}e^{-i\omega t S_z}\begin{bmatrix} 1 \\ 0 \\ 1 \end{bmatrix} = \frac{1}{\sqrt{2}}\begin{bmatrix} e^{-i\omega t} & 0 & 0 \\ 0 & 0 & 0 \\ 0 & 0 & e^{+i\omega t} \end{bmatrix}\begin{bmatrix} 1 \\ 0 \\ 1 \end{bmatrix}$$

$$|\psi(t)\rangle = \frac{1}{\sqrt{2}}\begin{bmatrix} e^{-i\omega t} \\ 0 \\ e^{+i\omega t} \end{bmatrix}$$

(b) The probability amplitude of finding the projection of the initial
state $|\psi(0)\rangle$ in the current state $|\psi(t)\rangle$ is found from the inner product
of the two states. The probability is the amplitude squared:

$$p(t) = |\langle\psi(0)|\psi(t)\rangle|^2 = \left|\frac{1}{2}\begin{bmatrix} 1 & 0 & 1 \end{bmatrix}\begin{bmatrix} e^{-i\omega t} \\ 0 \\ e^{+i\omega t} \end{bmatrix}\right|^2 = \frac{1 + \cos(2\omega t)}{2}$$

Note that every $t' = \frac{\pi}{\omega}$ seconds, the state returns to its initial state
at $t = 0$. That is, there is a periodic rotation of the initial state
$|\psi(t = 0)\rangle$ caused by this Hamiltonian.

Example 7. This example introduces you to *creation* and *annihilation* operators in matrix form.

Consider a Hilbert space with five orthonormal vectors,

$$\text{Hilbert space} = \{|0\rangle, |1\rangle, |2\rangle, |3\rangle, |4\rangle\} = \left\{ \begin{bmatrix} 1 \\ 0 \\ 0 \\ 0 \\ 0 \end{bmatrix}, \begin{bmatrix} 0 \\ 1 \\ 0 \\ 0 \\ 0 \end{bmatrix}, \begin{bmatrix} 0 \\ 0 \\ 1 \\ 0 \\ 0 \end{bmatrix}, \begin{bmatrix} 0 \\ 0 \\ 0 \\ 1 \\ 0 \end{bmatrix}, \begin{bmatrix} 0 \\ 0 \\ 0 \\ 0 \\ 1 \end{bmatrix} \right\}$$

represented using ket notation as $\{|0\rangle, |1\rangle, |2\rangle, |3\rangle, |4\rangle\}$. These are called the number states for reasons that will become clear later. We define two operators $\widehat{B}$ and its Hermitian conjugate $\widehat{B}^\dagger$ as follows:

$$\widehat{B} = \begin{bmatrix} 0 & \sqrt{1} & 0 & 0 & 0 \\ 0 & 0 & \sqrt{2} & 0 & 0 \\ 0 & 0 & 0 & \sqrt{3} & 0 \\ 0 & 0 & 0 & 0 & \sqrt{4} \\ 0 & 0 & 0 & 0 & 0 \end{bmatrix}, \quad \widehat{B}^\dagger = \begin{bmatrix} 0 & 0 & 0 & 0 & 0 \\ \sqrt{1} & 0 & 0 & 0 & 0 \\ 0 & \sqrt{2} & 0 & 0 & 0 \\ 0 & 0 & \sqrt{3} & 0 & 0 \\ 0 & 0 & 0 & \sqrt{4} & 0 \end{bmatrix}$$

(a) Calculate $\widehat{B}|0\rangle$, $\widehat{B}|1\rangle$, $\widehat{B}|2\rangle$, $\widehat{B}|3\rangle$, $\widehat{B}|4\rangle$ and from that deduce that $\widehat{B}|n\rangle = \sqrt{n}|n-1\rangle$.
(b) Calculate $\widehat{B}^\dagger|0\rangle$, $\widehat{B}^\dagger|1\rangle$, $B^\dagger|2\rangle$, $\widehat{B}^\dagger|3\rangle$ and from that deduce that $\widehat{B}^\dagger|n\rangle = \sqrt{n+1}|n+1\rangle$.
(c) Find $\widehat{B}^\dagger\widehat{B}|n\rangle$. What are the eigenvalue and eigenvector of $\widehat{B}^\dagger\widehat{B}$?
(d) If the Hamiltonian of a system is $H = \hbar\omega(\widehat{B}^\dagger\widehat{B} + \frac{I}{2})$, where I is the identity matrix, find the eigenvalues and eigenvectors of H.
(e) Do operators $\widehat{B}$ and $\widehat{B}^\dagger$ commute? If not, what is their commutation relationship, i.e., $[\widehat{B}, \widehat{B}^\dagger]$?

Solution: It is easy to check that the basis vectors in the given Hilbert space are orthonormal:

$$\langle 0|0\rangle = \langle 1|1\rangle = \langle 2|2\rangle = \langle 3|3\rangle = \langle 4|4\rangle = 1,$$

and inner products like $\langle 1|2\rangle = 0$.

We can write the following:

$$\langle n|n \rangle = 1 \quad \text{and} \quad \langle n|n+1 \rangle = 0 \text{ and } \langle n|n-1 \rangle = 0$$

(a) We now apply the operator B from the left side to ket vector:

$$\hat{B}|0\rangle = \begin{bmatrix} 0 & \sqrt{1} & 0 & 0 & 0 \\ 0 & 0 & \sqrt{2} & 0 & 0 \\ 0 & 0 & 0 & \sqrt{3} & 0 \\ 0 & 0 & 0 & 0 & \sqrt{4} \\ 0 & 0 & 0 & 0 & 0 \end{bmatrix} \begin{bmatrix} 1 \\ 0 \\ 0 \\ 0 \\ 0 \end{bmatrix} = \begin{bmatrix} 0 \\ 0 \\ 0 \\ 0 \\ 0 \end{bmatrix} = 0$$

as all components are zero. Note that this resulting vector does not belong to the Hilbert space anymore.

$$\hat{B}|1\rangle = \begin{bmatrix} 0 & \sqrt{1} & 0 & 0 & 0 \\ 0 & 0 & \sqrt{2} & 0 & 0 \\ 0 & 0 & 0 & \sqrt{3} & 0 \\ 0 & 0 & 0 & 0 & \sqrt{4} \\ 0 & 0 & 0 & 0 & 0 \end{bmatrix} \begin{bmatrix} 0 \\ 1 \\ 0 \\ 0 \\ 0 \end{bmatrix} = \begin{bmatrix} \sqrt{1} \\ 0 \\ 0 \\ 0 \\ 0 \end{bmatrix} = \sqrt{1} \begin{bmatrix} 1 \\ 0 \\ 0 \\ 0 \\ 0 \end{bmatrix} = \sqrt{1}|0\rangle$$

$$\hat{B}|2\rangle = \begin{bmatrix} 0 & \sqrt{1} & 0 & 0 & 0 \\ 0 & 0 & \sqrt{2} & 0 & 0 \\ 0 & 0 & 0 & \sqrt{3} & 0 \\ 0 & 0 & 0 & 0 & \sqrt{4} \\ 0 & 0 & 0 & 0 & 0 \end{bmatrix} \begin{bmatrix} 0 \\ 0 \\ 1 \\ 0 \\ 0 \end{bmatrix} = \begin{bmatrix} 0 \\ \sqrt{2} \\ 0 \\ 0 \\ 0 \end{bmatrix} = \sqrt{2} \begin{bmatrix} 0 \\ 1 \\ 0 \\ 0 \\ 0 \end{bmatrix} = \sqrt{2}|1\rangle$$

$$\hat{B}|3\rangle = \begin{bmatrix} 0 & \sqrt{1} & 0 & 0 & 0 \\ 0 & 0 & \sqrt{2} & 0 & 0 \\ 0 & 0 & 0 & \sqrt{3} & 0 \\ 0 & 0 & 0 & 0 & \sqrt{4} \\ 0 & 0 & 0 & 0 & 0 \end{bmatrix} \begin{bmatrix} 0 \\ 0 \\ 0 \\ 1 \\ 0 \end{bmatrix} = \begin{bmatrix} 0 \\ 0 \\ \sqrt{3} \\ 0 \\ 0 \end{bmatrix} = \sqrt{3} \begin{bmatrix} 0 \\ 0 \\ 1 \\ 0 \\ 0 \end{bmatrix} = \sqrt{3}|2\rangle$$

$$\widehat{B}|4\rangle = \begin{bmatrix} 0 & \sqrt{1} & 0 & 0 & 0 \\ 0 & 0 & \sqrt{2} & 0 & 0 \\ 0 & 0 & 0 & \sqrt{3} & 0 \\ 0 & 0 & 0 & 0 & \sqrt{4} \\ 0 & 0 & 0 & 0 & 0 \end{bmatrix} \begin{bmatrix} 0 \\ 0 \\ 0 \\ 0 \\ 1 \end{bmatrix} = \begin{bmatrix} 0 \\ 0 \\ 0 \\ \sqrt{4} \\ 0 \end{bmatrix} = \sqrt{4} \begin{bmatrix} 0 \\ 0 \\ 0 \\ 1 \\ 0 \end{bmatrix} = \sqrt{4}|3\rangle$$

As you see, the role of $\widehat{B}$ is to reduce the state's label by one unit and multiply it by the square root of its previous index. Hence, we may propose the following general form for the action of $\widehat{B}$ on a given state $|n\rangle$:

$$\widehat{B}|n\rangle = \sqrt{n}|n-1\rangle$$

$\widehat{B}$ is called the *lowering or annihilation* operator as it reduces the number index of the state (n) by one. These states are called the **number** or **Fock** states (named after the Russian scientist Vladimir A. Fock). They can stand for the number of photons in an electromagnetic field or LC oscillator or the number of phonons in an acoustic wave in a solid.

(b) We now apply the operator $\widehat{B}^\dagger$ from the left side to ket vector:

$$\widehat{B}^\dagger|0\rangle = \begin{bmatrix} 0 & 0 & 0 & 0 & 0 \\ \sqrt{1} & 0 & 0 & 0 & 0 \\ 0 & \sqrt{2} & 0 & 0 & 0 \\ 0 & 0 & \sqrt{3} & 0 & 0 \\ 0 & 0 & 0 & \sqrt{4} & 0 \end{bmatrix} \begin{bmatrix} 1 \\ 0 \\ 0 \\ 0 \\ 0 \end{bmatrix} = \begin{bmatrix} 0 \\ \sqrt{1} \\ 0 \\ 0 \\ 0 \end{bmatrix} = \sqrt{1} \begin{bmatrix} 0 \\ 1 \\ 0 \\ 0 \\ 0 \end{bmatrix} = \sqrt{1}|1\rangle$$

$$\widehat{B}^\dagger|1\rangle = \begin{bmatrix} 0 & 0 & 0 & 0 & 0 \\ \sqrt{1} & 0 & 0 & 0 & 0 \\ 0 & \sqrt{2} & 0 & 0 & 0 \\ 0 & 0 & \sqrt{3} & 0 & 0 \\ 0 & 0 & 0 & \sqrt{4} & 0 \end{bmatrix} \begin{bmatrix} 0 \\ 1 \\ 0 \\ 0 \\ 0 \end{bmatrix} = \begin{bmatrix} 0 \\ 0 \\ \sqrt{2} \\ 0 \\ 0 \end{bmatrix} = \sqrt{2} \begin{bmatrix} 0 \\ 0 \\ 1 \\ 0 \\ 0 \end{bmatrix} = \sqrt{2}|2\rangle$$

$$\widehat{B}^{\dagger}|2\rangle = \begin{bmatrix} 0 & 0 & 0 & 0 & 0 \\ \sqrt{1} & 0 & 0 & 0 & 0 \\ 0 & \sqrt{2} & 0 & 0 & 0 \\ 0 & 0 & \sqrt{3} & 0 & 0 \\ 0 & 0 & 0 & \sqrt{4} & 0 \end{bmatrix} \begin{bmatrix} 0 \\ 0 \\ 1 \\ 0 \\ 0 \end{bmatrix} = \begin{bmatrix} 0 \\ 0 \\ 0 \\ \sqrt{3} \\ 0 \end{bmatrix} = \sqrt{3} \begin{bmatrix} 0 \\ 0 \\ 0 \\ 1 \\ 0 \end{bmatrix} = \sqrt{3}|3\rangle$$

$$\widehat{B}^{\dagger}|3\rangle = \begin{bmatrix} 0 & 0 & 0 & 0 & 0 \\ \sqrt{1} & 0 & 0 & 0 & 0 \\ 0 & \sqrt{2} & 0 & 0 & 0 \\ 0 & 0 & \sqrt{3} & 0 & 0 \\ 0 & 0 & 0 & \sqrt{4} & 0 \end{bmatrix} \begin{bmatrix} 0 \\ 0 \\ 0 \\ 1 \\ 0 \end{bmatrix} = \begin{bmatrix} 0 \\ 0 \\ 0 \\ 0 \\ \sqrt{4} \end{bmatrix} = \sqrt{4} \begin{bmatrix} 0 \\ 0 \\ 0 \\ 0 \\ 1 \end{bmatrix} = \sqrt{4}|4\rangle$$

The previous equation suggests that the Hermitian conjugate of $\widehat{B}$, i.e., $\widehat{B}^{\dagger}$, *raises* the number (n) by one, and multiplies the state by the square root value of the new index. So, it is called the raising or creation operator,

$$\widehat{B}^{\dagger}|n\rangle = \sqrt{n+1}|n+1\rangle$$

Note that we did not calculate $\widehat{B}^{\dagger}|4\rangle$ because we don't have state $|5\rangle$ in the Hilbert space. The vector space's size must increase to six to accommodate this state. In general, the matrix size should be infinite to have any number state and matrices $\widehat{B}$ and $\widehat{B}^{\dagger}$ would grow to be of infinite size.

(c) To find $\widehat{B}^{\dagger}\widehat{B}|n\rangle$ we use their properties, which we discovered in parts (a) and (b) above:

$$\begin{aligned} \widehat{B}^{\dagger}B|n\rangle &= \widehat{B}^{\dagger}(\sqrt{n}|n-1\rangle) \\ &= \sqrt{n}\widehat{B}^{\dagger}(|n-1\rangle) \\ &= \sqrt{n}\sqrt{n-1+1}|n-1+1\rangle \\ &= n|n\rangle \end{aligned}$$

Therefore, we see that $\widehat{B}^\dagger \widehat{B}|n\rangle = n|n\rangle$, which means the number (n) is the eigenvalue of $\widehat{B}^\dagger \widehat{B}$, and its eigenvector is $|n\rangle$. $\widehat{B}^\dagger \widehat{B}$ is the **Number Operator** $(\widehat{N})$,

$$\widehat{N} = \widehat{B}^\dagger \widehat{B}, \quad \widehat{N}|n\rangle = n|n\rangle$$

(d) We apply $\widehat{H} = \hbar\omega\left(\widehat{B}^\dagger \widehat{B} + \frac{1}{2}\right)$ to $|n\rangle$ and use the above properties to find

$$\widehat{H}|n\rangle = \hbar\omega\left(\widehat{B}^\dagger \widehat{B} + \frac{I}{2}\right)|n\rangle$$

$$= \hbar\omega\left(\widehat{N} + \frac{I}{2}\right)|n\rangle$$

$$= \hbar\omega\left(\widehat{N}|n\rangle + \frac{I}{2}|n\rangle\right)$$

$$= \hbar\omega\left(n|n\rangle + \frac{1}{2}|n\rangle\right)$$

$$\widehat{H}|n\rangle = \hbar\omega\left(n + \frac{1}{2}\right)|n\rangle$$

Thus, the energy eigenvalues of this Hamiltonian are given as

$$E_n = \hbar\omega\left(n + \frac{1}{2}\right)$$

Recall that a simple harmonic oscillator's (SHO) energies are precisely the same as this, suggesting that the Hamiltonian in this example belongs to a SHO. Still, it is written as a 5×5 matrix, meaning that we are only interested in five initial states indexed by $n = 0, 1, 2, 3, 4$. The index n corresponds to the number of quanta of energy, where each quantum carries an energy of $\hbar\omega$. These could be photons in an LC oscillator circuit or phonons of acoustic oscillations in a crystal.

(e) No. These operators do not commute. We already showed that

$$\widehat{B}^\dagger \widehat{B}|n\rangle = n|n\rangle$$

On the other hand,

$$\begin{aligned}
\widehat{B}\widehat{B}^\dagger|n\rangle &= \widehat{B}(\widehat{B}^\dagger|n\rangle)\\
&= \widehat{B}(\sqrt{n+1}|n+1\rangle)\\
&= \sqrt{n+1}\widehat{B}|n+1\rangle\\
&= \sqrt{n+1}\sqrt{n+1}|n+1-1\rangle = (n+1)|n\rangle
\end{aligned}$$

which shows that

$$\widehat{B}^\dagger \widehat{B} \neq \widehat{B}\widehat{B}^\dagger$$

That is, they do not commute. The commutation relationship is,

$$[\widehat{B}, \widehat{B}^\dagger]|n\rangle = (\widehat{B}\widehat{B}^\dagger - \widehat{B}^\dagger \widehat{B})|n\rangle = \widehat{B}\widehat{B}^\dagger|n\rangle - \widehat{B}^\dagger \widehat{B}|n\rangle$$
$$= (n+1)|n\rangle - n|n\rangle = 1|n\rangle$$

Therefore,

$$[\widehat{B}, \widehat{B}^\dagger] = I \quad \text{or} \quad \widehat{B}\widehat{B}^\dagger = I + \widehat{B}^\dagger \widehat{B}$$

Example 8. Let the operators $\widehat{B}$ and $\widehat{B}^\dagger$ be defined in terms of position and momentum operators ($\hat{x}$ and $\hat{p}_x$) as follows:

$$\widehat{B} = \frac{1}{\sqrt{2m\hbar\omega}}(m\omega\hat{x} + i\hat{p}_x)$$

$$\widehat{B}^\dagger = \frac{1}{\sqrt{2m\hbar\omega}}(m\omega\hat{x} - i\hat{p}_x)$$

Recall that in quantum mechanics, operators are Hermitian. So, $\hat{x}^\dagger = \hat{x}$ and $\hat{p}^\dagger = \hat{p}$. Rewrite the Hamiltonian in the previous example $H = \hbar\omega\left(\widehat{B}^\dagger \widehat{B} + \frac{I}{2}\right)$ in terms of $\hat{x}$ and $\hat{p}_x$. What system does this Hamiltonian belong to? **Hint:** Recall that we showed in Chapter 3 that $[\hat{x}, \hat{p}_x] = \hat{x}\hat{p}_x - \hat{p}_x\hat{x} = i\hbar I$.

Solution: At first, we calculate $\widehat{B}^\dagger \widehat{B}$

$$\widehat{B}^\dagger \widehat{B} = \frac{1}{2m\hbar\omega}(m\omega\hat{x} - i\hat{p}_x)(m\omega\hat{x} + i\hat{p}_x)$$

$$= \frac{m^2\omega^2}{2m\hbar\omega}\hat{x}^2 - \frac{im\omega}{2m\hbar\omega}\hat{p}_x\hat{x} + \frac{im\omega}{2m\hbar\omega}\hat{x}\hat{p}_x + \frac{1}{2m\hbar\omega}\hat{p}_x^2$$

$$\widehat{B}^\dagger \widehat{B} = \frac{m\omega}{2\hbar}\hat{x}^2 + \frac{i}{2\hbar}(\hat{x}\hat{p}_x - \hat{p}_x\hat{x}) + \frac{1}{2m\hbar\omega}\hat{p}_x^2$$

Since $\hat{p}_x = -i\hbar\frac{d}{dx}$ and $[\hat{x}, \hat{p}_x] = i\hbar\widehat{I}$, $\widehat{B}^\dagger \widehat{B}$ can be rewritten as

$$\widehat{B}^\dagger \widehat{B} = \frac{m\omega}{2\hbar}\hat{x}^2 - \frac{\widehat{I}}{2} - \frac{\hbar}{2m\omega}\frac{d^2}{dx^2}$$

The above equation is then simplified to be

$$\hbar\omega\left(\widehat{B}^\dagger \widehat{B} + \frac{\widehat{I}}{2}\right) = -\frac{\hbar^2}{2m}\frac{d^2}{dx^2} + \frac{m\omega^2}{2}\hat{x}^2$$

The RHS looks familiar because it is the Hamiltonian of the SHO discussed in Chapter 1. In this example, by using a simple transformation, we showed that the Hamiltonian of the SHO can be written in terms of creation ($\widehat{B}^\dagger$) and annihilation ($\widehat{B}$) operators. Since $\widehat{B}^\dagger \widehat{B}$ is the number operator, $\widehat{N}$, the Hamiltonian can be written as

$$\widehat{H} = \hbar\omega\left(\widehat{N} + \frac{\widehat{I}}{2}\right)$$

With this notation, the eigenenergy of an SHO is found using $\widehat{B}^\dagger \widehat{B}|n\rangle = n|n\rangle$ and $\widehat{H}|n\rangle = E_n|n\rangle$ to be

$$E_n = \left(n + \frac{1}{2}\right)\hbar\omega$$

4.5 Problems

Section 4.1

(1) Prove Parseval's theorem (equation (4.8)) which states the energy content of signal in the x domain is the same as the energy in the k domain.

(2) Problem to help the reader understand quantum measurement: You have a million copies of a PiB in the state given by $(3\psi_1 + 4\psi_2)/5$, where ψ_1 and ψ_2 are the energy eigenfunctions in quantum numbers $n = 1$ and $n = 2$. If you measure the momentum of the particle, what values will you find? Assume that the PiB is of length L and the mass of the particle is M. Express your answer in terms L and M.

Section 4.2

(3) Evaluate $\langle 1|\hat{p}|1\rangle$, $\langle 2|\hat{p}|2\rangle$, $\langle 1|\hat{p}|2\rangle$, and $\langle 1|\hat{p}|3\rangle$ for a particle in a 1D particle in a box. Assume that the PiB is of length L and the mass of the particle is M. Express your answer in terms L and M.

You will learn later on that if the matrix element of the momentum operator $\hat{p}$ is nonzero between two states $\langle m|$ and $|n\rangle$, then light can be absorbed/emitted between these two states.

(4) For a PiB, express the (a) momentum and (b) energy operators in matrix form. Consider only quantum numbers $n = 1, 2$, and 3. Assume that the PiB is of length L and the mass of the particle is M. Express your answer in terms L and M.

(5) **(Closure theorem)** Show that the identity operator is $\hat{I} = \sum_n |n\rangle\langle n|$. The identity operator is defined by $\hat{I}|\phi\rangle = |\phi\rangle$ for any arbitrary vector $|\phi\rangle$. $\{|n\rangle\}$ is a complete set of basis vectors.

(6) You are given a state $|\text{init}\rangle = \frac{1}{\sqrt{N}}\{|\alpha_p\rangle + \sum_{\substack{i=1 \\ (i \neq p)}}^{N} |\alpha_i\rangle\}$. This state can be written as $|\text{init}\rangle = \frac{1}{\sqrt{N}}|\alpha_p\rangle + \frac{\sqrt{N-1}}{\sqrt{N}}|\bar{\alpha}_p\rangle$, where $|\bar{\alpha}_p\rangle = \frac{1}{\sqrt{N-1}}\sum_{\substack{i=1 \\ (i \neq p)}}^{N} |\alpha_i\rangle$.

You are also given an operator $\hat{W} = 2|\text{init}\rangle\langle\text{init}| - \hat{I}$, where $\hat{I}$ is the identity operator. Show that $\hat{W}|\alpha_p\rangle = \frac{2\sqrt{N-1}}{N}|\bar{\alpha}_p\rangle + \frac{2-N}{N}|\alpha_p\rangle$. We will use this problem when we discuss quantum information.

Section 4.3

(7) You are given a Hamiltonian, $H = \begin{pmatrix} 0 & 1 & 1 \\ 1 & 0 & 1 \\ 1 & 1 & 0 \end{pmatrix}$. Find the (a) eigenvalues and (b) eigenvectors, and (c) show that the eigenvectors are both orthogonal and orthonormal.

Section 4.4

(8) Show that the diagonal elements of $\hat{B} + \hat{B}^\dagger$ in Example 7 are zero. **Hint:** Use the properties of the operators and calculate $\langle n|\hat{B} + \hat{B}^\dagger|n\rangle$.

(9) Show that in Example 7, the Hamiltonian matrix's diagonal elements are n and nondiagonal elements are zero. **Hint:** Show that $\langle n|\hat{H}|n\rangle = n$, $\langle n|\hat{H}|n+1\rangle = 0$, and $\langle n|\hat{H}|n-1\rangle = 0$.

(10) You are given the voltage (V) and current (I) operators of an LC oscillator in terms of annihilation and creation operators ($\hat{B}$ and $\hat{B}^\dagger$),

$$\hat{V} = \sqrt{\frac{\hbar\omega}{2\mathbb{C}}}(\hat{B}^\dagger + \hat{B}) \ and \hat{I} = i\sqrt{\frac{\hbar\omega}{2\mathcal{L}}}(\hat{B}^\dagger - \hat{B})$$

Using them, rewrite the Hamiltonian of the LC oscillator in terms of $\hat{B}$ and $\hat{B}^\dagger$. $\mathcal{L}$ and $\mathbb{C}$ are values of inductance and capacitance, respectively. The frequency of oscillation is $\omega = 2\pi f = \frac{1}{\sqrt{\mathcal{L}\mathbb{C}}}$.

Hint: The Hamiltonian of an LC oscillator is a sum of inductive and capacitive energies in terms of I, V, $\mathcal{L}$ and $\mathbb{C}$.

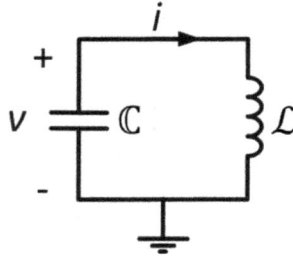

An LC resonator as an electrical model for SHO.

(11) The Hamiltonian matrix of a physical system in a 3D Hilbert space is

$$H = \begin{bmatrix} 1+\sqrt{5} & 0 & 1 \\ 0 & 1 & 0 \\ 1 & 0 & 1 \end{bmatrix} \ (eV)$$

(a) Find the energy eigenvalues and eigenvectors.

(b) If the system is in a superposition state $|\psi\rangle = \frac{1}{\sqrt{7}} \begin{bmatrix} -2 \\ 0 \\ \sqrt{3} \end{bmatrix}$, decompose the state into the energy eigenvectors of the Hamiltonian. Then find the expectation value or average energy of the system.

References

Dirac, P. A. M. (1958). *Principles of Quantum Mechanics*. Oxford University Press.

Reza, F. M. (1971). *Linear Space in Engineering*. Ginn, University of Michigan. This book presents Hilbert spaces from a general engineering point of view.

Schweber, S. S. (1994). *QED and the Men Who Made It: Dyson, Feynman, Schwinger, and Tomonaga*. Princeton University Press.

Steeb, H.-W. (1998). *Holbert Space, Wavelets, Generalized Functions and Modern Quantum mechanics*, Kluwer, Dordrecht. Chapter 2 of this book presents the mathematical physics aspects in depth with a historical background.

Zeidler, E. (1991). *Applied Functional Analysis: Applications to Mathematical Physics*, Chapter 2. Springer Verlag, New York. This book has many solved examples of Hilbert spaces as applied to quantum mechanics.

Chapter 5

UNCERTAINTY PRINCIPLE

Contents

Heisenberg derived the uncertainty relationship that relates the accuracies in the measurements of position and momentum to Planck's constant. It represents the fundamental limit to measuring both the momentum and position of a particle. Since this initial work, it has been realized that the uncertainty relationship holds for any two observables that do not commute. In this chapter, we will introduce the uncertainty relationship, briefly discuss the concept of preparation uncertainty using a Gaussian wave packet, and derive the generalized uncertainty relationship. A readable summary of recent work on extending the uncertainty relationship can be found in [Collings 2016].

In Chapter 3, we found that a particle in the eigenfunction of the momentum operator is,

$$\phi_k(x) = A\,e^{ikx} \tag{5.1}$$

where A is a normalization constant. Such a particle has a definite momentum $\hbar k$. By calculating the probability density to find the particle at location x (which is equal to $|A|^2$), we immediately observe that the particle is equally likely to be found at all points from $-\infty$ to ∞. So, while the momentum of the particle in this state is known to infinite accuracy, the position is completely uncertain. Let us now consider the wavefunction of a particle with a well-defined position,

$$\psi(x) = \delta(x - x_o) \tag{5.2}$$

This particle is located only at $x = x_o$. The wavefunction of this particle can be rewritten as,

$$\psi(x) = \delta(x - x_o) = \frac{1}{2\pi} \int_{-\infty}^{+\infty} e^{ikx} \, dk \tag{5.3}$$

Noting that the momentum eigenfunctions are given by equation (5.1), we immediately observe that $\delta(x-x_o)$ is an equal superposition of all components of momentum. So, if we measure the momentum of a wave function given by equation (5.2), we are equally likely to measure any Fourier component $\hbar k$. So, while the position is certain, the uncertainty in momentum is infinite. Since in the classical world, we are used to finding the position and momentum of a particle precisely, this may seem confusing. However, since quantum mechanics teaches us that the electron is also a wave, it is not too surprising. We know that a classical wave with a definite wave vector is spread over all space, and a spatially localized wave consists of many wave vector components in the Fourier domain. So, once we accept that the electron is a wave, and that a measurement leads to the collapse of the wavefunction, it is conceivable that the electron's position and momentum cannot be measured precisely.

One can extend the above discussion to find out to what accuracy the position and momentum of a quantum particle can be measured. This analysis resulted in the Heisenberg Uncertainty Principle, which states that the product of uncertainties in measuring the position and momentum of a particle can be no smaller than $\frac{\hbar}{2}$. The equation expressing this is:

$$\Delta x \, \Delta p \geq \frac{\hbar}{2} \tag{5.4}$$

where Δx and Δp are the standard deviations (uncertainties) in the measured values of position and momentum, respectively, over an ensemble of identically prepared quantum states. The above, called the *uncertainty relationship*, implies that if a particle's position is determined accurately (Δx is small), then the momentum is known with little accuracy (Δp is large). Conversely, if the momentum of a particle is determined accurately (Δp is small), then the position is known with little accuracy (Δx is large). It is straightforward to verify from equation (5.4) that for the wave function in equation (5.1) (equation (5.2)), where the momentum (position) is exactly known, the uncertainty in position (momentum) is infinite.

Δx and Δp are defined as follows:

$$(\Delta x)^2 = \langle \hat{x}^2 \rangle - \langle \hat{x} \rangle^2 \quad \text{and} \tag{5.5}$$

$$(\Delta p)^2 = \langle \hat{p}^2 \rangle - \langle \hat{p} \rangle^2 \tag{5.6}$$

$\langle \hat{x} \rangle$ and $\langle \hat{p} \rangle$ are the expectation/average values of position and momentum after measurements, while $\langle \hat{x}^2 \rangle$ and $\langle \hat{p}^2 \rangle$ are the average values of the respective squared quantities. The above equation is akin to *variance* in the values obtained after measuring position and momentum from a large sample of identically prepared quantum states. If the position and momentum are measured by repeating an experiment N times then $\langle \hat{x} \rangle^2$, $\langle \hat{x}^2 \rangle$, $\langle \hat{p} \rangle^2$, and $\langle \hat{p}^2 \rangle$ are defined by

$$\langle \hat{x} \rangle^2 = \left(\frac{1}{N} \sum_{i=1}^{N} x_i \right)^2, \quad \langle \hat{x}^2 \rangle = \frac{1}{N} \sum_{i=1}^{N} x_i^2, \quad \langle \hat{p} \rangle^2 = \left(\frac{1}{N} \sum_{i=1}^{N} p_i \right)^2,$$

$$\text{and} \quad \langle \hat{p}^2 \rangle = \frac{1}{N} \sum_{i=1}^{N} p_i^2$$

where the subscript refers to the observable measured from the i-th replica of the sample, and N is the total number of measurements.

5.1   The Uncertainty Relationship for a Gaussian Wave function

We will now work out a particular case of the uncertainty relationship by considering a particle in a Gaussian-shaped wave function given by

$$\psi(x) = \left[\frac{1}{2\pi\sigma^2}\right]^{\frac{1}{4}} \exp\left(-\frac{x^2}{4\sigma^2}\right) \tag{5.7}$$

The expectation value of position is

$$\langle \hat{x} \rangle = \int_{-\infty}^{+\infty} dx\, \psi(x)^* x \psi(x) = \left[\frac{1}{2\pi\sigma^2}\right]^{\frac{1}{2}} \int_{-\infty}^{+\infty} dx\, x \exp\left(-\frac{x^2}{2\sigma^2}\right) = 0 \tag{5.8}$$

The expectation value of x^2 is

$$\langle \hat{x}^2 \rangle = \int_{-\infty}^{+\infty} dx\, \psi(x)^* x^2 \psi(x) = \left[\frac{1}{2\pi\sigma^2}\right]^{\frac{1}{2}} \frac{1}{2}\sqrt{\pi}(2\sigma^2)^{\frac{3}{2}} = \sigma^2 \tag{5.9}$$

In finding the above expectation values, we have used the following integral identities:

$$\int_{-\infty}^{+\infty} dx\, \exp(-cx^2) = \sqrt{\frac{\pi}{c}} \tag{5.10}$$

$$\int_{-\infty}^{+\infty} dx\, x \exp(-cx^2) = 0 \tag{5.11}$$

because x and e^{-cx^2} are odd and even functions

$$\int_{-\infty}^{+\infty} dx\, x^2 \exp(-cx^2) = -\frac{\partial}{\partial c} \int_{-\infty}^{+\infty} dx\, e^{-cx^2} = -\frac{\partial}{\partial c}\sqrt{\frac{\pi}{c}} = \frac{1}{2}\frac{\sqrt{\pi}}{c^{\frac{3}{2}}} \tag{5.12}$$

The expectation value of momentum (by using its operator form) is

$$\langle \hat{p} \rangle = \int_{-\infty}^{+\infty} dx\, \psi(x)^* \frac{\hbar}{i}\frac{\partial}{\partial x}\psi(x)$$

Using the expression for the wave function, we have

$$\langle \hat{p} \rangle = \left[\frac{1}{2\pi\sigma^2}\right]^{\frac{1}{2}} \int_{-\infty}^{+\infty} dx\, \exp\left(-\frac{x^2}{4\sigma^2}\right)\frac{\hbar}{i}\frac{\partial}{\partial x}\exp\left(-\frac{x^2}{4\sigma^2}\right)$$

Because the integrand is an odd function of position x,

$$\langle \hat{p} \rangle = \left[\frac{1}{2\pi\sigma^2}\right]^{\frac{1}{2}} \int_{-\infty}^{+\infty} dx \exp\left(-\frac{x^2}{4\sigma^2}\right) \frac{\hbar}{i} \frac{x}{2\sigma^2} \exp\left(-\frac{x^2}{4\sigma^2}\right) = 0$$
(5.13)

The expectation value of p^2 is

$$\langle \hat{p}^2 \rangle = \int_{-\infty}^{+\infty} dx \, \psi(x)^* \hat{p}^2 \, \psi(x) = \int_{-\infty}^{+\infty} dx \, \psi(x)^* \left(\frac{\hbar}{i}\right)^2 \frac{\partial^2}{\partial x^2} \psi(x)$$

$$\langle \hat{p}^2 \rangle = -\hbar^2 \left[\frac{1}{2\pi\sigma^2}\right]^{\frac{1}{2}} \int_{-\infty}^{+\infty} dx \exp\left(-\frac{x^2}{4\sigma^2}\right) \frac{\partial^2}{\partial x^2} \exp\left(-\frac{x^2}{4\sigma^2}\right)$$

$$= -\hbar^2 \left[\frac{1}{2\pi\sigma^2}\right]^{\frac{1}{2}} \int_{-\infty}^{+\infty} dx \exp\left(-\frac{x^2}{4\sigma^2}\right) \left[\frac{1}{2\sigma^2} - \left(\frac{x}{2\sigma^2}\right)^2\right]$$

$$\times \exp\left(-\frac{x^2}{4\sigma^2}\right) = \hbar^2 \left[\frac{1}{2\sigma^2} - \frac{1}{4\sigma^2}\right] = \frac{\hbar^2}{4\sigma^2}$$

$$\langle \hat{p}^2 \rangle = \frac{\hbar^2}{4\sigma^2}$$
(5.14)

Using equations (5.8), (5.9), (5.13), and (5.14), we find that

$$(\Delta x)^2 (\Delta p)^2 = \frac{\hbar^2}{4}$$
(5.15)

That is,

$$\Delta x \, \Delta p = \frac{\hbar}{2}$$
(5.16)

This tells us that for a Gaussian wave function, the products of uncertainities in the measurement of position and momentum is precisely $\frac{\hbar}{2}$. No other quantum state (wave function) yields a smaller product of Δx and Δp.

In addition to the uncertainty in measurement discussed above, it should be noted that additional uncertainties due to other external factors in the measuring apparatus should be considered, which would only increase the uncertainty. For this reason, the uncertainty relation derived above for a Gaussian-shaped wave function is also called *preparation uncertainty*, meaning that a Gaussian-shaped wave function cannot be *prepared* to yield a value for $\Delta x \, \Delta p$ that is smaller than $\frac{\hbar}{2}$. See (Collings 2016), for a deeper discussion of uncertainty relationship.

> **Concept:** The position and momentum of a particle cannot be measured with infinite accuracy. The accuracy of each measurement is limited by $\Delta x \, \Delta p \geq \frac{\hbar}{2}$.

5.2 Energy–time Uncertainty Relationship

The uncertainty relation between energy and time is

$$\Delta E \, \Delta t \geq \frac{\hbar}{2} \tag{5.17}$$

While mathematically valid, equation (5.17) should be interpreted with care because time is not an observable or an operator. A typical application of this relationship is in the emission of photons accompanying an electron transitioning from a higher E_2 to a lower energy E_1. If the life-time in the excited state is Δt, then the photon's energy can be measured only to within a maximum accuracy of $(\hbar/2\Delta t)$ around $E_2 - E_1$. Even with the best equipment, the measured photon energy from many such transitions will have an energy spread of at least $\hbar/2\Delta t$.

> **Concept:** The energy of photon emission from an excited state cannot be measured with infinite accuracy.

5.3 Voltage–current Uncertainty Relationship

We defined the voltage $(\widehat{V})$ and current $(\widehat{J})$ operators for an LC oscillator in Problem 10 of Chapter 4. We recall here that they were

$$\widehat{V} = \sqrt{\frac{\hbar\omega}{2\mathbb{C}}} \left(\widehat{B}^\dagger + \widehat{B} \right) \quad \text{and} \quad \widehat{J} = i\sqrt{\frac{\hbar\omega}{2\mathcal{L}}} \left(\widehat{B}^\dagger - \widehat{B} \right)$$

where $\widehat{B}$ and $\widehat{B}^\dagger$ were the lowering and raising operators acting on number states, $|n\rangle$ of an SHO. We use the symbol $\widehat{J}$ for the current operator instead of $\widehat{I}$ because we reserve $\widehat{I}$ to represent the identity operator and I to represent the identity matrix. Recall that $\mathcal{L}$ and $\mathbb{C}$ are the inductance and capacitance, respectively.

In this section, $\widehat{J}$ represents the current operator to avoid confusion with the identity operator $\widehat{I}$, but it does not mean current density. We are now going to calculate the uncertainty product $\Delta V \Delta J$, just as we figured the $\Delta x \, \Delta p$ uncertainty product in Section 5.1.1. To calculate the uncertainty in measurement of voltage and current, we will calculate the variance just as we did for the position–momentum uncertainty relation by considering a state with n photons,

$$\Delta V^2 = \langle n|\widehat{V}^2|n\rangle - \langle n|\widehat{V}|n\rangle^2$$

$$\Delta J^2 = \langle n|\widehat{J}^2|n\rangle - \langle n|\widehat{J}|n\rangle^2$$

Let us now evaluate $\langle n|\widehat{V}|n\rangle$, $\langle n|\widehat{V}^2|n\rangle$, $\langle n|\widehat{J}|n\rangle$, and $\langle n|\widehat{J}^2|n\rangle$.

Recall that the properties of $\widehat{B}$ and $\widehat{B}^\dagger$ are

$$\widehat{B}|n\rangle = \sqrt{n}\,|n-1\rangle \quad \text{and} \quad \widehat{B}^\dagger|n\rangle = \sqrt{n+1}\,|n+1\rangle$$

Using the above, we have

$$\langle n|\widehat{V}|n\rangle = \sqrt{\frac{\hbar\omega}{2\mathbb{C}}}\langle n|\widehat{B}^\dagger + \widehat{B}|n\rangle = \frac{\hbar\omega}{2\mathbb{C}}\left\{\langle n|\widehat{B}^\dagger|n\rangle + \langle n|\widehat{B}|n\rangle\right\}$$

$$= \frac{\hbar\omega}{2\mathbb{C}}\left\{\sqrt{n+1}\langle n|n+1\rangle + \sqrt{n}\langle n|n-1\rangle\right\} = 0$$

$$\langle n|\widehat{V}^2|n\rangle = \frac{\hbar\omega}{2\mathbb{C}}\langle n|\widehat{B}^\dagger\widehat{B}^\dagger + \widehat{B}\widehat{B}^\dagger + \widehat{B}^\dagger\widehat{B} + \widehat{B}\widehat{B}|n\rangle$$

We have used the fact that states with a different number of excitations n are orthogonal, e.g., $\langle n|n+1\rangle = 0$. Since $\widehat{B}\widehat{B}^\dagger = \widehat{I} + \widehat{B}^\dagger\widehat{B}$, we can rewrite the previous equation as

$$\langle n|\widehat{V}^2|n\rangle = \frac{\hbar\omega}{2\mathbb{C}}\langle n|\widehat{B}^\dagger\widehat{B}^\dagger + \widehat{I} + 2\widehat{B}^\dagger\widehat{B} + \widehat{B}\widehat{B}|n\rangle$$

$$= \frac{\hbar\omega}{2\mathbb{C}}\left\{\langle n|2\widehat{B}^\dagger\widehat{B}|n\rangle + \langle n|\widehat{I}|n\rangle\right\} = \frac{\left(n+\frac{1}{2}\right)\hbar\omega}{\mathbb{C}}$$

Note that we have used $\langle n|\widehat{B}\widehat{B}|n\rangle = \sqrt{n}\sqrt{n-1}\langle n|n-2\rangle = 0$, and analogously we can prove that $\langle n|\widehat{B}^\dagger\widehat{B}^\dagger|n\rangle = 0$. Similarly, for the current operator, it can be verified that

$$\langle n|\widehat{J}|n\rangle = 0$$

$$\langle n|\widehat{J}^2|n\rangle = -\frac{\hbar\omega}{2\mathcal{L}}\langle n|\widehat{B}^\dagger\widehat{B}^\dagger - \widehat{B}\widehat{B}^\dagger - \widehat{B}^\dagger\widehat{B} + \widehat{B}\widehat{B}|n\rangle$$

$$= \frac{-\hbar\omega}{2\mathcal{L}}\left\{-\langle n|2\widehat{B}^\dagger\widehat{B}|n\rangle - \langle n|\widehat{I}|n\rangle\right\} = \frac{\left(n+\frac{1}{2}\right)\hbar\omega}{\mathcal{L}}$$

Using the above expressions, we have

$$\Delta V^2 = \langle n|\widehat{V}^2|n\rangle - \langle n|\widehat{V}|n\rangle = \frac{\left(n+\frac{1}{2}\right)\hbar\omega}{\mathbb{C}}$$

$$\Delta J^2 = \langle n|\widehat{J}^2|n\rangle - \langle n|\widehat{J}|n\rangle = \frac{\left(n+\frac{1}{2}\right)\hbar\omega}{\mathcal{L}}$$

$$\Delta V^2\Delta J^2 = \frac{\left(n+\frac{1}{2}\right)^2\hbar^2\omega^2}{\mathcal{L}\mathbb{C}}$$

Since $\omega = \frac{1}{\sqrt{\mathcal{L}\mathbb{C}}}$,

$$\Delta V\Delta J = \left(n+\frac{1}{2}\right)\hbar\omega^2$$

The above equation tells us that for $n \neq 0$, we have $\Delta V\Delta J \geq \frac{\hbar\omega^2}{2}$. Hence, both voltage and current in a harmonic oscillator cannot be measured with zero uncertainty.

If $n = 0$, for the oscillator state $|0\rangle$ (which is usually called the vacuum state because it has zero photons), we have

$$\Delta V\Delta J = \frac{\hbar\omega^2}{2}$$

This tells us that there is a minimum uncertainty in measuring voltage and current only if there is no photon in the LC oscillator.

5.4   Generalized Uncertainty Relationship

The uncertainty relationships discussed earlier are special cases of the uncertainty relationship between any two operators $\widehat{B}$ and $\widehat{C}$ that do not commute. That is, $\widehat{B}\widehat{C} \neq \widehat{C}\widehat{B}$ or $\widehat{B}\widehat{C} - \widehat{C}\widehat{B} \neq 0$. The commutator bracket $[\widehat{B}, \widehat{C}]$ is defined as:

$$[\widehat{B}, \widehat{C}] = \widehat{B}\widehat{C} - \widehat{C}\widehat{B}$$

The operator standing for deviation from the average value is

$$\Delta B = \widehat{B} - \langle \widehat{B} \rangle \tag{5.18}$$

$$\Delta C = \widehat{C} - \langle \widehat{C} \rangle \tag{5.19}$$

The standard deviation (σ) in measurement observables B and C are then

$$\sigma(B) = \sqrt{\langle \widehat{B^2} \rangle - \langle \widehat{B} \rangle^2}$$

$$\sigma(C) = \sqrt{\langle \widehat{C^2} \rangle - \langle \widehat{C} \rangle^2}$$

The generalized uncertainty principle can then be derived to be (the proof is quite simple)

$$\sigma(B)\sigma(C) \geq \frac{1}{2}|\langle \widehat{B}\widehat{C} - \widehat{C}\widehat{B} \rangle| \quad \text{or} \quad \Delta B\,\Delta C \geq \frac{1}{2}|\langle [\widehat{B}, \widehat{C}] \rangle| \tag{5.20}$$

The second form of the above equation is also commonly seen in the literature, where ΔB and ΔC are the standard deviations.

If two operators $\widehat{B}$ and $\widehat{C}$ commute, then $[\widehat{B}, \widehat{C}] = 0$, meaning that we can measure both operators' eigenvalues with infinite accuracy. However, when $[\widehat{B}, \widehat{C}] = \widehat{B}\widehat{C} - \widehat{C}\widehat{B} \neq 0$, the inequality of equation (5.20) says that the eigenvalues of $\widehat{B}$ and $\widehat{C}$ **cannot** be measured with infinite accuracy. The reader should verify that the position–momentum and time–energy uncertainty relationships are derivable from equation (5.20).

Concept: If two operators $\widehat{B}$ and $\widehat{C}$ do not commute, they cannot be measured simultaneously with infinite accuracy.

Example 1. Prove that two operators $\widehat{B}$ and $\widehat{C}$ commute if they have the same eigenstates. What is the uncertainty product $\Delta B\,\Delta C$ if $\widehat{B}$ and $\widehat{C}$ have the same eigenstates?

Solution: We are going to prove that $\widehat{B}\widehat{C} = \widehat{C}\widehat{B}$ or $[\widehat{B}, \widehat{C}] = 0$ if $\widehat{B}$ and $\widehat{C}$ have the same eigenstates. We assume that the eigenvalues and eigenstates of $\widehat{C}$ are c and $|\beta\rangle$, then we apply $\widehat{B}\widehat{C}$ to $|\beta\rangle$, from the left:

$$\widehat{B}\widehat{C}|\beta\rangle = \widehat{B}c|\beta\rangle = c\widehat{B}|\beta\rangle$$

If $|\beta\rangle$ is also an eigenstate of $\widehat{B}$ such that $\widehat{B}|\beta\rangle = b|\beta\rangle$, then we can write

$$\widehat{B}\widehat{C}|\beta\rangle = bc|\beta\rangle$$

Now we apply $\widehat{C}\widehat{B}$ to $|\beta\rangle$

$$\widehat{C}\widehat{B}|\beta\rangle = \widehat{C}b|\beta\rangle = bc|\beta\rangle$$

So, we have shown that $\widehat{B}\widehat{C} - \widehat{C}\widehat{B} = 0$ if operators $\widehat{B}$ and $\widehat{C}$ have the same eigenstate. Note that according to equation (5.20), measurements are possible with infinite accuracy. That is, because

$$\Delta B \, \Delta C \geq \frac{1}{2}|\langle \widehat{B}\widehat{C} - \widehat{C}\widehat{B}| = 0$$

This means that nothing prevents the uncertainty in both ΔB and ΔC from being zero. Finally, recall that position and momentum do not commute and that the Hamiltonian and time do not commute

$$[\hat{x}, \hat{p}] = i\hbar \Longrightarrow \Delta x \, \Delta p \geq \hbar/2$$
$$[\widehat{H}, t] = i\hbar \Longrightarrow \Delta E \, \Delta t \geq \hbar/2$$

where Δx, Δp, ΔE, and Δt are the standard deviations in the measurements of position, momentum, energy, and time, respectively.

Example 2. Using the generalized uncertainty principle, determine if the x and y components of the position vector can be measured to infinite accuracy.

Solution: To determine this, we set $\widehat{B} = x$ and $\widehat{C} = y$ in equation (5.20). Then,

$$[\widehat{B}\widehat{C} - \widehat{C}\widehat{B}]f(x,y) = [xy - yx]f(x,y) = 0$$

where $f(x,y)$ is an arbitrary function.

Substituting this in equation (5.20), we get

$$\Delta x \, \Delta y \geq 0$$

That is, position x and the y-component of position can be measured to infinite accuracy.

Example 3. Using the generalized uncertainty principle, determine $\Delta x \, \Delta p_x$.

Solution: To determine this, we set $\widehat{B} = x$ and $\widehat{C} = \hat{p}_x = \frac{\hbar}{i}\frac{\partial}{\partial x}$ in equation (5.20). Then, apply $\widehat{B}\widehat{C} - \widehat{C}\widehat{B}$ to an arbitrary wave function $f(x)$,

$$\frac{\hbar}{i}\left[x\frac{\partial}{\partial x} - \frac{\partial}{\partial x}x\right]f(x) = \frac{\hbar}{i}\left[x\frac{\partial f}{\partial x} - x\frac{\partial f}{\partial x} - f(x)\right] = \frac{\hbar}{i}f(x)$$

$$[x\hat{p}_x - \hat{p}_x x]\,f(x) = \frac{\hbar}{i}f(x)$$

Substituting this in equation (5.20), we immediately get the uncertainty principle discussed in Section 5.1.1:

$$\Delta x \, \Delta p_x \geq \frac{\hbar}{2}$$

That is, the x components of both momentum and position cannot be determined to infinite accuracy. The reader should verify that in comparison, the x-component of position and the y-component of momentum can be simultaneously determined.

5.5 Problems

(1) (a) Show that the x-component of position and the y-component of momentum commute. (b) Can the x-component of position and the y-component of momentum be determined to infinite accuracy? Why?
(2) Show that the x and z components of position can be simultaneously determined to infinite accuracy, using the generalized uncertainty relation.

References

Shankar, R. (2008). *Principles of Quantum Mechanics*, 2nd ed. Springer.

Collings, J. and Van Huele, J-F.S., Qualifying and Quantifying the Uncertainty in the Heisenberg Uncertainty Relations, J. Utah Academy of Sciences, vol 92, 223 (2016).

Chapter 6

PROJECTION OPERATORS

Contents

6.1 Measurement in terms of Projection Operators

Mathematically, there is an operator corresponding to every observable in quantum mechanics. A single measurement yields a single eigenvalue of the operator. A second operator, called the *projection operator*, represents the measurement process — it measures if the quantum system's wave function has a projection (or component) along an eigenfunction corresponding to the operator. Further, the projection operator extracts the probability amplitude corresponding to the eigenfunction.

Let us consider an electron in a system with a Hamiltonian $(\widehat{H})$ with eigenvalues (E_n) and eigenfunctions $(|\chi_n\rangle)$,

$$\widehat{H}|\chi_n\rangle = E_n|\chi_n\rangle \qquad (6.1)$$

For simplicity, we will assume the Hilbert space to be finite dimensional, $n = 1, 2, \ldots, N$. The wave function is in the following superposition of energy eigenfunctions before measurement,

$$\text{(Before measurement) } |\psi\rangle = \sum_{n=1}^{N} a_n|\chi_n\rangle \qquad (6.2)$$

The *projection operator* for the nth quantum number is defined by

$$\widehat{P}_n = |\chi_n\rangle\langle\chi_n| \tag{6.3}$$

As an example, to evaluate the probability of finding the electron in an energy E_3, we build P_3 using state χ_3,

$$P_3 = |\chi_3\rangle\langle\chi_3| \tag{6.4}$$

By applying it to the superposition state $|\psi\rangle$,

$$|\psi'\rangle = P_3|\psi\rangle = |\chi_3\rangle\langle\chi_3|\psi\rangle = a_3|\chi_3\rangle \tag{6.5}$$

The above equation shows the collapse of the wave function $|\psi\rangle$ to eigenfunction $|\chi_3\rangle$ upon measurement. The probability of finding the electron at energy E_3 is

$$\text{Probability } (E_3) = |a_3|^2 \tag{6.6}$$

We can state the above mathematically by saying that the probability of obtaining the eigenvalue E_3 upon measurement is

$$\text{Probability } (E_3) = \langle\psi|P_3|\psi\rangle = \langle\psi|\chi_3\rangle\langle\chi_3|\psi\rangle = a_3^*a_3 = |a_3|^2 \tag{6.7}$$

For any state, the probability of measuring energy E_n and state $|\chi_n\rangle$ is given by

$$\text{Probability } (E_n) = \langle\psi|P_n|\psi\rangle = |a_n|^2 \tag{6.8}$$

In general, applying $\widehat{P}_n$ to a superposition state yields a quantity proportional to the eigenfunction with quantum number n,

$$|\psi'_n\rangle = \widehat{P}_n|\psi\rangle = |\chi_n\rangle\langle\chi_n|\psi\rangle = a_n|\chi_n\rangle \tag{6.9}$$

Note that the above state $|\psi'_n\rangle$ is not a valid wave function after measurement because it is not normalized. We normalize $|\psi'_n\rangle$ by dividing it by its norm which is $|a_n|$:

$$|\psi'_n\rangle = \frac{\widehat{P}_n|\psi\rangle}{|a_n|} = \frac{\widehat{P}_n|\psi\rangle}{\sqrt{\langle\psi|\widehat{P}_n|\psi\rangle}} \tag{6.10}$$

As an exercise, prove that the norm of the wave function $|\psi'_n\rangle$ is unity.

Going back to the case of an electron being measured by P_3, we see that the final collapsed state after normalization is

$$|\psi_3'\rangle = \frac{P_3|\psi\rangle}{|a_3|} = |\chi_3\rangle \qquad (6.11)$$

Now, if we operate with the measurement operator P_3 again for a second time, the state remains unchanged as the system has already collapsed into $|\chi_3\rangle$ after the first measurement. The probability of getting $|\chi_3\rangle$ after the second measurement is 1, as seen in the following:

$$P_3|\psi_3'\rangle = P_3|\chi_3\rangle = |\chi_3\rangle\langle\chi_3|\chi_3\rangle = |\chi_3\rangle \qquad (6.12)$$

Here, $\langle\chi_3|\chi_3\rangle = 1$ because of orthonormality.

After measuring with P_3, we interrogate the system with a different projection operator P_5, and obtain

$$P_5|\psi_3'\rangle = P_5|\chi_3\rangle = |\chi_5\rangle\langle\chi_5|\chi_3\rangle = 0 \qquad (6.13)$$

This tells us that after the measurement and collapse of the state into $|\chi_3\rangle$, later measurements will not find the system in any other state. We again used the fact that eigenfunctions are mutually orthonormal, $\langle\chi_5|\chi_3\rangle = 0$.

You might have surmised that the projection operator has the following property:

$$\widehat{P}_m\widehat{P}_n = \delta_{mn}\widehat{P}_n \qquad (6.14)$$

That is, for a given Hamiltonian with N eigenfunctions, there is a set containing N projection operators:

$$\text{Set of projection operators} = \left\{\widehat{P}_1, P_2, P_3, \ldots, P_N\right\} \qquad (6.15)$$

The operators are mutually orthogonal,

$$\text{e.g.,} \quad P_3P_5 = 0 \quad \text{and} \quad P_3P_3 = P_3. \qquad (6.16)$$

If a matrix is such that $P_x P_x = P_x$, we call it *idempotent*. Note that the projection operator is Hermitian but not unitary, meaning that the Hermitian conjugate of the projection operator is not its inverse. Furthermore, applying the projection operator to a quantum state (wave function) changes the state's norm as we saw before. So, it is said that the projection operator is *not norm-conserving*.

> **Concept:** The projection operator helps treat the process of measurement mathematically. The definition of the projection operator is $\widehat{P}_n = |\chi_n\rangle\langle\chi_n|$, where $|\chi_n\rangle$ are the eigenfunctions of the quantity being measured.

6.1.1 *Summary of The Properties of Projection Operator*

We can write an arbitrary operator and its expectation value in terms of the projection operator. Let us consider the Hamiltonian as an example. From Chapters 2 and 3, the expectation value of energy is

$$\langle\widehat{H}\rangle = \sum_{n=1}^{N} |a_n|^2 E_n$$

Using equation (6.8), we can rewrite the above equation as

$$\langle\widehat{H}\rangle = \sum_{n=1}^{N} \langle\psi|P_n|\psi\rangle E_n$$

Using equation (6.3), which defines the projection operator, the above equation becomes

$$\langle\widehat{H}\rangle = \sum_{n=1}^{N} \langle\psi|\chi_n\rangle\langle\chi_n|\psi\rangle E_n = \langle\psi|\sum_{n=1}^{N}|\chi_n\rangle\langle\chi_n|E_n|\psi\rangle$$

Since $\langle\widehat{H}\rangle = \langle\psi|\widehat{H}|\psi\rangle$, from the above equation, we see that the Hamiltonian is

$$\widehat{H} = \sum_{n=1}^{N} E_n|\chi_n\rangle\langle\chi_n|$$

which can be rewritten as

$$\widehat{H} = \sum_{n=1}^{N} E_n\widehat{P}_n \tag{6.17}$$

The above equation shows that the Hamiltonian can be expanded using the projection operator. Finally, the closure theorem (Chapter 4) can be rewritten in terms of the projection operators as

$$\hat{I}_{N\times N} = \sum_{n=1}^{N} |\chi_n\rangle\langle\chi_n| = \sum_{n=1}^{N} \widehat{P}_n \qquad (6.18)$$

Table 6.1 summarizes the properties of the projection operator. As they are straightforward to verify, we will leave this as an exercise to the reader.

Example 1. This is a simple example of a quantum bit (qubit) and is also known as a **two-level system**.

Consider a quantum system with only two energy levels. This example is a PiB with only two eigenstates: $n = 1$ and $n = 2$. The lower energy state E_1 has a wave function $|0\rangle = \chi_1$, and the second, higher energy state E_2 has a wave function $|1\rangle = \chi_2$. The resulting Hilbert space of this two-level system is composed of two orthonormal basis states, which we represent as

$$\text{Hilbert space:} \left\{ |0\rangle = \begin{bmatrix} 1 \\ 0 \end{bmatrix}, \quad |1\rangle = \begin{bmatrix} 0 \\ 1 \end{bmatrix} \right\} \qquad (6.19)$$

Table 6.1. Summary of properties of the projection operator.

Definition	$\widehat{P}_n = |\chi_n\rangle\langle\chi_n|$
Idempotency/orthogonality	$\widehat{P}_m\widehat{P}_n = \delta_{mn}\widehat{P}_n$
	$= \begin{cases} \widehat{P}_n & \text{for } n = m \text{ (idempotency)} \\ 0 & \text{for } m \neq n \text{ (orthogonality)} \end{cases}$
Wave function normalization after measurement	$|\psi'_n\rangle = \dfrac{\widehat{P}_n|\psi\rangle}{\sqrt{\langle\psi|\widehat{P}_n|\psi\rangle}}$
Identity operator (closure theorem)	$\hat{I}_{N\times N} = \sum\limits_{n=1}^{N} \widehat{P}_n$
Probability of finding eigenfunction χ_n	$\left\langle \psi|\widehat{P}_n|\psi \right\rangle$
Operator $\hat{O}$ with eigenvalues a_n	$\hat{O} = \sum\limits_{n=1}^{N} a_n\widehat{P}_n$

An arbitrary state of the two-level system is a superposition of the above basis states,

$$|\psi\rangle = \alpha|0\rangle + \beta|1\rangle = \begin{bmatrix} \alpha \\ \beta \end{bmatrix}, \quad \text{where } |\alpha|^2 + |\beta|^2 = 1 \qquad (6.20)$$

Using the properties of the projection operator summarized in Table 6.1: (a) Write down the projection operators $\widehat{P}_n = |\chi_n\rangle\langle\chi_n|$ to measure energy in both braket and matrix notations. (b) Verify the idempotency and orthogonality properties of the projection operator, i.e., $\widehat{P}_m\widehat{P}_n = \delta_{mn}\widehat{P}_n$. (c) Show that the sum of the projection operators is the identity operator, i.e., $\widehat{I}_{N\times N} = \sum_{n=1}^{N} \widehat{P}_n$.

Solution: (a) As the Hilbert space is two dimensional, we need two projection operators corresponding to states $|0\rangle$ and $|1\rangle$. The probability of measuring E_1 is $|\alpha|^2$ and the probability of measuring E_2 is $|\beta|^2$. Using the definition of the projection operator in (6.3), we can immediately write down the projection operators in both braket and matrix forms,

$$\widehat{P}_0 = |0\rangle\langle0| = \begin{bmatrix} 1 \\ 0 \end{bmatrix} \begin{bmatrix} 1 & 0 \end{bmatrix} = \begin{bmatrix} 1 & 0 \\ 0 & 0 \end{bmatrix} \qquad (6.21)$$

$$\widehat{P}_1 = |1\rangle\langle1| = \begin{bmatrix} 0 \\ 1 \end{bmatrix} \begin{bmatrix} 0 & 1 \end{bmatrix} = \begin{bmatrix} 0 & 0 \\ 0 & 1 \end{bmatrix} \qquad (6.22)$$

By applying $\widehat{P}_0$ to $|\psi\rangle$, we interrogate if the state has projection on $|0\rangle$

$$\widehat{P}_0|\psi\rangle = |0\rangle\langle0|\psi\rangle = \alpha|0\rangle$$

$$\text{or in matrix form} \quad \begin{bmatrix} 1 & 0 \\ 0 & 0 \end{bmatrix} \begin{bmatrix} \alpha \\ \beta \end{bmatrix} = \alpha \begin{bmatrix} 1 \\ 0 \end{bmatrix} \qquad (6.23)$$

That is, the probability of finding the two-level system in state 0 is $|\alpha|^2$. Similarly, by applying $\widehat{P}_1$ to $|\psi\rangle$, we interrogate if the state has projection on $|1\rangle$:

$$\widehat{P}_1|\psi\rangle = |1\rangle\langle 1|\psi\rangle = \beta|1\rangle$$

$$\text{or in matrix form} \quad \begin{bmatrix} 0 & 0 \\ 0 & 1 \end{bmatrix} \begin{bmatrix} \alpha \\ \beta \end{bmatrix} = \beta \begin{bmatrix} 0 \\ 1 \end{bmatrix} \quad (6.24)$$

So, the probability of finding the system to be in state $|1\rangle$ is $|\beta|^2$. After the measurement, the state collapses to one of the basis vectors $|0\rangle$ or $|1\rangle$.

(b) To verify idempotency and orthogonality, $\widehat{P}_m\widehat{P}_n = \delta_{mn}\widehat{P}_n$, we multiply the matrices for $\widehat{P}_0$ and $\widehat{P}_1$,

$$\widehat{P}_0\widehat{P}_0 = \begin{bmatrix} 1 & 0 \\ 0 & 0 \end{bmatrix} \begin{bmatrix} 1 & 0 \\ 0 & 0 \end{bmatrix} = \begin{bmatrix} 1 & 0 \\ 0 & 0 \end{bmatrix} = \widehat{P}_0 \quad \text{and}$$

$$\widehat{P}_1\widehat{P}_1 = \begin{bmatrix} 0 & 0 \\ 0 & 1 \end{bmatrix} \begin{bmatrix} 0 & 0 \\ 0 & 1 \end{bmatrix} = \begin{bmatrix} 0 & 0 \\ 0 & 1 \end{bmatrix} = \widehat{P}_1$$

$$\widehat{P}_0\widehat{P}_1 = \begin{bmatrix} 1 & 0 \\ 0 & 0 \end{bmatrix} \begin{bmatrix} 0 & 0 \\ 0 & 1 \end{bmatrix} = \begin{bmatrix} 0 & 0 \\ 0 & 0 \end{bmatrix} = 0 \quad \text{and}$$

$$\widehat{P}_1\widehat{P}_0 = \begin{bmatrix} 0 & 0 \\ 0 & 1 \end{bmatrix} \begin{bmatrix} 1 & 0 \\ 0 & 0 \end{bmatrix} = \begin{bmatrix} 0 & 0 \\ 0 & 0 \end{bmatrix} = 0$$

(c) To demonstrate the identity operator property of the projection operator $I_{N \times N} = \sum_{n=1}^{N} \widehat{P}_n$, we simply add the matrices for $\widehat{P}_0$ and $\widehat{P}_1$,

$$\widehat{P}_0 + \widehat{P}_1 = \begin{bmatrix} 1 & 0 \\ 0 & 0 \end{bmatrix} + \begin{bmatrix} 0 & 0 \\ 0 & 1 \end{bmatrix} = \begin{bmatrix} 1 & 0 \\ 0 & 1 \end{bmatrix} = \widehat{I}$$

Example 2. Expectation value
In this example, we calculate the expectation value of energy using three different approaches:

(a) projection operator method,
(b) braket notation, and
(c) matrix method.

A particle is in a unique Harmonic potential with three energy levels, and its quantum state is given by

$$|\psi\rangle = \frac{2}{\sqrt{14}}|\chi_1\rangle + \frac{1}{\sqrt{14}}|\chi_2\rangle + \frac{3}{\sqrt{14}}|\chi_3\rangle$$

The eigenenergies and eigenstates are related by

$$\widehat{H}\,|\chi_n\rangle = \left(n + \frac{1}{2}\right)\varepsilon|\chi_n\rangle$$

where ε is a constant. Find the expectation value of energy using the three methods and verify that they yield the same value.

Solution: We first verify that the quantum state is normalized:

$$\langle\psi|\psi\rangle = \left\{\frac{2}{\sqrt{14}}|\chi_1\rangle + \frac{1}{\sqrt{14}}|\chi_2\rangle + \frac{3}{\sqrt{14}}|\chi_3\rangle\right\}^\dagger$$

$$\times \left\{\frac{2}{\sqrt{14}}|\chi_1\rangle + \frac{1}{\sqrt{14}}|\chi_2\rangle + \frac{3}{\sqrt{14}}|\chi_3\rangle\right\}$$

$$= \frac{4}{14}\langle\chi_1|\chi_1\rangle + \frac{1}{14}\langle\chi_2|\chi_2\rangle + \frac{9}{14}\langle\chi_3|\chi_3\rangle$$

$$= \frac{4}{14} + \frac{1}{14} + \frac{9}{14} = 1$$

To demonstrate normalization, note the use of orthonormality of the eigenstates, i.e., $\langle\chi_n|\chi_m\rangle = \delta_{nm}$.

Method (a): Projection operator method: The projection operators corresponding to the three energy eigenstates are

$$\widehat{P}_1 = |\chi_1\rangle\langle\chi_1|, \quad P_2 = |\chi_2\rangle\langle\chi_2|, \quad \text{and} \quad P_3 = |\chi_3\rangle\langle\chi_3| \qquad (6.25)$$

To find the expectation value, we use equation (6.17), where $\widehat{H} = \sum_{n=1}^{N} E_n \widehat{P}_n$.

$$\langle E\rangle = \langle\psi|\widehat{H}|\psi\rangle = \langle\psi|E_1\widehat{P}_1 + E_2 P_2 + E_3 P_3|\psi\rangle$$

$$\langle E\rangle = E_1\langle\psi|\widehat{P}_1|\psi\rangle + E_2\langle\psi|P_2|\psi\rangle + E_3\langle\psi|P_3|\psi\rangle \qquad (6.26)$$

Now, let us evaluate the three terms in equation (6.26) with the projection operator. Note that these terms correspond to the probability of measuring eigenenergies E_1, E_2, and E_3.

Probability of measuring $E_1 = \frac{\varepsilon}{2}$ is

$$p(E_1) = \left\langle \psi | \widehat{P}_1 | \psi \right\rangle = \langle \psi | \frac{2}{\sqrt{14}} | \chi_1 \rangle$$

$$= \left\{ \frac{2}{\sqrt{14}} | \chi_1 \rangle + \frac{1}{\sqrt{14}} | \chi_2 \rangle + \frac{3}{\sqrt{14}} | \chi_3 \rangle \right\}^{\dagger} \frac{2}{\sqrt{14}} | \chi_1 \rangle = \frac{4}{14}$$

Probability of measuring $E_2 = \frac{3\varepsilon}{2}$ is

$$p(E_2) = \langle \psi | P_2 | \psi \rangle = \langle \Psi | \frac{1}{\sqrt{14}} | \chi_2 \rangle$$

$$= \left\{ \frac{2}{\sqrt{14}} | \chi_1 \rangle + \frac{1}{\sqrt{14}} | \chi_2 \rangle + \frac{3}{\sqrt{14}} | \chi_3 \rangle \right\}^{\dagger} \frac{1}{\sqrt{14}} | \chi_2 \rangle = \frac{1}{14}$$

Probability of measuring $E_3 = \frac{5\varepsilon}{2}$ is

$$p(E_3) = \langle \psi | P_3 | \psi \rangle = \langle \Psi | \frac{3}{\sqrt{14}} | \chi_3 \rangle$$

$$= \left\{ \frac{2}{\sqrt{14}} | \chi_1 \rangle + \frac{1}{\sqrt{14}} | \chi_2 \rangle + \frac{3}{\sqrt{14}} | \chi_3 \rangle \right\}^{\dagger} \frac{3}{\sqrt{14}} | \chi_3 \rangle = \frac{9}{14}$$

Substituting the above three equations in equation (6.26), we have

$$\langle E \rangle = E_1 \langle \psi | \widehat{P}_1 | \psi \rangle + E_2 \langle \psi | P_2 | \psi \rangle + E_3 \langle \psi | P_3 | \psi \rangle$$

$$= \frac{4}{14} \left(\frac{\varepsilon}{2} \right) + \frac{1}{14} \left(\frac{3\varepsilon}{2} \right) + \frac{9}{14} \left(\frac{5\varepsilon}{2} \right) = \frac{13}{7} \varepsilon$$

Method (b): Braket notation: The expression for average energy with a normalized wave function is

$$\langle E \rangle = \langle \psi | \widehat{H} | \psi \rangle$$

Substituting for $|\psi\rangle$ in the above expression for $\langle E \rangle$, we have

$$\langle E \rangle = \langle \psi | \widehat{H} | \psi \rangle$$

$$= \langle \psi | \widehat{H} \left\{ \frac{2}{\sqrt{14}} |\chi_1\rangle + \frac{1}{\sqrt{14}} |\chi_2\rangle + \frac{3}{\sqrt{14}} |\chi_3\rangle \right\}$$

$$= \langle \psi | \left\{ \frac{2}{\sqrt{14}} \widehat{H} |\chi_1\rangle + \frac{1}{\sqrt{14}} \widehat{H} |\chi_2\rangle + \frac{3}{\sqrt{14}} \widehat{H} |\chi_3\rangle \right\}$$

Now using $\widehat{H}|\chi_n\rangle = E_n|\chi_n\rangle$, the above equation becomes

$$\langle E \rangle = \langle \psi | \left\{ \frac{2}{\sqrt{14}} E_1|\chi_1\rangle + \frac{1}{\sqrt{14}} E_2|\chi_2\rangle + \frac{3}{\sqrt{14}} E_3|\chi_3\rangle \right\}$$

$$= \frac{2}{\sqrt{14}} E_1 \langle \psi|\chi_1\rangle + \frac{1}{\sqrt{14}} E_2 \langle \psi|\chi_2\rangle + \frac{3}{\sqrt{14}} E_3 \langle \psi|\chi_3\rangle$$

Substituting for $\langle \psi |$, and using the orthonormality $\langle \chi_m | \chi_n \rangle = \delta_{m,n}$, we have

$$\langle E \rangle = \left| \frac{2}{\sqrt{14}} \right|^2 E_1 + \left| \frac{1}{\sqrt{14}} \right|^2 E_2 + \left| \frac{3}{\sqrt{14}} \right|^2 E_3 = \frac{13}{7} \varepsilon$$

Method (c): Matrix method: Here we represent the three eigenstates by

$$\{\chi_1, \chi_2, \chi_3\} = \left\{ \begin{bmatrix} 1 \\ 0 \\ 0 \end{bmatrix}, \begin{bmatrix} 0 \\ 1 \\ 0 \end{bmatrix}, \begin{bmatrix} 0 \\ 0 \\ 1 \end{bmatrix} \right\}$$

They form an orthonormal basis for this problem with only three eigenvalues. The Hamiltonian in this basis is a diagonal 3×3 matrix,

$$H = \begin{bmatrix} E_1 & 0 & 0 \\ 0 & E_2 & 0 \\ 0 & 0 & E_3 \end{bmatrix} = \begin{bmatrix} \left(1 + \frac{1}{2}\right)\varepsilon & 0 & 0 \\ 0 & \left(2 + \frac{1}{2}\right)\varepsilon & 0 \\ 0 & 0 & \left(3 + \frac{1}{2}\right)\varepsilon \end{bmatrix}$$

and the wave function can be written as

$$|\psi\rangle = \frac{2}{\sqrt{14}}|\chi_1\rangle + \frac{1}{\sqrt{14}}|\chi_2\rangle + \frac{3}{\sqrt{14}}|\chi_3\rangle = \frac{1}{\sqrt{14}}\begin{bmatrix} 2 \\ 1 \\ 3 \end{bmatrix}$$

Now let us recalculate the expectation value of energy using the matrix form.

$$\langle E\rangle = \langle\psi|H|\psi\rangle = \begin{bmatrix} \frac{2}{\sqrt{14}} & \frac{1}{\sqrt{14}} & \frac{3}{\sqrt{14}} \end{bmatrix}^*$$

$$\times \begin{bmatrix} \left(1+\frac{1}{2}\right)\varepsilon & 0 & 0 \\ 0 & \left(2+\frac{1}{2}\right)\varepsilon & 0 \\ 0 & 0 & \left(3+\frac{1}{2}\right)\varepsilon \end{bmatrix} \begin{bmatrix} \frac{2}{\sqrt{14}} \\ \frac{1}{\sqrt{14}} \\ \frac{3}{\sqrt{14}} \end{bmatrix}$$

$$\langle E\rangle = \langle\psi|H|\psi\rangle = \begin{bmatrix} \frac{2}{\sqrt{14}} & \frac{1}{\sqrt{14}} & \frac{3}{\sqrt{14}} \end{bmatrix}^*$$

$$\times \begin{bmatrix} \frac{2}{\sqrt{14}}\left(1+\frac{1}{2}\right)\varepsilon & 0 & 0 \\ 0 & \frac{1}{\sqrt{14}}\left(2+\frac{1}{2}\right)\varepsilon & 0 \\ 0 & 0 & \frac{3}{\sqrt{14}}\left(3+\frac{1}{2}\right)\varepsilon \end{bmatrix}$$

$$\langle E\rangle = \langle\psi|H|\psi\rangle = \left|\frac{2}{\sqrt{14}}\right|^2 E_1 + \left|\frac{1}{\sqrt{14}}\right|^2 E_2 + \left|\frac{3}{\sqrt{14}}\right|^2 E_3 = \frac{13}{7}\varepsilon$$

6.2 Problems

(1) For a three-level system, write the projection operators $\widehat{P}_1$, P_2, and P_3 in the matrix form and then show that $\hat{I}_{3\times3} = \sum_{n=1}^{3}\widehat{P}_n$.

(2) **Rotation of basis vectors:** In quantum mechanics and quantum information, we encounter situations where measurements can be made with different basis vectors, that represent the same physical problem. In the two-level system (a single qubit),

our original basis vectors are $|0\rangle = \begin{bmatrix} 0 \\ 1 \end{bmatrix}$ and $|1\rangle = \begin{bmatrix} 1 \\ 0 \end{bmatrix}$. Now, we generate a new basis, by superposition of $\{|0\rangle, |1\rangle\}$,

$$|+\rangle = \frac{|0\rangle + |1\rangle}{\sqrt{2}} = \frac{1}{\sqrt{2}} \begin{bmatrix} 1 \\ 1 \end{bmatrix}$$

$$|-\rangle = \frac{|0\rangle - |1\rangle}{\sqrt{2}} = \frac{1}{\sqrt{2}} \begin{bmatrix} 1 \\ -1 \end{bmatrix}$$

(a) Show that the new basis vectors $\{|+\rangle, |-\rangle\}$ are also orthonormal. Rewrite a superposition $|\psi\rangle = \alpha|0\rangle + \beta|1\rangle$ in terms of the new basis states.

(b) **Measuring in the new basis:** Build projection operators in the new basis and apply them to the old basis $\{|0\rangle, |1\rangle\}$. What is the result of the measurements?

Chapter 7

QUANTUM INFORMATION

Contents

7.1 Introduction

We saw in Chapter 1 that an electron inside a quantum dot (PIB) has a discrete set of allowed energy levels and corresponding wave functions. Let us consider the lowest two energy levels corresponding to eigenfunctions $|\chi_1\rangle$ and $|\chi_2\rangle$. In this chapter, we will refer to these states as $|0\rangle$ and $|1\rangle$. We will use these two energy levels as a basis for a new computational scheme. Recall that in classical computers, we assign a low voltage to logic state 0 and a high voltage to logic

Quantum Bit

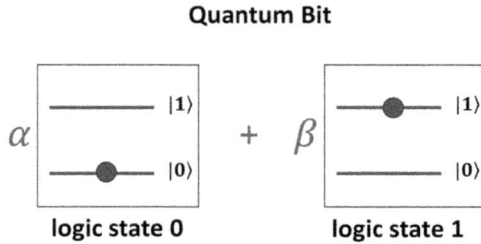

Figure 7.1. An electron in a single two-level system. Logic state 0 corresponds to an electron being in the state $|0\rangle$ and logic state 1 corresponds to an electron in the state $|1\rangle$. In quantum computing, the two-level system can also be in a superposition of logic states 0 and 1. $|0\rangle$ and $|1\rangle$ correspond to $|\chi_1\rangle$ and $|\chi_2\rangle$ respectively.

state 1. In analogy to classical logic states, we say that if an electron is in the (Figure 7.1):

- lower energy level (E_1) with wave function $|0\rangle$, then it is in logic state 0 and
- higher energy level (E_2) with wave function $|1\rangle$, then it is in logic state 1.

Let us further assume for the moment that we have designed the quantum dot (by materials engineering) in such a way that an electron cannot get to the energy E_3 and higher. We are only interested in having two distinct physical states in the quantum dot to build the computational paradigm.

In matrix form, these states are

$$|0\rangle = \begin{bmatrix}1\\0\end{bmatrix} \quad \text{(logic 0)}$$

$$|1\rangle = \begin{bmatrix}0\\1\end{bmatrix} \quad \text{(logic 1)}$$

In contrast to classical physics, the general state of the system is a superposition of logic 0 and 1 states:

$$|\psi\rangle = \alpha|0\rangle + \beta|1\rangle = \alpha\begin{bmatrix}1\\0\end{bmatrix} + \beta\begin{bmatrix}0\\1\end{bmatrix} = \begin{bmatrix}\alpha\\\beta\end{bmatrix} \tag{7.1}$$

The Hilbert space of this simplified system is two dimensional, with $\{|0\rangle, |1\rangle\}$ forming the orthonormal eigenfunctions. The coefficients α and β are complex numbers that should satisfy

$$|\alpha|^2 + |\beta|^2 = 1 \tag{7.2}$$

The above condition means the probabilities to find the electron in the lower ($|0\rangle$) and the higher energy ($|1\rangle$) add up to one. The two-level system described above to represent logic states 0 and 1 is a *quantum bit (qubit)*.

To implement the qubit above, the physical system was an electron in a quantum dot with two discrete energy levels. There are many other examples of qubits:

- **Photon polarization:** Two orthogonal polarizations (say x- and y-polarizations) of a single photon represent the $|0\rangle$ and $|1\rangle$ states of a qubit.
- **Spin:** An electron s up- and down-spins inside a quantum dot form the qubit. In a later chapter, we will discuss spin and its connection to equation (7.1).
- **Magnetic flux qubit:** A metallic loop of aluminum interrupted by a thin layer of an Aluminum Oxide (Josephson junction). When the temperature is under 1 K, aluminum becomes superconducting, forming a permanent current in the loop. The direction of current in the loop (clockwise or anticlockwise) corresponds to the two qubit states. We can say that the clockwise current corresponds to qubit state $|0\rangle$ and the counterclockwise current corresponds to qubit state $|1\rangle$.

In each case, the qubit's state is a superposition of the two basis states (equation (7.1)). To summarize, we will assume in this chapter that the vector space of a single qubit consists of states $|0\rangle$ and $|1\rangle$, associated with energy levels E_1 and E_2. While we ignore the higher energy levels ($E_3, E_4, \ldots$), in reality, excitations involving the higher energy levels cannot be neglected as they cause unwanted loss of information in the qubits. They are also beneficial but are not essential to describe the fundamentals of quantum computing.

Concept: A qubit is a two-level system represented by states $|0\rangle$ and $|1\rangle$. It can be put in a superposition, $|\psi\rangle = \alpha|0\rangle + \beta|1\rangle$.

7.2 Review of Quantum Measurement and Mathematical Machinery

The states (wave functions) of the two-level system $|0\rangle$ and $|1\rangle$ are normalized and orthogonal to each other. That is,

$$\langle 0|1\rangle = \langle 1|0\rangle = 0 \quad \text{and} \quad \langle 0|0\rangle = \langle 1|1\rangle = 1 \tag{7.3}$$

We can define a new set of orthonormal states which are linear combinations of $\{|0\rangle, |1\rangle\}$,

$$|+\rangle \equiv \frac{1}{\sqrt{2}}(|0\rangle + |1\rangle) \quad \text{and} \quad |-\rangle \equiv \frac{1}{\sqrt{2}}(|0\rangle - |1\rangle)) \tag{7.4}$$

where the states $|0\rangle$ and $|1\rangle$ can be represented by horizontal and vertical polarizations of a photon — in these orthogonal states, the horizontally (vertically) polarized light will not pass through vertical (horizontal) polarizers. The $|+\rangle$ and $|-\rangle$ states correspond to 45° and $-45°$ polarized photons, respectively (Figure 7.2).

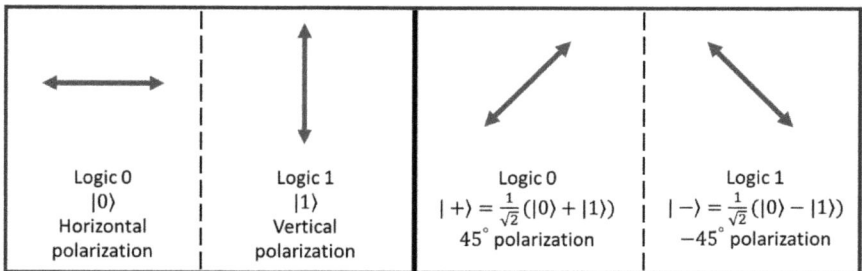

Logic 0	Logic 1	Logic 0	Logic 1
$|0\rangle$	$|1\rangle$	$|+\rangle = \frac{1}{\sqrt{2}}(|0\rangle + |1\rangle)$	$|-\rangle = \frac{1}{\sqrt{2}}(|0\rangle - |1\rangle)$
Horizontal polarization	Vertical polarization	45° polarization	$-45°$ polarization

Figure 7.2. Logic states 0 and 1 are represented by qubit states $|0\rangle$ and $|1\rangle$, respectively. Physically, $|0\rangle$ and $|1\rangle$ can be represented by horizontal and vertical polarization. Alternatively, the logic states 0 and 1 can be represented by qubit states $|+\rangle$ and $|-\rangle$ which correspond to 45° and $-45°$ polarized photons. See equations (7.4) and (7.6).

This orthogonality of the second set of basis states $\{|+\rangle, |-\rangle\}$ can be verified using the braket algebra,

$$\langle +|-\rangle = \langle -|+\rangle = 0 \quad \text{and} \quad \langle +|+\rangle = \langle -|-\rangle = 1 \tag{7.5}$$

and $\{|0\rangle, |1\rangle\}$ expressed in terms of $\{|+\rangle, |-\rangle\}$ is

$$|0\rangle = \frac{1}{\sqrt{2}}(|+\rangle + |-\rangle) \quad \text{and} \quad |1\rangle = \frac{1}{\sqrt{2}}(|+\rangle - |-\rangle) \tag{7.6}$$

That is, we can use these new basis vectors $\{(|+\rangle, |-\rangle)\}$ instead of the original basis vectors $\{|0\rangle, |1\rangle\}$ to represent the logic states.

Below, we will discuss quantum measurement in three scenarios, but before that, let us review the basics of measurement, collapse, and the projection operator discussed in Chapter 6.

Projection operator: A simple mathematical model for measurement is called the projection operator $(\widehat{P})$. For a vector space with basis vectors of $\{|0\rangle, |1\rangle\}$, the two projection operators are built as follows:

$$\widehat{P}_0 = |0\rangle\langle 0| = \begin{bmatrix} 1 \\ 0 \end{bmatrix} \begin{bmatrix} 1 \\ 0 \end{bmatrix}^+ = \begin{bmatrix} 1 & 0 \\ 0 & 0 \end{bmatrix} \tag{7.7}$$

$$\widehat{P}_1 = |1\rangle\langle 1| = \begin{bmatrix} 0 \\ 1 \end{bmatrix} \begin{bmatrix} 0 \\ 1 \end{bmatrix}^+ = \begin{bmatrix} 0 & 0 \\ 0 & 1 \end{bmatrix} \tag{7.8}$$

Operating $\widehat{P}_0$ on a qubit state $|u\rangle = \begin{bmatrix} \alpha \\ \beta \end{bmatrix}$ measures the projection along qubit $|0\rangle$,

$$\widehat{P}_0|u\rangle = |0\rangle\langle 0|u\rangle = \alpha|0\rangle$$

$$\text{or in matrix form} \quad \begin{bmatrix} 1 & 0 \\ 0 & 0 \end{bmatrix} \begin{bmatrix} \alpha \\ \beta \end{bmatrix} = \alpha \begin{bmatrix} 1 \\ 0 \end{bmatrix} \tag{7.9}$$

The above equation tells us that the probability of finding the electron in state $|0\rangle$ is $|\alpha|^2$. Note that the projection operator is not unitary as it does not preserve the norm of the vector that it acts on. That is,

$$\langle u|\widehat{P}_j^\dagger \widehat{P}_j|u\rangle \leq 1$$

with the equality sign holding only if $|u\rangle$ is a pure $|0\rangle$ or $|1\rangle$ state.

Similarly, to measure the projection along qubit state $|1\rangle$, we operate $|u\rangle$ by $\widehat{P}_1$,

$$\widehat{P}_1|u\rangle = |1\rangle\langle 1|u\rangle = \beta|1\rangle$$

or in matrix form $\quad \begin{bmatrix} 0 & 0 \\ 0 & 1 \end{bmatrix} \begin{bmatrix} \alpha \\ \beta \end{bmatrix} = \beta \begin{bmatrix} 0 \\ 1 \end{bmatrix}$ (7.10)

The probability of finding the system to be in state $|1\rangle$ is $|\beta|^2$. We see in equations (7.9) and (7.10) that measurement collapses state $|u\rangle$ to one of the basis vectors. It then follows that

$$\widehat{P}_0\widehat{P}_0|u\rangle = \widehat{P}_0\alpha \begin{bmatrix} 1 \\ 0 \end{bmatrix} = \alpha\widehat{P}_0 \begin{bmatrix} 1 \\ 0 \end{bmatrix} = \alpha \begin{bmatrix} 1 \\ 0 \end{bmatrix}$$ (7.11)

and

$$\widehat{P}_1\widehat{P}_1|u\rangle = \beta \begin{bmatrix} 0 \\ 1 \end{bmatrix}$$ (7.12)

In other words, we have just shown that repeated application of the same projection operator $\widehat{P}_j$ reveal no additional information after the first measurement by $\widehat{P}_j$. The mathematical implication of this is

$$\widehat{P}_j\widehat{P}_j = \widehat{P}_j, \quad \text{where } j = 0 \text{ or } 1$$ (7.13)

Example: Show that applying operator $\widehat{P}_1$ after applying $\widehat{P}_0$ returns 0. What does this mean in physical terms?

Solution: Equation (7.9) shows that operating $\widehat{P}_0$ on $|u\rangle$ tells us of the extent of the projection of the qubit along the first basis state $|0\rangle$. Applying $\widehat{P}_1$ after the collapse of $|u\rangle$ to $|0\rangle$ asks if the new state has a nonzero projection on the second basis state $|1\rangle$. As the bases $\{|0\rangle, |1\rangle\}$ are orthonormal, state $|0\rangle$ has no projection on $|1\rangle$. Mathematically, this is

$$\widehat{P}_1\widehat{P}_0|u\rangle = \widehat{P}_1\alpha|0\rangle = \alpha\widehat{P}_1|0\rangle = \alpha|1\rangle\langle 1|0\rangle = 0$$ (7.14)

Before moving to quantum computing, let's summarize measurement using three simple examples.

Case 1 (Measurement and collapse of state): We are given a two-level system in the state,

$$|\psi\rangle = \alpha|0\rangle + \beta|1\rangle, \quad \text{where } |\alpha|^2 + |\beta|^2 = 1 \qquad (7.15)$$

If we make measurements in the $\{|0\rangle, |1\rangle\}$ bases,

(a) What are the possible results from the first measurement?
(b) What is the system's quantum state after additional measurements on the state obtained in part (a)?

Solution: A single measurement can only yield one of the states $|0\rangle$ or $|1\rangle$.

If we measure state $|0\rangle$, the system will remain in state $|0\rangle$ after measurement. We say that the wave function has *collapsed* to state $|0\rangle$. A later measurement will find the system only in state $|0\rangle$.

Alternatively, if we measure state $|1\rangle$, the system will remain in state $|1\rangle$ after measurement. We say that the wave function has *collapsed* to state $|1\rangle$. A later measurement will find the system to be only in state $|1\rangle$.

If we repeat the above experiment many times with the same initial state $|\psi\rangle$, we measure qubits $|0\rangle$ and $|1\rangle$ with probabilities of $|\alpha|^2$ and $|\beta|^2$, respectively. For example, if $\alpha = \beta = \frac{1}{\sqrt{2}}$, and we repeat the experiment 1000 times in the same state $|\psi\rangle$, the measurement will collapse into states $|0\rangle$ and $|1\rangle$ approximately 500 times each.

Case 2 (Measuring in a different basis state): Put a qubit in state $|\psi\rangle = \alpha|0\rangle + \beta|1\rangle$. If we make measurements in the $\{|+\rangle, |-\rangle\}$ bases,

(a) What are the possible results from the first measurement?
(b) What is the system's quantum state after additional measurements on the state obtained in part (a)?

Solution: The state $|\psi\rangle = \alpha|0\rangle + \beta|1\rangle$ can be written in the $\{|+\rangle, |-\rangle\}$ basis as (using equation (7.6)),

$$|\psi\rangle = \left(\frac{\alpha + \beta}{\sqrt{2}}\right)|+\rangle + \left(\frac{\alpha - \beta}{\sqrt{2}}\right)|-\rangle \qquad (7.16)$$

Like Case 1, we can build two projection operators, $\widehat{P}_+$ and $\widehat{P}_-$, and apply them to the state given by equation (7.16) to find the measurement result.

$$\widehat{P}_+ = |+\rangle\langle+| \quad \text{and} \quad \widehat{P}_- = |-\rangle\langle-| \tag{7.17}$$

A single measurement can only yield the state $|+\rangle$ or the state $|-\rangle$. If we measure state $|+\rangle$ ($|-\rangle$), the system will remain in state $|+\rangle$ ($|-\rangle$) after the measurement unless it is disturbed. A later measurement will find the system only in state $|+\rangle$ ($|-\rangle$). The wave function collapses into the measured state.

If we repeat the above experiment many times with the same initial state $|\psi\rangle$, we see from equation (7.16) that we will obtain qubits $|+\rangle$ and $|-\rangle$ with probabilities of $\frac{|\alpha+\beta|^2}{2}$ and $\frac{|\alpha-\beta|^2}{2}$, respectively.

Case 3 (Repeated measurements on a different basis): Create a single copy of state $|0\rangle$. Make three consecutive measurements. Measurement 1 is in basis state $\{|+\rangle, |-\rangle\}$. Using the resultant state of Measurement 1, use the basis state $\{|0\rangle, |1\rangle\}$ for Measurement 2. Finally, using the resultant state of Measurement 2, the third measurement is in basis state $\{|+\rangle, |-\rangle\}$. Discuss the possible results.

Solution: Note from equation (7.6) that $|0\rangle$ is a linear combination of the $\{|+\rangle, |-\rangle\}$ states. So, Measurement 1 could yield either state $|+\rangle$ or $|-\rangle$. Measurement on many replicas of state $|0\rangle$ will yield $|+\rangle$ and $|-\rangle$ with equal probabilities of $\frac{1}{2}$ because the magnitude of their coefficients in equation (7.6) is equal to $\frac{1}{\sqrt{2}}$.

We note that both $|+\rangle$ and $|-\rangle$ states are superpositions of $\{|0\rangle, |1\rangle\}$ with coefficients of equal magnitudes (equation (7.4)). So, Measurement 2 will yield either $|0\rangle$ or $|1\rangle$ (with equal probability) independent of the result of Measurement 1.

Independent of if the wave function after Measurement 2 collapsed into state $|0\rangle$ or state $|1\rangle$, Measurement 3 will yield either state $|+\rangle$ or $|-\rangle$ with equal probability due to equation (7.6).

To summarize, if we collapse into a $|+\rangle$ in Measurement 1, Measurement 3 could result in either $|+\rangle$ or $|-\rangle$ states with an equal probability of $\frac{1}{2}$.

Summary of braket and matrix representations of a single qubit system:

Consider single qubit states $|\psi\rangle = \alpha|0\rangle + \beta|1\rangle$ and

$$|\chi\rangle = a|0\rangle + b|1\rangle \tag{7.18}$$

The inner product $\langle\psi|\chi\rangle$ is

$$\langle\psi|\chi\rangle = \alpha^*a\langle0|0\rangle + \alpha^*b\langle0|1\rangle + \beta^*a\langle1|0\rangle + \beta^*b\langle1|1\rangle$$
$$\langle\psi|\chi\rangle = \alpha^*a + \beta^*b \tag{7.19}$$

where we have used the orthonormality of states $|0\rangle$ and $|1\rangle$.

The matrix notation is not as compact as the bra and ket notation but is of pedagogic value. In the following, we present the matrix form of bra and ket notation, superposition, and inner product.

Matrix representation of a single qubit

The basis ket states $|0\rangle$ and $|1\rangle$ are represented by column vectors,

$$|0\rangle = \begin{bmatrix} 1 \\ 0 \end{bmatrix} \quad \text{and} \quad |1\rangle = \begin{bmatrix} 0 \\ 1 \end{bmatrix} \tag{7.20}$$

An arbitrary superposition of the $|0\rangle$ and $|1\rangle$ states is given by $|\chi\rangle = \alpha|0\rangle + \beta|1\rangle$. This can be written in matrix form as

$$|\chi\rangle = \alpha \begin{bmatrix} 1 \\ 0 \end{bmatrix} + \beta \begin{bmatrix} 0 \\ 1 \end{bmatrix} = \begin{bmatrix} \alpha \\ \beta \end{bmatrix} \tag{7.21}$$

The bra states corresponding to ket states $|0\rangle$ and $|1\rangle$ are represented by row vectors, which are the transpose and complex conjugate of the corresponding ket vectors,

$$\langle0| = \begin{bmatrix} 1 & 0 \end{bmatrix} \quad \text{and} \quad \langle1| = \begin{bmatrix} 0 & 1 \end{bmatrix} \tag{7.22}$$

The bra state corresponding to ket state $|\chi\rangle$ in equation (7.21) is

$$\langle\chi| = \alpha^*\begin{bmatrix} 1 & 0 \end{bmatrix} + \beta^*\begin{bmatrix} 0 & 1 \end{bmatrix} = \begin{bmatrix} \alpha^* & \beta^* \end{bmatrix} \tag{7.23}$$

Inner product of single qubit states in matrix form

The inner product of $\langle 0|0 \rangle$ is

$$\langle 0|0 \rangle = \begin{bmatrix} 1 & 0 \end{bmatrix} \begin{bmatrix} 1 \\ 0 \end{bmatrix} = 1 \qquad (7.24)$$

The inner product of $\langle 0|1 \rangle$ is

$$\langle 0|1 \rangle = \begin{bmatrix} 1 & 0 \end{bmatrix} \begin{bmatrix} 0 \\ 1 \end{bmatrix} = 0 \qquad (7.25)$$

The inner product of $|\psi\rangle = \alpha|0\rangle + \beta|1\rangle$ and $|\chi\rangle = a|0\rangle + b|1\rangle$ is

$$\langle \psi|\chi \rangle = \begin{bmatrix} \alpha^* & \beta^* \end{bmatrix} \begin{bmatrix} a \\ b \end{bmatrix} = \alpha^* a + \beta^* b \qquad (7.26)$$

Concept: A single qubit can be represented using either the braket notation $|0\rangle$ and $|1\rangle$ or matrices,

$$|0\rangle = \begin{bmatrix} 1 \\ 0 \end{bmatrix} \quad \text{and} \quad |1\rangle = \begin{bmatrix} 0 \\ 1 \end{bmatrix}$$

7.2.1 *Two qubits*

Recall that we have a shift register with m cells in classical logic circuits, and each cell stores a single bit. The juxtaposition of m bits gives us 2^m different m-bit words or states. For $m = 2$, we have four states, 00, 01, 10, and 11.

To have two qubits, we need two physical systems of the same kind. For example, in the context of an electron in a quantum dot or PiB, if we juxtapose two dots with an electron in each of them, the two-bit system can have four different configurations, as shown in Figure 7.3. Refer to the two qubits as qubit-a and qubit-b. Recall that the vector space of a single qubit has $\{|0\rangle, |1\rangle\}$ or equivalently $\left\{ \begin{bmatrix} 1 \\ 0 \end{bmatrix}, \begin{bmatrix} 0 \\ 1 \end{bmatrix} \right\}$ as the basis vectors. However, we need a four-dimensional Hilbert space to represent the four different states of a two-qubit system. An orthonormal basis of this four-dimensional Hilbert space

Basis Vector: $|00\rangle$, Energy: E_{00}		Basis Vector: $|10\rangle$, Energy: E_{10}		Basis Vector: $|01\rangle$, Energy: E_{01}		Basis Vector: $|11\rangle$, Energy: E_{11}	
qubit-a	qubit-b	qubit-a	qubit-b	qubit-a	qubit-b	qubit-a	qubit-b
logic state 0	logic state 0	logic state 1	logic state 0	logic state 0	logic state 1	logic state 1	logic state 1

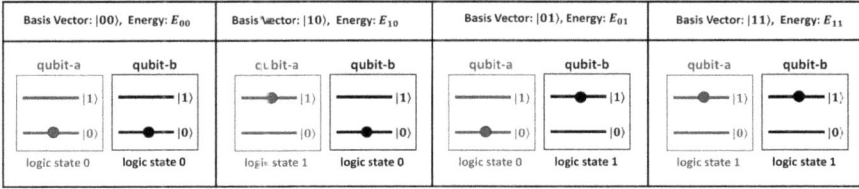

Figure 7.3. Two qubits, and their corresponding basis vectors, and energy levels.

is $\{|00\rangle, |01\rangle, |10\rangle$ and $|11\rangle\}$ in ket notation and

$$\left\{ \begin{bmatrix} 1 \\ 0 \\ 0 \\ 0 \end{bmatrix}, \begin{bmatrix} 0 \\ 1 \\ 0 \\ 0 \end{bmatrix}, \begin{bmatrix} 0 \\ 0 \\ 1 \\ 0 \end{bmatrix}, \begin{bmatrix} 0 \\ 0 \\ 0 \\ 1 \end{bmatrix} \right\} \quad \text{in matrix notation} \quad (7.27)$$

The two-qubit system's arbitrary state is a 4×1 vector in a superposition of the above vectors. In braket notation, the first number is the state of the first qubit (qubit-a in Figure 7.3), and the second number is the state of the second qubit (qubit-b in Figure 7.3). So, in the state $|01\rangle$, qubit-a is in state $|0\rangle$, and qubit-b is in state $|1\rangle$. The general state of a two-qubit system is

$$|\psi\rangle = \alpha|00\rangle + \beta|01\rangle + \gamma|10\rangle + \delta|11\rangle \quad (7.28)$$

As a result of normalization, $|\alpha|^2 + |\beta|^2 + |\gamma|^2 + |\delta|^2 = 1$. The two-qubit state has a component of each of the four logic states $00, 01, 10,$ and 11. A measurement of the two-qubit state $|\psi\rangle$ in the measurement basis $\{|00\rangle, |01\rangle, |10\rangle$ and $|11\rangle\}$ can, however, only yield one of the four basis states: $|00\rangle$ or $|01\rangle$ or $|10\rangle$ or $|11\rangle$.

Equation (7.28) can also be written explicitly as

$$|\psi\rangle = \alpha|0_a0_b\rangle + \beta|0_a1_b\rangle + \gamma|1_a0_b\rangle + \delta|1_a1_b\rangle \quad (7.29)$$

where the states of qubit-a and qubit-b are explicitly shown Another form for equations (7.28) and (7.29) is as follows:

$$|\psi\rangle = \alpha|0\rangle_a|0\rangle_b + \beta|0\rangle_a|1\rangle_b + \gamma|1\rangle_a|0\rangle_b + \delta|1\rangle_a|1\rangle_b \quad (7.30)$$

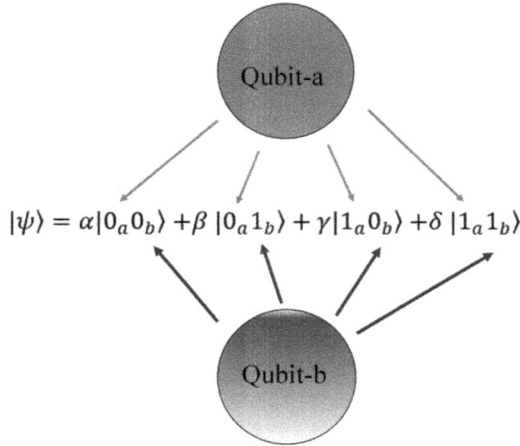

Figure 7.4. A two qubit system. Figure art by Mrunal Shenoy.

We will usually use equation (7.28) with the explicit understanding that the first number in each of the terms stands for qubit-a and the second number in each of the terms stands for qubit-b. See Figure 7.4.

Summary of braket and matrix representations of a two-qubit system

Multiplying each bra to its corresponding ket yields the inner product.

$$\langle \psi \phi | \delta \zeta \rangle = \langle \psi | \delta \rangle \langle \phi | \zeta \rangle \tag{7.31}$$

To clarify, equation (7.31) is equivalent to

$$((\langle \psi_a | \langle \phi_b |) | \delta_a \rangle | \zeta_b \rangle = \langle \psi_a | \delta_a \rangle \langle \phi_b | \zeta_b \rangle$$

We will now discuss a single qubit state's generalization to two qubit states, using the tensor (also known as Kronecker) product notation. In the box titled **Matrix representation of a two-qubit system**, we will show how to expand the Hilbert space using tensor products which generate the basis vectors for a higher number of qubits starting from the basis vectors of a single qubit.

Matrix representation of a two-qubit system

The basis states/kets $|00\rangle$, $|01\rangle$, $|10\rangle$, and $|11\rangle$ are represented by column vectors,

$$|00\rangle = \begin{bmatrix} 1 \\ 0 \\ 0 \\ 0 \end{bmatrix}, \quad |01\rangle = \begin{bmatrix} 0 \\ 1 \\ 0 \\ 0 \end{bmatrix}, \quad |10\rangle = \begin{bmatrix} 0 \\ 0 \\ 1 \\ 0 \end{bmatrix}, \quad \text{and} \quad |11\rangle = \begin{bmatrix} 0 \\ 0 \\ 0 \\ 1 \end{bmatrix}$$

$$(7.32)$$

To systematically derive this from the bases of a single qubit system, it is useful to discuss the *tensor product*, which is also called the Kronecker product. The tensor product of two matrices is defined as follows:

$$\begin{aligned} \text{kron}(A, B) = A \otimes B &= \begin{bmatrix} a_{11} & a_{12} \\ a_{21} & a_{22} \end{bmatrix} \begin{bmatrix} b_{11} & b_{12} \\ b_{21} & b_{22} \end{bmatrix} \\ &= \begin{bmatrix} a_{11} \begin{bmatrix} b_{11} & b_{12} \\ b_{21} & b_{22} \end{bmatrix} & a_{12} \begin{bmatrix} b_{11} & b_{12} \\ b_{21} & b_{22} \end{bmatrix} \\ a_{21} \begin{bmatrix} b_{11} & b_{12} \\ b_{21} & b_{22} \end{bmatrix} & a_{22} \begin{bmatrix} b_{11} & b_{12} \\ b_{21} & b_{22} \end{bmatrix} \end{bmatrix} \end{aligned}$$

The two matrices involved in the tensor product do not have to be square. The tensor product repeats the replicas of the second matrix, multiplied by the first matrix (vector) element. Both python and MATLAB have functions (for example, **kron**) which perform the tensor product.

To expand the Hilbert space of the single qubit to the Hilbert space of two qubits, we write

$$|00\rangle = |0\rangle \otimes |0\rangle = \begin{pmatrix} 1 \\ 0 \end{pmatrix} \otimes \begin{pmatrix} 1 \\ 0 \end{pmatrix} = \begin{bmatrix} 1\begin{pmatrix} 1 \\ 0 \end{pmatrix} \\ 0\begin{pmatrix} 1 \\ 0 \end{pmatrix} \end{bmatrix} = \begin{bmatrix} \begin{pmatrix} 1 \cdot 1 \\ 1 \cdot 0 \end{pmatrix} \\ \begin{pmatrix} 0 \cdot 1 \\ 0 \cdot 0 \end{pmatrix} \end{bmatrix} = \begin{bmatrix} 1 \\ 0 \\ 0 \\ 0 \end{bmatrix}$$

Similarly, the other state vectors of a two-qubit system are as:

$$|01\rangle = |0\rangle \otimes |1\rangle = \begin{pmatrix} 1 \\ 0 \end{pmatrix} \otimes \begin{pmatrix} 0 \\ 1 \end{pmatrix} = \begin{bmatrix} 1\begin{pmatrix} 0 \\ 1 \end{pmatrix} \\ 0\begin{pmatrix} 0 \\ 1 \end{pmatrix} \end{bmatrix} = \begin{bmatrix} \begin{pmatrix} 1 \cdot 0 \\ 1 \cdot 1 \end{pmatrix} \\ \begin{pmatrix} 0 \cdot 0 \\ 0 \cdot 1 \end{pmatrix} \end{bmatrix} = \begin{bmatrix} 0 \\ 1 \\ 0 \\ 0 \end{bmatrix}$$

$$|10\rangle = |1\rangle \otimes |0\rangle = \begin{pmatrix} 0 \\ 1 \end{pmatrix} \otimes \begin{pmatrix} 1 \\ 0 \end{pmatrix} = \begin{bmatrix} 0\begin{pmatrix} 1 \\ 0 \end{pmatrix} \\ 1\begin{pmatrix} 1 \\ 0 \end{pmatrix} \end{bmatrix} = \begin{bmatrix} \begin{pmatrix} 0 \cdot 1 \\ 0 \cdot 0 \end{pmatrix} \\ \begin{pmatrix} 1 \cdot 1 \\ 1 \cdot 0 \end{pmatrix} \end{bmatrix} = \begin{bmatrix} 0 \\ 0 \\ 1 \\ 0 \end{bmatrix}$$

$$|11\rangle = |1\rangle \otimes |1\rangle = \begin{pmatrix} 0 \\ 1 \end{pmatrix} \otimes \begin{pmatrix} 0 \\ 1 \end{pmatrix} = \begin{bmatrix} 0\begin{pmatrix} 0 \\ 1 \end{pmatrix} \\ 1\begin{pmatrix} 0 \\ 1 \end{pmatrix} \end{bmatrix} = \begin{bmatrix} \begin{pmatrix} 0 \cdot 0 \\ 0 \cdot 1 \end{pmatrix} \\ \begin{pmatrix} 1 \cdot 0 \\ 1 \cdot 1 \end{pmatrix} \end{bmatrix} = \begin{bmatrix} 0 \\ 0 \\ 0 \\ 1 \end{bmatrix}.$$

In some texts, the two-qubit state $|00\rangle$ is written as $|0\rangle|0\rangle$. This is to identify that there is a tensor product in between those vectors.

An arbitrary two-qubit state is a superposition of the $|00\rangle$, $|01\rangle$, $|10\rangle$, and $|11\rangle$ states given by $|\chi\rangle = \alpha|00\rangle + \beta|01\rangle + \gamma|10\rangle + \delta|11\rangle$,

$$|\chi\rangle = \alpha \begin{bmatrix} 1 \\ 0 \\ 0 \\ 0 \end{bmatrix} + \beta \begin{bmatrix} 0 \\ 1 \\ 0 \\ 0 \end{bmatrix} + \gamma \begin{bmatrix} 0 \\ 0 \\ 1 \\ 0 \end{bmatrix} + \delta \begin{bmatrix} 0 \\ 0 \\ 0 \\ 1 \end{bmatrix} = \begin{bmatrix} \alpha \\ \beta \\ \gamma \\ \delta \end{bmatrix}$$

The bra states corresponding to the above are row vectors

$$\langle 00| = \begin{bmatrix} 1 & 0 & 0 & 0 \end{bmatrix}$$
$$\langle 01| = \begin{bmatrix} 0 & 1 & 0 & 0 \end{bmatrix}$$
$$\langle 10| = \begin{bmatrix} 0 & 0 & 1 & 0 \end{bmatrix}$$
$$\langle 11| = \begin{bmatrix} 0 & 0 & 0 & 1 \end{bmatrix} \quad \text{and}$$
$$\langle \chi| = \begin{bmatrix} \alpha^* & \beta^* & \gamma^* & \delta^* \end{bmatrix}$$

Inner product of two-qubit states in matrix form

The inner product of $\langle 00|00\rangle$ in matrix form is

$$\langle 00|00\rangle = \begin{bmatrix} 1 & 0 & 0 & 0 \end{bmatrix} \begin{bmatrix} 1 \\ 0 \\ 0 \\ 0 \end{bmatrix} = 1 \tag{7.33}$$

The inner product of $\langle 00|01\rangle$ in matrix form is

$$\langle 00|01\rangle = \begin{bmatrix} 1 & 0 & 0 & 0 \end{bmatrix} \begin{bmatrix} 0 \\ 1 \\ 0 \\ 0 \end{bmatrix} = 0 \tag{7.34}$$

The inner product of

$$|\psi\rangle = \alpha|0\rangle_a|0\rangle_b + \beta|0\rangle_a|1\rangle_b + \gamma|1\rangle_a|0\rangle_b + \delta|1\rangle_a|1\rangle_b = \begin{bmatrix} \alpha \\ \beta \\ \gamma \\ \delta \end{bmatrix}$$

and

$$|\chi\rangle = a|0\rangle_a|0\rangle_b + b|0\rangle_a|1\rangle_b + c|1\rangle_a|0\rangle_b + d|1\rangle_a|1\rangle_b = \begin{bmatrix} a \\ b \\ c \\ d \end{bmatrix}$$

is,

$$\langle \psi|\chi\rangle = \begin{bmatrix} \alpha^* & \beta^* & \gamma^* & \delta^* \end{bmatrix} \begin{bmatrix} a \\ b \\ c \\ d \end{bmatrix} = \alpha^* a + \beta^* b + \gamma^* c + \delta^* d \tag{7.35}$$

7.2.2   *Three qubits*

Now consider a system with three qubits: qubit-a, qubit-b, and qubit-c. Recall that three classical bits can create eight different binary numbers:

Set of classical bits: $\{000, 001, 010, 011, 100, 101, 110, 111\}$

$$(7.36)$$

In equation (7.36), the first, second, and third numbers in every triplet stand for the state of bit-a, bit-b, and bit-c, respectively.

Quantum mechanically, the three-qubit system has eight ortho-normal basis vectors which we symbolically represent by eight vectors: $|000\rangle$, $|001\rangle$, $|010\rangle$, $|011\rangle$, $|100\rangle$, $|101\rangle$, $|110\rangle$, and $|111\rangle$. The three-qubit system can be in an arbitrary linear combination of these eight basis vectors,

$$|\psi\rangle = a_1|000\rangle + a_2|001\rangle + a_3|010\rangle + a_4|011\rangle + a_5|100\rangle + a_6|101\rangle$$
$$+ a_7|110\rangle + a_8|111\rangle \tag{7.37}$$

Equivalent forms of the above equation are as follows:

$$|\psi\rangle = a_1|0\rangle_a|0\rangle_b|0\rangle_c + a_2|0\rangle_a|0\rangle_b|1\rangle_c + a_3|0\rangle_a|1\rangle_b|0\rangle_c + a_4|0\rangle_a|1\rangle_b|1\rangle_c$$
$$+ a_5|1\rangle_a|0\rangle_b|0\rangle_c + a_6|1\rangle_a|0\rangle_b|1\rangle_c + a_7|1\rangle_a|1\rangle_b|0\rangle_c$$
$$+ a_8|1\rangle_a|1\rangle_b|1\rangle_c \tag{7.38}$$

and

$$|\psi\rangle = a_1|0_a0_b0_c\rangle + a_2|0_a0_b1_c\rangle + a_3|0_a1_b0_c\rangle + a_4|0_a1_b1_c\rangle$$
$$+ a_5|1_a0_b0_c\rangle + a_6|1_a0_b1_c\rangle + a_7|1_a1_b0_c\rangle + a_8|1_a1_b1_c\rangle. \tag{7.39}$$

Note that in equations (7.37), (7.38), and (7.39), $\sum_{j=1}^{8} |a_j|^2 = 1$ because $|\psi\rangle$ is normalized.

The reader will often see other notations to represent the above. For example, take $|110\rangle$. Other notations include $|1\rangle \otimes |1\rangle \otimes |0\rangle$ or simply $|1\rangle|1\rangle|0\rangle$, where the explicit tensor product symbol ($\otimes$) is not included for the sake of brevity. The box below shows how to use the tensor product notation for three qubits.

Like the earlier case for two qubits, multiplying each bra to its corresponding ket yields the inner product of the three qubit states. That is, the first bra to the first ket, the second bra to the second ket, and so on,

$$\langle \alpha\beta\gamma | \delta\zeta\eta \rangle = \langle \alpha|\delta\rangle \langle \beta|\zeta\rangle \langle \gamma|\eta\rangle \tag{7.40}$$

An arbitrary three-qubit state is a superposition of the eight basis states written above.

Matrix representation of a three-qubit system

Now we are going to generalize the previous concept of expanding the vector space to three qubits. The shorthand notations of basis states/kets for three-qubits are $|000\rangle$, $|001\rangle$, $|010\rangle$, $|100\rangle$, $|011\rangle$, $|101\rangle$, $|110\rangle$, and $|111\rangle$. They are represented by column vectors:

$$|000\rangle = \begin{bmatrix} 1 \\ 0 \\ 0 \\ 0 \\ 0 \\ 0 \\ 0 \\ 0 \end{bmatrix}, \quad |001\rangle = \begin{bmatrix} 0 \\ 1 \\ 0 \\ 0 \\ 0 \\ 0 \\ 0 \\ 0 \end{bmatrix}, \quad |010\rangle = \begin{bmatrix} 0 \\ 0 \\ 1 \\ 0 \\ 0 \\ 0 \\ 0 \\ 0 \end{bmatrix}, \quad |011\rangle = \begin{bmatrix} 0 \\ 0 \\ 0 \\ 1 \\ 0 \\ 0 \\ 0 \\ 0 \end{bmatrix}$$

$$|100\rangle = \begin{bmatrix} 0\\0\\0\\0\\1\\0\\0\\0 \end{bmatrix}, \quad |101\rangle = \begin{bmatrix} 0\\0\\0\\0\\0\\1\\0\\0 \end{bmatrix}, \quad |110\rangle = \begin{bmatrix} 0\\0\\0\\0\\0\\0\\1\\0 \end{bmatrix}, \quad \text{and} \quad |111\rangle = \begin{bmatrix} 0\\0\\0\\0\\0\\0\\0\\1 \end{bmatrix}$$

We show how this basis can be built from one and two qubit states using the tensor product. As an example, we work out the matrix representation of a three-qubit state $|000\rangle$ using the fact that the matrix representation of the two-qubit state $|00\rangle$ was

$$|00\rangle = \begin{bmatrix} 1\\0\\0\\0 \end{bmatrix}$$

Hence, we write $|000\rangle = |0\rangle \otimes |0\rangle \otimes |0\rangle = |0\rangle \otimes |00\rangle$

$$|000\rangle = \begin{pmatrix} 1\\0 \end{pmatrix} \otimes \begin{bmatrix} 1\\0\\0\\0 \end{bmatrix} = \begin{bmatrix} 1 \cdot \begin{bmatrix} 1\\0\\0\\0 \end{bmatrix} \\ 0 \cdot \begin{bmatrix} 1\\0\\0\\0 \end{bmatrix} \end{bmatrix} = \begin{bmatrix} 1\\0\\0\\0\\0\\0\\0\\0 \end{bmatrix}$$

This simply means that we repeat the two-qubit vector ($|00\rangle$) by as many components present in the first qubit ($|0\rangle$) while multiplying them by the zeros and ones (entries) of the first qubit.

As an exercise, repeat the above process and find the vector representation of other three-qubit states.

Matrix representation of a three-qubit system

With matrix notation, the state in equation (7.37) becomes

$$|\chi\rangle = \begin{bmatrix} a_1 \\ a_2 \\ a_3 \\ a_4 \\ a_5 \\ a_6 \\ a_7 \\ a_8 \end{bmatrix}$$

7.2.3 *m qubits*

Now consider a system with m qubits called qubit-1, qubit-2, and so on until qubit-m. There are 2^m classical states for this system,

$$\{00\ldots0, \quad 0\ldots01, \quad 0\ldots10, \quad \ldots, \quad 10\ldots.0,$$
$$0\ldots11, \quad \ldots, \quad 1\ldots.111\} \tag{7.41}$$

In each of the 2^m states in equation (7.41), the jth number is the state of qubit-j. For example, the first number is the state of qubit-1, the second number is the state of qubit-2, etc., and the m-th number is the state of qubit-m.

The general state of the m-qubit system is a superposition of all 2^m possible classical states,

$$|\psi\rangle = \alpha_1|000\ldots00\rangle + \alpha_2|100\ldots00\rangle + \cdots + \alpha_{2^m-1}|111\ldots10\rangle$$
$$+ \alpha_{2^m}|111\ldots11\rangle \tag{7.42}$$

where $\sum_{j=1}^{2^m} |\alpha_j|^2 = 1$ because $|\psi\rangle$ is normalized (the sum of the probabilities of finding the qubits in the 2^m basis states is one). Note that in equation (7.42), the m qubits store a superposition with components of all 2^m of the basis states. In a classical computer, the program or code runs over only one of the 2^m inputs at each time. On the other hand, a quantum computer evaluates all 2^m inputs simultaneously because the input is in a superposition of 2^m states. This feature is known as *quantum parallelism*, which we will discuss later.

Each of the vectors in equation (7.42) is a symbolic representation of $2^m \times 1$ basis vectors resulting from the tensor product of m single qubit states. For example, $|000\ldots00\rangle$ is

$$|00\ldots0\rangle = |0\rangle \otimes |0\rangle \otimes \ldots \otimes |0\rangle$$

$$= \begin{pmatrix} 1 \\ 0 \end{pmatrix} \otimes \begin{pmatrix} 1 \\ 0 \end{pmatrix} \otimes \ldots \otimes \begin{pmatrix} 1 \\ 0 \end{pmatrix} = \begin{bmatrix} 1 \\ 0 \\ \cdot \\ \cdot \\ 0 \end{bmatrix} \qquad (7.43)$$

Similarly, the basis vector corresponding to $|111\ldots11\rangle$ is

$$|11\ldots1\rangle = |1\rangle \otimes |1\rangle \otimes \ldots \otimes |1\rangle$$

$$= \begin{pmatrix} 0 \\ 1 \end{pmatrix} \otimes \begin{pmatrix} 0 \\ 1 \end{pmatrix} \otimes \ldots \otimes \begin{pmatrix} 0 \\ 1 \end{pmatrix} = \begin{bmatrix} 0 \\ 0 \\ \cdot \\ \cdot \\ 1 \end{bmatrix} \qquad (7.44)$$

Hence, the superposition in equation (7.42) can be rewritten as the weighted sum of $2^m \times 1$ vectors:

$$|\psi\rangle = \alpha_1 \begin{bmatrix} 1 \\ 0 \\ \cdot \\ \vdots \\ 0 \end{bmatrix} + \alpha_2 \begin{bmatrix} 0 \\ 1 \\ \cdot \\ \vdots \\ 0 \end{bmatrix} + \cdots + \alpha_{2^m-1} \begin{bmatrix} 0 \\ 0 \\ \cdot \\ \vdots \\ 1 \\ 0 \end{bmatrix} + \alpha_{2^m} \begin{bmatrix} 0 \\ 0 \\ \cdot \\ \vdots \\ 1 \end{bmatrix} \quad (7.45)$$

The matrix notation becomes infeasible when the number of qubits increases. We see that the vectors' length in matrix form for one-, two-, and three-qubits states are 2, 4, and 8 long. For a m-qubit state the vector is 2^m long. As m increases, the matrix form becomes cumbersome to use while the abstract braket notation is concise. The student should become familiar with the braket notation while knowing its connection to the matrix form (which is used in numerical calculations).

Concept: As the number of qubits becomes large, the matrix notation becomes unwieldy. The braket notation is more compact to use.

7.3   Product and Entangled States

For particular values of the coefficients in equation (7.42), the wave function of the m-qubit system can be written as an outer or Kronecker *product* of the wave function of qubit-1 times qubit-2 ... times qubit-m

$$|\psi\rangle = |\chi_1\rangle_{\text{qubit}-1} \otimes |\chi_2\rangle_{\text{qubit}-2} \otimes |\chi_3\rangle_{\text{qubit}-3}$$
$$\otimes \cdots \otimes |\chi_{m-1}\rangle_{\text{qubit}-(m-1)} \otimes |\chi_m\rangle_{\text{qubit}-m} \quad (7.46)$$

We refer to this as a **product state**.

Example: For $m = 2$,

$$|\psi\rangle = \frac{1}{2}|0\rangle_a|0\rangle_b + \frac{1}{2}|0\rangle_a|1\rangle_b + \frac{1}{2}|1\rangle_a|0\rangle_b + \frac{1}{2}|1\rangle_a|1\rangle_b \qquad (7.47)$$

is a product state because it *factors* out as an outer product of the wave functions of qubit-a times qubit-b

$$|\psi\rangle = \left(\frac{1}{\sqrt{2}}|0\rangle + \frac{1}{\sqrt{2}}|1\rangle\right)_{\text{qubit}-a} \otimes \left(\frac{1}{\sqrt{2}}|0\rangle + \frac{1}{\sqrt{2}}|1\rangle\right)_{\text{qubit}-b} \qquad (7.48)$$

For an arbitrary quantum state, the wave function $|\psi\rangle$ in equation (7.42) cannot be simplified into an outer product of wave functions, as shown in equation (7.46). A state that cannot be simplified into an outer product of the underlying qubits is called an **entangled state**.

Example: For $m = 2$,

$$|B_1\rangle = \frac{1}{\sqrt{2}}(|0\rangle_a|0\rangle_b + |1\rangle_a|1\rangle_b) \quad \text{(Bell state)} \qquad (7.49)$$

is an entangled state. You will not be able to write the above quantum state as a product state of qubits a and b,

$$|B_1\rangle \neq |\chi_0\rangle_{\text{qubit}-a} \otimes |\chi_1\rangle_{\text{qubit}-b}$$

$|B_1\rangle$ in equation (7.49) is called a Bell state, named after John Stewart Bell, famous for his work on Bell's inequality. The other two-qubit Bell states are:

$$|B_2\rangle = \frac{|01\rangle + |10\rangle}{\sqrt{2}}, \quad |B_3\rangle = \frac{|00\rangle - |11\rangle}{\sqrt{2}}, \quad \text{and} \quad |B_4\rangle = \frac{|01\rangle - |10\rangle}{\sqrt{2}}$$

While it is easy to see that equation (7.49) is not a product state, how can one find out if a general two-qubit state like $a|00\rangle + b|01\rangle + c|10\rangle + d|11\rangle$ is an entangled or product state?

Example: Show that $a|00\rangle + b|01\rangle + c|10\rangle + d|11\rangle$ is a product state if and only if $ad = bc$.

Solution: In matrix notation, the state $a|00\rangle + b|01\rangle + c|10\rangle + d|11\rangle$

is written as $\begin{bmatrix} a \\ b \\ c \\ d \end{bmatrix}$.

We start by assuming that the given state is a product state of two unknown single qubit states $|u\rangle$ and $|v\rangle$ as:

$$\begin{bmatrix} a \\ b \\ c \\ d \end{bmatrix} = |u\rangle \otimes |v\rangle = \begin{bmatrix} \alpha \\ \beta \end{bmatrix} \otimes \begin{bmatrix} \gamma \\ \delta \end{bmatrix} = \begin{bmatrix} \alpha\gamma \\ \alpha\delta \\ \beta\gamma \\ \beta\delta \end{bmatrix} \tag{7.50}$$

Hence, the above four equations' solution will yield α, β, γ, and δ. In other words,

$$\alpha\gamma = a, \quad \alpha\delta = b, \quad \beta\gamma = c, \quad \text{and} \quad \beta\delta = d$$

Substituting $\gamma = a/\alpha$ and $\beta = d/\delta$ in $\beta\gamma = c$, we get

$$\frac{d}{\delta} \times \frac{a}{\alpha} = \frac{ad}{b} = c \Rightarrow ad = bc$$

For the second part of this proof, we recall *if and only if* statements are proved both ways. So, we assume that $ad = bc$ and then show that $\begin{bmatrix} a \\ b \\ c \\ d \end{bmatrix}$ can be written in a product form. Since $\frac{b}{a} = \frac{d}{c}$, we factor a from the first two entries and factor c from the third and fourth entries of the state vector to write

$$\begin{bmatrix} a \\ b \\ c \\ d \end{bmatrix} = \begin{bmatrix} a \cdot \begin{bmatrix} 1 \\ \frac{b}{a} \end{bmatrix} \\ c \cdot \begin{bmatrix} 1 \\ \frac{d}{c} \end{bmatrix} \end{bmatrix} = \begin{bmatrix} a \\ c \end{bmatrix} \otimes \begin{bmatrix} 1 \\ \frac{d}{c} \end{bmatrix} \tag{7.51}$$

Hence, we showed that the 4×1 state vector is decomposable as the outer product of two single-qubit states. Therefore, the state is not entangled, and the proof is complete. Having seen the above example, you can easily show that the other Bell states described on the previous page are also entangled states.

We will now discuss the measurement of a two-qubit entangled state. Consider a system consisting of two noninteracting qubits a and b in the state,

$$|\psi\rangle = \alpha|0\rangle_a|0\rangle_b + \beta|1\rangle_a|1\rangle_b, \quad \text{where } |\alpha|^2 + |\beta|^2 = 1$$

What are the possible results if only qubit-a is measured? What will the wave function be after measurement?

$\rightarrow$ After measurement, we will find qubit-a in either

- $|0\rangle$ (with probability $|\alpha|^2$) or
- $|1\rangle$ (with probability $|\beta|^2$).

$\rightarrow$ If qubit-a is found to be in state $|0\rangle$, the collapsed state of the system after the measurement is $|0\rangle_a|0\rangle_b$. We will find qubit-b to be in state 0 with 100% guarantee.

$\rightarrow$ However, if qubit-a is found in state $|1\rangle$, the collapsed state of the system after the measurement is $|1\rangle_a|1\rangle_b$. From this, you can say that qubit-b will definitely be (100%) in state $|1\rangle_b$.

$\rightarrow$ Note that even though qubit-b is not measured, its quantum state gets determined by the measurement of qubit-a. There is no classical analogy to the observation that the measured state of qubit-b is correlated to the result of the measurement of qubit-a in state $|\psi\rangle$. This is called Einstein–Podolsky–Rosen paradox. They discussed the paradox because they were not at ease with qubit-a's measured state deciding the state of qubit-b. The strangeness of this is more apparent if one thinks of separating qubits a and b galaxies apart such that information transfer between them is not possible in one's lifetime. But this is how nature behaves, and this paradox is the difference between our expectations from classical nature and the tangible way nature works.

Projection operators are used to discuss measurement mathematically. We will now discuss a Bell state's measurement using projection operators to familiarize the reader with this.

We first decide to measure qubit-a using projection operator $\widehat{P}_{0a}$ on qubit-a and the second projection operator $\widehat{P}_{0b}$ acts on qubit-b. Hence,

$$|m1\rangle = \widehat{P}_{0a}[\alpha|00\rangle + \beta|11\rangle]$$
$$= \alpha\widehat{P}_{0a}|0\rangle|0\rangle + \beta\widehat{P}_{0a}|1\rangle|1\rangle = \alpha|0\rangle|0\rangle + \beta 0|1\rangle|1\rangle = \alpha|00\rangle \quad (7.52)$$

So, the measurement of the first qubit returned $|0\rangle$ with probability $|\alpha|^2$. The system has now collapsed to state $|00\rangle$. If we apply the projection operator $\widehat{P}_{0b}$ on the second qubit of $|00\rangle$, we get $|00\rangle$ with 100% probability. Applying $\widehat{P}_{1b}$ on the second qubit of $|00\rangle$ gives zero. These are summarized mathematically as:

$$|m2\rangle = \widehat{P}_{0b}|00\rangle = |C\rangle\widehat{P}_{0b}|0\rangle = 1|00\rangle \quad \rightarrow \quad 100\% \text{ probable} \quad (7.53)$$
$$|m3\rangle = P_{1b}|00\rangle = |C\rangle P_{1b}|0\rangle = 0 \quad \rightarrow \quad 0\% \text{ probable} \quad (7.54)$$

In contrast, measurement of qubits a and b in the product state given by $|\psi\rangle = \frac{1}{2}|0\rangle_a|0\rangle_b + \frac{1}{2}|0\rangle_a|1\rangle_b + \frac{1}{2}|1\rangle_a|0\rangle_b + \frac{1}{2}|1\rangle_a|1\rangle_b$ can be summarized as follows:

- If qubit-a is found in state $|0\rangle$, then qubit-b can be found in state $|0\rangle$ with 50% probability and state $|1\rangle$ with 50% probability.
- If qubit-a is found in state $|1\rangle$, then again qubit-b can be found in state $|0\rangle$ with 50% probability and state $|1\rangle$ with 50% probability.

Below, we will see that the measurement of qubit-a is *uncorrelated* to the measurement of qubit-b in the product state given by equation (7.47):

$$|m1\rangle = \widehat{P}_{0a}\frac{|00\rangle + |01\rangle + |10\rangle + |11\rangle}{2}$$
$$= \frac{1}{2}(\widehat{P}_{0a}|0\rangle|0\rangle + \widehat{P}_{0a}|0\rangle|1\rangle + \widehat{P}_{0a}|1\rangle|0\rangle + \widehat{P}_{0a}|1\rangle|1\rangle)$$

$$= \frac{1}{2}(|0\rangle|0\rangle + |0\rangle|1\rangle + 0|1\rangle|0\rangle + 0|1\rangle|1\rangle)$$

$$= \frac{1}{2}(|00\rangle + |01\rangle) = \frac{1}{2}|0\rangle(|0\rangle + |1\rangle) \tag{7.55}$$

The above equation implies that if we measure qubit-a, we will find it to be in state $|0\rangle$ with a 50% probability. We now take this system after measuring qubit-a, which is the normalized version of the above state $|m1\rangle$,

$$|m1\rangle = \frac{1}{\sqrt{2}}(|00\rangle + |01\rangle) \quad \text{(normalized version of equation (7.55))}$$

$$\tag{7.56}$$

Now we repeat measurements 2 and 3 on normalized $|m1\rangle$ as before:

$$|m2\rangle = \widehat{P}_{0b}|m1\rangle = \frac{1}{\sqrt{2}}\widehat{P}_{0b}(|00\rangle + |01\rangle)$$

$$= \frac{1}{\sqrt{2}}(|0\rangle\widehat{P}_{0b}|0\rangle + |0\rangle\widehat{P}_{0b}|1\rangle) = \frac{1}{\sqrt{2}}|0\rangle|0\rangle \tag{7.57}$$

That is, the probability of finding qubit-b in state $|0\rangle$ is 50%.

$$|m3\rangle = \widehat{P}_{1b}|m1\rangle = \frac{1}{\sqrt{2}}(|0\rangle\widehat{P}_{1b}|0\rangle + |0\rangle\widehat{P}_{1b}|1\rangle)$$

$$= \frac{1}{\sqrt{2}}(|0\rangle \cdot 0 + |0\rangle|1\rangle) = \frac{1}{\sqrt{2}}|01\rangle \tag{7.58}$$

That is, qubit-b is found in states $|0\rangle$ and $|1\rangle$, each with 50% probability. The measurement results of qubit-b are uncorrelated to the result of the first measurement on qubit-a.

Concept: Quantum states can be classified into *product states* and *entangled states*. Entangled states are unique to quantum mechanics. They are central to quantum computing.

7.4   Quantum Computing Basics

Students with a background in classical computing or logic circuits already know that any complex mathematical/logical operation is decomposable into a sequence of simpler gate operations. Some of

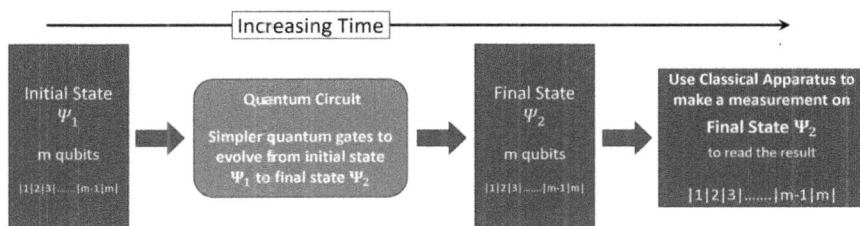

Figure 7.5. An overview of a simple model of quantum computing. The number of qubits stays the same over the computation as time progresses. To find the result of a computation, we make measurements on the qubits of the final state Ψ_2. The evolution from initial state Ψ_1 to final state Ψ_2 is broken down into simpler quantum gates as discussed in the text.

the basic gates in classical computing are NOT, AND, and OR gates. Similarly, any quantum computing task is decomposable as a sequential application of simpler quantum gates.

A quantum computer consists of m qubits which start at an initial state Ψ_1 (see Figure 7.5). After the computation (see orange rounded box), the number of qubits stays the same, but the final state of the qubits is Ψ_2. We would measure the qubits in state Ψ_2 to find the answer to our computation.

As opposed to classical logic circuits in which all logic functions are implementable with NAND gates, more than one gate is required for performing an arbitrary computation in quantum computing. A commonly considered set of quantum gates are as follows:

(1) controlled NOT gate,
(2) $\pi/8$ gate,
(3) Hadamard gate,
(4) phase gate.

The Hadamard, $\pi/8$, and phase gates operate only a single qubit, independent of the state of other qubits.[1] The controlled NOT gate

---

[1] The controlled NOT, $\pi/8$, and Hadamard gates are essential gates. The phase gate is nonessential (for reasons discussed later).

requires two qubits. The design of a quantum computer is inefficient with only the gates discussed earlier. Like classical computing, added gates are helpful for efficient computation. Some commonly discussed gates are:

(1) single qubit $\widehat{X}$, $\widehat{Y}$, and $\widehat{Z}$ gates,
(2) two qubit versions of the single qubit gates (1), where the gate operation affects a target qubit only if the control qubit is in logic state 1,
(3) N-qubit versions of the single qubit gates in (1), where the transformation is applied only if $N - 1$ controls have specific values,
(4) more generally, one thinks of a controlled U gate, where a unitary transform (matrix) $\widehat{U}$ is applied on the target qubit, only if the control qubit has a particular value (state).

In this chapter, we only name some important gates but they will be discussed in greater detail in the next chapter.

What is quantum parallelism?

One of the main advantages of quantum computing is access to massive parallelism, which does not have a counterpart in classical computation. Quantum parallelism is key to quantum computing. Let us say that we have $N = 2^m$ inputs, $|bases\rangle = \{|000\ldots.00\rangle,$ $|100\ldots.00\rangle, |111\ldots.10\rangle, |111\ldots.11\rangle\}$, corresponding to the m bits. Evaluating a function $f(|bases\rangle)$ corresponding to these inputs in a classical logic circuit will take $N = 2^m$ evaluations of $f(|bases\rangle)$. With classical logic circuits, each of the 2^m inputs would be given one at a time, meaning that the calculation of the output is sequential.

Quantum mechanics, however, allows us to create an input state which is an outer product of (a) superposition of all $N = 2^m$ input states, and (b) the state $|0\rangle$, which consists of p ancillary qubits,

$$|\text{input}\rangle = \frac{1}{\sqrt{N}}\{|000\ldots.00\rangle + |000\ldots.01\rangle + \cdots + |111\ldots.10\rangle$$

$$+ |111\ldots.11\rangle\}|0\rangle \qquad (7.59)$$

The $|0\rangle = |00..0..00\rangle$ state outside the curly bracket contains p ancillary qubits consisting only of $|0\rangle$s at the beginning of the computation, and contains the output after operation with f. A single operation of $f(|\text{bases}\rangle)$ on this input state creates an output state which is an entangled state of (a) the $N = 2^m$ input basis states and (b) the corresponding value of $f(|\text{bases}\rangle)$ as shown in the following:

$$|\text{output}\rangle = \frac{1}{\sqrt{N}}\{|000\ldots00\rangle|f(000\ldots00)\rangle$$
$$+ |000\ldots01\rangle|f(000\ldots01)\rangle$$
$$+ \cdots$$
$$+ |111\ldots10\rangle|f(111\ldots10)\rangle$$
$$+ |111\ldots11\rangle|f(111\ldots11)\rangle\} \qquad (7.60)$$

Note that once we measure the output state, we will collapse into one of these N output states and hence will have only one value of $f(|\text{bases}\rangle = $ collapsed state). The power of quantum computation arises from exploiting the above quantum parallelism to perform desired unitary transformations, which will create an output state that contains the answer to our mathematical problem. We seldom make measurements on a state with an equal superposition, such as the $|\text{output}\rangle$ above. Many more gate operations are involved. Grover's search algorithm discussed in the following illustrates a rich application of quantum parallelism, the evaluation of function (f) for all input variants with the aid of superposition. To probe further into quantum algorithms, please refer (McMahon, 2008) and (Nielsen, 2000).

Another common way to write a superposition of states is to use the summation notation. The states with m qubits above can be written as

$$|\text{input}\rangle = \frac{1}{\sqrt{N}} \sum_{j=0}^{2^m-1} |j\rangle|0\rangle$$

$$|\text{output}\rangle = \frac{1}{\sqrt{N}} \sum_{j=0}^{2^m-1} |j\rangle|f(j)\rangle$$

7.4.1   *Unitary operation on qubits*

The transformation of a single qubit state

$$|\chi\rangle = a_\chi|0\rangle + b_\chi|1\rangle = \begin{pmatrix} a_\chi \\ b_\chi \end{pmatrix}.$$

to a new state

$$|\varphi\rangle = a_\varphi|0\rangle + b_\varphi|1\rangle = \begin{pmatrix} a_\varphi \\ b_\varphi \end{pmatrix}$$

is a rotation in the vector space spanned by $|0\rangle$ and $|1\rangle$. The operator (or matrix) $\widehat{U}$ rotates $|\chi\rangle$ to $|\varphi\rangle$,

$$|\varphi\rangle = \widehat{U}|\chi\rangle \tag{7.61}$$

As wave functions are normalized, we know that

$$\langle\varphi|\varphi\rangle = 1 \tag{7.62}$$

Substituting equation (7.61) in equation (7.62), we have

$$\langle\varphi|U^+U|\varphi\rangle = 1 \quad \Rightarrow \quad U^+U = I \tag{7.63}$$

where $\widehat{I}$ is the identity operator. An operator with the property given by equation (7.63) is called a *unitary operator* (see Chapter 4). In quantum mechanics, all rotations are performed only by unitary transformations, independent of the number of qubits. Note that unitary transformations conserve the norm of the initial qubit state.

The same concept applies to a multi-(m) qubit state. As in Figure 7.5, consider the initial ($|\Psi_1\rangle$) and final ($|\Psi_2\rangle$) states of the m qubit quantum computer. The transformation from $|\Psi_1\rangle$ to $|\Psi_2\rangle$ is

$$|\Psi_2\rangle = \widehat{U}|\Psi_1\rangle$$

where $\widehat{U}$ is now a transformation that acts on the m-qubit state. The transformation $\widehat{U}$ is equivalent to consecutive operations by simpler gates that act on only one or two (of the m) qubits at a time. Mathematically, this statement is

$$|\Psi_2\rangle = \widehat{U}_p \widehat{U}_{p-1} \ldots \widehat{U}_2 \widehat{U}_1 |\Psi_1\rangle$$

where

$$\widehat{U} = \widehat{U}_p \widehat{U}_{p-1} \ldots \widehat{U}_2 \widehat{U}_1$$

Each of the transformations $\widehat{U}_1$ through $\widehat{U}_p$ represent one of the simple gates that act on either one or two of the m qubits at a time. In the above equation, $\widehat{U}_1$ acts on state $|\Psi_1\rangle$ first; $\widehat{U}_2$ follows this and so on until $\widehat{U}_p$. See Figure 7.6. The convention is identical to matrix multiplication, where the operator on $|\Psi_1\rangle$ acts sequentially (timewise) from right to left. The order of $\widehat{U}_j$ is not interchangeable, similar to matrix multiplication.

In the next section, we will discuss Grover's algorithm. A thorough discussion of algorithms based on quantum computing can be found in Nielsen and Chuang (2000).

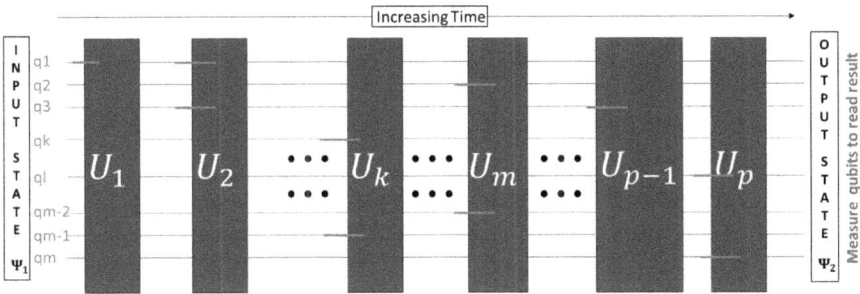

Figure 7.6. Sequential operation of one and two qubit gates in a quantum computer with m qubits. The horizontal blue lines stand for the qubits. The input state $|\Psi_1\rangle$ is at the left end (earliest time), and the output state $|\Psi_2\rangle$ is at the right end (end of computation). The boxes stand for the various one and two qubit operations. The orange lines stand for the qubits on which the gate to its right acts. So, U_1, a single qubit gate operation, which affects only qubit $q1$, acts on the input first. The second gate operation U_2 is a two-qubit gate operation, which affects only qubits $q1$ and $q3$ and so on until we reach the last gate operation U_p. The result of the computation is measured at the output.

7.5 Grover's Algorithm: For Database Search

Grover's algorithm uses quantum mechanics to perform an efficient database *search operation*. It is significantly more efficient than any known search algorithm on a classical computer. You have a phone number, and you want to find the name associated with it from the phonebook. In a phonebook with N entries, this search will take an average of $\frac{N}{2}$ operations. So classically, $O(N)$ steps[2] are required. Using quantum mechanics, Grover's algorithm provides a method to find the name associated with the phone number in $O(\sqrt{N})$ operations. The speedup is immense for a database (such as a phonebook) with many entries. If the phonebook has 10^{10} entries, it will take only 10^5 operations on a quantum computer but will take 5×10^9 operations on a classical computer. The flowchart in Figure 7.7 provides an overview of Grover's algorithm, and it will be helpful in keeping track of the steps involved in the following derivation.

Statement of the search problem: We have N inputs, which obey the property:

$$f(\alpha_j) = 1 \quad \text{only for one input} \tag{7.64}$$

$$f(\alpha_j) = 0 \quad \text{for all other inputs} \tag{7.65}$$

Devise an algorithm to find the input α_j that corresponds to $f(\alpha_j) = 1$.

We discussed before that the quantum computer takes an input quantum state and performs unitary transformations to reach the output quantum state (Figures 7.5 and 7.6). The quantum computer rotates the input vector to a final output vector, which represents the mathematical problem's solution. The challenges are to find:

(i) a mathematical algorithm,
(ii) a sequence of simpler unitary transformations (gates) to implement the algorithm on qubits, and
(iii) hardware implementation.

---

[2] $O(N)$ means N basic multiplications.

Given: N phone numbers and the associated names
Find: Given one of the N phone numbers, find the name associated with the phone number.
Classical Computational Cost: $N/2$
Quantum Computational Cost: $\sqrt{N}$

Given: $f(\alpha_j) = \begin{cases} 1 & \text{for } \alpha_j = \alpha_p \\ 0 & \text{for all other } j \end{cases}$ where $j \in \{1, 2, \ldots integer \ldots, N\}$
Find: Value of j for which $f(\alpha_j) = 1$
Given one of the N phone numbers, find the name associated with the phone number
Classical Computational Cost: $N/2$
Quantum Computational Cost: $\sqrt{N}$

$|init\rangle = \frac{1}{\sqrt{N}}\{|p_1\rangle|n_1\rangle + |p_2\rangle|n_2\rangle + \cdots + |p_p\rangle|n_p\rangle + \cdots + |p_{N-1}\rangle|n_{N-1}\rangle + |p_N\rangle|n_N\rangle\}$

$|init\rangle = \frac{1}{\sqrt{N}}|p_p\rangle|n_p\rangle + \frac{\sqrt{N-1}}{\sqrt{N}}\left[\frac{1}{\sqrt{N-1}}\{|p_1\rangle|n_1\rangle + |p_2\rangle|n_2\rangle + \cdots + |p_p\rangle|n_p\rangle + \cdots + |p_{N-1}\rangle|n_{N-1}\rangle + |p_N\rangle|n_N\rangle\}\right]$

$|init\rangle = \frac{1}{\sqrt{N}}|\alpha_p\rangle + \frac{\sqrt{N-1}}{\sqrt{N}}|\bar{\alpha}_p\rangle$

For large N, $|init\rangle$ is almost parallel to $|\bar{\alpha}_p\rangle$.
Rotation of $|init\rangle$ by close to $90°$ aligns it along $|\alpha_p\rangle$.
Grover's algorithm finds an efficient implementation of a rotation
Matrix to rotate $|init\rangle$ to $|\alpha_p\rangle$ in the N-dimensional Hilbert space.

Once the mathematical algorithm behind the rotation matrix has been determined, the next step is to break it up into simple 1-, 2- or few qubit operations.

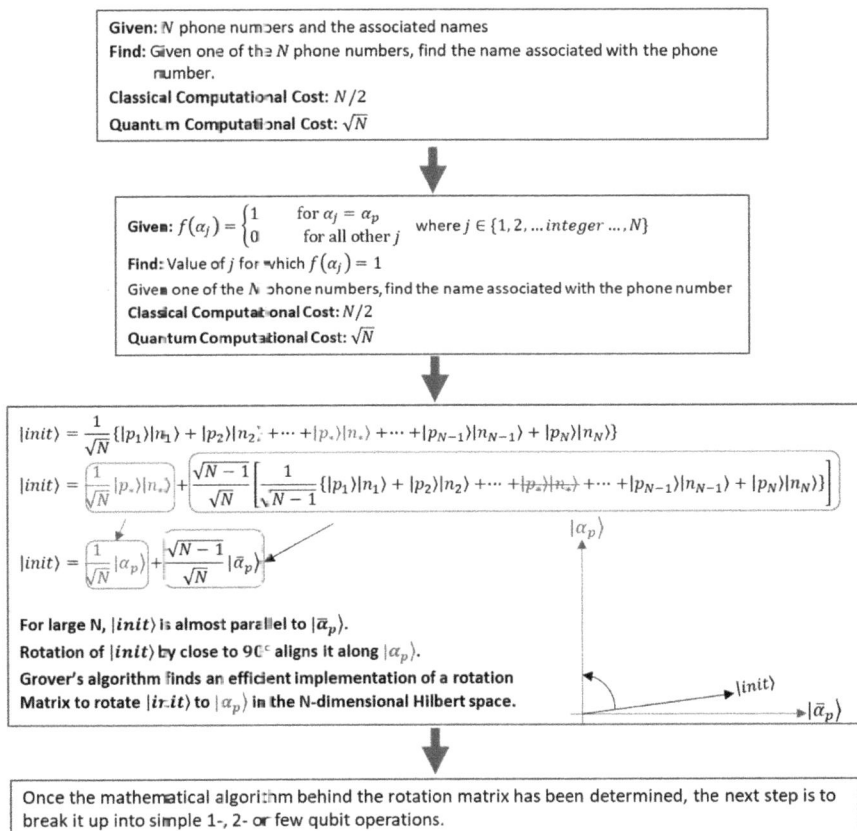

Figure 7.7. An overview of Grover's algorithm for database search.

Grover's algorithm takes care of step (i) by proposing the following (see flowchart in 7.7):

(a) Start with an m-qubit long input state init⟩, which is the equal superposition of all 2^m states,

$$|init\rangle = \frac{1}{\sqrt{N}}\{|000\ldots00\rangle + |100\ldots00\rangle + \cdots + |111\ldots10\rangle$$
$$+ |111\ldots11\rangle\} \tag{7.66}$$

Corresponding to one of these states, $\alpha_j = |\alpha_p\rangle$, $f(\alpha_j = |\alpha_p\rangle) = 1$. For the other $N - 1$ states, $f(\alpha_j \neq |\alpha_p\rangle) = 0$. We want to find the state $\alpha_j = |\alpha_p\rangle$.

(b) If we can find a unitary transformation that will rotate $|\text{init}\rangle$ to state $|\alpha_p\rangle$, we would have completed the search.

(c) The realization of the gates required to form the unitary transformations $\widehat{W}$ and $\widehat{V}$ is the last component. The reader can follow the examples discussed in Section 8.4.6 to work out a sequence of gates. The gates required to create an equal superposition are discussed in Section 8.4.5.

The input $|\text{init}\rangle$ in equation (7.66) can be rewritten as

$$|\text{init}\rangle = \frac{1}{\sqrt{N}} \left\{ |\alpha_p\rangle + \sum_{\substack{j=1 \\ (j \neq p)}}^{N} |\alpha_j\rangle \right\} \tag{7.67}$$

Note that the first and second terms inside the bracket are orthogonal. Further, the second term is a sum of $N - 1$ orthogonal basis states. Normalization of the second term gives

$$\sum_{j=1(j \neq p)}^{N} |\alpha_j\rangle \xrightarrow{\text{Normalization}} \frac{1}{\sqrt{N-1}} \sum_{j=1(j \neq p)}^{N} |\alpha_j\rangle \equiv |\bar{\alpha}_p\rangle \tag{7.68}$$

where $|\bar{\alpha}_p\rangle$ is an equal superposition of the $N - 1$ basis states orthogonal to $|\alpha_p\rangle$. Equations (7.66) and (7.67) can now be written as,

$$|\text{init}\rangle = \frac{1}{\sqrt{N}} |\alpha_p\rangle + \frac{\sqrt{N-1}}{\sqrt{N}} |\bar{\alpha}_p\rangle \tag{7.69}$$

$$|\text{init}\rangle = \sin(\theta) |\alpha_p\rangle + \cos(\theta) |\bar{\alpha}_p\rangle$$

where $\sin(\theta) = \frac{1}{\sqrt{N}}$ and $\cos(\theta) = \frac{\sqrt{N-1}}{\sqrt{N}}$. See Figure 7.8.

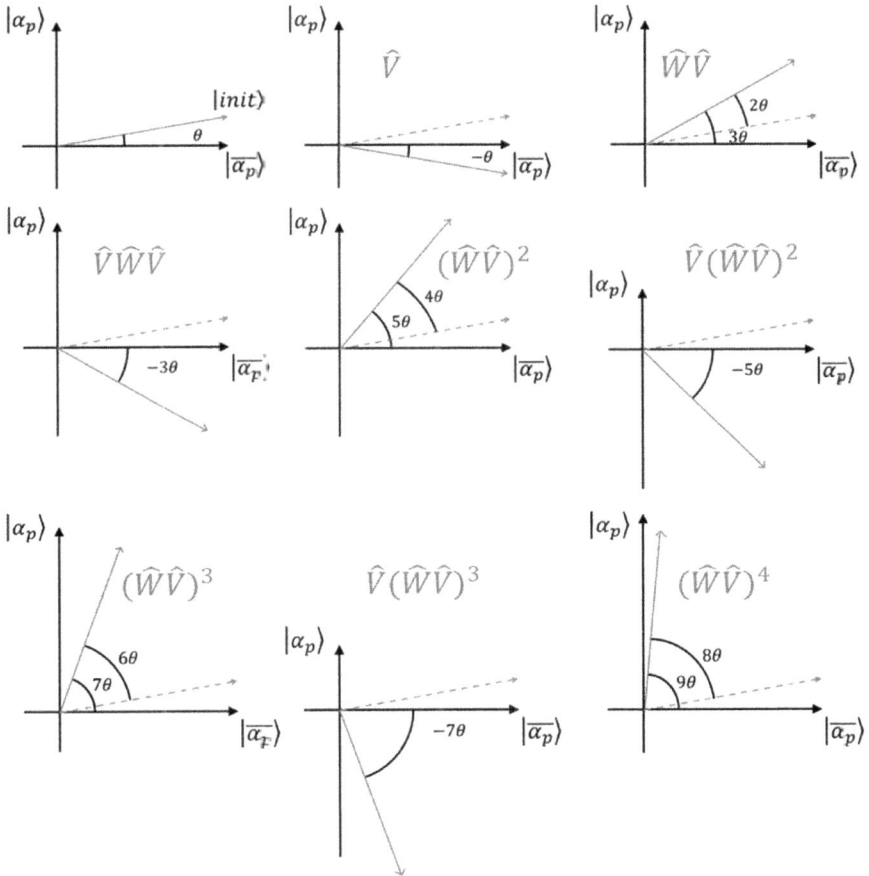

Figure 7.8. Interpretive figure for Grover's algorithm. Each action by WV causes a rotation by an angle of 2θ.

We define a unitary transformation $\widehat{V}$ that takes any state and performs the following operation:

$$\widehat{V}|\alpha_j\rangle = (-1)^{f((|\alpha_j\rangle))}|\alpha_j\rangle \qquad (7.70)$$

We know from the definition of f in equations (7.64) and (7.65) that

$$\widehat{V}|\alpha_p\rangle = -|\alpha_p\rangle \quad \text{and} \quad \widehat{V}|\bar{\alpha}_p\rangle = |\bar{\alpha}_p\rangle \qquad (7.71)$$

That is, operator $\widehat{V}$ flips the state $|\alpha_p\rangle$ to $-|\alpha_p\rangle$ while leaving $|\bar{\alpha}_p\rangle$ unchanged. See Figure 7.8(b) for an illustration. In matrix form, the above equation is

$$\widehat{V}\begin{pmatrix} |\bar{\alpha}_p\rangle \\ |\alpha_p\rangle \end{pmatrix} = \begin{pmatrix} 1 & 0 \\ 0 & -1 \end{pmatrix}\begin{pmatrix} |\bar{\alpha}_p\rangle \\ |\alpha_p\rangle \end{pmatrix} \tag{7.72}$$

An example of the realization of gates for unitary transformation $\widehat{V}$ is discussed in Section 8.4.6 (it flips the sign of only a selected basis vector).

The unitary transformation $\widehat{W}$ is defined by

$$\widehat{W} = 2|\text{init}\rangle\langle\text{init}| - \widehat{I} \tag{7.73}$$

We will show that $\widehat{W} = 2|\text{init}\rangle\langle\text{init}| - \widehat{I}$ leaves the component of $|\psi\rangle$ along $|\text{init}\rangle$ unchanged and inverts (puts a minus sign in front of) the component of $|\psi\rangle$ orthogonal to $|\text{init}\rangle$. We can illustrate this by writing $|\psi\rangle$ as a linear combination of $|\text{init}\rangle$ and the component orthogonal to it, $|\text{init}_\perp\rangle$,

$$|\psi\rangle = a|\text{init}\rangle + b|\text{init}_\perp\rangle$$

Then $\widehat{W}$ acting on $|\psi\rangle$ gives

$$\widehat{W}|\psi\rangle = [2|\text{init}\rangle\langle\text{init}| - \widehat{I}][a|\text{init}\rangle + b|\text{init}_\perp\rangle]$$
$$\widehat{W}|\psi\rangle = 2a|\text{init}\rangle - [a|\text{init}\rangle + b|\text{init}_\perp\rangle]$$
$$\widehat{W}|\psi\rangle = a|\text{init}\rangle - b|\text{init}_\perp\rangle$$

In $\widehat{W}|\psi\rangle$, the component of $|\psi\rangle$ along $|\text{init}\rangle$ is unchanged but the phase of the component orthogonal to $|\text{init}\rangle$ has changed by $e^{i\pi}$ (see Figure 7.9). An example of gates for the unitary transform $\widehat{W}$ is discussed in Chapter 8.

Operating $\widehat{W}$ on $|\alpha_p\rangle$,

$$\widehat{W}|\alpha_p\rangle = \frac{2\sqrt{N-1}}{N}|\bar{\alpha}_p\rangle + \frac{2-N}{N}|\alpha_p\rangle \quad \text{and}$$

$$\widehat{W}|\bar{\alpha}_p\rangle = -\frac{2-N}{N}|\bar{\alpha}_p\rangle + \frac{2\sqrt{N-1}}{N}|\alpha_p\rangle \tag{7.74}$$

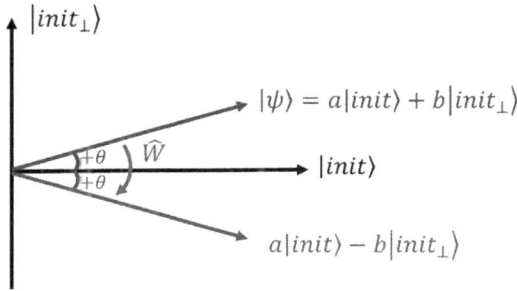

Figure 7.9. The application of $\widehat{W}$ on the superposition $|\psi\rangle = a|init\rangle + b|init_\perp\rangle$ leaves the component of $|\psi\rangle$ along the $|init\rangle$ unchanged while multiplying $|init_\perp\rangle$ by -1.

In matrix form, the above equation is

$$\widehat{W}\begin{pmatrix} |\bar{\alpha}_p\rangle \\ |\alpha_p\rangle \end{pmatrix} = \begin{pmatrix} \dfrac{N-2}{N} & \dfrac{2\sqrt{N-1}}{N} \\ \dfrac{2\sqrt{N-1}}{N} & -\dfrac{N-2}{N} \end{pmatrix} \begin{pmatrix} |\bar{\alpha}_p\rangle \\ |\alpha_p\rangle \end{pmatrix} \qquad (7.75)$$

From equations (7.72) and (7.75), by using matrix multiplication, we see that

$$\widehat{W}\widehat{V}\begin{pmatrix} |\bar{\alpha}_p\rangle \\ |\alpha_p\rangle \end{pmatrix} = \begin{pmatrix} \dfrac{N-2}{N} & -\dfrac{2\sqrt{N-1}}{N} \\ \dfrac{2\sqrt{N-1}}{N} & \dfrac{N-2}{N} \end{pmatrix} \begin{pmatrix} |\bar{\alpha}_p\rangle \\ |\alpha_p\rangle \end{pmatrix} \quad \text{or}$$

$$\widehat{W}\widehat{V}\begin{pmatrix} |\bar{\alpha}_p\rangle \\ |\alpha_p\rangle \end{pmatrix} = \begin{pmatrix} \cos(2\theta) & -\sin(2\theta) \\ \sin(2\theta) & \cos(2\theta) \end{pmatrix} \begin{pmatrix} |\bar{\alpha}_p\rangle \\ |\alpha_p\rangle \end{pmatrix} \qquad (7.76)$$

The 2×2 matrix in the above equation is a rotation matrix for angle 2θ,

$$\widehat{W}\widehat{V} = \begin{pmatrix} \dfrac{N-2}{N} & -\dfrac{2\sqrt{N-1}}{N} \\ \dfrac{2\sqrt{N-1}}{N} & \dfrac{N-2}{N} \end{pmatrix} = \begin{pmatrix} \cos(2\theta) & -\sin(2\theta) \\ \sin(2\theta) & \cos(2\theta) \end{pmatrix} \qquad (7.77)$$

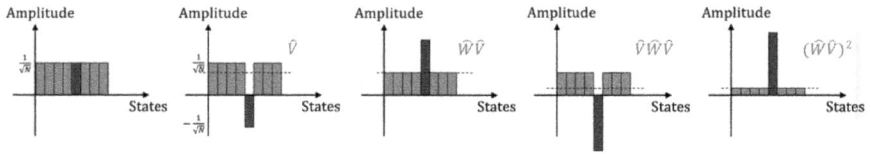

Figure 7.10. A second interpretive figure for Grover's algorithm.

where

$$2\theta = \arctan\left(\frac{2\sqrt{N-1}}{N-2}\right) \xrightarrow{\text{large } N} \arctan\left(\frac{2}{\sqrt{N}}\right) \qquad (7.78)$$

Figure 7.8 illustrates the rotation of $|init\rangle$ by 2θ. $\widehat{W}\widehat{V}$ rotates state $|\bar{\alpha}_p\rangle$ by an angle 2θ towards the desired state $|\alpha_p\rangle$. See Figure 7.10 for an alternative figure that explains the application of the $\widehat{W}$ and $\widehat{V}$ gates. Note that when $\widehat{W}\widehat{V}$ acts on $|\bar{\alpha}_p\rangle$, the amplitude of $|\bar{\alpha}_p\rangle$ has decreased and the amplitude of $|\alpha_p\rangle$ has increased,

$$\widehat{W}\widehat{V}|\bar{\alpha}_p\rangle = \frac{N-2}{N}|\bar{\alpha}_p\rangle - \frac{2\sqrt{N-1}}{\sqrt{N}}|\alpha_p\rangle \text{ (equation (7.76))}$$

Try working out $(\widehat{W}\widehat{V})^2|\bar{\alpha}_p\rangle$ and you will find that the amplitude of $|\bar{\alpha}_p\rangle$ has further decreased and the amplitude of $|\alpha_p\rangle$ has further increased. After applying $\widehat{W}\widehat{V}$ p times (see Figure 7.8), we have

$$(\widehat{W}\widehat{V})^p|\bar{\alpha}_p\rangle = \cos(2p\theta)|\bar{\alpha}_p\rangle - \sin(2p\theta)|\alpha_p\rangle \qquad (7.79)$$

From the above equation, we can see that the RHS is equal to the desired state (answer) $|\alpha_p\rangle$ when $2p\theta = \pi/2$. The state $|\bar{\alpha}_p\rangle$ has rotated to the desired state (answer) of the search problem, which is $|\alpha_p\rangle$. According to equation (7.79), this happens when $2p\theta = \pi/2$ or number of operations to rotate state $|\bar{\alpha}_p\rangle$ to $|\alpha_p\rangle$ is approximately

$$p = \frac{1}{\arctan\left(\frac{2\sqrt{N-1}}{N-2}\right)}\frac{\pi}{4}$$

For large N, this equation becomes

$$\lim_{N\to\infty} p = \lim_{N\to\infty} \frac{\pi}{8}\frac{N-2}{\sqrt{N-1}} = \frac{\pi}{8}\sqrt{N} = O(\sqrt{N}) \qquad (7.80)$$

The above result is not precise because the initial state $|\text{init}\rangle$ has a small component along $|\alpha_p\rangle$ while mostly being in the direction of $|\bar{\alpha}_p\rangle$ (equation (7.69)). Applying $(\widehat{W}\widehat{V})^p$ to $|\text{init}\rangle$, we have

$$(\widehat{W}\widehat{V})^p|\text{init}\rangle = \frac{1}{\sqrt{N}}[\sin(2p\theta)|\bar{\alpha}_p\rangle + \cos(2p\theta)|\alpha_p\rangle]$$

$$+ \frac{\sqrt{N-1}}{\sqrt{N}}[\cos(2p\theta)|\bar{\alpha}_p\rangle - \sin(2p\theta)|\alpha_p\rangle]$$

$$= \left[\frac{1}{\sqrt{N}}\sin(2p\theta) + \frac{\sqrt{N-1}}{\sqrt{N}}\cos(2p\theta)\right]|\bar{\alpha}_p\rangle$$

$$+ \left[\frac{1}{\sqrt{N}}\cos(2p\theta) - \frac{\sqrt{N-1}}{\sqrt{N}}\sin(2p\theta)\right]|\alpha_p\rangle \quad (7.81)$$

The state $|\text{init}\rangle$ rotates to $|\alpha_p\rangle$ when p satisfies

$$\frac{1}{\sqrt{N}}\sin(2p\theta) + \frac{\sqrt{N-1}}{\sqrt{N}}\cos(2p\theta) = 0 \quad (7.82)$$

$$p = \frac{\arctan(-\sqrt{N-1})}{2\arctan\left(\frac{2\sqrt{N-1}}{N-2}\right)} \quad (7.83)$$

In the above equations, p, which is the number of operations of $\widehat{W}\widehat{V}$ needed to rotate $|\text{init}\rangle$ to $|\alpha_p\rangle$ is not an integer. So, while we can get close to the answer $|\alpha_p\rangle$, there will usually be a tiny component along $|\bar{\alpha}_p\rangle$. As a result, we might have to perform the entire computation a few times to find the correct answer. However, since the incorrect answer can be easily verified, the verification process is not time consuming. This probabilistic nature of obtaining the correct answer is a feature of many quantum computing algorithms.

7.5.1 *Phonebook with four names*

Note that in our description of Grover's algorithm from equations (7.66) to (7.84), we did not explicitly write down the value of the phone numbers. In working out the concrete examples here, we will write the phone number explicitly.

Let the states $|00\rangle$, $|01\rangle$, $|10\rangle$ and $|11\rangle$ represent the four names, and let the states $|p_1\rangle$, $|p_2\rangle$, $|p_3\rangle$, and $|p_4\rangle$ represent the phone numbers p_1, p_2, p_3, and p_4 associated with these names. The input state $|\text{init}, p\rangle$ is the entangled state,

$$|\text{init}\rangle = \frac{1}{2}\{|00\rangle|p_1\rangle + |01\rangle|p_2\rangle + |10\rangle|p_3\rangle + |11\rangle|p_4\rangle\}$$

$$= \frac{1}{2}\{|00, p_1\rangle + |01, p_2\rangle + |10, p_3\rangle + |11, p_4\rangle\} \tag{7.84}$$

The entanglement is between the name state ($|00\rangle$, $|01\rangle$, $|10\rangle$, and $|11\rangle$) and their respective phone number state ($|p_1\rangle$, $|p_2\rangle$, $|p_3\rangle$, and $|p_4\rangle$).

To deal with a concrete example, let us assume that the phone number given to us is p_3, and we have to find the name $|10\rangle$ associated with it. The property of the function f in the algorithm (equations (7.64) and (7.65)) is as follows:

$$f(|p_3\rangle) = 1 \quad \text{and} \tag{7.85}$$

$$f(|p_1\rangle) = f(|p_2\rangle) = f(|p_4\rangle) = 0 \tag{7.86}$$

Note that the function f results in 1 only for the state corresponding to the phone number. Equation (7.84) is expressed in a form like equation (7.69) as

$$|\text{init}\rangle = \frac{1}{2}|10, p_3\rangle + \frac{\sqrt{3}}{2}|\overline{10, p}\rangle \tag{7.87}$$

where

$$|\overline{10, p}\rangle = \frac{1}{\sqrt{3}}(|00, p_1\rangle + |01, p_2\rangle + |11, p_4\rangle) \tag{7.88}$$

Equation (7.87), an equal superposition of the four states, is shown graphically in Figure 7.11(a).

Now that we have set up the initial state, let us start applying the operator $\widehat{W}\widehat{V}$ as required by equation (7.76) of Grover's algorithm.

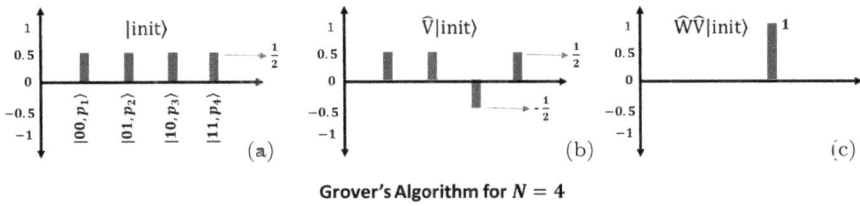

Grover's Algorithm for $N = 4$

Figure 7.11. Evolution of states for Grover's algorithm with $N = 4$, where we search for the phone number $|p_3\rangle$. (a) The initial state, where the vertical bars show the wave function amplitude corresponding to the basis states. (b) The application of the $\widehat{V}$ operator inverts the state corresponding to the phone number that we are searching for. (c) Each application of $\widehat{W}\widehat{V}$ amplifies the state with the phone number. For the $N = 2$ case, subfigures (a), (b), and (c) correspond to equations (7.84), (7.90), and (7.94), respectively.

First, let us apply $\widehat{V}$ on $|\text{init}, p\rangle$,

$$\widehat{V}|\text{init}\rangle = \frac{1}{2}(-1)^{f(|p_3\rangle)}|10, p_3\rangle$$

$$+ \frac{\sqrt{3}}{2}\left[\frac{1}{\sqrt{3}}((-1)^{f(|p_1\rangle)}|00, p_1\rangle + (-1)^{f(|p_2\rangle)}|01, p_2\rangle\right.$$

$$\left. + (-1)^{f(|p_4\rangle)}|11, p_4\rangle)\right] \tag{7.89}$$

Using equations (7.85) and (7.86) in equation (7.89) gives

$$\widehat{V}|\text{init}\rangle = -\frac{1}{2}|10, p_3\rangle + \frac{\sqrt{3}}{2}\left[\frac{1}{\sqrt{3}}(|00, p_1\rangle + |01, p_2\rangle + |11, p_4\rangle)\right]$$

$$\widehat{V}|\text{init}\rangle = -\frac{1}{2}|10, p_3\rangle + \frac{\sqrt{3}}{2}|\overline{10, p}\rangle \tag{7.90}$$

Equation (7.90) is shown graphically in Figure 7.11(b), where only the state containing the phone number has been inverted (has a negative sign). The operator $\widehat{W}$ is

$$\widehat{W} = 2|\text{init}\rangle\langle\text{init} - \widehat{I} = \frac{1}{2}|10, p_3\rangle\langle10, p_3| + \frac{\sqrt{3}}{2}|\overline{10, p}\rangle\langle10, p_3|$$

$$+ \frac{\sqrt{3}}{2}|10, p_3\rangle\langle\overline{10, p}| + \frac{3}{2}|\overline{10, p}\rangle\langle\overline{10, p}| - \widehat{I} \tag{7.91}$$

Applying $\widehat{W}\widehat{V}$ to $|10, p_3\rangle$, we have

$$\widehat{W}\widehat{V}|10, p_3\rangle = \frac{1}{2}|10, p_3\rangle - \frac{\sqrt{3}}{2}|\overline{10, p}\rangle \tag{7.92}$$

and, applying $\widehat{W}\widehat{V}$ to $|\overline{10, p}\rangle$, we have

$$\widehat{W}\widehat{V}|\overline{10, p}\rangle = \frac{\sqrt{3}}{2}|10, p_3\rangle + \frac{1}{2}|\overline{10, p}\rangle \tag{7.93}$$

Substituting equations (7.92) and (7.93) in $\widehat{W}\widehat{V}|\text{init}, p\rangle$ (see equation (7.87)) gives

$$\widehat{W}\widehat{V}|\text{init}\rangle = \frac{1}{2}\widehat{W}\widehat{V}|10, p_3\rangle + \frac{\sqrt{3}}{2}\widehat{W}\widehat{V}|\overline{10, p}\rangle = |10, p_3\rangle = |10\rangle|p_3\rangle \tag{7.94}$$

So, we applied $\widehat{W}\widehat{V}$ only once to find the correct name $|10\rangle$ associated with the phone number $|p_3\rangle$. Equation (7.94) is shown graphically in Figure 7.11(c), where the state with the correct phone number is amplified.

7.5.2 *Phonebook with eight names*

Our goal here is to show that applying $\widehat{W}\widehat{V}$ twice yields the correct phone number (p_1) with maximum probability; however, if you apply $\widehat{W}\widehat{V}$ a third time, the probability of finding the correct phone number decreases.

Let the eight names $|000\rangle$, $|001\rangle$, $|010\rangle$, $|011\rangle$, $|100\rangle$, $|101\rangle$, $|110\rangle$, and $|111\rangle$ be associated with phone numbers $|p_1\rangle$, $|p_2\rangle$, $|p_3\rangle$, $|p_4\rangle$, $|p_5\rangle$, $|p_6\rangle$, $|p_7\rangle$, and $|p_8\rangle$, respectively. We start with an initial state $|\text{init}, p\rangle$, which is an equal superposition of the $N = 2^3 = 8$ states (names) entangled with their respective phone number,

$$|\text{init}\rangle = \frac{1}{\sqrt{8}} \left\{ \begin{array}{l} |000, p_1\rangle + |001, p_2\rangle + |010, p_3\rangle + |011, p_4\rangle \\ + |100, p_5\rangle + |101, p_6\rangle + |110, p_7\rangle + |111, p_8\rangle \end{array} \right\} \tag{7.95}$$

To deal with a concrete example, we assume that we want to find the name associated with the phone number p_1. So, by application of Grover's algorithm, we must correctly identify the name $|000\rangle$. The properties of the function f in the algorithm (equations (7.64) and (7.65)) are as follows:

$$f(|p_1\rangle) = 1 \quad \text{and} \tag{7.96}$$

$$f(|p_2\rangle) = f(|p_3\rangle) = f(|p_4\rangle) = f(|p_5\rangle)$$
$$= f(|p_6\rangle) = f(|p_7\rangle) = f(|p_8\rangle) = 0 \tag{7.97}$$

Equation (7.95) can be rewritten as

$$|\text{init}\rangle = \frac{1}{\sqrt{8}}|000, p_1\rangle + \frac{\sqrt{7}}{\sqrt{8}}|\overline{000, p}\rangle \quad \text{where} \tag{7.98}$$

$$|\overline{000, p}\rangle = \frac{1}{\sqrt{7}}\{|001, p_2\rangle + |010, p_3\rangle + |011, p_4\rangle + |100, p_5\rangle$$
$$+ |101, p_6\rangle + |110, p_7\rangle + |111, p_8\rangle\} \tag{7.99}$$

Equation (7.99) is shown graphically in Figure 7.12(a), an equal superposition of all eight states. We can now start by applying $\widehat{W}\widehat{V}$ as required by Grover's algorithm. First, let us apply $\widehat{V}$ on $|\text{init}\rangle$. Using equations (7.96) and (7.97),

$$\widehat{V}|000, p_1\rangle = (-1)^{f(|p_1\rangle)}|000\rangle|p_1\rangle = -|000, p_1\rangle \quad \text{and} \tag{7.100}$$

$$\widehat{V}|\overline{000, p}\rangle = \frac{1}{\sqrt{7}}\left\{(-1)^{f(|p_2\rangle)}|001\rangle|p_2\rangle + (-1)^{f(|p_3\rangle)}|010\rangle|p_3\rangle\right.$$
$$+ (-1)^{f(|p_4\rangle)}|011\rangle|p_4\rangle + (-1)^{f(|p_5\rangle)}|100\rangle|p_5\rangle$$
$$+ (-1)^{f(|p_6\rangle)}|101\rangle|p_6\rangle + (-1)^{f(|p_7\rangle)}|110\rangle|p_7\rangle$$
$$\left. + (-1)^{f(|p_8\rangle)}|111\rangle|p_8\rangle\right\} \tag{7.101}$$

$$\widehat{V}|\overline{000, p}\rangle = |\overline{000, p}\rangle \tag{7.102}$$

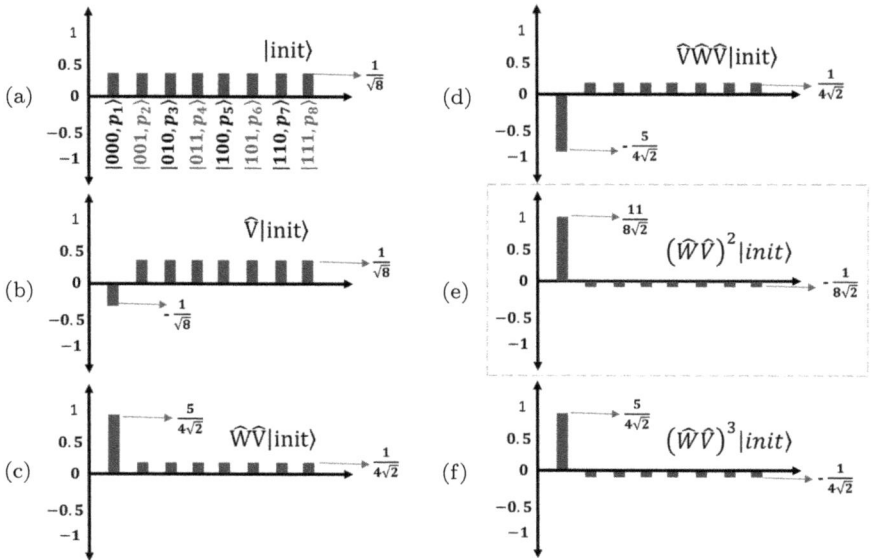

Figure 7.12. Evolution of states for Grover's algorithm with $N = 8$, where we search for the phone number $|p_1\rangle$. The vertical bars show the wave function amplitude corresponding to the basis states in subfigure (a). Application of the $\widehat{V}$ operator inverts the state corresponding to the phone number that we are searching for (subfigures (b) and (d)). Each application of $\widehat{W}\widehat{V}$ amplifies the state with the desired phone number (subfigures (c) and (e)). Subfigures (a), (b), (c), (e), and (f) correspond to equations (7.98), (7.103), (7.107), (7.110), and (7.113), respectively.

From equations (7.100) and (7.102), we have

$$\widehat{V}|\text{init}\rangle = -\frac{1}{\sqrt{8}}|000, p_1\rangle + \frac{\sqrt{7}}{\sqrt{8}}|\overline{000, p}\rangle \qquad (7.103)$$

Equation (7.103) is shown graphically in Figure 7.12(b), where only the state containing the phone number has been inverted (has a negative sign). The next step in the algorithm is to evaluate $\widehat{W}\widehat{V}|\text{init}, p\rangle$. We first note that $\widehat{W} = 2|\text{init}\rangle\langle\text{init}| - \widehat{I}$ can be expanded as

$$\widehat{W} = \frac{1}{4}|000, p_1\rangle\langle 000, p_1| + \frac{\sqrt{7}}{4}|\overline{000, p}\rangle\langle 000, p_1|$$

$$+ \frac{\sqrt{7}}{4}|000, p_1\rangle\langle\overline{000, p}| + \frac{7}{4}|\overline{000, p}\rangle\langle\overline{000, p}| - \widehat{I} \quad (7.104)$$

Using equations (7.100), (7.102), and (7.104), we have

$$\widehat{W}\widehat{V}|000, p_1\rangle = -\frac{1}{4}|000, p_1\rangle - \frac{\sqrt{7}}{4}|\overline{000, p}\rangle + |000, p_1\rangle$$

$$\widehat{W}\widehat{V}|000, p_1\rangle = \frac{3}{4}|000, p_1\rangle - \frac{\sqrt{7}}{4}|\overline{000, p}\rangle \quad \text{and} \quad (7.105)$$

$$\widehat{W}\widehat{V}|\overline{000, p}\rangle = \frac{\sqrt{7}}{4}|000, p_1\rangle + \frac{7}{4}|\overline{000, p}\rangle - |\overline{000, p}\rangle$$

$$\widehat{W}\widehat{V}|\overline{000, p}\rangle = \frac{\sqrt{7}}{4}|000, p_1\rangle + \frac{3}{4}|\overline{000, p}\rangle \quad (7.106)$$

Using equations (7.105) and (7.106), we have

$$\widehat{W}\widehat{V}|\text{init}, p\rangle = [2|\text{init}\rangle\langle\text{init}| - \widehat{I}]\widehat{V}|\text{init}, p\rangle$$

$$= \frac{5}{4\sqrt{2}}|000, p_1\rangle + \frac{\sqrt{7}}{4\sqrt{2}}|\overline{000, p}\rangle \quad (7.107)$$

We see from equation (7.107) that after a single operation on $|\text{init}, p\rangle$ by $\widehat{W}\widehat{V}$, the probability of getting the correct phone number ($|p_1\rangle$) and name ($000\rangle$) is $\left|\frac{5}{4\sqrt{2}}\right|^2 = 0.781$. The probability of getting any other state (name) is $\left|\frac{1}{4\sqrt{2}}\right|^2 = 0.031$. So, the probability of getting the wrong phone number and name is $7 \times \left|\frac{1}{4\sqrt{2}}\right|^2 = 0.219$. Equation (7.107), shown graphically in Figure 7.12(c), demonstrates the amplification of the state with the correct phone number.

The next step is to find $(\widehat{W}\widehat{V})^2|\text{init}, p\rangle$. Applying $\widehat{W}\widehat{V}$ to equation (7.107),

$$(\widehat{W}\widehat{V})^2|\text{init}, p\rangle = \frac{5}{4\sqrt{2}}\widehat{W}\widehat{V}|000, p_1\rangle + \frac{\sqrt{7}}{4\sqrt{2}}\widehat{W}\widehat{V}|\overline{000, p}\rangle \quad (7.108)$$

Using equations (7.105) and (7.106) in the previous equation gives

$$(\widehat{W}\widehat{V})^2|\text{init}, p\rangle = \frac{5}{4\sqrt{2}}\left(\frac{3}{4}|000, p_1\rangle - \frac{\sqrt{7}}{4}|\overline{000}, p\rangle\right)$$

$$+ \frac{\sqrt{7}}{4\sqrt{2}}\left(\frac{\sqrt{7}}{4}|000, p_1\rangle + \frac{3}{4}|\overline{000}, p\rangle\right) \quad (7.109)$$

$$(\widehat{W}\widehat{V})^2|\text{init}, p\rangle = \frac{11}{8\sqrt{2}}|000, p_1\rangle - \frac{\sqrt{7}}{8\sqrt{2}}|\overline{000}, p\rangle \quad (7.110)$$

If we make a measurement now, the probability of getting the correct phone number ($|p_1\rangle$) and name ($|000\rangle$) is $\left|\frac{11}{8\sqrt{2}}\right|^2 = 0.945$. Equation (7.110), shown graphically in Figure 7.12(e), shows further amplification of the state with the correct phone number.

Let us apply $\widehat{W}\widehat{V}$ for the third time to see whether the probability to find the correct answer increases or decreases. Applying $\widehat{W}\widehat{V}$ to equation (7.110),

$$(\widehat{W}\widehat{V})^3|\text{init}, p\rangle = \frac{11}{8\sqrt{2}}\widehat{W}\widehat{V}|000, p_1\rangle - \frac{\sqrt{7}}{8\sqrt{2}}\widehat{W}\widehat{V}|\overline{000}, p\rangle \quad (7.111)$$

Using equations (7.105) and (7.106) in the previous equation gives

$$(\widehat{W}\widehat{V})^3|\text{init}, p\rangle = \frac{11}{8\sqrt{2}}\left(\frac{3}{4}|000, p_1\rangle - \frac{\sqrt{7}}{4}|\overline{000}, p\rangle\right)$$

$$+ \frac{\sqrt{7}}{8\sqrt{2}}\left(\frac{\sqrt{7}}{4}|000, p_1\rangle + \frac{3}{4}|\overline{000}, p\rangle\right) \quad (7.112)$$

$$(\widehat{W}\widehat{V})^3|\text{init}, p\rangle = \frac{40}{32\sqrt{2}}|000, p_1\rangle - \frac{8\sqrt{7}}{32\sqrt{2}}|\overline{000}, p\rangle$$

$$= \frac{5}{4\sqrt{2}}|000, p_1\rangle - \frac{\sqrt{7}}{4\sqrt{2}}|\overline{000}, p\rangle \quad (7.113)$$

After three rotations by $\widehat{W}\widehat{V}$, the probability of getting the correct phone number ($|p_1\rangle$) and name ($|000\rangle$) decreases to $\left|\frac{5}{4\sqrt{2}}\right|^2 = 0.781$. So, the optimum number of rotations for the search problem with $N = 8$ is two. Equation (7.113) is shown graphically in Figure 7.12(e).

The sequence of basic gates to realize the $\widehat{V}$ and $\widehat{W}$ operators are given in Chapter 8.

7.6 Public Key Distribution

One of the promising applications of quantum computing is harvesting a quantum state's properties to distribute the encryption key. The impossibility of copying a quantum state, and irreversibility of the measurement process make the quantum key distribution (QKD) algorithms resilient to eavesdropping and attack. We assume that two parties A and B (Alice and Bob), intend to send a secret message through a classical channel. Alice can scramble her bit sequence, for example, by XOR-ing the sequence with a long pseudo-random bit sequence called the Key. Before communication, Alice and Bob "somehow" share this long Key. Once Bob receives the scrambled message, he can descramble it by XOR-ing the received bits with the Key to retrieve the message. Now the third party, Eve (eavesdropper), can copy Alice's message and try many different pseudo-random bit sequences and XOR gates to retrieve the secret message. The more Eve listens to the channel and copies the encrypted message, she will finally decipher the Key through statistical methods. Even if Alice and Bob find out that they have been hacked, there is no way to change the Key because they are physically far away from each other, and the channel is not safe anymore to transfer the Key.

Different classical encryption algorithms exist to make things harder for Eve and make it easier and safer for Alice and Bob to exchange the key and guarantee the channel's security. Quantum mechanics provides a protocol for Alice and Bob to share a secret key. This method does not let Eve copy the transmitted message from the channel, and most importantly, if Eve measures the states, both Alice and Bob will notice it. There are different quantum key distribution algorithms (QKD) like BB84, B92, and E91, which are named after the inventors and the year of their introduction. Here we explain the simplified version of Bennet and Brassard in 1984 (BB84).

Alice and Bob can encode their qubit states as the polarization of photons. The photons are transferred by a fiber-optic cable or a laser link between two satellites. Linear (up and down) polarizations are states $\{|0\rangle, |1\rangle\}$; diagonal polarization can be encoded as $\{|+\rangle, |-\rangle\}$. However, at the beginning of the communication, the receiver (Bob) has no idea which polarizer Alice will use to send her photons (up/down or diagonal).

(1) There are two sets of bases $\{|0\rangle, |1\rangle\}$ and $\{|+\rangle, |-\rangle\}$. Alice generates N bits using one of these bases. We recall that

$$|+\rangle \equiv \frac{1}{\sqrt{2}}(|0\rangle + |1\rangle)$$

$$|-\rangle \equiv \frac{1}{\sqrt{2}}(|0\rangle - |1\rangle)$$

From the above equations,

$$|0\rangle = \frac{1}{\sqrt{2}}(|+\rangle + |-\rangle) \tag{7.114}$$

$$|1\rangle = \frac{1}{\sqrt{2}}(|+\rangle - |-\rangle) \tag{7.115}$$

Recall that measuring states $|+\rangle$ and $|-\rangle$ using the $\{|0\rangle, |1\rangle\}$ basis yields $|0\rangle$ and $|1\rangle$ each with 50% probability. For example, suppose you receive a diagonally polarized photon but use a vertically polarized film to measure the polarization. In that case, you will measure the photon with up or down polarization, each with a 50% probability. Similarly, measuring the states $|0\rangle$ and $|1\rangle$ using the $\{|+\rangle, |-\rangle\}$ basis yields $|+\rangle$ and $|-\rangle$ each with 50% probability. In these cases, you also have a 50% probability for measuring each of the diagonally polarized photons.

(2) Alice transmits a string of N qubits to Bob.

(3) Bob does not know the basis used by Alice to generate the qubits. He randomly chooses the bases to make the measurement and then measures the qubit state.

(4) Alice then announces all N values of the basis state she used publicly.

(5) On an average, Bob would have used the same basis as Alice in approximately $N/2$ bits. Bob discards those bits where he used a basis different from Alice and announced which $N/2$ qubits are discarded. Out of the remaining $N/2$ measurements, where Bob used the same bases as Alice, he publicly announces the M values of the states he measured (we will call this State$_{\text{Bob}}$). Alice compares to see if all M values announced by Bob match the qubit state generated by her (we call these State$_{\text{Alice}}$). That is, she checks if State$_{\text{Bob}}$ = State$_{\text{Alice}}$ for all M results announced by Bob.

If Alice and Bob agree on all M qubit values, it is safe to assume that Eve did not eavesdrop. These M publicly announced states are discarded because everybody knows these qubit values. The remaining $(N/2 - M)$ values of the qubits form the key.

(6) If Eve eavesdropped, she would intercept the qubits transmitted by Alice, make measurements, and then retransmit them to Bob. This clearly will lead to Bob receiving states that Alice did not generate. Note that Eve measures before Bob and Alice make their public announcements. So, Eve has to choose the basis for her measurements randomly. For example, let us suppose that Alice had prepared a state in the $\{|+\rangle, |-\rangle\}$ basis. If Eve measured in the $\{|+\rangle, |-\rangle\}$ bases, Bob would measure the correct state generated by Alice, and the eavesdropping will go undetected. The probability of this occurring is $1/2$ (probability of Eve using the $\{|+\rangle, |-\rangle\}$ basis). If Eve had instead measured in the $\{|0\rangle, |1\rangle\}$ basis, then the probability that Bob measured the correct state is 50%. This is the case because

$$|0\rangle \equiv \frac{1}{\sqrt{2}}(|+\rangle + \ -\rangle)$$

$\rightarrow$ probability that Bob got the correct answer is 50%

$$|1\rangle \equiv \frac{1}{\sqrt{2}}(|+\rangle - |-\rangle)$$

$\rightarrow$ probability that Bob got the correct answer is 50%

The probability that Eve intercepted but Bob and Alice still agreed with their result and did not detect the eavesdropping is $75\% = 50\% + 25\%$.

The probability of 50% corresponds to event #1 when both Alice and Bob have the same basis for measuring their qubits. Eve has a 50% chance to use the same basis for the measurement and send the result to Bob, in which case they all agree on the value of the qubit.

The probability of 25% corresponds to the less probable event # 2. This is when Alice sends her qubit in one basis, e.g., $\{|+\rangle, |-\rangle\}$, and Eve uses (with 50% probability) the wrong basis $\{|0\rangle, |1\rangle\}$ to measure what Alice has sent. This will collapse the state into her basis $\{|0\rangle, |1\rangle\}$. Then, Bob ends up measuring something in agreement with Alice only 50% of the time in event # 2.

So, on average, the probability that Bob will measure the correct state after an eavesdropping event (either event 1 or event 2) is $0.5 + 0.25 = \left(\frac{3}{4}\right)$. Now, if Bob and Alice compare their results for all M values of the qubit, the probability that they would agree 100% of the time (i.e., being lured by Eve) is minuscule because this chance is $\left[\frac{3}{4} \times \frac{3}{4} \times \cdots \times \frac{3}{4} = \left(\frac{3}{4}\right)^M\right]$. If $M = 50$, the probability that they agree with all 50 values is $\left(\frac{3}{4}\right)^{50} \sim 5.7 \times 10^{-5}$. So, if Bob's value of the qubit does not agree with Alice's for the M cases, they discard the key and regenerate a new key because there is a good chance that Eve has eavesdropped.

Note that this protocol does not use entanglement but uses only the collapse of the wave function and the inability to make copies of unknown quantum mechanical states to ensure security. The E91 algorithm is a QKD method based on entangled Bell states invented by Artur Ekert in 1991 (Ekert, 1991). For an interesting description of encryption (from classical to quantum), the reader is referred to the popular book (Singh, 2000).

7.7 Problems

Section 7.1

(1) **Sample set A:** You are given a million copies of a quantum state $\frac{1}{\sqrt{2}}(|0\rangle - |1\rangle)$.
Sample set B: You are given half a million copies each of states $|0\rangle$ and $|1\rangle$.

Assume that there are no defective states. Can you differentiate between the two sample sets? If yes, suggest an experiment that will help you to do so. Be detailed in your answer. [You are free to design your own crazy experiment.]

Sections 7.1 and 7.3

(2) Consider the state $|\chi\rangle = \frac{1}{\sqrt{3}}\{|00\rangle + |10\rangle + |01\rangle\}$. If a measurement of both qubits is made in the $\{|+\rangle, |-\rangle\}$ states, what is the probability of finding the $|+\rangle$ and $|-\rangle$ states for each qubit?

(3) Consider the state $|\chi\rangle = \frac{1}{\sqrt{2}}\{|000\rangle - |111\rangle\}$. What states will you find when you measure $|\chi\rangle$ in a perfect "error-free" world and what are the corresponding probabilities? The bases for the measurement is $\{|000\rangle, |001\rangle, |010\rangle, |100\rangle, |011\rangle, |101\rangle, |110\rangle, |111\rangle\}$.

Section 7.5

(4) Using $\widehat{W} = 2|\text{init}\rangle\langle\text{init}| - \widehat{I}$, show that $\widehat{W}|\alpha_p\rangle = \frac{2\sqrt{N-1}}{N}|\bar{\alpha}_p\rangle + \frac{2-N}{N}|\alpha_p\rangle$ and $\widehat{W}\,\bar{\alpha}_p\rangle = -\frac{2-N}{N}|\bar{\alpha}_p\rangle + \frac{2\sqrt{N-1}}{N}|\alpha_p\rangle$.

(5) Calculate $\exp(-i\theta\widehat{Y})$, where $\widehat{Y} = \begin{pmatrix} 0 & -i \\ i & 0 \end{pmatrix}$. Use the expansion of $\exp(x)$ as a series and then add the terms in the summation. You will find a cute answer.

(6) Calculate $(\widehat{W}\widehat{V})^2|\bar{\alpha}_p\rangle$. If $2\theta = \arctan\left(\frac{2\sqrt{N-1}}{N-2}\right)$ determine the angle vector $|\bar{\alpha}_p\rangle$ rotates by.

(7) Grover's algorithm demonstrates a significant speedup for the search problem. Identify the critical step(s) in the algorithm that

requires a quantum computer. Why can these step(s) not be carried out on a digital computer with the same efficiency?

References

Ekert, A. (1991). Quantum cryptography based on Bell's theorem. *Physical Review Letters*, **67**, 661.

McMahon, D. (2007). *Quantum Computing Explained*. Wiley-IEEE Computer Society.

Nielsen, M. and Chuang, I. L. (2000). *Quantum Computation and Quantum Information*. Cambridge University Press, Cambridge.

Singh, S. (2000). *The Code Book: Science of Secrecy from Ancient Egypt to Quantum Cryptography*. Anchor Publishing.

Chapter 8

QUANTUM GATES

Contents

8.1 Introduction

An overview of the need to break up a complex quantum computation in terms of simpler one and two qubit gates was provided in Section 7.4. While we mentioned the names of some important quantum gates, their mathematical operation was not defined in Chapter 7. In this chapter, we will discuss quantum gates in greater detail. We will also discuss gates that perform the V and W unitary transformations used in Grover's algorithm. Finally, we discuss quantum teleportation, which uses some of the gates defined here.

Before discussing the gates, we discuss the concept of a Bloch sphere, which provides a useful pictorial representation of a single qubit in an otherwise mathematical topic.

8.2   Visualization of a Single Qubit on a Sphere

We discussed in Section 7.1 that a **qubit** can be in an arbitrary superposition of 0 and 1 logic states. This corresponds to the physical system being in a linear combination of $|0\rangle$ and $|1\rangle$ states,

$$|\psi\rangle = \alpha|0\rangle + \beta|1\rangle \tag{8.1}$$

where $|\alpha|^2 + |\beta|^2 = 1$ due to normalization. As α and β are complex numbers, their representation in terms of amplitude and phase factor are

$$\alpha = e^{i\phi_1}|\alpha| \quad \text{and} \quad \beta = e^{i\phi_2}|\beta|$$

and $|\psi\rangle$ can be rewritten as

$$|\psi\rangle = e^{i\phi_1}(|\alpha||0\rangle + e^{i\phi}|\beta||1\rangle) \tag{8.2}$$

where $\phi = \phi_2 - \phi_1$. Within an overall phase factor $e^{i\phi_1}$, the arbitrary state of a qubit is

$$|\alpha||0\rangle + e^{i\phi}|\beta||1\rangle$$

where,

$$0 \le |\alpha| \le 1, \quad 0 \le |\beta| \le 1, \quad \text{and} \quad 0 \le \phi \le 2\pi \tag{8.3}$$

Because the wave function is normalized, $|\beta| = \sqrt{1 - |\alpha|^2}$. For a fixed value of $|\alpha|$ and $|\beta|$. each value of ϕ from 0 to 2π results in a different state of the qubit. Similarly, for a given value of ϕ, the values of $|\alpha|$ and $|\beta|$ can also be varied to result in different qubit states. The normalized qubit state can be represented on the surface of a unit sphere by

$$|\alpha| = \cos\left(\frac{\theta}{2}\right) \quad \text{and} \quad |\beta| = \sin\left(\frac{\theta}{2}\right), \quad \text{where } 0 \leq \theta \leq \pi \qquad (8.4)$$

The reader should convince themself of the above statement. Substituting equation (8.4) in equation (8.1), the arbitrary state of a qubit is

$$|\psi\rangle = \cos\left(\frac{\theta}{2}\right)|0\rangle + e^{i\phi}\sin\left(\frac{\theta}{2}\right)|1\rangle \quad \text{(up to an overall phase factor)} \qquad (8.5)$$

where $0 \leq \theta \leq \pi$ and $0 \leq \phi \leq 2\pi$. Values of θ and ϕ outside this range correspond to points on the Bloch sphere that map into points arising from θ and ϕ inside this range. An overall phase factor that multiplies equation (8.5) cannot be measured.

Equation (8.5) stands for a general qubit state $|\psi\rangle$, which has a unit norm and is represented by a point on the Bloch sphere (see Figure 8.1). Some specific points on the Bloch sphere are:

- $+z$ axis $(\theta = 0)$ corresponds to $|0\rangle$
- $+x$ axis $\left(\theta = \frac{\pi}{2}, \phi = 0\right)$ corresponds to $\frac{|0\rangle + |1\rangle}{\sqrt{2}}$
- $+y$ axis $\left(\theta = \frac{\pi}{2}, \phi = \frac{\pi}{2}\right)$ corresponds to $\frac{|0\rangle + i|1\rangle}{\sqrt{2}}$
- $-z$ axis $(\theta = \pi)$ corresponds to $|1\rangle$
- $-x$ axis $\left(\theta = \frac{\pi}{2}, \phi = \pi\right)$ corresponds to $\frac{|0\rangle - |1\rangle}{\sqrt{2}}$
- $-y$ axis $\left(\theta = \frac{\pi}{2}, \phi = -\frac{\pi}{2}\right)$ corresponds to $\frac{|0\rangle - i|1\rangle}{\sqrt{2}}$

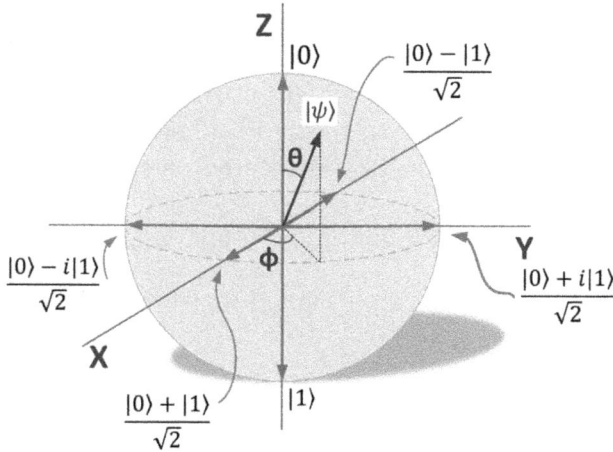

Figure 8.1. Points on the Bloch sphere represent the set of all allowed single-qubit states. Each state has a unique set of angles θ and ϕ. Remember that even though $|0\rangle$ and $|1\rangle$ are orthogonal vectors in Hilbert space, they are diametrically opposite in the Bloch sphere. In general, any two vectors on the Bloch sphere that are diametrically opposite to each are orthogonal to each other.

8.3   Single Qubit Gates

In this section, we will discuss common gates with a single input and output qubit. They are the Hadamard (H), X, Y, Z, S, and T gates. The operation of these single qubit gates can also be visualized on the Bloch sphere as discussed in Figure 8.2.

8.3.1   *Hadamard Gate ($\hat{H}$)*

The **Hadamard gate** ($\hat{H}$) operates on a single qubit to perform the following rotation:

$$\hat{H}|0\rangle = \frac{1}{\sqrt{2}}(|0\rangle + |1\rangle) = |+\rangle \tag{8.6}$$

$$\hat{H}|1\rangle = \frac{1}{\sqrt{2}}(|0\rangle - |1\rangle) = |-\rangle \tag{8.7}$$

Input	Gate	Output		Input	Gate	Output
$\lvert 0\rangle$ / $\lvert 1\rangle$	X	$\lvert 1\rangle$ / $\lvert 0\rangle$		$\lvert 0\rangle$ / $\lvert 1\rangle$	H	$\lvert +\rangle = \frac{1}{\sqrt{2}}(\lvert 0\rangle + \lvert 1\rangle)$ / $\lvert -\rangle = \frac{1}{\sqrt{2}}(\lvert 0\rangle - \lvert 1\rangle)$
$\lvert 0\rangle$ / $\lvert 1\rangle$	Y	$e^{i\pi/2}\lvert 1\rangle$ / $e^{-i\pi/2}\lvert 0\rangle$		$\lvert 0\rangle$ / $\lvert 1\rangle$	S	$\lvert 0\rangle$ / $i\lvert 1\rangle = e^{i\pi/2}\lvert 1\rangle$
$\lvert 0\rangle$ / $\lvert 1\rangle$	Z	$\lvert 0\rangle$ / $-\lvert 1\rangle = e^{i\pi}\lvert 1\rangle$		$\lvert 0\rangle$ / $\lvert 1\rangle$	T	$\lvert 0\rangle$ / $e^{i\pi/4}\lvert 1\rangle$
				$\lvert 0\rangle$ / $\lvert 1\rangle$	T†	$\lvert 0\rangle$ / $e^{-i\pi/4}\lvert 1\rangle$

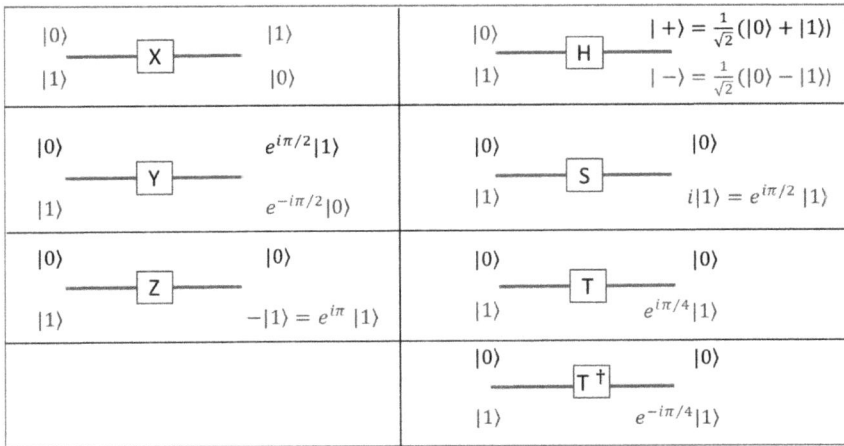

Figure 8.2. Operation of X, Y, Z, Hadamard (H), phase (S), and T gates. States on the left are the input to the gates and the states on the right are the output. Time increases from left to right in the sub figures.

The resulting orthogonal states $\{\lvert +\rangle, \lvert -\rangle\}$, called the *diagonal* or *conjugate* bases, are often used in quantum computing discussions,

$$\lvert +\rangle \equiv \frac{1}{\sqrt{2}}(\lvert 0\rangle + \lvert 1\rangle) \tag{8.8}$$

$$\lvert -\rangle \equiv \frac{1}{\sqrt{2}}(\lvert 0\rangle - \lvert 1\rangle) \tag{8.9}$$

In matrix form, the Hadamard gate is

$$\hat{H} = \frac{1}{\sqrt{2}}\begin{pmatrix} 1 & 1 \\ 1 & -1 \end{pmatrix} \tag{8.10}$$

Since $\lvert 0\rangle = \begin{bmatrix} 1 \\ 0 \end{bmatrix}$ and $\lvert 1\rangle = \begin{bmatrix} 0 \\ 1 \end{bmatrix}$, in matrix form, equations (8.6) and (8.7) are

$$\frac{1}{\sqrt{2}}\begin{pmatrix} 1 & 1 \\ 1 & -1 \end{pmatrix}\begin{bmatrix} 1 \\ 0 \end{bmatrix} = \frac{1}{\sqrt{2}}\begin{bmatrix} 1 \\ 1 \end{bmatrix} = \lvert +\rangle \quad \text{and}$$

$$\frac{1}{\sqrt{2}}\begin{pmatrix} 1 & 1 \\ 1 & -1 \end{pmatrix}\begin{bmatrix} 0 \\ 1 \end{bmatrix} = \frac{1}{\sqrt{2}}\begin{bmatrix} 1 \\ -1 \end{bmatrix} = \lvert -\rangle$$

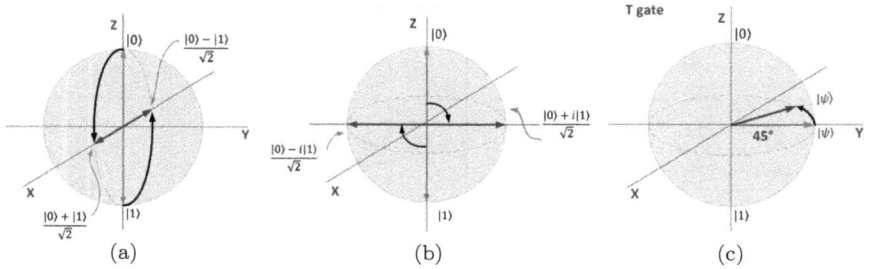

Figure 8.3. (a) Applying the Hadamard gate to state $|1\rangle$ is equivalent to rotating it by 90 degrees around the y-axis to bring it along the $-x$-axis while rotating $|0\rangle$ to be aligned along the $+x$-axis. (b) Applying the phase gate after the Hadamard gate moves $|0\rangle$ to be along $+y$-axis and $|1\rangle$ to be along the $-y$-axis. (c) The action of T-gate on a general qubit $|\psi\rangle$ leads to 45 degrees rotation around the z-axis. This figure shows a particular case of $|\psi\rangle$ along the y axis.

Note the fact that when the Hadamard gate operates on $|0\rangle$ (or $|1\rangle$), it rotates the qubit around the y-axis by 90 degrees to create an equal superposition of $|0\rangle$ and $|1\rangle$. See Figure 8.3(a).

8.3.2   *T gate ($\pi/8$ gate)*

The **T gate** ($\hat{T}$) (also called $\pi/8$ gate) leaves the state $|0\rangle$ intact but multiplies $|1\rangle$ by an extra phase factor.

$$\hat{T}|0\rangle = |0\rangle$$
$$\hat{T}|1\rangle = e^{i\pi/4}|1\rangle$$

That is, $\hat{T}$ acting on an arbitrary qubit rotates the angle ϕ in Figure 8.3(c) by an angle of $\frac{\pi}{4}$ about the z-axis,

$$\hat{T}[\alpha|0\rangle + e^{i\phi}\beta|1\rangle] = \alpha|0\rangle + e^{i(\phi+\frac{\pi}{4})}\beta|1\rangle$$

In matrix form, the $\hat{T}$ gate is

$$\hat{T} = \begin{pmatrix} 1 & 0 \\ 0 & e^{i\pi/4} \end{pmatrix} \tag{8.11}$$

Given that both the real and imaginary parts of $e^{i\pi/4}$ are the irrational number $\frac{1}{\sqrt{2}}$, any arbitrary single qubit superposition is realizable by combining the Hadamard operator and the T gate. We can create any rotation in the Bloch sphere from $|0\rangle$ or $|1\rangle$ or any other single qubit state. While we do not discuss the proof here, the reader should try to convince themselves of the validity of the above claim.

8.3.3 *Phase gate* $(\hat{S})$

The **phase gate** $(\hat{S})$ performs the following operation:

$$\hat{S}|0\rangle = |0\rangle$$
$$\hat{S}|1\rangle = i|1\rangle$$

That is, $\hat{S}$ acting on an arbitrary qubit rotates the angle ϕ in the Bloch sphere of Figure 8.1 by $\frac{\pi}{2}$ (rotation about the z-axis by an angle of $\frac{\pi}{2}$),

$$\hat{S}\left[\cos\left(\frac{\theta}{2}\right)|0\rangle + e^{i\phi}\sin\left(\frac{\theta}{2}\right)|1\rangle\right]$$

$$= \cos\left(\frac{\theta}{2}\right)|0\rangle + e^{i\left(\phi+\frac{\pi}{2}\right)}\sin\left(\frac{\theta}{2}\right)|1\rangle$$

$$\hat{S}[\alpha|0\rangle + e^{i\phi}\beta|1\rangle] = \alpha|0\rangle + e^{i\left(\phi+\frac{\pi}{2}\right)}\beta|1\rangle$$

See Figure 8.4. In matrix form, the phase gate performs the following rotation around the z-axis:

$$\hat{S} = \begin{pmatrix} 1 & 0 \\ 0 & i \end{pmatrix}$$

From the above definition of the phase and T gates, it is easy to verify that the phase gate is equivalent to operating sequentially with the

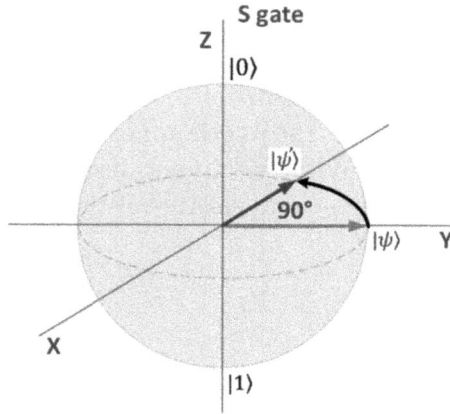

Figure 8.4. Action of S gate on the qubit is equivalent to applying T gate twice. I.e., it result in a 90-degrees rotation around the z-axis.

T gate twice:

$$\hat{S} = \hat{T}^2 \quad \text{or} \quad \hat{S} = \begin{pmatrix} 1 & 0 \\ 0 & i \end{pmatrix} = \begin{pmatrix} 1 & 0 \\ 0 & e^{i\pi/4} \end{pmatrix} \begin{pmatrix} 1 & 0 \\ 0 & e^{i\pi/4} \end{pmatrix}$$

8.3.4 $\hat{X}, \hat{Y},$ *and* $\hat{Z}$ *gates/operators*

The **X**, **Y**, and **Z gates** are also common in quantum computing. In matrix form, they are:

$$\hat{X} = \begin{pmatrix} 0 & 1 \\ 1 & 0 \end{pmatrix}, \quad \hat{Y} = \begin{pmatrix} 0 & -i \\ i & 0 \end{pmatrix}, \quad \text{and} \quad \hat{Z} = \begin{pmatrix} 1 & 0 \\ 0 & -1 \end{pmatrix} \quad (8.12)$$

Note that these gates only operate on a single qubit, independent of of other qubits. We can see from equations (8.12) that

$$\hat{X}|0\rangle = |1\rangle \quad \text{and} \quad \hat{X}|1\rangle = |0\rangle \quad (8.13)$$

In matrix form, the above equation reads

$$\begin{pmatrix} 0 & 1 \\ 1 & 0 \end{pmatrix} \begin{bmatrix} 1 \\ 0 \end{bmatrix} = \begin{bmatrix} 0 \\ 1 \end{bmatrix} \quad \text{and} \quad \begin{pmatrix} 0 & 1 \\ 1 & 0 \end{pmatrix} \begin{bmatrix} 0 \\ 1 \end{bmatrix} = \begin{bmatrix} 1 \\ 0 \end{bmatrix}$$

The X gate *flips* the qubit state $|0\rangle$ to $|1\rangle$ and vice versa (like a classical NOT gate). This is equivalent to rotating a single qubit around x-axis by 180 degrees on the Bloch sphere. That is, $\hat{X}$ acting on an arbitrary qubit causes a rotation $\phi \to 2\pi - \phi$ and $\theta \to \pi - \theta$ in the Bloch sphere of Figure 8.1,

$$\hat{X}\left[\cos\left(\frac{\theta}{2}\right)|0\rangle + e^{i\phi}\sin\left(\frac{\theta}{2}\right)|1\rangle\right]$$

$$= \cos\left(\frac{\pi-\theta}{2}\right)|0\rangle + e^{i(2\pi-\phi)}\sin\left(\frac{\pi-\theta}{2}\right)|1\rangle$$

The Y gate performs the following rotation:

$$\hat{Y}|0\rangle = i|1\rangle = e^{i\pi/2}|1\rangle \quad \text{and} \quad \hat{Y}|1\rangle = -i|0\rangle = e^{-i\pi/2}|0\rangle \quad (8.14)$$

which in matrix form is

$$\begin{pmatrix} 0 & -i \\ i & 0 \end{pmatrix}\begin{bmatrix} 1 \\ 0 \end{bmatrix} = i\begin{bmatrix} 0 \\ 1 \end{bmatrix} \quad \text{and} \quad \begin{pmatrix} 0 & -i \\ i & 0 \end{pmatrix}\begin{bmatrix} 0 \\ 1 \end{bmatrix} = -i\begin{bmatrix} 1 \\ 0 \end{bmatrix}$$

The Y gate rotates state $|0\rangle$ to $e^{i\pi/2}|1\rangle$ and rotates state $|1\rangle$ to $e^{-i\pi/2}|0\rangle$. This is equivalent to rotating a single qubit around the y axis by 180 degrees on the Bloch sphere or $\phi \to \pi - \phi$ and $\theta \to \pi - \theta$ in the Bloch sphere of Figure 8.1,

$$\hat{Y}\left[\cos\left(\frac{\theta}{2}\right)|0\rangle + e^{i\phi}\sin\left(\frac{\theta}{2}\right)|1\rangle\right]$$

$$= \cos\left(\frac{\pi-\theta}{2}\right)|0\rangle + e^{i(\pi-\phi)}\sin\left(\frac{\pi-\theta}{2}\right)|1\rangle$$

Finally, the Z gate, works in a similar way:

$$\hat{Z}|0\rangle = |0\rangle \quad \text{and} \quad \hat{Z}|1\rangle = -|1\rangle = e^{-i\pi}|1\rangle \quad (8.15)$$

or

$$\begin{pmatrix} 1 & 0 \\ 0 & -1 \end{pmatrix}\begin{bmatrix} 1 \\ 0 \end{bmatrix} = \begin{bmatrix} 1 \\ 0 \end{bmatrix} \quad \text{and} \quad \begin{pmatrix} 1 & 0 \\ 0 & -1 \end{pmatrix}\begin{bmatrix} 0 \\ 1 \end{bmatrix} = -\begin{bmatrix} 0 \\ 1 \end{bmatrix}$$

The Z gate leaves state $|0\rangle$ as is but shifts the phase of $|1\rangle$ by π radian S (that is, it rotates $|1\rangle$ to $-|1\rangle$). This is equivalent to rotating the single qubit around z-axis by 180 degrees on the Bloch sphere ($\phi \to \pi + \phi$ and θ):

$$\hat{Z}\left[\cos\left(\frac{\theta}{2}\right)|0\rangle + e^{i\phi}\sin\left(\frac{\theta}{2}\right)|1\rangle\right]$$

$$= \cos\left(\frac{\theta}{2}\right)|0\rangle + e^{i(\pi+\phi)}\sin\left(\frac{\theta}{2}\right)|1\rangle$$

Figure 8.5 visualizes the operations of X, Y, and Z gates on the Bloch sphere.

8.3.5   *Notation for Operations on a Multi-Qubit State*

Let us review the mathematical manipulation of some two-qubit states using the matrix format. Consider a two-qubit operator where the X gate only operates on the first qubit (qubit-a) and the Z gate operates only on the second qubit (qubit-b). As two qubits span a vector space of dimension equal to $2 \times 2 = 4$, the matrix representation of the gates is also four dimensional. Consider the

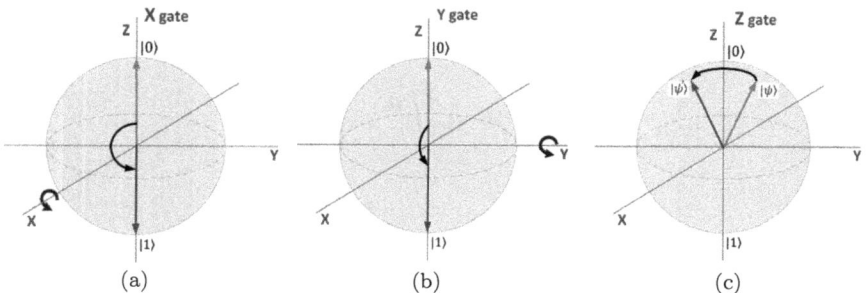

Figure 8.5. (a) Action of X gate on a qubit is equivalent to rotation around x-axis by 180 degrees. (b) Action of Y gate is equivalent to a rotation of qubit by 180 degrees around y–axis. (c) Z-gate rotates the qubit around z-axis by 180 degrees.

tensor product of X and Z gates,

$$\hat{Q} = \hat{X} \otimes \hat{Z} = \begin{bmatrix} 0 & 1 \\ 1 & 0 \end{bmatrix} \otimes \hat{Z} = \begin{bmatrix} 0 \cdot \hat{Z} & \hat{Z} \\ \hat{Z} & 0 \cdot \hat{Z} \end{bmatrix} = \begin{bmatrix} 0 & 0 & 1 & 0 \\ 0 & 0 & 0 & -1 \\ 1 & 0 & 0 & 0 \\ 0 & -1 & 0 & 0 \end{bmatrix},$$

which can be written as

$$\hat{Q} = \hat{X} \otimes \hat{Z} = \begin{bmatrix} 0 & 1 \\ 1 & 0 \end{bmatrix} \otimes \begin{bmatrix} 1 & 0 \\ 0 & -1 \end{bmatrix} \tag{8.16}$$

$$\hat{Q} = \begin{bmatrix} 0 \cdot \begin{bmatrix} 1 & 0 \\ 0 & -1 \end{bmatrix} & 1 \cdot \begin{bmatrix} 1 & 0 \\ 0 & -1 \end{bmatrix} \\ 1 \cdot \begin{bmatrix} 1 & 0 \\ 0 & -1 \end{bmatrix} & 0 \cdot \begin{bmatrix} 1 & 0 \\ 0 & -1 \end{bmatrix} \end{bmatrix} = \begin{bmatrix} 0 & 0 & 1 & 0 \\ 0 & 0 & 0 & -1 \\ 1 & 0 & 0 & 0 \\ 0 & -1 & 0 & 0 \end{bmatrix} \tag{8.17}$$

So, applying the two-qubit gate $\hat{Q}$ to $|01\rangle$ means applying $\hat{X}$ to the first qubit ($|0\rangle$) and $\hat{Z}$ to the second qubit ($|1\rangle$). This corresponds to left-multiplying the 4×1 vector representation of $|01\rangle$ by the above 4×4 matrix representing $\hat{Q}$ to obtain

$$\hat{Q}|01\rangle = \hat{X} \otimes \hat{Z}|01\rangle$$

$$= \begin{bmatrix} 0 & 0 & 1 & 0 \\ 0 & 0 & 0 & -1 \\ 1 & 0 & 0 & 0 \\ 0 & -1 & 0 & 0 \end{bmatrix} \begin{bmatrix} 0 \\ 1 \\ 0 \\ 0 \end{bmatrix} = \begin{bmatrix} 0 \\ 0 \\ 0 \\ -1 \end{bmatrix} = -\begin{bmatrix} 0 \\ 0 \\ 0 \\ 1 \end{bmatrix} = -|11\rangle \tag{8.18}$$

In the braket notation, the above equation is

$$\hat{Q}|01\rangle = \hat{X} \otimes \hat{Z}|01\rangle = \hat{X} \otimes \hat{Z}(|0\rangle \otimes |1\rangle)$$
$$= (\hat{X}|0\rangle) \otimes (\hat{Z}|1\rangle) = (|1\rangle) \otimes (-|1\rangle) = -|11\rangle \tag{8.19}$$

In the previous line, we split the operators to work on their corresponding ket vectors and take their tensor product.

Let us now discuss the matrix representation of the three-qubit operator $\hat{Q} = \hat{X} \otimes \hat{X} \otimes \hat{Z}$ and apply it to the state $|010\rangle$. We will also show that this is simply equal to applying each operator on its corresponding single qubit and taking the three resulting vectors' tensor product. We saw the matrix form of $\hat{X} \otimes \hat{Z}$ in equation (8.17); hence for the three-qubit operator $\hat{Q}$,

$$\hat{Q} = \hat{X} \otimes \hat{X} \otimes \hat{Z}$$

$$= \hat{X} \otimes \begin{bmatrix} 0 & 0 & 1 & 0 \\ 0 & 0 & 0 & -1 \\ 1 & 0 & 0 & 0 \\ 0 & -1 & 0 & 0 \end{bmatrix} = \begin{bmatrix} 0 & 1 \\ 1 & 0 \end{bmatrix} \otimes \begin{bmatrix} 0 & 0 & 1 & 0 \\ 0 & 0 & 0 & -1 \\ 1 & 0 & 0 & 0 \\ 0 & -1 & 0 & 0 \end{bmatrix} \tag{8.20}$$

$$\hat{Q} = \begin{bmatrix} 0 \cdot \begin{bmatrix} 0 & 0 & 1 & 0 \\ 0 & 0 & 0 & -1 \\ 1 & 0 & 0 & 0 \\ 0 & -1 & 0 & 0 \end{bmatrix} & 1 \cdot \begin{bmatrix} 0 & 0 & 1 & 0 \\ 0 & 0 & 0 & -1 \\ 1 & 0 & 0 & 0 \\ 0 & -1 & 0 & 0 \end{bmatrix} \\ 1 \cdot \begin{bmatrix} 0 & 0 & 1 & 0 \\ 0 & 0 & 0 & -1 \\ 1 & 0 & 0 & 0 \\ 0 & -1 & 0 & 0 \end{bmatrix} & 0 \cdot \begin{bmatrix} 0 & 0 & 1 & 0 \\ 0 & 0 & 0 & -1 \\ 1 & 0 & 0 & 0 \\ 0 & -1 & 0 & 0 \end{bmatrix} \end{bmatrix} \tag{8.21}$$

$$\hat{Q} = \begin{bmatrix} 0 & 0 & 0 & 0 & 0 & 0 & 1 & 0 \\ 0 & 0 & 0 & 0 & 0 & 0 & 0 & -1 \\ 0 & 0 & 0 & 0 & 1 & 0 & 0 & 0 \\ 0 & 0 & 0 & 0 & 0 & -1 & 0 & 0 \\ 0 & 0 & 1 & 0 & 0 & 0 & 0 & 0 \\ 0 & 0 & 0 & -1 & 0 & 0 & 0 & 0 \\ 1 & 0 & 0 & 0 & 0 & 0 & 0 & 0 \\ 0 & -1 & 0 & 0 & 0 & 0 & 0 & 0 \end{bmatrix} \tag{8.22}$$

Also, note that if we multiply the matrix $\hat{Q}$ by the matrix representation of $|010\rangle$, we get

$$\hat{Q}|010\rangle = \begin{bmatrix} 0 & 0 & 0 & 0 & 0 & 0 & 1 & 0 \\ 0 & 0 & 0 & 0 & 0 & 0 & 0 & -1 \\ 0 & 0 & 0 & 0 & 1 & 0 & 0 & 0 \\ 0 & 0 & 0 & 0 & 0 & -1 & 0 & 0 \\ 0 & 0 & 1 & 0 & 0 & 0 & 0 & 0 \\ 0 & 0 & 0 & -1 & 0 & 0 & 0 & 0 \\ 1 & 0 & 0 & 0 & 0 & 0 & 0 & 0 \\ 0 & -1 & 0 & 0 & 0 & 0 & 0 & 0 \end{bmatrix} \begin{bmatrix} 0 \\ 0 \\ 1 \\ 0 \\ 0 \\ 0 \\ 0 \\ 0 \end{bmatrix} = \begin{bmatrix} 0 \\ 0 \\ 0 \\ 1 \\ 0 \\ 0 \\ 0 \\ 0 \end{bmatrix} = -|100\rangle \tag{8.23}$$

This example, however, is a lot simpler in the bra-ket notation if we only record multiplying each operator by its corresponding ket and write the results in compact form by removing the Kronecker product signs,

$$\hat{Q}|010\rangle = \hat{X} \otimes \hat{X} \otimes \hat{Z}(|0\rangle \otimes |1\rangle \otimes |0\rangle) = (\hat{X}|0\rangle) \otimes (\hat{X}|1\rangle) \otimes (\hat{Z}|0\rangle)$$
$$= (|1\rangle) \otimes (|0\rangle) \otimes (|0\rangle) = |100\rangle \tag{8.24}$$

Note: If the tensor product sign is missing and there is no caveat regarding the size of the vector space/number of qubits, then $\hat{Q} = \hat{X}\hat{X}\hat{Z}$ is simply a 2×2 matrix, and $\hat{Q}|0\rangle$ means something very different,

$$\hat{Q}|0\rangle = \hat{X}\hat{X}\hat{Z}|0\rangle = \hat{X}\hat{X}|0\rangle = \hat{X}|1\rangle = |0\rangle \tag{8.25}$$

That is, $\hat{Q}$ simply acts on a single $|0\rangle$ qubit.

While the above gates help with a single qubit rotation, conditional operations/rotations that depend on a second qubit's state are central to computation. The two-qubit CX gate defined below is essential to understand conditional operations.

8.4   Gates with Two or More Qubits

This section discusses some of the classic two-qubit gates (CX, CY, CZ, and CH) and applications involving multi-qubit gates. We will discuss cases involving the generation of an entangled state, an equal superposition state, and the gates used in Grover's algorithm.

8.4.1 *Controlled X gate*

The two qubits of a controlled X (CX) gate at the input are the control (A) and target (B) qubits. The CX gate flips the target bit/input B ($0 \rightarrow 1$ and $1 \rightarrow 0$) at the output terminal (D) only if the control qubit (A) is 1. If the control qubit is in state 0, the target qubit is unchanged at the output. The gate and truth table are shown in Figure 8.6. In summary, the CX gate changes 10 to 11 and 11 to 10, where the first qubit is the control, and the second qubit is the target. The CX gate is also called the controlled NOT or CNOT gate.

We will show later that if the control is in a superposition state, the relative phase factors of the control qubit can be altered, a phenomenon known as the *phase kickback*.

In the braket notation, the CX gate performs the following operation on the two-qubit states:

$$\widehat{CX}|00\rangle = |00\rangle$$
$$\widehat{CX}|01\rangle = |01\rangle$$

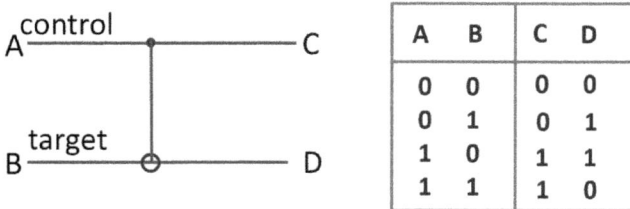

A	B	C	D
0	0	0	0
0	1	0	1
1	0	1	1
1	1	1	0

Figure 8.6. CX gate: If the control A = 1, the target input (B) gets flipped at terminal D. If the control A = 0, then D = B. C = A always (independent of the value of B). The truth table is shown on the right. The solid dot and open circle represent the control and target qubits, respectively.

$$\widehat{CX}|10\rangle = |11\rangle$$
$$\widehat{CX}|11\rangle = |10\rangle$$

It is easy to verify that the following matrix operator gives the CX gate:

$$\widehat{CX} = \begin{pmatrix} 1 & 0 & 0 & 0 \\ 0 & 1 & 0 & 0 \\ 0 & 0 & 0 & 1 \\ 0 & 0 & 1 & 0 \end{pmatrix}$$

The CX gate on the input $|00\rangle = \begin{bmatrix} 1 \\ 0 \\ 0 \\ 0 \end{bmatrix}$ gives an output identical to the input. That is,

$$\begin{pmatrix} 1 & 0 & 0 & 0 \\ 0 & 1 & 0 & 0 \\ 0 & 0 & 0 & 1 \\ 0 & 0 & 1 & 0 \end{pmatrix} \begin{bmatrix} 1 \\ 0 \\ 0 \\ 0 \end{bmatrix} = \begin{bmatrix} 1 \\ 0 \\ 0 \\ 0 \end{bmatrix}$$

Similarly, the CX gate on $|01\rangle = \begin{bmatrix} 0 \\ 1 \\ 0 \\ 0 \end{bmatrix}$ leaves the same output as the input.

Conversely, CX operating on $|10\rangle = \begin{bmatrix} 0 \\ 0 \\ 1 \\ 0 \end{bmatrix}$ gives an output $|11\rangle = \begin{bmatrix} 0 \\ 0 \\ 0 \\ 1 \end{bmatrix}$ as shown in the following:

$$\begin{pmatrix} 1 & 0 & 0 & 0 \\ 0 & 1 & 0 & 0 \\ 0 & 0 & 0 & 1 \\ 0 & 0 & 1 & 0 \end{pmatrix} \begin{bmatrix} 0 \\ 0 \\ 1 \\ 0 \end{bmatrix} = \begin{bmatrix} 0 \\ 0 \\ 0 \\ 1 \end{bmatrix}$$

or CX operating on $|11\rangle = \begin{bmatrix} 0 \\ 0 \\ 0 \\ 1 \end{bmatrix}$ gives $|10\rangle = \begin{bmatrix} 0 \\ 0 \\ 1 \\ 0 \end{bmatrix}$. When the CX gate operates on a superposition, it acts independently on each term of the superposition (like any unitary transformation or linear operator)

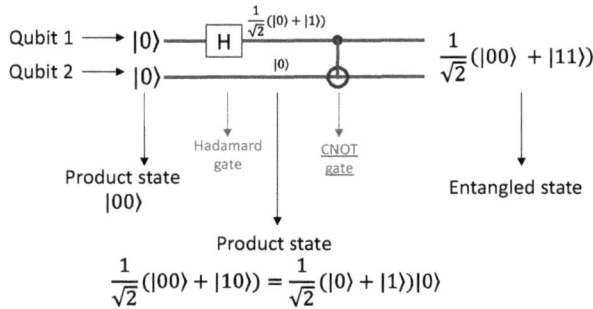

Figure 8.7. The sequential operation of gates to transform a product state $|00\rangle$ to a Bell state (entangled state).

following the truth table in Figure 8.6. This can be mathematically written as

$$\widehat{CX}[\alpha|00\rangle + \beta|01\rangle + \gamma|10\rangle + \delta|11\rangle] = \alpha|00\rangle + \beta|01\rangle + \gamma|11\rangle + \delta|10\rangle$$

where only the last two terms have changed in the superposition (red = control, blue = target).

Example 1. Transforming $|00\rangle$ to an entangled state

Using the gates defined above, we can take $|00\rangle$ to the entangled state $|B_1\rangle = \frac{1}{\sqrt{2}}(|00\rangle + |11\rangle)$ (referred to as a Bell state) through a unitary transformation. In Figure 8.7, the two horizontal lines represent qubits and time progresses from left to right. We start with a product state $|00\rangle$ and apply a Hadamard gate to qubit 1. Then, we apply a CX gate with qubit 1 as the *control bit* and qubit 2 as the *target bit* to create a Bell state. Figure 8.8 shows the creation of a two-qubit entangled state using a CX gate implemented using optical pulses. The underlying qubit is the two-level system introduced in Figures 7.1 and 7.2.

Example 2. Exchanging the control and target qubits of a CX gate

In the CX gate (Figure 8.6), the control is qubit 1, and the target is qubit 2. This example shows how to make the control qubit 2 and the target qubit 1. See Figure 8.9. It is equivalent to applying the Hadamard gate on both qubits, then applying the CX gate, followed

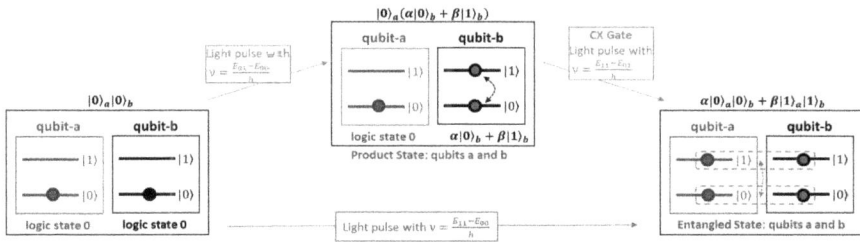

Figure 8.8. The creation of an entangled state from a product state $|00\rangle$. The entangled state $\alpha|00\rangle+\beta|11\rangle$ (right) can be created by applying appropriate gates. These gates can be realized by applying light pulses of appropriate frequencies and timings. The red and green paths above show some possible ways to create the entangled state. The first step of the green path is a single qubit operation where $|0\rangle_b$ is converted to $\alpha|0\rangle_b+\beta|1\rangle_b$. The second step of the green path involves a CX gate.

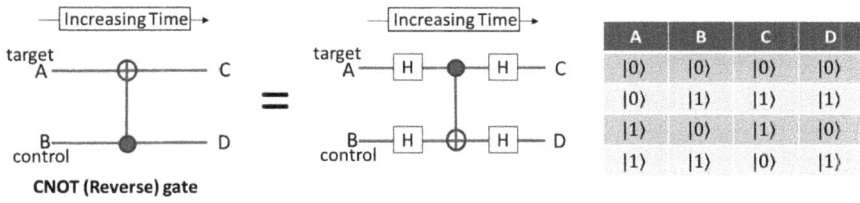

Figure 8.9. Circuit to exchange the control and target qubits of a CX gate using four Hadamard gates. If the control $B = 1$, the target (A) input gets flipped at terminal C. If the control $B = 0$, then $C = A$. $D = B$ regardless of the value of B. The truth table for this is included on the right.

by the re-application of the Hadamard gate to both qubits. All that is needed to show the equivalence of the two gates is to multiply gate (rotation) matrices of each operation from left to right and show that both circuits have the same matrix.

8.4.2 *Swap Gate*

This gate swaps (interchanges) the states of two qubits, i.e., if the input is 01, the output is 10, and vice versa. Note that 00 and 11 are unchanged after swapping. Figure 8.10 shows the truth table and circuit using the sequential application of CX gates.

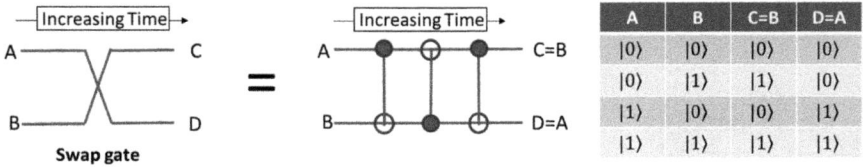

A	B	C=B	D=A
$\lvert 0\rangle$	$\lvert 0\rangle$	$\lvert 0\rangle$	$\lvert 0\rangle$
$\lvert 0\rangle$	$\lvert 1\rangle$	$\lvert 1\rangle$	$\lvert 0\rangle$
$\lvert 1\rangle$	$\lvert 0\rangle$	$\lvert 0\rangle$	$\lvert 1\rangle$
$\lvert 1\rangle$	$\lvert 1\rangle$	$\lvert 1\rangle$	$\lvert 1\rangle$

Figure 8.10. Swap gate interchanges the two bits such that $C = B$ and $D = A$.

8.4.3 *Controlled Y, Z, H, S, and CCX gates*

A controlled U gate is a straightforward extension of the CX gate. It is a two-qubit gate where the unitary transformation $\hat{U}$ is applied to the target qubit *only if* the control is in logic state 1. If the control qubit is in logic state 0, then the target qubit is left unchanged. When a controlled U gate is applied to an input state that is a superposition of many terms, it leaves the 0 or 1 state of the control qubit unchanged in each of the terms. However, note that this is not the same as saying the state of the control qubit is unchanged — in fact, the phase factor of the control qubit state can change as shown in the section discussing phase kickback. Common operations used in quantum computing are $\hat{U} = \hat{X}, \hat{Y}, \hat{Z}, \hat{H}, \hat{S}$ gates, shown in Figure 8.11. The controlled gates operate on two qubits, and it is easy to verify that in matrix form, the operators for these gates are as follows:

$$\text{Control} - X = \widehat{CX} = \begin{pmatrix} 1 & 0 & 0 & 0 \\ 0 & 1 & 0 & 0 \\ 0 & 0 & 0 & 1 \\ 0 & 0 & 1 & 0 \end{pmatrix}$$

$$\text{Control} - S = \widehat{CS} = \begin{pmatrix} 1 & 0 & 0 & 0 \\ 0 & 1 & 0 & 0 \\ 0 & 0 & 1 & 0 \\ 0 & 0 & 0 & e^{i\pi/4} \end{pmatrix}$$

$$\text{Control} - Y = \widehat{CY} = \begin{pmatrix} 1 & 0 & 0 & 0 \\ 0 & 1 & 0 & 0 \\ 0 & 0 & 0 & i \\ 0 & 0 & -i & 0 \end{pmatrix}$$

$$\text{Control} - H = \widehat{CH} = \begin{pmatrix} 1 & 0 & 0 & 0 \\ 0 & 1 & 0 & 0 \\ 0 & 0 & \dfrac{1}{\sqrt{2}} & \dfrac{1}{\sqrt{2}} \\ 0 & 0 & \dfrac{1}{\sqrt{2}} & -\dfrac{1}{\sqrt{2}} \end{pmatrix}$$

$$\text{Control} - Z = \widehat{CZ} = \begin{pmatrix} 1 & 0 & 0 & 0 \\ 0 & 1 & 0 & 0 \\ 0 & 0 & 0 & -i \\ 0 & 0 & i & 0 \end{pmatrix}$$

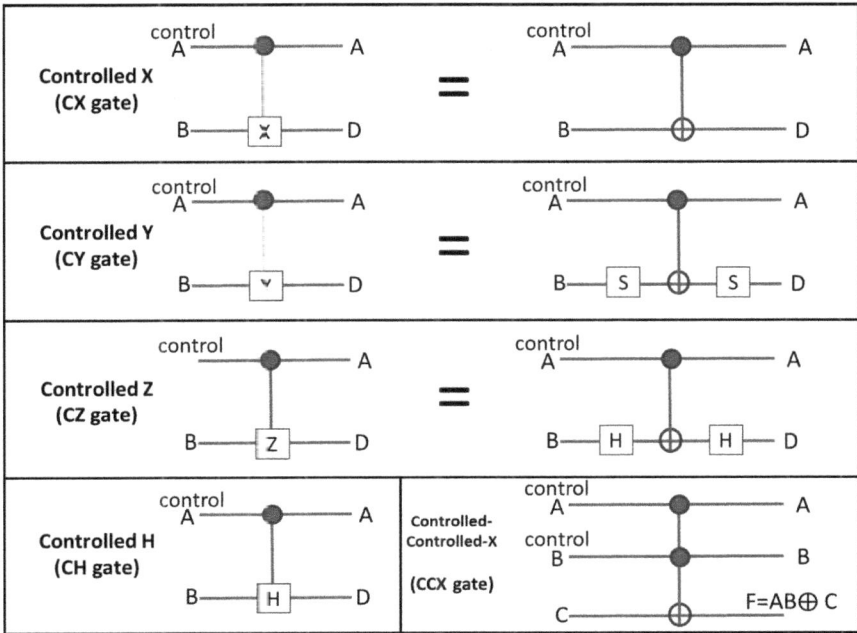

Figure 8.11. Controlled X, Y, Z (CX, CY, CZ), Hadamard (CH) gates, and the controlled-controlled-X (CCX) gate. The CCX gate is a three-terminal gate, while the others are two-terminal gates. The solid dot and open circle represent the control and target qubits, respectively.

In bra-ket notation, the operation by the control-H gate yields

$$\widehat{CH}|00\rangle = |00\rangle$$

$$\widehat{CH}|01\rangle = |01\rangle$$

$$\widehat{CH}|10\rangle = \frac{1}{\sqrt{2}}[|10\rangle + |11\rangle]$$

$$\widehat{CH}|11\rangle = \frac{1}{\sqrt{2}}[|10\rangle - |11\rangle]$$

In matrix form, the above equations correspond to

$$\begin{pmatrix} 1 & 0 & 0 & 0 \\ 0 & 1 & 0 & 0 \\ 0 & 0 & \frac{1}{\sqrt{2}} & \frac{1}{\sqrt{2}} \\ 0 & 0 & \frac{1}{\sqrt{2}} & -\frac{1}{\sqrt{2}} \end{pmatrix} \begin{bmatrix} 1 \\ 0 \\ 0 \\ 0 \end{bmatrix} = \begin{bmatrix} 1 \\ 0 \\ 0 \\ 0 \end{bmatrix} \quad\bigg|\quad \begin{pmatrix} 1 & 0 & 0 & 0 \\ 0 & 1 & 0 & 0 \\ 0 & 0 & \frac{1}{\sqrt{2}} & \frac{1}{\sqrt{2}} \\ 0 & 0 & \frac{1}{\sqrt{2}} & -\frac{1}{\sqrt{2}} \end{pmatrix} \begin{bmatrix} 0 \\ 0 \\ 1 \\ 0 \end{bmatrix} = \frac{1}{\sqrt{2}} \left(\begin{bmatrix} 0 \\ 0 \\ 1 \\ 0 \end{bmatrix} + \begin{bmatrix} 0 \\ 0 \\ 0 \\ 1 \end{bmatrix} \right)$$

$$\begin{pmatrix} 1 & 0 & 0 & 0 \\ 0 & 1 & 0 & 0 \\ 0 & 0 & \frac{1}{\sqrt{2}} & \frac{1}{\sqrt{2}} \\ 0 & 0 & \frac{1}{\sqrt{2}} & -\frac{1}{\sqrt{2}} \end{pmatrix} \begin{bmatrix} 0 \\ 1 \\ 0 \\ 0 \end{bmatrix} = \begin{bmatrix} 0 \\ 1 \\ 0 \\ 0 \end{bmatrix} \quad\bigg|\quad \begin{pmatrix} 1 & 0 & 0 & 0 \\ 0 & 1 & 0 & 0 \\ 0 & 0 & \frac{1}{\sqrt{2}} & \frac{1}{\sqrt{2}} \\ 0 & 0 & \frac{1}{\sqrt{2}} & -\frac{1}{\sqrt{2}} \end{pmatrix} \begin{bmatrix} 0 \\ 0 \\ 0 \\ 1 \end{bmatrix} = \frac{1}{\sqrt{2}} \left(\begin{bmatrix} 0 \\ 0 \\ 1 \\ 0 \end{bmatrix} - \begin{bmatrix} 0 \\ 0 \\ 0 \\ 1 \end{bmatrix} \right)$$

Figure 8.11 also shows the three qubit controlled-controlled-X gate, the output of which is (A-and-B)-XOR-(C)

$$\hat{F} = AB \oplus C$$

The output of an XOR gate is logic state 1 if and only if exactly one of the two inputs is in logic state 1. The CCX gate is also called the controlled-controlled NOT (CCNOT) or CCN or Toffoli gate. The control in the CCX gate can be generalized to construct a controlled f gate (controlled function). The output D is the XOR operation of the target and a function f of the control bits A and B. Figure 8.12 shows the symbol for controlled f gates with two and three inputs.

The $\widehat{CCZ}$ gate only applies the $\hat{Z}$ gate to the target if both controls are in state 1. So, the $\widehat{CCZ}$ gate will flip $|111\rangle$ to $-|111\rangle$ while leaving the other seven inputs unchanged. Figure 8.13 shows the construction of a $\widehat{CCZ}$ gate in terms of the Hadamard and the

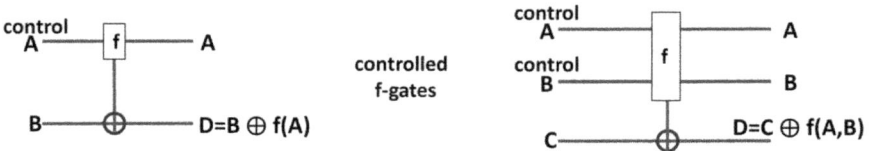

Figure 8.12. Controlled f gates with (left) two and (right) three inputs.

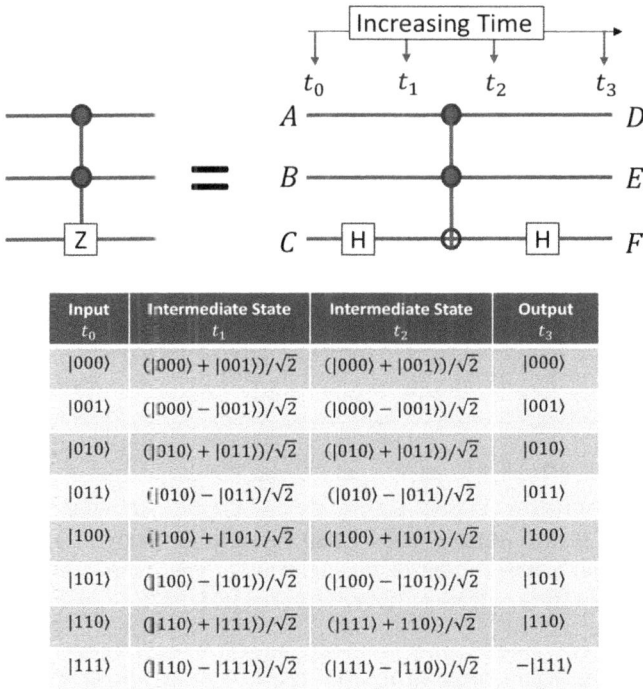

Figure 8.13. The CCZ gate symbol and decomposition in terms of Hadamard and CCX gates. The bottom figure shows the truth table, where at time t_0, we have the various input states, and at time t_3, we have the corresponding output states. The state of the qubits at times t_1 and t_2 before and after the CCX gate are shown.

$\widehat{CCX}$ gate. The truth table also describes the evolution of the input with time by showing the intermediate states at times t_1 and t_2. The reader should work these out.

8.4.4 *Phase Kickback*

Consider a CZ gate with the control in the superposition state $\frac{1}{\sqrt{2}}[|0\rangle + |1\rangle]$, and the target in state $|1\rangle$ (Figure 8.14). We see that at the output, the target is unchanged but the relative phase of the control has changed, putting the control to be in a state $\frac{1}{\sqrt{2}}[|0\rangle - |1\rangle]$. This process where the control develops a new phase based on the

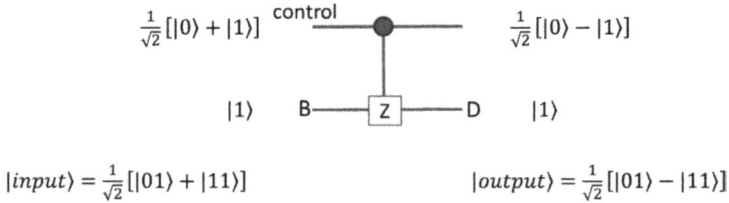

$|input\rangle = \frac{1}{\sqrt{2}}[|01\rangle + |11\rangle]$ $|output\rangle = \frac{1}{\sqrt{2}}[|01\rangle - |11\rangle]$

Figure 8.14. Illustration of phase kickback using a CZ gate.

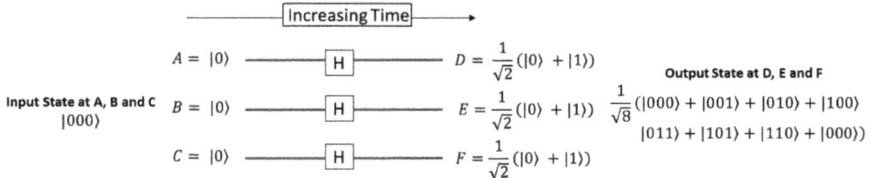

Figure 8.15. Transforming a three-qubit input state $|000\rangle$ to an output state which is an equal superposition of the eight underlying basis states $\frac{1}{\sqrt{8}}\{|000\rangle + |001\rangle + |010\rangle + |011\rangle + |100\rangle + |101\rangle + |110\rangle + |111\rangle\}$.

state of the target is sometimes referred to as a *phase kickback*. This concept is used in some quantum algorithms.

8.4.5 *Creating an Equal Superposition State*

Let us imagine how to transform a state $|000\rangle$ to $\frac{1}{\sqrt{8}}\{|000\rangle + |001\rangle + |010\rangle + |011\rangle + |100\rangle + |101\rangle + |110\rangle + |111\rangle\}$, which is an equal superposition of the eight basis states of the three-qubit system.

The output is simply the product state obtained by transforming each of the three qubits from $|0\rangle$ to $\frac{1}{\sqrt{2}}(|0\rangle + |1\rangle)$:

$$\frac{1}{\sqrt{2}}(|0\rangle + |1\rangle)\frac{1}{\sqrt{2}}(|0\rangle + |1\rangle)\frac{1}{\sqrt{2}}(|0\rangle + |1\rangle)$$

$$= \frac{1}{\sqrt{8}}\{|000\rangle + |001\rangle + |010\rangle + |011\rangle + |100\rangle + |101\rangle + |110\rangle + |111\rangle\}$$

To create the above state, the Hadamard gate operates independently on the three input qubits $|000\rangle$ (Figure 8.15).

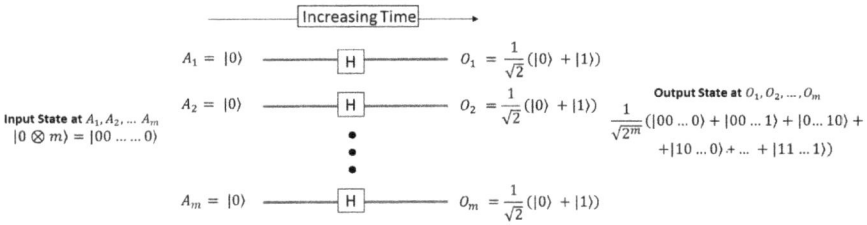

Figure 8.16. Transforming an m-qubit state $|00\ldots0\rangle$ to an equal superposition of the underlying $N = 2^n$ basis states.

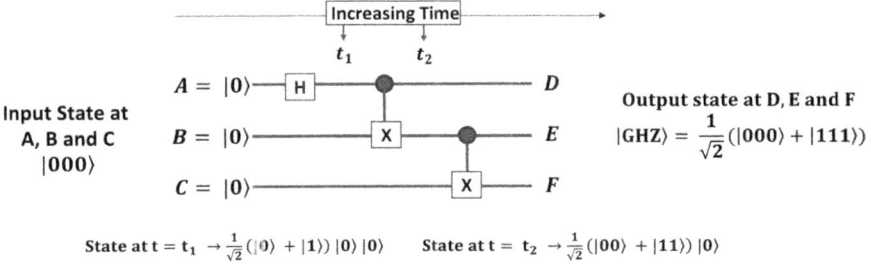

Figure 8.17. Gate sequence for generating the GHZ state from $|000\rangle$ using one and two qubit gates.

A common m-qubit input state in many quantum computing algorithms is an equal superposition of the $N = 2^m$ states. Such a state is a product state because it is created by transforming each qubit of the m-qubits from $|0\rangle$ to $\frac{1}{\sqrt{2}}(|0\rangle + |1\rangle)$. The creation of such a qubit is shown in Figure 8.16.

Example 3. Transforming $|000\rangle$ to a GHZ state, which is a three-qubit Bell-like state defined by

$$|GHZ\rangle = \frac{1}{\sqrt{2}}(|000\rangle + |111\rangle)$$

Figure 8.17 shows the one and two qubit gate operations that rotate an input state $|000\rangle$ to the $|GHZ\rangle$ state.

8.4.6   *Gates for Grover's Algorithm*

First, consider the $\hat{V}$ gate, which changes the phase of one term in a superposition (Section 7.5). To be concrete, we will see how to transform $\frac{1}{\sqrt{8}}\{|000\rangle + |001\rangle + |010\rangle + |011\rangle + |100\rangle + |101\rangle + |110\rangle + |111\rangle\}$ to $\frac{1}{\sqrt{8}}\{-|000\rangle + |001\rangle + |010\rangle + |011\rangle + |100\rangle + |101\rangle + |110\rangle + |111\rangle\}$.

Since the only difference between the input and output is the sign before $|000\rangle$ we can apply the X-gate to all bits so that $|000\rangle$ becomes a $|111\rangle$. Then, we can apply the CCZ gate to change the phase of the $|111\rangle$ state to $-|111\rangle$, while keeping all other states in the superposition unchanged. Finally re-apply X-gates to each qubit to flip it back to the original state. You should use the X-gate and CCZ gate properties to convince yourself that the circuit in Figure 8.18 implements the transformation.

Next, let us construct a gate for the $\hat{W}$ transform from Grover's algorithm. To recall, the W-gate was defined to be $\hat{W} = 2|\text{init}\rangle\langle\text{init}| - \hat{I}$, where $|\text{init}\rangle = \frac{1}{\sqrt{8}}\{|000\rangle + |001\rangle + |010\rangle + |011\rangle + |100\rangle + |101\rangle + |110\rangle + |111\rangle\}$.

In a manner similar to the previous cases, the reader should convince themselves that the operator $2|0 \otimes m\rangle\langle 0 \otimes m| - \hat{I}$ acting on $|\psi\rangle$ creates a vector with the same component along $|0 \otimes m\rangle$. However, the phase of the component of $|\psi\rangle$ perpendicular to $|0 \otimes m\rangle$ is multiplied by $e^{i\pi}$. Here $|0 \otimes m\rangle$ is a vector where all m-qubits are in state $|0\rangle$.

We saw in Section 8.4.5 that the product state $|\text{init}\rangle$ is generated from $|0 \otimes m\rangle$ by applying a Hadamard operator on each of the m

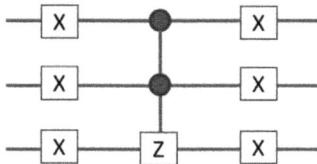

Figure 8.18. A gate that flips only the sign of $|000\rangle$ to $-|000\rangle$ while leaving the other seven basis states unchanged.

qubits. That is,

$$|\text{init}\rangle = \hat{H}^{\otimes m}|0 \otimes m\rangle$$

We also know that $\hat{H}\hat{H} = I$, the 2×2 identity matrix, and $\hat{H}^{\otimes m}\hat{H}^{\otimes m} = I_{2^m}$, the 2^m-dimensional identity matrix. Then, we can write $\hat{W} = 2|\text{init}\rangle\langle\text{init}| - \hat{I}$ as

$$\hat{W} = \hat{H}^{\otimes m}[2|0 \otimes m\rangle\langle 0 \otimes m| - \hat{I}]\hat{H}^{\otimes m}$$

Note that $[2|0 \otimes m\rangle\langle 0 \otimes m| - \hat{I}]$, when acting on a vector, keeps the sign of the component of $|\psi\rangle$ along $|0 \otimes m\rangle$ unchanged but inverts the sign of the component perpendicular to $|0 \otimes m\rangle$. Using the m-qubit version of Figure 8.18, we have realized the W-gate up to an overall phase factor of (-1) as shown in Figure 8.19.[1] The three-qubit version of this gate is a special case where $m = 3$.

Concept: Quantum gates are unitary. That is, there is a one-to-one mapping between the input and output states.

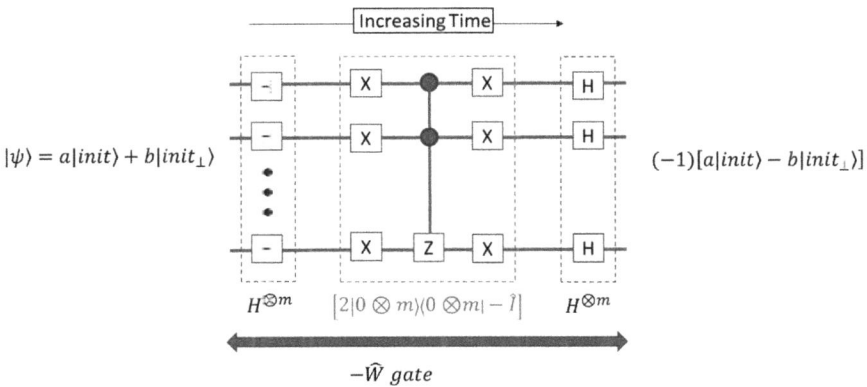

Figure 8.19. W gate multiplied by (-1).

---

[1]Note that we do not care about the overall phase factor of a quantum state but the relative phase factors between two terms of a quantum wave function are of paramount importance.

8.5   Quantum Teleportation

Let us ask first what *classical teleportation* is. If you have watched science fiction movies, information or matter can magically travel between two locations, even without physically traveling between them. There has so far been no experimental demonstration of this — do not bet your money on it either!

On the other hand, the quantum teleportation that we describe here is real. Teleportation over thousands of kilometers between a ground station and a satellite was demonstrated in 2017. Researchers expect that quantum teleportation will be important in building future quantum communication networks, quantum internet, and distributed quantum computing. Quantum repeaters utilize quantum teleportation. A repeater is a device used to receive communication signals and deliver corresponding amplified ones. It is important to realize that quantum teleportation does not involve teleportation of matter, unlike in science fiction movies. Rather, it means the teleportation of an unknown qubit state, after a step involving the classical transfer of matter. So, quantum teleportation will only result in the teleportation of information and not matter.

This section shows a simple example in which Alice (who is on earth) tries to teleport a qubit state in her possession to Bob (who is far away, say the moon). Alice's qubit $|\chi\rangle = a|0_{A1}\rangle + b|1_{A1}\rangle$ can be teleported to Bob such that Bob has a qubit $a|0_B\rangle + b|1_B\rangle$. You may ask why Alice does not simply measure her state and tell Bob the values of a and b using a classical communication channel. The problem is that if Alice does a measurement in her basis $\{|0_{A1}\rangle, |1_{A1}\rangle\}$, she either gets $|0_{A1}\rangle$ or $|1_{A1}\rangle$, and no information about a and b.

For teleportation, Alice and Bob make a two-qubit entangled state. Bob travels from earth to moon with his qubit of the entangled state. Alice has a second qubit in the state $|\chi\rangle = a|0_{A1}\rangle + b|1_{A1}\rangle$. Now Alice and Bob can communicate classically between their locations. (Note that this requires at least the time for light to propagate between earth and moon. So, a state cannot be teleported faster than the speed of light.) We will show that after a classical communication, the quantum teleportation protocol will transfer Alice's state to Bob such that Bob has the state $a|0_B\rangle + b|1_B\rangle$. It is important to note that Alice loses her state $a|0_{A1}\rangle + b|1_{A1}\rangle$ (because copying of an unknown state is not allowed quantum mechanically according to the no-cloning theorem).

Alice has a single copy of a quantum state

$$|\chi\rangle = a|0_{A1}\rangle + b|1_{A1}\rangle \tag{8.26}$$

that she wants to send to Bob. Alice can send her qubit by shipping her whole laboratory while keeping the cryogenic fridge, power supplies, and electronic peripherals, e.g., by keeping her superconducting qubit or spin-based qubit state inside a quantum dot on a silicon chip still alive and in superposition state! But is there a faster and more efficient method? The answer is yes, if quantum entanglement is used. We describe this process now.

While both Alice and Bob were on earth, they used Alice's second qubit $A2$ with a qubit B of Bob to create the entangled state,

$$\psi\rangle = \frac{1}{\sqrt{2}}(|0_{A2}0_B\rangle + |1_{A2}1_B\rangle) \tag{8.27}$$

The three qubit system is in state:

$$|\Theta\rangle = |\chi\rangle \otimes |\psi\rangle = [a|0_{A1}\rangle + b|1_{A1}\rangle] \otimes \frac{1}{\sqrt{2}}(|0_{A2}0_B\rangle + |1_{A2}1_B\rangle) \tag{8.28}$$

$$|\Theta\rangle = \frac{a(|0_{A1}0_{A2}0_B\rangle + |0_{A1}1_{A2}1_B\rangle) + b(|1_{A1}0_{A2}0_B\rangle + |1_{A1}1_{A2}1_{B_i}\rangle)}{\sqrt{2}} \tag{8.29}$$

Bob travels to moon while Alice stays on earth.

Alice then applies a CX operation to her qubits, with $A1$ being the *control* and $A2$ being the *target*. So whenever $A1$ is $|0\rangle$, there is no change in $A2$, and if $A1$ is $|1\rangle$, the value of qubit $A2$ flips. This gives

$$|\Theta_1\rangle = \widehat{CX}(A1, A2)|\Theta\rangle$$

$$= \frac{a(|0_{A1}0_{A2}0_B\rangle + |0_{A1}1_{A2}1_B\rangle) + b(|1_{A1}1_{A2}0_B\rangle + |1_{A1}0_{A2}1_B\rangle)}{\sqrt{2}} \tag{8.30}$$

Next Alice applies a Hadamard gate to the first qubit $A1$. Recalling that applying a Hadamard gate is given by equations (8.6) and (8.7),

$$|\Theta_2\rangle = H_{A1}|\Theta_1\rangle$$
$$= \frac{H_{A1}a|0_{A1}\rangle(|0_{A2}0_B\rangle + |1_{A2}1_B\rangle) + H_{A1}b|1_{A1}\rangle(|1_{A2}0_B\rangle + |0_{A2}1_B\rangle)}{\sqrt{2}}$$

$$(8.31)$$

The above equation gives

$$|\Theta_2\rangle = a\left(\frac{|0_{A1}\rangle + |1_{A1}\rangle}{\sqrt{2}}\right)\left(\frac{|0_{A2}0_B\rangle + |1_{A2}1_B\rangle}{\sqrt{2}}\right)$$
$$+ b\left(\frac{|0_{A1}\rangle - |1_{A1}\rangle}{\sqrt{2}}\right)\left(\frac{|1_{A2}0_B\rangle + |0_{A2}1_B\rangle}{\sqrt{2}}\right) \quad (8.32)$$

Factoring out the two qubits of Alice's system gives

$$|\Theta_2\rangle = H_{A1}|\Theta_1\rangle = \frac{1}{2}\{|0_{A1}0_{A2}\rangle[a|0_B\rangle + b|1_B\rangle]$$
$$+ |1_{A1}0_{A2}\rangle[a|0_B\rangle - b|1_B\rangle] + |0_{A1}1_{A2}\rangle[a|1_B\rangle + b|0_B\rangle]$$
$$+ |1_{A1}1_{A2}\rangle[a|1_B\rangle - b|0_B\rangle]\} \quad (8.33)$$

In the next step, Alice measures her two qubits. She will get one of the four answers in the table below, each with a probability of $1/4$. Each of the results collapses Bob's state to the one shown in the second column of the table. Alice then communicates the values she measured for bits $A1$ and $A2$ to Bob using a classical communication channel. Depending on what Alice conveys to Bob, he performs the operation prescribed in the third column of the table on his qubit and always ends up having $[a|0_B\rangle + b|1_B\rangle]$. Note that this was the original state for teleportation to Bob. So, the teleportation is done! And Alice is in the classical state she collapsed into (left column of the table).

Note that as the final state is a single qubit state, $\hat{Z}_B\hat{X}_B$ means applying two single qubit gates in sequence, from right to left. Recall that $\hat{X}_B$ flips the qubit and $\hat{Z}_B|0_B\rangle = |0_B\rangle$ and $\hat{Z}_B|1_B\rangle = -|1_B\rangle$.

$$\hat{Z}_B\hat{X}_B[a|1_B\rangle - b|0_B\rangle] = a\hat{Z}_B(|0_B\rangle) - b\hat{Z}_B(|1_B\rangle) = a|0_B\rangle + b|1_B\rangle$$

which is the state Alice wanted to send to Bob.

Alice measures	Bob's qubit will be in state	Depending on the qubit values Alice measured for $A1$ and $A2$ and conveyed to Bob, he performs the following operation
$\lvert 0_{A1} 0_{A2}\rangle$	$a\lvert 0_B\rangle + b\lvert 1_B\rangle$	I (doing nothing)
$\lvert 1_{A1} 0_{A2}\rangle$	$a\lvert 0_B\rangle - b\lvert 1_B\rangle$	Z_B
$\lvert 0_{A1} 1_{A2}\rangle$	$a\lvert 1_B\rangle + b\lvert 0_B\rangle$	X_B
$\lvert 1_{A1} 1_{A2}\rangle$	$a\lvert 1_B\rangle - b\lvert 0_B\rangle$	$Z_B X_B$

8.6 Problems

Sections 8.4

(1) (a) Draw the circuit to realize the state $\lvert\chi\rangle = \frac{1}{\sqrt{2}}\{\lvert 000\rangle - \lvert 111\rangle\}$ from the initial state $\lvert 000\rangle$.

 (b) If a real experiment is done to build the circuit in (a), what do you expect if the experiment is done a hundred times. You can also try this using a quantum simulator.

(2) A two-qubit Hadamard gate acts on the state $\lvert 00\rangle$. Determine if the resulting state is an entangled or product state.

Hint: Apply $\hat{H} \otimes \hat{H}$ on the state $\lvert 00\rangle$ and write the resulting state in vector format, i.e., $[a\ b\ c\ d]$. Is $ad = bc$ for this state?

(3) Design a gate to transform $\frac{1}{\sqrt{8}}\{\lvert 000\rangle + \lvert 001\rangle + \lvert 010\rangle + \lvert 100\rangle + \lvert 011\rangle + \lvert 101\rangle + \lvert 110\rangle + \lvert 111\rangle +\}$ to $\frac{1}{\sqrt{8}}\{-\lvert 000\rangle + \lvert 001\rangle + \lvert 010\rangle + \lvert 011\rangle - \lvert 100\rangle + \lvert 101\rangle + \lvert 110\rangle + \lvert 111\rangle\}$.

Sections 8.5

(4) Show that if Bob applies $\hat{Z}_B \hat{X}_B$ to $a\lvert 1_B\rangle - b\lvert 0_B\rangle$, the result is $a\lvert 0_B\rangle + b\lvert 1_B\rangle$.

Chapter 9

TWO-LEVEL SYSTEM IN
A TIME-DEPENDENT FIELD

Contents

In Chapters 7 and 8, we learned that a physical system with two distinct energy levels can be used to represent a qubit. However, leaving the qubit on its own in its stationary state has no practical application. To put the qubit to use, we must utilize time-varying signals such as laser beams or microwave pulses to control and change the qubit's state. That is, we need to "perturb" the quantum system in question. To engineer the response of the quantum system to our desired "perturbation", we use Schrödinger's equation to predict how the system will evolve over time.

In this chapter, we will study a two-level system in the presence of a time-dependent potential (Figure 9.1). Our plan is to calculate the time evolution of the wave function of a **two-level system**. We will learn how the population of each state evolves with time in response to an electromagnetic field, e.g., a microwave pulse. Based on this, we will see examples on how to design microwave pulses to implement NOT and Hadamard gates if we use the two-level system as a qubit.

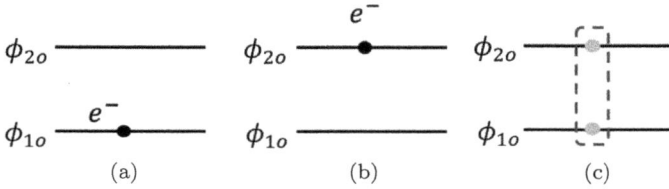

Figure 9.1. A two-level system consists of only two distinct energy levels. The electron is in (a) the lower energy ϕ_{1o}, (b) the higher energy ϕ_{2o} and (c) a superposition of ϕ_{1o} and ϕ_{2o}.

9.1 Time-Dependent Perturbation Theory

This section will discuss how an electron transitions between the states of a two-level system in response to a time-dependent electromagnetic potential. The reference states in our discussion will refer to the states of the unperturbed or stationary system.

The time-dependent perturbation, modeled as a potential energy $U(\bar{r}, t)$ is added to a time-independent Hamiltonian H_o,

$$\widehat{H} = \widehat{H}_o + \widehat{U}(\bar{r}, t) \tag{9.1}$$

The eigenenergies and eigenfunctions of $\widehat{H}_o$ are assumed to be known and denoted by (E_{1o}, ϕ_{1o}) and (E_{2o}, ϕ_{2o}). The time-dependent potential $U(\bar{r}, t)$ is also known. For brevity, we will drop the *hat* over the time-dependent potential. Then, the wave function of the time-dependent Schrödinger equation with Hamiltonian $\widehat{H}$ can be obtained by solving (Chapter 1)

$$i\hbar\frac{\partial \psi(\bar{r}, t)}{\partial t} = \widehat{H}\psi(\bar{r}, t) \tag{9.2}$$

We will expand the solution to the above differential equation as a superposition of unperturbed eigenfunctions,

$$\psi = a_1(t)e^{-\frac{iE_{1o}t}{\hbar}}\phi_{1o} + a_2(t)e^{-\frac{iE_{2o}t}{\hbar}}\phi_{2o} \tag{9.3}$$

Note that if there was no perturbation, the coefficients a_1 and a_2 would be constants. In the presence of a time-dependent perturbation, the coefficients will become time-dependent. To find the time dependence of the coefficients, we form and solve coupled differential equations that govern these coefficients. Substituting equation (9.3) into equation (9.2) gives us the equations for the time evolution of the coefficients as follows:

$$i\hbar e^{-\frac{iE_{1o}t}{\hbar}}\phi_{1o}\frac{\partial a_1(t)}{\partial t} + i\hbar e^{-\frac{iE_{2o}t}{\hbar}}\phi_{2o}\frac{\partial a_2(t)}{\partial t}$$

$$= a_1(t)e^{-\frac{iE_{1o}t}{\hbar}}U\phi_{1o} + a_2(t)e^{-\frac{iE_{2o}t}{\hbar}}U\phi_{2o} \qquad (9.4)$$

Applying the operator $\int dv\, \phi_{1o}^*$ to both sides of equation (9.4) from the left side gives us

$$i\hbar e^{-\frac{iE_{1o}t}{\hbar}}\left[\int dv\, \phi_{1o}^*\phi_{1o}\right]\frac{\partial a_1(t)}{\partial t} + i\hbar e^{-\frac{iE_{2o}t}{\hbar}}\left[\int dv\, \phi_{1o}^*\phi_{2o}\right]\frac{\partial a_2(t)}{\partial t}$$

$$= a_1(t)e^{-\frac{iE_{1o}t}{\hbar}}\left[\int dv\, \phi_{1o}^*U\phi_{1o}\right] + a_2(t)e^{-\frac{iE_{2o}t}{\hbar}}\left[\int dv\, \phi_{1o}^*U\phi_{2o}\right]$$

$$(9.5)$$

Using the orthonormality of the wave functions, $\int dv\, \phi_{po}^*\phi_{no} = \delta_{pn}$. Since only the first term is nonzero on the LHS, equation (9.5) becomes

$$i\hbar\frac{\partial a_1(t)}{\partial t} = a_1(t)\left[\int dv\, \phi_{1o}^*U\phi_{1o}\right] + a_2(t)e^{\frac{i(E_{1o}-E_{2o})t}{\hbar}}\left[\int dv\, \phi_{1o}^*U\phi_{2o}\right]$$

$$(9.6)$$

Similarly, applying $\int dv\, \phi_{2o}^*$ to both sides of equation (9.4) from the left side gives,

$$i\hbar\frac{\partial a_2(t)}{\partial t} = a_1(t)e^{\frac{i(E_{2o}-E_{1o})t}{\hbar}}\left[\int dv\, \phi_{2o}^*U\phi_{1o}\right] + a_2(t)\left[\int dv\, \phi_{2o}^*U\phi_{2o}\right]$$

$$(9.7)$$

The integrals on the RHS of the above equations called the matrix element $U_{pn}(t)$ are as follows:

$$U_{11}(t) = \int dv\, \phi_{1o}^*(\bar{r})U(\bar{r},t)\phi_{1o}(\bar{r}) = \langle\phi_{1o}|U|\phi_{1o}\rangle \qquad (9.8)$$

$$U_{22}(t) = \int dv\, \phi_{2o}^*(\bar{r})U(\bar{r},t)\phi_{2o}(\bar{r}) = \langle\phi_{2o}|U|\phi_{2o}\rangle \qquad (9.9)$$

$$U_{12}(t) = \int dv\, \phi_{1o}^*(\bar{r})U(\bar{r},t)\phi_{2o}(\bar{r}) = \langle\phi_{1o}|U|\phi_{2o}\rangle \qquad (9.10)$$

$$U_{21}(t) = \int dv\, \phi_{2o}^*(\bar{r})U(\bar{r},t)\phi_{1o}(\bar{r}) = \langle\phi_{2o}|U|\phi_{1o}\rangle = U_{12}(t)^* \qquad (9.11)$$

Note that the perturbation $U(\bar{r}, t)$ is a real (as opposed to a complex) quantity. The **resonant frequency** ω_{21} is defined as the difference of the unperturbed energy levels divided by Planck's constant,

$$\omega_{21} = \frac{E_{2o} - E_{1o}}{\hbar} \tag{9.12}$$

With this substitution, equations (9.6) and (9.7) can be rewritten as

$$i\hbar \frac{\partial a_1(t)}{\partial t} = U_{11}(t)a_1(t) + e^{i\omega_{12}t}U_{12}(t)a_2(t) \quad \text{and} \tag{9.13}$$

$$i\hbar \frac{\partial a_2(t)}{\partial t} = e^{i\omega_{21}t}U_{21}(t)a_1(t) + U_{22}(t)a_2(t) \tag{9.14}$$

Equations (9.13) and (9.14) represent a set of coupled linear differential equations which can be solved to find the coefficients $a_1(t)$ and $a_2(t)$, from which the wave function in equation (9.3) can be constructed. Note that the matrix elements U_{11}, U_{22} and U_{12}, and the resonant frequency ω_{21} are known because we already know the unperturbed eigenenergies and eigenstates (those with a subscript o) and the form of perturbation $U(\bar{r}, t)$ on the system.

Let us consider a special case with only two energy levels in the presence of an AC electric potential (see Figure 9.1). The energy eigenvalues and wave functions are denoted by (E_{1o}, ϕ_{1o}) and (E_{2o}, ϕ_{2o}). The perturbing time-dependent Hamiltonian is written as a traveling wave with frequency ω and wave vector $|\bar{k}| = \frac{2\pi}{\lambda}$,

$$U(\bar{r}, t) = u(\bar{r}) \cos\left(\bar{k} \cdot \bar{r} - \omega t\right) = \frac{1}{2}\left[u(\bar{r})e^{+i\bar{k}\cdot\bar{r}}e^{-i\omega t} + u(\bar{r})e^{-i\bar{k}\cdot\bar{r}}e^{+i\omega t}\right] \tag{9.15}$$

where $u(\bar{r})$ is real and represents the amplitude of the perturbing potential.

$$u(\bar{r}) = q\bar{\mathcal{E}} \cdot \bar{r} \tag{9.16}$$

where $\bar{\mathcal{E}}$ is a constant electric field, $\bar{r}$ is the position, and q is the electronic charge. Note that in equation (9.16), the perturbing electric potential energy $u(\bar{r})$ is caused by the perturbing electric field $\bar{\mathcal{E}}$.

Both terms *perturbing field* and *perturbing potential* are used in literature because the concept of *field* and *potential* are closely related.

Using equations (9.8)–(9.15), we have

$$U_{ij}(t) = \int dv\, \phi_{io}^*(\bar{r})\, U(\bar{r}, t)\phi_{jo}(\bar{r})$$

$$= \left[e^{-i\omega t} u_{ij} + e^{+i\omega t} \tilde{u}_{ij} \right], \quad (i, j \in 1, 2) \tag{9.17}$$

where u_{ij} and $\tilde{u}_{ij}$ are short-hand notations for the following matrix elements:

$$u_{ij} = \frac{1}{2} \int dv\, \phi_{io}^*(\bar{r}) u(\bar{r}) e^{+i\bar{k}\cdot\bar{r}} \phi_{jo}(\bar{r}) \tag{9.18}$$

and

$$\tilde{u}_{ij} = \frac{1}{2} \int dv\, \phi_{io}^*(\bar{r}) u(\bar{r}) e^{-i\bar{k}\cdot\bar{r}} \phi_{jo}(\bar{r}) \tag{9.19}$$

It is now easy to see that

$$\tilde{u}_{ij} = u_{ji}^* \tag{9.20}$$

Next, we assume:

Assumption 1. The physical dimension of the two-level system (qubit) is much smaller than the wavelength. Therefore, it is a good approximation to note that the region dv at location $\bar{r}$ which contributes to the integral in equations (9.18) and (9.19) is much smaller than the wavelength. As $|\bar{r}| \ll \lambda$, we can assume that $|\bar{k} \cdot \bar{r}| = \frac{2\pi}{\lambda}|\bar{r}| \approx 0$ which means $e^{+i\bar{k}\cdot\bar{r}} \approx 1$. This assumption allows us to write

$$\tilde{u}_{ij} = u_{ij} \tag{9.21}$$

Assumption 2. The perturbing potential $u(\bar{r})$ has an odd symmetry when $u(\bar{r}) = q\bar{\mathcal{E}}\cdot\bar{r}$, where the quantum system is placed at the origin. Thus, an integral such as $\int dv\, q\bar{\mathcal{E}} \cdot \bar{r}|\phi_{1o}(\bar{r})|^2 = 0$ because $|\phi_{1o}(\bar{r})|^2$ has even symmetry and $q\bar{\mathcal{E}} \cdot \bar{r}$ has odd symmetry.

Using assumptions 1 and 2, we can simplify u_{11} and u_{22} to

$$u_{11}(t) = \int dv\, \phi_{1o}^*(\bar{r}) u(\bar{r}) e^{+i\bar{k}\cdot\bar{r}} \phi_{1o}(\bar{r}) = \int dv\, q\bar{\mathcal{E}} \cdot \bar{r} \, |\phi_{1o}(\bar{r})|^2 = 0$$

$$\to U_{11} = 0 \tag{9.22}$$

$$u_{22}(t) = \int dv\, \phi_{2o}^*(\bar{r}) u(\bar{r}) e^{+i\bar{k}\cdot\bar{r}} \phi_{2o}(\bar{r}) = \int dv\, q\bar{\mathcal{E}} \cdot \bar{r} \, |\phi_{2o}(\bar{r})|^2 = 0$$

$$\to U_{22} = 0 \tag{9.23}$$

Using equations (9.17)–(9.23), equations (9.13) and (9.14) can be written as

$$i\hbar \frac{\partial a_1(t)}{\partial t} = \left[e^{i(\omega_{12}-\omega)t} u_{12} + e^{i(\omega_{12}+\omega)t} u_{12} \right] a_2(t) \tag{9.24}$$

$$i\hbar \frac{\partial a_2(t)}{\partial t} = \left[e^{i(\omega_{21}-\omega)t} u_{21} + e^{i(\omega_{21}+\omega)t} u_{21} \right] a_1(t) \tag{9.25}$$

Now, we will further assume that the AC perturbation has a frequency ω close to ω_{21}. Then, the solutions to equations (9.24) and (9.25) have a term with a low-frequency component that depends on $|\omega - \omega_{21}|$ and a term with a high frequency component that depends on $|\omega + \omega_{21}|$. The terms that depend on $|\omega - \omega_{21}|$ vary slowly with time compared to the terms that depend on $|\omega + \omega_{21}|$. Here, we are only interested in the slow-varying component because the rapidly changing part will effectively average out to be zero in a time scale t that follows $2\pi|\omega + \omega_{21}|^{-1} < t < 2\pi|\omega - \omega_{21}|^{-1}$. This special case is called the *rotating wave approximation* (RWA). Dropping the high-frequency components in equations (9.24) and (9.25), the resulting low-frequency parts governing the time evolution are

$$i\hbar \frac{\partial a_1(t)}{\partial t} = e^{i\Delta t} u_{12} a_2(t) \tag{9.26}$$

$$i\hbar \frac{\partial a_2(t)}{\partial t} = e^{-i\Delta t} u_{21} a_1(t) \tag{9.27}$$

The **detuning** of the system (difference between *incident* and *resonant/natural* frequency) is

$$\Delta = \omega - \omega_{21} \quad \textbf{(detuning frequency)} \tag{9.28}$$

Equations (9.26) and (9.27) are two coupled differential equations in which we are solving for a_1 and a_2 for which we are solving. To solve this differential equation, we must eliminate one of the unknowns (say a_1) to obtain the second-order differential equation in terms of a_2. Taking the time derivative of equation (9.27),

$$i\hbar\frac{\partial^2 a_2}{\partial t^2} = e^{-i\Delta t}u_{21}\frac{\partial a_1}{\partial t} - i\Delta e^{-i\Delta t}u_{21}a_1 \tag{9.29}$$

Now, using equation (9.20) and substituting for $\frac{\partial a_1}{\partial t}$ and a_1 from equations (9.26) and (9.27) into equation (9.29), we have

$$\frac{\partial^2 a_2}{\partial t^2} + i\Delta\frac{\partial a_2}{\partial t} + \frac{|u_{12}|^2}{\hbar^2}a_2 = 0 \tag{9.30}$$

where $u_{12} = u_{21}^*$ gives us $u_{12}u_{21} = |u_{12}|^2$. The above equation is a linear second-order differential equation with constant coefficients in time. Hence, the trial solution is an oscillatory solution with frequency B, i.e., $a_2(t) = Ce^{iBt}$, where B and C are constants. Replacing this in equation (9.30), we get a quadratic equation for B:

$$a_2(t)\left\{B^2 + \Delta B - \frac{|u_{12}|^2}{\hbar^2}\right\} = 0. \tag{9.31}$$

The roots of this equation are

$$B = \frac{-\Delta \pm \sqrt{4\Omega^2 + \Delta^2}}{2} \tag{9.32}$$

where

$$\Omega = \frac{|u_{12}|}{\hbar}. \quad \textbf{(Rabi frequency)} \tag{9.33}$$

Ω is called the *Rabi frequency*. It depends on the matrix element of the perturbing potential between the two unperturbed states ϕ_{1o} and ϕ_{2o} in equation (9.18). Note that in writing the matrix element u_{12}, we used Assumption 1 to approximate that $e^{+i\bar{k}\cdot\bar{r}} \approx 1$, That is,

$$\Omega = \frac{|\langle\phi_{1o}|q\bar{\mathcal{E}}\cdot\bar{r}|\phi_{2o}\rangle|}{\hbar} \tag{9.34}$$

If we define,

$$\Omega' = \sqrt{\Omega^2 + \frac{\Delta^2}{4}} \tag{9.35}$$

the roots can be rewritten more compactly as

$$B_{1,2} = -\frac{\Delta}{2} \pm \Omega' \tag{9.36}$$

As the equation has two roots, the total solution is a linear combination of two oscillatory solutions for each frequency found in equation (9.36). The two terms in $a_2(t)$ are solutions,

$$a_2(t) = e^{-\frac{i\Delta t}{2}} \left[C_1 e^{i\Omega't} + C_2 e^{-i\Omega't} \right] \tag{9.37}$$

To find $a_1(t)$, we substitute equation (9.37) into equation (9.27) and this results in

$$\frac{\hbar\Delta}{2} \left[C_1 e^{i\Omega't} + C_2 e^{-i\Omega't} \right] - \hbar\Omega' e^{-\frac{i\Delta t}{2}} \left[C_1 e^{i\Omega't} - C_2 e^{-i\Omega't} \right]$$

$$= e^{-i\Delta t} u_{21} a_1(t) \tag{9.38}$$

$$a_1(t) = \frac{\hbar}{u_{21}} e^{\frac{i\Delta t}{2}} \left[C_1 \left(\frac{\Delta}{2} - \Omega' \right) e^{i\Omega't} + C_2 \left(\frac{\Delta}{2} + \Omega' \right) e^{-i\Omega't} \right] \tag{9.39}$$

Since the system is in the unperturbed (stationary) state ϕ_{1o} at time $t = 0$, we can set the boundary condition as

$$a_1(t = 0) = 1 \quad \text{(to within a phase factor)} \tag{9.40}$$

$$a_2(t = 0) = 0 \tag{9.41}$$

Substituting equation (9.41) into equation (9.37), we get

$$C_1 = -C_2 \tag{9.42}$$

Equation (9.37) then becomes

$$a_2(t) = C_1 e^{-\frac{i\Delta t}{2}} \left[e^{i\Omega't} - e^{-i\Omega't} \right] \tag{9.43}$$

Using equations (9.40) and (9.42) in equation (9.39) gives

$$1 = \frac{\hbar}{u_{21}} C_1 \left[\left(\frac{\Delta}{2} - \Omega' \right) - \left(\frac{\Delta}{2} + \Omega' \right) \right] \tag{9.44}$$

$$\Rightarrow C_1 = -\frac{u_{21}}{\hbar} \frac{1}{2\Omega'} \tag{9.45}$$

After substituting for C_1 and C_2 from equations (9.42) and (9.45) in equations (9.39) and (9.43),

$$a_1(t) = -\frac{e^{\frac{i\Delta t}{2}}}{2\Omega'} \left[\left(\frac{\Delta}{2} - \Omega' \right) e^{i\Omega' t} - \left(\frac{\Delta}{2} + \Omega' \right) e^{-i\Omega' t} \right] \tag{9.46}$$

$$a_2(t) = -\frac{u_{21}}{\hbar} \frac{1}{2\Omega'} e^{-\frac{i\Delta t}{2}} \left[e^{i\Omega' t} - e^{-i\Omega' t} \right] \tag{9.47}$$

By replacing exponentials with trigonometric functions using Euler's identity, we arrive at the equations for the coefficients of the two levels in the system:

$$a_1(t) = +e^{\frac{i\Delta t}{2}} \left[\cos \left(\Omega' t \right) - \frac{i\Delta}{2\Omega'} \sin \left(\Omega' t \right) \right] \tag{9.48}$$

$$a_2(t) = -\frac{iu_{21}}{\hbar} \frac{1}{\Omega'} e^{-\frac{i\Delta t}{2}} \sin \left(\Omega' t \right) \tag{9.49}$$

Substituting equations (9.48) and (9.49) into equation (9.3), at last, we find that the time-dependent wave function ψ is

$$\psi = e^{\frac{i\Delta t}{2}} \left[\cos \left(\Omega' t \right) - \frac{i\Delta}{2\Omega'} \sin \left(\Omega' t \right) \right] e^{-\frac{iE_{1o}t}{\hbar}} \phi_{1o}$$

$$- \frac{iu_{21}}{\hbar} \frac{1}{\Omega'} e^{-\frac{i\Delta t}{2}} \left[\sin \left(\Omega' t \right) \right] e^{-\frac{iE_{2o}t}{\hbar}} \phi_{2o} \tag{9.50}$$

This solution tells us that applying a periodic time-dependent electric potential will result in a solution that oscillates between ϕ_{1o} and ϕ_{2o}. The rate of oscillation between the two coefficients $a_1(t)$ and $a_2(t)$ in equations (9.48) and (9.49) depends on the Rabi frequency and detuning. In general, the solution is a superposition of both eigenfunctions.

This is better understood if we calculate the probabilities of the electron being in each eigenstate of the unperturbed Hamiltonian separately. The probability $P_1(t)$ of finding the electron in state ϕ_{1o} at any time is found using the coefficient a_1,

$$P_1(t) = |a_1(t)|^2 = \cos^2(\Omega't) + \frac{\Delta^2}{4\Omega'^2}\sin^2(\Omega't) \tag{9.51}$$

and the probability $P_2(t)$ of finding the electron in state ϕ_{2o} is found using the coefficient a_2,

$$P_2(t) = |a_2(t)|^2 = \frac{\Omega^2}{\Omega'^2}\sin^2(\Omega't) \tag{9.52}$$

From these equations, we can reach a few conclusions about how the probability of finding an electron in states ϕ_{1o} and ϕ_{2o} changes with time:

- Equation (9.52) implies that if the electron is in state ϕ_{1o} at $t = 0$, then the probability of finding the electron in state ϕ_{2o} at a later time varies between zero and $\frac{\Omega^2}{\Omega'^2}$. The probability of finding the electron in state ϕ_{2o} never reaches unity unless the detuning is zero, i.e., $\Delta = \omega - \omega_{21} = 0$.
- Similarly, the probability of finding the electron in state ϕ_{1o} oscillates between $\frac{\Delta^2}{4\Omega'^2}$ and one, and never reaches zero unless the detuning is zero. When the detuning is zero, the driving AC frequency is equal to the resonant frequency of the two-level system; that is, $\omega = \omega_{21}$.
- The frequency Ω' with which the probabilities $P_1(t)$ and $P_2(t)$ oscillate depends on both the Rabi frequency (Ω) and detuning (Δ); $\Omega' = \sqrt{\Omega^2 + \frac{\Delta^2}{4}}$.
- The sum of the probabilities of finding the electron in either state ϕ_{1o} or ϕ_{2o} at any time t is one. It is easy to verify from equations (9.51) and (9.52) that

$$P_1(t) + P_2(t) = 1 \tag{9.53}$$

- When the detuning is zero, the AC driving frequency is equal to the natural or resonant frequency, $\omega = \omega_{21}$, then

$$P_1(t) = \cos^2(\Omega t) \quad \text{and} \tag{9.54}$$

$$P_2(t) = \sin^2(\Omega t) \tag{9.55}$$

In this case, we say the two-level system (atom or qubit) resonates with the external **driving** signal. Only when the detuning is equal to zero, the probability of finding the electron in states ϕ_{1o} and ϕ_{2o} oscillates the full range between zero and one with time t. Then, the frequency of oscillation, or *swapping*, between these states is precisely the Rabi frequency.

Figure 9.2 shows a two-level system where the difference between the ground (E_{1o}) and excited (E_{2o}) energy states are $E_{21} = E_{2o} - E_{1o} = 16.5\,\mu eV$, which corresponds to $\omega_{21} = 2\pi \times 4\,\text{GHz}$. A coherent electromagnetic wave that is detuned from the two-level system by $\Delta = 2\pi \times 10\,\text{MHz}$ drives the system. In response to this perturbation, the probabilities P_1 and P_2 for the electron to occupy states ϕ_{1o} and ϕ_{2o} oscillate according to Figure 9.2(a). When detuning is zero, i.e., when the resonant frequency matches the frequency of the electromagnetic wave, there is a full swing of probabilities P_1 and P_2 between the minimum possible value of zero and the maximum possible value of one, at Rabi frequency $\Omega = 2\pi \times 3.34\,\text{MHz}$, as shown in Figure 9.2(b).

If the system under study is an atom and the eigenstates ϕ_{1o} and ϕ_{2o} are s and p orbitals, respectively, then the total state ψ in equation (9.50) is an orbital that wobbles between a spherical (s) and dumbbell (p) shape at Rabi frequency.

In the example below, we use several different notations to represent the states in the two-level system. These notations are used interchangably in various books and journal papers. See table.

	Notation/Terminology Used in Various Sources			
Logic state	Quantum computing (Chapters 7 and 8)	This chapter for the two-level system		Miscellaneous literature
0	$|0\rangle$	ϕ_{1o}	$|g\rangle$	Ground state
1	$|1\rangle$	ϕ_{2o}	$|e\rangle$	Excited state

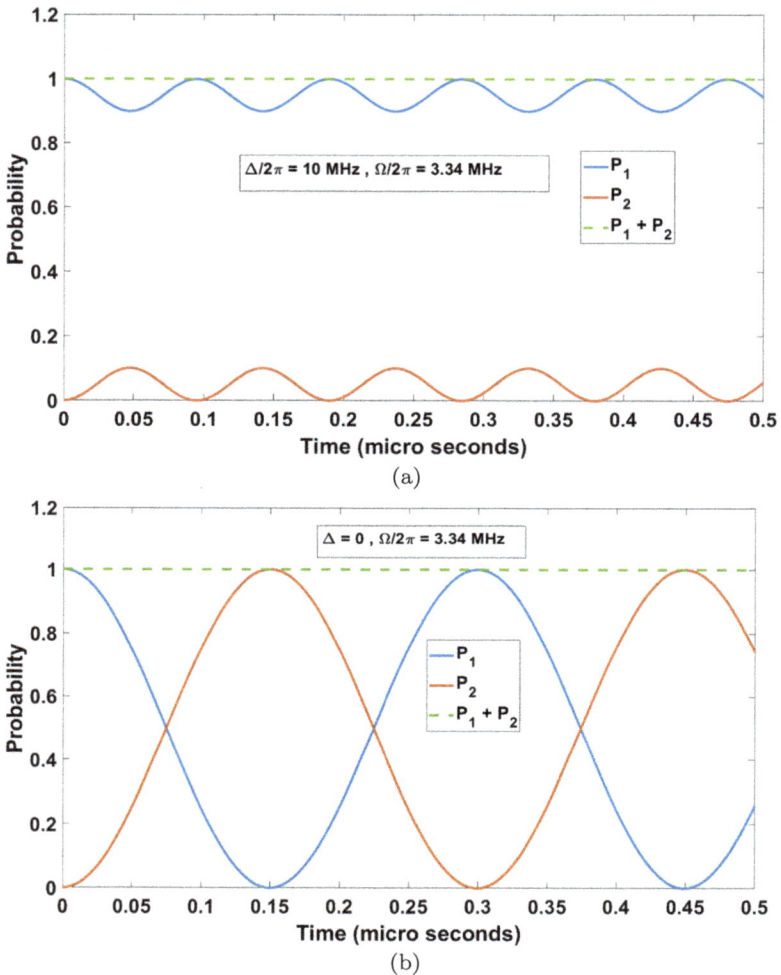

Figure 9.2. The probability oscillates sinusoidally when an AC potential interacts with the two-level system. We have chosen $\omega_{21} = 2\pi\,(4\,\text{GHz})$. Detuning (a) $\frac{\Delta}{2\pi} = 10\,\text{MHz}$ and (b) $\Delta = 0$. (a) shows that the oscillation of probabilities has a higher frequency when the detuning is larger. Also, with nonzero detuning, the probability to find the state ϕ_{2o} is always smaller than 1. We have assumed that at $t = 0$, the electron is only in state ϕ_{1o} (100% probability).

9.2   Examples

Example 1 (Superconducting qubit). A transmon qubit is an LC oscillator that uses an inductor (Josephson junction) to create two distinct energy levels to represent logic states 0 and 1. The difference between the ground ($|g\rangle$ or ϕ_{1o}) and excited ($|e\rangle$ or ϕ_{2o}) energy levels in this qubit is $4\,\mathrm{GHz}$. Assume that the Rabi frequency is $\Omega = 2\pi \times 3.34\,\mathrm{MHz}$ and the probability to find the two-level system in states $|g\rangle$ and $|e\rangle$ at time $t = 0$ are one and zero, respectively. In terms of the equations discussed in this section, $|g\rangle$ and $|e\rangle$ will correspond to the previously used notation ϕ_{1o} and ϕ_{2o}, respectively.

(a) **X-Gate:** For the qubit in Figure 9.3(a) and (b), design an appropriate microwave pulse applied to the XY-line that will rotate the qubit around X-axis from the ground to the excited state on the Bloch sphere (by 180 degrees). Assume that the pulse is in resonance with the qubit, e.g., it has the same frequency as the qubit, i.e., detuning is zero ($\Delta = 0$).

(b) **Hadamard Gate:** What change must be made to the XY-line pulse to build a Hadamard gate (i.e., rotating the qubit by 90 degrees)?

Solution: (a) As the detuning is zero, the probability of finding the system in the ground $|g\rangle$ and excited $|e\rangle$ states can fully oscillate between zero and one with Rabi frequency of $3.34\,\mathrm{MHz}$, as shown in Figure 9.2(b). At instances where P_g is equal to one, P_e becomes zero, and vice versa. The first instance the probabilities are reversed from their initial conditions is at $t_\pi = 150\,\mathrm{ns}$. This process can also be described as the qubit in ground state $|g\rangle$ rotating by 180 degrees on the Bloch sphere and landing on the excited state $|e\rangle$, as depicted in Figure 9.3(c). Therefore, the pulse length on the XY-line for the X-gate should be $150\,\mathrm{ns}$. The subscript π is to remind us that this is the required duration of the pulse to rotate the qubit by π. Recalling Chapter 8, we see that this is precisely the action of the X-gate or Pauli matrix σ_x.

(b) Recall from Chapter 8 that rotating a qubit around the X-axis by 90 degrees brings the qubit to the equatorial plane of the Bloch sphere

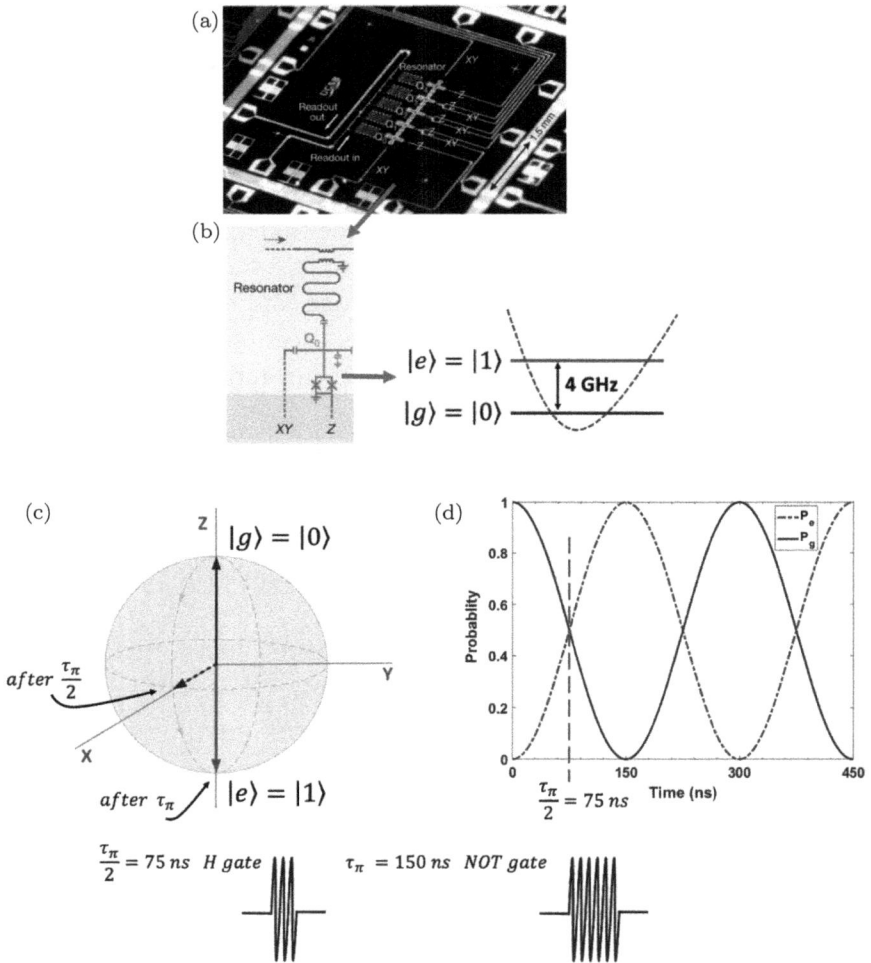

Figure 9.3. (a) A microphotograph of a 5-qubit processor (Barends, 2014). The qubits are nonlinear LC resonators called transmons. Each transmon has two states separated by an energy in the 4–5 GHz range. All qubits are coupled to each other capacitively. A common transmission line is coupled to the resonator of each qubit to read them. Each qubit has a flux line (shown by Z) to tune its energy separation, i.e., qubit frequency. The XY-line carries a microwave pulse close to each qubit to rotate the qubits around the X- or Y-axis on the Bloch sphere. (b) Circuit diagram of the first qubit in the array (Q_0). The qubit's energy levels are tunable by the magnetic flux created by the current in the Z-line.

←——

Figure 9.3. (*Continued*) Q_0 is coupled to the neighboring qubits by a capacitor, and it is coupled to its adjacent resonator (meandering type) for readout. (c) The application of a 4 GHz pulse to the XY-line swaps the probability to be in the ground or excited states with a Rabi frequency of 3.34 MHz. When the qubit at time $t = 0$ is in the $|0\rangle$ state, cutting the pulse at 75 ns leaves the qubit in the equator (Hadamard-gate). Turning off the pulse at 150 ns leaves the qubit in state $|1\rangle$, (a 180 degree rotation), which is a NOT gate (Adapted with permission from John Martinis).

and creates a 50%–50% superposition of ground and excited ($|g\rangle$ and $|e\rangle$) states when the initial state is $|g\rangle$. This rotation is enabled by the Hadamard gate (Chapter 8). To implement the Hadamard gate by a microwave pulse on the XY-line, we require a 4 GHz pulse for a time $t_{\pi/2} = 75$ ns for the proper rotation. At the end of this pulse, the occupation probabilities of $|g\rangle$ and $|e\rangle$ states are both 50% (see Figure 9.3(d)).

Example 2. For a two-level system with a resonant frequency $\frac{\omega_{21}}{2\pi} = 4$ GHz, what is the consequence of increasing the Rabi frequency? How can the Rabi frequency be increased? Plot the probabilities $P_1(t)$ and $P_2(t)$ in equations (9.51) and (9.52) versus the Rabi frequency Ω. Assume that detuning is zero and use a *fixed* pulse length of 100 ns.

Solution: The Rabi frequency determines the period at which the probability oscillates between the states of a two-level system (see Figure 9.2). We know from equation (9.33) that the Rabi frequency is proportional to the matrix element of the applied potential energy $q\vec{\mathcal{E}} \cdot \vec{r}$ between the ground and excited states.

We start with the condition that the system is only in state ϕ_{1o} at time $t = 0$ ($P_1 = 1$ and $P_2 = 0$). In Figure 9.4, we show the state of the system at $t = 100$ ns as the Rabi frequency (electric field amplitude) increases. At specific values of the Rabi frequency $\Omega = (2n - 1) \times 2.5$ MHz ($n = 1, 2, 3, \ldots$), there is zero probability of finding the system in the ground state and 100% probability in the excited state ($P_1 = 0$ and $P_2 = 1$). Finally, with $\hbar\Omega = |u_{12}|$ from equation (9.34), we find that the matrix element of $q\vec{\mathcal{E}} \cdot \vec{r}$ is equal to 1.65 neV when the Rabi frequency is 2.5 MHz in Figure 9.4.

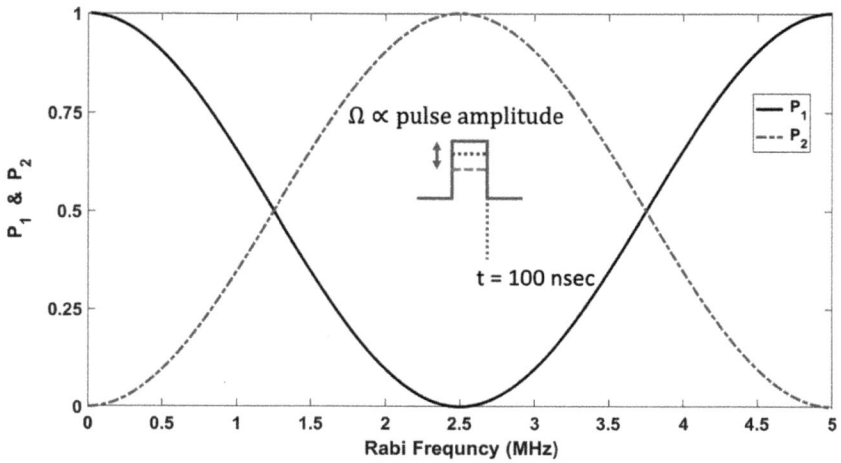

Figure 9.4. The swapping of occupation probabilities by changing the amplitude of the 4 GHz pulse with a fixed length (100 ns). The required pulse amplitude (voltage) is extracted from the Rabi frequency at which swapping occurs (2.5 MHz here).

Technologically, there are two ways to rotate a qubit. We can change the duration of the pulse length and keep the magnitude of the electric field fixed. Alternatively, we can keep the microwave pulse length fixed and change the amplitude of the pulse as done in this example.

Example 3. In this example, we plot the Chevron pattern of a qubit gate. Note that ϕ_{1o} and ϕ_{2o} represent the qubits $|g\rangle$ or $|0\rangle$ and $|e\rangle$ or $|1\rangle$.

(a) Plot $P_1(t)$ as a function of time and detuning. Assume detuning Δ varies from -50 MHz to $+50$ MHz and the time period of $P_2(t)$ is $T = 2\pi/\Omega = 300$ ns.

(b) Show which instants of time correspond to $\sqrt{\text{NOT}}$ gate and NOT gate (also known as the X-gate)? Why?

Solution: (a) We plot equation (9.51) in 2D as a function of time and detuning. When detuning is zero, maximum swapping between zero and one for $P_1(t)$ can be seen in Figure 9.5(a). Additionally,

(a)

(b)

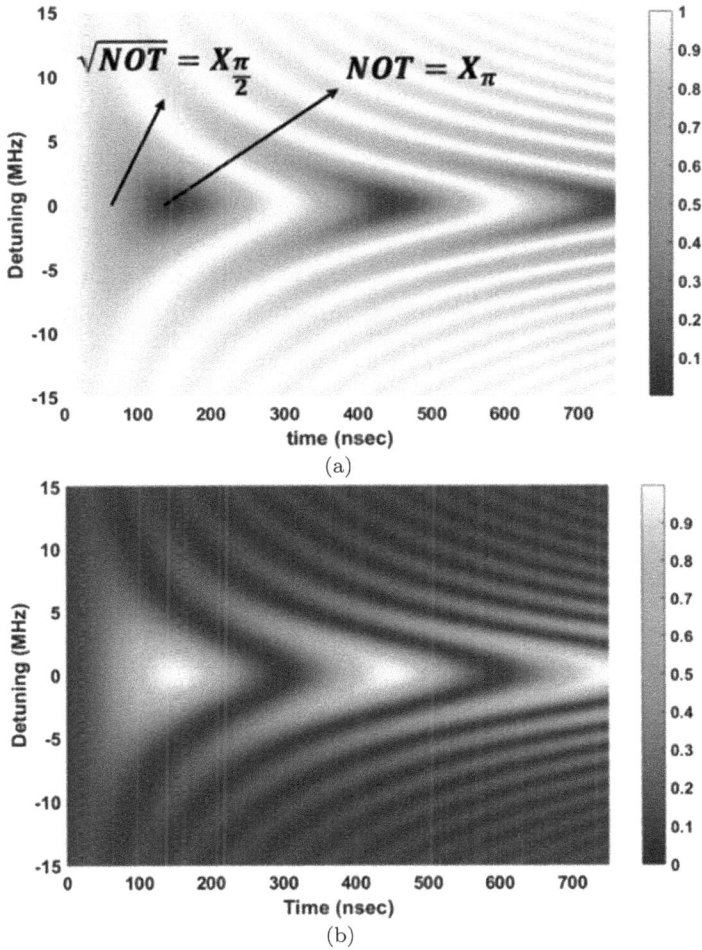

Figure 9.5. 2D plots of the probability of the system being in its (a) ground state $P_1(t)$ and (b) excited state $P_2(t)$, as a function of time and frequency detuning. The Rabi frequency is 3.34 MHz, and the qubit frequency is 4 GHz.

$P_2(t) = 1 - P_1(t)$ can be observed by comparing the two subfigures in Figure 9.5. The population of qubits in the ground state (ϕ_{1o}) and excited state (ϕ_{2o}) oscillates with the Rabi frequency, which is $\frac{\Omega}{2\pi} = 3.34$ MHz. The time required for a pulse to rotate the qubit from 0 to 1 or vice versa (the function of a X or NOT gate) is $t_\pi = 150$ ns.

A pulse with half of this period rotates the qubit by $\pi/2$, hence $t_{\frac{\pi}{2}} = 75\,\text{ns}$ is the required time to implement the $\sqrt{\text{NOT}}$ gate.

The following table summarizes the important frequencies that are used in studying the two-level system perturbed by a time-dependent field. Indices 1 and 2 are sometimes called $|g\rangle$ (ground) and $|e\rangle$ (excited) states depending on the context.

Frequency	Symbol	Meaning	Equation
Driving frequency	ω	Driving frequency of applied ac field	$\omega = 2\pi f$
Natural frequency	ω_{21}	Resonant/natural frequency of two-level system	$\omega_{21} = \dfrac{E_{2o} - E_{1o}}{\hbar}$
Rabi Frequency	Ω	Frequency of transition between the two levels when $\omega = \omega_{21}$	$\Omega = \dfrac{|u_{12}|}{\hbar}$
Detuning	Δ	Difference between driving and resonant frequencies	$\Delta = \omega - \omega_{21}$

Consider a two-level system with energy eigenvalues E_{1o} and E_{2o} and corresponding eigenfunctions ϕ_{1o} and ϕ_{2o}. If a time-dependent perturbation $U(\bar{r}, t) = u(\bar{r}) \cos(\bar{k} \cdot \bar{r} - \omega t)$ is applied, the probability to find the particle in states ϕ_{1o} and ϕ_{2o} will sinusoidally vary with time.

The probability oscillates with a frequency, $\Omega' = \sqrt{\Omega^2 + \frac{\Delta^2}{4}}$ where

$\Omega = \frac{|u_{12}|}{\hbar}$ (Rabi frequency, which depends on the strength of the ac field)

$\Delta = \omega - \omega_{21}$ (Detuning frequency, which is the difference between the driving and resonant frequencies)

The probability oscillates between the minimum value (zero) and maximum value (one) only when the detuning is zero.

9.3 Problems

(1) Prove equations (9.17) and (9.20).
(2) Show that the sum of probabilities in equations (9.51) and (9.52) is one.
(3) Consider a two-level system with energy level spacing equal to 2 eV. At time $t = 0$, the electron is in state 1. Plot the probability to find the electron as a function of time when the frequency of the applied AC potential is (a) 2.05 eV, (b) 2 eV and (c) 1.99995 eV. Assume that the matrix element $u_{12} = 6.5 \times 10^{-6}$ meV.

Reference

Barends, R., Kelly, J., Megrant, A. *et al.* (2014). Superconducting quantum circuits at the surface code threshold for fault tolerance. *Nature* **508**, 500–503.

Chapter 10

TUNNELING

Contents

10.1 Introduction

Tunneling is a key quantum mechanical phenomenon because of
which an electron can travel through a classically forbidden region
(barrier), where its kinetic energy (KE) is smaller than the potential
energy (U). A simple example of hardware to create a potential

(a)

(b)

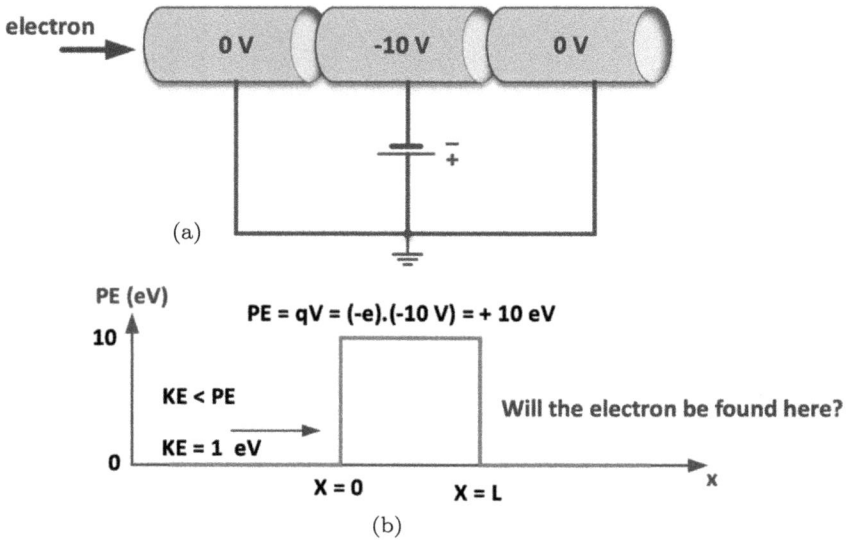

Figure 10.1. (a) A set of three tubes are kept at different electrostatic potentials. The left and right tubes are kept at an electrostatic potential of zero and the middle tube is kept at an electrostatic potential of $U_o/(-q)$, where $-q$ is the charge of the electron. For the electron to feel a repulsive potential U_o, the central tube's potential will be a negative number, such as $-10\,\text{V}$. We can assume that the left and right tubes are infinitely long while the middle tube has a finite width of L, and the transition between the tubes are abrupt. (b) The figure shows a left region $(x < 0)$ where the potential energy (U) is zero, followed by a potential energy barrier from $x = 0$ to L with a constant $U = U_o$, which is then followed by a right region $(x > L)$ with $U = 0$. An electron with kinetic energy less than U_o is incident from the left. The question is, will the electron cross the barrier to be found on the right side where $x > L$?

barrier using three co-axial metal tubes is shown in Figure 10.1(a) and the corresponding electrostatic potential that an electron feels along the axis is shown in Figure 10.1(b). We will assume that the electron travels along the tube's axis and calculate the probability to tunnel in the following sections.

Tunneling has many applications in science and engineering. One typical example in electrical devices is the *tunneling of*

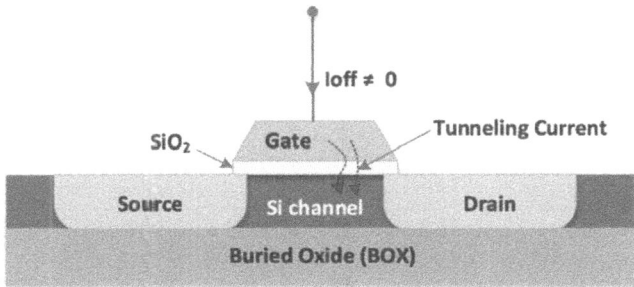

Figure 10.2. In a MOSFET in its OFF state, the voltage on the source and gate are zero, and the drain can be connected to a positive voltage. If the insulating layer, (silicon oxide or hafnium oxide) is thin, electrons can tunnel from the gate to the channel. This can cause current to flow between the drain and gate terminal, leading to heat dissipation in the OFF state.

electrons through the metal-oxide semiconductor field-effect transistor's (MOSFET) gate oxide.[1] This unwanted tunneling gives rise to power dissipation in the OFF state, meaning a decrease in the I_{on}/I_{off} ratio (ON to OFF state currents), an important figure of merit for digital logic gates. As the oxide has become thinner with MOSFETs' miniaturization, electrons are more likely to tunnel from the gate metal into the silicon through the gate oxide layer (Figure 10.2), which is an insulator. As a result, undesirable leakage current flows in the gate contact. For a circuit (for example, a CPU chip) with millions of transistors, the leakage current leads to battery drainage and heat generation.

The second example is *Fowler–Nordheim* tunneling, which describes the device physics of field emission from sharp tips and ionization of atoms in strong electric fields. Free electrons in a solid

---

[1]Metal-oxide semiconductor field-effect transistor (MOSFET) is an electrical switch in digital circuits. The gate oxide is an insulator and is critical to operation. A desirable property of a MOSFET is that at dc, current should flow only between the source and drain as the gate oxide is an insulator. But in reality, there is an undesirable small tunnel current.

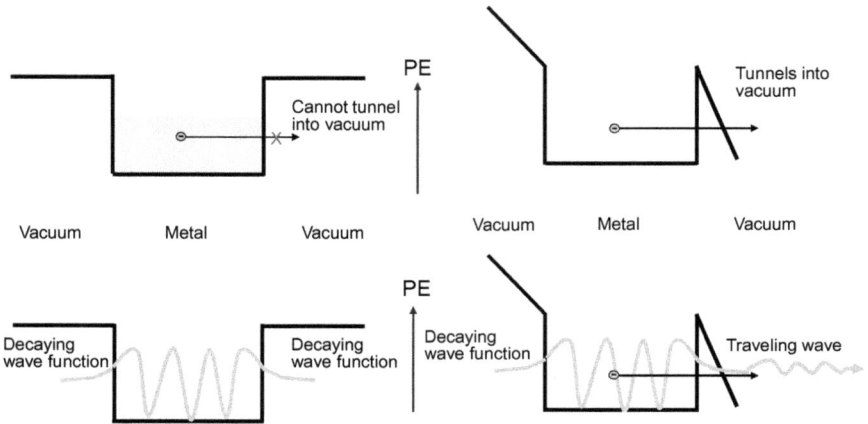

Figure 10.3. Figure shows electrons inside a metal (yellow). The black line represents the potential felt by the electron without (left) and with (right) an electric field. Applying a positive voltage in the vicinity of the metal tip changes the potential such that electrons can tunnel into vacuum as shown (right). The bottom figures show the electron's wave function, which decays in the barrier region but are traveling waves in the metal and far to the right in the presence of an electric field.

are bound by the nuclei that make up the lattice. As a result, when an electron tries to escape from a solid, it feels high potential energy in the vacuum outside and is pulled back into the solid (Figure 10.3). However, when a large electric field is applied, the potential energy outside the metal can become lower than the KE of the electron inside the metal (see Figure 10.3, right). Now Electrons can tunnel through the barrier into the vacuum. This phenomenon is known as Fowler–Nordheim tunneling.

One of the applications of this effect is in field emission displays (FED). They were a promising technology in the 1990s due to their brightness and thin physical structure compared to voluminous cathode ray tube (CRT) displays. A two-dimensional array of sharp nanoscale metal tips is placed very close to a screen biased with a positive voltage and covered with a phosphorescent material. Addressing the tips with image data (coded as voltages pulses) recreates it on the screen. The metal tips biased with high voltage emit electrons due to tunneling out of the tip (Figure 10.3, right).

These electrons create a bright spot on the screen upon colliding with it. The metal tips with insufficient bias do not emit an electron and leave a dark spot on the screen. However, with the advent of other low-cost alternatives like LCD, plasma panel display, and OLED TV, the FED technology became obsolete.

The third example is *Zener tunneling*, where an electron tunnels from a semiconductor's valence to its conduction band, causing a reverse breakdown in PN junction diodes. This event occurs at a well-defined reverse bias voltage (called *Zener voltage*) when a large amount of current can flow through the diode. Both voltage regulators and protection circuits that limit the applied voltage in a sensitive circuit use the Zener diode. The voltage across a device connected in parallel with the diode is thus limited, and the device is protected.

The fourth example is a *double barrier resonant tunneling diode (RTD)*, which consists of two potential barriers in series (Figure 10.4). The RTD is a device that has a negative differential resistance. Thus, it is a component which can be used in high-frequency oscillators. The negative resistance compensates for the Ohmic loss of the LC tank in the oscillator circuit and keeps the oscillation undamped. The RTD also has the property that electrons transmit across the device only at specific energy windows – an electron filter. We will discuss the RTD in greater detail in Section 10.5.

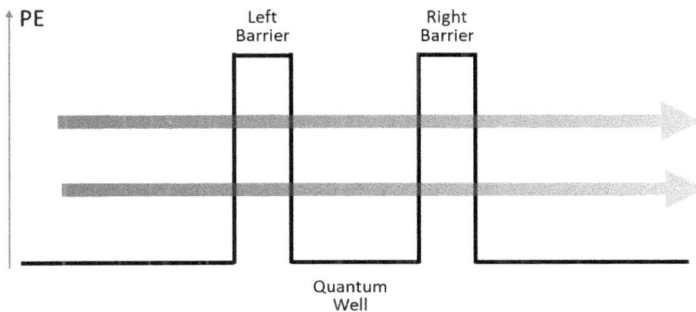

Figure 10.4. A double barrier resonant tunneling diode consists of two barriers connected in series. The region between the left and right barriers is called a quantum well. Electrons transmit across the device only at certain energy windows (red bands). Electrons incident at other energies are primarily reflected.

Another exciting technology in which tunneling plays a vital role is **scanning tunneling microscopy (STM)**, which led to the 1986 Nobel Prize for its inventors, Gerd Binnig and Heinrich Rohrer of the IBM Research Center at Zürich. With this microscope, it is possible to see the surface of conducting solids with atomic-scale resolution. This will also be discussed later.

Concept: Tunneling is an essential quantum phenomenon, where a particle can travel through classically forbidden regions, where the potential energy is larger than the kinetic energy.

Tunneling is seen in various applications (transistors, field emission devices, and scanning tunneling microscopes, etc).

10.2   Tunneling through Single Barriers

We will now derive the expression for transmission probability of electrons incident across a single barrier shown in Figure 10.5. Let us solve Schrödinger equation in three regions: (1) left of the barrier $(x < 0)$, (2) inside the barrier $(0 < x < L)$, and (3) right side of the barrier $(x > L)$. The solution gives us the wave function, from which we can calculate the transmission and reflection coefficients.

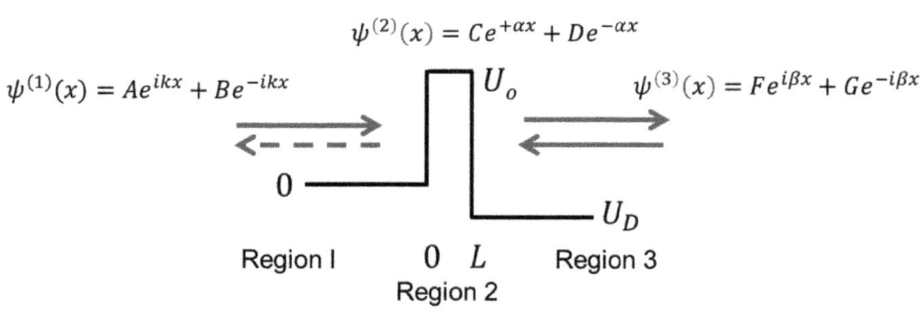

$$\psi^{(2)}(x) = Ce^{+\alpha x} + De^{-\alpha x}$$

$$\psi^{(1)}(x) = Ae^{ikx} + Be^{-ikx} \qquad U_o \qquad \psi^{(3)}(x) = Fe^{i\beta x} + Ge^{-i\beta x}$$

$$U_D$$

Region I   0 L   Region 3

Region 2

Figure 10.5. A single barrier, where the potential in Regions 1 and 3 are not the same. Region 2 $(0 < x < L)$ is the barrier. $\psi^{(1)}$, $\psi^{(2)}$, and $\psi^{(3)}$ represent a single wave function at a given energy.

The potential energy (PE) as a function of position x is

$$U(x) = \begin{cases} 0 & x < 0 \quad \text{(Region 1)} \\ U_o & 0 < x < L \quad \text{(Region 2)} \\ U_D & x > L \quad \text{(Region 3)} \end{cases} \qquad (10.1)$$

where L is the barrier width, U_o is the barrier height, and q is the absolute value of the charge of an electron ($q = +1.6 \times 10^{-19}$ C). Our strategy to find the wave function is as follows:

- The PE, $U(x)$ within each of these three regions does not change. We know how to solve Schrödinger equation within each region and write down the wave functions' mathematical forms in the three regions.
- We assume that an electron wave is incident from the left side of Region 1.
- Since the wave function and its first derivative are continuous at all points, we will match their values at $x = 0$ and $x = L$, which are the boundaries between the three regions. Reviewing the properties of the wave function discussed in Chapter 1 will be useful.

Region 1 ($-\infty < x < 0$)

For this region, Schrödinger equation is,

$$-\frac{\hbar^2}{2m}\frac{d^2\psi}{dx^2} = E\psi \qquad (10.2)$$

we write the wave function as

$$\psi^{(1)}(x) = Ae^{ikx} + Be^{-ikx} \qquad (10.3)$$

where the wave vector is

$$k = \frac{2\pi}{\lambda} = \sqrt{\frac{2mE}{\hbar^2}} \qquad (10.4)$$

In $\psi^{(1)}(x)$, Ae^{ikx} is the incident wave propagating along $+x$ direction and Be^{-ikx} is the wave reflected from the barrier which is propagating along $-x$ direction.

Region 2 $(0 < x < L)$

For this region, Schrödinger equation is

$$-\frac{\hbar^2}{2m}\frac{d^2\psi}{dx^2} + U_o\psi = E\psi \tag{10.5}$$

and we write the wave function as

$$\psi^{(2)}(x) = Ce^{+\alpha x} + De^{-\alpha x} \tag{10.6}$$

$$\text{where } \alpha = \sqrt{\frac{2m(U_o - E)}{\hbar^2}} \tag{10.7}$$

When $E < U_o$, the value of α is a real number. The first term of $\psi^{(2)}$ is an exponentially growing term, and the second term is an exponentially decaying term, as x increases.

When $E > U_o$, $\alpha = i\beta'$ is purely imaginary. The first term of $\psi^{(2)}$ is a traveling wave moving in the $+x$ direction $(Ce^{+i\beta' x})$ and the second term is a traveling wave in the $-x$ directions $(De^{-i\beta' x})$.

Region 3 $(L < x < \infty)$

For this region, Schrödinger equation is

$$-\frac{\hbar^2}{2m}\frac{d^2\psi}{dx^2} - U_D\psi = E\psi \tag{10.8}$$

and we write the wave function as

$$\psi^{(3)}(x) = Fe^{i\beta x} + Ge^{-i\beta x} \tag{10.9}$$

$$\text{where } \beta = \sqrt{\frac{2m(E - U_D)}{\hbar^2}} \tag{10.10}$$

When $E > U_D$, the first term of $\psi^{(3)}$ is a traveling wave along $+x$ direction $(Fe^{+i\beta x})$ and the second term is a traveling wave in the $-x$ directions $(Ge^{-i\beta x})$.

If a wave is incident from Region 1 toward the barrier, there can only be a transmitted wave propagating to the right in Region 3. As a result, the amplitude G is equal to zero. The wave function in

Region 3 is

$$\psi^{(3)}(x) = Fe^{i\beta x} \tag{10.11}$$

The wave functions $\psi^{(1)}$, $\psi^{(2)}$, and $\psi^{(3)}$ satisfy Schrödinger equation separately in each of the three regions; they do not represent the full solution until the constants A, B, C, D, and F are determined. Note that one can think of A, B, C, D, and F as amplitudes of the waves. The continuity of the wave function and its first derivative applied at boundaries $x = 0$ and $x = L$ yield the following equations:

$$\psi^{(1)}(x = 0^-) = \psi^{(2)}(x = 0^+) \Rightarrow A + B = C + D \tag{10.12}$$

$$\left.\frac{d\psi^{(1)}(x)}{dx}\right|_{x=0^-} = \left.\frac{d\psi^{(2)}(x)}{dx}\right|_{x=0^+} \Rightarrow ik(A - B) = \alpha(C - D) \tag{10.13}$$

$$\psi^{(2)}(x = L^-) = \psi^{(3)}(x = L^+) \Rightarrow Ce^{\alpha L} + De^{-\alpha L} = Fe^{i\beta L} \tag{10.14}$$

$$\left.\frac{d\psi^{(2)}(x)}{dx}\right|_{x=L^-} = \left.\frac{d\psi^{(3)}(x)}{dx}\right|_{x=L^+} \Rightarrow \alpha(Ce^{\alpha L} - De^{-\alpha L}) = i\beta Fe^{i\beta L} \tag{10.15}$$

The four linear equations (10.12) to (10.15) currently have five unknowns, but we can express amplitudes B, C, D, and F in terms of the incident amplitude A. Using equations (10.12) and (10.13), we find C and D in terms of A and B. Substituting these expressions for C and D in equations (10.14) and (10.15), we express B and F in terms of A as

$$\begin{bmatrix} B \\ F \end{bmatrix} = \begin{bmatrix} \alpha - ik & -(\alpha + i\beta)\,e^{-\alpha L}e^{i\beta L} \\ \alpha + ik & -(\alpha - i\beta)\,e^{\alpha L}e^{i\beta L} \end{bmatrix}^{-1} \begin{bmatrix} \alpha + ik \\ \alpha - ik \end{bmatrix} A \tag{10.16}$$

In equation (10.3), A is the amplitude of the incident (right moving) wave and B is amplitude of the reflected (left moving) wave. Similarly, in equation (10.11), F is the amplitude of the transmitted wave.

From the above equation, we have

$$\frac{B}{A} = \frac{(\alpha + i\beta)(\alpha - ik)e^{-\alpha L} - (\alpha - i\beta)(\alpha + ik)e^{\alpha L}}{(\alpha + i\beta)(\alpha + ik)e^{-\alpha L} - (\alpha - i\beta)(\alpha - ik)e^{\alpha L}} e^{-i\beta L} \quad (10.17)$$

$$\frac{F}{A} = \frac{-4i\alpha k}{(\alpha + ik)(\alpha + i\beta)e^{-\alpha L} - (\alpha - ik)(\alpha - i\beta)e^{\alpha L}} e^{-i\beta L} \quad (10.18)$$

Using the expression for the probability current density in section 10.8, the incident, transmitted and reflected probability current densities are $\frac{\hbar k}{m}|A|^2$, $\frac{\hbar \beta}{m}|F|^2$ and $\frac{\hbar k}{m}|B|^2$. As the incident probability current density should equal the sum of the transmitted and reflected components, the reflection probability (R), which is the ratio of the reflected to incident probability current densities is,

$$\textbf{(Reflection probability)} \qquad R = \left|\frac{B}{A}\right|^2 \qquad (10.19)$$

Similarly, the transmission probability (T), which is the ratio of the transmitted to incident probability current densities is,

$$\textbf{(Transmission probability)} \quad T = \frac{\beta}{k}\left|\frac{F}{A}\right|^2 \qquad (10.20)$$

It can be verified that the sum of the reflection and transmission probabilities, as expected, should add up to one, $T + R = 1$. Figure 10.6 shows the transmission (T) and reflection (R) probabilities as a function of energy for a barrier height equal to $U_o = 1\,\text{eV}$ and width equal to $L = 2\,\text{nm}$. The transmission and reflection probabilities have surprising features if the electron is considered a classical particle. (However, if we accept that the electron is a wave, these features are not very surprising.)

When KE > maximum of PE $(E > U_o)$, we have the following:

- Classically, the probability for particle transmission probability over the barrier is one. But quantum mechanically, the transmission probability is less than/equal to one and oscillates with energy. The value of energy where $T = 1$ corresponds to a resonance condition $kL = n\pi$, where $k = \sqrt{2m(E - U_o)/\hbar^2}$. We can derive this from equation (10.18).

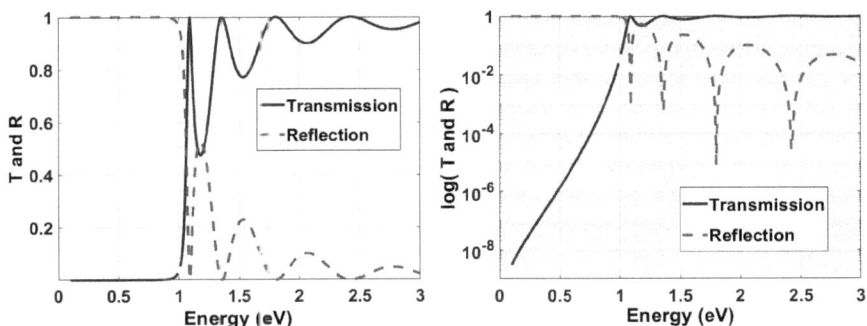

Figure 10.6. Transmission (T) and reflection (R) probability as a function of energy. (Left) Linear scale. For $E < U_o$, T is very small but nonzero. For $E > U_o$, T oscillates and is less than one at most energies. (right) Log version of plot on the left. Here $\log(T)$ increases exponentially with energy and is proportional to $\sqrt{U_o - E}$, at energies much smaller than the barrier height. Free electron mass is assumed, barrier height $U_o = 1\,\mathrm{eV}$, $U_D = 0$, and barrier width $L = 2\,\mathrm{nm}$.

- A classical analogy is water waves in a shallow channel, where the wave's velocity is depth dependent. Hence, when the wave passes over an area with an underwater barrier (or ditch), the sudden depth change causes a velocity change. As a result, water waves reflect at the border even though the water's surface is above both the riverbed and the barrier/ditch.

When KE < maximum of PE $(E < U_o)$, we have the following:

- Classically, the reflection probability is one and the transmission probability is correspondingly zero. But quantum mechanically, the reflection probability can be smaller than unity, meaning that transmission is possible or penetration of the particle through the barrier is possible.
- Transmission probability monotonically increases with energy.
- When $E < 0$, there are no solutions which are traveling waves in both Regions 1 and 3.

The phenomenon of tunneling also has an analogy with electromagnetic waves traveling through media with a spatially varying refractive index (hence different wave velocity). Reflection also occurs when a microwave signal travels in a transmission line with a discontinuity (for example, a width change). Here the change of impedance (or group velocity of the microwave along the line) leads to reflection.

As an exercise, you should calculate the tunneling probability for a classical particle with a mass of $1\,\mathrm{mg}$, $L = 0.02\,\mathrm{mm}$, and barrier height of $1\,\mathrm{eV}$, and observe the differences. This will help you realize why tunneling is not observed in the classical world (see Problem 4).

10.2.1 *Wide–tall barriers*

The expression for the transmission probability T has a more appealing form in the limit of a wide/tall barrier, which mathematically corresponds to

$$\alpha L \gg 1 \Rightarrow \sqrt{\frac{2m\,(U_o - E)}{\hbar^2}}\,L \gg 1 \tag{10.21}$$

Note that $\alpha L \gg 1$, when L is large and the potential energy U_o is (much) larger than the energy E.

For a wide/tall barrier, assuming $U_D = 0$ and $\beta = k$,

$$T = \left|\frac{F}{A}\right|^2 \approx \frac{16\alpha^2 k^2 e^{-2\alpha L}}{|(\alpha - ik)\,(\alpha - ik)|^2} \tag{10.22}$$

Equation (10.22) implies that for a wide/tall barrier, the transmission probability decreases:

- exponentially with an increase in barrier width L (see Figure 10.7),
- exponentially with an increase in $\sqrt{U_o - E}$ (see Figure 10.8).

(Note that the other factors in the above equation for transmission probability depend only algebraically on $U_o - E$, which is a much weaker dependence than exponential dependence.)

Concept: When a particle with kinetic energy smaller than the potential energy of the barrier is incident, the tunneling probability decreases exponentially with increase in both (a) barrier width L and (b) difference in energy between the barrier height and kinetic energy $(U_o - E)$.

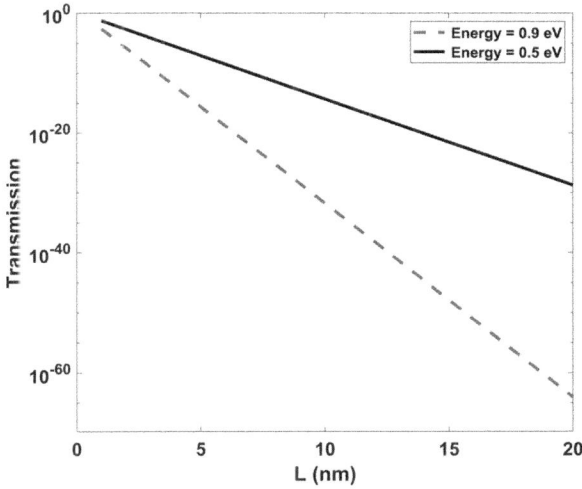

Figure 10.7. The Logarithm of transmission probability decreases nearly linearly with increase in barrier width L (or in other words, T decreases exponentially with an increase in barrier width). Electron energy is 0.5 or 0.9 eV and the barrier height is $U_o = 1$ eV. The mass corresponds to that of a free electron.

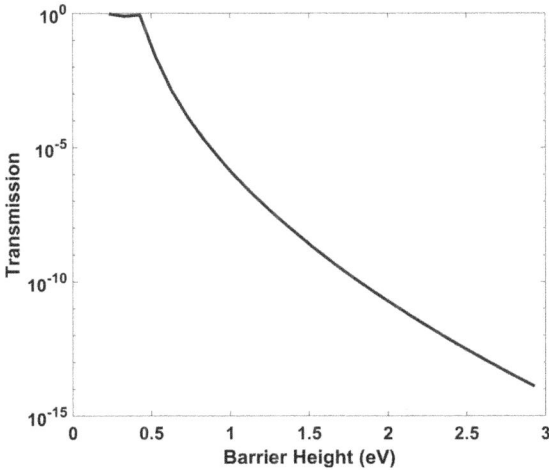

Figure 10.8. For a given energy, the logarithm of transmission probability decreases with an increase in barrier height and is proportional to $-\sqrt{U_o - E}$. The barrier width is $L = 2$ nm and electron energy is 0.5 eV. The mass corresponds to that of a free electron.

10.3   Tunneling Time through a Single Rectangular Barrier

Classically, the time it takes for an electron to travel over a barrier of width L is

$$t = \frac{L}{\text{velocity}} = \frac{L}{\sqrt{\frac{2(E-U_o)}{m}}} \tag{10.23}$$

where $E > U_o$ and $\text{KE} = E - U_o = \frac{mv^2}{2}$. At energies $E < U_o$, the question "What is the time for an electron with $E < U_o$ to tunnel through the barrier?" is valid in quantum mechanics, even when an electron cannot classically travel through the barrier. This time would set the limit for the speed at which devices based on tunneling operate. Another interesting question is, "Does the time it takes an incident electron to get reflected and transmitted differ?".

The answer to these questions on tunneling time is not settled, but we will discuss two different literature viewpoints. The first viewpoint is that the relevant timescale is the *dwell time* (t_{dwell}). The second viewpoint is a method to calculate the times for tunneling and reflection which utilizes a small signal AC potential, following the work of Marcus Büttiker (Büttiker, 1983). Consider an electron incident from the left. The current density carried by this electron would depend on the electron's transmission probability. Intuitively, the easier it is for the electron to tunnel through the barrier, the shorter the amount of time the electron would spend inside the barrier. We can also calculate the probability of finding the electron in the barrier. The *dwell time* is the ratio between the probability to find the electron in the barrier and the probability current density.

$$
t_{\text{dwell}} = \frac{\begin{array}{c}\text{Probability per unit area to find the electron in the}\\ \text{barrier region of length } L\end{array}}{\text{Current density of the electron wave}}
$$

$$
= \frac{P(x = 0 \text{ to } L)/\text{Area}}{J} \tag{10.24}
$$

Using equation (10.6),

$$\frac{P\left(x = 0 \text{ to } L\right)}{\text{Area}} = \frac{1}{\text{Area}} \int_0^L dx \left|\psi^{(2)}(x)\right|^2$$

$$= \frac{1}{\text{Area}} \int_0^L dx \left|Ce^{\alpha x} + De^{-\alpha x}\right|^2 \qquad (10.25)$$

We can calculate the current density (J) using the equation we will derive in Section 10.8. As the probability current density is the same at every point x in steady state, we can use the wave function either to the left, right, or barrier region. If the probability current density is not identical at every point x, the probability will either grow or decay in some regions of space, which is not possible in steady-state conditions. We will evaluate the current density using the expression for the wave function to the barrier's right for convenience. Using equation (10.11) (note $\beta = k$ in equation (10.11) because we assume that the potential both to the left and right of the barrier is 0),

$$J = \frac{i\hbar}{m} \left[\psi^{(3)*} \frac{d\psi^{(3)}}{dx} - \psi^{(3)} \frac{d\psi^{(3)*}}{dx}\right] = -\frac{\hbar k}{2m}|F|^2 \qquad (10.26)$$

Remember from equations (10.12) to (10.15), the amplitudes C, D, and F can be expressed in terms of the amplitude of the incident wave A. So, the dwell time will depend only on the energy and parameters of the barrier,

$$t_{\text{dwell}} = \frac{1}{T} \frac{mk}{\hbar \alpha} \frac{2\alpha L \left(\alpha^2 - k^2\right) + k_o^2 \sinh(2\alpha L)}{4\alpha^2 k^2 + k_o^4 \sinh^2(\alpha L)} \qquad (10.27)$$

where $T = \left|\frac{F}{A}\right|^2$ is transmission, $k = \frac{\sqrt{2mE}}{\hbar}$, $\alpha = \frac{\sqrt{2m(U_o - E)}}{\hbar}$, and $k_o = \frac{\sqrt{2mU_o}}{\hbar}$. Figure 10.9 shows the dwell time (t_{dwell}) versus energy (black solid line). The blue dashed line is the amount of time it takes for an electron with energy E to travel a distance L. The blue dashed line shows that as the energy of the electron increases, the time it takes to travel over the barrier decreases. When $E < U_o$, the black solid line is however counterintuitive because the dwell time decreases with decrease in energy of the incident electron. When $E > U_o$, the

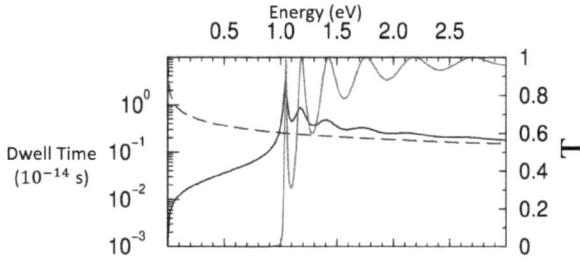

Figure 10.9. The solid black line is the dwell time and the blue dashed line corresponds to barrier width over velocity (L/velocity, equation (10.23)). The transmission probability T is shown in red (Patiño, 2015).

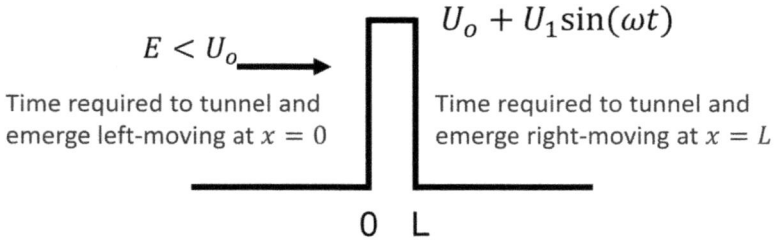

Figure 10.10. An electron is incident at energy E. What is the time required for the incident electron to emerge at $x = L$ and at $x = 0$?

dwell time oscillates with increase in energy but approaches the classical value (blue dashed line) at high energies.

Note that the dwell time does not differentiate between the time required for transmission and reflection. Büttiker suggested an alternate definition for the time necessary for an electron to tunnel through the barrier. This formalism further differentiated between the time required for transmission and reflection. The idea consists of imposing a tiny AC voltage with potential energy $U_1\sin(\omega t)$ on the barrier such that the barrier potential energy changes with time as $U_o + U_1\sin(\omega t)$ (see Figure 10.10). Note that $U_o = qV_o$ and $U_1 = qV_1$, where V_o and V_1 are the dc potential and amplitude of ac potential, respectively. If the AC potential changes slowly compared to the transit time, then the transmission probability through the barrier at any given time simply depends on the value of the potential energy

$U_o + U_1 \sin(\omega t)$ at that time. This is known as the quasi-static limit. As the oscillation frequency increases, the transmission and reflection probabilities through the barrier begin to deviate from the quasi-static limit. The timescale (inverse of AC frequency) at which this deviation begins to occur defines the time required for transmission and reflection. We will not derive these expressions here but will simply provide the final result for an opaque barrier (where the transmission probability is small) when the energy of the tunneling particle $E < U_o$ and when $\hbar\omega \ll E$:

$$\text{Time for transmission, } \tau_T = \frac{L}{(\hbar\alpha/m)} = \frac{L}{\sqrt{\frac{2(U_o-E)}{m}}} \qquad (10.28)$$

$$\text{Time for reflection, } \tau_R = \frac{\hbar}{U_o}\frac{k}{\alpha} = \frac{\hbar}{U_o}\sqrt{\frac{E}{U_o - E}} \qquad (10.29)$$

The reader should compare the denominators of equations (10.23) and (10.28). The former equation's denominator is the velocity of an electron and is valid only when $E > U_o$, while the latter equation's denominator is a velocity-like term that depends on $U_o - E$ and is valid only when $E < U_o$.

Note that τ_T is equal to the width of the barrier divided by the velocity of a particle with energy $U_o - E$. However, there is no particle here with velocity $\hbar\alpha/m$. The reflection times τ_R, on the other hand, are independent of the barrier width and depend only on the barrier height and energy.

The hope is that experiments will determine if the transit time through a barrier is closer to the dwell or transmission time, or something else.

10.4 Tunneling through a Double Barrier Resonant Tunneling Structure

The double barrier resonant tunneling structure (DBRTS) consists of a quantum well (PiB) in between two barriers (Figure 10.11). We will assume that an electron is incident from the left and then calculate the probability of finding the electron on the right ($x > L + b_2$). A specific DBRTS example consists of a quantum well made with

GaAs and barriers made of $Al_xGa_{1-x}As$, where x is the fraction of gallium atoms replaced by aluminum atoms. The barrier height depends on x. This is because the bandgap (the difference between the minimum conduction band energy and maximum valence band energy, i.e., $E_g = E_c - E_v$) changes with Al content. As Al is a smaller atom, it has a higher electronegativity than its neighbor Ga in the periodic table. Hence, it does not lose an electron as easily as Ga, which means that $Al_xGa_{1-x}As$ has a higher bandgap than GaAs. Therefore, increasing the Al percentage (x) increases E_g. The multiple layers of GaAs and AlGaAs, as shown in Figure 10.11, result in a potential energy change along the x-axis for the electron.

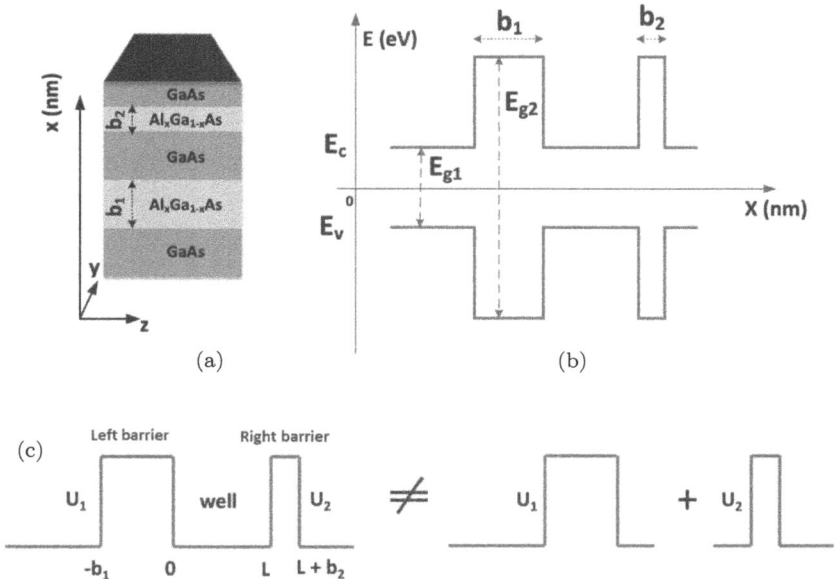

Figure 10.11. (a) Five layers of materials, alternating between GaAs and $Al_xGa_{1-x}As$, form the heterostructure. (b) The conduction and valence bands corresponding to these five layers. We will neglect the valence band for the discussion. (c) Double barrier resonant tunneling structure. The widths of the left and right barriers are b_1 and b_2, respectively, and potential energies of the barriers (barrier heights) are U_1 and U_2, respectively. The width of the quantum well is L. Transmission probability through two barriers in series is not equal to the product of the transmission probability through each barrier individually.

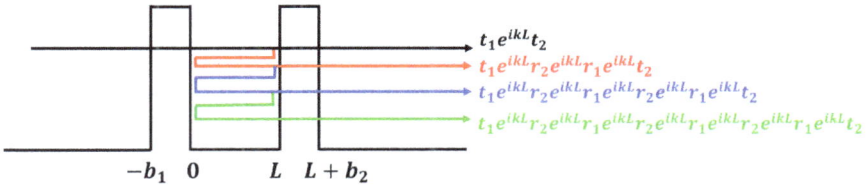

$t_1 e^{ikL} t_2$

$t_1 e^{ikL} r_2 e^{ikL} r_1 e^{ikL} t_2$

$t_1 e^{ikL} r_2 e^{ikL} r_1 e^{ikL} r_2 e^{ikL} r_1 e^{ikL} t_2$

$t_1 e^{ikL} r_2 e^{ikL} r_1 e^{ikL} r_2 e^{ikL} r_1 e^{ikL} r_2 e^{ikL} r_1 e^{ikL} t_2$

$-b_1 \quad 0 \qquad L \quad L+b_2$

Figure 10.12. The propagation of an electron wave that transmits through the left and right barriers can be broken down into a number of paths. Shown in black is direct transmission through the two barriers. The red path involves reflection at the right barrier followed by a reflection at the left barrier before the wave is transmitted to the right, and so on. The waves reflected back to the left are not shown but each of the black, red, blue, and green lines has a component moving to the left of $x = -b_1$. t_1 and t_2 (r_1 and r_2) are the transmission (reflection) amplitudes at the left and right barriers, respectively. They are calculated using equations (10.17) and (10.18).

The potential energy of the DBRTS is

$$U(x) = \begin{cases} 0 & x < -b_1 & \text{(Region 1)} \\ U_1 & -b_1 < x < 0 & \text{(Region 2)} \\ 0 & 0 < x < L & \text{(Region 3)} \\ U_2 & L < x < L+b_2 & \text{(Region 4)} \\ 0 & L+b_2 < x & \text{(Region 5)} \end{cases} \qquad (10.30)$$

We can derive an expression for tunneling through this structure by matching the wave functions and its first derivative at the interfaces, $x = -b_1, 0, L, L + b_2$, using the continuity and differentiability properties of wave functions. Instead of what we did with single rectangular barriers, we will calculate the tunneling probability using a method inspired by electromagnetic wave propagation through layered dielectric media (Datta, 1988).

Consider a wave incident from the left, as shown by the black line in Figure 10.12. The electron waves partially reflect at the left barrier and partially transmit through the left barrier to the well. The right moving wave (black line) in the quantum well partially reflects at the right barrier (upper red line) and partially transmits across the right barrier toward the right side (black line). The amplitude of the transmitted component is $t_1 e^{ikL} t_2$, where t_1 is the transmission amplitude across the left barrier, e^{ikL} is the phase factor accumulated in the quantum well, and t_2 is the transmission amplitude across the right barrier. The reflection amplitudes at the left and right barriers are represented by r_1 and r_2, respectively. The partial wave moving to the left after the reflection at the right barrier reaches the left barrier, where it is partially transmitted (not shown) and partially reflected (red line); at the right barrier, the wave is then partially reflected toward the left barrier (blue line) and partially transmitted across the right barrier (red line, transmission amplitude of $t_1 e^{ikL} r_2 e^{ikL} r_1 e^{ikL} t_2$). This process continues *ad infinitum* as shown by the right-moving partial waves to the right of $L + b_2$.

The total transmission amplitude (t) to the right is the sum of the infinite series of partial waves transmitted,

$$t = t_1 e^{ikL} t_2 + t_1 e^{ikL} r_2 e^{ikL} r_1 e^{ikL} t_2 + t_1 e^{ikL} r_2 e^{ikL} r_1 e^{ikL} r_2 e^{ikL} r_1 e^{ikL} t_2$$

$$+ t_1 e^{ikL} r_2 e^{ikL} r_1 e^{ikL} r_2 e^{ikL} r_1 e^{ikL} r_2 e^{ikL} r_1 e^{ikL} t_2 + \cdots \qquad (10.31)$$

We can rewrite the above equation as

$$t = t_1 e^{ikL} t_2 (1 + r_2 e^{ikL} r_1 e^{ikL} + r_2 e^{ikL} r_1 e^{ikL} r_2 e^{ikL} r_1 e^{ikL}$$

$$+ r_2 e^{ikL} r_1 e^{ikL} r_2 e^{ikL} r_1 e^{ikL} r_2 e^{ikL} r_1 e^{ikL} + \cdots) \qquad (10.32)$$

The term in the parentheses above is a geometric series:

$$t = t_1 e^{ikL} t_2 [1 + (r_2 e^{ikL} r_1 e^{ikL}) + (r_2 e^{ikL} r_1 e^{ikL})^2$$

$$+ (r_2 e^{ikL} r_1 e^{ikL})^3 + \cdots] \qquad (10.33)$$

As the term $u = r_2 e^{ikL} r_1 e^{ikL}$ has a magnitude that is smaller than one, the above geometric series is convergent. Remembering that when $u < 1$, $1 + u + u^2 + u^3 + \cdots = \frac{1}{1-u}$, we can write the transmission amplitude t as

$$t = \frac{t_1 e^{ikL} t_2}{1 - r_2 r_1 e^{i2kL}} \qquad (10.34)$$

The transmission probability (T) is

$$T = |t|^2 = \left| \frac{t_1 t_2}{1 - r_2 r_1 e^{i2kL}} \right|^2 \qquad (10.35)$$

where we have used the fact that $\left|e^{ikL}\right| = 1$ in the numerator. To calculate the transmission probability T through the DBRTS, the transmission and reflection amplitudes through the left and right barriers are required. The expressions for these are in equations (10.17) and (10.18).

When the transmission amplitudes t_1 and t_2 are much smaller than one, at energies close to $kL = n\pi$, equation (10.35) simplifies to

$$T = |t|^2 = \frac{\Gamma_1 \Gamma_2}{(E - E_1)^2 + \left(\frac{\Gamma_1 + \Gamma_2}{2}\right)^2} \tag{10.36}$$

$$\text{where } \Gamma_1 = \frac{\hbar}{\tau}|t_1|^2 \quad \text{and} \quad \Gamma_2 = \frac{\hbar}{\tau}|t_2|^2 \tag{10.37}$$

In the above equation, $\tau = \frac{2L}{\text{velocity}} = \frac{2L}{(\hbar k/m)}$ is the time required for a free particle to traverse the length of the quantum well twice (back and forth). E_1 is the energy level of a resonance (note that there are many such resonances at energies that satisfy $kL = n\pi$). Equation (10.36) is sometimes known as the Breit–Wigner formula, which we will prove later in this section.

Let us now discuss the key features of *transmission versus energy* qualitatively by assuming that t_1 and t_2 are energy independent. For small values of t_1 and t_2, the DBRTS has transmission peaks at energies that approximately satisfy the PiB quantization condition $kL = n\pi$. One would expect this intuitively because the wave function becomes small at the well's boundaries (but not exactly zero as in the PiB). In Figure 10.13, we have plotted the transmission as a function of energy by assuming that $t_1 = t_2$. Assuming well width $L = 7\,\text{nm}$ and electron mass $m = 0.6\,m_0$, the energy levels of the resonances are

$$E_n = \frac{\hbar^2}{2m}\left(\frac{n\pi}{L}\right)^2 = \frac{\left(1.04 \times 10^{-34}\,\text{Js}\right)^2}{2 \times 0.6 \times 9.1 \times 10^{-31}\,\text{kg}}\left(\frac{n\pi}{7 \times 10^{-9}\,\text{m}}\right)^2 \tag{10.38}$$

which gives $E_1 = 12.47\,\text{meV}$ and $E_2 = 49.88\,\text{meV}$ and so on. Note that the transmission peaks in Figure 10.13 occur at the above resonant energies determined by the PiB condition. Also note that the double-barrier heterostructure behaves as an electron

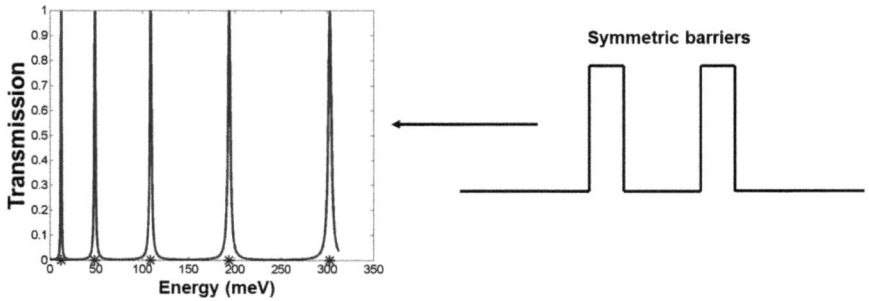

Figure 10.13. The transmission probability versus energy through symmetric barriers (equation (10.35)). The transmission amplitudes of both barriers are assumed to be energy independent for simplicity. The other parameters are $t_1 = t_2 = 0.3$, well width $L = 7\,\text{nm}$, and electron mass $m = 0.6\,m_0$. The transmission probability peaks occur at the PiB resonant energies as discussed in the text.

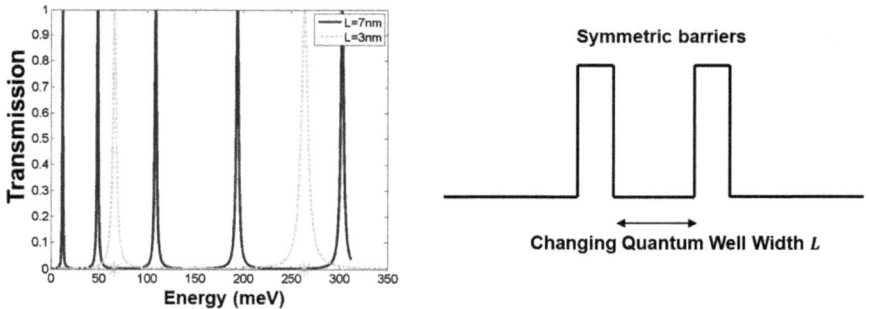

Figure 10.14. The peaks in transmission probability shift to lower energies as the well width increases. The other parameters are $t_1 = t_2 = 0.3$ and $m = 0.6\,m_0$.

filter — only electrons with an energy close to the PiB's resonant energy transmit with high probability.

Figure 10.14 shows the transmission probability versus energy for two different well widths. One can see that making the well width larger decreases the energy of the resonances in a PiB. Further, *symmetric barriers have a maximum transmission probability of one.* While this is true for arbitrary barrier strengths (the barrier strength is determined by both the barrier height and width), we show this to be the case for barriers with small values of t_1 and t_2.

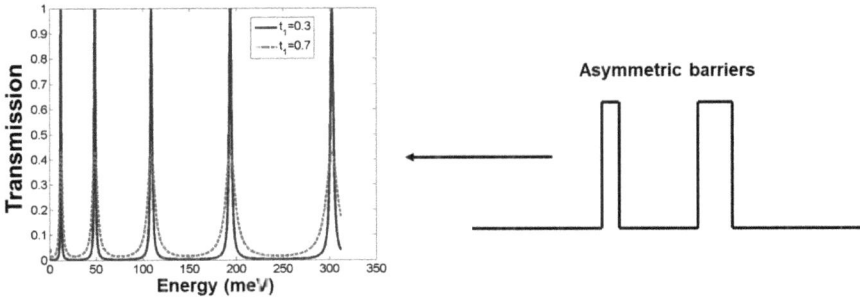

Figure 10.15. The transmission probability versus energy for symmetric (blue) and asymmetric (red) barriers. For the symmetric barrier, $t_1 = t_2 = 0.3$, and for the asymmetric barrier, $t_1 = 0.7$ and $t_2 = 0.3$. The other parameters are well width $L = 7\,$nm and electron mass $m = 0.6\,m_0$.

The transmission probability given by equation (10.36) has its maximum value g

$$\max{(T)} = \frac{\Gamma_1\Gamma_2}{\left(\frac{\Gamma_1+\Gamma_2}{2}\right)^2} \tag{10.39}$$

when $E = E_1$. Also, $\max(T) = 1$ when $\Gamma_1 = \Gamma_2$ and $\max(T) < 1$ when $\Gamma_1 \neq \Gamma_2$. Note that $\Gamma_1 = \Gamma_2$ implies that the transmission probabilities of both barriers are identical (equation (10.37)). Figure 10.15 shows the transmission for symmetric and asymmetric barriers. The transmission is smaller than one for asymmetric barriers.

Concept: The transmission probability through two barriers in series has resonant peaks. For symmetric barriers, the maximum transmission through a double barrier structure is one, many orders of magnitude larger than the transmission through a single barrier.

10.5 Resonant Tunneling Diode

The DBRTS is the basis for the resonant tunneling diode, which shows a negative differential resistance (NDR). NDR means dI/dV is negative at some bias ranges (current decreases with an increase in voltage). Devices with NDR are used in building high-frequency oscillators.

To explain the operation of an RTD, we will simplify the transmission versus energy $[T(E)]$ curve to consist of just the two lowest resonant energy levels (Figure 10.16). When we apply a bias voltage between the device's left and right ends, the electrostatic potential drops across the device. For simplicity, assume that this potential only drops between the two barriers. Electrons from the left and right contacts are incident onto the device at energies below the red dashed line (called the Fermi energy). At tiny applied voltages, there are no electrons at the energy of the transmission resonance, and only a small amount of current can flow through the lower tail (low energy tail) of the transmission resonance (Figure 10.16(a) and 10.16(b)).

As the applied voltage increases, the current through the tail of the transmission resonance increases. The current flow becomes a maximum when electrons flow through the transmission resonance (Figure 10.16(c)). This happens when the first quantum well resonance is usually between the Fermi energy of the left and right contacts. Further increase in the applied voltage eventually makes the resonant energy level in the well fall below the lowest energy on the device's left side (Figure 10.16(d)). Now, current can only flow through the upper part of the transmission peak (which has a low transmission), and the magnitude of the current drops (Figure 10.16(d)). Drop of current with an increasing applied voltage leads to a negative differential resistance. In the current–voltage characteristics of Figure 10.16 (right), there is a further increase in current with voltage (point D) because electrons enter from the left contact into the second resonant level in the quantum well and flow into the right contact.

Another interesting point to note is that the I–V plot's derivative, i.e., dI/dV, resembles the transmission $T(E)$. The peaks of $\frac{dI}{dV}$ correspond to energy levels of the quantum well, i.e., E_1 and E_2 (Figure 10.16). Therefore, more generally, the energy levels (for example, the bandgap) and density of electronic states can be calculated from the I–V plot, as briefly discussed in the following section.

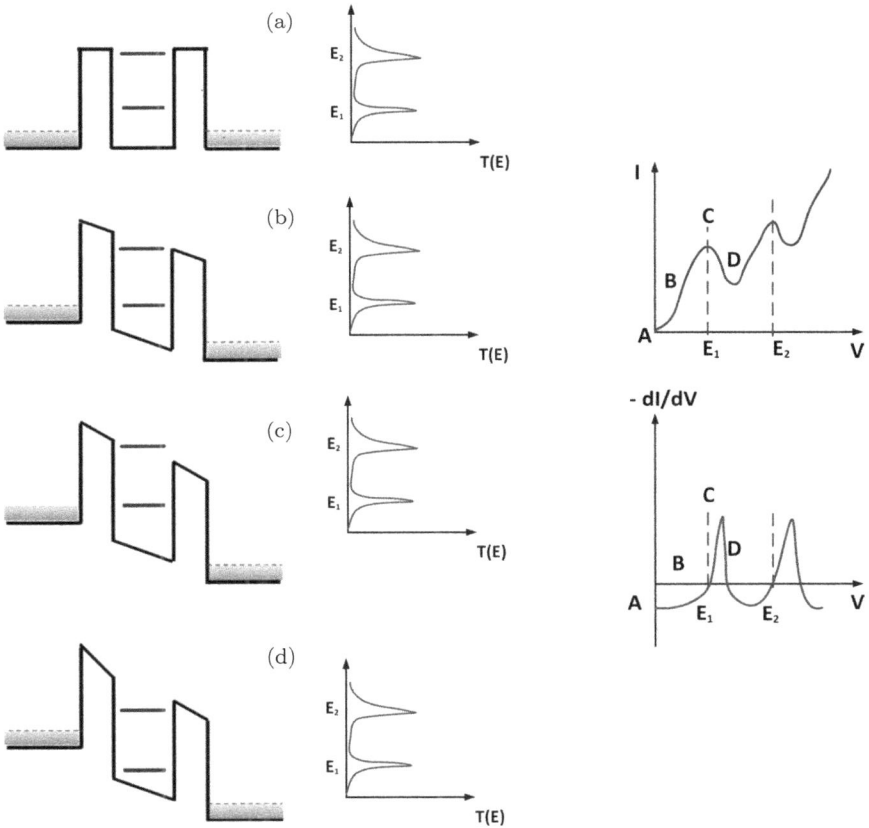

Figure 10.16. (a)–(d) show the evolution of the resonant energy level with applied bias. The top-most plot on the right (red curve) shows how the current changes with bias corresponding to (A) through (D). The derivative of the I–V plot resembles the $T(E)$ or LDOS (the curve on the bottom right). The peaks correspond to the energies where $T(E)$ also peaks, i.e., E_1 and E_2 levels in the quantum well. The energy levels E_1 and E_2 change with applied bias.

10.5.1 Scanning tunneling microscopy and spectroscopy

In this section, we will discuss two more applications of tunneling. The first application is scanning tunneling microscopy, which makes it possible to see the surface of a sample with atomic-scale resolution.

Examples of samples are graphene, boron nitride, carbon nanotubes, nanowires, and Cu surfaces. The sample to be observed is on a conducting plate connected to a sensitive current to voltage converting amplifier (transimpedance amplifier). A very sharp metallic tip with a curvature smaller than 50 nm is brought into proximity with the surface. If the tip and sample are biased (say 0.1–1 V) and the shortest distance between the tip and sample is larger than 2 nm, then there is no current passing between the tip and sample because of the large vacuum barrier between them. However, if the tip is very close to the sample (of the order of 0.5 nm), the electrons in the tip (sample) see a narrower potential barrier, and they can tunnel through the barrier to the sample (tip). Hence, a tunnel current flows through the narrow gap to complete the circuit. This tunnel current, converted to a voltage, provides a feedback control mechanism that tries to keep the current constant by applying signals back to piezoelectric motors, which move the tip up and down along z direction (Figure 10.17). Two other X and Y motors move the tip in x and y directions. If the tip gets very close to the surface during the movement, the feedback loop applies a voltage to pull up the tip so that the tunnel current is unchanged. On the other hand, if the tip-to-surface distance increases due to a sudden discontinuity or ditch on the sample surface, the current drop is sensed. To keep a constant current, the feedback controller applies a voltage to the z-motor to push the tip slightly toward the surface.

Recording all these fluctuations (ups and downs) of the control voltage and plotting them along x and y directions reveal the sample's image with atomic-scale resolution. Figure 10.18 shows the STM image taken from a single layer of graphene, a 2D hexagonal lattice of carbon atoms. The bright spots are carbon atoms, which form a hexagonal ring with dark centers. Note that optical microscopes do not have the resolution to see atomic-scale features because of the optical diffraction limit.

The second application is scanning tunneling spectroscopy (STS), which makes it possible to gain information about the energy spectrum of a sample placed on a metallic/semiconducting substrate. This method reveals the local density of states (LDOS) of a material.

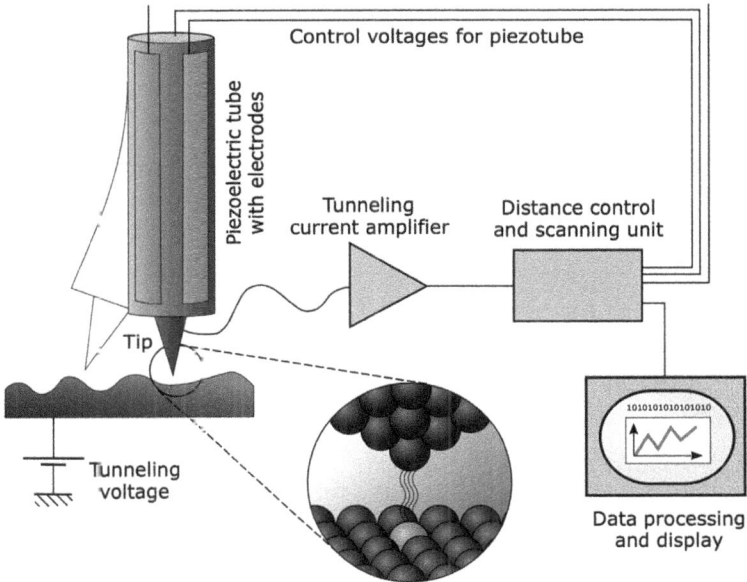

Figure 10.17. Scanning tunneling microscope. The feedback loop using an amplifier tries to keep the tunneling current between the tip and sample constant. As a result, the recorded voltage variation shows the topography of the surface with atomic resolution. Image courtesy: Michael Schmid (TU Wien).

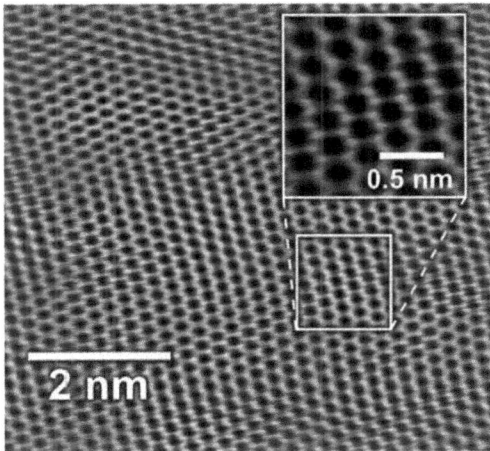

Figure 10.18. STM image taken from a single layer graphene surface. Image provided by Hannes C. Schniepp's laboratory (Schniepp Lab).

LDOS is the number of quantum states per unit volume per unit energy at any point and is discussed in detail in Chapter 12. Like in STS, a sharp metallic tip located approximately a nanometer from the sample is used to apply a voltage (similar to the STM introduced in Figure 10.17). Assume you have a nanowire sample at hand, and you want to find out its bandgap or eigenenergies (Figure 10.19(a)). The nanowire is placed on a substrate, e.g., silicon substrate, and the metallic tip approaches the sample. A small (vacuum) barrier separates the metallic tip and the sample (nanowire). In addition, there might be a barrier between the sample and the substrate. The barrier between the tip and the sample is the larger of the two barriers. A voltage applied between the tip and the substrate causes current to flow through the sample's energy levels. The transmission probability for electrons to tunnel from the tip to the sample at any location depends on the sample's LDOS.

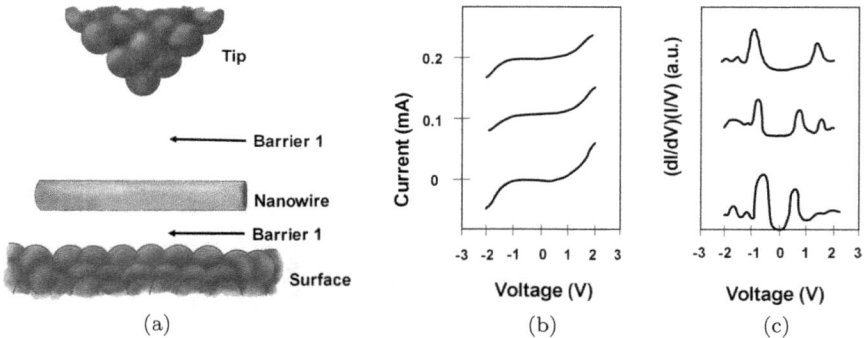

Figure 10.19. Scanning tunneling spectroscopy (STS): (a) A graphic showing an STM tip separated from a nanowire by a large barrier (Barrier 1). In some experiments, a smaller barrier (Barrier 2) may exist between the nanowire and the substrate (called surface here). The differential conductance (dI/dV) depends on the local density of states of the nanowire when it interacts with the substrate. (b) Experimental current (I) versus voltage (V) measured for silicon nanowire with diameters ranging from 1.3 to 7 nm (labeled 1 and 6, respectively) using STS. (c) $\left(\frac{\partial I}{\partial V}\right)/\left(\frac{I}{V}\right)$ shows the bandgap (E_g) increases with decreasing diameter of the nanowire. This shows the physics of a particle in a box or quantum confinement in a nanowire. Adapted from (Ma, 2003). Figure drawing by Mrunal Shenoy.

Figure 10.19(b) is an example of a recorded current versus applied voltage. Taking the derivatives of the current versus voltage $\left(\frac{\partial I}{\partial V}\right)$ shows peaks bearing a close resemblance to the transmission, $T(E)$, or the local density of states (LDOS) as shown in Figure 10.19(c). Figure 10.19 shows STS microscopy results for silicon nanowires of different thicknesses. Due to the nanowire's confined cross-section, the energy difference between eigenenergies (conduction and valence bands) becomes smaller as the nanowire diameter increases. You see this in Figure 10.19(c), which shows the bandgap or difference between transmission peaks corresponding to the conduction and valence band edges increases as the diameter decreases from 7 to 1.3 nm revealing the effect of quantum confinement. The horizontal axis is in terms of voltage, but they can be multiplied by electron charge and expressed as energy (eV).

10.6 Derivation of Breit–Wigner Formula

In this section, we will derive equation (10.36), the Breit–Wigner formula. The reader should first verify that when an electron reflects from an infinitely tall barrier, the phase of the reflected amplitude changes by π compared to the incident amplitude.[2] In the limit of a finite but tall and thick barrier, the reflection amplitudes at the two barriers of the DBRTS are

$$r_1 \sim -\sqrt{1 - |t_1|^2} \quad \text{and} \quad r_2 \sim -\sqrt{1 - |t_2|^2} \tag{10.40}$$

Using equation (10.40), the transmission probability in equation (10.35) is

$$T \approx \frac{|t_1|^2 |t_2|^2}{\left|1 - \sqrt{1 - |t_1|^2}\sqrt{1 - |t_2|^2}e^{i2kL}\right|^2} \tag{10.41}$$

---

[2]This is similar to the reflection of an electromagnetic wave from a metal surface. As the tangential electric field in the metal must be zero, the reflected electromagnetic wave has to be 180° out of phase.

Noting that $\sqrt{1 - |t_i|^2} = 1 - \frac{|t_i|^2}{2} + \frac{|t_i|^4}{4} - \cdots$, we can neglect $|t_i|^4$ and higher-order terms when $|t_i|^2 \ll 1$ for $i = 1$ or 2, equation (10.41) simplifies to

$$T \approx \frac{|t_1|^2 |t_2|^2}{\left| 1 - \left[1 - \frac{|t_1|^2}{2} \right] \left[1 - \frac{|t_2|^2}{2} \right] e^{i2kL} \right|^2} \tag{10.42}$$

The denominator of equation (10.42) can be expanded and rewritten as

$$\left| 1 - \left[1 - \frac{|t_1|^2}{2} \right] \left[1 - \frac{|t_2|^2}{2} \right] e^{i2kL} \right|^2 \sim \left| 1 - e^{i2kL} + \frac{|t_1|^2 + |t_2|^2}{2} e^{i2kL} \right|^2 \tag{10.43}$$

by neglecting terms of the order of $|t_1|^2 |t_2|^2$ because they are small. Equation (10.43) becomes

$$T \approx \frac{|t_1|^2 |t_2|^2}{\left| 1 - e^{i2kL} + \frac{|t_1|^2 + |t_2|^2}{2} e^{i2kL} \right|^2} \tag{10.44}$$

At energies $E_n + \Delta E$ close to the energy levels of the PiB, the wave vector is

$$k = \sqrt{\frac{2m(E_n + \Delta E)}{\hbar^2}} \sim \sqrt{\frac{2mE_n}{\hbar^2}} \left(1 + \frac{\Delta E}{2E_n} \right) = k_n \left(1 + \frac{\Delta E}{2E_n} \right) \tag{10.45}$$

Using the above equation, e^{i2kL} is approximated by

$$e^{i2kL} \sim e^{i2k_n L} e^{ik_n L \frac{\Delta E}{E_n}} \tag{10.46}$$

At the energy levels of a PiB, we know from Chapter 1 that $k_n L = n\pi$. As $e^{i2k_n L} = 1$, we can simplify the above equation,

$$e^{i2kL} \sim e^{ik_n L \frac{\Delta E}{E_n}} \qquad (10.47)$$

At energy level E_n in the well, the classical velocity is given by

$$v = \frac{\hbar k_n}{m} \qquad (10.48)$$

$$E_n = \frac{1}{2}mv^2 \qquad (10.49)$$

Using these expressions, we can rewrite e^{i2kL} (equation (10.47)) close to resonance to be

$$e^{i2kL} = e^{i\frac{\Delta E}{\hbar} \times \frac{2L}{v}} \qquad (10.50)$$

Defining $\tau = \frac{2L}{v}$, the time needed for an electron to traverse the length of the well twice (back and forth), the exponential factor is then $\frac{\Delta E}{\hbar}\frac{2L}{v} = \frac{\Delta E}{(\hbar/\tau)}$. If ΔE is a small number compared to $\hbar/\tau$, a Taylor series expansion gives us

$$e^{i2kL} = 1 + i\frac{\Delta E}{(\hbar/\tau)} \qquad (10.51)$$

Substituting equation (10.51) in the denominator for the expression for the transmission T, we have to leading order in $|t_1|^2$ and $|t_1|^2$ (terms of order $|t_1|^4$ and $|t_1|^4$ are smaller and hence neglected)

$$1 - e^{i2kL} + \frac{|t_1|^2 + |t_2|^2}{2}e^{i2kL} = -i\frac{\Delta E}{(\hbar/\tau)} + \frac{|t_1|^2 + |t_2|^2}{2} \qquad (10.52)$$

Using equation (10.52) in (10.44) gives

$$T \sim \frac{|t_1|^2 |t_2|^2}{\left| -i\frac{\Delta E}{(\hbar/\tau)} + \frac{|t_1|^2 + |t_2|^2}{2} \right|^2} \qquad (10.53)$$

Using $\tau = \frac{2L}{v}$, equation (10.53) can be written as

$$T \sim \frac{\frac{\hbar}{\tau}|t_1|^2 \frac{\hbar}{\tau}|t_2|^2}{\left|\Delta E + i\frac{\frac{\hbar}{\tau}|t_1|^2 + \frac{\hbar}{\tau}|t_2|^2}{2}\right|^2} \sim \frac{\Gamma_1 \Gamma_2}{(E - E_n)^2 + \left(\frac{\Gamma_1 + \Gamma_2}{2}\right)^2} \qquad (10.54)$$

close to any energy resonance E_n. The energies Γ_1 and Γ_2 are

$$\Gamma_1 = \frac{\hbar}{\tau}|t_1|^2 \quad \text{and} \quad \Gamma_2 = \frac{\hbar}{\tau}|t_2|^2 \qquad (10.55)$$

and represent the strength of coupling to the left and right semi-infinite regions. This proves equation (10.36).

10.7   Tunneling through Multiple Barriers

The method presented next is analogous to a procedure used to calculate the transmission and reflection of electromagnetic waves in a cascade of microwave circuits (see periodic structures chapter in (Collin, 2001)). We are interested in calculating electron tunneling probabilities through multiple barriers because it is applicable to real-world problems. The design and simulation of lasers and detectors in infrared and terahertz ranges use semiconductor layers, as shown in Figure 10.11(a).

In the previous two sections of this chapter, we discussed tunneling through a single barrier (Figure 10.5) and double barrier (Figure 10.11) structures. We matched the wave function and its first derivative at $x = 0$ and $x = L$ to derive the tunneling probability expression through a single barrier. We could have followed the same procedure to derive the expression for tunneling through the double barrier structure. But, instead, we utilized a different scheme, which consisted of adding the transmission amplitudes to the right of the right barrier at $x = L + b_2$ to obtain the total transmission amplitude. The two methods are equivalent.

We will now address the transmission amplitude calculation through a system consisting of multiple barriers, such as in Figure 10.20, using the intuitive and efficient technique of scattering matrices. The multiple barrier structure consists of $N + 2$ regions, labeled $0, 1, 2, \ldots, N + 1$. Regions 0 and $N + 1$ have flat potentials

Regions 0   1   2   3   4   5   6   7   8

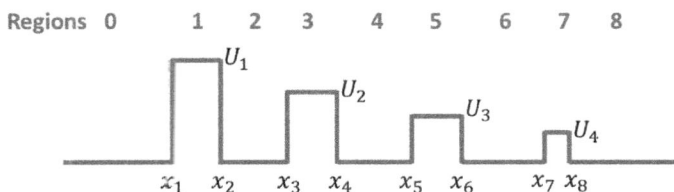

Figure 10.20. Transmission through a system consisting of multiple barriers.

that are semi-infinitely long. Regions 0 and $N+1$ extend to $x = -\infty$ and $x = +\infty$, respectively. By calculating the scattering matrices through N independent regions and then cascading them, we can find the transmission amplitude through a multiple-barrier structure. In Figure 10.20, there are seven regions $(N = 7)$.

The scattering matrix M_n of region n is defined by

$$\begin{pmatrix} A_{n-1}e^{+ikx_{n-1}} \\ B_{n-1}e^{-ikx_{n-1}} \end{pmatrix} = M_{n-1} \begin{pmatrix} A_n e^{+ikx_n} \\ B_n e^{-ikx_n} \end{pmatrix} \tag{10.56}$$

Here, x_n is the point that separates region n from region $n + 1$. Equation (10.56) relates the wave functions at $x = x_{n-1}$ to those at $x = x_n$.

For the potential profile in Figure 10.20, by cascading the individual scattering matrices across the seven regions, the total scattering matrix M can be written as

$$\begin{pmatrix} A_1 e^{+ikx_1} \\ B_1 e^{-ikx_1} \end{pmatrix} = M \begin{pmatrix} A_8 e^{+ikx_8} \\ B_8 e^{-ikx_8} \end{pmatrix}$$

$$= M_1 M_2 M_3 M_4 M_5 M_6 M_7 \begin{pmatrix} A_8 e^{+ikx_8} \\ B_8 e^{-ikx_8} \end{pmatrix} \tag{10.57}$$

$$M = M_1 M_2 M_3 M_4 M_5 M_6 M_7 \tag{10.58}$$

The above equation relates the wave function incident on Region 1 to the wave function exiting Region 7. More generally, for a system with $N + 2$ regions including semi-infinite regions 0 and $N + 1$,

$$\begin{pmatrix} A_1 e^{+ikx_1} \\ B_1 e^{-ikx_1} \end{pmatrix} = M \begin{pmatrix} A_{N+1} e^{+ikx_{N+1}} \\ B_{N+1} e^{-ikx_{N+1}} \end{pmatrix}$$

$$= M_1 M_2 \cdots M_{N-1} M_N \begin{pmatrix} A_{N+1} e^{+ikx_{N+1}} \\ B_{N+1} e^{-ikx_{N+1}} \end{pmatrix} \qquad (10.59)$$

The above equation relates the wave function incident on Region 1 to the wave function exiting region $N + 1$. In Figure 10.21, M_1, M_3, M_5, and M_7 represent the scattering matrices of barriers, and M_2, M_4, and M_6 represent the scattering matrices of the regions with constant potential energy where the wave function is a plane wave.

An expression for scattering matrix M through a single barrier: Consider a single barrier shown in Figure 10.22. The wave function to the left, right, and inside the barrier are also depicted in the figure. One can express the amplitudes A_n and B_n of the

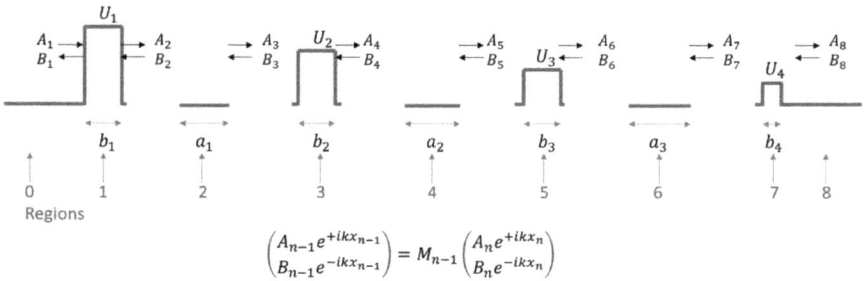

$$\begin{pmatrix} A_{n-1} e^{+ikx_{n-1}} \\ B_{n-1} e^{-ikx_{n-1}} \end{pmatrix} = M_{n-1} \begin{pmatrix} A_n e^{+ikx_n} \\ B_n e^{-ikx_n} \end{pmatrix}$$

Figure 10.21. The transmission through a system consisting of multiple barriers can be decomposed into transmission through the seven regions shown above. This multiple-barrier system consists of four barrier regions and three regions in between the barriers where the potential energy does not change with position. The wave functions on the left of each of these seven regions, can be written as $A_n e^{ikx} + B_n e^{-ikx}$. The wave function to the right of the fourth barrier is $A_8 e^{ikx} + B_8 e^{-ikx}$. The wave functions in the barrier regions may either be plane waves or decaying/growing waves depending on the relative energy of the electron with respect to the barrier height. The wave vectors in the various regions are generally different.

$$\psi^{(2)}(x) = Ce^{+ax} + De^{-ax}$$

$$\psi^{(1)}(x) = A_{n-1}e^{ikx} + B_{n-1}e^{-ikx}$$

U_0

$$\psi^{(3)}(x) = A_n e^{i\beta x} + B_n e^{-i\beta x}$$

0

U_D

x_n

x_{n-1}

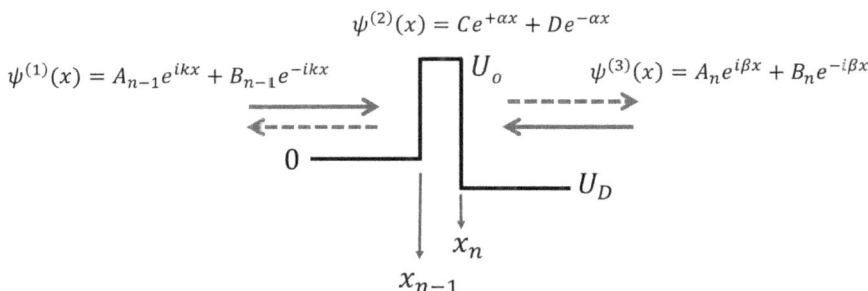

Figure 10.22. Wave function to the left, right, and inside the barrier. The wave functions and the first derivatives can be matched at x_{n-1} and x_n to derive an expression for the scattering matrix M of this barrier.

wave functions on the right-hand side of the barrier in terms of the amplitudes A_{n-1} and B_{n-1} of the wave functions on the left-hand side of the barrier by matching the wave functions and their derivatives as described in the following:

Matching the wave function and its first derivative at $x = x_{n-1}^-$ and $x = x_{n-1}^+$ (choosing $x_1 = 0$), we have

$$A_{n-1}e^{ikx_{n-1}} + B_{n-1}e^{-ikx_{n-1}} = Ce^{\alpha x_{n-1}} + De^{-\alpha x_{n-1}}$$

$$(10.60)$$

$$ik(A_{n-1}e^{ikx_{n-1}} - B_{n-1}e^{-ikx_{n-1}}) = \alpha(Ce^{\alpha x_{n-1}} - De^{-\alpha x_{n-1}})$$

$$(10.61)$$

Analogously, at $x = x_n^-$ and $x = x_n^+$, we have

$$Ce^{\alpha x_n} + De^{-\alpha x_n} = A_n e^{i\beta x_n} + B_n e^{-i\beta x_n} \qquad (10.62)$$

$$\alpha(Ce^{\alpha x_n} - De^{-\alpha x_n}) = i\beta(A_n e^{i\beta x_n} - B_n e^{-i\beta x_n}) \qquad (10.63)$$

Defining

$$\begin{pmatrix} \tilde{A}_{n-1} \\ \tilde{B}_{n-1} \end{pmatrix} = \begin{pmatrix} A_{n-1}e^{+ikx_{n-1}} \\ B_{n-1}e^{-ikx_{n-1}} \end{pmatrix}$$

(this is the wave function at the left of the barrier, $x = x_{n-1}$)

$$(10.64)$$

$$\begin{pmatrix} \tilde{C} \\ \tilde{D} \end{pmatrix} = \begin{pmatrix} Ce^{\alpha x_{n-1}} \\ De^{-\alpha x_{n-1}} \end{pmatrix} \tag{10.65}$$

$$\begin{pmatrix} \tilde{A}_n \\ \tilde{B}_n \end{pmatrix} = \begin{pmatrix} A_n e^{+i\beta x_n} \\ B_n e^{-i\beta x_n} \end{pmatrix}$$

(this is the wave function at the right of the barrier $x = x_n$)
(10.66)

and, substituting equations (10.64) to (10.66) in equations (10.60) to (10.63), we have

$$\tilde{A}_{n-1} + \tilde{B}_{n-1} = \tilde{C} + \tilde{D} \tag{10.67}$$

$$ik(\tilde{A}_{n-1} - \tilde{B}_{n-1}) = \alpha(\tilde{C} - \tilde{D}) \tag{10.68}$$

$$\tilde{C}e^{\alpha L_n} + \tilde{D}e^{-\alpha L_n} = \tilde{A}_n + \tilde{B}_n \tag{10.69}$$

$$\alpha(\tilde{C}e^{\alpha L_n} - \tilde{D}e^{-\alpha L_n}) = i\beta(\tilde{A}_n - \tilde{B}_n) \tag{10.70}$$

where

$$L_n = x_n - x_{n-1} \tag{10.71}$$

is the width of the barrier.

Equations (10.67) and (10.68) in matrix form are

$$\begin{pmatrix} 1 & 1 \\ 1 & -1 \end{pmatrix} \begin{pmatrix} \tilde{A}_{n-1} \\ \tilde{B}_{n-1} \end{pmatrix} = \begin{pmatrix} 1 & 1 \\ \alpha/ik & -\alpha/ik \end{pmatrix} \begin{pmatrix} \tilde{C} \\ \tilde{D} \end{pmatrix} \tag{10.72}$$

Pre-multiplying by

$$\text{inv} \begin{pmatrix} 1 & 1 \\ 1 & -1 \end{pmatrix} = \frac{1}{2} \begin{pmatrix} 1 & 1 \\ 1 & -1 \end{pmatrix}$$

equation (10.72) becomes

$$\begin{pmatrix} \tilde{A}_{n-1} \\ \tilde{B}_{n-1} \end{pmatrix} = \frac{1}{2} \begin{pmatrix} 1 & 1 \\ 1 & -1 \end{pmatrix} \begin{pmatrix} 1 & 1 \\ \alpha/ik & -\alpha/ik \end{pmatrix} \begin{pmatrix} \tilde{C} \\ \tilde{D} \end{pmatrix} \quad \text{or} \tag{10.73}$$

$$\begin{pmatrix} \tilde{A}_{n-1} \\ \tilde{B}_{n-1} \end{pmatrix} = \frac{1}{2ik} \begin{pmatrix} ik+\alpha & ik-\alpha \\ ik-\alpha & ik+\alpha \end{pmatrix} \begin{pmatrix} \tilde{C} \\ \tilde{D} \end{pmatrix} \tag{10.74}$$

Similarly, equations (10.69) and (10.70) in matrix form are

$$\begin{pmatrix} e^{\alpha L_n} & e^{-\alpha L_n} \\ e^{\alpha L_n} & -e^{-\alpha L_n} \end{pmatrix} \begin{pmatrix} \tilde{C} \\ \tilde{D} \end{pmatrix} = \begin{pmatrix} 1 & 1 \\ \frac{i\beta}{\alpha} & -\frac{i\beta}{\alpha} \end{pmatrix} \begin{pmatrix} \tilde{A}_n \\ \tilde{B}_n \end{pmatrix} \qquad (10.75)$$

Pre-multiplying by

$$\begin{pmatrix} e^{\alpha L_n} & e^{-\alpha L_n} \\ e^{\alpha L_n} & -e^{-\alpha L_n} \end{pmatrix}^{-1} = \frac{1}{2} \begin{pmatrix} e^{-\alpha L_n} & e^{-\alpha L_n} \\ e^{\alpha L_n} & -e^{\alpha L_n} \end{pmatrix}$$

equation (10.75) becomes

$$\begin{pmatrix} \tilde{C} \\ \tilde{D} \end{pmatrix} = \frac{1}{2} \begin{pmatrix} e^{-\alpha L_n} & e^{-\alpha L_n} \\ e^{\alpha L_n} & -e^{\alpha L_n} \end{pmatrix} \begin{pmatrix} 1 & 1 \\ \frac{i\beta}{\alpha} & -\frac{i\beta}{\alpha} \end{pmatrix} \begin{pmatrix} \tilde{A}_n \\ \tilde{B}_n \end{pmatrix} \quad \text{or} \quad (10.76)$$

$$\begin{pmatrix} \tilde{C} \\ \tilde{D} \end{pmatrix} = \frac{1}{2\alpha} \begin{pmatrix} (\alpha + i\beta)e^{-\alpha L_n} & (\alpha - i\beta)e^{-\alpha L_n} \\ (\alpha - i\beta)e^{+\alpha L_n} & (\alpha + i\beta)e^{+\alpha L_n} \end{pmatrix} \begin{pmatrix} \tilde{A}_n \\ \tilde{B}_n \end{pmatrix} \qquad (10.77)$$

Substituting equation (10.77) in equation (10.74), the components of the wave function on the right side of the barrier $\begin{pmatrix} \tilde{A}_n \\ \tilde{B}_n \end{pmatrix}$ can be written in terms of components of wave function on the left side of the barrier $\begin{pmatrix} \tilde{A}_{n-1} \\ \tilde{B}_{n-1} \end{pmatrix}$ as

$$\begin{pmatrix} \tilde{A}_{n-1} \\ \tilde{B}_{n-1} \end{pmatrix} = \frac{1}{2ik} \begin{pmatrix} ik + \alpha & ik - \alpha \\ ik - \alpha & ik + \alpha \end{pmatrix}$$

$$\times \frac{1}{2\alpha} \begin{pmatrix} (\alpha + i\beta)e^{-\alpha L_n} & (\alpha - i\beta)e^{-\alpha L_n} \\ (\alpha - i\beta)e^{+\alpha L_n} & (\alpha + i\beta)e^{+\alpha L_n} \end{pmatrix} \begin{pmatrix} \tilde{A}_n \\ \tilde{B}_n \end{pmatrix}$$

$$(10.78)$$

Equation (10.78) has the form

$$\begin{pmatrix} \tilde{A}_{n-1} \\ \tilde{B}_{n-1} \end{pmatrix} = M \begin{pmatrix} \tilde{A}_n \\ \tilde{B}_n \end{pmatrix} \qquad (10.79)$$

where

$$M = \frac{1}{2ik} \begin{pmatrix} ik + \alpha & ik - \alpha \\ ik - \alpha & ik + \alpha \end{pmatrix} \frac{1}{2\alpha} \begin{pmatrix} (\alpha + i\beta)e^{-\alpha L_n} & (\alpha - i\beta)e^{-\alpha L_n} \\ (\alpha - i\beta)e^{+\alpha L_n} & (\alpha + i\beta)e^{+\alpha L_n} \end{pmatrix}$$

$$(10.30)$$

The incident wave propagates without reflection in a region of width a, where the potential energy (U) does not change with position and the KE is greater than U. The M matrix for such a region with a flat potential is simply

$$M = \begin{pmatrix} e^{-ika} & 0 \\ 0 & e^{+ika} \end{pmatrix} \tag{10.81}$$

where k is the wave vector. This can be verified easily from equation (10.80) by setting $L_n = a$ and $\alpha = ik = i\beta$ (that is, there is no barrier, and the potential is flat). The M in equation (10.79) corresponds to the M_{n-1} in equation (10.56).

Note: In the derivation here, A_{n-1}, B_{n-1}, and A_n (Figure 10.22) correspond to A, B, and F of a single barrier (Figure 10.5). In Figure 10.5, a wave incident from the right-hand side, corresponding to B_n in Figure 10.22, does not exist. Recall that we neglected G in equation (10.9).

The calculation of the transmission and reflection probabilities through the structure in Figure 10.20 is straightforward. If a wave is incident from the left, part of the wave transmits to the right, and the rest reflects to the left. There can be no left-moving wave in Region 8 because the wave was incident from the left. That is, $B_8 = 0$. Equation (10.57) now reads

$$\begin{pmatrix} A_1 e^{+ikx_1} \\ B_1 e^{-ikx_1} \end{pmatrix} = M \begin{pmatrix} A_8 e^{+ikx_8} \\ 0 \end{pmatrix} \tag{10.82}$$

Solving the above equation, the transmitted and reflected waves are

$$A_1 e^{+ikx_1} = M_{11} A_8 e^{+ikx_8} \tag{10.83}$$

$$B_1 e^{-ikx_1} = M_{21} A_8 e^{+ikx_8} \tag{10.84}$$

M_{11} and M_{21} correspond to the $(1,1)$ and $(2,1)$ elements of the 2×2 matrix M, where M is the scattering matrix in equation (10.58).

Solving equations (10.83) and (10.84), the transmission and reflection probabilities are

$$T = \left| \frac{A_8 e^{+ikx_8}}{A_1 e^{+ikx_1}} \right|^2 = \left| \frac{1}{M_{11}} \right|^2 \tag{10.85}$$

$$R = \left| \frac{B_1 e^{-ikx_1}}{A_1 e^{+ikx_1}} \right|^2 = \left| \frac{M_{21}}{M_{11}} \right|^2 \tag{10.86}$$

10.7.1 *Tunneling Through Arbitrary Potentials*

There are many technological examples where one would need to calculate the transmission through nonrectangular potential profiles, such as in Figure 10.23. A silicon nanowire surface may be rough due to the fabrication process's imperfection. The same can happen if the interface between the oxide layer and channel under the MOSFET gate is not perfect. The problem of electron transport through materials should account for a nonuniform change in cross-section. This is equivalent to assuming that there are many different quantum wells of different heights and widths attached after each other along the electron path. Hence, the simple problem of an electron traveling through an ideal nanowire or channel has become a more complex problem involving multiple reflections and transmissions.

To calculate the transmission probability and wave function, the reader should treat the irregular barrier (blue) in Figure 10.23 by the array of rectangular barriers (red) shown. The rectangular barriers

Original potential   Discretized version

Figure 10.23. (Left) Original potential profile (blue) through which we want to calculate the transmission probability and wave function. (Right) Discretized version of the potential on the left (red). The dashed blue line is the same potential as the left. We can use the scattering matrix approach of this section on the rectangular barriers (red) to calculate the transmission and wave function.

in Figure 10.23 are separated by flat regions of zero width (Regions 2, 4, and 6) to yield a barrier of arbitrary shape.

In another analogy, if a microstrip line's fabrication process on a PCB or semiconductor substrate suffers from imperfection, i.e., imprecise etching and lithography, you may end up having a microstrip with dented edges. This leads to multiple random reflections of microwave signals passing through the line and manifesting as degradation in transmission and increased reflection. The microstrip is usually discretized into many serial pieces of different widths and lengths to model the problem. The total T and R are calculated by multiplying the scattering matrices of each section called $ABCD$ matrices. See (Collin, 2001)

10.8 Continuity Equation and Probability Current Density

Continuity equations pervade all of engineering and science. A generic form of the continuity equation relates the change of density with time to the divergence of the flux at the same point. The continuity equation for electrons in semiconductors is

$$\frac{\partial n(\bar{r}, t)}{\partial t} = -\nabla \cdot J_P(\bar{r}, t)$$

where $n(\bar{r}, t)$ is the electron density and $J_P(\bar{r}, t)$ is the electron flux (number of particles flowing per unit area per unit time at position $\bar{r}$ at time t). Flux is the same as current density (current per unit area per unit time), a vector. If we include generation and recombination processes, then the above equation is generalized to

$$\frac{\partial n(\bar{r}, t)}{\partial t} = -\nabla \cdot J_P(\bar{r}, t) + G - R$$

G is the generation rate for creating electrons by shining light on a semiconductor, and R is the recombination rate for losing electrons due to electron–hole recombination.

In quantum mechanics, the density of interest is the *probability density* related to the *probability flux* or *probability current density*. The *probability current density* forms the basis to derive the Büttiker formula for the *electric current* in nanoscale devices.

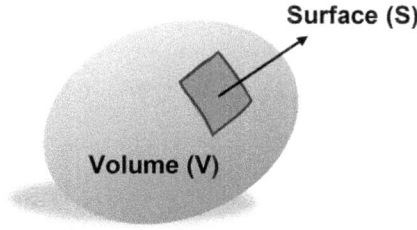

Figure 10.24. The time rate of change of probability density at a point is equal to the negative of the divergence of probability current density at that point. That is, the rate of change of probability with respect to time inside volume V depends on the flux of probability current density leaving the surface S enclosing the volume V.

Consider a volume V in a quantum object. The time rate of change of the quantum mechanical probability in volume V depends on the probability of current density flowing out from surface S that bounds V.

The probability density at $(\bar{r}, t)$ is defined as

$$P(\bar{r}, t) = \Psi^*(\bar{r}, t)\Psi(\bar{r}, t) \tag{10.87}$$

The rate of change of probability density can be obtained by taking the partial derivative of equation (10.87) as shown below,

$$\frac{\partial P(\bar{r}, t)}{\partial t} = \left[\frac{\partial \Psi^*(\bar{r}, t)}{\partial t}\right]\Psi(\bar{r}, t) + \Psi^*(\bar{r}, t)\left[\frac{\partial \Psi(\bar{r}, t)}{\partial t}\right] \tag{10.88}$$

For the time derivatives in equation (10.88), we are going to substitute Schrödinger equation and its complex conjugate,

$$i\hbar\frac{\partial \Psi(\bar{r}, t)}{\partial t} = \left[-\frac{\hbar^2}{2m}\nabla^2 + U(\bar{r}, t)\right]\Psi(\bar{r}, t) \tag{10.89}$$

$$i\hbar\frac{\partial \Psi^*(\bar{r}, t)}{\partial t} = -\left[-\frac{\hbar^2}{2m}\nabla^2 + U(\bar{r}, t)\right]\Psi^*(\bar{r}, t) \tag{10.90}$$

Multiplying equation (10.89) by Ψ^* and equation (10.90) by Ψ, and adding the two resulting equations, we obtain

$$\frac{\partial P}{\partial t} = \frac{i\hbar}{2m}\left[\Psi^*\nabla^2\Psi - \Psi\nabla^2\Psi^*\right] = \frac{i\hbar}{2m}\nabla\cdot\left[\Psi^*\nabla\Psi - \Psi\nabla\Psi^*\right] \tag{10.91}$$

By defining $J_P(\bar{r}, t)$ to be

$$J_P(\bar{r}, t) = -\frac{i\hbar}{2m} \left[\Psi^* \nabla \Psi - \Psi \nabla \Psi^* \right] \tag{10.92}$$

equation (10.91) can be written as

$$\frac{\partial P(\bar{r}, t)}{\partial t} = -\nabla \cdot J_P(\bar{r}, t) \tag{10.93}$$

The interpretation of the above equation is the rate of change of probability density with time at any spatial location $\bar{r}$, is equal to the divergence of the probability current density J_P. Verify that the dimension of $J_P(\bar{r}, t)$ is 1/second.

Probability density, $P(\bar{r}, t) = \Psi^*(\bar{r}, t)\Psi(\bar{r}, t)$

Probability current density, $J_P(\bar{r}, t) = -\frac{i\hbar}{2m} \left[\Psi^* \nabla \Psi - \Psi \nabla \Psi^* \right]$

Continuity equation for probability density, $\frac{\partial P(\bar{r}, t)}{\partial t} = -\nabla \cdot J_P(\bar{r}, t)$

Example 1. Find the probability current density of a particle in an eigenfunction of a free particle.

Solution: The eigenfunction of a free particle is $\psi = \frac{1}{\sqrt{L}} e^{ikx}$ (see Chapter 1). Substituting the wave function in the expression for current density, equation (10.92), we find that the probability current density is

$$J_P(\bar{r}, t) = -\frac{i\hbar}{2m} \left[\Psi^* \nabla \Psi - \Psi \nabla \Psi^* \right] = \frac{\hbar k}{m} \cdot \frac{1}{L}.$$

The above equation means that the particle moves along the x direction with wave vector k.

Example 2. Find the probability current density of a particle in an eigenfunction of a PiB.

Solution: The eigenfunction of a PiB is $\psi = \sqrt{\frac{2}{L}} \sin\left(\frac{n\pi}{L} x\right)$. Substituting the wave function in the expression for current density, equation (10.92), we find that the probability current density is

$$J_P(\bar{r},t) = -\frac{i\hbar}{2m}\left[\Psi^*\nabla\Psi - \Psi\nabla\Psi^*\right]$$

$$= -\frac{i\hbar(n\pi/L)}{2m}\left[\sin\left(\frac{n\pi}{L}x\right)\cos\left(\frac{n\pi}{L}x\right)\right.$$

$$\left. - \sin\left(\frac{n\pi}{L}x\right)\cos\left(\frac{n\pi}{L}x\right)\right] = 0,$$

which makes sense! The probability current density of a standing wave is zero. The electron is confined in the potential well and going nowhere.

10.9 Problems

Section 10.2

(1) Calculate the transmission and reflection probabilities of an electron incident at a barrier that is 3 nm thick with a barrier height of 200 meV. Plot the transmission (T) and reflection (R) probabilities in the energy window 0.1 meV $< E <$ 600 meV. Draw plots where T and R are both in linear and logarithmic scales while the energy axis is only in linear scale. Assume the mass to be the rest mass of a free electron (9.1×10^{-31} kg).

(2) Consider a heterostructure consisting of three layers Si–SiO$_2$–Si. The central Sio$_2$ layer is a barrier (insulator). Silicon, which is a semiconductor, is on the two sides of the central silicon dioxide layer. The barrier for electrons in the conduction band of silicon that tunnel through the silicon dioxide to the silicon on the other side is approximately 3.4 eV. The silicon dioxide region is 1.4 nm thick. Assume that the potential energy in both silicon layers is the same. This problem is representative of the device physics associated with tunneling through the insulator in MOSFETs.

(a) Calculate the tunneling probability for electrons to tunnel through the silicon dioxide as a function of energy. Make a plot, and mark all the differences between a quantum and a classical barrier.

(b) Calculate the tunneling probability for a barrier that is 6 nm thick. All other parameters are the same as part (a). In a single graph, compare the tunneling probability of parts (a) and (b). How does the tunneling probability change as a function of barrier thickness?

(Focus on representing the plot-data in an easy-to-read manner. Use logarithmic scales when it is challenging to read values.)

(3) Assume an electron is moving to the right from $x < 0$, and it is facing a potential trough with a depth of $-|U_o|$, that is, PE $= -|U_o|$. (a) Show that there will be a reflection at the trough and calculate the reflection (R) and transmission (T) probabilities. (b) Plot the reflection and transmission probabilities in terms of KE$/|U_o|$, where KE is the kinetic energy of the incident electron. Note that this problem is like the case of a potential barrier when KE $>$ PE, discussed in this chapter. The only difference is that you can simply replace U_o with $-|U_o|$ in the corresponding equations for T and R.

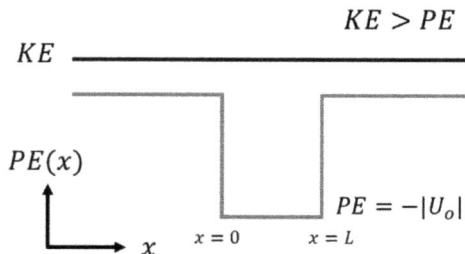

The water waves in shallow channels follow a nonlinear wave equation, as a result of which the velocity of the wave is depth dependent. Hence, when the wave passes over an area with an underwater ditch, the sudden depth change causes a sudden velocity change, causing a reflection of waves, even though the water's surface is above the riverbed.

(4) To appreciate why quantum mechanical tunneling is not observed in the classical world as experienced every day, assume that the particle's mass is 1 mg, $L = 0.02$ mm, and the barrier height is 1 eV. Calculate T and R and plot them versus energy.

What do you conclude about the tunneling probability of this heavy particle?

(5) Show that the sum of the reflection and transmission probabilities for the single barrier is equal to one.

Section 10.3

(6) Consider a rectangular barrier with a barrier height U_o and barrier width L. The dwell time in the barrier for an electron incident with energy $E < U_o$ is defined to be,

$$t_{\text{dwell}} = \frac{\text{Probability per unit area to find the electron in the barrier region of length } L}{\text{Current density of the electron wave}}$$

Show that the dwell time is equal to

$$t_{\text{dwell}} = \frac{1}{T}\frac{mk}{\hbar\alpha}\frac{2\alpha L\left(\alpha^2 - k^2\right) + k_o^2\sinh(2\alpha L)}{4\alpha^2 k^2 + k_o^4\sinh^2(\alpha L)}$$

where $k = \frac{\sqrt{2mE}}{\hbar}$, $\alpha = \frac{\sqrt{2m(U_o - E)}}{\hbar}$, $k_o = \frac{\sqrt{2mU_o}}{\hbar}$, and T is transmission.

Section 10.4

(7) Double barrier resonant tunneling structures are nanostructures that have a negative differential resistance, which is useful to make oscillators. Consider the barrier regions to be made of AlGaAs and the other three regions (left, well, and right) to be made of GaAs.

(a) Perform a literature search to find the barrier height at the AlGaAs–GaAs interface. Does this change with the percentage of aluminum?

(b) Assume that the left and right barriers are identical and have both t_1 and t_2 equal to 0.2, and a well width of 50 Å. The transmission amplitude of the two barriers being independent of energy is assumed for simplicity. Plot the transmission probability as a function of energy.

(c) At what energies do the peaks of the transmission occur? Compare the energy of the first transmission maxima to the lowest energy of a particle in a box with the same well width and infinite barrier thickness.

(8) (a) Calculate and plot the transmission probability versus energy through the double barrier structure shown below. On the same figure, plot t_1 versus energy. Calculate t_1, t_2, r_1, and r_2 using the expressions for a single barrier. What is the maximum value of transmission in your plots? What is the maximum theoretical value of transmission? Assume that $L = 50\,\text{Å}$, $L_B = 10\,\text{Å}$, and $U_o = 400\,\text{meV}$.

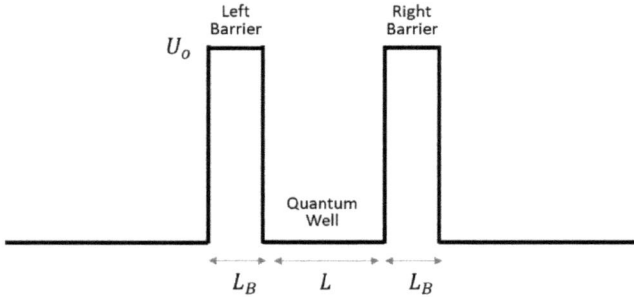

(b) Plot and discuss the effects of making both barriers thinner, $L_B = 6\,\text{Å}$.

Section 10.6

(9) For a 1D double barrier resonant tunneling barrier, show that the transmission probability as a function of energy near a resonant energy level E_1 in the well is

$$T = \frac{\Gamma_1 \Gamma_2}{(E - E_1)^2 + \left(\frac{\Gamma_1 + \Gamma_2}{2}\right)^2}$$

Assume the transmission probabilities through the individual barriers are both smaller than one.

(a) Express Γ_1 and Γ_2 in terms of the length of the quantum well L, mass m, and the transmission amplitude through the two barriers. What are the units of Γ_1 and Γ_2?

(b) What physical quantity does $\frac{\hbar}{\Gamma_1 + \Gamma_2}$ represent?

(c) Using the expression for the transmission probability from equation (10.36), find the maximum possible transmission probability through the double barrier structure. Give an example of the device's barrier heights and widths that will have the highest possible transmission probability. (You should not have to calculate anything significant for this part.)

Section 10.7

(10) Calculate the transmission probability of a triangular barrier as shown in the following figure for energies between $10\,\mathrm{meV}$ and $1\,\mathrm{eV}$. The values of the parameters are $b = 5\,\mathrm{nm}$, $U_1 = 800\,\mathrm{meV}$, and $U_2 = 200\,\mathrm{meV}$. Hint: Break this down into many rectangular barriers with varying widths and heights.

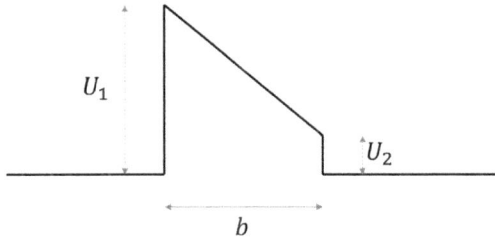

(11) In the following figure, $U_o = 400\,\mathrm{meV}$, $L = 50\,\text{Å}$, and $L_B = 5\,\text{Å}$. Calculate the transmission probability for energies from 1 to $350\,\mathrm{meV}$.

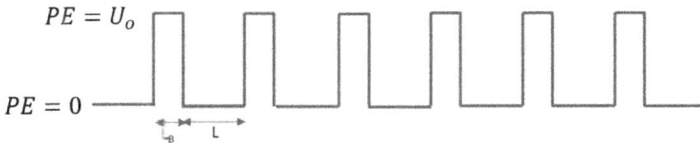

References

Büttiker, M. (1983). Larmor precession and the traversal time for tunneling. *Physical Review B* **27**, 6179.

Collin, R. E. (2001). *Foundations for Microwave Engineering*, 2nd edn. IEEE-Wiley Press.

Datta, S. (1988). *Quantum Phenomena, Modular Series on Solid State Devices*, Vol. 8. Addison-Wesley.

Ma, D. D. D. *et al.* (2003). Small-Diameter Silicon Nanowire Surfaces. *Science* **299**, 1874.

Patiño, E. J. and Kelkar, N. G. (2015). Experimental determination of tunneling characteristics and dwell times from temperature dependence of $Al/Al_2O_3/Al$ junctions. *Applied Physics Letters* **107**, 253502.

Chapter 11

QUANTUM DOTS, WELLS, AND NANOWIRES: SEPARATION OF VARIABLES

Contents

11.1 Introduction

Our goal in this chapter is to demonstrate the role of quantization in zero-. one-, and two-dimensional (0D, 1D, and 2D) nanostructures using the separation of variables. We emphasize solving Schrödinger equation analytically, with boundary conditions based on the physical system's geometry, to find the eigenvalues and eigenfunctions.

We start from an electron in an infinite solid (which we call *bulk*) and consider three cases: (I) a two-dimensional solid (quantum well), (II) a one-dimensional solid (nanowire or nanotube), and

(III) a zero-dimensional solid (quantum dot). They are all relevant to semiconducting nanodevices. Examples include single-electron transistors and nanowire transistors quantum well photodetectors, THz sensors, and laser diodes. Our emphasis is on interpreting the solutions of the Schrödinger equation. In later chapters, we will learn how to use these results to calculate the conductance, absorption, and other properties.

11.1.1   *Free electron*

The Schrödinger equation for an electron in vacuum is

$$\hat{H}\psi = \left\{ -\frac{\hbar^2}{2m_o}\nabla^2\Psi + U(\bar{r}) \right\}\psi = E\psi \qquad (11.1)$$

where $U(\bar{r})$ is a constant potential energy that we set to 0 everywhere. The Laplacian operator ∇^2 is defined as $\frac{\partial^2}{\partial x^2} + \frac{\partial^2}{\partial y^2} + \frac{\partial^2}{\partial z^2}$. The solution of the above equation is a plane wave,

$$\psi(\bar{r}) = Ae^{+i\bar{k}\cdot\bar{r}}, \text{ where } \bar{k}\cdot\bar{r} = k_x x + k_y y + k_z z \qquad (11.2)$$

The inner product of the wave vector and position vector, $\bar{k}\cdot\bar{r}$, is the phase of the wave function. The name *plane wave* comes from the concept that the wave function is constant at all points on a plane perpendicular to the propagation direction (Figure 11.1).

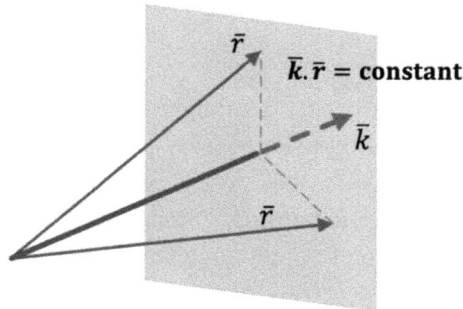

Figure 11.1. A rectangular plane (yellow) on which the phase of the wave $\bar{k}\cdot\bar{r}$ is constant because the dot product of $\bar{k}$ and any point $\bar{r}$ on this plane does not change.

The energy eigenvalues are

$$E(\bar{k}) = \frac{\hbar^2}{2m_o}\left(k_x^2 + k_y^2 + k_z^2\right) = \frac{\hbar^2}{2m_o}|\bar{k}|^2 \qquad (11.3)$$

where we replaced the sum of the wave vector components squared by its norm squared, $|\bar{k}|^2$, or $\bar{k}\cdot\bar{k}$, the inner product of the wave vector with itself. Equation (11.3) shows that the relationship between energy and wave vector (called the *energy dispersion*) is parabolic for a free electron. The curvature of this parabolic shape is inversely proportional to the electron mass m_o. Since energy is a function of three variables (k_x, k_y, k_z), plotting it as $E(k_x, k_y, k_z)$ requires four dimensions, which is not easy to imagine since we are familiar with three-dimensional entities. However, it is possible to draw constant energy contours (surfaces). We can draw all points in 3D k-space (k_x, k_y, k_z), which satisfy $E(k_x, k_y, k_z) = E_1$. As you may have surmised, this is the equation of a sphere whose radius $|\bar{k}|$ is constant and is determined as follows:

$$E_1 = \frac{\hbar^2}{2m_o}|\bar{k}|^2 \rightarrow |\bar{k}| = \sqrt{\frac{2m_o E_1}{\hbar^2}} \qquad (11.4)$$

Figure 11.2 shows the constant energy surface or contour corresponding to $E = 1\,\mathrm{eV}$ for a free electron.

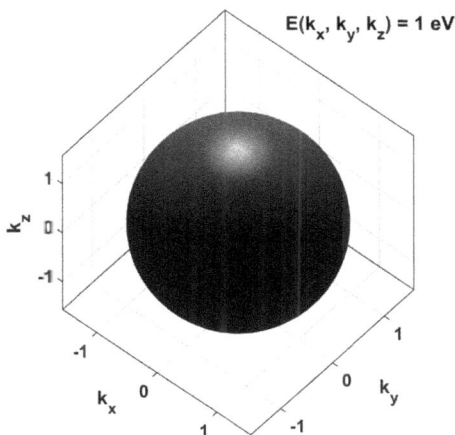

Figure 11.2. The constant energy surface for a free electron having $E_1 = 1\,\mathrm{eV}$ is a sphere with radius $|\bar{k}| = \sqrt{\frac{2m_o E_1}{\hbar^2}} = \frac{1}{2.7\,\text{Å}}$.

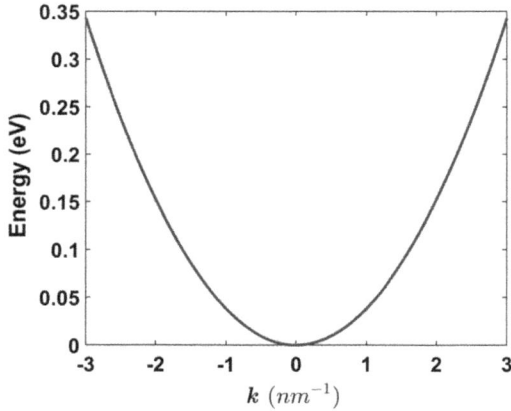

Figure 11.3. Energy versus wave vector for a free electron is a parabola, following equation (11.3).

For each energy value, there is a sphere in k-space. The constant energy contours for a free electron are infinite, and they can be imagined as concentric spheres with a common center at $(k_x, k_y, k_z) = (0, 0, 0)$.

Equation (11.3) can also be visualized as a one-dimensional plot if we are careful and keep in mind that $|\bar{k}|$ is the norm of (k_x, k_y, k_z). The curvature of this plot is inversely proportional to the mass of the electron. Figure 11.3 is a plot of equation (11.3).

11.1.2 *Effective mass and electrons in solids*

As we saw in the previous section, the relationship between E and $|\bar{k}|$ for a free electron is a parabola. If we take the second derivative of energy $E(\bar{k})$ with respect to k in equation (11.3), we get the curvature,

$$\text{Curvature} = \frac{\partial^2 E(k_x, k_y, k_z)}{\partial k^2} = \frac{\hbar^2}{m_o} \tag{11.5}$$

which tells us that the mass of the electron is inversely proportional to the curvature, i.e.,

$$m_o = \frac{\hbar^2}{\frac{\partial^2 E(k_x,k_y,k_z)}{\partial k^2}} \tag{11.6}$$

For a free electron, its mass m_o is a fundamental constant.

In Chapter 14, we will learn that an electron in an infinitely large (bulk) solid moves under the influence of periodic electric potential created by the infinite array of atomic nuclei. Hence, modeling an electron's environment by a spatially independent $U(\bar{r})$ (as in the case of a free electron) is not enough. Later, we will learn how to include this periodic potential and solve Schrödinger equation to find the energy dispersion for solids (metals, semiconductors, and insulators). However, there are cases where the $E(\bar{k})$ for solids resembles a parabola as if the electron is still free but with a different mass than the free electron mass m_o. This mass (m^*), which we call the *effective mass*, is found from the curvature of the $E(\bar{k})$. In the simple case of Figure 11.2, the constant energy surfaces are spheres in k-space, and the curvatures along k_x, k_y, and k_z are the same. In this case, the effective mass in the three different directions is equal,

$$m_{xx}^* = m_{yy}^* = m_{zz}^* = \frac{\hbar^2}{\frac{\partial^2 E(k_x,k_y,k_z)}{\partial k_{ii}^2}}, \quad i = x, y, z \tag{11.7}$$

If the constant energy surfaces in k-space are not spherical the curvature depends on the direction in k-space. Therefore, we must discriminate between these directions as the electron possesses different effective masses in different directions. Mathematically put, the effective mass is a tensor, which is an array of nine numbers corresponding to nine alternatives,

$$m_{ii}^* = \frac{\hbar^2}{\frac{\partial^2 E(k_x,k_y,k_z)}{\partial k_{ii}^2}} \text{ for equal indices } ii = xx, yy, zz \tag{11.8}$$

$$m_{ij}^* = \frac{\hbar^2}{\frac{\partial^2 E(k_x, k_y, k_z)}{\partial k_i \partial k_j}} \text{ for unequal indices } ij = xy, xz, yx, yz, zx, zy$$

$$(11.9)$$

The above terms can be arranged in a matrix called the **effective mass tensor** (M^*):

$$M^* = \begin{bmatrix} m_{xx}^* & m_{xy}^* & m_{xz}^* \\ m_{yx}^* & m_{yy}^* & m_{yz}^* \\ m_{zx}^* & m_{zy}^* & m_{zz}^* \end{bmatrix} \tag{11.10}$$

For a spherical constant energy surface, only diagonal terms are nonzero and equal. This type of dispersion $E(\bar{k})$ is called **isotropic**, since it does not matter in which direction the electron is moving, and it always has the same effective mass. However, in general, this is not the case, and we have to calculate other elements of the effective mass tensor from the dispersion relation.

The strengths of the electric potential of the nuclei and periodicity of the atoms in a solid determine if the effective mass tensor is diagonal or not. For a given energy, a higher effective mass means the electrons have a low velocity (recall that velocity is momentum divided by mass, $v = \hbar k / m^*$) as they move through the solid; correspondingly, a smaller effective mass means the electrons have a higher velocity in the solid.

To simplify our discussion in this chapter, we will replace the mass m_o in the free electron Schrödinger equation (equation (11.1)) by a single effective mass m^*:

$$\hat{H}\psi = \left\{ -\frac{\hbar^2}{2m^*}\nabla^2 + U(\bar{r}) \right\} \psi = E\psi \tag{11.11}$$

We assume that m^*, the effective mass, is either available from experiments or from prior theoretical work. $U(\bar{r})$ consists of two parts: (1) the potential we apply by an external voltage and (2) the potential imposed by the geometric boundary of the device.

Before moving to the following section, it is necessary to distinguish semiconductors from metals and insulators. We provide a simplified view of this here. A more precise definition will be given in

Chapter 14 when we calculate the *energy dispersion* or *band structure* of electrons in infinitely periodic solids.

11.1.3   *Conceptual introduction to energy bands*

When atoms in a periodic solid interact with electrons, solving Schrödinger's equation gives us the energy levels of the electrons. These energy levels form bands of allowed energies as shown in Figure 11.4. Each band has a different quantum number called the **band index**. The second quantum number is the wave vector k.

Periodic metals, semiconductors, and insulators all have bands. In a **metal**, electrons are filled up to an energy that lies inside a band. As a result, the energy difference between the highest occupied energy and the lowest unoccupied energy level is zero (Figure 11.4). At zero temperature, electrons are filled until the **Fermi energy**, which is represented by E_F.

In a **semiconductor**, electrons are filled up to an energy that fully occupies a band but leaves the next higher energy band fully empty. As a result, the energy difference between the highest occupied energy and lowest unoccupied energy at zero temperature is larger than zero. This energy difference is called the **bandgap** of the material, which ranges from tens of meV to a few eV in semiconductors.

For a semiconductor at zero temperature, the Fermi energy lies in the bandgap, but its precise energetic location is not unique. The

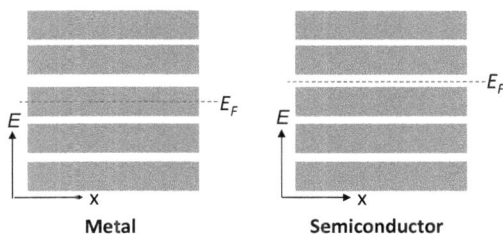

Figure 11.4. The main difference between a metal and semiconductor is the energy until which electrons are filled at zero kelvin. The x-axis is spatial location, and the y-axis is energy.

location can be anywhere in the bandgap depending on the density of added dopants atoms which change the density of electrons and from holes at finite temperature.

We summarize a few conclusions from studying semiconductor (see Figure 11.5):

- The highest occupied band at zero temperature is referred to as the **valence band** and its maximum energy is represented by E_v (see Figure 11.5 middle).
- The lowest unoccupied band at zero temperature is referred to as the **conduction band** and its minimum energy is represented by E_c (see Figure 11.5 middle).
- Usually, only the highest occupied (valence) and lowest unoccupied (conduction) bands at zero temperature are drawn (see Figure 11.5 right).

Bandgap = lowest unoccupied energy − Highest occupied energy

$$E_g = E_c - E_v$$

In Figure 11.5, E_c and E_v are independent of position because the electrostatic potential is periodic, and no external bias is applied.

The energy dispersion shown in Figure 11.5 has its conduction band minimum and valence band maximum at the same wave vector k. It can be represented mathematically in a manner akin to the free

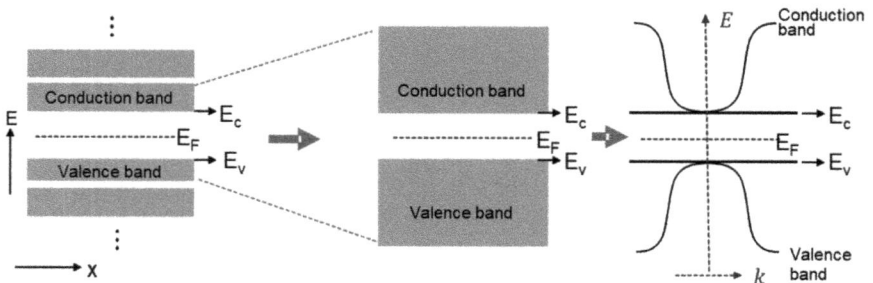

Figure 11.5. Conventional representation of how conduction and valence bands are used in semiconductor devices. The middle figure is a magnified version of the left figure.

electrons in equation (11.3):

$$\text{Conduction band: } E = E_c + \frac{\hbar^2 k^2}{2m_c} \tag{11.12}$$

$$\text{Valence band: } \quad E = E_v - \frac{\hbar^2 k^2}{2m_v} \tag{11.13}$$

In these equations, the conduction band minimum and valence band maximum are both at $k = 0$. m_c and $-m_v$ are the effective masses of electrons in the conduction and valence bands, respectively. The reader should note that the hole represents the absence of an electron in the valence bands — its effective mass is m_v. Note that the effective masses of electrons and holes in the valence band have opposite signs.

Figure 11.5 shows the dispersion for conduction and valence band states in a bulk semiconductor. The term *bulk* means that the material is still an infinite solid, and it is not yet cut or shaped into a nanoscale object. The energy dispersion relationships of the conduction and valence bands of the isotropic (properties do not change with direction) semiconductor shown in equations (11.12) and (11.13) can be obtained from the following Hamiltonians separately for the conduction and valence bands:

$$\hat{H} = -\frac{\hbar^2}{2m_c}\nabla^2 + E_c$$

$$= -\frac{\hbar^2}{2m_c}\frac{d^2}{dx^2} - \frac{\hbar^2}{2m_c}\frac{d^2}{dy^2} - \frac{\hbar^2}{2m_c}\frac{d^2}{dz^2} + E_c \quad \text{(Conduction band)}$$

$$\hat{H} = \frac{\hbar^2}{2m_v}\nabla^2 + E_v$$

$$= \frac{\hbar^2}{2m_v}\frac{d^2}{dx^2} + \frac{\hbar^2}{2m_v}\frac{d^2}{dy^2} + \frac{\hbar^2}{2m_v}\frac{d^2}{dz^2} + E_v \quad \text{(Valence band)}$$

Note: There is no allowed energy level within the bandgap. For semiconductors, E_g is variable from $38\,\text{meV}$ (for bismuth) to $5\,\text{eV}$ (for diamond — carbon) (Madelung, 1996). This is like the concept of a forbidden band in photonic crystals or acoustic crystals, where there is no k-state (mode) available for a range of optical or acoustic frequencies to propagate.

Concept: An electron in a semiconductor interacts with the nuclei and other electrons. To a good approximation, the electron behaves like a free particle near the conduction band bottom and valence band top, with an effective mass m_c and $-m_v$, respectively. The effective mass of holes in the valence band is m_v, which has an opposite sign to the effective mass of an electron.

11.2   Quantum Wells, Nanowires, and Quantum Dots

In *highly symmetric structures*, a method called the *separation of variables* is useful in obtaining analytical solutions of partial differential equations (Brown, 2011). In this method, one starts with Schrödinger's equation in 3D and finds a way to decompose it into three differential equations, each in a single variable x, y, and z.

Nanowires, quantum wells, and quantum dots can be made either out of metal, semiconductors, or insulators. Our discussion here will focus on semiconducting nanostructures of simple rectangular forms, as shown in Figure 11.6. Note that for other structures, e.g., spherical quantum dots or cylindrical nanowires, Schrödinger equation is solved in spherical or cylindrical coordinates, respectively. Different complex cross-sections or forms can always be solved numerically. Our starting point will be the bulk conduction and valence bands of a semiconductor. We will solve Schrödinger equation for electrons and holes separately. The potential energy of an electron in the conduction band of the nanostructure is assumed to be

$$U(x,y,z) = \begin{cases} E_c \text{ inside the semiconductor nanostructure} \\ \infty \text{ outside the semiconductor nanostructure} \end{cases}$$

Similarly, the potential energy of a hole in the valence band is assumed to be

$$U(x,y,z) = \begin{cases} E_v \text{ inside the semiconductor nanostructure} \\ -\infty \text{ outside the semiconductor nanostructure} \end{cases}$$

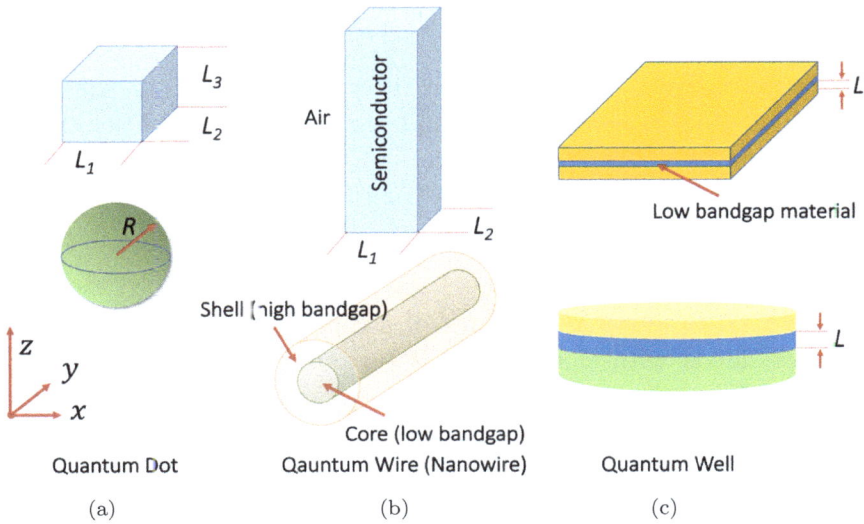

Figure 11.6. A quantum dot, nanowire, and quantum well are depicted. The potential energy for an electron is assumed to be the bulk value of the conduction band minimum E_c inside the nanostructure and a very high value or infinite outside. The potential energy for holes is the valence band maximum E_v inside the nanostructure and the potential energy outside is minus infinity. (a) A rectangular cuboid and a spherical quantum dot surrounded by vacuum. (b) A nanowire of a rectangular cross-section. A nanowire made of a core material surrounded by a high bandgap (potential) shell. It has a circular cross-section and is infinitely long along the y-axis. (c) A quantum well (blue) is made by sandwiching a low bandgap semiconductor layer of width L (e.g., silicon) between two high bandgap materials (e.g., silicon dioxide, yellow). The electron is free in the xy plane and is confined in the z direction. Ideally, the well extends from minus infinity to plus infinity in xy plane. A circular quantum well (blue) in between two oxides (yellow and green).

Note again that E_c is the bottom of the bulk conduction band and E_v is top of the bulk valence band. The main steps to calculate the energy levels of nanostructures in this chapter are:

(1) Define the Hamiltonian in the conduction band.
(2) Solve the time-independent Schrödinger equation with appropriate boundary conditions and use the separation of variables to

find the wave functions and energy levels for an electron in the conduction band.

(3) Repeat steps 1 and 2 above for the valence band.
(4) Calculate the conduction band minimum and valence band maximum of the nanostructure.
(5) Find the bandgap of the nanostructure.

11.2.1   *Quantum well*

Figure 11.7(a) shows an example of a quantum well (QW); a thin layer of silicon with thickness L, is surrounded by two thick SiO_2 layers. SiO_2 possesses a large energy barrier for electrons to enter from Si. Hence, its effect can be modeled as a barrier with a practically infinite potential energy. The electron movement is confined along the layering direction (z-axis), and free in the xy plane.

The Hamiltonian for an electron in the conduction band of a QW is

$$\hat{H} = -\frac{\hbar^2}{2m_c}\frac{d^2}{dx^2} - \frac{\hbar^2}{2m_c}\frac{d^2}{dy^2} - \frac{\hbar^2}{2m_c}\frac{d^2}{dz^2} + U(x,y,z) \quad (11.14)$$

$$\text{where,} \quad U(x,y,z) = \begin{cases} E_c & \text{for } 0 < z < L \\ \infty & \text{otherwise} \end{cases} \quad (11.15)$$

That is, the potential energy for an electron in the conduction band is E_c (of bulk) inside the QW and ∞ outside the QW. In reality, the potential outside the well is finite but immense. We assume an infinite potential for simplicity. The structure is of indefinite length in the x and y directions, but the Si layer has a finite width L along the z direction. Schrödinger equation for the system is

$$\hat{H}\chi = E\chi \Rightarrow \{\widehat{KE}_x + \widehat{KE}_y + \widehat{KE}_z + E_c\}\chi = E\chi \quad (11.16)$$

$$\text{where } \widehat{KE}_x = -\frac{\hbar^2}{2m_c}\frac{d^2}{dx^2}, \quad \widehat{KE}_y = -\frac{\hbar^2}{2m_c}\frac{d^2}{dy^2}, \quad \widehat{KE}_z = -\frac{\hbar^2}{2m_c}\frac{d^2}{dz^2}$$

$$(11.17)$$

We try a trial solution to the above differential equation,

$$\chi(x,y,z) = X(x)Y(y)Z(z) \quad (11.18)$$

which is a product of three functions that individually depend only on one variable, x, y, or z. If we substitute equation (11.18) into equation (11.16) to obtain,

$$\{\widehat{KE}_x + \widehat{KE}_y + \widehat{KE}_z + E_c\}X(x)Y(y)Z(z) = EX(x)Y(y)Z(z)$$

Pre-multiplying both the LHS and RHS by $\frac{1}{X(x)Y(y)Z(z)}$, we obtain

$$\frac{1}{X(x)}\widehat{KE}_x X(x) + \frac{1}{Y(y)}\widehat{KE}_y Y(y) + \frac{1}{Z(z)}\widehat{KE}_z Z(z) = E - E_c \quad (11.19)$$

Remember that in equation (11.19), each of the three terms on the left-hand side is a function of only one variable. Equation (11.19) can be rewritten as

$$\frac{1}{X(x)}\widehat{KE}_x X(x) = E - E_c - \frac{1}{Y(y)}\widehat{KE}_y Y(y) - \frac{1}{Z(z)}\widehat{KE}_z Z(z) \quad (11.20)$$

As the left-hand side depends only on variable x and the right-hand side is independent of x, the only way we can get them to be equal

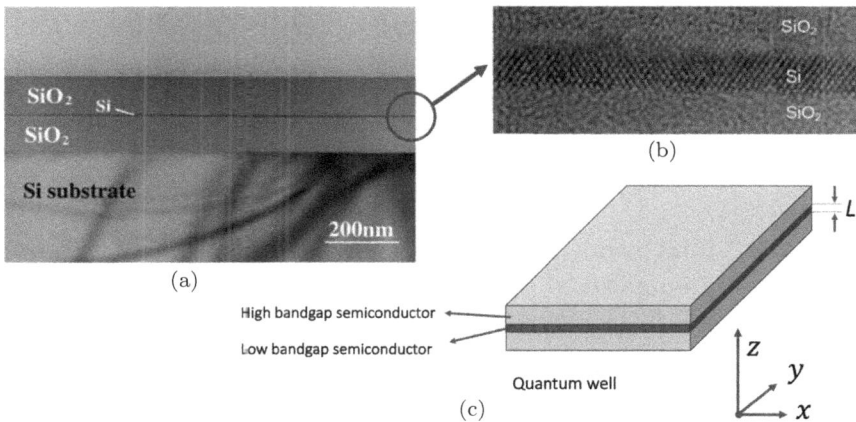

(a)

(b)

(c)

Figure 11.7. (a) An example of a quantum well made by confining a silicon layer between two layers of silicon dioxide. Note that the electron is free to move in the silicon layer horizontally (toward and away from the reader, toward the left and right directions), but it is confined like a particle in a box along the vertical direction (z-axis). (b) A zoomed-in view of the Si quantum well in (a) shows a silicon layer that is merely seven atomic layers thick. The pictures were taken using a transmission electron microscopy (Cho, 2007). (c) The simple model of a quantum well in which the electron is free to move in the low bandgap layer along with xy plane but is confined in the z direction by the high bandgap material.

to each other is if each equals the same constant ε_1. This gives us

$$-\frac{\hbar^2}{2m_c}\frac{d^2X(x)}{dx^2} = \varepsilon_1 X(x) \qquad \text{for} \quad -\infty < x < \infty \qquad (11.21)$$

$$\frac{\hbar^2}{2m_c}\frac{1}{Y(y)}\frac{d^2Y(y)}{dy^2} + \frac{\hbar^2}{2m_c}\frac{1}{Z(z)}\frac{d^2Z(z)}{dz^2} + (E - E_c) = \varepsilon_1$$

Rearranging the terms we have,

$$\frac{\hbar^2}{2m_c}\frac{1}{Y(y)}\frac{d^2Y(y)}{dy^2} = \varepsilon_1 - \left(+\frac{\hbar^2}{2m_c}\frac{1}{Z(z)}\frac{d^2Z(z)}{dz^2} + (E - E_c)\right)$$

As the left-hand side depends only on variable y and the right-hand side is independent of y, the only way we can get them to be equal is if each is equal to the same constant $-\varepsilon_2$. This gives us

$$-\frac{\hbar^2}{2m_c}\frac{d^2Y(y)}{dy^2} = \varepsilon_2 Y(y) \quad \text{for} \quad -\infty < y < \infty \qquad (11.22)$$

Similarly, we can get that

$$-\frac{\hbar^2}{2m_c}\frac{d^2Z(z)}{dz^2} = \varepsilon_3 Z(z) \quad \text{for} \quad 0 < z < L \qquad (11.23)$$

$$\text{where } E - E_c = \varepsilon_1 + \varepsilon_2 + \varepsilon_3 \qquad (11.24)$$

Equations (11.21) and (11.22) correspond to free particles because the QW is unbounded in the xy plane, and equation (11.23) is the one for a particle in a box (PiB) due to confinement along the z direction. The solutions of equations (11.21)–(11.23) give the wave functions and energy levels for electrons in the conduction band,

$$X(x) = \sqrt{\frac{1}{L_x}}e^{ik_x x} \quad \text{and} \quad \varepsilon_1 = \frac{\hbar^2 k_x^2}{2m_c} \qquad (11.25)$$

$$Y(y) = \sqrt{\frac{1}{L_y}}e^{ik_y y} \quad \text{and} \quad \varepsilon_2 = \frac{\hbar^2 k_y^2}{2m_c} \qquad (11.26)$$

$$Z(z) = \sqrt{\frac{2}{L}}\sin\left(\frac{n\pi}{L}z\right) \quad \text{and} \quad \varepsilon_3 = \frac{\hbar^2}{2m_c}\left(\frac{n\pi}{L}\right)^2 \qquad (11.27)$$

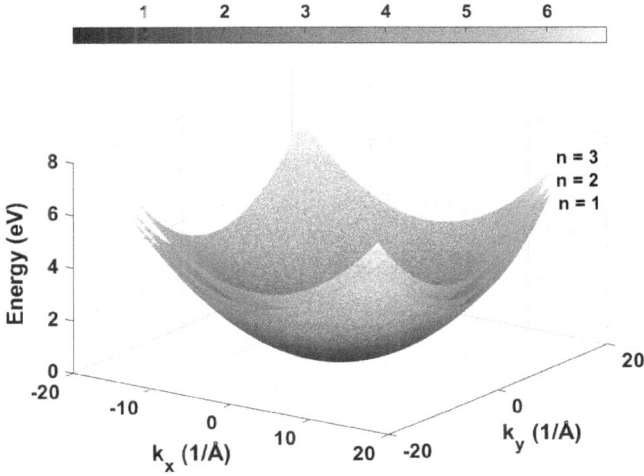

Figure 11.8. The parabolic energy surfaces, $E(k_x, k_y)$, for the first three eigenenergies in conduction band indexed by $n = 1, 2, 3$. The vertical axis unit is eV, and units of k_x and k_y are $1/\text{Å}$, $L = 70\,\text{Å}$, and $m_c^* = 0.067\,m_0$.

The energy eigenvalue is the sum of three individual terms in equation (11.24):

$$E_{k_x, k_y, n} = E_c + \varepsilon_n + \frac{\hbar^2}{2m_c}\left(k_x^2 + k_y^2\right) \tag{11.28}$$

$$\text{where} \quad \varepsilon_n = \frac{\hbar^2}{2m_c}\left(\frac{n\pi}{L}\right)^2 \tag{11.29}$$

Note that the eigenenergy is a function of (n, k_x, k_y). Corresponding to each value of n (integer quantum number), the energy dispersion relationship is a 2D parabolic function of k_x and k_y. Figure 11.8 shows the dispersion plots for $n = 1, 2$, and 3. This indicates that the electron behaves as a free-electron wave along the x and y directions, however along z it acts like a confined particle in a potential well with a discrete set of energy eigenvalues as we saw for PiB.

The wave function of an electron in the QW is the product of three different solutions, $X(x)$, $Y(y)$, and $Z(z)$:

$$\chi_{k_x, k_y, n}(x, y, z) = \frac{1}{\sqrt{L_x L_y}}\sqrt{\frac{2}{L}}\,e^{ik_x x}e^{ik_y y}\sin\left(\frac{n\pi}{L}z\right) \tag{11.30}$$

n is a positive integer $(1, 2, 3, \ldots)$ and $-\infty < k_x$ and $k_y < +\infty$. Note that in equations (11.28) and (11.30), we have added the quantum numbers as subscripts to the wave functions and energy values. The energy difference between the minimum of each conduction sub-band is in the meV range corresponding to very long wavelength infrared (IR) or tetrahertz (THz) photons. Therefore, engineering the energy levels within quantum wells is a way of designing IR and THz lasers and detectors that have useful applications in security scanning and detection of explosive materials.

Similarly, for holes in the valence band, the eigenfunction is a product of traveling waves in the xy plane and a standing wave in the z direction,

$$X(x) = \sqrt{\frac{1}{L_x}} \, e^{ik_x x} \quad \text{and} \quad \varepsilon_1 = \frac{\hbar^2 k_x^2}{2m_v} \tag{11.31}$$

$$Y(y) = \sqrt{\frac{1}{L_y}} \, e^{ik_y y} \quad \text{and} \quad \varepsilon_2 = \frac{\hbar^2 k_y^2}{2m_v} \tag{11.32}$$

$$Z(z) = \sqrt{\frac{2}{L}} \, sin \left(\frac{n'\pi}{L} z \right) \quad \text{and} \quad \varepsilon_3 = \frac{\hbar^2}{2m_v} \left(\frac{n'\pi}{L} \right)^2 \tag{11.33}$$

$$E_{k_x, k_y, n'} = E_v - \varepsilon_{n'} - \frac{\hbar^2}{2m_v} \left(k_x^2 + k_y^2 \right)$$

$$= E_v - \frac{\hbar^2}{2m_v} \left(k_x^2 + k_y^2 + \left(\frac{n'\pi}{L} \right)^2 \right) \tag{11.34}$$

where

$$\varepsilon_{n'} = \frac{\hbar^2}{2m_v} \left(\frac{n'\pi}{L} \right)^2 \tag{11.35}$$

The energy dispersion is a set of parabolas in (k_x, k_y) plane as shown in Figure 11.8, but inverted due to the negative sign in equation (11.34). See Figure 11.9. The minimum energy of holes corresponds to $k_x = k_y = 0$ and $n' = 1$. The holes behave like bubbles in water — they rise to the highest point in the energy dispersion (valence band maximum in Figure 11.9) which is their minimum energy point.

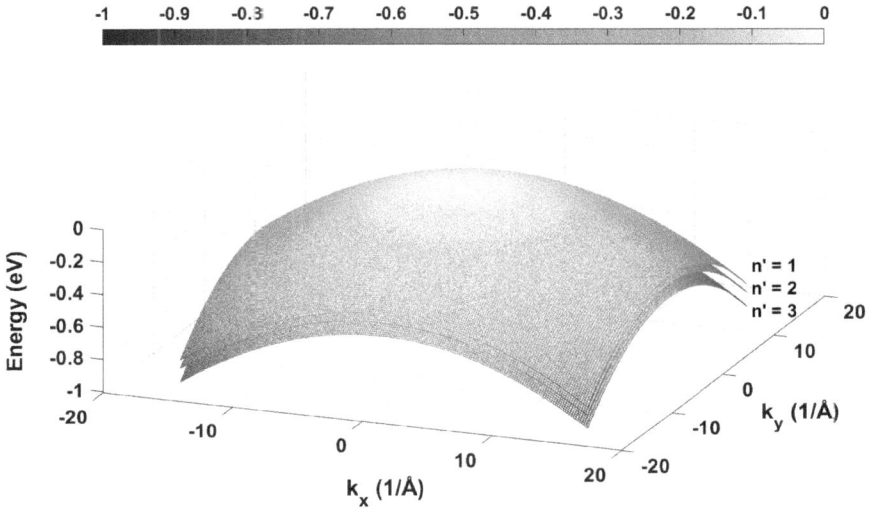

Figure 11.9. The parabolic energy surfaces, $E(k_x, k_y)$, for the highest three eigenenergies in the valence band indexed by $n' = 1, 2, 3$. Energy is given in eV, and units of k_x and k_y are $1/\text{Å}$. $L = 70\,\text{Å}$ and $m_v^* = 0.45\,m_0$.

The wave function of holes in the valence band of the QW is

$$\chi_{k_x, k_y, n'}(x, y, z) = \frac{1}{\sqrt{L_x L_y}} \sqrt{\frac{2}{L}} \, e^{ik_x x} e^{ik_y y} \sin\left(\frac{n'\pi}{L} z\right) \quad (11.36)$$

n' is a positive integer and $-\infty < k_x$ and $k_y < +\infty$.

Summary: There is a dispersion relationship $E(k_x, k_y)$ in the conduction band for each value of quantum number n, referred to as the nth *conduction sub-band* (see Figure 11.10). The sub-band minimum has an energy equal to $E_c + \varepsilon_n$ which, varies with n. The dispersion relationship $E(k_x, k_y)$ in the valence band for each value of quantum number n' is referred to as the n'th *valence sub-band*.

Conduction band minimum, valence band maximum, and bandgap: The conduction band minimum of the QW is obtained by substituting $k_x = k_y = 0$ and $n = 1$ in equation (11.28) is,

$$E_c(\text{QW}) = E_c + \frac{\hbar^2}{2m_c}\left(\frac{\pi}{L}\right)^2 \quad (11.37)$$

Bulk   **Quantum Well**

$$E = E_c + \frac{\hbar^2}{2m_c}\left(k_x^2 + k_y^2 + \left(\frac{n\pi}{L}\right)^2\right)$$

$$E = E_c + \frac{\hbar^2 k^2}{2m_c}$$

$$E_c(\text{quantum well}) = E_c + \frac{\hbar^2}{2m_c}\left(\frac{\pi}{L}\right)^2$$

$$E_v(\text{quantum well}) = E_c - \frac{\hbar^2}{2m_v}\left(\frac{\pi}{L}\right)^2$$

$$E = E_v - \frac{\hbar^2 k^2}{2m_v}$$

$$E = E_v - \frac{\hbar^2}{2m_v}\left(k_x^2 + k_y^2 + \left(\frac{n'\pi}{L}\right)^2\right)$$

$$E_g(\text{bulk}) = E_c - E_v$$

$$E_g(QW) = E_g(\text{bulk}) + \frac{\hbar^2}{2}\left(\frac{1}{m_c} + \frac{1}{m_v}\right)\left(\frac{\pi}{L}\right)^2$$

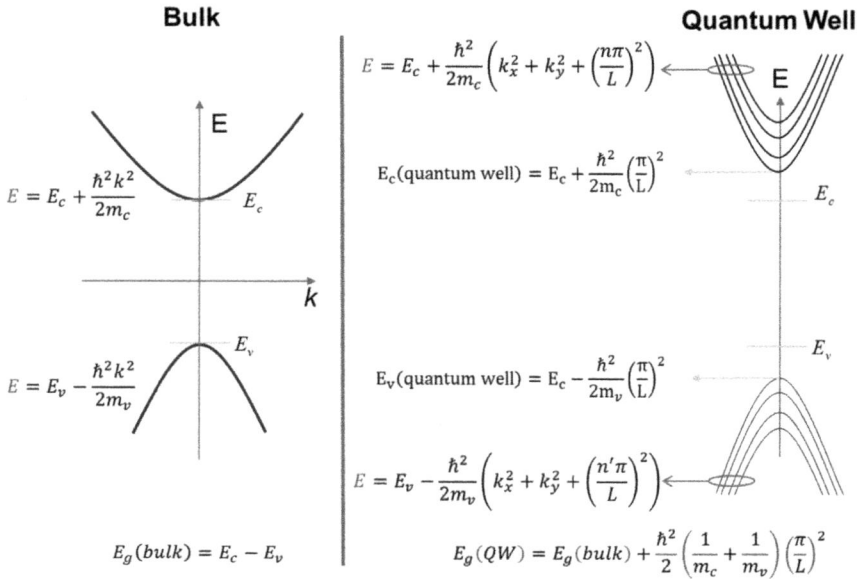

Figure 11.10. The left-hand side subfigure corresponds to a bulk semiconductor. A 2D cross-sectional view of the 3D $E(k_x, k_y)$ for a quantum well is shown in the right-hand side subfigure. The y-axis is energy, and the x-axis is wave vector (k). The different sub-bands in the quantum well correspond to quantum numbers n (conduction band) and n' (valence band). A quantum well's bandgap increases as the well width L decreases. Photons can be either emitted or absorbed when an electron transitions between states of the conduction and valence bands, and within states of the conduction or valence band.

Similarly, the valence band maximum obtained by substituting $k_x = k_y = 0$ and $n' = 1$ is,

$$E_v(QW) = E_v - \frac{\hbar^2}{2m_v}\left(\frac{\pi}{L}\right)^2 \tag{11.38}$$

From equations (11.37) and (11.38), the bandgap of the QW is

$$E_g(QW) = E_c(QW) - E_v(QW)$$

$$E_g(QW) = E_g(\text{bulk}) + \frac{\hbar^2}{2}\left(\frac{1}{m_c} + \frac{1}{m_v}\right)\left(\frac{\pi}{L}\right)^2 \tag{11.39}$$

We see from equation (11.39) that the bandgap of the QW minus the bandgap of bulk decreases inversely proportional to the QW width-squared $\left(\frac{1}{L^2}\right)$.

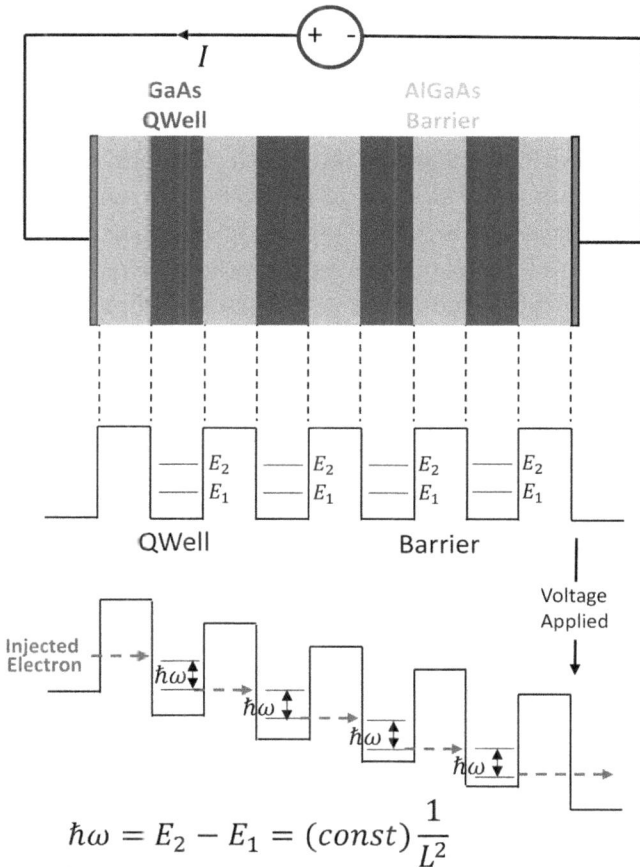

$$\hbar\omega = E_2 - E_1 = (const)\frac{1}{L^2}$$

Figure 11.11. A terahertz source based on a multiple quantum well structure. (Top) The orange regions represent the barriers (for example, made of AlGaAs) and the blue regions represent the well (made out of GaAs). (Middle) The potential energy of the structure as a function of position at zero bias. (Bottom) The potential energy of the structure at a finite bias causes electrons to flow through the structure (red dashed) lines. Photons are emitted as electrons transition from energy E_2 to E_1.

Quantum wells are used to generate terahertz radiation for security detectors and terahertz lasers. For example, in Figure 11.11, an electron can be injected into the $n = 2$ conduction sub-band in the left most quantum well (orange). This electron will emit a photon of frequency $f = \frac{E_2 - E_1}{\hbar}$ and go to the $n = 1$ conduction sub-band of the same quantum well. The electron will then tunnel to the $n = 2$ conduction sub-band of the second quantum well and the process continues, thereby emitting more photons with a frequency of $f = \frac{E_2 - E_1}{h}$. The process continues as the electron propagates to the right. We can engineer the width and barrier heights such that the quantum well can emit in a wavelength range that is larger than 4000 nm, where it is difficult to use conduction to valence band transitions in bulk semiconductors for practical reasons (there are not many stable semiconductors with corresponding bandgaps).

Concept: Electrons in quantum wells are traveling waves along the plane of the well and standing waves in the direction perpendicular to the well.

The bandgap (E_g) of a quantum well is larger than that of the bulk semiconductor due to quantization of the wave vector perpendicular to the well.

$$E_g(\text{QW}) = E_g(\text{bulk}) + \frac{\hbar^2}{2} \left(\frac{1}{m_c} + \frac{1}{m_v} \right) \left(\frac{\pi}{L} \right)^2$$

11.2.2 *Nanowire*

A nanowire (NW) is a narrow cross-section semiconductor surrounded by either a higher energy bandgap material or a vacuum. For example, a long pillar of silicon surrounded by air or silicon dioxide is a nanowire.

Figure 11.12(a) shows an array of indium antimonide nanowires grown on a silicon substrate using a bottom-up growth mechanism called the vapor–liquid–solid (VLS) method. In the beginning, a silicon wafer is covered by a silicon nitride (Si_xN_y) mask. The mask is then etched to create an array of holes. The holes are filled with gold (Au) nanodroplets. After that, In and Sb enter the reaction chamber in gaseous form. The Au droplets catalyze and accelerate the solidification process of InSb. That is, wherever a gold droplet exists, the InSb solidifies. Figure 11.12(b) and (c) illustrate these processes.

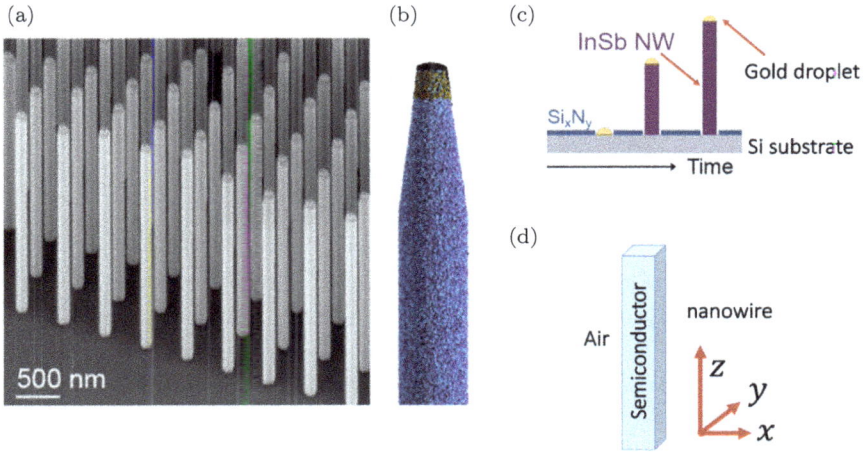

Figure 11.12. (a) An array of the indium antimonide (InSb) nanowires grown on silicon substrate using the vapor–liquid–solid (VLS) process. (b) The yellow-colored top is the liquid gold droplet that catalyzes (enhances) the solidification of InSb (blue) from the vapor phase. (c) The VLS process is pictured as time passes. The silicon nitride layer works as a mask to let the nanowires grow wherever there is a gold nanodroplet in the mask. (d) A simple model of a semiconductor nanowire with the axis along the z direction and surrounded by a higher bandgap material or vacuum. The figure is adapted with permission from (Badawy, 2019).

In nanofabrication technologies, selective etching of a silicon wafer's surface leaves silicon columns, which an oxide layer covers. Repetitive etching can reduce the diameter of the silicon column to nanoscale wires with a cross-section as small as 3 nm. An example of a silicon nanowire array grown by this top-down method is shown in Figure 11.15(a). In a nanowire, electrons move freely as plane waves along the nanowire axis (z-axis) in the limit of an infinitely long nanowire but are confined in the cross-sectional plane, i.e., the xy plane.

The derivation of the eigenfunctions and energy levels is similar to that of the quantum well. So, we will discuss only the main steps here. The Hamiltonian for an electron in the conduction band is,

$$\widehat{H} = -\frac{\hbar^2}{2m_c}\frac{d^2}{dx^2} - \frac{\hbar^2}{2m_c}\frac{d^2}{dy^2} - \frac{\hbar^2}{2m_c}\frac{d^2}{dz^2} + U(x,y,z) \qquad (11.40)$$

where the potential energy is

$$U(x, y, z) = \begin{cases} E_c & \text{for } 0 < x < L_1, \ 0 < y < L_2, \ -\infty < z < +\infty \\ \infty & \text{otherwise} \end{cases}$$

$$(11.41)$$

The potential energy for an electron in the conduction band is E_c inside the nanowire and ∞ outside the nanowire. In a manner identical to the quantum well, we solve Schrödinger equation,

$$\widehat{H}\chi = E\chi \Rightarrow \left\{ \widehat{KE_x} + \widehat{KE_y} + \widehat{KE_z} + U(r) \right\} \chi = E\chi \qquad (11.42)$$

with a trial solution that is separable in the three dimensions,

$$\chi(x, y, z) = X(x)Y(y)Z(z) \qquad (11.43)$$

This gives us three independent differential equations in variables x, y, and z from which functions $X(x)$, $Y(y)$, and $Z(z)$ can be determined:

$$-\frac{\hbar^2}{2m_c}\frac{d^2 X(x)}{dx^2} = \varepsilon_1 X(x) \quad \text{for} \quad 0 < x < L_1 \qquad (11.44)$$

$$-\frac{\hbar^2}{2m_c}\frac{d^2 Y(y)}{dy^2} = \varepsilon_2 Y(y) \quad \text{for} \quad 0 < y < L_2 \qquad (11.45)$$

$$-\frac{\hbar^2}{2m_c}\frac{d^2 Z(z)}{dz^2} = \varepsilon_3 Z(z) \quad \text{for} \quad -\infty < y < \infty \qquad (11.46)$$

$$\text{and} \quad E - E_c = \varepsilon_1 + \varepsilon_2 + \varepsilon_3 \qquad (11.47)$$

The solution in the x and y directions correspond to a PiB, whose solution we know to be,

$$X(x) = A_1 \sin\left(\frac{l\pi}{L_1}x\right) \quad \text{and} \quad \varepsilon_1 = \frac{\hbar^2}{2m_c}\left(\frac{l\pi}{L_1}\right)^2 \qquad (11.48)$$

$$Y(y) = A_2 \sin\left(\frac{n\pi}{L_2}y\right) \quad \text{and} \quad \varepsilon_2 = \frac{\hbar^2}{2m_c}\left(\frac{n\pi}{L_2}\right)^2 \qquad (11.49)$$

The solution in the z direction is a plane wave,

$$Z(z) = A_3 \, e^{ik_z z} \quad \text{and} \quad \varepsilon_3 = \frac{\hbar^2 k_z^2}{2m_c} \tag{11.50}$$

Substituting equations (11.48), (11.49), and (11.50) in equation (11.43), the eigenfunction is,

$$\chi_{l,m,k}(x, y, z) = A \sin\left(\frac{l\pi}{L_1}x\right) \sin\left(\frac{m\pi}{L_2}y\right) e^{ik_z z} \tag{11.51}$$

The eigenfunction has two components. One part is the free-electron wave, which propagates along the nanowire axis; the second part is standing waves in the nanowire's confined directions.

The energy eigenvalues are

$$E_{l,m,k} = E_c + \varepsilon_{l,m} + \frac{\hbar^2 k_z^2}{2m_c} = E_c + \frac{\hbar^2}{2m_c}\left[\left(\frac{l\pi}{L_1}\right)^2 + \left(\frac{m\pi}{L_2}\right)^2 + k_z^2\right] \tag{11.52}$$

where

$$\varepsilon_{l,r} = \frac{\hbar^2}{2m_c}\left(\frac{l\pi}{L_1}\right)^2 + \frac{\hbar^2}{2m_c}\left(\frac{m\pi}{L_2}\right)^2 \tag{11.53}$$

l and m are positive integers and $-\infty < k_z < +\infty$.

The conduction band's energy dispersion consists of a parabola for each allowed value of l and m (see Figure 11.13(a)). Each of these parabolas is called an (l, m) conduction sub-band. Figure 11.13(a) shows a set of parabolas, i.e., $E(k_z)$ for different combinations of the integers l and m.

In the valence band of the nanowire, the Hamiltonian is,

$$\widehat{H} = -\frac{\hbar^2}{2m_v}\frac{d^2}{dx^2} - \frac{\hbar^2}{2m_v}\frac{d^2}{dy^2} - \frac{\hbar^2}{2m_v}\frac{d^2}{dz^2} + U(x, y, z) \tag{11.54}$$

$$U(x, y, z) = \begin{cases} E_v & \text{for } 0 < x < L_1, \ 0 < y < L_2, \ -\infty < z < +\infty \\ -\infty & \text{otherwise} \end{cases} \tag{11.55}$$

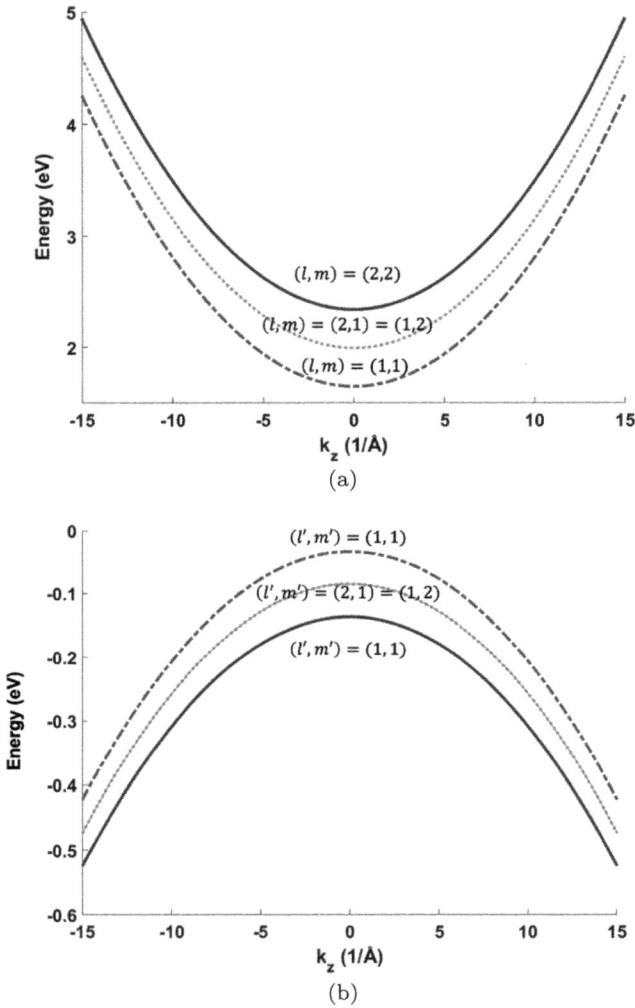

Figure 11.13. (a) Eigenenergies of a GaAs nanowire showing different parabolas for each pair of quantum numbers (l, m). Note that the eigenenergies for $(l, m) = (1, 2)$ and $(2, 1)$ are equal. (b) The valence sub-bands for the same nanowire. The minimum conduction and maximum valence energies of bulk GaAs used in this plot are $1.42 \, \text{eV}$ and $0 \, \text{eV}$, respectively. Note that for $(l', m') = (1, 2)$ and $(2, 1)$ the sub-bands have the same energy.

That is, the potential energy of a hole in the valence band is E_v inside the nanowire and $-\infty$ outside the nanowire. Now, in a manner identical to the conduction band, the energy eigenfunctions and eigenvalues for a hole in the valence band are

$$\chi(x, y, z) = X(x)Y(y)Z(z) \tag{11.56}$$

$$X(x) = A_1 \sin\left(\frac{l'\pi}{L_1}x\right) \quad \text{and} \quad \varepsilon_1 = \frac{\hbar^2}{2m_v}\left(\frac{l'\pi}{L_1}\right)^2 \tag{11.57}$$

$$Y(y) = A_2 \sin\left(\frac{m'\pi}{L_2}y\right) \quad \text{and} \quad \varepsilon_2 = \frac{\hbar^2}{2m_v}\left(\frac{m'\pi}{L_2}\right)^2 \tag{11.58}$$

The solution in the z direction is a plane wave,

$$Z(z) = A_3 e^{ik_z z} \quad \& \quad \varepsilon_3 = \frac{\hbar^2 k_z^2}{2m_v} \tag{11.59}$$

Substituting equations (11.57), (11.58), and (11.59) into equation (11.56) gives that the wave function is

$$\chi_{l',m'\,k}(x, y, z) = A \sin\left(\frac{l'\pi}{L_1}x\right) \sin\left(\frac{m'\pi}{L_2}y\right) e^{ik_z z} \tag{11.60}$$

The energy eigenvalues are,

$$E_{l',m',k} = E_v - \varepsilon_{l',m'} - \frac{\hbar^2 k_z^2}{2m_v} = E_v - \frac{\hbar^2}{2m_v}\left[\left(\frac{l'\pi}{L_1}\right)^2 + \left(\frac{m'\pi}{L_2}\right)^2 + k_z^2\right] \tag{11.61}$$

where,

$$\varepsilon_{l',m'} = \frac{\hbar^2}{2m_v}\left(\frac{l'\pi}{L_1}\right)^2 + \frac{\hbar^2}{2m_v}\left(\frac{m'\pi}{L_2}\right)^2 \tag{11.62}$$

l' and m' are positive integers and $-\infty < k_z < +\infty$.

The energy dispersion $[E(k)]$ in the valence band consists of an inverted parabola for each allowed value of l' and m' (see Figure 11.13). Each of these parabolas is called a (l', m') *valence sub-band.*

Conduction band minimum, valence band maximum, and bandgap: From equation (11.51), we find that the lowest energy conduction band state, which corresponds to $l = m = 1$ and $k = 0$ is,

$$E_c(\text{NW}) = E_c + \frac{\hbar^2}{2m_c}\left[\left(\frac{\pi}{L_1}\right)^2 + \left(\frac{\pi}{L_2}\right)^2\right] \tag{11.63}$$

From equation (11.60), we find that the highest valence band state, which corresponds to $l' = m' = 1$ and $k = 0$ is

$$E_v(\text{NW}) = E_v - \frac{\hbar^2}{2m_v}\left[\left(\frac{\pi}{L_1}\right)^2 + \left(\frac{\pi}{L_2}\right)^2\right] \tag{11.64}$$

From equations (11.63) and (11.64), the bandgap of the nanowire is,

$$E_g(\text{NW}) = E_c(\text{NW}) - E_v(\text{NW})$$

$$E_g(\text{NW}) = E_g(\text{bulk}) + \frac{\hbar^2}{2}\left(\frac{1}{m_c} + \frac{1}{m_v}\right)\left[\left(\frac{\pi}{L_1}\right)^2 + \left(\frac{\pi}{L_2}\right)^2\right] \tag{11.65}$$

Summary: The properties of nanowire band structure and its comparison with a bulk semiconductor are summarized in Figure 11.14.

When a nanowire is fabricated from a bulk semiconductor, its bandgap, $E_g(\text{NW}) > E_g(\text{bulk})$. The increase in bandgap is contained in the second term in equation (11.65). This effect is observed in photoluminescence spectroscopy of silicon nanowires. When silicon nanowires are exposed to UV light, electrons absorb a large amount of energy. After losing some of their excess energy as heat, electrons end up being at the bottom of the lowest parabola in the conduction band. Then, they recombine with a hole in the valence band and the reduction in the electron's energy results in a photon emission with energy,

$$\hbar\omega = h \cdot \frac{c}{\lambda} = E_g(\text{NW})$$

Bulk

Nanowire

$$E = E_c + \frac{\hbar^2}{2m_c}\left[\left(\frac{l\pi}{L_1}\right)^2 + \left(\frac{m\pi}{L_2}\right)^2 + k_z^2\right]$$

$$E_c(\text{Nanowire}) = E_c + \frac{\hbar^2}{2m_c}\left[\left(\frac{\pi}{L_1}\right)^2 + \left(\frac{\pi}{L_2}\right)^2\right]$$

E_c

$$E = E_c + \frac{\hbar^2 k^2}{2m_c}$$

E_c

$$E_v(\text{Nanowire}) = E_v - \frac{\hbar^2}{2m_v}\left[\left(\frac{\pi}{L_1}\right)^2 + \left(\frac{\pi}{L_2}\right)^2\right]$$

E_v

$$E = E_v - \frac{\hbar^2 k^2}{2m_v}$$

$$E = E_v - \frac{\hbar^2}{2m_v}\left[\left(\frac{l'\pi}{L_1}\right)^2 + \left(\frac{m'\pi}{L_2}\right)^2 + k_z^2\right]$$

$$E_g(bulk) = E_c - E_v$$

$$E_g(\text{NW}) = E_g(bulk) + \frac{\hbar^2}{2}\left(\frac{1}{m_c} + \frac{1}{m_v}\right)\left[\left(\frac{\pi}{L_1}\right)^2 + \left(\frac{\pi}{L_2}\right)^2\right]$$

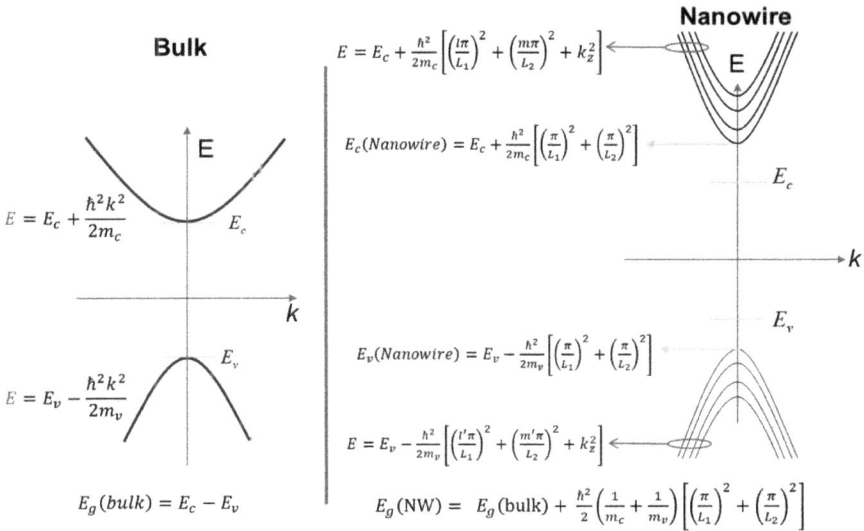

Figure 11.14. The left-hand side subfigure shows the energy levels of a bulk semiconductor. The energy levels of the nanowire vary with the wave vector along the nanowire axis and is shown in the right-hand side subfigure. The various sub-bands correspond to different quantum numbers (l, m). The nanowire bandgap increases as the nanowire cross-section decreases. Photons can be either emitted or absorbed when an electron transitions between states of the conduction and valence bands, or within states of the conduction or valence band.

where, ω and λ are the frequency and wavelength of the emitted photon. We showed above that the bandgap increases as the nanowire diameter decreases. As the nanowire diameter becomes smaller, Figure 11.15 shows that the wavelength moves to smaller values because the bandgap increases.

Concept: Electrons in a nanowire are traveling waves along its axis and standing waves in the cross-sectional direction.

The bandgap (E_g) of a nanowire is larger than that of the bulk semiconductor it is made from due to quantization of the wave vectors along the cross-section.

$$E_g(\text{NW}) = E_g(\text{bulk}) + \frac{\hbar^2}{2}\left(\frac{1}{m_c} + \frac{1}{m_v}\right)\left[\left(\frac{\pi}{L_1}\right)^2 + \left(\frac{\pi}{L_2}\right)^2\right]$$

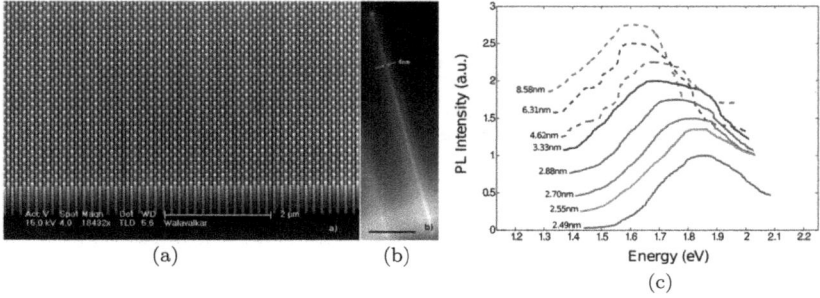

(a)   (b)

(c)

Figure 11.15. Visible photoluminescence from silicon nanowires when they are excited by UV light. (a) An array of nanowires made from silicon by repetitive oxidation and etching of oxides using HF to narrow the diameter down to (b) 4 nm. (c) The shift of the photoluminescence (PL) peak intensity is proof of increasing the bandgap by reducing the nanowire diameter (Walavalkar, 2010). The x-axis is the energy of the emitted photon. There is a broad range of emitted photon energies because of nonuniformities in NW diameter and defects.

11.2.3   *Quantum dots*

Imagine that a solid semiconductor is cut in all three directions to have a finite size of a few nanometers. For example, a small rectangular cuboid of silicon surrounded by air or silicon dioxide is a quantum dot. Although the structure is still three dimensional, a quantum dot is referred to as a zero-dimensional structure to show that the electron's confinement is in all three directions instead of nanowire/well, where the confinement is in one/two directions.

Consider a quantum dot made of a semiconductor. The Hamiltonian for an electron in the conduction band is

$$\widehat{H} = \widehat{\mathrm{KE}} + \widehat{\mathrm{PE}} \tag{11.66}$$

$$\widehat{H} = -\frac{\hbar^2}{2m_c}\frac{d^2}{dx^2} - \frac{\hbar^2}{2m_c}\frac{d^2}{dy^2} - \frac{\hbar^2}{2m_c}\frac{d^2}{dz^2} + U(x,y,z) \tag{11.67}$$

The potential energy of an electron in the conduction band of the quantum dot is

$$U(x,y,z) = \begin{cases} E_c & \text{for } 0 < x < L_1, \ 0 < y < L_2, \ 0 < z < L_3 \\ \infty & \text{otherwise} \end{cases}$$

$$\tag{11.68}$$

where L_1, L_2, and L_3 are the dimensions of the quantum dot in the x, y, and z directions (recall Figure 11.16(a)). The potential energy is the bulk value of the conduction band minima E_c inside the quantum dot and ∞ outside the quantum dot. Schrödinger equation inside the quantum dot can be solved in a similar way to the previous cases. The difference is that the electron in the z direction is also confined. We will have three PiB-like solutions in x, y and z directions. The Hamiltonian is

$$\hat{H}\chi = E\chi \Rightarrow \left\{ \widehat{KE}_x + \widehat{KE}_y + \widehat{KE}_z + U(r) \right\} \chi = E\chi \qquad (11.69)$$

$$\widehat{KE}_x = -\frac{\hbar^2}{2m_c}\frac{d^2}{dx^2}, \ \widehat{KE}_y = -\frac{\hbar^2}{2m_c}\frac{d^2}{dy^2}, \ \widehat{KE}_z = -\frac{\hbar^2}{2m_c}\frac{d^2}{dz^2}$$

$$(11.70)$$

We try a trial solution to the above differential equation,

$$\chi(x, y, z) = X(x)Y(y)Z(z) \qquad (11.71)$$

The solution consists of the product of three terms, $X(x)$, $Y(y)$, and $Z(z)$, which are each function of only x, y, and z, respectively. Substituting $X(x)Y(y)Z(z)$ into equation (11.69) and pre-multiplying by $\frac{1}{X(x)Y(y)Z(z)}$, we get

$$\frac{1}{X(x)}\widehat{KE}_x X(x) + \frac{1}{Y(y)}\widehat{KE}_y Y(y) + \frac{1}{Z(z)}\widehat{KE}_z Z(z) = E - E_c \quad (11.72)$$

We note that in equation (11.72), each of the three terms on the left-hand side is a function of only one variable. Equation (11.72) can be rewritten as

$$\frac{1}{X(x)}\widehat{KE}_x X(x) = -\frac{1}{Y(y)}\widehat{KE}_y Y(y) - \frac{1}{Z(z)}\widehat{KE}_z Z(z) + (E - E_c)$$

$$(11.73)$$

The validities of equations (11.71), (11.72), and (11.73) are inside the quantum dot. Outside the quantum dot, the potential energy is infinite, and the wave function is *zero*. We note that the left-hand side of equation (11.73) depends only on variable x, while the right-hand side is independent of variable x (it depends only on variables

y and z). For the LHS to be equal to the RHS, each one of them has to be equal to the same constant. That is, equation (11.73) can be broken down into two independent equations,

$$\frac{1}{X(x)}\widehat{KE}_xX(x) = \varepsilon_1 \implies -\frac{\hbar^2}{2m_c}\frac{d^2X(x)}{dx^2} = \varepsilon_1X(x) \quad \text{for } 0 < x < L_1 \quad \text{and}$$
(11.74)

$$-\frac{\hbar^2}{2m_c}\frac{1}{Y(y)}\frac{d^2Y(y)}{dy^2} - \frac{\hbar^2}{2m_c}\frac{1}{Z(z)}\frac{d^2Z(z)}{dz^2} = (E - E_c) - \varepsilon_1 \quad (11.75)$$

where ε_1 is the constant. Now, one can rewrite equation (11.75) as

$$-\frac{\hbar^2}{2m_c}\frac{1}{Y(y)}\frac{d^2Y(y)}{dy^2} = \frac{\hbar^2}{2m_c}\frac{1}{Z(z)}\frac{d^2Z(z)}{dz^2} + (E - E_c) - \varepsilon_1 \quad (11.76)$$

We note that the left-hand side of equation (11.76) depends only on variable y, while the right-hand side is independent of variable y and depends only on variables z. The only way we know for the LHS to be equal to the RHS is for each of them to be equal to the same constant. Equation (11.76) is then equivalent to two independent equations,

$$-\frac{\hbar^2}{2m_c}\frac{d^2Y(y)}{dy^2} = \varepsilon_2Y(y) \quad \text{for } 0 < y < L_2 \quad (11.77)$$

$$-\frac{\hbar^2}{2m_c}\frac{d^2Z(z)}{dz^2} = \varepsilon_3Z(z) \quad \text{for } 0 < z < L_3, \quad \text{where} \quad (11.78)$$

$$E - E_c = \varepsilon_1 + \varepsilon_2 + \varepsilon_3 \quad (11.79)$$

In summary, we have shown above that solving Schrödinger equation for the rectangular cuboid quantum dot shown in Figure 11.6(a) is equivalent to solving three equations in a single variable, namely equations (11.74), (11.77), and (11.78). Each of these equations corresponds to a one-dimensional particle in a box (PiB) in the x, y, and z directions, respectively. Their solution is well known to us, and we will write it down here

$$X(x) = \sqrt{\frac{2}{L_1}}\sin\left(\frac{l\pi}{L_1}x\right) \quad \text{and} \quad \varepsilon_1 = \frac{\hbar^2}{2m_c}\left(\frac{l\pi}{L_1}\right)^2 \quad (11.80)$$

$$Y(y) = \sqrt{\frac{2}{L_2}} \sin\left(\frac{m\pi}{L_2}y\right) \quad \text{and} \quad \varepsilon_2 = \frac{\hbar^2}{2m_c}\left(\frac{m\pi}{L_2}\right)^2 \quad (11.81)$$

$$Z(z) = \sqrt{\frac{2}{L_3}} \sin\left(\frac{n\pi}{L_3}z\right) \quad \text{and} \quad \varepsilon_3 = \frac{\hbar^2}{2m_c}\left(\frac{n\pi}{L_3}\right)^2 \quad (11.82)$$

where l, m, and n are positive integers. The wave function of an electron in the conduction band of the QD is confined in all directions:

$$\chi_{l,m,n}(x, y, z) = \sqrt{\frac{8}{L_1 L_2 L_3}} \sin\left(\frac{l\pi}{L_1}x\right) \sin\left(\frac{m\pi}{L_2}y\right) \sin\left(\frac{n\pi}{L_3}z\right)$$
$$(11.83)$$

where $A = \sqrt{\frac{8}{L_1 L_2 L_3}}$ is the normalization constant.

Substituting the values of ε_1, ε_2, and ε_3, and from equations (11.80), (11.81), and (11.82) in equation (11.79), the energy eigenvalues corresponding to quantum numbers (l, m, n) are

$$E_{l,m,n} = E_c + \frac{\hbar^2}{2m_c}\left(\frac{l\pi}{L_1}\right)^2 + \frac{\hbar^2}{2m_c}\left(\frac{m\pi}{L_2}\right)^2 + \frac{\hbar^2}{2m_c}\left(\frac{n\pi}{L_3}\right)^2 \quad (11.84)$$

In this case, in contrast to the 2D and 1D cases, there is no dispersion, i.e., E is not a function of the continuous quantum number k (wave vector). The allowed quantum numbers are a set of discrete positive integers (l, m, n). l, m, and n are all positive integers $(1,2,3,\dots)$.

Figure 11.16 shows the discrete energy levels in the conduction band of the quantum dot instead of the continuum in bulk. The minimum of the conduction band has shifted up from the bulk conduction band by,

$$\frac{\hbar^2}{2m_c}\left(\frac{\pi}{L_1}\right)^2 + \frac{\hbar^2}{2m_c}\left(\frac{\pi}{L_2}\right)^2 + \frac{\hbar^2}{2m_c}\left(\frac{\pi}{L_3}\right)^2$$

The Hamiltonian for a hole in the valence band of the quantum dot is,

$$\hat{H} = -\frac{\hbar^2}{2m_v}\frac{d^2}{dx^2} - \frac{\hbar^2}{2m_v}\frac{d^2}{dy^2} - \frac{\hbar^2}{2m_v}\frac{d^2}{dz^2} + U(x, y, z) \quad (11.85)$$

Bulk

Quantum Dot

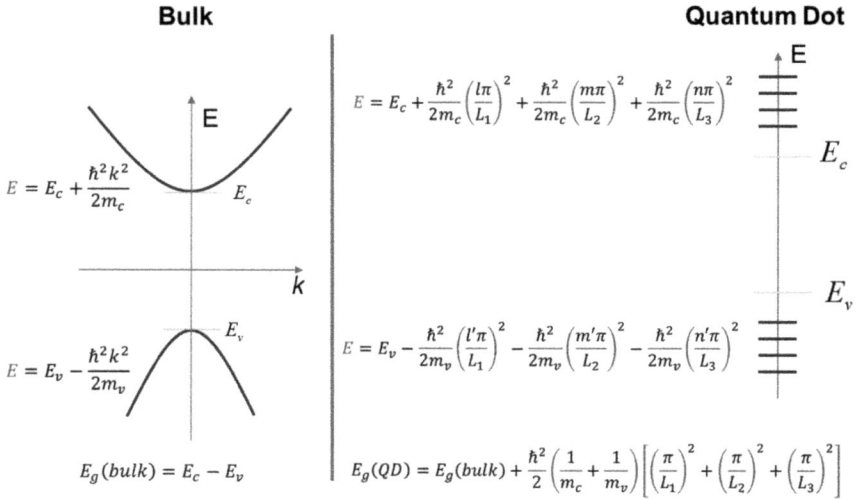

$$E = E_c + \frac{\hbar^2 k^2}{2m_c}$$

$$E = E_v - \frac{\hbar^2 k^2}{2m_v}$$

$$E = E_c + \frac{\hbar^2}{2m_c}\left(\frac{l\pi}{L_1}\right)^2 + \frac{\hbar^2}{2m_c}\left(\frac{m\pi}{L_2}\right)^2 + \frac{\hbar^2}{2m_c}\left(\frac{n\pi}{L_3}\right)^2$$

$$E = E_v - \frac{\hbar^2}{2m_v}\left(\frac{l'\pi}{L_1}\right)^2 - \frac{\hbar^2}{2m_v}\left(\frac{m'\pi}{L_2}\right)^2 - \frac{\hbar^2}{2m_v}\left(\frac{n'\pi}{L_3}\right)^2$$

$$E_g(bulk) = E_c - E_v$$

$$E_g(QD) = E_g(bulk) + \frac{\hbar^2}{2}\left(\frac{1}{m_c}+\frac{1}{m_v}\right)\left[\left(\frac{\pi}{L_1}\right)^2 + \left(\frac{\pi}{L_2}\right)^2 + \left(\frac{\pi}{L_3}\right)^2\right]$$

Figure 11.16. The quantum dot consists only of discrete energy levels due to quantization in all three dimensions. A quantum dot's bandgap increases as the dimensions L_1, L_2 and L_3 decreases. Photons can be either emitted or absorbed when an electron transitions between states of the conduction and valence bands, or within states of the conduction or valence band.

where m_v is the effective mass of a hole in the valence band. The potential energy of a hole in the valence band is,

$$U(x,y,z) = \begin{cases} E_v & \text{for } 0 < x < L_1,\ 0 < y < L_2,\ 0 < z < L_3 \\ -\infty & \text{otherwise} \end{cases}$$

$$(11.86)$$

That is, the potential energy of a hole in the valence band is the bulk value of the valence band maxima E_v inside the quantum dot and $-\infty$ outside the quantum dot. In a manner identical to the derivation for the conduction band, the wave function and energy levels in the valence band can be derived to be,

$$X(x) = \sqrt{\frac{2}{L_1}}\sin\left(\frac{l'\pi}{L_1}x\right) \quad \text{and} \quad \varepsilon_1 = \frac{\hbar^2}{2m_v}\left(\frac{l'\pi}{L_1}\right)^2 \quad (11.87)$$

$$Y(y) = \sqrt{\frac{2}{L_2}}\sin\left(\frac{m'\pi}{L_2}y\right) \quad \text{and} \quad \varepsilon_2 = \frac{\hbar^2}{2m_v}\left(\frac{m'\pi}{L_2}\right)^2 \quad (11.88)$$

$$Z(z) = \sqrt{\frac{2}{L_3}} \sin\left(\frac{n'\pi}{L_3}z\right) \quad \text{and} \quad \varepsilon_3 = \frac{\hbar^2}{2m_v}\left(\frac{n'\pi}{L_3}\right)^2 \quad (11.89)$$

where l', m', and n' are positive integers. The wave function of an electron in the valence band of the QD is,

$$\chi_{l',m',n'}(x,y,z) = X(x)Y(y)Z(z)$$

$$= \sqrt{\frac{8}{L_1 L_2 L_3}} \sin\left(\frac{l'\pi}{L_1}x\right) \sin\left(\frac{m'\pi}{L_2}y\right) \sin\left(\frac{n'\pi}{L_3}z\right)$$

$$(11.90)$$

where $A = \sqrt{\frac{\varepsilon}{L_1 L_2 L_3}}$ is the normalization constant. The energy eigenvalues in the valence band corresponding to quantum numbers (l', m', n') are,

$$E_{l',m',n'} = E_v - \frac{\hbar^2}{2m_v}\left(\frac{l'\pi}{L_1}\right)^2 - \frac{\hbar^2}{2m_v}\left(\frac{m'\pi}{L_2}\right)^2 - \frac{\hbar^2}{2m_v}\left(\frac{n'\pi}{L_3}\right)^2$$

$$(11.91)$$

l', m', and n' are all positive integers.

Note that in equations (11.90) and (11.91), we have added the quantum numbers as subscripts to the wave functions and energy eigenvalues.

Again, as shown in Figure 11.16, the valence band's energy eigenvalues are quantized in a quantum dot as opposed to a continuum of allowed values in bulk. The maximum energy of the valence band of the quantum dot has shifted down from the valence band of the corresponding bulk material by,

$$\frac{\hbar^2}{2m_v}\left(\frac{l'\pi}{L_1}\right)^2 + \frac{\hbar^2}{2m_v}\left(\frac{m'\pi}{L_2}\right)^2 + \frac{\hbar^2}{2m_v}\left(\frac{n'\pi}{L_3}\right)^2$$

Equations (11.84) and (11.91) tell us that the energy levels in the conduction and valence bands of a semiconductor can take only quantized values. The lowest-lying conduction band energy of the quantum dot $E_c(\text{QD})$ has $l = m = n = 1$ in equation (11.84), and its energy is

$$E_c(\text{QD}) = E_c + \frac{\hbar^2}{2m_c}\left(\frac{\pi}{L_1}\right)^2 + \frac{\hbar^2}{2m_c}\left(\frac{\pi}{L_2}\right)^2 + \frac{\hbar^2}{2m_c}\left(\frac{\pi}{L_3}\right)^2 \quad (11.92)$$

The highest lying energy in the valence band of the quantum dot $E_v(\text{QD})$ corresponds to $l' = m' = n' = 1$ in equation (11.91) and is given by

$$E_v(\text{QD}) = E_v - \frac{\hbar^2}{2m_v}\left(\frac{\pi}{L_1}\right)^2 - \frac{\hbar^2}{2m_v}\left(\frac{\pi}{L_2}\right)^2 - \frac{\hbar^2}{2m_v}\left(\frac{\pi}{L_3}\right)^2 \quad (11.93)$$

The **bandgap of the quantum dot** is defined by

$$\text{Bandgap (QD)} = E_c(\text{QD}) - E_v(\text{QD})$$

$$E_g(\text{QD}) = E_g(\text{bulk}) + \frac{\hbar^2}{2}\left(\frac{1}{m_c} + \frac{1}{m_v}\right)\left[\left(\frac{\pi}{L_1}\right)^2 + \left(\frac{\pi}{L_2}\right)^2 + \left(\frac{\pi}{L_3}\right)^2\right]$$

$$(11.94)$$

where $E_c - E_v$ is the bandgap of the bulk material. We see from equation (11.94) that the bandgap of the QD minus the bandgap of bulk scales inversely as the square of the QD dimensions. The energy of photons emitted by a quantum dot is approximately $E_g(\text{QD})$, which corresponds to a photon with wavelength equal to $\lambda = \frac{hc}{E_g(\text{QD})}$.

Example 1. GaAs has a bandgap of 1.4 eV. As a result, it emits light at wavelengths of 881 nm and lower. Design a GaAs quantum dot that will emit light with a wavelength of 700 nm and lower but will not emit light at wavelengths higher than 700 nm.

Solution: One way to design such a quantum dot is to choose the dimensions $L_1 = L_2 = L_3 = L$ so that the bandgap of the quantum dot corresponds to a wavelength of 700 nm. The relationship between the bandgap of the quantum dot and wavelength is

$$\lambda = \frac{hc}{E_g(\text{QD})} \quad (11.95)$$

Substituting the values of Planck's constant h and the speed of light c, we have,

$$\lambda(\text{m}) = \frac{6.6267 \times 10^{-34}\,\text{J}\cdot\text{s} \cdot 3 \times 10^8\,\text{m}\cdot\text{s}^{-1}}{E_g(\text{QD}) \text{ in J}}$$

$$= \frac{1.9875 \times 10^{-25}\,\text{J}\cdot\text{m}}{E_g(\text{QD}) \text{ in J}} \sim \frac{1234 \times 10^{-9}\,\text{m}}{E_g(\text{QD}) \text{ in eV}}$$

The above equation can be written as,

$$\lambda(\text{in terms of nm}) \sim \frac{1234}{E_g(\text{QD}) \text{ in eV}} \tag{11.96}$$

As our GaAs quantum dot should emit light only of wavelength 700 nm and smaller, using equation (11.96), we find that,

$$E_g\,(\text{eV}) = \frac{1234}{700\,\text{nm}} = 1.76\,\text{eV}$$

Now using equation (11.94) with the bandgap of bulk GaAs $(E_g(\text{bulk}) = 1.42\,\text{eV})$, and effective mass of electrons $(m_c = 0.067\,m_0)$ and holes $(m_v = 0.45\,m_0)$ in GaAs, we have for a quantum dot with $L_1 = L_2 = L_3 = L$,

$$E_g(\text{QD}) = E_g(\text{bulk}) + \frac{\hbar^2}{2}\left(\frac{1}{m_c} + \frac{1}{m_v}\right)\left\{\left(\frac{\pi}{L}\right)^2 + \left(\frac{\pi}{L}\right)^2 + \left(\frac{\pi}{L}\right)^2\right\}$$

Using that the bandgap is 1.76 eV, we can find the size of the quantum dot,

$$1.76\,\text{eV} = 1.42\,\text{eV} + \frac{3\hbar^2\pi^2}{2}\left(\frac{1}{m_c} + \frac{1}{m_v}\right)\frac{1}{L^2}$$

$$L \sim 7.54 \times 10^{-9}\,\text{m} = 7.54\,\text{nm}$$

As we saw before, the change of bandgap from bulk is inversely proportional to L^2, where L is the quantum dot's dimension. When the quantum dots are excited by high-energy UV light or by injecting an electric current, the high-energy electrons lose their energy as heat until they reach the minimum conduction band energy, $E_c(\text{QD})$.

Figure 11.17. Quantum dots dispersed in organic solutions emit light under excitation by UV light. Each color, i.e., the emitted light's wavelength, is determined by the quantum dot's size as per equation (11.94). The size of the quantum dots increases from left to right. Picture courtesy of PlasmaChem GmbH.

After that, they recombine with a hole in the valence band and emit a photon whose energy corresponds to $E_g(\text{QD})$. The wavelength of the emitted light is inversely proportional to this bandgap. Hence, larger quantum dots emit light of a more reddish color, and the smaller ones emit shorter wavelengths (bluish colors) (Figure 11.17). This is the principle behind quantum dot-based LED television displays called QLED TVs. Recent technologies of growing PbSe quantum dots in different sizes led to the fabrication of IR sensors covering different wavelengths (Palomaki, 2020). Figure 11.18 shows that in contrast to silicon and GaAs with a continuous absorption spectrum, the absorption spectra of an IR camera can be tuned by the size of the embedded quantum dots to detect only the IR wavelengths in the desired range.

Concept: Electrons in a quantum dot have discrete energy levels. The quantum dot can be visualized to be the equivalent of a PiB in all three dimensions. The bandgap (E_g) of a quantum dot is

$$E_g(\text{QD}) = E_g(\text{bulk}) + \frac{\hbar^2}{2}\left(\frac{1}{m_c} + \frac{1}{m_v}\right)\left[\left(\frac{\pi}{L_1}\right)^2 + \left(\frac{\pi}{L_2}\right)^2 + \left(\frac{\pi}{L_3}\right)^2\right]$$

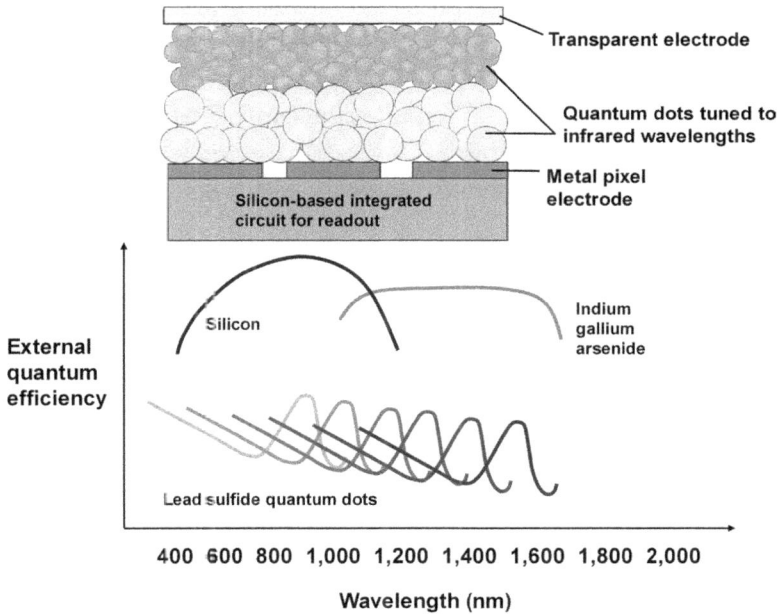

Figure 11.18. Quantum dots applied in IR image sensor for security applications. Adapted from (Palomaki, 2020). Technical art by Mrunal Shenoy.

11.3 Problems

(1) Derive expressions for the energy levels and wave functions for a particle of effective mass M on an infinitesimally thin cylindrical shell (an example is an ultrathin aluminum foil rolled up as a cylinder) with radius R and length L. The potential energy of the particle on the cylinder is a constant U_o. The potential energy outside the cylinder is infinity.

(2) Derive expressions for the energy levels and wave functions for a particle of effective mass M on an infinitesimally thin cylindrical shell of radius R and infinite length. The potential energy of the particle on the cylinder is a constant U_o. The potential energy outside the cylinder is infinity.

(3) Consider a bulk semiconductor with a bandgap of 0.7 eV. Assume that the effective mass of both electrons and holes is 9.1×10^{-31} kg. Furthermore, this semiconductor is known to be a

poor emitter of light at both 500 and 8000 nm. How would you engineer this structure so that it emits light with a wavelength of (a) 500 nm and (b) 8000 nm? Find the nanostructure dimensions to achieve (a) and (b). Your design can be based on quantum wells, dots, or nanowires.

(4) Consider a gallium arsenide nanowire with a square cross-section.

  (a) Plot the bandgap as a function of the edge dimension L. Vary L from 1 to 15 nm in increments of 1 nm. In the y-axis of the plot, indicate both the numerical value of the bandgap and the wavelength/color of light that the bandgap corresponds to.

  (b) Write down the wave functions and plot the energy bands for the two lowest sub-band energies in the conduction band and the two highest sub-band energies in the valence band. Use $L = 2$ nm.

Assume that the effective masses of electrons and holes are $0.067\,m_o$ and $0.47\,m_o$, respectively, where m_o is the free electron mass.

References

Badawy, G. *et al.* (2019). High mobility stemless InSb nanowires. *Nano Letters* **19**(6), 3575–3582.

Brown, J. and Churchill, R. (2011). *Fourier Series and Boundary Value Problems*, 8th edn. McGraw-Hill Education.

Cho, E. C. *et al.* (2007). Photoluminescence in crystalline silicon quantum wells. *Journal of Applied Physics* **101**, 024321.

Madelung, O. (Ed.) (1996). *Semiconductors: Basic Data*. Springer Verlag, Berlin, Heidelberg.

Palomaki, P. and Keuleyan, S. (2020). Move over, CMOS: Here come snapshots by quantum dots. *IEEE Spectrum*.

Walavalkar, S. S. *et al.* (2010). Tunable visible and near-IR emission from sub-10 nm etched single-crystal Si nanopillars. *Nano Letters* **10**(11), 4423–4428.

Chapter 12

DENSITY OF STATES

Contents

Density of states is used in the calculation of electron density, current (Chapter 13), scattering rates, and optical absorption in nanomaterials (Chapter 18).

12.1 Density of States

Before mathematically defining density of states (DOS), let us explain it conceptually. Consider a tower that has nonuniformly spaced rooms at different elevations indexed by H_i (Figure 12.1(a)). Let us assume that each room has one person occupying it. The potential energy (E) of each room (or of a person occupying it) is proportional to H ($E = mgH$). Hence, we can replace H by E. If we are asked to count the number of people occupying this tower, all we need is to build a histogram using bars of unit length at each $H_i(E_i)$ and plot it along E (or H). Then the total number of people in the tower is merely counting the discrete vertical bars as we increase E_i.

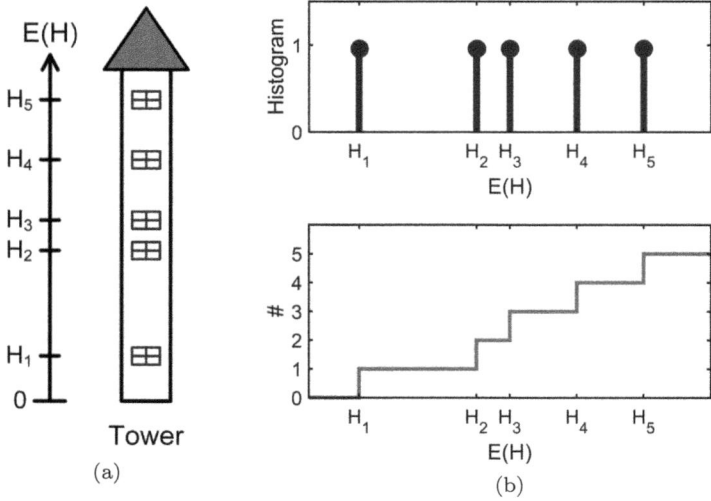

Figure 12.1. (a) A tower with five nonuniformly spaced rooms. (b) Distribution (density) histogram of rooms (top) and the total number of rooms as the sum of unit bars (bottom). It is assumed that the probability of a person occupying each room is always 1.

Figure 12.1(b) shows the histogram and the total number of people in the tower found by adding up the bars from left to right.

However, to make things mathematically sound, we need to add more information to Figure 12.1. Firstly, the density should have the unit of person per unit height (or per unit potential energy). Hence, if we replace the discrete unit-height bars with a continuous spike-like distribution whose integral is equal to one, the job is done. Secondly, as there is no room between two consecutive heights, i.e., H_i and H_{i+1}, the density must be zero between them. The Dirac delta function $\delta(H)$ or $\delta(E)$ satisfies both these necessities. Figure 12.2(a) shows the updated density of rooms, which is a set of delta functions peaking at each $H_i(E_i)$. This can be written as

$$\text{Density of rooms} = \sum_i \delta(H - H_i) \quad \text{or} \quad \sum_i \delta(E - E_i) \qquad (12.1)$$

Now, if the people living in the tower decide to show up or disappear randomly, then the height of each delta-function must be weighted by p_i, which is the probability of a person showing up in their assigned room (H_i). To account for this, equation (12.1) is updated as

$$\text{Density of people} = \sum_i p_i \delta(E - E_i) \qquad (12.2)$$

If you are asked how many people are living in the tower, add up the numbers or calculate the area in Figure 12.2(a) to find,

$$\text{No. of people} = \int_{E=0}^{E>E_5} \sum_i p_i \delta(E - E_i) dE \qquad (12.3)$$

Recall that integration of a Dirac delta function is a step function,

$$\int_{-\infty}^{+\infty} \delta(E - E_i) dE = \int_{E<E_i}^{E>E_i} \delta(E - E_i) dE = 1 \cdot u(E - E_i) \quad (12.4)$$

The term u is called the step function, which is defined by

$$u(E - E_i) = \begin{cases} 1 & \text{for } E > E_i \\ 0 & \text{for } E < E_i \end{cases} \qquad (12.5)$$

Therefore, equation (12.3) can be reduced to a sum of shifted step functions of unit step weighted by probability p_i:

$$\text{No. of people as a function of } E = \int_0^E \sum_i p_i \delta\left(E' - E_i\right) dE'$$

$$= \sum_i p_i \, u(E - E_i) \qquad (12.6)$$

See Figure 12.2(b) for a plot of equation (12.6).

Similar to equation (12.2), the density of (electronic) states or DOS measures the number of available states per unit volume per unit energy at spatial location r and energy E. The DOS depends on energy eigenvalues and their corresponding eigenfunctions and is

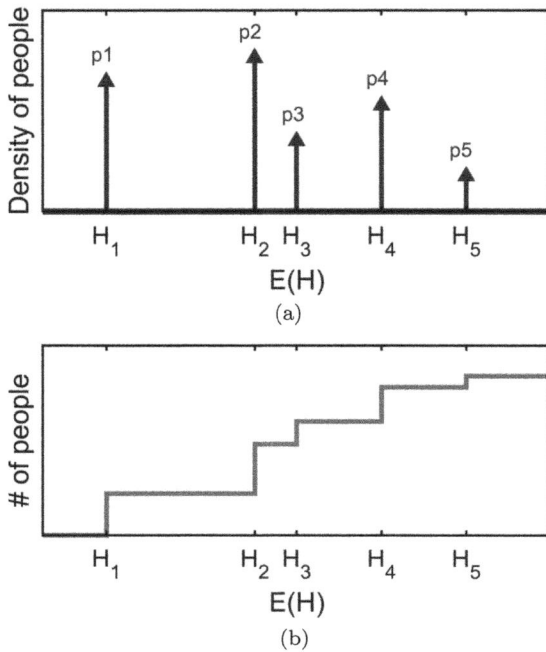

Figure 12.2. (a) The Density of people in a tower as a series of weighted delta functions that peak at each energy (height) of E_i (or H_i). (b) The area under the curve in (a). The cumulative sum (integral) of the density is used to find out the number of people.

defined by

$$\text{DOS}(\bar{r}, E) = \sum_n |\psi_n(\bar{r})|^2 \, \delta(E - E_n) \qquad (12.7)$$

where n is the set of all quantum numbers. The term $|\psi_n(\bar{r})|^2$ has units of inverse volume (in 3D), and it determines the probability of occupying state n at various spatial points. It is multiplied by delta functions, which peak at the energy eigenvalues E_n. The unit of the above delta function is per unit energy. DOS units in perfectly 1D, 2D, and 3D cases are $m^{-1}J^{-1}$, $m^{-2}J^{-1}$, and $m^{-3}J^{-1}$, respectively. You can see this by noting that the dimensions of the eigenfunctions in 1D, 2D, and 3D are $m^{-\frac{1}{2}}$, m^{-1}, and $m^{-3/2}$, respectively.

Definition of delta function: The following properties define the delta function:

$$\delta(E - E_n) = \begin{cases} \infty & \text{at } E = E_n \\ 0 & \text{at other energies} \end{cases} \qquad (12.8)$$

and

$$\int_{E_n - \delta E}^{E_n + \delta E} dE\, \delta\,(E - E_n) = 1 \qquad (12.9)$$

These equations mean that even though the delta function's maximum value is infinity, the area under the delta function is finite and equal to 1. Since the integration is around E_n, only those peaks centered on E_n will give a nonzero value. Equation (12.9) is called the sifting property of the Delta function.

From the definition of the DOS (equation (12.7)) and the properties of the Dirac delta function, we will show that for a system with discrete energy levels,

$$\int_{\text{all space}} dv \int_{E_n - \delta E}^{E_n + \delta E} \text{DOS}\,(r, E)$$
$$= \text{Number of energy levels with } E = E_n \qquad (12.10)$$

The number of energy levels with $E = E_n$ can be larger than one if more than one quantum number corresponds to the same energy value. For example, in the quantum dot discussed in Chapter 11, the three sets of quantum numbers $(l, m, n) = (1, 1, 2)$, $(1, 2, 1)$, and $(2, 1, 1)$ all have the same energy if $L_1 = L_2 = L_3$. The energy level is called three-fold *degenerate*. This is like having a single room in the tower occupied by three people. So, the RHS of equation (12.10) will be 3 for energy value $E_{1,1,2}$. On the other hand, the energy value $E_{1,1,1}$ is single-fold degenerate. The terminology for the number of quantum numbers corresponding to a given energy value is known as **degeneracy**.

Equation (12.10) can be verified by noting that

$$\int_{\text{all space}} dv \int_{E_n-\delta E}^{E_n+\delta E} dE \ \text{DOS}(\bar{r}, E)$$

$$= \int_{\text{all space}} dv \int_{E_n-\delta E}^{E_n+\delta E} dE \sum_m |\psi_m(\bar{r})|^2 \delta(E - E_m)$$

$$= \sum_m \int_{\text{all space}} dv |\psi_m(\bar{r})|^2 \int_{E_n-\delta E}^{E_n+\delta E} dE \ \delta(E - E_m) \qquad (12.11)$$

Noting that the normalization of the wave function implies $\int_{\text{all space}} dv |\psi_m(\bar{r})|^2 = 1$ and since $\int_{E_n-\delta E}^{E_n+\delta E} dE \ \delta(E - E_m) = 1$, the above equation becomes

$$\int_{\text{all space}} dv \int_{E_m-\delta E}^{E_m+\delta E} dE \ \text{DOS}(\bar{r}, E)$$

$$= \sum_m \int_{E_m-\delta E}^{E_m+\delta E} \delta(E - E_m)$$

$$= \text{Number of energy levels with } E = E_m$$

Note: In the literature, you will see the density of optical modes in an optical waveguide, the density of acoustic modes in a sound cavity or musical instrument, or the density of phonon modes (quantum of vibrations). Their mathematical definition is similar to the DOS defined in equation (12.7).

We will now discuss the DOS for various nanostructures that we have discussed previously in Chapter 11.

Concept: The density of states (DOS) is the number of quantum states per unit energy per unit volume at spatial location $\bar{r}$ and energy E.

$$\text{DOS}(\bar{r}, E) = \sum_n |\psi_n(\bar{r})|^2 \delta(E - E_n)$$

Definition of step function: The step function is defined by

$$u(\gamma - \gamma_o) = \begin{cases} 1 & \text{for } \gamma > \gamma_o \\ 0 & \text{for } \gamma < \gamma_o \end{cases} \quad \text{(definition of step function)}$$

The step function u does not have any dimensions. If its argument $\gamma > \gamma_o$, then $u = 1$ and if $\gamma < \gamma_o$, then $u = 0$. The step function is discontinuous when $\gamma = \gamma_o$. See the following figure for an illustration of the step function.

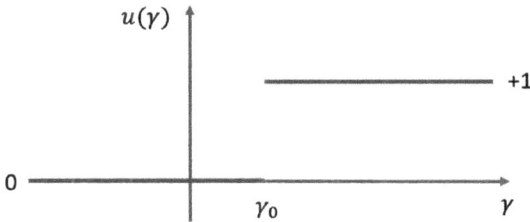

The derivative of the step function is a delta function at $\gamma = \gamma_o$ as shown in the following figure.

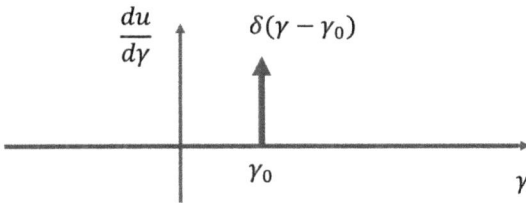

12.1.1 *Density of states for a particle in a 1D box*

The energy eigenfunctions and eigenvalues of the PiB are (Chapter 1) as follows:

$$\chi_n(x) = \sqrt{\frac{2}{L}} \sin\left(\frac{n\pi}{L}x\right) \quad \text{and} \quad E_n = \frac{\hbar^2}{2m_c}\left(\frac{n\pi}{L}\right)^2 + U$$

where we have used the effective mass m_c of the electron as the PiB may be made of a semiconductor. In vacuum, we would simply use the vacuum mass of the electron (m_o). Substituting these in equation (12.7), the DOS is

$$\text{DOS}(x, E) = \frac{2}{L} \sum_n \sin^2\left(\frac{n\pi}{L}x\right) \delta(E - E_n) \tag{12.12}$$

Note that DOS is a two-dimensional function. At a constant x value, along the E (energy) axis, there are several delta functions. At $E = E_n$, the peaks follow the form of the $\sin^2\left(\frac{n\pi}{L}x\right)$ function along the x direction.

12.1.2   DOS of a 1D free particle

In contrast to the PiB, where we had a discrete set of quantum numbers and corresponding energy levels (E_n), the energy of a free particle is a continuous function of its wave vector (quantum number), i.e., $E(\bar{k})$. The Hamiltonian of a free particle in 1D is

$$\hat{H} = -\frac{\hbar^2}{2m_c}\frac{\partial^2}{\partial x^2} + E_c$$

where m_c is the effective mass of the electron. The wave functions and energy eigenvalues of the 1D free particle are (as we learned in Chapter 1)

$$\chi_k(x) = \sqrt{\frac{1}{L}}e^{ikx} \quad \text{and} \quad E_k = \frac{\hbar^2 k^2}{2m_c} + E_c \tag{12.13}$$

Since L is the length of the 1D box, the $\sqrt{\frac{1}{L}}$ factor normalizes the free particle wave function in 1D space; Further, we will set L to be infinity. E_c is a constant. Substituting equation (12.13) in equation (12.7), we find that the DOS is

$$\text{DOS}(x, E) = \frac{1}{L} \sum_k \delta(E - E_k) = \frac{1}{L} \sum_k \delta\left(E - E_c - \frac{\hbar^2 k^2}{2m_c}\right) \tag{12.14}$$

In equation (12.14), the quantum number k of the free particle is a continuous variable. The summation over wave vector k is not straightforward to evaluate with basic calculus. We will adopt an intuitive and direct approach based on a counting argument. We will see how many allowed k values are found within an interval of energy lying between E and $E + \Delta E$. We will assume that the free particle is over a wire of length L and apply the periodic (Born–von Karman) boundary condition:

$$\chi_n(x) = \chi_n(x + L) \rightarrow e^{ikx} = e^{ik(x+L)} \tag{12.15}$$

We will finally set L equal to infinity at the end of the calculation.

The boundary condition in equation (12.15) implies that the allowed values of the wave vector (k) are

$$k = \frac{2\pi n}{L} \quad \text{(allowed values of wave vector } k) \tag{12.16}$$

where n is an integer. The length in wave vector space occupied by a single k-point is

$$\Delta k = \frac{2\pi}{L} \tag{12.17}$$

The k space is a one-dimensional array of points that are $2\pi/L$ apart, meaning that the larger the L, the smaller the distance between neighboring k-points (Δk).

The electron waves are incident from both (a) left to right $(+x$ direction) and (b) right to left $(-x$ direction) in the energy range from E to $E + dE$, as shown in Figure 12.3. They occupy a length dk_L and dk_R in wave vector space. dk_L (dk_R) is the length of k-space for waves incident in $-x$ direction $(+x$ direction) in energy window from E to $E + dE$. The total number of states incident from the left (N_L) and right (N_R) in the energy window from E to $E + dE$ in Figure 12.3 is

$$N_L = \frac{dk_L}{\Delta k} = \frac{dk_L}{2\pi/L} \tag{12.18}$$

$$N_R = \frac{dk_R}{\Delta k} = \frac{dk_R}{2\pi/L} \tag{12.19}$$

The total number of states in the energy window from E to $E + dE$ is

$$N = N_L + N_R \tag{12.20}$$

From the definition of DOS, and equations (12.18), (12.19), and (12.20), we have

$$\text{DOS} = \frac{N}{L\,dE} = \frac{N_L + N_R}{L\,dE} = \frac{1}{2\pi} \frac{dk_L + dk_R}{dE} \tag{12.21}$$

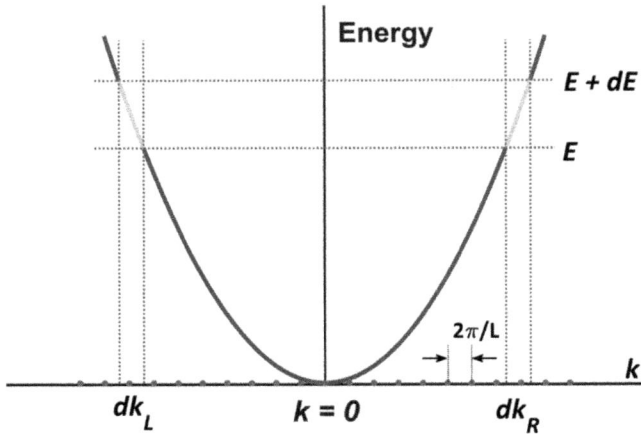

Figure 12.3. $E(k)$ dispersion relationship of a free particle. dk_L and dk_R are the lengths in wave vector space corresponding to the energy window from E to $E + dE$.

The above equation can be simplified further by using the dispersion relation, $E = E_c + \frac{\hbar^2 k^2}{2m_c}$, which gives

$$\frac{dk_L}{dE} = \frac{1}{\hbar}\sqrt{\frac{m_c}{2(E - E_c)}} \quad \text{and} \quad \frac{dk_R}{dE} = \frac{1}{\hbar}\sqrt{\frac{m_c}{2(E - E_c)}} \qquad (12.22)$$

Substituting equation (12.22) in (12.21), we find that the DOS is

$$\text{DOS}(E) = \text{DOS}_L(E) + \text{DOS}_R(E) \qquad (12.23)$$

where each term for one kind of electron spin is

$$\text{DOS}_L(x, E) = \text{DOS}_R(x, E) = \frac{1}{2\pi\hbar}\sqrt{\frac{m_c}{2(E - E_c)}} \qquad (12.24)$$

Using the fact that $E - E_c = \frac{\hbar^2 k^2}{2m_c}$, we can write equation (12.24) as

$$\text{DOS}_L(x, E) = \text{DOS}_R(x, E) = \frac{1}{2\pi\hbar}\frac{m_c}{\hbar|k|} \qquad (12.25)$$

Substituting equation (12.24) in equation (12.23), we have shown that the 1D DOS per degree of spin is given by,

$$\text{DOS}(x, E) = \frac{1}{L}\sum_{k} \delta\left(E - E_k\right) = \frac{1}{\pi\hbar}\sqrt{\frac{m_c}{2\left(E - E_c\right)}} u(E - E_c)$$

(12.26)

"Per degree of spin" simply means that we have not included a multiplicative factor of 2 on the RHS of equation (12.26) to account for up and down spin electrons.[1]

The step function in equation (12.26) reminds us that when $E < E_c$, DOS $= 0$. The 1D DOS given by equation (12.26) has a singularity at $E = E_c$ which is called the van Hove singularity. In experiments with nanostructures, scattering mechanisms (such as electron–phonon interaction) make the sharp peaks of the van Hove singularity be lowered and rounded, a process called broadening.

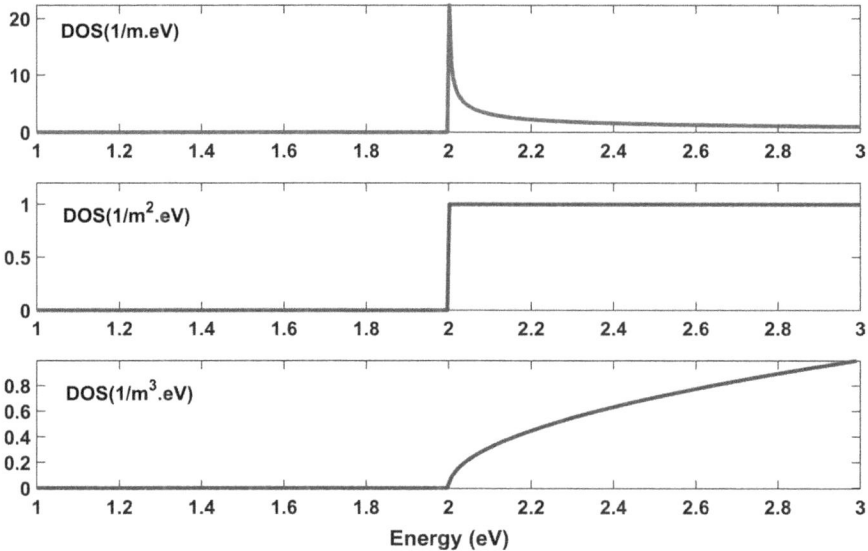

Figure 12.4. The DOS of a free electron in (top) 1D, (middle) 2D, and (bottom) 3D case when $E_c = 2\,\text{eV}$ and effective mass $m_c = 9.1 \times 10^{-31}\text{kg}$.

---

[1]In ferromagnetic materials, the DOS for spin-up and spin-down electrons can be different, and the total density of states is not necessarily one of them multiplied by 2.

Numerical computations also adopt a broadening to prevent the DOS from becoming infinite.

Figure 12.4 shows a plot of DOS versus energy obtained from equation (12.26), which decreases with energy, proportional to $\frac{1}{\sqrt{E-E_c}}$.

12.1.3 DOS of a 2D free particle

The Hamiltonian of a purely 2D sheet with a constant potential is given by

$$\widehat{H} = -\frac{\hbar^2}{2m_c}\nabla^2_{xy} + E_c \tag{12.27}$$

The energy eigenfunctions and eigenvalues are

$$\psi_{k_x,k_y}(x,y) = \frac{1}{\sqrt{L_xL_y}}e^{ik_xx}e^{ik_yy} \quad \text{and} \quad E_{k_x,k_y} = E_c + \frac{\hbar^2}{2m_c}\left(k_x^2 + k_y^2\right) \tag{12.28}$$

Like the 1D case, we assume that the wave function is periodic in the x and y directions. The periodicity of the box is now L_x and L_y, which we will later set to infinity,

$$\psi_{k_x,k_y}(x + L_x, y + L_y) = \psi_{k_x,k_y}(x,y) \tag{12.29}$$

From this, we find allowed discrete set of values for k_x and k_y which are as follows (see Figure 12.5):

$$k_x = \frac{2\pi n}{L_x} \quad \text{and} \quad k_y = \frac{2\pi m}{L_y} \quad (m,n) \in \mathbb{Z} \tag{12.30}$$

This means that the distance between two consecutive k points in 2D k-space is $\frac{2\pi}{L_x}$ and $\frac{2\pi}{L_y}$ along the k_x and k_y directions, respectively. The element of the area occupied by a single k-point in 2D k-space is a rectangular area $\Delta A = \frac{4\pi^2}{L_xL_y}$ (see Figure 12.5).

We will calculate the number of k-points between energies E and $E + dE$. To find this, we first calculate the number of k-points in the energy window dE, which is shown by the annular shadow of the

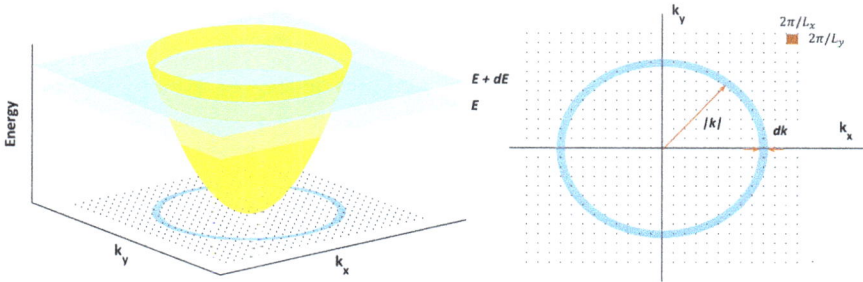

Figure 12.5. (Left) Energy dispersion for a free electron in 2D. The k-points corresponding to energies between E and $E + dE$ are on the annular disk on the (k_x, k_y) plane. The inner radius of the disk corresponds to energy E and the outer radius corresponds to energy $E + dE$. (Right). The inner radius of the annular disk is $|k|$ and its thickness is dk. The top right shows that the area occupied by a single k-point on the (k_x, k_y) plane is $\Delta A = \frac{4\pi^2}{L_x L_y}$.

$E(k)$ parabola on the 2D k plane shown in Figure 12.5. The number of (k_x, k_y) points inside this annular disk with width dk is equal to the area of the annular disk divided by the area occupied by a single k-point,

$$N_k = \frac{2\pi \,|k|\, dk}{\Delta A} = \frac{2\pi \,|k|\, dk}{\frac{4\pi^2}{L_x L_y}} = \frac{L_x L_y}{2\pi} |k|\, dk \qquad (12.31)$$

From equation (12.28), we see that the differential dE is related to the magnitude of the k vector $|k|$ via

$$dE = \frac{\hbar^2}{m_c} |k|\, dk \qquad (12.32)$$

From the definition of DOS, and equations (12.31) and (12.32), we have

$$\text{DOS} = \frac{N_k}{L_x L_y dE} = \frac{L_x L_y}{L_x L_y 2\pi} \frac{|k|\, dk}{dE} = \frac{m_c}{2\pi \hbar^2} \qquad (12.33)$$

We multiply the previous equation with a step function to emphasize that the DOS is nonzero only when the energy is larger than E_c (see Figure 12.4), hence

$$\text{DOS}(x, y, E) = \frac{m_c}{2\pi\hbar^2} u(E - E_c) \quad \text{(per degree of spin)} \qquad (12.34)$$

The DOS expression does not depend on the z-coordinate because we assume that we are in a perfectly 2D (x, y) system.

12.1.4   DOS of a 3D free particle

The Hamiltonian, energy eigenfunctions, and eigenvalue for a free particle in 3D are,

$$\widehat{H} = -\frac{\hbar^2}{2m_c} \nabla^2 + E_c \qquad (12.35)$$

$$\psi_{k_x, k_y, k_z}(x, y, z) = \frac{1}{\sqrt{L_x L_y L_z}} e^{ik_x x} e^{ik_y y} e^{ik_z z} \quad \text{and}$$

$$E_{k_x, k_y, k_z} = E_c + \frac{\hbar^2}{2m_c} \left(k_x^2 + k_y^2 + k_z^2 \right) \qquad (12.36)$$

We assume the electron is within a 3D box with dimensions L_x, L_y, and L_z and impose the following periodic boundary conditions:

$$\psi_{k_x, k_y, k_z}(x + L_x, y + L_y, z + L_z) = \psi_{k_x, k_y, k_z}(x, y, z)$$

Later, we will increase the box size to be infinity. The allowed discrete values for k_x, k_y, and k_z due to periodic boundary conditions are,

$$k_x = \frac{2\pi n}{L_x} \quad \text{and} \quad k_y = \frac{2\pi m}{L_y} \quad \text{and} \quad k_z = \frac{2\pi p}{L_z} \qquad (12.37)$$

where m, n, and p are integers. The distances between two consecutive k values is $\frac{2\pi}{L_x}$, $\frac{2\pi}{L_y}$, and $\frac{2\pi}{L_z}$ in the x, y, and z directions, respectively. The element of volume in the 3D k-space occupied by a single k-point is then

$$\Delta V = \frac{8\pi^3}{L_x L_y L_z} \qquad (12.38)$$

The number of available k states between energies E and $E + dE$ is the number of k points in the spherical shell of thickness dE in

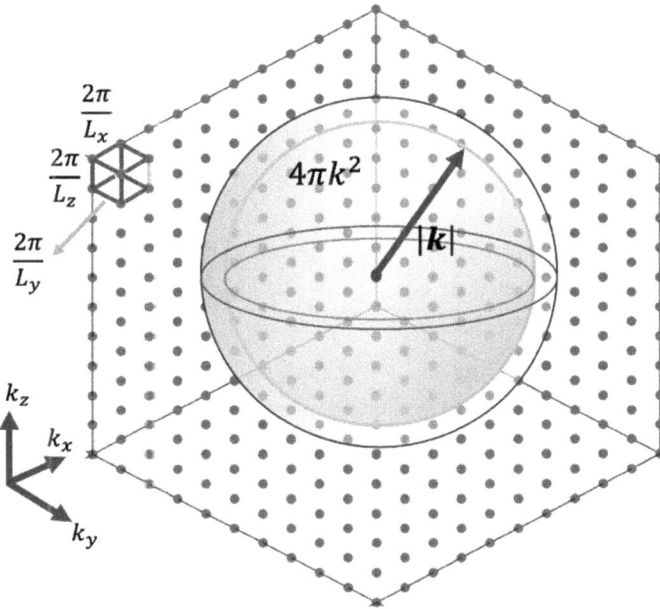

Figure 12.6. 3D k space with a discrete set of available k-states. The shell (green) is enclosed by two constant energy spherical surfaces at energies of E and $E + dE$. The volume of the shell in k-space is $4\pi|k|^2 dk$. The small cube on the top left shows the volume occupied by a single k-point on the (k_x, k_y, k_z) space is $\frac{8\pi^3}{L_x L_y L_z}$.

Figure 12.6. This is found by dividing the volume of the shell by the volume dV occupied by a single k-point in 3D wave vector space:

$$N_k = \frac{4\pi \, |k|^2 \, dk}{\Delta V} = \frac{4\pi \, |k|^2 \, dk}{\frac{8\pi^3}{L_x L_y L_z}} = \frac{L_x L_y L_z}{2\pi^2} |k|^2 \, dk \qquad (12.39)$$

Finding the differential of $E(dE)$ from equation (12.36), and substituting it into the expression for the DOS, we have

$$\text{DOS} = \frac{N_k}{L_x L_y L_z dE} = \frac{L_x L_y L_z}{L_x L_y L_z 2\pi^2} \frac{|k|^2 \, dk}{dE}$$

$$= \frac{m_c}{2\pi^2 \hbar^2} |k| = \frac{m_c}{2\pi^2 \hbar^3} \sqrt{2m_c(E - E_c)}$$

We multiply the previous equation with a step function to emphasize that the DOS is nonzero only when the energy is larger than E_c (see Figure 12.4), hence

$$\text{DOS}(x, y, z, E) = \frac{m_c}{2\pi^2\hbar^3}\sqrt{2m_c(E - E_c)}u(E - E_c) \quad \text{(per degree of spin)} \tag{12.40}$$

Summary: The Hamiltonian and density of states for free particles in 1D, 2D, and 3D are in the following table:

	Hamiltonian	Free-particle DOS per spin
1D	$H = -\frac{\hbar^2}{2m_c}\frac{\partial^2}{\partial x^2} + E_c$	$\frac{1}{\pi\hbar}\sqrt{\frac{m_c}{2(E - E_c)}}u(E - E_c)$
2D	$H = -\frac{\hbar^2}{2m_c}\nabla^2_{xy} + E_c$	$\frac{m_c}{2\pi\hbar^2}u(E - E_c)$
3D	$H = -\frac{\hbar^2}{2m_c}\nabla^2 + E_c$	$\frac{m_c}{2\pi^2\hbar^3}\sqrt{2m_c(E - E_c)}u(E - E_c)$

12.1.5   *Nanowire with a finite cross-section*

Consider a nanowire with a rectangular cross-section of $L_1 \times L_2$ in the xy plane and an infinite length along the z direction. The energy eigenfunctions and eigenvalues for the nanowire were derived in Chapter 11 to be

$$\chi_{l,m,k}(x, y, z) = \sqrt{\frac{4}{L_1 L_2}}\sqrt{\frac{1}{L_z}}\sin\left(\frac{l\pi}{L_1}x\right)\sin\left(\frac{m\pi}{L_2}y\right)e^{ik_z z} \tag{12.41}$$

$$E_{l,m,k} = E_c + \frac{\hbar^2}{2m_c}\left(\frac{l\pi}{L_1}\right)^2 + \frac{\hbar^2}{2m_c}\left(\frac{m\pi}{L_2}\right)^2 + \frac{\hbar^2 k_z^2}{2m_c}$$

$$= E_c + \varepsilon_{l,m} + \frac{\hbar^2 k_z^2}{2m_c} \tag{12.42}$$

where L_z is the length of the nanowire (L_z will be set to infinity at the end of the calculation). L_z is added to normalize the free-electron wave function along the z direction; but it will disappear during the integration process over k-space.

This shows that the band structure is composed of many parabolas in terms of k_z starting at different energy values determined by a pair of quantum numbers (l, m),

$$\varepsilon_{l,m} = \frac{\hbar^2}{2m_c}\left(\frac{l\pi}{L_1}\right)^2 + \frac{\hbar^2}{2m_c}\left(\frac{m\pi}{L_2}\right)^2 \tag{12.43}$$

Substituting equations (12.41), (12.42), and (12.43) in equation (12.7) gives us

$$\text{DOS}(\bar{r}, E) = \sum_{l,m,k} |\chi_{l,m,k_z}(r)|^2 \delta(E - E_{l,m,k_z}) \tag{12.44}$$

$$\text{DOS}(x, y, z, E) = \frac{4}{L_1 L_2 L_z} \sum_{l,m} \sin^2\left(\frac{l\pi}{L_1}x\right)\sin^2\left(\frac{m\pi}{L_2}y\right)$$

$$\times \sum_{k_z} \delta\left(E - E_c - \varepsilon_{l,m} - \frac{\hbar^2 k_z^2}{2m_c}\right) \tag{12.45}$$

The summation over k_z in equation (12.45) is identical to equation (12.14) of the 1D free particle except that E_c is replaced by the bottom of each subband $E_c + \varepsilon_{l,m}$. Using methods similar to the derivation for 1D DOS (equations (12.13)–(12.26)), one can show that,

$$\frac{1}{L_z} \sum_{k_z} \delta(E - E_c - \varepsilon_{l,m} - \frac{\hbar^2 k_z^2}{2m_c})$$

$$= \frac{1}{\pi\hbar}\sqrt{\frac{m_c}{2(E - E_c - \varepsilon_{l,m})}} u(E - E_c - \varepsilon_{l,m}) \tag{12.46}$$

where the conduction band bottom is at $E_c + \varepsilon_{l,m}$. Substituting equation (12.46) into (12.45), the DOS for the nanowire is

$$\text{DOS}(x, y, z, E) = \frac{4}{L_1 L_2}\frac{1}{\pi\hbar} \sum_{l,m} \sin^2\left(\frac{l\pi}{L_1}x\right)\sin^2\left(\frac{m\pi}{L_2}y\right)$$

$$\times \sqrt{\frac{m_c}{2(E - E_c - \varepsilon_{l,m})}} u(E - E_c - \varepsilon_{l,m}) \tag{12.47}$$

per degree of spin. The meaning of equation (12.47) is that the nanowire DOS at the location (x, y) is the sum of the 1D free particle

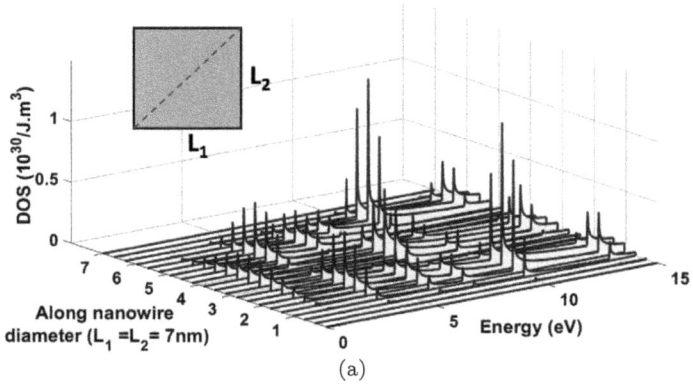

(a)

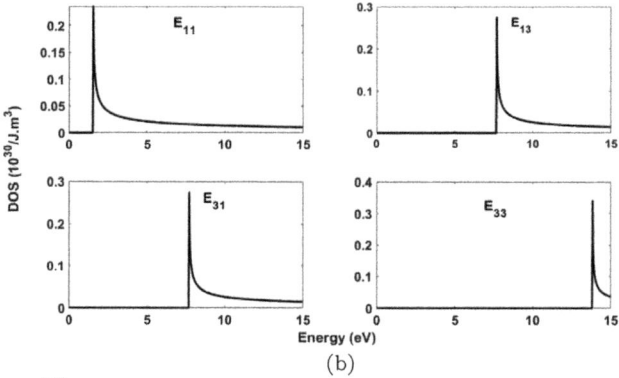

(b)

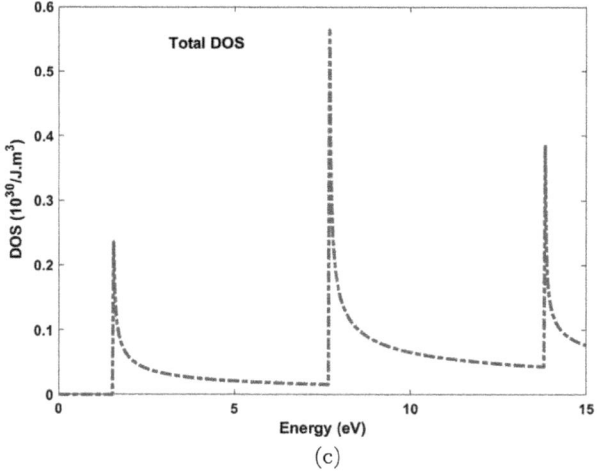

(c)

Figure 12.7. (a) Conduction band DOS versus energy of a nanowire at various points along the line $x = y$, which runs diagonally along the nanowire's square cross-section. (b) DOS versus energy at the point $x = y = 3.5$ nm inside the nanowire. There is a singularity in the DOS at the bottom of each new sub-band along the energy axis. The partial DOS due to a sub-band (l, m) starts at the minimum sub-band energy $\varepsilon_{l,m}$. (c) The sum of the partial DOS from the sub-bands in subfigure (b). Note that for some quantum number combinations, the energy eigenvalues are degenerate, i.e., $E_{1,2} = E_{2,1}$ and $E_{1,3} = E_{3,1}$. This is because both sides of the nanowire are equal, i.e., $L_1 = L_2$. Note that in these calculations, $E_c = 2$ eV, the effective mass of an electron is $m_c = 0.01\, m_o$, and the nanowire dimensions are $L_1 = L_2 = 7$ nm.

DOS (equation (12.26)) with different subband energies $E_c + \varepsilon_{l,m}$, weighted by the eigenfunction magnitude squared corresponding to subband (l,m), which is

$$\frac{4}{L_1 L_2} \sin^2\left(\frac{l\pi}{L_1}x\right) \sin^2\left(\frac{m\pi}{L_2}y\right)$$

Figure 12.7 shows the DOS for a nanowire with a square cross-section. The van Hove singularity when a new subband appears in the summation in equation (12.47) should be apparent. The DOS corresponding to $(l, m) = (1, 2)$ and $(2, 1)$ lie on top of each other because these two subbands are degenerate (meaning, have the same energy). As a reminder, E_c is the conduction band of the bulk semiconductor, while the conduction band of the nanowire is $E_c + \varepsilon_{l,m}$. The partial DOS versus energy for the lowest four sub-bands at $x = 5$ nm and $y = 2.5$ nm is shown in Figure 12.8.

Similar to the derivation for DOS in the conduction band, the DOS in the valence band can be derived using the maximum energy of the valence band (E_v) and effective mass of holes (m_v). The DOS in the valence band is then given by

$$\text{DOS}(x, y, z, E) = \frac{4}{L_1 L_2} \frac{1}{\pi\hbar} \sum_{l',m'} \sin^2\left(\frac{l'\pi}{L_1}x\right) \sin^2\left(\frac{m'\pi}{L_2}y\right)$$

$$\times \sqrt{\frac{m_v}{2\left(E_v - \varepsilon_{l',m'} - E\right)}}\, u(E_v - \varepsilon_{l',m'} - E) \quad (12.48)$$

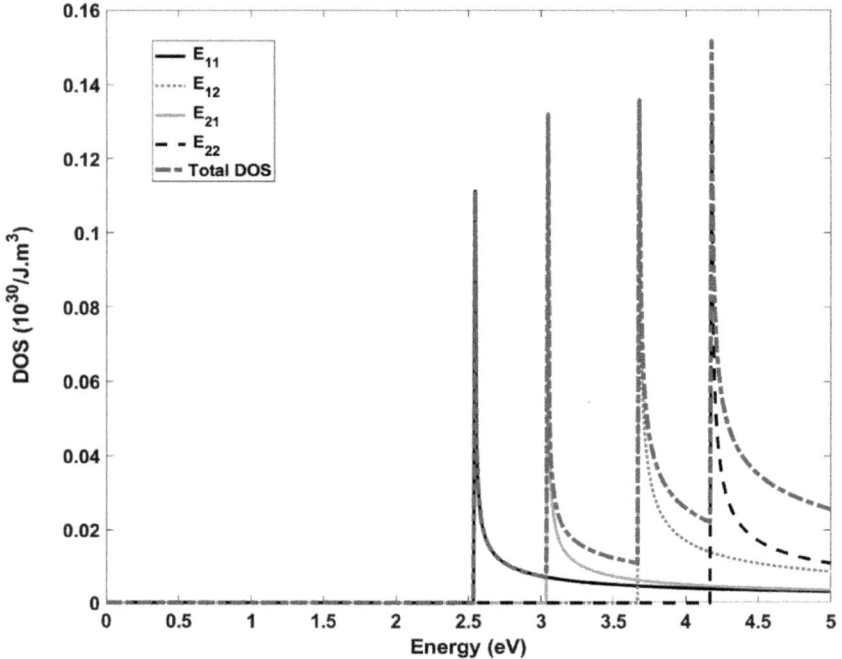

Figure 12.8. DOS of the lowest four conduction subbands for a nanowire with $L_1 = 15\,\text{nm}$ and $L_2 = 10\,\text{nm}$, and the total DOS versus energy at the point $x = 5\,\text{nm}$ and $y = 2.5\,\text{nm}$.

Example 1. Calculate and then plot the DOS versus energy in the valence band of a nanowire which has $L_1 = 10\,\text{nm}$ and $L_2 = 5\,\text{nm}$. Assume that effective mass of holes and maximum of the bulk valence band energy are $m_v = 0.1\ m_o$ and $E_v = -1\,\text{eV}$. Find the partial DOS and total DOS versus energy at $x = 5\,\text{nm}$ and $y = 2.5\,\text{nm}$ from the lowest four valence subbands (the subbands closest to the bulk valence band). Assume that the energy of the holes is changing from $-5\,\text{eV}$ to $0\,\text{ev}$.

Solution: Within this range, we calculate $\text{DOS}(x = 5\,\text{nm}, y = 2.5\,\text{nm}, E)$ by sweeping the integer indices, l' and m', over 1, 2, and 3. Note that for even values of l', the wave function is zero at $x = 5\,\text{nm}$. Therefore only odd indices l' equal to 1 and 3 have nonzero contributions to the total DOS.

As $\varepsilon_{l'=1,m'=1} = -0.188\,\text{eV}$, the first peak of the DOS is at $E = E_v - E_{11} = -1.188\,\text{eV}$, as shown in Figure 12.9.

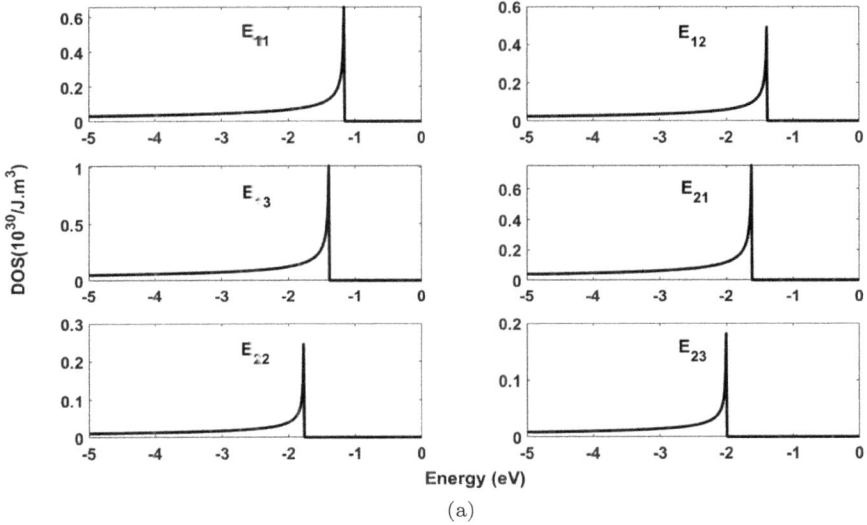

(a)

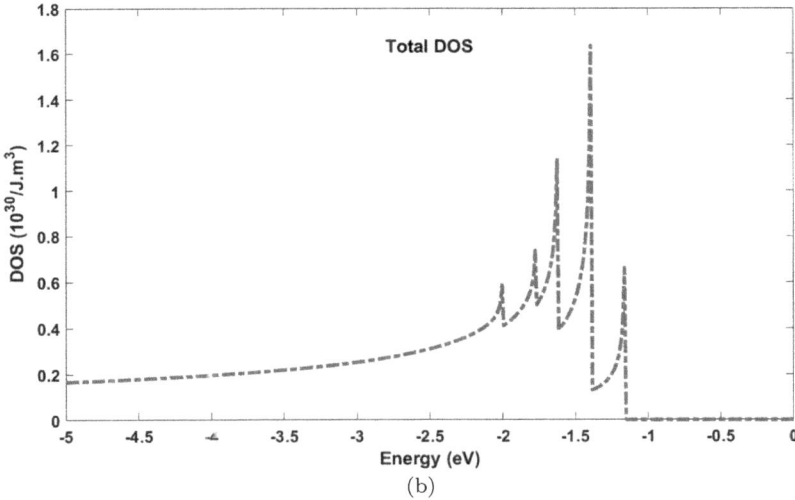

(b)

Figure 12.9. Valence band DOS versus energy of a nanowire with $L_1 = 10\,\text{nm}$ and $L_2 = 5\,\text{nm}$ plotted at the point ($x = 5\,\text{nm}$, $y = 2.5\,\text{nm}$). Note that $E_v = -1\,\text{eV}$, and the effective mass of the hole is $m_v/m_0 = 0.1$. The red dashed line is the sum of four partial DOS.

12.2   Quantum Well

Recall from Chapter 11 that in a quantum well, the electron is confined along the z direction and is free to move in the xy plane. Hence, the eigenfunction consists of a free propagating wave in the xy plane and a standing wave in the z direction. We found the energy eigenfunctions and eigenvalues in Chapter 11 to be

$$\chi_{k_x,k_y,n}(x,y,z) = X(x)Y(y)Z(z) = \frac{1}{\sqrt{L_x L_y}}\sqrt{\frac{2}{L}}e^{ik_x x}e^{ik_y y}\sin\left(\frac{n\pi}{L}z\right)$$
(12.49)

$$E_{k_x,k_y,n} = E_c + \frac{\hbar^2}{2m_c}\left[k_x^2 + k_y^2 + \left(\frac{n\pi}{L}\right)^2\right] = E_c + \varepsilon_n + \frac{\hbar^2}{2m_c}(k_x^2 + k_y^2)$$
(12.50)

The bottom of each parabolic subband is at a quantized energy $E_c + \varepsilon_n$ where,

$$\varepsilon_n = \frac{\hbar^2}{2m_c}\left(\frac{n\pi}{L}\right)^2$$
(12.51)

k_x and k_y are the wave vectors in the xy plane and n is the quantum number due to confinement in the z direction within the well.

To calculate the DOS, we substitute equations (12.49), (12.50), and (12.51) in equation (12.7) to find

$$DOS(x,y,z,E) = \sum_{n,k_x,k_y}\left|\chi_{n,k_x,k_y}(\bar{r})\right|^2\delta(E - E_{n,k_x,k_y})$$
(12.52)

$$DOS(x,y,z,E)$$

$$= \frac{2}{L_x L_y L}\sum_n \sin^2\left(\frac{n\pi}{L}z\right)\sum_{k_x,k_y}\delta\left(E - E_c - \varepsilon_n - \frac{\hbar^2(k_x^2 + k_y^2)}{2m_c}\right)$$
(12.53)

In the above equation, the term $\frac{1}{L_x L_y}\sum_{k_x,k_y}\delta\left(E - E_c - \varepsilon_n - \frac{\hbar^2(k_x^2+k_y^2)}{2m_c}\right)$ is the DOS of a 2D free particle derived in equation (12.34). Now, using equation (12.34), we have

$$DOS(x,y,z,E) = \frac{1}{L}\frac{m_c}{\pi\hbar^2}\sum_n \sin^2\left(\frac{n\pi}{L}z\right)u(E - E_c - \varepsilon_n)$$
(12.54)

An example of DOS for a quantum well is shown in Figure 12.10. The DOS consists of a set of steps, where a new step is added at every subband minima $E_c + \varepsilon_n$. The meaning of equation (12.54) is that the quantum well DOS at location z is the sum of the 2D free particle DOS with step height equal to $\frac{m_c}{2\pi\hbar^2}$ (equation (12.34)) weighted by the eigenfunction magnitude squared corresponding to subband n, which is $\frac{2}{L}\sin^2\left(\frac{n\pi}{L}z\right)$. Figure 12.10 shows the DOS of a quantum well with a width of 7 nm. With increase in energy, when a new subband opens, the DOS shows a step increase, where the height of the step depends on the spatial location along the width of the quantum well.

In a manner analogous to the conduction band, the DOS in the valence band is

$$\mathrm{DOS}(x, y, z, E) = \frac{1}{L}\frac{m_v}{\pi\hbar^2}\sum_{n'}\sin^2\left(\frac{n'\pi}{L}z\right)u(E_v - \varepsilon_{n'} - E) \quad (12.55)$$

Example 2. Consider a quantum well with width $L = 1\,\mathrm{nm}$. Assume that the holes have a high effective mass $m_v/m_0 = 0.6$ and valence band maximum is located at $E_v = -1\,\mathrm{eV}$. Calculate the first four partial DOS and the total DOS as a function of energy in the valence band at the center of the quantum well, $z = 0.5\,\mathrm{nm}$.

Solution: The lowest energy in the valence band corresponds to the quantum number $n' = 1$. The subband energies are given by

$$E_{n'} = -(E_v + \varepsilon_{n'}), \quad \text{where,} \quad \varepsilon_{n'} = -\frac{\hbar^2}{2m_v}\left(\frac{n'\pi}{L}\right)^2$$

For $n' = 1$, we have $\varepsilon_{n'} = -0.62675\,\mathrm{eV}$, and hence, $E_1 = -(E_V + \varepsilon_{n'}) = -1.62675\,\mathrm{eV}$. Using the DOS equation in (12.55), we note that the Heaviside function (step function) should begin at $E_1 = -1.62675\,\mathrm{eV}$, as shown in Figure 12.11. The DOS corresponding to $n' = 1, 3$, and 5 are also shown. As the effective mass is 60 times higher than that of the electron in the case corresponding to Figure 12.10(b) and the well width is 7 times narrower, the DOS here is larger by about 420 times compared to Figure 12.10(b). The total DOS, which is plotted in Figure 12.11, is the sum of the four partial DOS.

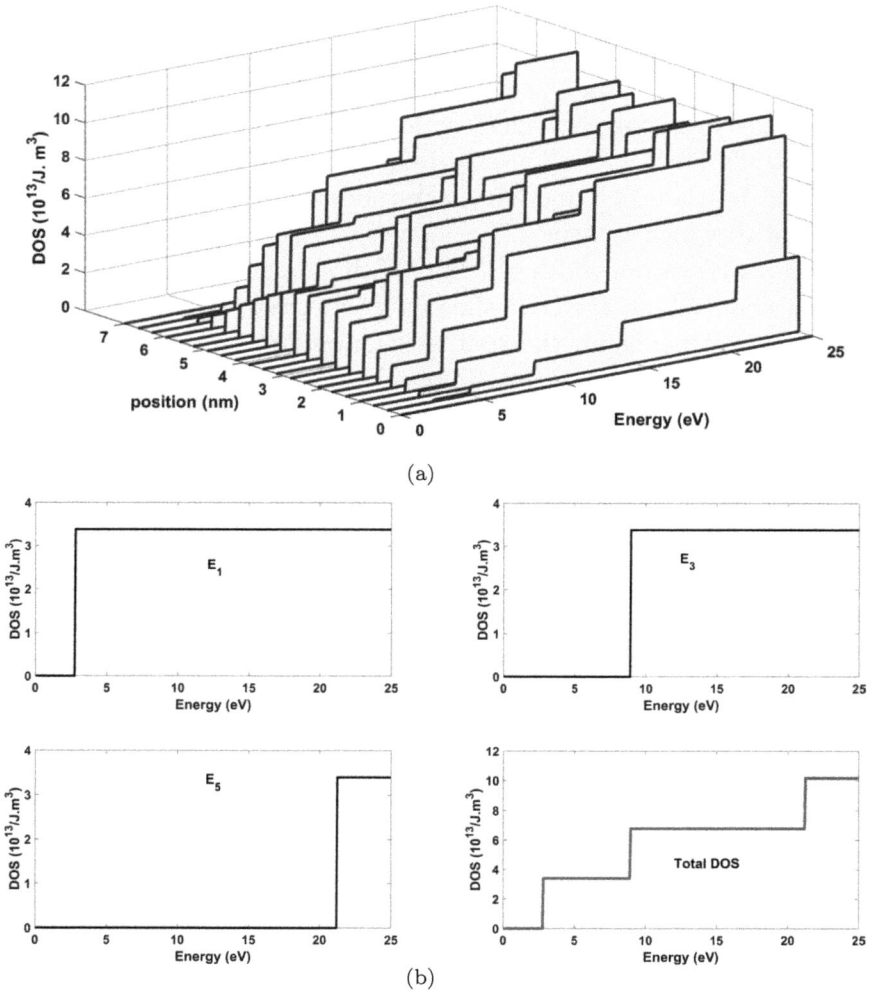

Figure 12.10. (a) DOS of a quantum well as a function of energy at various points along the well width (z direction). We have chosen well width $L = 7\,\text{nm}$, $E_c = 2\,\text{eV}$, and effective mass of electron $m_c/m_0 = 0.01$. (b) DOS as a function of energy at $z = \frac{L}{2} = 3.5\,\text{nm}$. The three black lines corresponding to the DOS for quantum numbers $n = 1, 3$, and 5, which are nonzero for energies larger than the corresponding subband minima $E_c + \varepsilon_n$ at $L/2$. The red line is the total DOS which is the sum of the three partial DOS. The DOS at the $n = 2$ and 4 subbands are zero at $L/2$ because the wave function is zero.

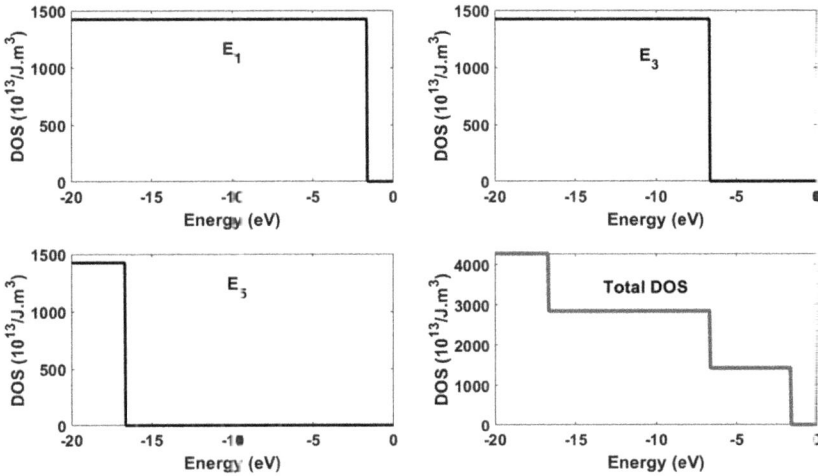

Figure 12.11. Valence band DOS versus energy at $z = 0.5\,\text{nm}$ in a quantum well with width $L = 1\,\text{nm}$. Note that $E_v = -1\,eV$, and hole effective mass is $m_v/m_0 = 0.6$. Each partial DOS is a step function which is nonzero when $E < E_v + \varepsilon_{n'}$. The total, which is the sum of the partial DOS corresponding to $n' = 1, 3$, and 5, is the red line.

Summary: Density of states in nanostructures studied in this chapter.

DOS $(x,\ y,\ z,\ E)$ Units: $\text{m}^{-3}\text{J}^{-1}$

Nanowire valence band	$\dfrac{4}{L_1 L_2}\dfrac{1}{\pi\hbar}\sum_{l',m'}\sin^2\left(\dfrac{l'\pi}{L_1}x\right)\sin^2\left(\dfrac{m'\pi}{L_2}y\right)$ $\times\sqrt{\dfrac{m_v}{2(E_v-\varepsilon_{l',m'}-E)}}\,u(E_v - \varepsilon_{l',m'} - E)$
Nanowire conduction band	$\dfrac{4}{L_1 L_2}\dfrac{1}{\pi\hbar}\sum_{l,m}\sin^2\left(\dfrac{l\pi}{L_1}x\right)\sin^2\left(\dfrac{m\pi}{L_2}y\right)$ $\times\sqrt{\dfrac{m_c}{2(E-E_c-\varepsilon_{l,m})}}\,u(E - E_c - \varepsilon_{l,m})$
Quantum well valence band	$\dfrac{1}{L}\dfrac{m_v}{\pi\hbar^2}\sum_{n'}\sin^2\left(\dfrac{n'\pi}{L}z\right)u(E_v - \varepsilon_{n'} - E)$
Quantum well conduction band	$\dfrac{1}{L}\dfrac{m_c}{\pi\hbar^2}\sum_{n}\sin^2\left(\dfrac{n\pi}{L}z\right)u(E - E_c - \varepsilon_n)$

12.3   Problems

(1) Plot the DOS in the conduction and valence bands of a 3D semiconductor with $E_c = 0.56\,\text{eV}$ and $E_v = -0.56\,\text{eV}$. Plot from -1.5 to $+1.5\,\text{eV}$. The effective mass of an electron in the conduction band is $0.06\,m_o$ and the effective mass of the hole in the valence band is $-m_o$. m_o is the rest mass of an electron.

(2) Plot the density of states as a function of energy for a GaAs quantum well with a width of $6\,\text{nm}$. In plotting the DOS, choose a point located at the center of the well. Show the DOS contribution due to each subband and the total DOS. Include at least the three lowest subbands in the conduction band and the three highest subbands in the valence band. Assume that the effective masses of electrons and holes are $0.067\,m_o$ and $0.47\,m_o$, respectively.

(3) Plot the density of states as a function of energy for a GaAs nanowire with a $6\,\text{nm} \times 6\,\text{nm}$ square cross-section. In plotting the DOS, choose a point located at the center of the square cross-section. Show the DOS contribution due to each subband and the total DOS. Include at least the three lowest subbands in the conduction band and the three highest subbands in the valence band. Assume that the effective masses of electrons and holes are $0.067\,m_o$ and $0.47\,m_o$, respectively.

(4) Plot the density of states integrated over the entire cross-sectional area as a function of energy for a GaAs wire with a $6\,\text{nm} \times 6\,\text{nm}$ square cross-section. Include at least the three lowest subbands in the conduction band and the three highest subbands in the valence band. Assume that the effective masses of electrons and holes are $0.067\,m_o$ and $0.47\,m_o$, respectively.

Chapter 13

ELECTRON DENSITY AND CURRENT

Contents

Imagine a transistor that is designed with a nanoscale channel using either top-down lithography or emerging materials, such as nanowires/carbon nanotubes. Such a channel is connected to terminals called source and drain metal electrodes. The electrostatic potential landscape of the channel is modulated using a third electrode called a gate. The gate is deposited on top of the channel, and it is separated from it by a thin insulating layer. Let us further assume that the transit time of the electron in this transistor is smaller than the scattering time such that we are operating in the ballistic limit. How would one calculate the current flowing in this transistor at various biases?

We will learn about the formulae used to calculate the electron density and the current in such devices using quantum mechanics. We will derive the Landauer–Büttiker equation to calculate the current in nanostructures.

13.1 Electron Density

The electron and hole densities in nanostructures depend on the density of states which we have discussed in the previous chapter. Recall that the DOS is the number of states per unit energy per unit dimension (length, area, or volume). If the DOS is nonzero at a specific energy and spatial location, it will contribute to the electron density as long as the probability to find an electron at that energy is nonzero. The electron density is the weighted sum of the $DOS(r, E)$ and the probability to find an electron at energy E, which is represented by $f(E)$. In the conduction band of a semiconductor, the electron density is

$$n(r) = \int_{E_c}^{E_{cmax}} DOS_c(r, E) f(E) dE \qquad (13.1)$$

The subscript index c stands for "conduction band." The lower limit of the integral E_c begins at the bottom of the conduction band. For example, in a nanowire (Chapter 11), E_c is the energy of the lowest parabola corresponding to quantum numbers $(n, m) = (1, 1)$. E_{cmax} is infinity. But in practical computations, we take E_{cmax} to be a large energy at which the probability of finding an electron has become so minuscule that it does not contribute to the integral. The function $f(E)$ is the probability of finding an electron in equilibrium at energy E, and it is called the Fermi–Dirac or Fermi function. We are only discussing the expression for electron density at equilibrium (without an applied bias) in equation (13.1).

In summary, if you consider a nanostructure at equilibrium, the number of electrons per unit volume per unit energy at location r and energy E depends on

(1) the probability of finding an electron at energy E,
(2) the number of available states at energy E at location r.

At equilibrium, the Fermi function is,

$$f(E) = \frac{1}{1 + \exp\left(\frac{E - E_F}{k_B T}\right)} \qquad (13.2)$$

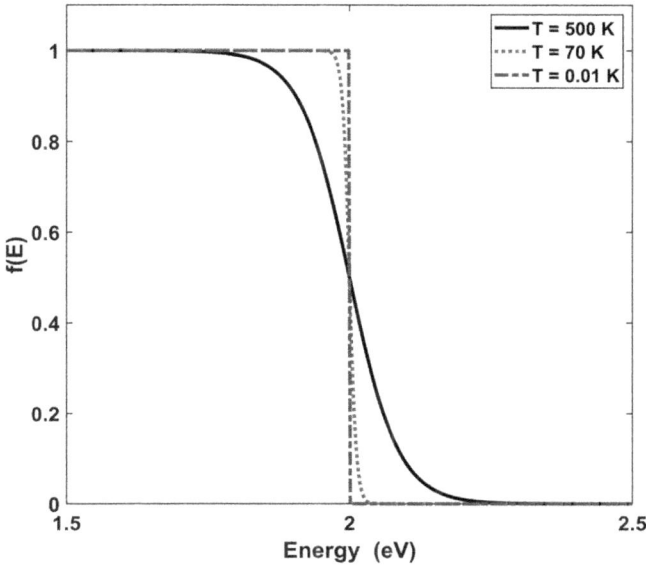

Figure 13.1. The Fermi function $f(E)$ for three different temperatures. As the temperature decreases, the change in the Fermi function progressively becomes more abrupt near $E = E_F = 2\,\mathrm{eV}$.

where k_B is the Boltzmann constant, T is the absolute temperature (in Kelvin), and E_F is the Fermi energy. A plot of the Fermi function is shown in Figure 13.1 for different temperatures. The properties of the Fermi function are

$$f(E) \to 1 \quad \text{when} \quad E - E_F \ll 3k_BT$$

$$f(E) \to 0 \quad \text{when} \quad E - E_F \gg 3k_BT \quad \text{and}$$

$$\text{at } E = E_F, \text{ it is } f(E = E_F) = \frac{1}{2}.$$

When $E - E_F > 3k_BT$, the Fermi function can be approximated by

$$f(E) \sim \exp\left[-\left(\frac{E - E_F}{k_BT}\right)\right] \tag{13.3}$$

because in the denominator of equation 13.2, the $\exp\left(\frac{E-E_F}{k_BT}\right)$ term is much larger than 1. The probability of finding a hole[1] at energy E in

---

[1] A hole at energy E is the absence of an electron in the valence band at energy E.

the valence band is $1 - f(E)$, or in other words, the probability that an electron does not occupy energy state E. Like equation (13.1), the density of holes in a semiconductor is given by

$$p(r) = \int_{E_{vmin}}^{E_{vmax}} DOS_v(r, E)[1 - f(E)]dE \qquad (13.4)$$

The subscript v refers to the valence band.

Note that the energy of holes increases as E becomes more negative. E_{vmin} is replaced by minus infinity $(-\infty)$ and E_{vmax} is the energy of the top of the valence band. In semiconductors, we usually keep track of the electron density in the conduction band and hole density in the valence band. But it is important to keep in mind that though there are many electrons in the valence band, equation (13.1) only gives the electrons in the conduction band.

The electron density in a metal is calculated using

$$n(r) = \int_{-\infty}^{+\infty} DOS(r, E)f(E)dE$$

The integrand goes to zero for energies larger than the Fermi energy because the Fermi function decreases exponentially with increase in energy. Calculation of the electrical current in nanodevices at very low applied voltages (also called linear response) benefits from the derivative of the Fermi function, which is

$$-\frac{\partial f(E)}{\partial E} = \frac{1}{k_B T} \frac{\exp\left(\frac{E - E_F}{k_B T}\right)}{\left[1 + \exp\left(\frac{E - E_F}{k_B T}\right)\right]^2} \qquad (13.5)$$

As shown in Figure 13.2, the derivative resembles a Dirac delta function at very low temperatures and becomes the delta function at $T \to 0\,\mathrm{K}$.

Example 1. Show that the integral of equation (13.5) is one at any temperature.

Solution:

$$\int_{-\infty}^{\infty} -\frac{\partial f(E)}{\partial E} dE = -[f(E = \infty) - f(E = -\infty)] = -(0 - 1) = 1$$

$$(13.6)$$

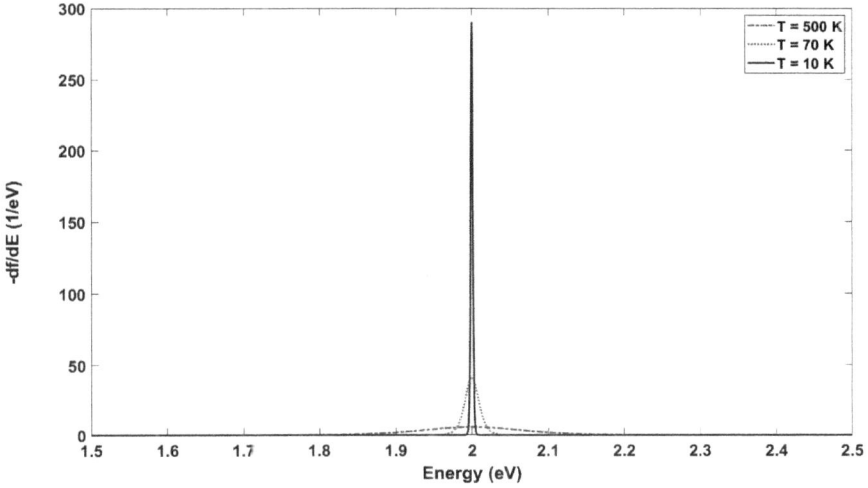

Figure 13.2. The derivative of the Fermi function with respect to energy multiplied by -1 has a peak at the Fermi energy, which is chosen to be $E_f = 2\,\text{eV}$. The width of the peak scales with $k_B T$.

13.1.1 *Electron Density in 3D Semiconductor*

We will now derive the expressions for the electron density in the conduction band of a semiconductor,

$$n = \frac{2}{\sqrt{\pi}} N_c F_{1/2}(\eta_c) \quad \text{and} \tag{13.7}$$

$$n = N_c \exp\left(\frac{E_F - E_c}{k_B T}\right) \tag{13.8}$$

E_c is the conduction band minimum.

Substituting the equation for 3D DOS from equation (12.40) into the expression for electron density in equation (13.1), we have ($m = m_c$, the conduction band's effective mass, and a multiplicative factor of 2 for spin has been included upfront)

$$n = 2 \times \frac{m_c \sqrt{2 m_c}}{2 \pi^2 \hbar^3} \int_{E_c}^{\infty} dE \sqrt{(E - E_c)} \frac{1}{1 + \exp\left(\frac{E - E_F}{k_B T}\right)} \tag{13.9}$$

Changing the variables to

$$\eta = \frac{E - E_c}{k_B T} \quad \text{and} \quad \eta_c = \frac{E_F - E_c}{k_B T} \tag{13.10}$$

equation (13.9) can be rewritten as,

$$n = \frac{m_c \sqrt{2 m_c} (k_B T)^{3/2}}{\pi^2 \hbar^3} \int_0^\infty d\eta \sqrt{\eta} \frac{1}{1 + e^{(\eta - \eta_c)}}$$

$$n = \frac{2}{\sqrt{\pi}} N_c F_{1/2}(\eta_c) \quad \text{(same as 13.9)}$$

The first term N_c is called the effective density,

$$N_c = 2 \left[\frac{m_c k_B T}{2 \pi \hbar^2} \right]^{\frac{3}{2}} \tag{13.11}$$

and the second term is called the **Fermi half** function, which is defined as,

$$F_{1/2}(\eta_c) = \int_0^\infty d\eta \frac{\sqrt{\eta}}{1 + e^{(\eta - \eta_c)}} \tag{13.12}$$

There is no closed-form analytical expression for the Fermi half function, so it must be computed numerically. However, when the conduction band is at least $3 k_B T$ above the Fermi energy, equation (13.3) can be used to approximate equation (13.12) to

$$F_{1/2}(\eta_c) \sim \int_0^\infty d\eta \sqrt{\eta} e^{-(\eta - \eta_c)} = \frac{\sqrt{\pi}}{2} e^{\eta_c} = \frac{\sqrt{\pi}}{2} e^{(E_F - E_c)/k_B T} \tag{13.13}$$

Substituting equation (13.13) in equation (13.9), we get equation (13.8),

$$n = N_c \exp\left(\frac{E_F - E_c}{k_B T} \right)$$

Note that as the temperature increases, N_c increases as $T^{\frac{3}{2}}$ but the other term increases exponentially because $E_F - E_c$ is negative. The change in carrier density with variation in temperature is important in determining the conductivity of a semiconductor.

13.2 Current

Consider a nanoscale device, as shown in Figure 13.3, which is composed of a nanostructure biased between two conductors called source and drain. The third terminal, the gate, controls the electrostatic potential landscape inside the device and affects the transport of charge between source and drain. The gate electrode is electrically separated from the nanodevice (which we also call a channel) with an insulating layer. To study the current quantum mechanically, we need

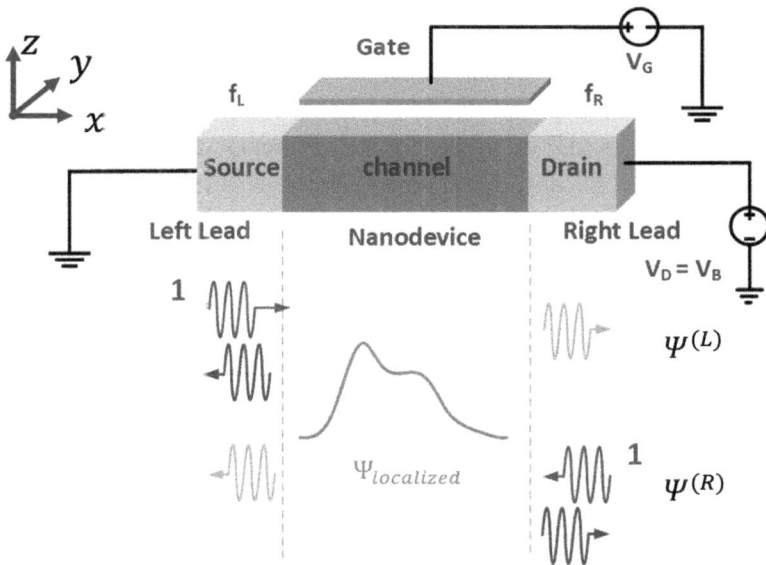

Figure 13.3. A device (channel) connected to two contacts (source and drain) through which DC current flows. The device is also connected electrostatically to a gate. The wave function consists of electron waves incident from the left, incident from the right, and localized waves. The number "1" to the left of $\Psi^{(L)}$ (to the right of $\Psi^{(R)}$) is added to indicate the wave is incident from the left (right) with unit amplitude; part of the wave is transmitted and the other part is reflected.

to know the electron wave functions. Three types of wave functions exist in the devices:

(a) electron waves incident from the left contact $\Psi^{(L)}_{n,k_L}$,
(b) electron waves incident from the right contact $\Psi^{(R)}_{m,k_R}$,
(c) localized waves in the device (they do not propagate either to the left or right contacts).

The source and drain are so large relative to the nanodevice that we can assume they are in local equilibrium. When the bias voltage is zero, a single Fermi energy defines the system (nanodevice and two contacts). If a bias voltage of V_B is applied to the right contact with respect to the left contact, the relationship between the Fermi energy in the two contacts and the bias voltage is:

$$E_{FL} - E_{FR} = qV_B \tag{13.14}$$

E_{FL} and E_{FR} are the Fermi energies in the left (source) and right (drain) contacts, respectively. If V_B is positive, the Fermi energy of the left contact is higher than the Fermi energy of the right contact, $E_{FL} > E_{FR}$. $f_L(E)$ and $f_R(E)$ stand for the Fermi functions in the left and right contacts.

Let us derive the celebrated Büttiker formula for current in a nanodevice. We will assume that electrons are occupied only in a single subband in the two leads of Figure 13.3. This makes the derivation easier to understand, but the reader can generalize this to the case with electrons being present in multiple subbands. We will drop the indices n, k_L (m, k_R) from $\psi^{(L)}_{n,k_L}$ $(\psi^{(R)}_{m,k_R})$ and simply write it as $\psi^{(L)}$ $(\psi^{(R)})$.

We derived earlier (Chapter 10) that the *probability current density* $(J^{(L)}_P)$ due to an electron in a wave function $\psi^{(L)}$ is

$$J^{(L)}_P(x,y,z,E) = \frac{\hbar}{2m_c i}\left[\psi^{(L)*}(x,y,z)\frac{d\psi^{(L)}(x,y,z)}{dx} - c.c.\right] \tag{13.15}$$

$$J^{(R)}_P(x,y,z,E) = \frac{\hbar}{2m_c i}\left[\psi^{(R)*}(x,y,z)\frac{d\psi^{(R)}(x,y,z)}{dx} - c.c.\right] \tag{13.16}$$

where the waves are incident from the left contact with wave vector k_L at energy E. The probability for a wave to be incident from the left contact depends on the contact's Fermi function. So, the *electrical current density* due to an electron incident from the left contact at energy E is

$$j^{(L)}(x, y, z, E) = -qJ_P^{(L)}(x, y, z, E)f_L(E) \qquad (13.17)$$

Note that $j^{(L)}$ is used to denote the electric current density due to wave function $\psi^{(L)}$, which is the probability current density in equation (13.15) multiplied by the probability $f_L(E)$ that an electron wave is present at energy E and the charge of an electron $-q$. Similarly, the electrical current density due to an electron incident from the right contact is

$$j^{(R)}(x, y, z, E) = -qJ_P^{(R)}(x, y, z, E)f_R(E) \qquad (13.18)$$

The net current density at (x, y, z) is the sum of the current densities due to left and right incident waves at all energies, which is

$$j(x, y, z) = \sum_{k_L} j^{(L)}(x, y, z, E) + \sum_{k_R} j^{(R)}(x, y, z, E) \qquad (13.19)$$

The electric current at any cross-section x at energy E is the integral of equation (13.19) over the cross-section (y, z),

$$j(x) = \int dy \int dz \; j(x, y, z)$$

$$= \int dy \int dz \left[\sum_{k_L} j^{(L)}(x, y, z, E) + \sum_{k_R} j^{(R)}(x, y, z, E) \right]$$
$$(13.20)$$

The wave function of the electrons incident from the left lead is

$$\psi^{(L)} = A_{k_L} \times \begin{cases} e^{ik_L x}\chi(y, z) + r_{L\leftarrow L}e^{-ik_L x}\chi(y, z) & \text{(in the left lead)} \\ t_{R\leftarrow L}e^{+ik_R x}\chi(y, z) & \text{(in the right lead)} \end{cases}$$
$$(13.21)$$

where k_L and k_R are the wave vectors in the left and right leads, respectively. $\chi(y, z)$ is the wave function in the cross-sectional direction of the lead. $t_{R\leftarrow L}$ is the transmission amplitude for a wave incident in the left contact to be transmitted to the right contact. $t_{L\leftarrow R}$ is the transmission amplitude for a wave incident in the right contact to be transmitted to the left contact. $r_{L\leftarrow L}$ and $r_{R\leftarrow R}$ are the reflection probabilities for waves incident from the left and right contacts, respectively. Note that we have assumed that the wave function is separable along the x and (y, z) directions as in Chapter 11. The wave function of the electrons incident from the right lead is

$$\psi^{(R)} = A_{k_R} \times \begin{cases} e^{-ik_R x}\chi(y, z) + r_{R\leftarrow R}\, e^{+ik_R x}\chi(y, z) & \text{(in the right lead)} \\ t_{L\leftarrow R}\, e^{-ik_L x}\chi(y, z) & \text{(in the left lead)} \end{cases}$$

$$(13.22)$$

where A_{k_L} and A_{k_R} are normalization constants. Substituting equation (13.22) in the second term of equation (13.20), we have

$$\int dy \int dz \sum_{k_R} j^{(R)}(x, y, z, E)$$

$$= \frac{-q\hbar}{m_c i} \sum_{k_R} |A_{k_R}|^2 (-k_L)\, t_{L\leftarrow R}^*\, t_{L\leftarrow R}\, f_R(E) \int dy \int dz\, \chi^*(y, z)\chi(y, z)$$

$$= \frac{-q\hbar}{m_c} \sum_{k_R} |A_{k_R}|^2 (-k_L)|t_{L\leftarrow R}|^2 f_R(E) \qquad (13.23)$$

We have used the fact that normalization demands that the integral along the cross-section $\int dy \int dz\, \chi^*(y, z)\chi(y, z) = 1$. Substituting the normalization $|A_{k_R}|^2 = \frac{1}{L}$ in equation (13.23),

$$\int dy \int dz \sum_{k_R} j^{(R)}(x, y, z, E) = \frac{1}{L}\frac{q\hbar}{m_c} \sum_{k_R} k_L\, |t_{L\leftarrow R}|^2 f_R(E) \quad (13.24)$$

Because the summation over k_R is over a continuous variable, we simplify this by converting it to an integral. We can convert the above summation over a wave vector to an integral over energy. To see this, the following two steps should be understood:

(i) Recall that each k-point occupies a space, $\Delta k = 2\pi/L$. So,

$$\sum_{k_R} A(k_R) = \sum_{k_R} A(k_R)\frac{\Delta k_R}{2\pi/L}$$

$$= \frac{L}{2\pi} \sum_{k_R} A(k_R)\Delta k_R$$

$$= \frac{L}{2\pi} \int A(k_R)dk_R$$

where $A(k_R)$ is an arbitrary quantity that depends on k_R.

(ii) We can write dk_R as $\left|\frac{dk_R}{dE}\right| dE$ and then write $A(E)$ instead of $A(k_R)$ because every value of $A(k_R)$ corresponds to an $A(E)$. Recall from Chapter 11 that energy is a function of the wave vector. Therefore, we can write the above equation as

$$\sum_{k_R} A(k_R) = \frac{L}{2\pi} \int A(E) \left|\frac{dk_R}{dE}\right| dE \qquad (13.25)$$

In the previous equation, we note that,

$$\left|\frac{\left(\frac{dk_R}{2\pi/L}\right)}{dE}\right| = \frac{\text{Number of states incident from the right contact in energy interval } dE}{dE} = L \cdot DOS_R(E)$$

$$(13.26)$$

where DOS_R (the number of states per unit energy per unit length in the right contact due to waves incident from the right contact into the device) was defined in equation (12.24). Substituting equation (13.26) into equation (13.25), we can convert the summation to the following integral,

$$\frac{1}{L}\sum_{k_R} A(k_R) = \int A(E) DOS_R(E) dE \qquad (13.27)$$

It is important to note that we must set $A(k_R) = (-k_L)|t_{L\leftarrow R}|^2 f_R(E)$ in equation (13.27). This gives us

$$\frac{1}{L}\sum_{k_R} k_L |t_{L\leftarrow R}|^2 f_R(E) = \int k_L |t_{L\leftarrow R}|^2 f_R(E) DOS_R(E) dE$$

Using the above equation in equation (13.24) gives,

$$\int dy \int dz \sum_{k_R} j^{(R)}(x,y,z,E)$$

$$= \frac{q\hbar}{m_c} \int k_L |t_{L\leftarrow R}|^2 f_R(E) DOS_R(E) dE \qquad (13.28)$$

To simplify the above equation further, we note that the DOS due to waves incident from the right contact was derived to be (equation (12.24)),

$$DOS_R(x, E) = \frac{1}{2\pi\hbar} \frac{m_c}{\hbar k_R}$$

Substituting this in equation (13.28), we have,

$$\int dy \int dz \sum_{k_R} j^{(R)}(x, y, z, E) = \frac{q}{h} \int dE \, \frac{k_L}{k_R} |t_{L\leftarrow R}|^2 f_R(E)$$

$$(13.29)$$

The transmission probability from the right contact to the left contact is defined by,

$$T_{L\leftarrow R}(E) = \left(\frac{k_L}{k_R}\right) |t_{L\leftarrow R}|^2 \qquad (13.30)$$

As the Fermi function $f_R(E)$ depends only on the contact Fermi function and total energy E, we finally arrive at the equation for the right-going part of the current density in our device,

$$\int dy \int dz \sum_{k_R} j^{(R)}(x, y, z, E) = \frac{q}{h} \int dE \, T_{L\leftarrow R}(E) f_R(E) \quad (13.31)$$

Similarly, for the left-going part of the current density, we write,

$$\int dy \int dz \sum_{k_L} j^{(L)}(x, y, z, E) = -\frac{q}{h} \int dE \, T_{R\leftarrow L}(E) f_L(E) \quad (13.32)$$

where

$$T_{R\leftarrow L}(E) = \left(\frac{k_R}{k_L}\right) |t_{R\leftarrow L}|^2 \qquad (13.33)$$

Substituting equations (13.31) and (13.32) in (13.20), we obtain the total current density,

$$j = -\frac{q}{h} \int dE \, [T_{R\leftarrow L}(E) f_L(E) - T_{L\leftarrow R}(E) f_R(E)] \qquad (13.34)$$

The condition that the current in a two-terminal device must be zero at equilibrium when $f_L(E) = f_R(E) = f(E)$ implies that

$$j = -\frac{q}{h} \int dE \; [T_{R \leftarrow L}(E) - T_{L \leftarrow R}(E)] \, f(E) = 0 \qquad (13.35)$$

This condition is satisfied if the left-to-right and right-to-left transmissions are equal,

$$T_{R \leftarrow L}(E) = T_{L \leftarrow R}(E) = T(E)$$

We call their common value $T(E)$. The electrical current is then,

$$j = -\frac{q}{h} \int dE \; T(E) \, [f_L(E) - f_R(E)] \qquad (13.36)$$

If we include a multiplicative factor of 2 on the RHS because both up and down spin electrons contribute to the total current, equation (13.34) becomes,

$$j = -\frac{2q}{h} \int dE \; T(E) \, [f_L(E) - f_R(E)] \qquad (13.37)$$

Equation (13.37) is known as the **Büttiker** formula. While this formula was derived above for a device with a single subband, it is applicable to devices with an arbitrary number of subbands.

It is important to mention that transmission coefficients are always equal in two-terminal devices (Equation 13.35). Otherwise, there would be nonzero current going from one terminal to the other even at zero voltage bias. The reciprocity in transmission between contacts breaks down if there are more than two contacts to the device and time-reversal symmetry is broken. For example, Figure 13.4 shows how a magnetic field causes the transmission of electron from terminal 1 to 2 ($T_{1 \rightarrow 2}$) to be different from the transmission from terminal 2 to 1 ($T_{2 \rightarrow 1}$). An electron incident from terminal 2 will go to terminal 3 due to the Lorentz force, instead of going to terminal 1 while an electron incident from terminal 1 will go to terminal 2 due to the Lorentz force.

Note: The reader should try to derive the Büttiker formula for the case with multiple subbands. In the following, we identify the

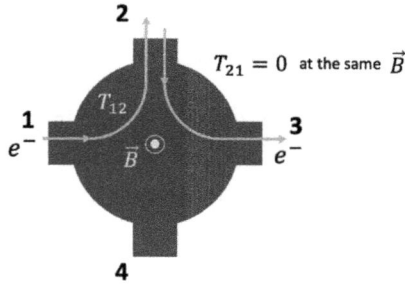

Figure 13.4. Breaking the time-reversal symmetry (or reciprocity) by applying a perpendicular magnetic field to a device, which causes $T_{2\to 1} \neq T_{1\to 2}$. $T_{2\to 1}$ is zero because the electron ends up going to terminal 3 instead of 1, due to the Lorentz force caused by the magnetic field pointing out of the plane of the page.

central equations of such a derivation. The expression for currents with many subbands are

$$J_{P,n}^{(L)}(x, y, z, E) = \frac{\hbar}{2m_c i}\left[\psi_{n,k_L}^{(L)*}(x, y, z)\frac{d\psi_{n,k_L}^{(L)}(x, y, z)}{dx} - c.c.\right]$$

(13.38)

$$J_{P,m}^{(R)}(x, y, z, E) = \frac{\hbar}{2m_c i}\left[\psi_{m,k_R}^{(R)*}(x, y, z)\frac{d\psi_{m,k_R}^{(R)}(x, y, z)}{dx} - c.c.\right]$$

(13.39)

where n (m) represents the subband quantum numbers of the waves incident from the left (right) contact with wave vector $k_L(k_R)$ at energy E.

The wave function of electrons incident from the left lead is

$$\psi_{n,k_L}^{(L)} = A_{n,k_L} \cdot \begin{cases} e^{ik_{n,L}x}\chi_n(y, z) + \sum_{n'} r_{Ln' \leftarrow Ln}\, e^{-ik_{n',L}x}\chi_{n'}(y, z) \\ \text{(in the left lead)} \\ \sum_{m'} t_{Rm' \leftarrow Ln}\, e^{+ik_{m',R}x}\chi_{m'}(y, z) \\ \text{(component transmitted to the right lead)} \end{cases}$$

(13.40)

and the wave function of electrons incident from the right lead is

$$\psi^{(R)}_{m,k_R} = A_{m,k_R} \cdot \begin{cases} e^{-ik_{m,R}x}\chi_m(y,z) + \sum_{m'} r_{Rm'\leftarrow Rm}\, e^{+ik_{m',R}x}\, \chi_{m'}(y,z) \\ \quad \text{(in the right lead)} \\ \sum_{n'} t_{Ln'\leftarrow Rm}\, e^{-ik_{n',L}x}\, \chi_{n'}(y,z) \\ \quad \text{(component transmitted to the left lead)} \end{cases}$$

(13.41)

where A_{n,k_L} and A_{m,k_R} are the normalization constants. n and m are the subband indices in the leads. An electron incident from the left (right) lead in subband n (m) has a wave vector $k_{n,L}$ ($k_{m,R}$). Note that the wave vector depends on the band index apart from the total energy because each subband has a different minimum energy (subband bottom).

The expressions for $T_{L\leftarrow R}$ and $T_{R\leftarrow L}$ become

$$T_{L\leftarrow R}(E) = \sum_{n',m} T_{Ln'\leftarrow Rm}(E) \tag{13.42}$$

$$T_{R\leftarrow L}(E) = \sum_{n,m'} T_{Rm'\leftarrow Ln}(E) \tag{13.43}$$

where

$$T_{Ln'\leftarrow Rm}(E) = \left(\frac{k_{n',L}}{k_{m,R}}\right) |t_{Ln'\leftarrow Rm}|^2 \tag{13.44}$$

$$T_{Rm'\leftarrow Ln}(E) = \left(\frac{k_{m',R}}{k_{n,L}}\right) |t_{Rm'\leftarrow Ln}|^2 \tag{13.45}$$

Note that like in the single subband case, reciprocity holds i.e.,

$$T_{R\leftarrow L}(E) = T_{L\leftarrow R}(E) = T(E)$$

In Section 13.2.2, we will discuss the calculation of $T(E)$ in a structure with multiple subbands when there is no scattering.

13.2.1 *Small Bias Conductance*

We will now derive an expression for small bias conductance using equation 13.37. Here, *small* refers to biases much smaller than the thermal energy $k_B T$. Because the small bias V_B is applied to the right contact (Figure 13.3), the Fermi energy of the right contact changes from E_F to $E_F - qV_B$, where E_F is the Fermi energy of the left contact,

$$E_{FL} - E_{FR} = qV_B \qquad (13.46)$$

As a result, the Fermi functions of the left and right contacts are:

$$f_L(E) = \frac{1}{1 + \exp\left(\frac{E - E_F}{k_B T}\right)} \quad \text{and}$$

$$f_R(E) = f(E) = \frac{1}{1 + \exp\left(\frac{E - E_F + qV_B}{k_B T}\right)} \qquad (13.47)$$

In the limit that $\frac{qV_B}{k_B T} \ll 1$, we can write $f_R(E)$ to leading order using the Taylor series expansion of the Fermi function in which qV_B is small,

$$f_R(E) = \frac{1}{1 + \exp\left(\frac{E - E_F + qV_B}{k_B T}\right)} \sim \frac{1}{1 + \exp\left(\frac{E - E_F}{k_B T}\right)}$$

$$+ qV_B \frac{\partial}{\partial E} \frac{1}{1 + \exp\left(\frac{E - E_F}{k_B T}\right)} = f_L(E) + qV_B \frac{\partial f}{\partial E}$$

$$(13.48)$$

Substituting equation 13.48 in equation 13.37, the small bias current is,

$$j \simeq \frac{2q^2}{h} V_B \int dE \, T(E) \left(\frac{\partial f}{\partial E}\right) \quad \text{(Small bias formula)} \qquad (13.49)$$

As the positive bias applied to the right contact, this current flow from the right to the left contact. The small bias conductance at finite temperature is,

$$G = \left|\frac{j}{V_B}\right| = \frac{2q^2}{h} \int dE \, T(E) \left|\frac{\partial f}{\partial E}\right| \qquad (13.50)$$

As the temperature decreases, the derivative of the Fermi function peaks around the Fermi energy and is nearly zero everywhere else. Assuming that $T(E) \sim T(E_F)$ in the energy window where $\left|\frac{\partial f}{\partial E}\right|$ is appreciable, we can rewrite the above equation as,

$$G = \left|\frac{j}{V_B}\right| = \frac{2q^2}{h}T(E_F)\int dE \left|\frac{\partial f}{\partial E}\right|$$

Now using equation 13.6,

$$G = \frac{j}{V_B} = \frac{2q^2}{h}T(E_F) \quad \text{(Small bias formula at } T=0\text{)} \quad (13.51)$$

At higher bias voltages, the Büttiker formula given in equation 13.37 should be used.

Example 2 (Minimum resistance of a nanowire with a single subband at low temperatures). We will now apply equation (13.37) to find the minimum resistance of a defect-free nanowire with a single subband. Defect-free means that an electron does not suffer any scattering event (loss of energy or momentum) during transport inside the channel. This mode of transport is called ballistic. Let there be a voltage V_B applied across the nanowire. We note that the minimum resistance would correspond to the maximum transmission $T(E)$ from the left to right contact. The maximum value of $T(E)$ for a wire with a single subband is 1.

At very low temperatures, $f_L(E)$ and $f_R(E)$ approach step functions. Then noting that $E_{FL} - E_{FR} = qV_B$, it is easy to see that the main contribution to the following integral comes from the shaded area in Figure 13.5. Then,

$$\int dE \, 1 \cdot [f_L(E) - f_R(E)] \sim qV_B$$

Using this in equation (13.37), the current is

$$j = \frac{2q^2}{h}V_B$$

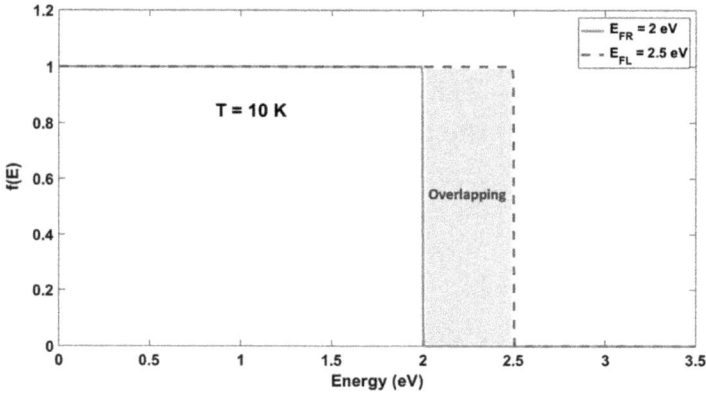

Figure 13.5. The low-temperature Fermi functions in the left and right contacts when $E_{FL} - E_{FR} = qV_B = 0.5\,\text{eV}$. At zero temperature, the Fermi function is a perfect step function.

The conductance of a wire with a single subband with no defects/scattering is

$$G_Q = \left|\frac{j}{V_B}\right| = \frac{2q^2}{h} = \frac{2 \times 1.6 \times 10^{-19}C \times 1.6 \times 10^{-19}C}{6.625 \times 10^{-34}\text{Js}} = 77.47\,\mu\text{S}$$

(13.52)

This is known as the *quantum of conductance*, which is the maximum possible conductance of a wire with a single subband available to carry current. The quantum of conductance only depends on the fundamental constants q and h. The resistance corresponding to the quantum of conductance is the *minimum resistance* possible for a wire with a single subband

$$R_{min} = \frac{1}{G_Q} = \left|\frac{V_B}{j}\right| = \frac{h}{2q^2} = 12.909\,\text{k}\Omega$$

(13.53)

Interestingly, it is not zero! The only way to reduce this resistance is to have multiple subbands by increasing the wire's diameter. Transport of electrons then occurs via parallel channels, and the total resistance decreases by a factor of M, which is the number of subbands at the Fermi energy $E = E_F$. In the following section, we will discuss the case of multiple subbands.

13.2.2 Calculation of T(E) in the Absence of Scattering

We will continue to assume that the Fermi energy of the left contact is higher than that of the right contact and zero temperature. For a nanostructure with multiple subbands, $T(E)$ is found by counting the subbands in the channel with electron velocity directed from left to right between Fermi energies E_{FL} and E_{FR}. The slope of the dispersion $E(k)$ is proportional to the group velocity of the electron (this is discussed in Chapter 14),

$$v_g = \frac{1}{\hbar} \frac{\partial E}{\partial k}$$

When the channel has multiple subbands, the $T(E)$ is found by counting the subbands with a *positive group velocity* (moving from left to right). After this, $T(E)$ as a function of energy is inserted in equation (13.36) or (13.50) to calculate the electric current or conductance. Figure 13.6 shows the band structure of a silicon nanowire and the plot of $T(E)$. The latter is found by starting at $E = 0$ and counting the number of subbands with a positive group velocity.

In the example shown in Figure 13.6, let us assume that the Fermi energy is slightly above 5 eV. If we apply a small bias, the conductance is three times larger than equation (13.52) because $T(E > 5\,\text{eV}) = 3$. The corresponding resistance is smaller than R_{min} (equation (13.53)) by three times as we have three parallel channels for the transport of electrons. This is the smallest resistance possible for a nanostructure where three subbands carry the current. Scattering of electrons can only increase this resistance.

The reader can verify from the energy dispersion relation (Chapter 11) that as the cross-section of a nanowire increases, the number of subbands at a particular energy increases proportionally with area. So, the resistance will decrease as 1/area.

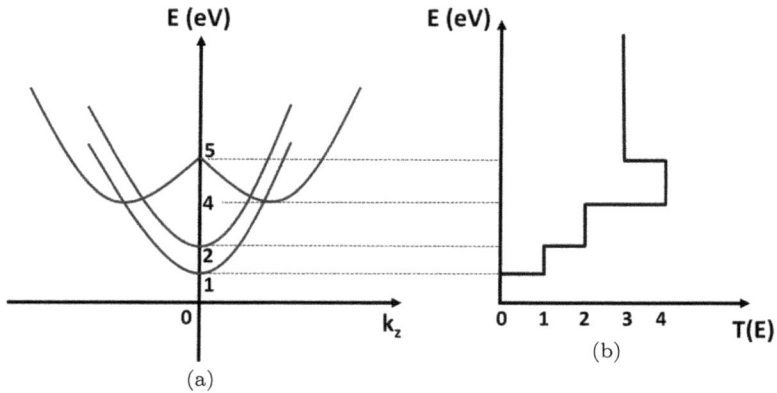

Figure 13.6. (a) Band structure of a nanowire and (b) $T(E)$ as a function of energy. Every time the horizontal line crosses a new subband, $T(E)$ increases by one, and if it loses a subband, $T(E)$ decreases by one. The topmost horizontal line at $5\,\mathrm{eV}$ shows that there are three subbands available for transport when $E > 5\,\mathrm{eV}$. Hence $T(E) = 3$ for $E > 5\,\mathrm{eV}$.

Concept: The Büttiker formula to calculate the current in nano-structures depends on the transmission $T(E)$ for electrons to flow between the Left and Right contacts. The formula is,

$$j = -\frac{2q}{h} \int dE \ T(E) \ [f_L(E) - f_R(E)]$$

The largest possible value of transmission for a device with a single conducting subband (or mode) is $T(E_F) = 1$. Using this, the maximum conductance of a single subband device is,

$$G_Q = \frac{2q^2}{h} = 77.47 \,\mu\mathrm{S}$$

This value depends only on fundamental constants, the charge of an electron and Planck's constant. It is known as the **quantum of conductance**.

The smallest possible resistance of a device with a single conducting subband (or mode) is

$$R_{min} = \frac{1}{G_Q} = \frac{h}{2q^2} = 12.909 \,\mathrm{k\Omega}$$

13.3 Qualitative Analogy to Heat Current

It is instructive to discuss the formal similarity between the equations for electric current and heat current. Heat current is the energy per unit time passing through a conductor due to the temperature difference between its two ends. Consider a conductor connected to left and right thermal baths kept at temperatures of T_L and T_R, respectively. We will completely neglect the flow of electrons here and consider only the flow of lattice vibrations or phonons. Phonons are quanta of lattice vibrations of the crystal lattice and are important to carrying heat in solids. The heat energy flowing per unit time (J_Q) between the two ends of a conductor is governed by an equation similar to equation (13.35),

$$J_Q = \frac{1}{h} \int E \times [T_{R \leftarrow L}(E) n_L(E) - T_{L \leftarrow R}(E) n_R(E)] dE \quad (13.54)$$

$T_{L \leftarrow R}(E)$ and $T_{R \leftarrow L}(E)$ are the transmission functions for phonons from right to left and left to right thermal baths, respectively. As heat is carried by phonons, $E = \hbar \omega$. In the above equation, instead of the Fermi function, we have the Bose–Einstein distribution function of the left and right thermal baths, that is, $n_L(E)$ and $n_R(E)$. Phonons, which are Bosons follow the Bose–Einstein statistics. At a given temperature and phonon energy $\hbar\omega$, the number of phonons is given by

$$n_L(E) = \frac{1}{\exp\left(\frac{\hbar\omega}{k_B T_L}\right) - 1} \quad \text{and} \quad n_R(E) = \frac{1}{\exp\left(\frac{\hbar\omega}{k_B T_R}\right) - 1} \quad (13.55)$$

Example 3 (Quantum of heat conductance). Show that in the presence of a tiny temperature difference $(T_L = T + \Delta T, T_R = T)$ between the two ends of a heat conductor, the maximum heat conductance (quantum of heat conductance) is equal to

$$G_Q = \frac{\pi^2 k_B^2 T}{3h} \quad \text{(quantum of heat conductance)} \quad (13.56)$$

Note that the quantum of heat conductance at $T = 1\,\mathrm{K}$ is $1\,\mathrm{pW/K}$.

Solution: The difference between the Bose–Einstein distributions in the left and right thermal baths can be approximated by

$$n_L - n_R = -\frac{\partial n}{\partial T}\Delta T \quad (13.57)$$

Using this equation and assuming we have only one acoustic mode traveling through the material to carry heat, i.e., $T_{L \leftarrow R} = T_{R \leftarrow L} = 1$, we can show that

$$J_Q = \frac{\pi^2 k_B^2 T}{3h} \Delta T = G_Q \Delta T \qquad (13.58)$$

13.4   Problems

Section 13.2

(1) Calculate the electron and hole densities in a 1D nanowire with a bandgap of 1 eV. The effective mass of both the conduction and valence bands is the free electron mass m_o. By assuming that there is only a single subband in the nanowire's conduction and valence bands, (a) write down the expression for calculating the electron and hole densities and simplify it as much as possible and (b) calculate numerical values for the electron and hole densities by assuming that the Fermi energy is 200 meV below the nanowire's conduction band. Assume the temperature is $T = 300$ K.

(2) Calculate the electron and hole densities in a 2D quantum well with a bandgap of 1 eV. The effective mass of the conduction and valence bands is the free electron mass m_o. By assuming that there is only a single subband in the quantum well's conduction and valence bands, (a) write down the expression for calculating the electron and hole densities and simplify it as much as possible and (b) calculate a numerical value for the electron and hole densities by assuming that the Fermi energy is 200 meV above the quantum well's valence band. Assume the temperature is $T = 300$ K.

(3) Consider a 1D nanowire with a single subband as shown in the following figure. Plot the following by assuming a (i) temperature of 116 K and (ii) Fermi energy of 10 meV above the band bottom:

(a) Transmission versus energy. Assume that there is no reflection/scattering.
(b) Fermi function versus energy.
(c) Product of the transmission and Fermi function versus energy.

Energy range is from -30 to $+30\,\mathrm{meV}$. Note that the band bottom is the zero of energy.

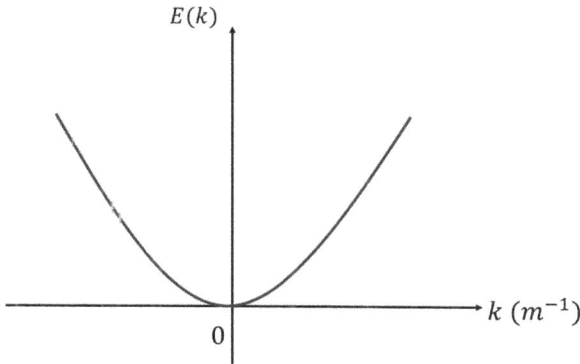

(4) Consider a 1D nanowire with a single subband, as shown in the above figure. What is the minimum possible small bias resistance at zero Kelvin when the Fermi energy is $10\,\mathrm{meV}$ (a) below and (b) above the band bottom?

(5) Consider a gallium arsenide nanowire with a square cross-section with edge length $L = 5\,\mathrm{nm}$. What is the *minimum possible small bias* resistance if the Fermi energy is (a) $50\,\mathrm{meV}$ above and (b) $50\,\mathrm{meV}$ below the conduction band minimum of the nanowire? Assume that there is no scattering, and the temperature is $300\,\mathrm{K}$. Assume that the effective masses of electrons and holes are both $0.067m_o$.

Note: The answer is not zero because electrons are occupied in the conduction band due to the tail in the Fermi factor. This problem is relatively easy once you estimate the energy levels.

(6) Consider a three-terminal device (e.g., transistor) with terminals 1, 2, and 3. Derive the formula to calculate the current in every terminal of the device. Assume the following:

(a) Total transmission from terminal i to terminal j is represented by $T_{i \rightarrow j}$.

(b) Electrons are incident from the terminals with the Fermi functions f_1, f_2, and f_3.

Assume that only a single subband is occupied in each of terminals 1, 2, and 3.

Chapter 14

PERIODIC SOLIDS

Contents

Chapter 11 discussed the energy levels of nanostructures when our starting point was the effective mass Hamiltonian. The effective mass depends on the material system (element type) and the atom's arrangement in the solid. This chapter will discuss the energy levels of nanostructures starting from atoms as the building blocks of materials. We will learn the following:

- the essential features of energy levels and orbitals of atoms,
- how orbitals of atoms interact to determine the energy levels of nanostructures,
- application of Bloch's theorem to determine the energy levels of a periodic solid,
- electron's effective mass derived from the band structure,
- what determines if a material is a semiconductor or metal,
- graphene and carbon nanotubes.

14.1   Orbitals and Energy Levels of Atoms

An atom consists of a positively charged nucleus made of protons and neutrons. The nucleus is so small that it is assumed to be a point charge. Electrons in an atom surround the nucleus and give atoms their finite size. In Chapter 1, we derived that the radius of an electron orbiting the nucleus of a hydrogen atom is the Bohr radius, which is 0.54 Å. The derivation used Bohr's quantization principle, which predates Schrödinger equation.

The Schrödinger equation for a hydrogen atom is,

$$\left[-\frac{\hbar^2}{2m_o}\nabla^2 - \frac{q^2}{4\pi\varepsilon_0 r} \right] \psi(\bar{r}) = E\psi(\bar{r}) \tag{14.1}$$

where m_o is the rest mass of an electron and $\bar{r}$ is the vector representing the location of the electron from the nucleus, which is located at $r = 0$. The nucleus is much heavier than the electron, so it is assumed to be fixed in space. The potential energy $-\frac{q^2}{4\pi\varepsilon_0 r}$ is spherically symmetric. Solving Schrödinger equation in the three spatial dimensions results in three quantum numbers:

- n the principal quantum number, which takes values of integers 1, 2, 3, and so on,
- l the orbital quantum number, which takes integer values 0, 1, ..., $n - 1$, for every principal quantum number n,
- m the magnetic quantum number, which takes the following values for each l:

$$-l, -(l - 1), -(l - 2), ..., 0, ..., (l - 2), (l - 1), l$$

Important facts about the wave functions and energy levels of electrons in atoms are:

- The eigenfunctions defined by the set of quantum numbers (n, l, m) are referred to as *orbitals*.
- The energy levels of electrons in a hydrogen atom whose Hamiltonian is given by equation (14.1) depends only on the principal quantum number n (and not on l and m),

$$E_n = -\frac{E_0}{n^2} \tag{14.2}$$

where $E_0 = \dfrac{q^2}{8\pi\varepsilon_0 a_0} = 13.6\,\text{eV}$ and $\tag{14.3}$

$$a_o = \frac{4\pi\varepsilon_0\hbar^2}{q^2 m_o} = 0.53\,\text{Å is the Bohr radius} \tag{14.4}$$

- The average radius of the orbitals increases with the principal quantum number n.
- The shape of $\psi_{n,l,m}$ depends only on the quantum numbers l and m. The letters s, p, d, and f are used to represent the quantum numbers $l = 0, 1, 2, 3, 4$. A subscript is also added to orbitals p and d to indicate the corresponding quantum number m of the orbitals, as shown in table 14.1. For example, p_x is a dumbbell-shaped orbital pointing in the x direction and corresponds to $m = -1$ (see Figure 14.1).

Table 14.1

n	l	m	Common Names (n, l, m)	Shapes
1	0	0	$1s$	Spherical
2	0	0	$2s$	Spherical
		-1	$2p_x$	Dumbbell-shaped along x-axis
	1	0	$2p_y$	Dumbbell-shaped along y-axis
		$+1$	$2p_z$	Dumbbell-shaped along z-axis
3	0	0	$3s$	Spherical
		-1	$3p_x$	Dumbbell-shaped along x-axis
	1	0	$3p_y$	Dumbbell-shaped along y-axis
		$+1$	$3p_z$	Dumbbell-shaped along z-axis
		-2	$3d_{yz}$	Cross-shaped in yz plane
		-1	$3d_{xz}$	Cross-shaped in xz plane
	2	0	$3d_{xy}$	Cross-shaped in xy plane
		$+1$	$3d_{x^2-y^2}$	Rotated cross in xy plane
		$+2$	$3d_{z^2}$	Dumbbell-shaped with a ring around the z-axis

The eigenfunctions of an electron in an atom at distance r from the nucleus with atomic number Z are

$$1s \text{ orbital, } \psi_{1s} = 2 \left(\frac{Z}{a_o} \right)^{3/2} e^{-\frac{\rho}{2}} \tag{14.5}$$

$$2s \text{ orbital, } \psi_{2s} = \frac{1}{2\sqrt{2}} \left(\frac{Z}{a_o} \right)^{3/2} \left(2 - \frac{\rho}{2} \right) e^{-\frac{\rho}{4}} \tag{14.6}$$

$$2p \text{ orbital, } \psi_{2p} = \frac{1}{2\sqrt{6}} \left(\frac{Z}{a_o} \right)^{3/2} \rho e^{-\frac{\rho}{4}}$$

(radial part of all p-orbitals) $\tag{14.7}$

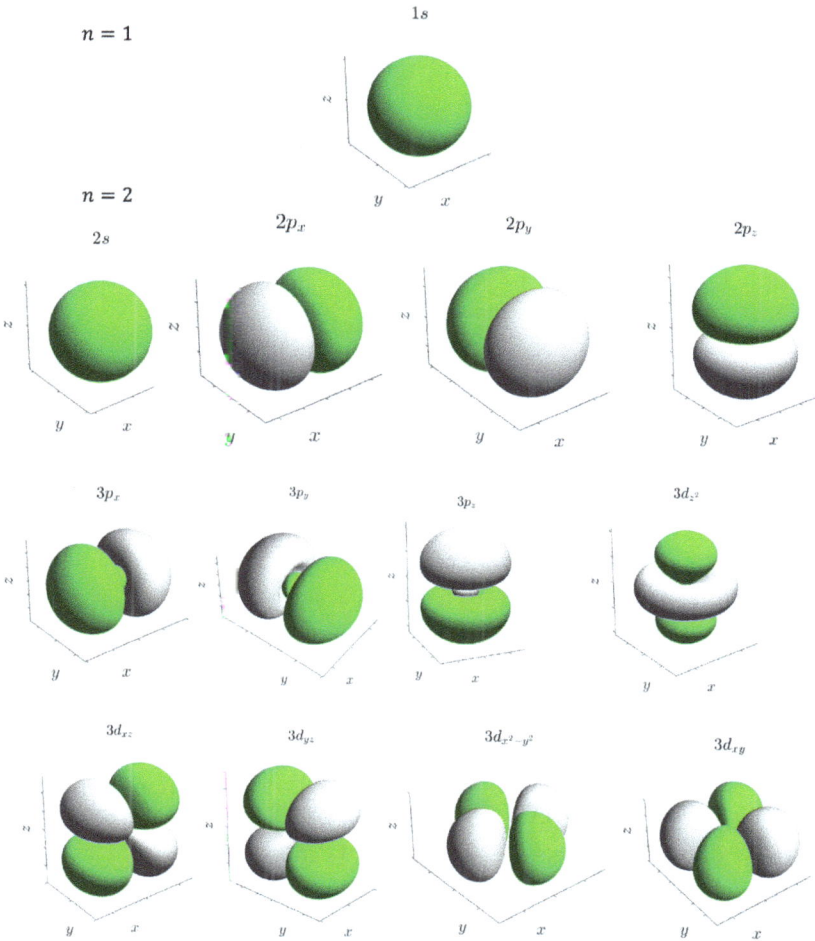

Figure 14.1. The shapes of s, p, and d orbitals are shown. It is not clear from the figures, but it should be noted that the larger the principal quantum number n, the larger the wave function's spatial extent from the nucleus.

where

$$\rho = Z \frac{r}{a_o} \qquad (14.8)$$

and a_o is the Bohr radius.

14.2   Energy Levels and Wave Functions of a Solid

Consider a cluster with N atoms and M electrons. The Hamiltonian in its full glory is

$$\widehat{H} = -\frac{\hbar^2}{2m_0} \sum_{i=1}^{M} \nabla_i^2 - q \sum_{i=1}^{M} \sum_{\alpha=1}^{N} V(\bar{r}_i - \bar{R}_\alpha)$$
$$+ \frac{1}{4\pi\varepsilon_0} \sum_{i=1}^{M} \sum_{j=1(i \neq j)}^{M} \frac{q^2}{|\bar{r}_i - \bar{r}_j|} \tag{14.9}$$

Here, the first term on the RHS is the kinetic energy of the M electrons, the second term is the electron–nuclei interaction, and the third term is the potential energy due to electron–electron repulsion. The M electron Schrödinger equation is

$$\widehat{H}\psi(\bar{r}_1, \bar{r}_2 \cdots \bar{r}_M) = E\psi(\bar{r}_1, \bar{r}_2 \cdots \bar{r}_M) \tag{14.10}$$

Note that the wave function depends on the coordinates of all M electrons. This equation has $3M$ degrees of freedom [(x, y, z) coordinates of each of the M electrons]. The $M = 1$ (one electron) case can be solved analytically. While an analytical solution is not known for values $M > 1$, even computationally solving this equation is very challenging because of the $\frac{q^2}{|\bar{r}_i - \bar{r}_j|}$ term. Hence, for higher values of M, we should rely on approximate numerical methods based on solving equations with a much smaller number of degrees of freedom. The density functional theory (DFT) and tight-binding approaches, which are examples of such methods, are discussed next briefly.

14.2.1   *Density functional theory*

The difficulty in solving equation (14.10) arises primarily from the electrical repulsion term between electrons. An important development in providing a systematic framework to solve these equations was made possible by Hohenberg and Kohn who proved that all ground-state properties of a material are a functional of the electron density. We will not provide a detailed discussion of the

Hohenberg–Kohn theorem but will outline an important development called the Kohn–Sham orbital equations that made calculations of electronic properties practical in material science and engineering applications.

Kohn and Sham showed that the multi-electron Schrödinger equation (equation (14.10)) can be replaced by an equation for a single electron experiencing an effective potential called $V_{DFT}(r)$, which is a functional of the local electron density. The single particle Hamiltonian is

$$\widehat{H} = -\frac{\hbar^2}{2m_0}\nabla_r^2 - q\sum_{\alpha=1}^{N} V\left(\bar{r} - \bar{R}_\alpha\right) - qV_{DFT}(\bar{r}) \quad (14.11)$$

$$V_{DFT}\left(\bar{r}\right) = \frac{1}{4\pi\varepsilon_0}\int dv\,\frac{\rho\left(\bar{r}'\right)}{|\bar{r} - \bar{r}'|} + V_{xc}(\bar{r}) \quad (14.12)$$

The first term in V_{DFT} is the classical Coulomb potential felt by an electron at $\bar{r}$ due to the electron density ρ at all other points $\bar{r}'$. V_{xc} which is called the exchange-correlation potential is the component due to the antisymmetry of the wave function and all other correlations that exist in the original system (equation (14.10)). The second term V_{xc} is a *functional* of density $\rho(\bar{r})$,

$$V_{xc}(\bar{r}) = \boldsymbol{F}\left[\rho(\bar{r})\right] \quad (14.13)$$

This observation is a direct result of Hohenberg–Kohn theorem which showed that all ground-state properties of a material are a functional of the electron density. Hence, the name *density functional theory*. The single-particle (electron) Schrödinger equation that is solved is

$$\widehat{H}\psi_\alpha(\bar{r}) = E_\alpha\psi_\alpha(\bar{r}) \quad (14.14)$$

We can find the single-particle eigenfunctions by solving equation (14.14) numerically. Based on these eigenfunctions, we can populate the energy levels with the M electrons in the system and then calculate the electron density ρ at zero temperature by using

$$\rho(\bar{r}) = \sum_\alpha |\psi_\alpha(\bar{r})|^2 \quad (14.15)$$

where α is the set of all occupied energy levels. The procedure to solve equations (14.11)–(14.15) involves starting with an assumed initial guess for the density. The potential energy term is calculated using equations (14.12) and (14.15), after which the Hamiltonian in equation (14.11) is updated. This equation yields a new wave function, from which a new density and V_{DFT} are obtained and substituted back into equation (14.11). This procedure is recursively continued until convergence, when consecutive iterations give the same result within an error threshold. See Figure 14.2 for a flowchart of the iterative procedure to solve the Kohn–Sham equations.

The functional dependence in $V_{xc}(\bar{r})$ on the density ($\boldsymbol{F}\left[\rho(\bar{r})\right]$) is actively being studied and developed by DFT specialists. Several functionals exist, and the particular choice in practical calculations depends on the material system and properties simulated. The number of atoms in a structure simulated with DFT packages has increased with the increase in computational power; as of 2023, DFT software packages can handle around a thousand atoms. Walter Kohn and John Pople were awarded the 1998 Nobel Prize in

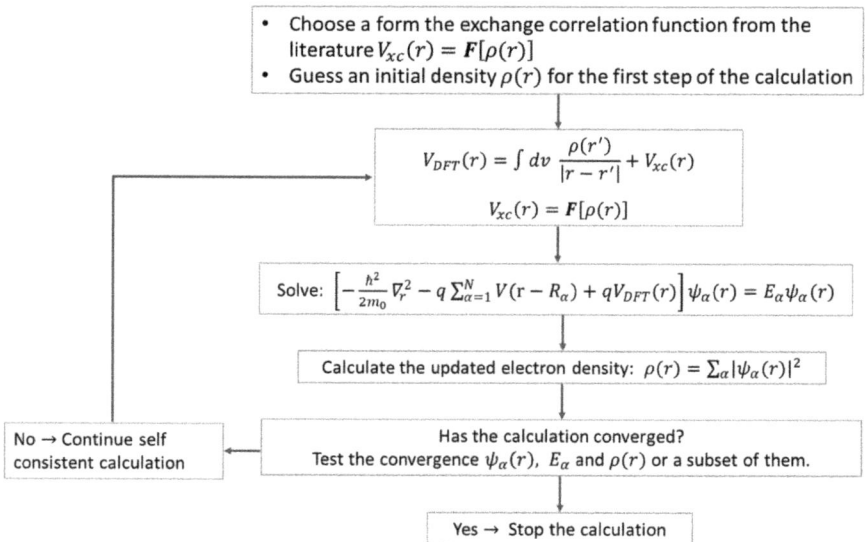

- Choose a form the exchange correlation function from the literature $V_{xc}(r) = \boldsymbol{F}[\rho(r)]$
- Guess an initial density $\rho(r)$ for the first step of the calculation

$$V_{DFT}(r) = \int dv\, \frac{\rho(r')}{|r - r'|} + V_{xc}(r)$$

$$V_{xc}(r) = \boldsymbol{F}[\rho(r)]$$

Solve: $\left[-\frac{\hbar^2}{2m_0}\nabla_r^2 - q\sum_{\alpha=1}^{N} V(r - R_\alpha) + qV_{DFT}(r)\right]\psi_\alpha(r) = E_\alpha\psi_\alpha(r)$

Calculate the updated electron density: $\rho(r) = \sum_\alpha |\psi_\alpha(r)|^2$

No → Continue self consistent calculation

Has the calculation converged?
Test the convergence $\psi_\alpha(r)$, E_α and $\rho(r)$ or a subset of them.

Yes → Stop the calculation

Figure 14.2. Flowchart of the iterative process involved in solving the Kohn–Sham equations.

chemistry for their work in creating methods to calculate material properties, which included the DFT.

14.2.2 *Tight-binding method*

A second, less rigorous approach to calculate the material properties is the tight-binding method. Consider a scenario where electrons are close to their mother atoms but can hop to neighboring atoms because of their proximity (Figure 14.3). The trial wave function of an electron in the solid is assumed to be a linear combination of atomic orbitals (LCAO) or wave functions centered on every atom i. The wave function for the entire system is assumed to be of the following form:

$$\psi(\bar{r}) = \sum_{n,i} c_{i,n}\phi_{i,n}(\bar{r}) \tag{14.16}$$

where $\phi_{i,n}(\bar{r})$ is the nth orbital centered on atom i. The orbital index n corresponds to orbitals $1s, 2s, 2p, 3s, 3p, 3d$ etc. In equation (14.16), the orbital wave functions $\phi_{i,n}(\bar{r})$ are known, and are used to solve for the coefficients $c_{i,n}$. Note that the trial wave function is a sum over all the orbital functions. In order to decrease computational complexity, only a small subset of orbitals, depending on the material system, are included in practical applications of the method.

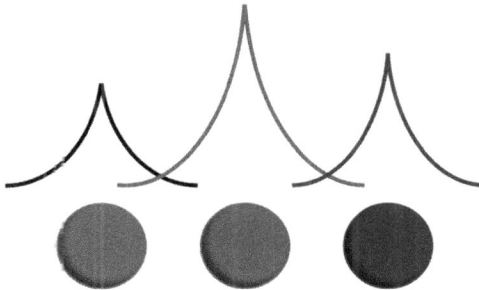

Figure 14.3. Atoms with one orbital centered on their nucleus interacting with each other. The solid circles represent the atoms. The peaks represent functions centered on each atom (an example would be atomic eigenfunctions).

We will first apply the tight-binding method to a system with two atoms, each of which contains a single orbital. The trial wave function of an electron in the two atom system is assumed to be an LCAO wave function centered on every atom. The wave function is,

$$\psi(\bar{r}) = c_1\phi_1(\bar{r}) + c_2\phi_2(\bar{r})$$

where ϕ_1 and ϕ_2 are orbitals centered on atoms 1 and 2, respectively. The expectation value of the energy corresponding to this wave function is

$$E = \frac{\int dv\, \psi^*(\bar{r})\,\widehat{H}\psi(\bar{r})}{\int dv\, \psi^*(\bar{r})\psi(\bar{r})}$$

$$E = \frac{|c_1|^2\, H(1,1) + c_1^* c_2 H(1,2) + c_2^* c_1 H(2,1) + |c_2|^2\, H(2,2)}{|c_1|^2\, S(1,1) + c_1^* c_2 S(1,2) + c_2^* c_1 S(2,1) + |c_2|^2\, S(2,2)}$$

where

$$H(i,j) = \int dv\, \phi_i^*(\bar{r})\widehat{H}\phi_j(\bar{r}) \quad \text{and}$$

$$S(i,j) = \int dv\, \phi_i^*(\bar{r})\,\phi_j(\bar{r})$$

for i and j equal to 1 and 2. $H(i,j)$ is called the *transfer integral* or *hopping integral* and is the integral of Hamiltonian between orbitals centered on atoms i and j. $S(i,j)$ is called the *overlap integral* and is the integral of the product of orbitals centered on atoms i and j. The values of c_1 and c_2 that minimize the energy E are determined by the minimization condition learned in calculus,

$$\frac{\partial E}{\partial c_1^*} = 0 \quad \text{and} \quad \frac{\partial E}{\partial c_2^*} = 0$$

The minimization with respect to c_1^* is

$$\frac{\partial E}{\partial c_1^*} = \frac{c_1 H(1,1) + c_2 H(1,2)}{|c_1|^2\, S(1,1) + c_1^* c_2 S(1,2) + c_2^* c_1 S(2,1) + |c_2|^2\, S(2,2)}$$

$$- \frac{\left[|c_1|^2\, H(1,1) + c_1^* c_2 H(1,2) + c_2^* c_1 H(2,1) + |c_2|^2\, H(2,2)\right] \times [c_1 S(1,1) + c_2 S(1,2)]}{\left[|c_1|^2\, S(1,1) + c_1^* c_2 S(1,2) + c_2^* c_1 S(2,1) + |c_2|^2\, S(2,2)\right]^2} = 0$$

$$\frac{\partial E}{\partial c_1^*} = \frac{c_1 H\left(1,1\right) + c_2 H(1,2)}{\left|c_1\right|^2 S\left(1,1\right) + c_1^* c_2 S(1,2) + c_2^* c_1 S(2,1) + \left|c_2\right|^2 S(2,2)}$$
$$- E\frac{\left[c_1 S\left(1,1\right) + c_2 S(1,2)\right]}{\left|c_1\right|^2 S\left(1,-\right) + c_1^* c_2 S(1,2) + c_2^* c_1 S(2,1) + \left|c_2\right|^2 S(2,2)} = 0$$

For the above equality to hold in general, the numerator should be zero. This gives us

$$\left[ES\left(1,1\right) - H\left(1,1\right)\right] c_1 + \left[ES\left(1,2\right) - H\left(1,2\right)\right] c_2 = 0$$

Similarly, the minimization with respect to c_2^*, $\frac{\partial E}{\partial c_2^*} = 0$ gives

$$\left[ES\left(2,1\right) - H\left(2,1\right)\right] c_1 + \left[ES\left(2,1\right) - H\left(2,2\right)\right] c_2 = 0$$

The previous two equations define the eigenvalue problem to find the energy eigenvalues E and the eigenfunction coefficients c_1 and c_2,

$$\begin{bmatrix} ES\left(1,1\right) - H\left(1,1\right) & ES\left(1,2\right) - H\left(1,2\right) \\ ES\left(2,1\right) - H\left(2,1\right) & ES\left(2,1\right) - H\left(2,2\right) \end{bmatrix} \begin{bmatrix} c_1 \\ c_2 \end{bmatrix} = 0$$

General case: Going back to the case when there is more than one orbital per atom, as given by the wave function in equation (14.16), the normalized wave function is

$$\psi(\bar{r}) = \frac{1}{\int dv\, \psi^*(\bar{r})\psi(\bar{r})} \sum_{n,i} c_{i,n} \phi_{i,n}\left(\bar{r}\right) \tag{14.17}$$

The energy corresponding to the wave function ψ is the expectation value of the Hamiltonian $(\hat{H})$ normalized by the norm of the wave function,

$$E = \frac{\int dv\, \psi^*\left(\bar{r}\right) \widehat{H}\psi(\bar{r})}{\int dv\, \psi^*(\bar{r})\psi(\bar{r})} \tag{14.18}$$

The Hamiltonian $\widehat{H}$ is given by equation (14.11). Substituting equation (14.17) in (14.18) gives

$$E = \frac{\displaystyle\sum_{n,i,m,j} c_{n,i}^* c_{m,j} H_{n,m}(i,j)}{\displaystyle\sum_{n,i,m,j} c_{n,i}^* c_{m,j} S_{n,m}(i,j)} \tag{14.19}$$

where

$$H_{n,m}\left(i,j\right) = \int dv\, \phi_{i,n}^{*}\left(\bar{r}\right) H \phi_{j,m}\left(\bar{r}\right) \text{ and} \qquad (14.20)$$

$$S_{n,m}\left(i,j\right) = \int dv\, \phi_{i,n}^{*}\left(\bar{r}\right) \phi_{j,m}\left(\bar{r}\right) \qquad (14.21)$$

$H_{n,m}\left(i,j\right)$ is related to the role of Hamiltonian H in transferring electrons between orbital n on atom i and orbital m on atom j. $H_{n,m}$ is known by a variety of names such as the *hopping integral energy/parameter* or *tight-binding parameter energy* or *tunneling integral* as it quantifies the movement/tunneling rate of an electron between orbitals n and m. If $i = j$, then it is called *onsite energy* as the term involves only a single atom. $S_{n,m}\left(i,j\right)$ is the overlap integral between the wave functions $\phi_{i,n}\left(\bar{r}\right)$ and $\phi_{j,m}(\bar{r})$. The unknown coefficients in equation (14.19), $c_{i,n}$, must minimize the energy of the system. i.e., they should minimize E. This method, called the variational method is used in many fields. An example is its use in optical and microwave engineering to find eigenfrequencies and eigenfunctions of waveguides. For applications in resonance cavities and waveguides, see (Sadiku, 2000) for further detail.

To find the values of the coefficients, we set the partial differential with respect to coefficients, $c_{k,p}^{*}$ equal to zero,

$$\frac{\partial E}{\partial c_{k,p}^{*}} = 0 \qquad (14.22)$$

Taking the derivative of equation (14.19),

$$\frac{\partial E}{\partial c_{k,p}^{*}} = \frac{\sum_{n,i,m,j} c_{m,j} H_{n,m}\left(i,j\right) \delta_{np}\delta_{ik}}{\sum_{n,i,m,j} c_{n,i}^{*} c_{m,j} S_{n,m}(i,j)}$$
$$- \left[\sum_{n,i,m,j} c_{n,i}^{*} c_{m,j} H_{n,m}(i,j) \right]$$

$$\times \frac{\sum_{n,i,m,j} c_{m,j} S_{n,m}(i,j) \, \delta_{np}\delta_{ik}}{\left[\sum_{n,i,m,j} c_{n,i}^* c_{m,j} S_{n,m}(i,j)\right]^2} = 0$$

We can rewrite the above equation by using equation (14.19) in the second term above to give

$$\frac{\partial E}{\partial c_{p,k}^*} = \frac{\sum_{m,j} c_{m,j} H_{p,m}(k,j)}{\sum_{n,i,m,j} c_{n,i}^* c_{m,j} S_{n,m}(i,j)}$$

$$- E \frac{\sum_{m,j} c_{m,j} S_{p,m}(k,j)}{\sum_{n,i,m,j} c_{n,i}^* c_{m,j} S_{n,m}(i,j)} = 0$$

As the denominator is common to both terms, the above equation is equivalent to the matrix equation,

$$\sum_{m,j} [E S_{p,m}(k,j) - H_{p,m}(k,j)] c_{m,j} = 0 \qquad (14.23)$$

As a reminder, the summation in the above equation is over indices m and j. Each value of (p,k) yields a different linear equation. More compactly, equation (14.23) is

$$[ES - H]c = 0 \qquad (14.24)$$

The number of unknown coefficients $c_{i,n}$ in equation (14.23) is equal to the total number of orbitals in the system with N atoms. By solving equation (14.23), we know both the energy eigenvalues E and the eigenfunctions by using the values of $c_{i,n}$ in equation (14.17). We know that the orbital wave function on an atom exponentially decays with increasing radial distance from the nuclei. As a result, it is only necessary to know the values of $H_{n,m}(i,j)$ and $S_{n,m}(i,j)$ between an atom and its nearest neighbors up to a cut off distance (typically less than 5 Å), as beyond that the values are inconsequential to the sum.

The procedure discussed above seems straightforward, but a few points make its implementation difficult. Neither the atomic wave functions $\phi_{i,n}(r)$ nor the potential $V_{DFT}(r)$ are not fully known. Using the best-known values for them provides useful guidelines, but ultimately, the values of $S_{n,m}(i,j)$ and $H_{n,m}(i,j)$ must be tweaked to match critical experimental parameters (such as effective mass and bandgap).

In the next section, we will assume that we are bringing two simplified hydrogen atoms (called H-atoms) together to build a H_2 molecule. To illustrate what we have learned in this section, we will assume that each hydrogen atom has a $1s$ orbital only (the other orbitals have higher energy and are neglected for simplicity). Section 14.3 uses the tight-binding approach to find the energy levels of simple 1D, 2D, and 3D periodic solids.

14.2.3   *Two interacting atoms*

This section will consider two identical atoms interacting with each other using the notation in Section 14.2.2. Each atom consists of a single s-orbital and a single electron (we will refer to this as the H-atom). If we simplify the notation for the diagonal elements of the matrix in equation (14.20), the onsite energy of an isolated atom is,

$$H_{s,s}(1,1) = \int dv\, \phi_{1,s}^*(\bar{r})\, \widehat{H}\phi_{1,s}(\bar{r}) = \varepsilon$$

$$H_{s,s}(2,2) = \int dv\, \phi_{2,s}^*(\bar{r})\, \widehat{H}\phi_{2,s}(\bar{r}) = \varepsilon$$

The off-diagonal matrix elements between the two atoms in equation (14.20) are called the hopping energies, and they are

$$H_{s,s}(1,2) = \int dv\, \phi_{1,s}^*(\bar{r})\, \widehat{H}\phi_{2,s}(\bar{r}) = t$$

$$H_{s,s}(2,1) = \int dv\, \phi_{2,s}^*(\bar{r})\, \widehat{H}\phi_{1,s}(\bar{r}) = t$$

For simplicity, assume that the orbitals on one atom are orthogonal to the second atom's orbitals. While this is not entirely true, the approximation simplifies our calculations. This condition in terms of the overlap matrix elements of equation (14.21) is as follows:

$$S_{s,s}(1,1) = \int dv\, \phi_{1,s}^*(\bar{r})\, \phi_{1,s}(\bar{r}) = 1$$

$$S_{s,s}(2,2) = \int dv\, \phi_{2,s}^*(\bar{r})\, \phi_{2,s}(\bar{r}) = 1$$

$$S_{s,s}(1,2) = \int dv\, \phi_{1,s}^*(\bar{r})\, \phi_{2,s}(\bar{r}) = 0$$

$$S_{s,\varepsilon}(2,1) = \int dv\, \phi_{2,s}^*(\bar{r})\, \phi_{1,s}(\bar{r}) = 0$$

Using the above notation involving ε and t, equation (14.23), which is the matrix equation needed to find the eigenvalues and wave functions, can be written as

$$\begin{pmatrix} \varepsilon & t \\ t & \varepsilon \end{pmatrix} \begin{pmatrix} c_1 \\ c_2 \end{pmatrix} = E \begin{pmatrix} c_1 \\ c_2 \end{pmatrix} \tag{14.25}$$

It is easy to verify that there are two solutions to the above eigenvalue equation. The first solution corresponds to

$$E_1 = \varepsilon - |t| \tag{14.26}$$

$$\begin{pmatrix} c_1 \\ c_2 \end{pmatrix} = \frac{1}{\sqrt{2}} \begin{pmatrix} 1 \\ -\mathrm{sign}(t) \end{pmatrix} \tag{14.27}$$

The corresponding wave function is

$$\psi_1(r) = \frac{1}{\sqrt{2}} [\phi_{1,s}(r) - \mathrm{sign}(t) \cdot \phi_{2,s}(r)] \tag{14.28}$$

The second solution corresponds to

$$E_2 = \varepsilon + |t| \tag{14.29}$$

$$\begin{pmatrix} c_1 \\ c_2 \end{pmatrix} = \frac{1}{\sqrt{2}} \begin{pmatrix} 1 \\ \mathrm{sign}(t) \end{pmatrix} \tag{14.30}$$

Quantum Mechanics for Engineers and Material Scientists

The corresponding wave function is

$$\psi_2(r) = \frac{1}{\sqrt{2}} [\phi_{1,s}(\bar{r}) + \text{sign}(t) \cdot \phi_{2,s}(\bar{r})] \qquad (14.31)$$

where $\text{sign}(t)$ corresponds to the sign of t. The splitting of the energy level ε of the isolated hydrogen atoms to $\varepsilon \pm |t|$, when two identical atoms are brought together, is shown in Figure 14.4. The magnitude of the energy level splitting is $2|t|$. If each atom brought in only one electron, the lower energy $\varepsilon - |t|$ will have one up-spin and one down-spin electron and the energy $\varepsilon + |t|$ will be empty.

Note that in solids, many ($\sim 10^{23}$) atoms are brought together. Hence, there are many filled bands similar to Figure 14.5, which are so close that they build a continuum of filled states called the valence band. The continuum of empty states that lie at energies higher than that of the filled states is called the conduction band. Depending on the details of the material system, there may be a large (insulators), a small (semiconductors), or no (metals) *energy gap* between the occupied and unoccupied bands. We will see a concrete example of this in Section 14.3.2.

14.2.3.1 Analogy to two coupled electrical resonators

In the previous section, we saw that when two identical atoms are brought close together, their energy levels split into two. The amount of splitting depends on the hopping energy t between the two atoms.

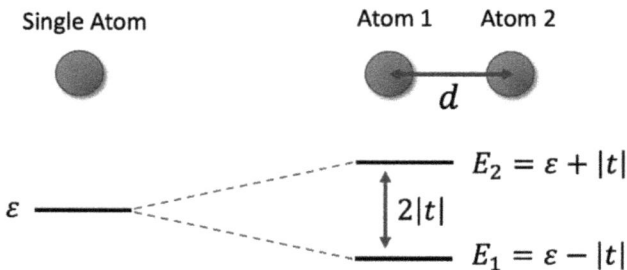

Figure 14.4. (Left) The energy levels of two identical H-atoms with energy level ε split as they come close to each other. The energy difference between the new energy levels $\varepsilon - |t|$ and $\varepsilon + |t|$ increases as the atoms approach each other.

Single Atom

Two-atom molecule

Four-Atom molecule

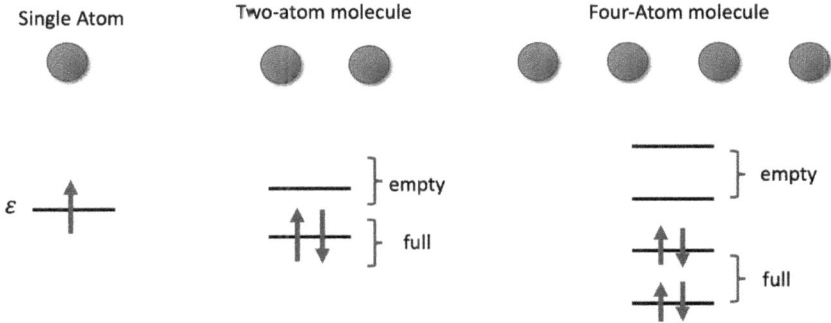

Figure 14.5. Four identical interacting atoms, each with a single energy level on each atom. The energy levels split into four when the atoms are brought close to each other. If each isolated atom has only one electron, when the atoms are brought together, the lowest two energy levels will be occupied, and the higher two energy levels will be empty.

The resulting molecule has two wave functions: one corresponding to the lower energy level and the second corresponding to the higher energy level (recall equations (14.26) and (14.29)).

This section shows a similar phenomenon when two identical LC resonators are brought close together (Figure 14.6). The resonant frequency of an isolated resonator is

$$\omega_0 = \frac{1}{\sqrt{\mathcal{L}\mathbb{C}}} \qquad (14.32)$$

If the resonators are brought close together, they interact with each other through a mutual inductance (modeled as M). To find the new resonant frequencies, we write Kirchhoff voltage law (KVL) in each loop. Note that we assume that the resonators have steady-state sinusoidal solutions; hence using phasors and KVL, we have that the currents in the two loops i_1 and i_2 are related by

$$\begin{cases} i_1 \left(\frac{1}{j\omega\mathbb{C}} + j\omega\mathcal{L} \right) - j\omega\mathcal{M}i_2 = 0 \\ -j\omega\mathcal{M}i_1 + i_2 \left(\frac{1}{j\omega\mathbb{C}} + j\omega\mathcal{L} \right) = 0 \end{cases} \qquad (14.33)$$

or

$$\begin{bmatrix} \mathcal{L}\mathbb{C} & -\mathcal{M}\mathbb{C} \\ -\mathcal{M}\mathbb{C} & \mathcal{L}\mathbb{C} \end{bmatrix} \begin{bmatrix} i_1 \\ i_2 \end{bmatrix} = \frac{1}{\omega^2} \begin{bmatrix} i_1 \\ i_2 \end{bmatrix} \qquad (14.34)$$

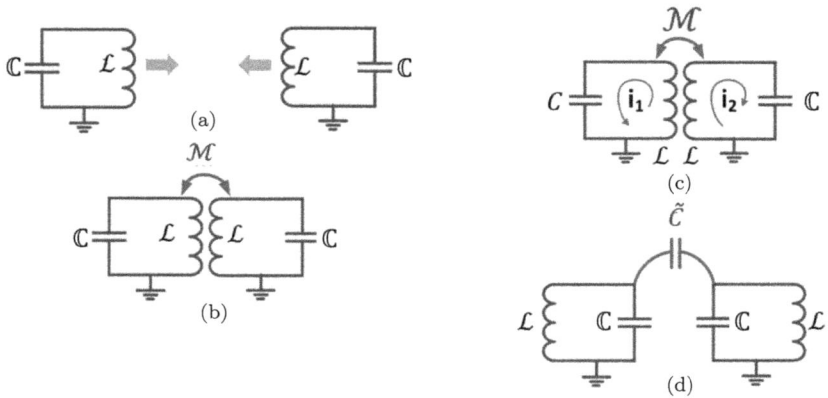

Figure 14.6. (a) Two separate LC resonators. (b and c) Inductively coupled LC resonators. (d) Capacitively coupled LC resonators.

The above equation has a mathematical form that is similar to equation (14.25) for two interacting atoms. The solutions to the above eigenvalue problem are

$$\omega_1 = \frac{1}{\sqrt{C(\mathcal{L} - \mathcal{M})}} \quad \text{and} \quad \omega_2 = \frac{1}{\sqrt{C(\mathcal{L} + \mathcal{M})}} \tag{14.35}$$

If we assume that the mutual inductance ($\mathcal{M}$) is much smaller than the inductance ($\mathcal{L}$), a Taylor series expansion of equation (14.35) gives

$$\omega_1 = \frac{1}{\sqrt{C\mathcal{L}\left(1 - \frac{\mathcal{M}}{\mathcal{L}}\right)}} \approx \omega_0 \left(1 + \frac{\mathcal{M}}{2\mathcal{L}}\right) \quad \text{and} \tag{14.36}$$

$$\omega_2 = \frac{1}{\sqrt{C\mathcal{L}\left(1 + \frac{\mathcal{M}}{\mathcal{L}}\right)}} \approx \omega_0 \left(1 - \frac{\mathcal{M}}{2\mathcal{L}}\right) \tag{14.37}$$

The mutual interaction between the two resonators lifts the degeneracy in resonant frequencies of the isolated resonators, with one resonant frequency being below ω_0 and the other being above ω_0 by $\frac{\mathcal{M}\omega_0}{2\mathcal{L}}$. Readers can show that if the resonators are coupled capacitively (e.g., through parasitic capacitance between the resonators, see

Figure 14.6), the resonant frequencies split as a function of the coupling capacitance.

As a final note, reconsider equation (14.34). The diagonal terms $(\mathcal{L}\mathbb{C})$ resemble the onsite energy (ε) in equation (14.25), and the off-diagonal terms $(\mathcal{M}\mathbb{C})$ resemble the hopping or tunneling between the two atoms, i.e., t in equation (14.25).

14.3 Periodic Systems

Consider an infinite 1D array of atoms where each atom is separated from its two nearest neighbors by precisely the same distance a (Figure 14.7(a)). When translated by all integer multiples of vector $\bar{a}$, the red dashed cell generates the entire 1D array of atoms along the horizontal axis. The 1D periodic system that is shown in Figure 14.7(b) also has a periodicity of length a, but the basic unit (red dashed cell) that repeats itself has five atoms. The smallest cell required to generate the periodic lattice of all atoms is called the primitive cell. The vector $\bar{a}$, which defines the primitive cell, is called the primitive vector. Note that the primitive cell is in general not unique for a given periodic lattice. The atoms in a primitive

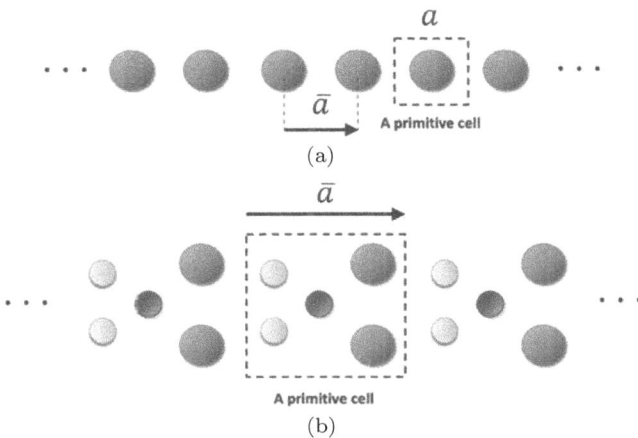

Figure 14.7. Two examples of 1D periodic systems. The example in (a) has one atom per primitive cell, and the example in (b) has five atoms per primitive cell. The colored circles represent the different atoms.

cell are known as the basis of the cell. Figure 14.7(a) is a primitive cell with one atom in its basis, while Figure 14.7(b) has five atoms. Examples of 1D periodic systems are polyethylene $(-C_2H_2-)$ and carbon nanotubes.

Moving over to two dimensions, Figure 14.8 presents two examples of a 2D lattice, a square, and a hexagonal lattice. There are now two primitive vectors, $\bar{a}$ and $\bar{b}$, necessary to define the 2D primitive cell. Translation by the rectangular shaded primitive cell defined by integer multiples of $\bar{a}$ and $\bar{b}$ generates the entire 2D lattice of atoms. Examples of 2D periodic solids are graphene, a semimetal, and boron nitride, a semiconductor.

A simple example of a 3D lattice is shown in Figure 14.9. In 3D, there are three primitive vectors $(\bar{a}, \bar{b}$ and $\bar{c})$, and the primitive cell occupies a 3D volume. Translation by the rectangular shaded cubic primitive cell by all integer multiples of $\bar{a}, \bar{b}$ and $\bar{c}$ generates the entire 3D lattice of atoms.

Primitive cell: The smallest cell that generates the periodic lattice of atoms without any overlap or gap, when translated by the set of all integer multiples of the primitive vector.

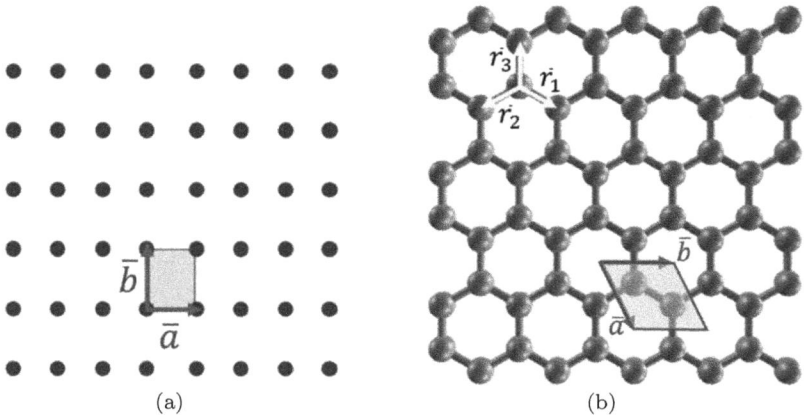

Figure 14.8. Rectangular lattice of atoms. The primitive vectors $\bar{a}$ and $\bar{b}$ and the primitive cell are shown. The basis has only one atom per primitive cell. (b) A hexagonal lattice of atoms, e.g., graphene. The primitive vectors $\bar{a}$ and $\bar{b}$ are not orthogonal to each other. The basis has two atoms per primitive cell.

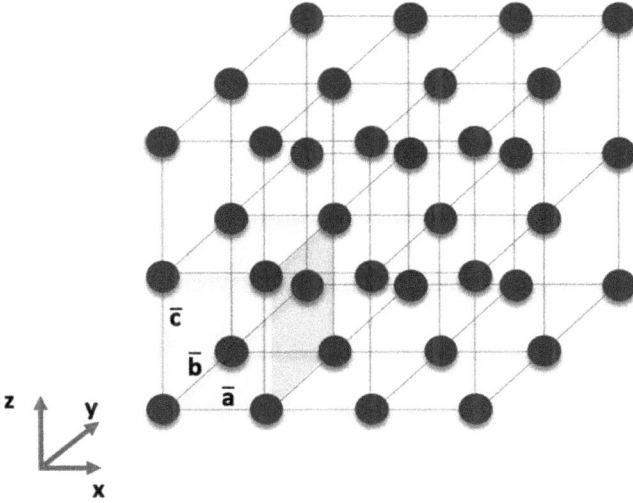

Figure 14.9. The primitive vectors $\bar{a}$, $\bar{b}$, and $\bar{c}$ and orthorhombic primitive cell are shown. The basis has only one atom. Note that each atom is shared by eight primitive cells.

The spatial location of an arbitrary primitive cell in a **1D periodic structure** is given by

$$x_{unit-cell}(n) = n\bar{a} \tag{14.38}$$

where n is the set of *all* integers and $\bar{a}$ is the primitive vector.

In a **2D periodic structure**, the spatial location of an arbitrary primitive cell is given by

$$\bar{r}^{(2D)}_{unit-cell}(n_1, n_2) = n_1\bar{a} + n_2\bar{b} \tag{14.39}$$

n_1 and n_2 are the set of all integers and $\bar{a}$ and $\bar{b}$ are the primitive vectors. By spanning n_1 and n_2 over the set of all integers, the entire 2D periodic system is generated without gaps or overlaps.

In a **3D periodic structure**, the spatial location of an arbitrary primitive cell is given by

$$\bar{r}^{(3D)}_{unit-cell}(n_1, n_2, n_3) = n_1\bar{a} + n_2\bar{b} + n_3\bar{c} \tag{14.40}$$

where $(\bar{a}, \bar{b}, \bar{c})$ are the three primitive vectors. By spanning (n_1, n_2, n_3) over all integers, the entire 3D periodic structure is generated without gaps or overlaps. Note that the unit vectors need not be orthogonal to each other in both 2D and 3D periodic structures.

Unit cell: Like the primitive cell, the unit cell generates the entire lattice without any overlap or gaps when translated periodically by the unit cell vectors. However, the primitive cell is the smallest possible unit cell. We will not use the unit cell concept in this book, but refer the reader to a book on solid-state physics for further information (Blakemore, 1985).

14.3.1 *Bloch theorem*

The Hamiltonian of a periodic system is also periodic. It obeys the mathematical property,

$$H\left(\bar{r}\right) = H\left(\bar{r} + n_1\bar{a} + n_2\bar{b} + n_3\bar{c}\right) \tag{14.41}$$

We can solve for the energy eigenvalues and eigenfunctions using equation (14.23). However, the system is infinitely large. So, a blind computation would be infeasible even if we knew all the elements of H and S. Luckily, solving for the energy eigenvalues and eigenfunctions of an infinite periodic system (if H and S are known) becomes simple by using the *Bloch theorem*.

The Bloch theorem states that the energy eigenfunctions of the periodic system are given by

$$\psi(\bar{r}) = e^{i\bar{k}\cdot\bar{r}}u_{\bar{k}}(\bar{r}) \tag{14.42}$$

where

$$u_{\bar{k}}\left(\bar{r}\right) = u_{\bar{k}}\left(\bar{r} + n_1\bar{a} + n_2\bar{b} + n_3\bar{c}\right) \tag{14.43}$$

has the same periodicity as the underlying lattice. So,

$$\psi(\bar{r} + n_1\bar{a} + n_2\bar{b} + n_3\bar{c}) = e^{i\bar{k}\cdot(n_1\bar{a}+n_2\bar{b}+n_3\bar{c})}\psi(\bar{r}) \tag{14.44}$$

In summary, the eigenfunction is a product of (i) a traveling wave $e^{i\bar{k}\cdot\bar{r}}$ and (ii) a function $u_{\bar{k}}(\bar{r})$ which has the same periodicity as the Hamiltonian.

14.3.2 A periodic 1D array with one atom per primitive cell

We assume that the atoms are identical and have only one orbital, represented by orbital index s. In discussing the wave function of orbitals, we learned that the wave function decreases exponentially as the distance from the nucleus increases.

We will further assume that an electron in atom n interacts only with its nearest neighbor atoms to its left $(n-1)$ and right $(n+1)$ (see Figure 14.10). We neglect the direct interaction of orbitals at atoms n and $n \pm 2$. That is, in terms of the notation in Section 14.2.2, the diagonal element of equations (14.20) is

$$\text{Onsite}: H(n,n) = \int dv \, \phi_n^*(\bar{r}) \, \widehat{H} \phi_n(\bar{r}) = \varepsilon$$

The above integral does not depend on the atom number n because all atoms are identical. The only nonzero off-diagonal matrix elements between two atoms are the nearest neighbor interactions in equations (14.20), which are the hopping energies given by

$$(\text{Hopping}) \ H(n,n+1) = \int dv \, \phi_n^*(\bar{r}) \widehat{H} \phi_{n+1}(\bar{r}) = t$$

$$H(n+1,n) = \int dv \, \phi_{n+1}^*(\bar{r}) \, \widehat{H} \phi_n(\bar{r}) = t^*$$

Recall that H is a Hermitian operator $(H = H^\dagger)$. So, $H(n,n+1) = H(n+1,n)^*$. We will assume that t is real. Again, for simplicity, we assume that the orbitals on one atom are orthogonal to the orbitals

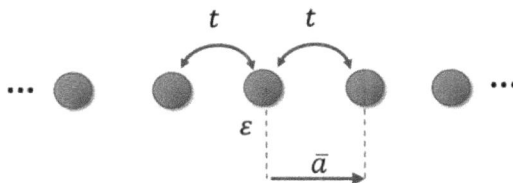

Figure 14.10. An infinitely long 1D array of atoms where the distance between consecutive atoms in the array is a. Each atom has a nonzero hopping energy with only its first neighbor atom to its left and right.

on the other atom. This condition in terms of equation (14.21) is

$$\text{(Overlap)} \quad S(n, n) = \int dv \, \phi_n^*(\bar{r}) \phi_n(\bar{r}) = 1$$

$$S(n, m \neq n) = \int dv \, \phi_n^*(\bar{r}) \phi_m(\bar{r}) = 0$$

While this condition is not entirely accurate, it simplifies our discussion without a loss of generality. Using the above notation involving the elements of $\widehat{H}$ and S (ε and t), equation (14.23), which is the matrix equation to find the energy eigenvalues and eigenfunctions can be written as

$$\widehat{H}\psi = E\psi$$

where the Hamiltonian matrix H and eigenfunction (or eigenvector) ψ of the periodic array are

$$\widehat{H} = \begin{pmatrix} \ddots & & & & \\ & \ddots & & & \\ & & \ddots & & \\ & t & \varepsilon & t & \\ & & t & \varepsilon & t \\ & & & t & \varepsilon & t \\ & & & & t & \varepsilon & t \\ & & & & & \ddots \\ & & & & & & \ddots \end{pmatrix} \quad \text{and} \quad \psi = \begin{bmatrix} \vdots \\ \psi(n-1) \\ \psi(n) \\ \psi(n+1) \\ \vdots \end{bmatrix}$$

(14.45)

Substituting equation (14.45) into $H\psi = E\psi$, the following equation is obtained for the nth row

$$\varepsilon\psi(n) + t\psi(n+1) + t\psi(n-1) = E\psi(n) \tag{14.46}$$

Bloch's theorem tells us that the wave function is of the form

$$\psi(n) = e^{ikna}u \tag{14.47}$$

where na denotes the location of atom n. Note that $\bar{r}$ in equation (14.42) is $\bar{r} = n\bar{a}$ and $\bar{k} \cdot \bar{r} = kna$. In the above equation, e^{ikna} is the traveling wave component and u is the periodic component of equation (14.42). Substituting equation (14.47) in equation (14.46) yields

$$te^{ik(n-1)a}u + \varepsilon e^{ikna}u + te^{ik(n+1)a}u = Ee^{ikna}u \qquad (14.48)$$

$$[\varepsilon + te^{ika} + te^{-ika}]u = Eu \qquad (14.49)$$

$$E = \varepsilon - 2t\cos(ka) \qquad (14.50)$$

The wave function and energy levels are given by equations (14.47) and (14.50). They depend on the quantum number k. Equation (14.50), which relates the energy of an electron to wave vector (k), is called the *band structure* or *band dispersion*. To determine the range of allowed quantum numbers, we note that the eigenfunction should take a unique value for each distinct quantum number. Note that distinct eigenfunctions are orthogonal to one another, as discussed in Chapters 1 and 2. The eigenfunctions are orthogonal only for k values lying in the range,

$$\text{1D Brillouin zone (BZ)}: \quad -\frac{\pi}{a} < k < \frac{\pi}{a} \qquad (14.51)$$

This range, which defines the unique set of all quantum numbers possible, is called the **1D BZ** of the corresponding 1D periodic solid. Note that no two quantum numbers can have the same wave function, but they can have the same energy. For example, consider the wave functions at $+|k|$ and $-|k|$, which have the same energy but orthogonal wave functions.

To see that the orthogonal eigenfunctions only lie in the window shown in equation (14.51), note that for any two values of the wave vector k separated by $\frac{2\pi}{a}$, the following holds true

$$e^{ika} = e^{i(k-\frac{2\pi}{a})a} \qquad (14.52)$$

As a result, k which lies in the range $-\frac{\pi}{a} < k < \frac{\pi}{a}$ and $k - \frac{2\pi}{a}$ which lies outside the range both correspond to the same eigenfunction e^{ika} (equation (14.47)). That is, k and $k - \frac{2\pi}{a}$ represent the same quantum number.

The energy eigenvalues (equation (14.50)) are plotted in Figure 14.11 as a function of the wave vector k. We saw in Figure 14.4

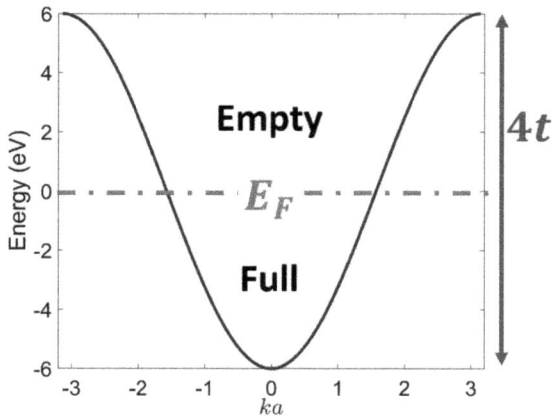

Figure 14.11. Energy band structure of a 1D lattice with one atom per primitive cell. Each atom has only one orbital. The x-axis is wave vector times primitive cell length. The hopping integral between an atom and its neighbors on the left and right is t. In the plot, the diagonal element of the Hamiltonian $\varepsilon = 0$ and $t = -3\,\text{eV}$.

that if two atoms, each with a single orbital, are brought next to each other, the energy eigenvalues split into two energy levels. If each atom has only one electron, the lower energy level fills with one up-spin and one down-spin electron, and the higher energy level is empty. Similarly, if four atoms, each with a single orbital, are brought next to each other, the energy eigenvalues split into four energy levels (Figure 14.5). Again, if the atoms have only one electron each, then each of the lower two energy levels fills with an up-spin and one down-spin electron. The upper two energy levels are empty. So again, in the four-atom case, only half the energy levels are filled. Similarly, in the case of the infinite chain of single-orbital atoms, each with only one electron, the energy levels will be half-filled at zero temperature, as shown in Figure 14.11. At zero temperature, the energy until which electrons are filled is called the Fermi energy (E_F). As the energy separation between the occupied and unoccupied states at zero temperature is zero, this 1D periodic solid is a *metal*.

Before closing this section, it is noteworthy to mention that the Hamiltonian matrix in equation (14.46) is formally similar to the finite difference form of Schrödinger equation in Section 1.6. See equation (1.111).

14.3.3 An infinite 1D array with two atoms per primitive cell

We now consider a slightly more complicated case consisting of a 1D string with two atoms per primitive cell (Figure 14.12). The system and Hamiltonian are defined as follows: (i) each atom has one orbital and (ii) the diagonal (onsite) energy of both the small red and big blue atoms are $\varepsilon = 0$. The hopping energy between nearest-neighbor atoms in neighboring primitive cells is t_2. The transfer integral between atoms inside the same primitive cell is t_1.

The Hamiltonian of primitive cell n is

$$H_{n,n} = \begin{pmatrix} \varepsilon & t_1 \\ t_1 & \varepsilon \end{pmatrix} \tag{14.53}$$

The Hamiltonian couplings of primitive cell n with $n+1$ and $n-1$ are

$$H_{n,n+1} = \begin{pmatrix} 0 & 0 \\ t_2 & 0 \end{pmatrix} \quad \text{and} \tag{14.54}$$

$$H_{n,n-1} = \begin{pmatrix} 0 & t_2 \\ 0 & 0 \end{pmatrix} \tag{14.55}$$

t_2 in equation (14.54) represents the coupling of the big blue (second) atom in primitive cell n to the small red (first) atom of primitive cell $n+1$, and t_2 in equation (14.55) represents the coupling of the red (first) atom in primitive cell n to the big blue (second) atom of primitive cell $n-1$.

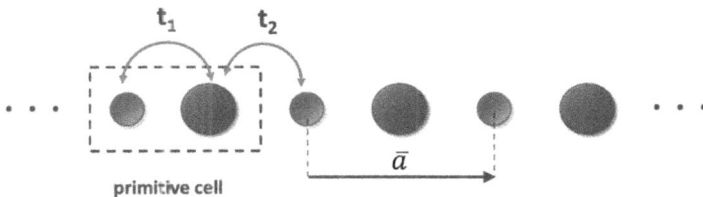

Figure 14.12. An infinitely long 1D array of atoms with each primitive cell containing two atoms, a small red and a big blue atom. The hopping energy between the atoms are t_1 and t_2 as shown above. The length of the primitive cell is $\bar{a}$.

Substituting the wave function and the Hamiltonian terms (equations (14.53)–(14.55)) into Schrödinger equation, we have (similar to equation (14.46))

$$H_{n,n}\psi_n + H_{n,n+1}\psi_{n+1} + H_{n,n-1}\psi_{n-1} = E\psi_n \tag{14.56}$$

Using Bloch theorem (equation (14.42)), the eigenfunction in any primitive cell is

$$\psi_n = e^{ikna}\begin{pmatrix} u_1 \\ u_2 \end{pmatrix} \tag{14.57}$$

where n is the primitive cell number and u_1 and u_2 are the components of the eigenfunction at the small-red and big-blue atoms, respectively. Substituting equation (14.57) into (14.56), we have

$$e^{ikna}\begin{pmatrix} \varepsilon & t_1 \\ t_1 & \varepsilon \end{pmatrix}\begin{pmatrix} u_1 \\ u_2 \end{pmatrix} + e^{ik(n+1)a}\begin{pmatrix} 0 & 0 \\ t_2 & 0 \end{pmatrix}\begin{pmatrix} u_1 \\ u_2 \end{pmatrix}$$

$$+ e^{ik(n-1)a}\begin{pmatrix} 0 & t_2 \\ 0 & 0 \end{pmatrix}\begin{pmatrix} u_1 \\ u_2 \end{pmatrix} = e^{ikna}\begin{pmatrix} E & 0 \\ 0 & E \end{pmatrix}\begin{pmatrix} u_1 \\ u_2 \end{pmatrix} \tag{14.58}$$

The above equation can be simplified to

$$\begin{pmatrix} \varepsilon & t_1 \\ t_1 & \varepsilon \end{pmatrix}\begin{pmatrix} u_1 \\ u_2 \end{pmatrix} + e^{ika}\begin{pmatrix} 0 & 0 \\ t_2 & 0 \end{pmatrix}\begin{pmatrix} u_1 \\ u_2 \end{pmatrix}$$

$$+ e^{-ika}\begin{pmatrix} 0 & t_2 \\ 0 & 0 \end{pmatrix}\begin{pmatrix} u_1 \\ u_2 \end{pmatrix} = \begin{pmatrix} E & 0 \\ 0 & E \end{pmatrix}\begin{pmatrix} u_1 \\ u_2 \end{pmatrix} \tag{14.59}$$

$$\begin{bmatrix} \varepsilon & t_1+t_2e^{-ika} \\ t_1 + t_2e^{ika} & \varepsilon \end{bmatrix}\begin{pmatrix} u_1 \\ u_2 \end{pmatrix} = E\begin{pmatrix} u_1 \\ u_2 \end{pmatrix} \tag{14.60}$$

Solving for the eigenvalues of the 2×2 matrix above gives

$$E = \varepsilon \pm \left| t_1 + t_2e^{ika} \right| \tag{14.61}$$

$$E = \varepsilon \pm \sqrt{(t_1 + t_2\cos(ka))^2 + t_2^2\sin^2(ka)} \tag{14.62}$$

$$E = \varepsilon \pm \sqrt{t_1^2 + t_2^2 + 2t_1t_2\cos(ka)} \tag{14.63}$$

The solution shown in equation (14.63) consists of *two energy bands* (see Figure 14.13):

$$E = \varepsilon + \sqrt{t_1^2 + t_2^2 + 2t_1 t_2 \cos(ka)} \tag{14.64}$$

$$E = \varepsilon - \sqrt{t_1^2 + t_2^2 + 2t_1 t_2 \cos(ka)} \tag{14.65}$$

The minimum (maximum) energy in the higher (lower) energy band given by equation (14.64) (equation (14.65)) occurs at $ka = \pi$. The bandgap, which is defined to be the *minima of upper energy band* minus the *maxima of lower energy band* is (see Figure 14.13)

$$Bandgap = 2\,|t_1 - t_2| \tag{14.66}$$

We now assume that each atom in the 1D string of atoms with two atoms per primitive cell has only one electron per atom. In a manner analogous to the previous case (single atom per primitive cell), we find that when there is only one orbital per atom and one electron per atom, the energy levels are half filled and half empty. That is, the lower band is full, and the upper empty band is empty at zero

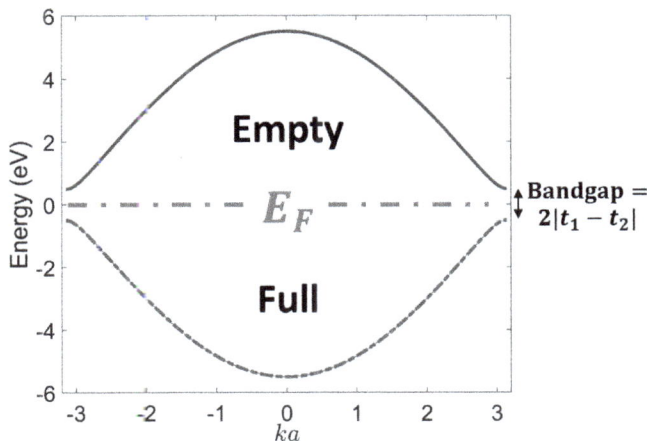

Figure 14.13. Energy band structure for the case with two atoms per primitive cell. The x-axis is wave vector times primitive cell length. We have chosen $\varepsilon = 0$, $t_1 = -3\,\text{eV}$, and $t_2 = -2.5\,\text{eV}$. The Fermi energy's location lies in the bandgap if each atom has only one electron. The reader should convince themselves that when $t_1 = t_2 = -3\,\text{eV}$, the energy eigenvalues match the case of Figure 14.11.

temperature. This corresponds to a semiconductor, where the lower band (full) is the valence band, and the upper band (empty) is the conduction band. It takes an energy of $2|t_1 - t_2|$ to excite an electron from the valence band to the conduction band.

Points to ponder:

- What would happen if each atom had two electrons?
- What happens if the small red atom has two electrons and the big blue atom has one electron?
- What happens when $t_1 = t_2$?

14.3.4   *3D solid (simple orthogonal crystal)*

To appreciate Bloch theorem's usefulness, we generalize the previous discussion to a 3D periodic arrangement of atoms in a cubic crystal/lattice (Figure 14.14). After solving for the energy levels and wave functions, we illustrate the (i) concept of the BZ and (ii) method of representing the band structure using *unique cuts* along the lines of high symmetry within the 3D BZ.

Three integers (n_1, n_2, n_3) define each atom in this orthorhombic lattice corresponding to the (x, y, z) directions, respectively. Further,

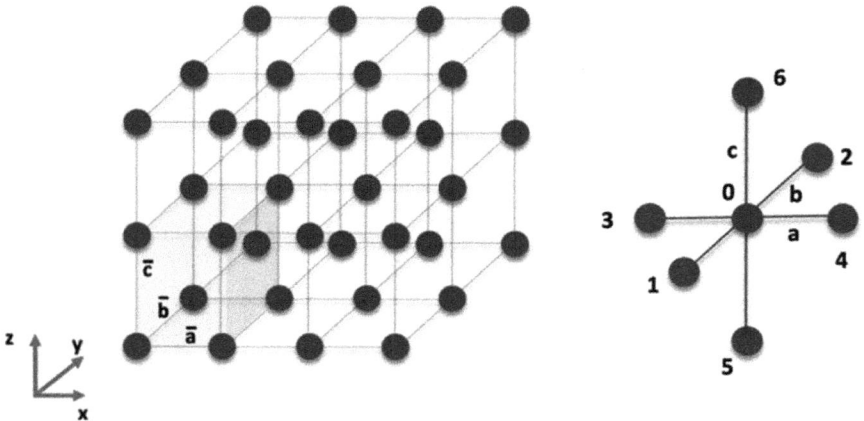

Figure 14.14. (Left) An orthorhombic crystal with orthogonal primitive vectors. Periodicity along x, y, and z directions are $\bar{a}$, $\bar{b}$, and $\bar{c}$, respectively. The circles represent the atoms. (Right) Each atom (numbered 0) interacts only with the six nearest neighbor atoms.

like in the 1D example of Section 14.3.2, we assume that each atom (n_1, n_2, n_3) can interact only with its six nearest neighbors along the x, y, and z directions. The only nonzero elements of H are,

$$\text{Onsite potential: } H\left(n_1, n_2, n_3; n_1, n_2, n_3\right) = \varepsilon \qquad (14.67)$$

Nearest neighbor hopping energy:

$$H\left(n_1, n_2, n_3; n_1 \pm 1, n_2, n_3\right) = H\left(n_1 \pm 1, n_2, n_3; n_1, n_2, n_3\right) = t_x$$
$$(14.68)$$

$$H\left(n_1, n_2, n_3; n_1, n_2 \pm 1, n_3\right) = H\left(n_1, n_2 \pm 1, n_3; n_1, n_2, n_3\right) = t_y$$
$$(14.69)$$

$$H\left(n_1, n_2, n_3; n_1, n_2, n_3 \pm 1\right) = H\left(n_1, n_2, n_3 \pm 1; n_1, n_2, n_3\right) = t_z$$
$$(14.70)$$

The overlap between different orbitals, for simplicity, is assumed to be zero. That is, the only nonzero elements of the overlap matrix are along the diagonal and equal to one,

$$\text{(Overlap) } S\left(n_1, n_2, n_3; n_1, n_2, n_3\right) = 1 \qquad (14.71)$$

Substituting the above nonzero terms into Schrödinger equation $\hat{H}\psi = E\psi$, we get

$$H\left(n_1, n_2, n_3; n_1, n_2, n_3\right) \psi\left(n_1, n_2, n_3\right)$$
$$+ H\left(n_1, n_2, n_3; n_1 + 1, n_2, n_3\right) \psi\left(n_1 + 1, n_2, n_3\right)$$
$$+ H\left(n_1, n_2, n_3; n_1 - 1, n_2, n_3\right) \psi\left(n_1 - 1, n_2, n_3\right)$$
$$+ H\left(n_1, n_2, n_3; n_1, n_2 + 1, n_3\right) \psi\left(n_1, n_2 + 1, n_3\right)$$
$$+ H\left(n_1, n_2, n_3; n_1, n_2 - 1, n_3\right) \psi\left(n_1, n_2 - 1, n_3\right)$$
$$+ H\left(n_1, n_2, n_3; n_1, n_2, n_3 + 1\right) \psi\left(n_1, n_2, n_3 + 1\right)$$
$$+ H\left(n_1, n_2, n_3; n_1, n_2, n_3 - 1\right) \psi\left(n_1, n_2, n_3 - 1\right)$$
$$= E\psi\left(n_1, n_2, n_3\right) \qquad (14.72)$$

Using Bloch's theorem, the wave function can be written as

$$\psi\left(n_1, n_2, n_3\right) = e^{i(k_x n_1 a + k_y n_2 b + k_z n_3 c)} u_{\vec{k}} \qquad (14.73)$$

where $u_{\vec{k}}\left(\vec{r}\right) = u_{\vec{k}}\left(\vec{r} + n_1\vec{a} + n_2\vec{b} + n_3\vec{c}\right)$

Substituting equation (14.73) into equation (14.72), we have

$$\varepsilon\, e^{i(k_x n_1 a + k_y n_2 b + k_z n_3 c)} u_{\bar{k}}$$

$$+ t_x e^{i\left(k_x(n_1+1)a + k_y n_2 b + k_z n_3 c\right)} u_{\bar{k}} + t_x e^{i\left(k_x(n_1-1)a + k_y n_2 b + k_z n_3 c\right)} u_{\bar{k}}$$

$$+ t_y e^{i\left(k_x n_1 a + k_y(n_2+1)b + k_z n_3 c\right)} u_{\bar{k}} + t_y e^{i\left(k_x n_1 a + k_y(n_2-1)b + k_z n_3 c\right)} u_{\bar{k}}$$

$$+ t_z e^{i\left(k_x n_1 a + k_y n_2 b + k_z(n_3+1)c\right)} u_{\bar{k}} + t_z e^{i\left(k_x n_1 a + k_y n_2 b + k_z(n_3-1)c\right)} u_{\bar{k}}$$

$$= E e^{i\left(k_x n_1 a + k_y n_2 b + k_z n_3 c\right)} u_{\bar{k}} \qquad (14.74)$$

Multiplying both sides of the above equation by $e^{-i(k_x n_1 a + k_y n_2 b + k_z n_3 c)}$, we have

$$\varepsilon u_{\bar{k}} + t_x e^{ik_x a} u_{\bar{k}} + t_x e^{-ik_x a} u_{\bar{k}}$$

$$+ t_y e^{ik_y b} u_{\bar{k}} + t_y e^{-ik_y b} u_{\bar{k}}$$

$$+ t_z e^{ik_z c} u_{\bar{k}} + t_z e^{-ik_z c} u_{\bar{k}} = E u_{\bar{k}} \qquad (14.75)$$

Simplifying, we find that the energy dispersion is given by

$$E = \varepsilon + 2t_x \cos(k_x a) + 2t_y \cos(k_y b) + 2t_z \cos(k_z c) \qquad (14.76)$$

The wave function is then

$$\psi(n_1, n_2, n_3) = e^{i\left(k_x n_1 a + k_y n_2 b + k_z n_3 c\right)} u \qquad (14.77)$$

where $u_{\bar{k}}$ is simply a constant u. The values of the three quantum numbers $(k_x k_y k_z)$ are unique as long as they yield unique wave functions (two unique wave functions are orthogonal to each other). Because $e^{ika} = e^{i(k - \frac{2\pi}{a})a}$, the ranges of (k_x, k_y, k_z) that yield unique wave functions are

$$\text{3D BZ}: -\frac{\pi}{a} < k_x < \frac{\pi}{a} \quad -\frac{\pi}{b} < k_y < \frac{\pi}{b} \quad -\frac{\pi}{c} < k_z < \frac{\pi}{c} \qquad (14.78)$$

The volume in k-space (wave vector space) that is limited to this region, which is defined by equation (14.78), contains all the unique quantum numbers (k-points) and is called the Brillouin Zone (BZ).

As the energy dispersion (equation (14.76)) is a function of (k_x, k_y, k_z), it cannot be visualized in 3D k-space unless we use constant energy surfaces, i.e., $E(k_x, k_y, k_z) = const$. Recall that in Chapter 11 we used such a method to visualize the 3D free

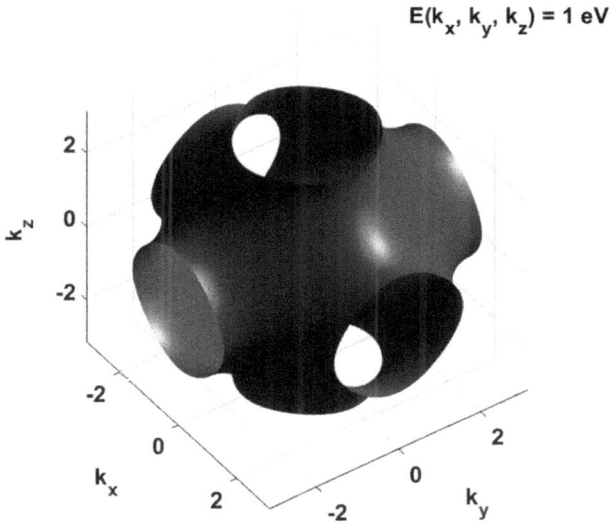

$E(k_x, k_y, k_z) = 1 \text{ eV}$

Figure 14.15. Constant energy contour (surface) corresponding to equation (14.76) for $E(k_x, k_y, k_z) = 1\,\text{eV}$ over the BZ.

electron dispersion (band structure), which led to spherical surfaces of different radii in 3D k-space to represent different constant energy surfaces. Figure 14.15 shows the constant energy contour of $E(k_x, k_y, k_z) = 1\,\text{eV}$ for $t_x = t_y = t_z = -1\,\text{eV}$ and $\varepsilon = 1\,\text{eV}$. $a = b = c = 1$ has been assumed.

The 3D BZ corresponding to the periodic orthorhombic lattice is shown in Figure 14.16. Note that the real space and wave vector space are Fourier conjugates of each other. If the real space period along the z axis (c) is largest among the three primitive vectors, then the corresponding k_z range in momentum space is the smallest $\left(-\frac{\pi}{c} < k_z < \frac{\pi}{c}\right)$ and vice versa. This should be reminiscent of Fourier transform properties for signals in time and frequency domains. If a signal has a large time span, it has a tighter bandwidth in the frequency domain and vice versa.

Visualizing the band structure: It is common to visualize the band structure $E(k)$ along the high symmetry lines within the BZ (e.g., red lines and designated points on them in Figure 14.16). These points reflect the symmetry of the crystal. For example, rotating the

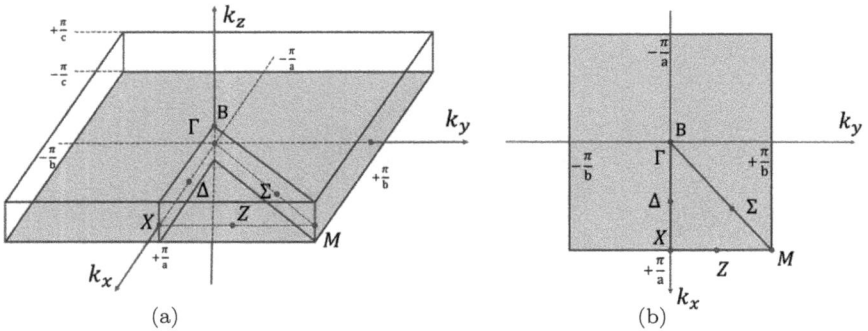

Figure 14.16. (a) 3D BZ for a 3D periodic solid with periodicities of $\bar{a}$, $\bar{b}$, and $\bar{c}$ along x, y, and z directions. (b) Top view of (a) showing the points and directions of high symmetry.

original crystal around the z axis by 90 degrees leaves the crystal unchanged. This is reflected in the band structure by seeing that rotating the BZ around the k_z axis at the gamma (Γ) point leaves the band structure the same because the crystal is symmetric upon a 90-degree rotation.

Example 1. Consider the 3D periodic solid with periodicities of a, b, and c along the x, y, and z directions, respectively. The onsite and hopping energies are chosen to be $\varepsilon = 1\,\mathrm{eV}$ and $t_x = t_y = t_z = -0.5\,\mathrm{eV}$. We will now calculate the expressions for $E(k_x, k_y, k_z)$ along the red lines of the minimum irreducible wedge within the BZ shown in Figure 14.16 using equation (14.76).

Along the Γ–X direction:

$$E\left(k_x, k_y = 0, k_z = 0\right) = 1\,\mathrm{eV} - 1\cos\left(k_x a\right) - 1\cos(0 \cdot b) - 1\cos\left(0 \cdot c\right)$$
$$= -1 - \cos(k_x a)$$

Along the X–M direction:

$$E\left(k_x = \frac{\pi}{a}, k_y, k_z = 0\right) = 1\,\mathrm{eV} - 1\cos\left(\pi\right) - 1\cos\left(k_y b\right) - 1\cos(0 \cdot c)$$
$$= 1 - \cos(k_y b)$$

Along the M–Γ direction (k_x and k_y are related by $k_y = \frac{a}{b}k_x$ along the $M\Gamma$ line):

$$E\left(k_x, k_y, k_z = 0\right) = 1\,\mathrm{eV} - 1\cos\left(k_x a\right) - 1\cos\left(\frac{a}{b}k_x b\right) - 1\cos\left(0\cdot c\right)$$

$$= 0 - 2\cos(k_x a)$$

Along the Γ–B direction (k_x and k_y are 0, k_z changed from 0 to $\frac{\pi}{c}$):

$$E\left(k_x = 0, k_y = 0, k_z\right) = 1\,\mathrm{eV} - 1 - 1 - 1\cos\left(k_z c\right) = -1 - \cos(k_z c)$$

We plot the above 1D functions of (k_x, k_y, k_z) along the four directions within the wedge and juxtapose this information besides each other like a map, as shown in Figure 14.17. This is a common way of visualizing the band structures of a 3D periodic solid because imagining $E(k_x, k_y, k_z)$ needs 4D space.

14.3.5 *Free electron and effective mass*

The band structure for the 3D cubic lattice has a minimum at $k_x = k_y = k_z = 0$. For very small values of $k_x, k_y,$ and k_z, using $\cos\left(\theta\right) \sim 1 - \frac{\theta^2}{2}$ (for small θ), the band structure in equation (14.76) can be approximated by

$$E\left(k_x, k_y, k_z\right) \sim E_1 - t_x k_x^2 a^2 - t_y k_y^2 b^2 - t_z k_z^2 c^2 \tag{14.79}$$

where $E_1 = \varepsilon + 2t_x + 2t_y + 2t_z$ is the minimum energy located at $k_x = k_y = k_z = 0$. The above equation is a parabola. Comparing it with the equation of a free electron (Chapter 11), we can rewrite

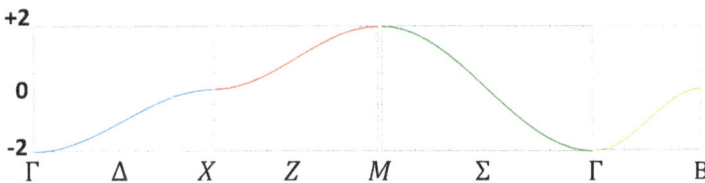

Figure 14.17. Plots of $E\left(k_x, k_y, k_z\right)$ along different high symmetry directions within the irreducible wedge in the BZ.

equation (14.79) as

$$E\left(k_x, k_y, k_z\right) \sim E_1 + \frac{\hbar^2 k_x^2}{2m_{xx}^*} + \frac{\hbar^2 k_y^2}{2m_{yy}^*} + \frac{\hbar^2 k_z^2}{2m_{zz}^*}$$

where

$$m_{xx}^* = \left(\frac{\partial^2 E(k)}{\partial k_x \partial k_x}\right)^{-1} = \left(\frac{\partial^2 E(k)}{\partial^2 k_x}\right)^{-1} = -\frac{\hbar^2}{2t_x a^2}$$

$$m_{yy}^* = \left(\frac{\partial^2 E(k)}{\partial^2 k_y}\right)^{-1} = -\frac{\hbar^2}{2t_y b^2}$$

$$m_{zz}^* = \left(\frac{\partial^2 E(k)}{\partial^2 k_z}\right)^{-1} = -\frac{\hbar^2}{2t_z c^2}$$

are the effective masses in the x, y, and z directions. The effective mass is inversely proportional to the strength of the hopping energy. If the effective mass along a particular direction is small (the hopping energy is large), it is easy for the electron to travel along that direction. Finally, the constant energy surface corresponding to equation (14.79) is an ellipsoid as shown in Figure 14.18.

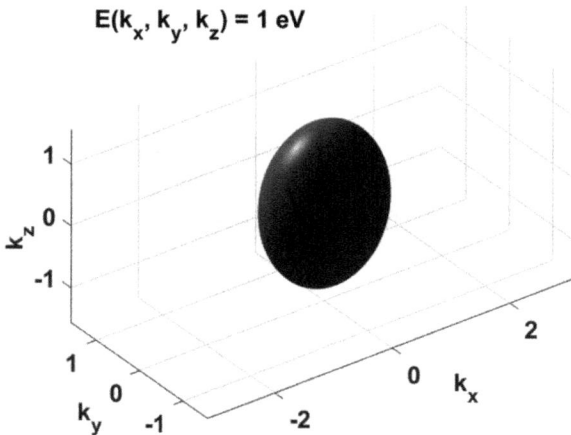

Figure 14.18. Constant energy ellipsoid for $E\left(k_x, k_y, k_z\right) = k_x^2 + 2k_y^2 + 0.6k_z^2$, corresponding to $-t_x a^2 = 1$, $-t_y b^2 = 2$, $-t_z c^2 = 0.6$ and $E_1 = 0$ in equation (14.79).

In equation (14.79), the energy dispersion relation for $E(k)$ only has terms that depend on k_x^2, k_y^2, and k_z^2. In general, there could be terms that depend on $k_x k_y$, $k_y k_z$, and $k_z k_z$. Then the effective mass is a tensor given by

$$
m^* = \begin{bmatrix} m_{xx}^* & m_{xy}^* & m_{xz}^* \\ m_{yx}^* & m_{yy}^* & m_{yz}^* \\ m_{zx}^* & m_{zy}^* & m_{zz}^* \end{bmatrix}
$$

where

$$
m_{xy}^* = \left(\frac{\partial^2 E(\bar{k})}{\partial k_x \partial k_y} \right)^{-1}
$$

For a homogenous band structure where the constant energy surfaces have a spherical shape, the diagonal terms are all equal, and the off-diagonal terms are zero.

For a 1D periodic lattice in Figure 14.10, the band structure is $E = \varepsilon + 2t\cos(ka)$. For negative values of the hopping energy $t = -|t|$, the energy minima for the band structure occurs at $ka = 0$. Expanding about this point, the reader can easily verify that

$$
E = E_C + \left(|t|a^2 \right) k^2 \tag{14.80}
$$

where

$$
E_c = \varepsilon - 2|t| \tag{14.81}
$$

Comparing the free particle expression $E = E_c + \frac{\hbar^2 k^2}{2m^*}$ and equation (14.80), we can identify

$$
\frac{\hbar^2}{2m^*} = |t|a^2 \tag{14.82}
$$

That is, the effective mass m^* is

$$
\frac{1}{m^*} = \frac{2|t|a^2}{\hbar^2} \quad \text{or} \quad m^* = \frac{\hbar^2}{2|t|a^2} \tag{14.83}
$$

This equation shows that the effective mass of an electron at the minimum of the energy band decreases inversely with an increase in the hopping energy t. This has an intuitive consequence in that the velocity of the electron at the bottom of the band $v = \frac{\hbar k}{m}$ increases with an increase in the value of the hopping energy t, which is a measure of the strength of electron hopping/transfer from atom to atom as indicated in the discussion of equation (14.20).

14.4   Graphene and Carbon Nanotube

Graphene consists of a hexagonal lattice of carbon atoms, where each carbon atom has three neighboring carbon atoms (Figure 14.19).

There are two carbon atoms per primitive cell, and the two *primitive vectors* $\bar{a}_1$ and $\bar{a}_2$ are equal in length. The shape of the primitive cell is that of a parallelogram. The *bond vectors* ($\bar{r}_1$, $\bar{r}_2$, and $\bar{r}_3$) have a length of $a_{cc} = 1.42\,\text{Å}$, which is the distance between carbon atoms. The angle between the bond vectors is $120°$ (see Figure 14.20). The primitive and bond vectors are related by,

$$\bar{a}_1 = \bar{r}_1 - \bar{r}_3 \quad \text{and} \quad \bar{a}_2 = \bar{r}_1 - \bar{r}_2 \tag{14.84}$$

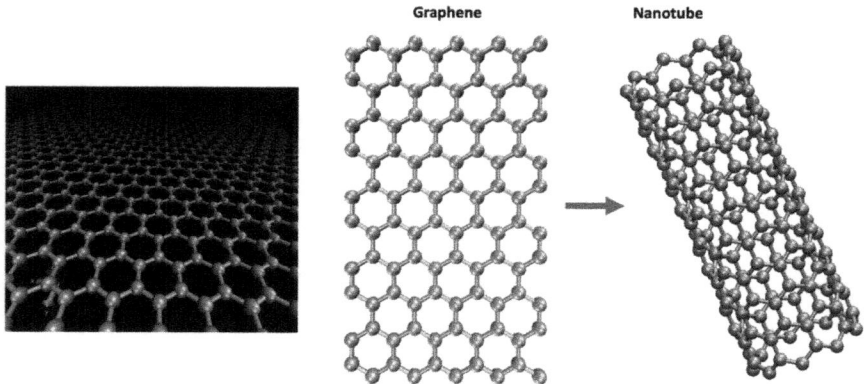

Figure 14.19. (Left) Graphene is a single layer of carbon atoms in a hexagonal lattice (each ball is a carbon atom). (Right) A nanotube is a graphene strip rolled up to form a hollow cylinder with a single layer of carbon atoms such that there are no dangling bonds.

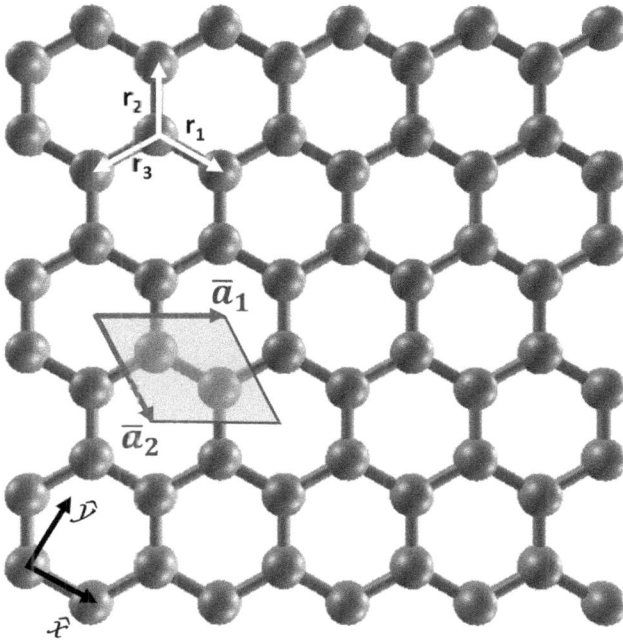

Figure 14.20. The shaded area in blue is a primitive cell of graphene. It contains two carbon atoms. The primitive vectors are $\bar{a}_1$ and $\bar{a}_2$. The bond vectors are $\bar{r}_1$, $\bar{r}_2$, and $\bar{r}_3$. Each carbon atom has three atoms that are equidistant from it with a bond length of 1.42 Å. The angle between the bond vectors is 120°. The x-axis is along $\bar{r}_1$, in this figure.

The lengths of the bond vector (distance between nearest-neighbor carbon atoms) are

$$|\bar{r}_1| = |\bar{r}_2| = |\bar{r}_3| = \frac{a}{\sqrt{3}} \qquad (14.85)$$

where

$$|\bar{a}_1| = |\bar{a}_2| = a = \sqrt{3}a_{cc} \qquad (14.86)$$

Noting that the angle between the bond vectors is 120°, it is easy to see that the bond vectors are given by

$$\bar{r}_1 = \frac{a}{\sqrt{3}}\hat{x}, \quad \bar{r}_2 = -\frac{a}{2\sqrt{3}}\hat{x} + \frac{a}{2}\hat{y}, \quad \text{and} \quad \bar{r}_3 = -\frac{a}{2\sqrt{3}}\hat{x} - \frac{a}{2}\hat{y} \quad (14.87)$$

The primitive vectors given by equation (14.84) are

$$\bar{a}_1 = \frac{\sqrt{3}a}{2}\hat{x} + \frac{a}{2}\hat{y} \quad \text{and} \quad \bar{a}_2 = \frac{\sqrt{3}a}{2}\hat{x} - \frac{a}{2}\hat{y} \qquad (14.88)$$

A carbon atom has six electrons. The $1s$ orbital has two electrons and the $2s$ and $2p$ orbitals share four electrons. Since the energies of the $2s$ and $2p$ orbitals are the same, the electrons do not prefer one unless interactions lift the degeneracy. The electrons in the $1s$, $2s$, and $2p$ orbitals of an isolated carbon atom are

$$\psi_{1s} = \frac{1}{\sqrt{\pi}}\left(\frac{6}{a_o}\right)^{3/2} e^{-6r/a_o} \qquad (14.89)$$

$$\psi_{2s} = \frac{1}{\sqrt{32\pi}}\left(\frac{6}{a_o}\right)^{3/2}\left(2 - \frac{6r}{a_o}\right) e^{-3r/a_o} \qquad (14.90)$$

$$\psi_{2px} = \frac{1}{\sqrt{32\pi}}\left(\frac{6}{a_o}\right)^{3/2}\left(\frac{6r}{a_o}\right) e^{-3r/a_o}\sin\theta\cos\phi \qquad (14.91)$$

$$\psi_{2py} = \frac{1}{\sqrt{32\pi}}\left(\frac{6}{a_o}\right)^{3/2}\left(\frac{6r}{a_o}\right) e^{-3r/a_o}\sin\theta\sin\phi \qquad (14.92)$$

$$\psi_{2pz} = \frac{1}{\sqrt{32\pi}}\left(\frac{6}{a_o}\right)^{3/2}\left(\frac{6r}{a_o}\right) e^{-3r/a_o}\cos\theta \qquad (14.93)$$

where a_o is the Bohr radius. At a distance of 1.42 A° from a carbon atom, the $1s$ wave function is smaller than the $2s$ and $2p$ wave functions. As a result, the $1s$ orbital of a carbon atom does not interact significantly with other carbon atoms. So, in our analysis, only the more diffuse $2s$ and $2p$ orbitals are included. Note that 1.42 A° is the distance between nearest neighbor carbon atoms in graphene.

Hybrid orbitals: The bonds in the plane of a graphene sheet consist of three electrons in $2s$, $2p_x$, and $2p_y$ orbitals on each carbon atom (Figure 14.21). With the covalent bonds between the carbon atoms, it is conventional to create new orbitals ϕ_1, ϕ_2, and

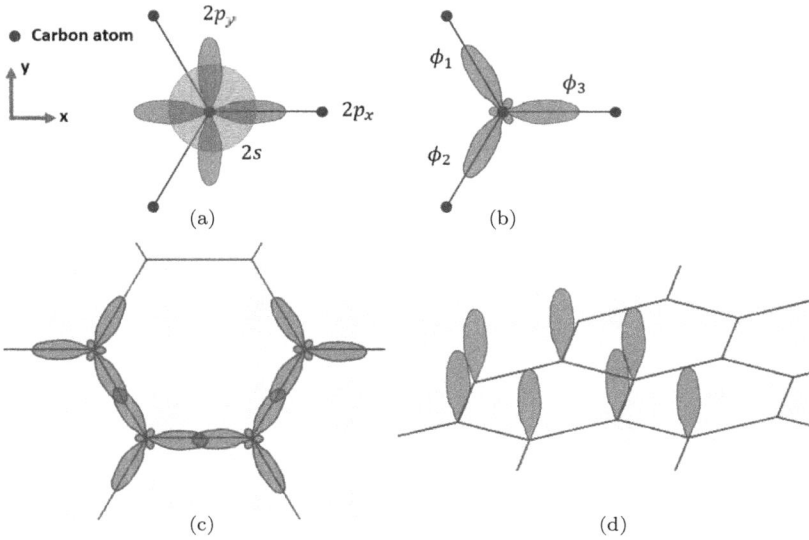

Figure 14.21. Four carbon atoms on the plane of graphene and the orbitals on them. (a) $2s$, $2p_x$, and $2p_y$ orbitals centered on a carbon atom. (b) Three electrons can be considered to be shared by the hybridized orbitals centered on a carbon atom. The fourth electron is in the $2p_z$ orbital perpendicular to the plane of graphene. (c) Hybridized orbitals in the plane of graphene and the (d) $2p_z$ orbitals perpendicular to the plane of graphene.

ϕ_3 pointing between the carbon atoms as shown in Figure 14.21. We can find the transformation between (ϕ_1, ϕ_2, and ϕ_3) and ($2s$, $2p_x$, and $2p_y$) using trigonometry,

$$\psi_{2s} = A_1(\phi_1 + \phi_2 + \phi_3) \tag{14.94}$$

$$\psi_{2p_x} = A_2(-\cos(60°)\,\phi_1 - \cos(60°)\,\phi_2 + \phi_3) \tag{14.95}$$

$$\psi_{2p_y} = A_3(\cos(30°)\,\phi_1 - \cos(30°)\,\phi_2) \tag{14.96}$$

where A_1, A_2 and A_3 are normalization constants. Assuming that ϕ_1, ϕ_2, and ϕ_3 are orthonormal and that $2s$, $2p_x$, and $2p_y$ orbitals are orthonormal give us the value of A_1, A_2 and A_3,

$$A_1 = \frac{1}{\sqrt{3}}, \quad A_2 = \sqrt{\frac{2}{3}} \quad \text{and} \quad A_3 = \sqrt{\frac{2}{3}} \tag{14.97}$$

Substituting equation (14.97) in equations (14.94)–(14.96) gives us

$$
\begin{pmatrix} \psi_{2s} \\ \psi_{2p_x} \\ \psi_{2p_y} \end{pmatrix} = \begin{bmatrix} \frac{1}{\sqrt{3}} & \frac{1}{\sqrt{3}} & \frac{1}{\sqrt{3}} \\ -\frac{1}{\sqrt{6}} & -\frac{1}{\sqrt{6}} & \frac{\sqrt{2}}{\sqrt{3}} \\ \frac{1}{\sqrt{2}} & -\frac{1}{\sqrt{2}} & 0 \end{bmatrix} \begin{pmatrix} \phi_1 \\ \phi_2 \\ \phi_3 \end{pmatrix}
$$

Multiplying the above equation by the inverse of the matrix in the square bracket, we obtain

$$
\begin{pmatrix} \phi_1 \\ \phi_2 \\ \phi_3 \end{pmatrix} = \begin{bmatrix} \frac{1}{\sqrt{3}} & -\frac{1}{\sqrt{6}} & \frac{1}{\sqrt{2}} \\ \frac{1}{\sqrt{3}} & -\frac{1}{\sqrt{6}} & -\frac{1}{\sqrt{2}} \\ \frac{1}{\sqrt{3}} & \frac{\sqrt{2}}{\sqrt{3}} & 0 \end{bmatrix} \begin{pmatrix} \psi_{2s} \\ \psi_{2p_x} \\ \psi_{2p_y} \end{pmatrix}
$$

from which ϕ_1, ϕ_2, and ϕ_3 can be written in terms of $2s$ and $2p$ orbitals,

$$\phi_1 = \frac{1}{\sqrt{3}}\psi_{2s} - \frac{1}{\sqrt{6}}\psi_{2p_x} + \frac{1}{\sqrt{2}}\psi_{2p_y} \qquad (14.98)$$

$$\phi_2 = \frac{1}{\sqrt{3}}\psi_{2s} - \frac{1}{\sqrt{6}}\psi_{2p_x} - \frac{1}{\sqrt{2}}\psi_{2p_y} \qquad (14.99)$$

$$\phi_3 = \frac{1}{\sqrt{3}}\psi_{2s} + \sqrt{\frac{2}{3}}\psi_{2p_x} \qquad (14.100)$$

ϕ_1, ϕ_2, and ϕ_3 are called *hybridized* orbitals, they point between two neighboring carbon atoms. Electrons in the hybridized orbitals of adjacent carbon atoms covalently bond in the plane of graphene. The bonds are known as the sp^2 bonds because they involve $2s$, $2p_x$, and $2p_y$ orbitals. Each carbon atom shares three of its electrons via the ϕ_1, ϕ_2, and ϕ_3 orbitals. The fourth electron of each carbon atom occupies the $2p_z$ orbital, perpendicular to the graphene sheet.

14.4.1   *Electronic properties of graphene*

To calculate the electronic properties, we will assume:

- The hopping energy is nonzero only between a carbon atom and its three nearest neighbors.

- The $2p_z$ orbitals primarily account for the electronic properties around the Fermi energy.
- The overlap matrix S is the identity matrix.

The Hamiltonian of the black primitive cell (D) and the transfer Hamiltonians (like transfer integrals or hopping energies) to its neighbors, the blue unit cells L, R, T and B are (Figure 14.22)

$$H_D = \begin{pmatrix} 0 & t \\ t & 0 \end{pmatrix}, \quad H_R = \begin{pmatrix} 0 & 0 \\ t & 0 \end{pmatrix}, \quad H_L = \begin{pmatrix} 0 & t \\ 0 & 0 \end{pmatrix},$$

$$H_T = \begin{pmatrix} 0 & t \\ 0 & 0 \end{pmatrix} \quad \text{and} \quad H_B = \begin{pmatrix} 0 & 0 \\ t & 0 \end{pmatrix} \tag{14.101}$$

We will now solve the Schrödinger equation, $H\psi = E\psi$. The wave function according to Bloch theorem is

$$\psi = e^{i\bar{k}\cdot\bar{r}} \begin{pmatrix} u_1 \\ u_2 \end{pmatrix} \tag{14.102}$$

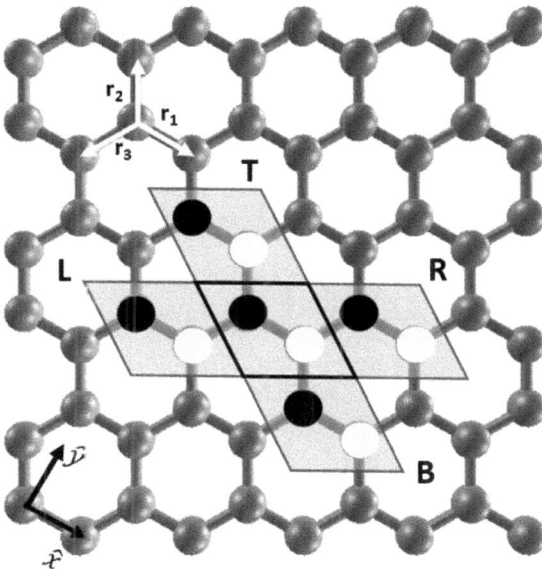

Figure 14.22. The hopping energy is nonzero only between the black primitive cell and its nearest neighbors, the blue primitive cells. The blue primitive cells to the left, right, top, and bottom are referred to as L, R, T, and B in the text. The x-axis is along $\bar{r}_1$, in this figure.

where the periodic parts of the wave function at the two atoms in a primitive cell are u_1 and u_2. $e^{i\bar{k}\cdot\bar{r}}$ is the traveling wave part of the wave function. Noting that the black primitive cell in Figure 14.22 has a nonzero hopping energy with the four blue primitive cells surrounding it, Schrödinger equation gives

$$H_D\psi_D + H_R\psi_R + H_L\psi_L + H_T\psi_T + H_B\psi_B = E\psi_D \qquad (14.103)$$

Substituting equation (14.102) into equation (14.103), we get

$$H_D e^{i\bar{k}\cdot 0}\begin{pmatrix} u_1 \\ u_2 \end{pmatrix} + H_R e^{i\bar{k}\cdot\bar{a}_1}\begin{pmatrix} u_1 \\ u_2 \end{pmatrix} + H_L e^{-i\bar{k}\cdot\bar{a}_1}\begin{pmatrix} u_1 \\ u_2 \end{pmatrix}$$

$$+ H_T e^{-i\bar{k}\cdot\bar{a}_2}\begin{pmatrix} u_1 \\ u_2 \end{pmatrix} + H_B e^{i\bar{k}\cdot\bar{a}_2}\begin{pmatrix} u_1 \\ u_2 \end{pmatrix} = E e^{i\bar{k}\cdot 0}\begin{pmatrix} u_1 \\ u_2 \end{pmatrix} (14.104)$$

Now substituting the various Hamiltonian terms from equation (14.101) in equation (14.104), we get

$$\left[\begin{pmatrix} 0 & t \\ t & 0 \end{pmatrix} + \begin{pmatrix} 0 & 0 \\ t & 0 \end{pmatrix}e^{i\bar{k}\cdot\bar{a}_1} + \begin{pmatrix} 0 & t \\ 0 & 0 \end{pmatrix}e^{-i\bar{k}\cdot\bar{a}_1}\right.$$

$$\left. + \begin{pmatrix} 0 & t \\ 0 & 0 \end{pmatrix}e^{-i\bar{k}\cdot\bar{a}_2} + \begin{pmatrix} 0 & 0 \\ t & 0 \end{pmatrix}e^{i\bar{k}\cdot\bar{a}_2}\right]\begin{pmatrix} u_1 \\ u_2 \end{pmatrix}$$

$$= E\begin{pmatrix} 1 & 0 \\ 0 & 1 \end{pmatrix}\begin{pmatrix} u_1 \\ u_2 \end{pmatrix} \qquad (14.105)$$

$$\left[\begin{matrix} 0 & t + te^{-i\bar{k}\cdot\bar{a}_2} + te^{-\bar{k}\cdot\bar{a}_1} \\ t + te^{i\bar{k}\cdot\bar{a}_2} + te^{i\bar{k}\cdot\bar{a}_1} & 0 \end{matrix}\right]\begin{pmatrix} u_1 \\ u_2 \end{pmatrix}$$

$$= \begin{pmatrix} E & 0 \\ 0 & E \end{pmatrix}\begin{pmatrix} u_1 \\ u_2 \end{pmatrix} \qquad (14.106)$$

$$E = \pm\left|t + te^{i\bar{k}\cdot\bar{a}_2} + te^{i\bar{k}\cdot\bar{a}_1}\right|$$

$$= \pm\left|te^{-i\bar{k}\cdot\bar{r}_1} + te^{-i\bar{k}\cdot\bar{r}_2} + te^{-i\bar{k}\cdot\bar{r}_3}\right| \qquad (14.107)$$

Using the values of the bond vectors $\bar{r}_1$, $\bar{r}_2$, and $\bar{r}_3$ in equation (14.87), equation (14.107) can be rewritten as

$$E = \pm |t| \left[1 + 4\cos^2 \left(\frac{k_y a}{2} \right) + 4\cos \left(\frac{\sqrt{3} k_x a}{2} \right) \cos \left(\frac{k_y a}{2} \right) \right]^{\frac{1}{2}} \quad (14.108)$$

The plot of equation (14.108) as a function of the wave vectors in the x and y directions is shown in Figure 14.23 by assuming $t = -3\,\mathrm{eV}$. The band with $E < 0$ is the valence band. It touches the conduction band $(E > 0)$ at $E = 0$. So the bandgap is zero. Graphene is called a semimetal because the Fermi energy is neither in a bandgap as for semiconductors nor within a band as for metal. The red hexagon in Figure 14.23 is the BZ of graphene which

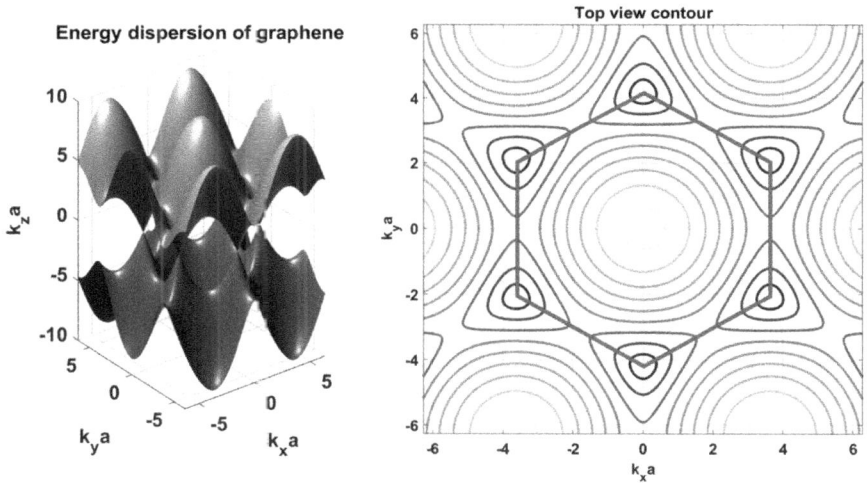

Figure 14.23. (a) $E(k_x, k_y)$, energy dispersion or band structure of graphene. The states with $E < 0$ are filled with electrons and the states with $E > 0$ are empty at zero temperature. The valence band $(E < 0)$ touches the conduction band states $(E > 0)$ at $E = 0$, at six points in the (k_x, k_y) plane, which are called Dirac points (see the cones). Out of these six points, only two correspond to unique quantum numbers. (b) A contour plot of $E(k_x, k_y)$ on the (k_x, k_y) plane. The points inside the red hexagon denote unique values of the quantum numbers (k_x, k_y) and define the BZ. The corners of the red hexagon are where conduction and valence bands touch at the tips of the cones at $E = 0\,\mathrm{eV}$ and they are called Dirac points. Hopping energy of $t = -3\,\mathrm{eV}$ has been assumed.

contains all unique values of the quantum numbers (k_x, k_y). Values of (k_x, k_y) outside the BZ correspond to the same wave function already accounted for inside the BZ.

The in-plane low-field mobility of electrons in graphene is high, with measured values larger than 20,000 cm^2/Vs. In comparison, the mobility of electrons in bulk silicon is 1,500 cm^2/Vs. As a result of the high mobility which translates to high-speed electronic devices, there have been several proposals for building both transistors and interconnects out of graphene. One of the most critical measures of a transistor is the ratio between its ON (operating state) current and OFF (inactive state) current. In an ideal case, the OFF current should be zero, meaning that the material making the transistor has a nonzero bandgap. To make graphene an electronic material for transistors, modifying graphene to have a finite bandgap is valuable. To achieve this, various options are being tried. Examples include forming a narrow graphene strip to induce a bandgap due to confinement, and using bilayer graphene with an electric field perpendicular to the planes to create a bandgap (Figure 14.24).

14.4.2 *Nanotube*

A carbon nanotube is topologically equivalent to rolling a graphene strip into a cylinder, where each carbon atom bonds to its three nearest neighbors. An infinitely long strip orthogonal to the vector

$$\bar{C} = m\bar{a}_1 + n\bar{a}_2 \qquad (14.109)$$

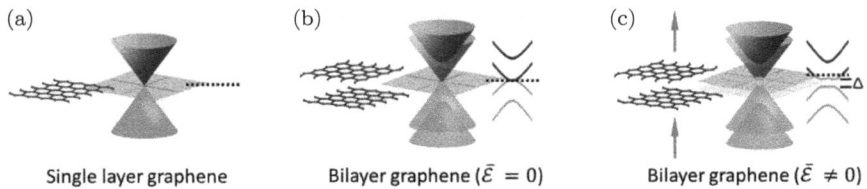

(a) Single layer graphene (b) Bilayer graphene ($\bar{\mathcal{E}} = 0$) (c) Bilayer graphene ($\bar{\mathcal{E}} \neq 0$)

Figure 14.24. Energy band structure of (a) single and (b) bilayer graphene. $\bar{\mathcal{E}}$ is the electric field. (c) A bilayer graphene with a nonzero electric field perpendicular to the graphene sheet has a nonzero bandgap Δ. These cones are similar to those in Figure 14.23(a). Permission to use figures was obtained from Feng Wang of Lawrence Berkeley Laboratory.

creates a nanotube. A nanotube generated by $\bar{C}$ is called an (m, n) nanotube — it is generated by translating m times along $\bar{a}_1$ and n times along $\bar{a}_2$ as shown in Figure 14.25.

Substituting the values of $\bar{a}_1$ and $\bar{a}_2$ from equation (14.88) in equation (14.109), $\bar{C}$ is given by

$$\bar{C} = \frac{\sqrt{3}a}{2}(m + n)\hat{x} + \frac{a}{2}(m - n)\hat{y} \qquad (14.110)$$

The circumference of the nanotube is equal to the length of $\bar{C}$,

$$\text{Circumference} = |\bar{C}| = \sqrt{m^2 + n^2 + mn}a \qquad (14.111)$$

Note that while graphene is an example of a 2D periodic lattice, the nanotube is an example of a 1D periodic lattice. The nanotube's primitive cell is shown in Figure 14.25(a) and (b). The primitive vector is the edge length of the box which, when translated along the direction perpendicular to $\bar{C}$, generates the entire nanotube. For the family of $(n, 0)$ nanotubes known as zigzag nanotubes, using geometry, you can verify that the primitive cell length is $\sqrt{3}a$.

For arbitrary integer values of (m, n), the nanotube is obtained by translating m-times along $\bar{a}_1$ and n-times along $\bar{a}_2$. Side and front views of a $(5,10)$ chiral nanotube of 1 nm length is shown in Figure 14.26(a). While the bonding is precisely sp^2 for graphene, the bonding along the nanotube cylinder is not precisely sp^2, because of the curvature of the nanotube surface. The fourth electron is in the $2p_z$ orbital which is perpendicular to the nanotube surface [see Figure 14.26(b)].

14.4.3 *Electronic properties of nanotubes*

There are two ways to calculate the electronic properties of nanotubes. One method is to apply Bloch's theorem to the 1D structure. Here, we will adopt a second method, which consists of starting with graphene's energy dispersion (equation (14.108)) and using the condition that the wave vector along the circumference must be quantized. As the nanotube is a cylinder, the single valuedness of

(a)

(b)

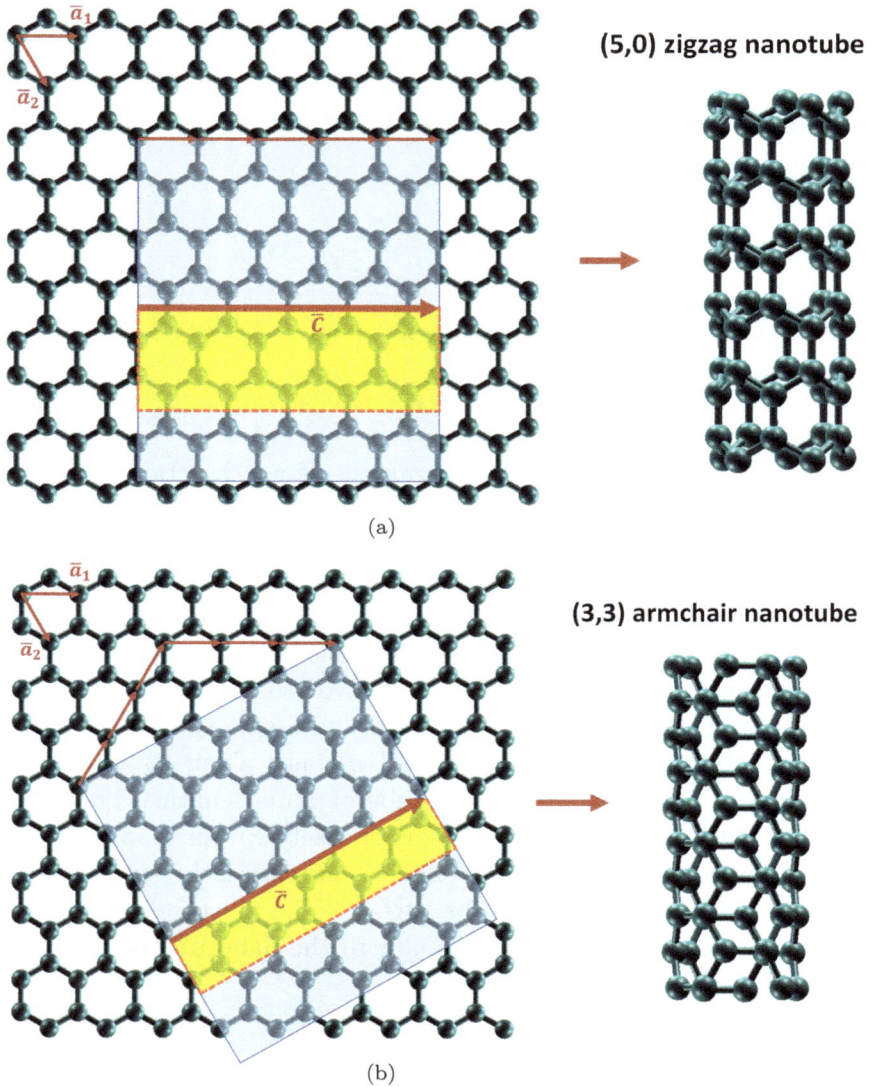

Figure 14.25. (a) A (5,0) zigzag nanotube, which is obtained by translating five times along $\bar{a}_1$ and zero times along $\bar{a}_2$, has 20 atoms per primitive cell. See the big blue rectangular box. The primitive cell length is $\sqrt{3}a$. (b) A (3,3) armchair nanotube, which is obtained by translating three times along $\bar{a}_1$ and three times along $\bar{a}_2$, has 12 atoms per primitive cell. The small yellow rectangular box in (a) and (b) is the 1D primitive cell.

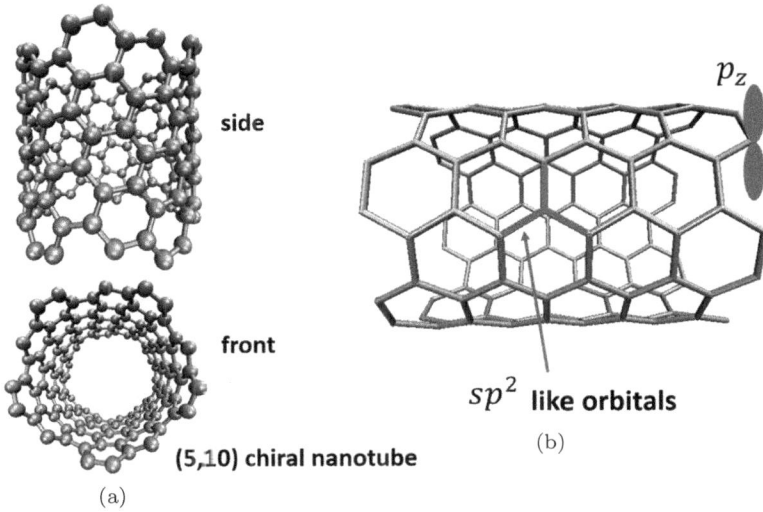

Figure 14.26. (a) An arbitrary (m, n) chiral nanotube is obtained by translating m-times along $\bar{a}_1$ and n-times along $\bar{a}_2$. (b) The bonding along the nanotube cylinder is primarily sp^2 (it is not precisely sp^2 because of the curvature of the nanotube surface; graphene is precisely sp^2). The fourth electron is in the orbital perpendicular to the nanotube cylinder.

the wave function demands

$$\psi (R, z, \phi) = \psi (R, z, \phi + 2\pi) \tag{14.112}$$

where R is the radius of the nanotube, ϕ is the angle in cylindrical coordinates, and z is the position along the length of the nanotube. The single valuedness given by the above equation implies that

$$e^{ik_c R\phi} = e^{ik_c R(\phi + 2\pi)} \tag{14.113}$$

where k_c is the wave vector along the circumference of the nanotube. Equation (14.113) implies that

$$k_c 2\pi R = k_c \left| \bar{C} \right| = 2\pi p$$

$$k_c = \frac{2\pi}{\left| \bar{C} \right|} p \tag{14.114}$$

where p is an integer.

Band structure of a $(N, 0)$ zigzag nanotube: From equation (14.111), we saw that for a $(N, 0)$ zigzag nanotube, $|\bar{C}| = Na$. As the circumferential wave vector k_y is quantized, we substitute

$$k_y = \frac{2\pi}{|\bar{C}|}p = \frac{2\pi}{Na}p \tag{14.115}$$

into equation (14.108) to get the energy dispersion relation of a $(N, 0)$ zigzag nanotube,

$$E = \pm |t| \left[1 + 4\cos^2\left(\frac{p\pi}{N}\right) + 4\cos\left(\frac{\sqrt{3}k_x a}{2}\right)\cos\left(\frac{p\pi}{N}\right)\right]^{\frac{1}{2}} \tag{14.116}$$

k_x is the wave vector along the axis of the nanotube. This 1D band structure is plotted in Figure 14.27 for $(6, 0)$ and $(5, 0)$ carbon nanotubes. Note that this is as if the 2D band structure of Figure 14.23(a) is chopped along $k_y = \frac{2\pi}{Na}p$ lines for different p values. For each value of integer p, there are two bands: one with $E > 0$ and the other with $E < 0$. If the solutions with $E > 0$ and $E < 0$ do not meet, then the nanotube is a semiconductor, but if the bands meet, the nanotube is a metal.

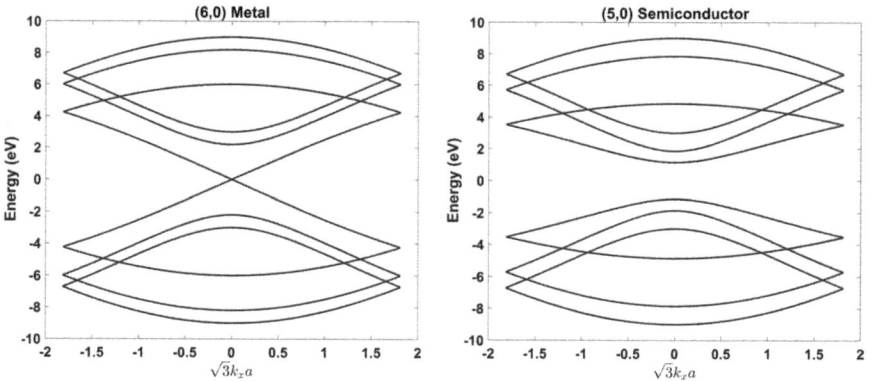

Figure 14.27. Energy band structure of (Left) $(6, 0)$ zigzag nanotube is metallic, and (right) $(5, 0)$ zigzag nanotube is semiconducting. The Fermi energy is at $E = 0$. A hopping energy of $t = -3\,\text{eV}$ has been assumed, the range of the x-axis is $-\pi < \sqrt{3}K_x a < \pi$, and the primitive cell length is $\sqrt{3}a$ for zigzag nanotubes.

Minimizing equation (14.116) with respect to k_x, we find that the minimum (for $E > 0$, equation with a positive sign in equation (14.116)) and maximum (for $E < 0$) of the energy dispersion for every value of p occur at $k_x = 0$. The equation that describes the minimum and maximum is obtained by setting $k_x = 0$ in equation (14.116):

$$\text{Min and Max value of } E = \pm t \left| 1 + 2 \cos \left(\frac{p\pi}{N} \right) \right| \tag{14.117}$$

The bands meet at $E = 0$ only when

$$\cos \left(\frac{p\pi}{N} \right) = -\frac{1}{2} \tag{14.118}$$

This can happen only if

$$\frac{p\pi}{N} = \frac{2\pi}{3} \Rightarrow p = \frac{2N}{3} \tag{14.119}$$

Recall that p must be an integer due to the quantization of the circumferential wave vector (equation (14.114)). So, a $(N, 0)$ zigzag nanotube can be a metal only if N is an integer multiple of 3. On the other hand, if N is not an integer multiple of 3, the bands do not meet at $E = 0$, and the nanotube is a semiconductor.

Band structure of a (N, N) armchair nanotube: From equation (14.111), we see that for a (N, N) zigzag nanotube, $|\bar{C}| = \sqrt{3} N a$. We also note that k_x must be quantized. Then, we substitute

$$k_x = \frac{2\pi}{|\bar{C}|} p = \frac{2\pi}{\sqrt{3} N a} p \tag{14.120}$$

in equation (14.108) to give the energy dispersion relation of a (N, N) armchair nanotube,

$$E = \pm |t| \left[1 + 4 \cos^2 \left(\frac{k_y a}{2} \right) + 4 \cos \left(\frac{p\pi}{N} \right) \cos \left(\frac{k_y a}{2} \right) \right]^{\frac{1}{2}} \tag{14.121}$$

where k_y is the wave vector along the axis of the nanotube. This is equivalent to chopping the 2D band structure in Figure 14.23(a) along $k_x = \frac{2\pi}{\sqrt{3}Na}p$ (p = integer). To obtain, a set of 1D band structure lines as a function of k_y. The minimum (for $E > 0$) and maximum (for $E < 0$) energies occur when $p = N$

$$E = \pm |t| \left| 1 - 2\cos\left(\frac{k_y a}{2}\right) \right| \qquad (14.122)$$

From the above equation, we can immediately see that for any (N, N) nanotube, the bandgap disappears when $k_y a = \pm 2\pi/3$.

14.5   Velocity, Force, and Effective Mass

We will now discuss velocity, the force felt by an electron, and effective mass, using the dispersion relationships discussed in this chapter. Our discussion will closely follow (Blakemore, 1985).

14.5.1   *Velocity*

The group velocity of any wave (electromagnetism, quantum mechanics, sound, etc.) in 1D is found from the dispersion relation by,

$$v_g = \frac{\partial \omega}{\partial k} = \frac{1}{\hbar}\frac{\partial E(k)}{\partial k} \qquad (14.123)$$

The dispersion $E(k)$ gives the relation between energy E or ω as a function of k. In three dimensions, the group velocity is given by the gradient of $E(\bar{k})$,

$$\bar{v}_g = \frac{1}{\hbar}\nabla_k E(\bar{k}) \qquad (14.124)$$

Example 2. Consider electrons in the conduction and valence bands at three different values of k (*zero*, $+|k|$, and $-|k|$) as shown in Figure 14.28. What is the velocity of an electron in the CB and VB at these three different k-points?

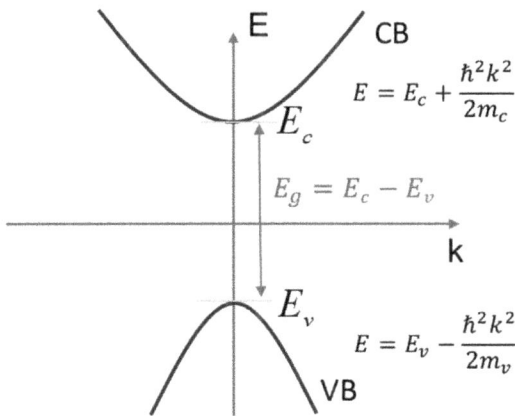

Figure 14.28. Band structure of a semiconductor with bandgap E_g.

Solution: The velocity is calculated and tabulated using equation (14.124) as follows:

k-point	0	$-|k|$	$+|k|$
Conduction band	$v_g = 0$	$v_g = -\frac{\hbar|k|}{m_c} < 0$	$v_g = \frac{\hbar|k|}{m_c} > 0$
Valence band	$v_g = 0$	$v_g = \frac{\hbar|k|}{m_v} > 0$	$v_g = -\frac{\hbar|k|}{m_v} < 0$
Comment	Electron is at rest	Electrons in CB and VB move along $-x$ and $+x$ directions, respectively	Electrons in CB and VB move along $+x$ and $-x$ directions, respectively

Note that an electron with $k < 0$ moves from left to right (in VB) or right to left (in CB) depending on the slope of $\frac{\partial E(k)}{\partial k}$. The direction of the electron wave propagation is determined by the slope of $\frac{\partial E}{\partial k}$ for all wave vectors.

14.5.2 *Force*

Consider the effect of a force ($\bar{F}$) on an electron due to an external field. If the electron traveled distance $d\bar{r}$ in time dt, the gain in energy (dE) is

$$dE = \bar{F} \cdot d\bar{r} \tag{14.125}$$

The rate of change of energy with time is

$$\frac{dE}{dt} = \bar{F} \cdot \frac{d\bar{r}}{dt} \tag{14.126}$$

The velocity of the electron is defined by

$$\bar{v}_g = \frac{d\bar{r}}{dt} \tag{14.127}$$

So

$$\frac{dE}{dt} = \bar{F} \cdot \bar{v}_g \tag{14.128}$$

Using the chain rule in differentiation, $\frac{dE}{dt}$ can also be written as

$$\frac{dE}{dt} = \frac{\partial E}{\partial k} \frac{\partial k}{\partial t} \tag{14.129}$$

Noting from equation (14.123), $v_g = \frac{\partial \omega}{\partial k} = \frac{1}{\hbar} \frac{\partial E(k)}{\partial k}$, we can write equation (14.129) as

$$\frac{dE}{dt} = v_g \hbar \frac{\partial k}{\partial t} \tag{14.130}$$

Comparing equations (14.128) and (14.130), we find that the force is found from the band structure as:

$$\bar{F} = \hbar \frac{d\bar{k}}{dt} \tag{14.131}$$

14.5.3 *Effective mass*

Acceleration ($\bar{a}$) is defined as

$$\bar{a} = \frac{d\bar{v}_g}{dt} \tag{14.132}$$

By using the definition of velocity in equation 14.124, equation 14.132 can be written as,

$$a = \frac{1}{\hbar} \frac{d}{dt} \frac{\partial E}{\partial k} \quad \left(\bar{a} = \frac{1}{\hbar} \frac{d}{dt} \nabla_k E(k) \quad in \quad 3D \right) \tag{14.133}$$

Using equations 14.124 and 14.128, the previous equation can be written as,

$$\bar{a} = \frac{1}{\hbar^2} \nabla_k (\bar{F} \cdot \nabla_k E(\bar{k})) \tag{14.134}$$

Assuming that force is constant, equation 14.134 can be written as,

$$\begin{pmatrix} a_x \\ a_y \\ a_z \end{pmatrix} = \frac{1}{\hbar^2} \begin{pmatrix} \dfrac{\partial^2 E}{\partial k^2 x} & \dfrac{\partial^2 E}{\partial k_x k_y} & \dfrac{\partial^2 E}{\partial k_x k_z} \\[2mm] \dfrac{\partial^2 E}{\partial k_y k_x} & \dfrac{\partial^2 E}{\partial k^2 y} & \dfrac{\partial^2 E}{\partial y k_z} \\[2mm] \dfrac{\partial^2 E}{\partial k_z k_x} & \dfrac{\partial^2 E}{\partial k_z k_y} & \dfrac{\partial^2 E}{\partial k^2 z} \end{pmatrix} \begin{pmatrix} F_x \\ F_y \\ F_z \end{pmatrix} \tag{14.135}$$

Noting that $\bar{F} = m\bar{a}$, we can identify the effective mass to have nine components defined by,

$$m^{-1} = \frac{1}{\hbar^2} \begin{pmatrix} \dfrac{\partial^2 E}{\partial k^2 x} & \dfrac{\partial^2 E}{\partial k_x k_y} & \dfrac{\partial^2 E}{\partial k_x k_z} \\[2mm] \dfrac{\partial^2 E}{\partial k_y k_x} & \dfrac{\partial^2 E}{\partial k^2 y} & \dfrac{\partial^2 E}{\partial y k_z} \\[2mm] \dfrac{\partial^2 E}{\partial k_z k_x} & \dfrac{\partial^2 E}{\partial k_z k_y} & \dfrac{\partial^2 E}{\partial k^2 z} \end{pmatrix} \tag{14.136}$$

From equation 14.135, the x-component of acceleration is,

$$a_x = m_{xx}^{-1} F_x + m_{xy}^{-1} F_y + m_{xz}^{-1} F_z \tag{14.137}$$

where,

$$m_{xx}^{-1} = \frac{1}{\hbar^2} \frac{\partial^2 E}{\partial k^2 x}, \quad m_{xy}^{-1} = \frac{1}{\hbar^2} \frac{\partial^2 E}{\partial k_x k_y}, \quad and \quad m_{xz}^{-1} = \frac{1}{\hbar^2} \frac{\partial^2 E}{\partial k_x k_z}$$

Example 3. Consider an electron in the conduction and valence bands of a semiconductor as shown in Figure 14.28. Using the expression for effective mass, calculate the effective mass in both the conduction and valence bands at $k = 0$. What is the sign of the effective mass in the conduction and valence bands? Note that m_c and m_v are both positive numbers.

Solution: The 1D version of equation (14.136) is

$$m^{-1} = \frac{1}{\hbar^2} \frac{\partial^2 E(k)}{\partial k^2}$$

Using this equation, it is easy to verify that the effective masses in the conduction and valence bands at $k = 0$ are m_c and $-m_v$, respectively. We must calculate the effective mass of an electron in the CB (VB) and show that it is positive (negative). The rest mass of an electron is always positive and a fundamental constant.

Example 4. In a 1D solid consisting of one atomic orbital per primitive cell discussed in Figures 14.10 and Figure 14.11, find the effective mass of an electron at various k-points.

Solution: The $E(k)$ relationship for this 1D solid was derived to be (equation (14.50))

$$E(k) = \varepsilon + 2t \cos(ka)$$

Using equation (14.136), the k-dependent effective mass is,

$$m = -\frac{1}{2ta^2 \cos(ka)}$$

If $t < 0$, the effective mass is,

$$m = \frac{1}{2 |t| a^2 \cos(ka)}$$

Then, the effective mass at $k = 0$ is $\frac{-1}{2ta^2}$ which is a positive number. The effective mass is negative for $|ka| > \frac{\pi}{2}$ which is expected as the dispersion is curved downward. The effective mass is infinite for $ka = \frac{\pi}{2}$.

14.6 Appendix: Proof of Bloch Theorem

To prove Bloch theorem, we consider a 1D case with a primitive vector $\bar{a}$. Then, we will show the steps leading to the generalization in three dimensions. We define the translation operator $\widehat{T}_{n_1\mathbf{a}}$, which acts on a function $\bar{F}(\bar{r})$ and translates it by $n_1\bar{a}$,

$$\widehat{T}_{n_1\mathbf{a}} F(\bar{r}) = F(\bar{r} + n_1\bar{a}) \tag{14.138}$$

Using this definition of the translation operator, we see that the eigenfunctions $\psi(\bar{r})$ obeys

$$\widehat{T}_{n_1\mathbf{a}}\psi(\bar{r}) = \psi(\bar{r} + n_1\bar{a}) \tag{14.139}$$

Schrödinger equation for the periodic solid is no different from that of an aperiodic solid,

$$\widehat{H}(\bar{r})\,\psi(\bar{r}) = E\psi(\bar{r}) \tag{14.140}$$

Now operating the LHS and RHS of the above equation by $T_{n_1\bar{a}}$, we have

$$\widehat{T}_{n_1\bar{a}}\left[H(\bar{r})\,\psi(\bar{r})\right] = \widehat{T}_{n_1\bar{a}}\left[E\psi(\bar{r})\right] \tag{14.141}$$

$$\widehat{T}_{n_1\bar{a}}\left[H(\bar{r})\,\psi(\bar{r})\right] = E\widehat{T}_{n_1\bar{a}}\psi(\bar{r}) \tag{14.142}$$

Also, from equation (14.138), we have that $T_{n_1\bar{a}}\left[H(\bar{r})\,\psi(\bar{r})\right]$ can be written as

$$\widehat{H}(\bar{r} + n_1\bar{a})\,\psi(\bar{r} + n_1\bar{a}) = E\psi(\bar{r} + n_1\bar{a}) \tag{14.143}$$

As the Hamiltonian has the same periodicity as the lattice,

$$\widehat{H}(\bar{r} + n_1\bar{a}) = \widehat{H}(\bar{r}) \tag{14.144}$$

equation (14.143) becomes,

$$\widehat{H}(\bar{r})\,\psi(\bar{r} + n_1\bar{a}) = E\psi(\bar{r} + n_1\bar{a}) \tag{14.145}$$

The above equation can be rewritten using equation (14.139) as

$$\widehat{H}(\bar{r})\,\widehat{T}_{n_1\bar{a}}\psi(\bar{r}) = E\widehat{T}_{n_1\bar{a}}\psi(\bar{r}) \tag{14.146}$$

By comparing equations (14.142) and (14.146), we notice that the Hamiltonian $\widehat{H}$ and the translation operator T commute,

$$\widehat{H}(\bar{r})\widehat{T}_{n_1\bar{a}} = \widehat{T}_{n_1\bar{a}}\widehat{H}(\bar{r}) \tag{14.147}$$

We know from Chapter 5 that a pair of commuting operators can share the same set of eigenfunctions. That is, the eigenfunctions of the Hamiltonian are also the eigenfunctions of the translation operator,

$$\widehat{T}_{n_1\bar{a}}\,\psi(\bar{r}) = \lambda_{n_1\bar{a}}\psi(\bar{r}) \text{ and} \tag{14.148}$$

$$\widehat{H}(\bar{r})\,\psi(\bar{r}) = E\psi(\bar{r}) \tag{14.149}$$

By operating on equation (14.148) with the second translation operator $T_{n_2\mathbf{a}}$, we can see that

$$\widehat{T}_{n_2\bar{a}}\widehat{T}_{n_1\bar{a}}\,\psi(\bar{r}) = \lambda_{n_2\bar{a}}\lambda_{n_1\bar{a}}\psi(\bar{r}) \tag{14.150}$$

$\lambda_{n_1\bar{a}}$ and $\lambda_{n_2\bar{a}}$ are the eigenvalues of the translation operators $\widehat{T}_{n_1\bar{a}}$ and $\widehat{T}_{n_2\bar{a}}$, respectively. Two consecutive translations by $n_1\bar{a}$ and $n_2\bar{a}$ (order does not matter) are equivalent to a single translation by $(n_1+n_2)\bar{a}$

$$\widehat{T}_{n_2\bar{a}}\widehat{T}_{n_1\bar{a}}\,\psi(\bar{r}) = \widehat{T}_{(n_1+n_2)\bar{a}}\psi(\bar{r}) = \lambda_{(n_1+n_2)\bar{a}}\psi(\bar{r}) \tag{14.151}$$

From equations (14.150) and (14.151),

$$\lambda_{(n_1+n_2)\bar{a}} = \lambda_{n_2\bar{a}}\lambda_{n_1\bar{a}} \tag{14.152}$$

Setting $n_2 = -n_1$ (which means reversing the shift by $n_1\bar{a}$), we expect to get $\psi(\bar{r})$ back,

$$\widehat{T}_{-n_1\bar{a}}\widehat{T}_{n_1\bar{a}}\,\psi(\bar{r}) = \lambda_{-n_1\bar{a}}\lambda_{n_1\bar{a}}\psi(\bar{r}) \tag{14.153}$$

But we also know from equation (14.139),

$$\widehat{T}_{-n_1\bar{a}}\widehat{T}_{n_1\bar{a}}\,\psi(\bar{r}) = \widehat{T}_{-n_1\bar{a}}\,\psi(\bar{r} + n_1\bar{a}) \tag{14.154}$$

$$= \psi(\bar{r} + n_1\bar{a} - n_1\bar{a}) \tag{14.155}$$

$$= \psi(\bar{r}) \tag{14.156}$$

So,

$$\lambda_{-n_1\bar{a}}\lambda_{r_{\_}\bar{a}} = 1 \quad \text{or} \quad \lambda_{-n_1\bar{a}} = 1/\lambda_{n_1\bar{a}} \tag{14.157}$$

A function for which equations (14.152) and (14.157) hold good and the eigenvalues $\lambda_{n_1\bar{a}}$ remain finite for any value of n_1 is,

$$\lambda_{n_1\bar{a}} = e^{i\bar{k}\cdot n_1\bar{a}} \tag{14.158}$$

where $\bar{k}$ is a real vector. Substituting the value of $\lambda_{n_1\bar{a}}$ in equation (14.148) and using equation (14.139), we have

$$\psi(\bar{r} + n_1\bar{a}) = e^{i\bar{k}\cdot n_1\bar{a}}\psi(\bar{r}) \tag{14.159}$$

From this, we see that the wave function $\psi(\bar{r})$ is

$$\psi(\bar{r}) = e^{i\bar{k}\cdot\bar{r}}u_{\bar{k}}(\bar{r}) \tag{14.160}$$

where $u_{\bar{k}}(\bar{r}) = u_{\bar{k}}(\bar{r} + n_1\bar{a})$ satisfies the equation.

We have so far proven Bloch theorem only for the 1D case. Establishing it in 3D by including the unit vectors $\bar{b}$ and $\bar{c}$ is straightforward. We will give the main equations to prove the theorem in 3D below:

$$\widehat{T}_{n_1\bar{a}+n_2\bar{b}+n_3\bar{c}}F(\bar{r}) = F(\bar{r} + n_1\bar{a} + n_2\bar{b} + n_3\bar{c}) \tag{14.161}$$

$$H(\bar{r})\,\widehat{T}_{n_1\bar{a}+n_2\bar{b}+n_3\bar{c}} = \widehat{T}_{n_1\bar{a}+n_2\bar{b}+n_3\bar{c}}H(\bar{r}) \tag{14.162}$$

$$\widehat{T}_{n_1\bar{a}-n_2\bar{b}+n_3\bar{c}}\psi(\bar{r}) = \lambda_{n_1\bar{a}+n_2\bar{b}+n_3\bar{c}}\psi(\bar{r}) \tag{14.163}$$

$$\lambda_{(n_1\bar{a}+n_2\bar{b}+n_3\bar{c})+(m_{\_}\bar{a}+m_2\bar{b}+m_3\bar{c})} = \lambda_{n_1\bar{a}+n_2\bar{b}+n_3\bar{c}}\,\lambda_{m_1\bar{a}+m_2\bar{b}+m_3\bar{c}} \tag{14.164}$$

$$\lambda_{-(n_1\bar{a}+n_2\bar{b}+n_3\bar{c})} = 1/\lambda_{n_1\bar{a}+n_2\bar{b}+n_3\bar{c}} \tag{14.165}$$

$$\lambda_{n_1\bar{a}+n_2\bar{b}+n_3\bar{c}} = e^{i\bar{k}\cdot n_1\bar{a}}e^{i\bar{k}\cdot n_2\bar{b}}e^{i\bar{k}\cdot n_3\bar{c}} \tag{14.166}$$

$$\psi(\bar{r} + n_1\bar{a} + n_2\bar{b} + n_3\bar{c}) = e^{i\bar{k}\cdot(n_1\bar{a}+n_2\bar{b}+n_3\bar{c})}\,\psi(\bar{r}) \tag{14.167}$$

$$\psi(\bar{r}) = e^{i\bar{k}\cdot\bar{r}}u_{\bar{k}}(\bar{r}) \tag{14.168}$$

where $u_{\bar{k}}(\bar{r}) = u_{\bar{k}}(\bar{r} + n_1\bar{a}+n_2\bar{b}+n_3\bar{c})$

14.7   Problems

Section 14.1

(1) Let us get to know atomic wave functions better as follows:

  (a) Write down the probability density corresponding to the $1s$ and $2s$ wave functions of the hydrogen atom. Plot the probability densities as a function of radius.

  (b) Draw a contour plot of the probability density associated with the (i) $2p_x$ wave function along the xy and xz planes, (ii) $2p_y$ wave function along the y–z and x–y planes, and (iii) $2p_z$ wave function along the xz and yz planes.

  (c) Plot the wave function of the $2p_x$ wave function along the x-axis. What is the sign of the $2p_x$ wave function for $x > 0$ and $x < 0$?

Section 14.3

(2) For the graphene lattice discussed in Section 14.3, identify a unit cell different from the primitive cell and write down the lattice vectors (not primitive vectors) that span real space.

(3) In Section 14.3.3, assume $t_1 = 1\,\text{eV}$ and $t_2 = 2\,\text{eV}$. Determine if the following cases correspond to a metal or semiconductor:

  (a) Atoms 1 and 2 have one electron.

  (b) Atom 1 has two electrons and atom 2 has one electron.

  (c) Atom 1 has one electron and atom 2 has two electrons.

Draw the band structure and the location of the Fermi energy in each of the above cases. You should not have to perform any numerical or analytical calculations for this problem.

(4) (a) Draw a (3,3) carbon nanotube. Show a primitive cell in your drawing.

  (b) How many atoms are there in a primitive cell of a (3, 3) carbon nanotube?

  (c) How many $E(k)$ dispersion relations are there in the wave vector range $-\frac{\pi}{T} < k < \frac{\pi}{T}$, where T is the primitive cell length along the axis of the nanotube. (You do not have to solve these dispersion relations.)

(5) Definition of (n, m) nanotube: A tube where the circumference is defined on a graphene sheet by a vector corresponding to translation along $\bar{a}$ and $\bar{b}$ by n and m times, respectively.

   (a) Draw the graphene strip that gives you $(4, 4)$, $(5, 1)$, and $(6, 0)$ nanotubes.
   (b) Derive an expression for the band structure of a (n, m) nanotube from the band structure of graphene. Plot the band structure for a $(5, 1)$ nanotube.

Assume that (i) the hopping energy between a carbon atom and its three nearest neighbors is $-3.1\,\text{eV}$, (ii) the onsite potential (diagonal components of the Hamiltonian) for carbon is $0\,\text{eV}$, and (iii) the overlap matrix is the identity matrix.

(6) Apply a uniform axial strain to a $(6, 0)$ nanotube. Find the bandgap as the strain changes from -3% (compression) to 0%–$+3\%$ (stretching). For simplicity, assume the following:

   • Cross-section of the nanotube does not change with strain.
   • The hopping energy for electrons between neighboring p_z orbitals changes as $= t_0(r_0/r)^2$, where r_0 is the equilibrium bond length and r is the bond length in the presence of strain. Neglect the curvature of the nanotube in this problem and assume $r_0 = 1.42\,\text{Å}$ for all bonds, just like in graphene, and $t_0 = -3.1\,\text{eV}$

References

Blakemore, J. S. (1985). *Solid State Physics*, 1st edn. Cambridge University Press.

Sadiku, M. N. O. (2000). *Numerical Techniques in Computational Electromagnetic*, 2nd edn. CRC Press.

Chapter 15

SPIN

Contents

Review of Magnetic Moments

Consider a tiny wire loop carrying the electric current which has a magnetic dipole moment ($\bar{\mu}$) located at $\bar{r}$ in a magnetic field ($\bar{B}(\bar{r})$). The **potential energy** U_E of the loop depends on the angle between its magnetic moment and magnetic field vectors and is given by

$$U_E(\bar{r}) = -\bar{\mu} \cdot \bar{B}(\bar{r}) \tag{A}$$

The potential energy is the lowest if the magnetic moment is aligned parallel to the magnetic field. Otherwise, the loop has higher potential energy, which is at its maximum when the magnetic moment and the field are anti-parallel.

The **force** experienced by the magnetic moment in the magnetic field is given by

$$\bar{F}(\bar{r}) = -\nabla(U_E) = \nabla(\bar{\mu} \cdot \bar{B}(\bar{r})) \tag{B}$$

Note that this force is associated purely with the dipole moment and that the charge of the particle can be zero.

The **torque** experienced by the magnetic moment in the magnetic field is given by

$$\bar{\tau} = \bar{\mu} \times \bar{B} \tag{C}$$

As a result of this torque, the magnetic moment precesses around the magnetic field axis if the magnetic moment and field are not collinear.

Based on equations (A), (B), and (C), it is easy to conclude the following:

1. In a *nonuniform magnetic field*, the force is nonzero and is dependent on the magnitude and direction of the magnetic field at location $\bar{r}$.

2. In a *uniform magnetic field*, while the torque experienced by a magnetic moment is at a maximum when $\bar{\mu}$ and $\bar{B}$ are perpendicular (equation (C)), the force experienced by the magnetic moment is zero (equation (B)).

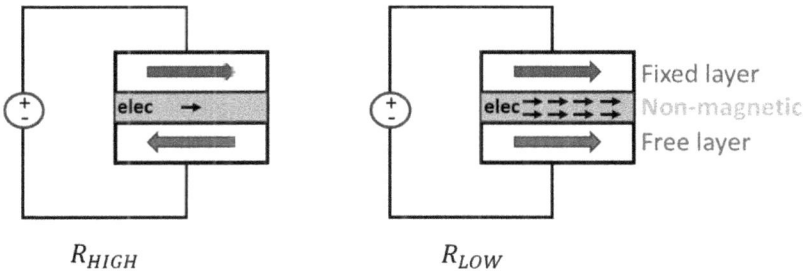

Figure 15.1. Operating principle of magnetic memory.

Electron spin is a quantum mechanical phenomenon that is of both fundamental and applied importance. The fundamental concept of electron spin was first proposed more than a hundred years ago, and its discoverers were awarded a Nobel Prize. Materials where the electron spin is controlled has entered technology. Magnetic memory is a prime example, it uses the giant magneto resistance (GMR) effect shown in Figure 15.1. The fixed and free layers are magnetic materials. When the magnetic polarities of the fixed and free layers are antiparallel, an electron injected from the fixed to the free layer is blocked because the fixed layer injects only electrons with right-pointing spins, but the free layer can primarily accept only electrons that have left-pointing spins. This results in a high resistance. When the magnetic polarities of the fixed and free layers are parallel, electrons can freely travel from the fixed to the free layer. This results in low resistance. These high (R_{HIGH}) and low (R_{LOW}) resistance states form the logic 0 and 1 states.

After qualitatively describing experiments leading to the discovery of spin, we will derive the operator for *spin angular momentum* (usually referred to as *spin*). We will then discuss the impossibility of measuring two different components of spin simultaneously and then proceed to discuss the spin–orbit interaction. One application involving the electron spin discussed is the Datta–Das transistor. We conclude with a brief discussion of fermions and bosons.

15.1   Stern–Gerlach Experiment and the Discovery of Spin

15.1.1   *Angular momentum of a charged particle leads to a magnetic moment*

It is important to review the relationship between magnetic moment and orbital motion of an electron before discussing the quantum mechanical concept of *spin*. The magnetic moment is a vector measure of magnetic strength and orientation. A particle with positive charge Q and mass m_o is moving counterclockwise in a circular loop with radius r and constant uniform angular frequency ω. The resulting current flowing in the counterclockwise direction has a magnitude equal to

$$I = \frac{Q}{T} = \frac{Q\omega}{2\pi} \tag{15.1}$$

where $T = 2\pi/\omega$ is the rotation period. The resulting **magnetic moment** $\bar{\mu}$ due to this circulating charge is a vector pointing up as shown in Figure 15.2(a) with magnitude and direction given by (Knight, 2016)

$$\bar{\mu} = IA\,\hat{k} \tag{15.2}$$

where A is the area of the loop $\hat{k}$ is a unit vector in the z-direction. Substituting for the current and area of the loop, $A = \pi r^2$, we have

$$\bar{\mu} = \frac{Q\omega}{2\pi}\pi r^2\,\hat{k} = \frac{1}{2}Q\omega r^2\,\hat{k} \tag{15.3}$$

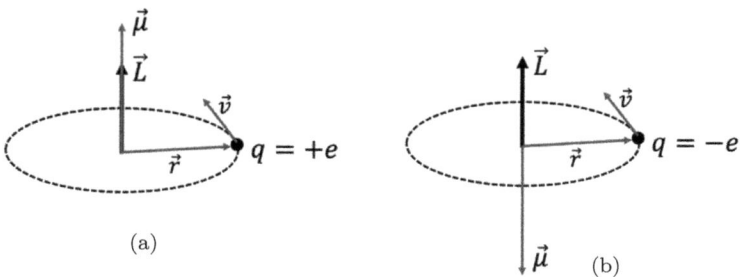

Figure 15.2. Orbital angular momentum and magnetic moment of a positive (a) and a negative (b) charge rotating around a center. Note that for the electron with $Q = -|q|$, the $\vec{L}$ and $\bar{\mu}$ are anti-parallel.

The **orbital angular momentum** $(\bar{L})$ of this positive charge rotating in the counterclockwise direction with speed $v = \omega r$ is a vector parallel to the magnetic moment and is given by

$$\bar{L} = m_o v r \,\hat{k} = m_o \omega r^2 \,\hat{k} \qquad (15.4)$$

Generalizing to a positive or negative charge $\pm Q$, the relationship between the orbital angular momentum and magnetic dipole moment is

$$\bar{\mu} = \frac{Q}{2m_o}\bar{L} = \begin{cases} \dfrac{+|q|}{2m_o}\bar{L} & \text{for positive charge, } Q = +|q| \\[2mm] \dfrac{-|q|}{2m_o}\bar{L} & \begin{array}{l}\text{for a negative charge such as}\\ \text{an electron, } Q = -|q|\end{array} \end{cases} \qquad (15.5)$$

where the magnitude of the charge of an electron is $q = +1.6 \times 10^{-19}$ C. This means that for an electron, the magnetic dipole moment and orbital angular momentum point in opposite directions as shown in Figure 15.2(b). Note that m_o is the electron rest mass. The above equation is true for both classical physics and quantum mechanics.

15.1.2 *Stern–Gerlach experiment*

In 1921, Otto Stern and Walter Gerlach at the University of Hamburg performed an exciting experiment in atomic physics that revealed the existence of electron spin and angular momentum. The simplified experimental setup is shown in Figure 15.3. A beam of neutral silver atoms is evaporated at a high temperature $(T > 2000°C)$ from the bottom part of a vacuum glass tube (note that the velocity-dependent Lorentz force in a magnetic field is zero because the silver atoms are neutral). The atoms are collimated to form a beam that travels through a nonuniform magnetic field and are detected by the screen at the top of the glass tube. The magnetic field is generated by two asymmetric poles, N and S, which are magnetized by coils, as shown in the figure. The resulting magnetic field is nonuniform in the z direction due to the shape of the poles. However, it is still a good approximation to assume that the magnetic field only points in

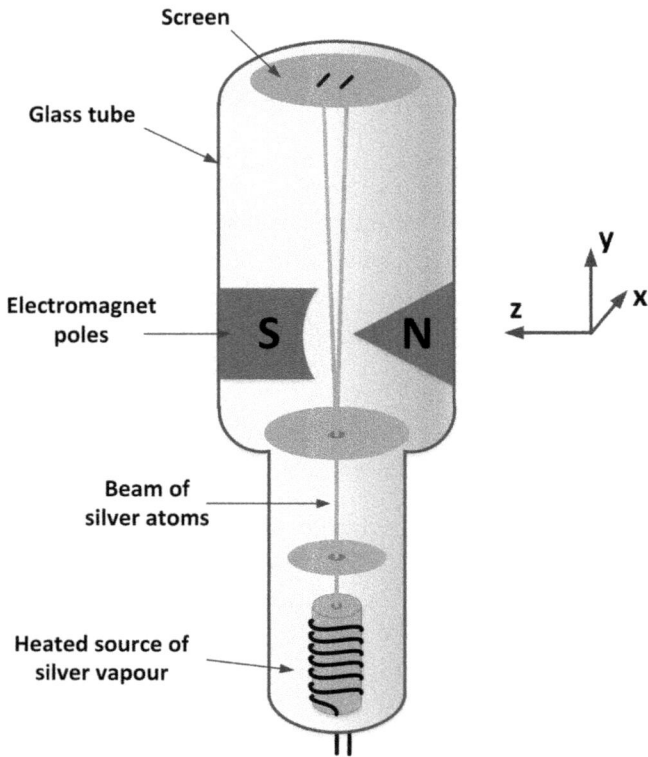

Figure 15.3. The apparatus used by Stern and Gerlach demonstrates the quantization of direction using a nonuniform magnetic field and a beam of vaporized neutral silver atoms.

the z direction in the region between the poles at the location of the beam (see Figure 15.4). The magnetic field is

$$\bar{B}(\text{beam}) \sim B_z(z)\,\hat{k} \tag{15.6}$$

and varies primarily with the z-coordinate. We recall that only in the presence of a nonuniform magnetic field created by the asymmetric magnetic poles will a magnetic moment experience a force given by equation (B),

$$\bar{F} = \nabla(\bar{\mu} \cdot \bar{B}(\bar{r})) \sim \mu_z \frac{\partial B_z}{\partial z}\hat{k} \tag{15.7}$$

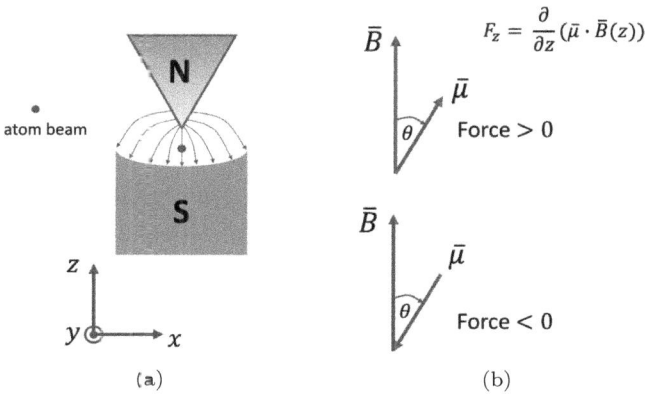

Figure 15.4. (a) Cross-section of the nonuniform magnetic field and (b) the resultant force on the atom beam is dependent on the gradient of the magnetic energy $E = -\bar{\mu} \cdot \bar{B}$. At the location of the beam, the magnetic field is mainly a function of the z-coordinate.

where μ_z and B_z are the z components of the magnetic moment and field respectively. Now, you may argue that since an atom was not biased to have a preferred magnetic moment direction when emitted from the silver source, all magnetic moment directions are equally probable. Hence, the expectation is that atoms arriving at the screen should leave a single wide spot instead of a narrow image due to forces that would exist for randomly oriented spins. Instead, the original beam split into two separate beams upon entering the nonuniform magnetic field, which resulted in two spots on the screen (top section of the glass tube in Figure 15.3). This was unexpected because a neutral silver atom's orbital angular momentum $(\bar{L})$ is zero, and hence its magnetic moment $(\bar{\mu})$ from the orbital angular momentum as per equation (15.5) is zero. Note that silver atoms have 47 electrons, of which 46 electrons fully fill the $1s$, $2s$, $2p$, $3s$, $3p$, $3d$, $4s$, $4p$, $4d$ shells and do not carry a net orbital angular momentum. The 47th electron is in the $5s$ orbital, and so it does not carry an orbital angular momentum either. It has been demonstrated that the splitting of the beam occurs for hydrogen and other atoms as well. Also, the z-axis in the figures does not have any particular significance. One could have chosen any other direction.

The magnetic field in Figure 15.3 can be arranged to be directed along x- or y-axis and it will still result in the splitting of the beam on the screen.

It was then proposed that the magnetic moment arises from another component of angular momentum, or, put differently, the intrinsic angular momentum of an electron (recall the single electron in the last shell of silver atom). In classical objects, angular momentum also arises from spinning around an axis passing through the object's center of mass. Perhaps, it is this *spinning* angular momentum of an electron which give rise to the magnetic moment of silver atoms. While it is tempting to think so, this is incorrect because electrons are regarded to be almost point-like objects, and calculations show that a large enough spin angular momentum cannot be generated by physically allowed velocities of rotation. Nevertheless, it was hypothesized that an electron possesses a nonzero spin angular momentum, whose origin is not based on classical physics. The two distinct spots (on the screen) in the Stern–Gerlach (SG) experiment show that the magnetic moment along the z direction, μ_z, can only have two distinct values of equal magnitude because the two beams diverge equally from the incident beam. These two distinct values are denoted by $+m$ and $-m$.

Let's do a thought experiment. Assume one of the deflected beams (let us take the $+z$ deflected one in Figure 15.3) is sent through another SG apparatus with a magnetic field along the $+z$ (see Figure 15.5(b)). From now on, we will refer to an SG experimental set up with a nonuniform magnetic field along directions x, y and z as SG_x, SG_y and SG_z, respectively. What will happen?

As shown in Figure 15.5(b), the beam still deflects toward $+z$ and exits the SG_z apparatus. Similarly, if we separate the $-z$ deflected beam and send it through SG_z apparatus, it ends up being deflected only towards $-z$ (Figure 15.5(c)). That is, if the nonuniform magnetic field is in the $\pm z$ direction, there will be no further separation of beams. We can then say that $+m$ and $-m$ are the eigenvalues of the magnetic moment operator along the z direction. The eigenvectors are represented by $|\uparrow z\rangle$ and $|\downarrow z\rangle$ (up and down directed magnetic moment). As the $+z$-directed magnetic moments

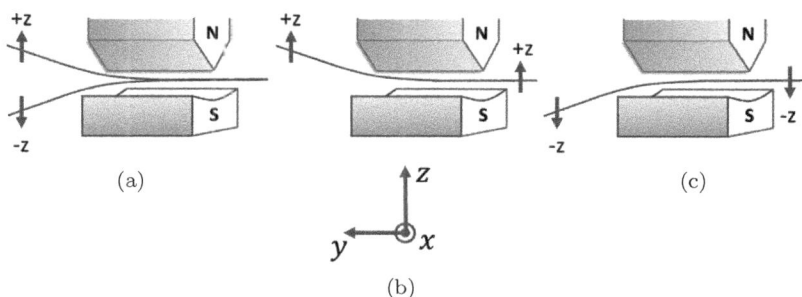

Figure 15.5. (a) A simplified diagram of a SG experiment where the nonuniform magnetic field is along the local z direction. If the $+z$- and $-z$-directed beams are interrogated by another set of SG apparatuses as shown in (b) and (c), the beams no longer split when they pass through the magnet. In (b), the up spin is always deflected toward the $+z$ direction and in (c), the down spin is always deflected toward the $-z$ direction.

remain $+z$-directed when they pass through the SG_z apparatus, we can say that $|\uparrow z\rangle$ and $|\downarrow z\rangle$ must be orthonormal to each other. They are eigenvectors (eigenfunctions) of the magnetic moment operator along the z direction. The SG magnet is modeled as a matrix operator S_z — which will be found later (equation (15.10)).

It is postulated that the same proportionality described in equation (15.5) exists between the magnetic dipole moment and the **spin angular momentum** $(\bar{S})$,

$$\bar{\mu} = -g_s \left(\frac{q}{2m_o} \right) \bar{S}$$

where g_s is a proportionality constant called the *gyromagnetic ratio*, $q = +1.6 \times 10^{-19}$ C and m_o is the mass of the electron. We multiply and divide the above equation by $\hbar$ and change the vector $\bar{S}$ to an operator symbol to obtain the magnetic moment operator,

$$\hat{\mu} = - \left(\frac{g_s \mu_B}{\hbar} \right) \hat{S} \tag{15.8}$$

where

$$\mu_B = \frac{q\hbar}{2m_o}$$

is called the *Bohr magneton*. From the above equation, the magnetic moment operator due to spin along any direction is

$$\hat{\mu}_j = -\left(\frac{g_s \mu_B}{\hbar}\right)\hat{S}_j, \quad \text{where } j = x, y, z \tag{15.9}$$

The experimentally determined eigenvalues of the magnetic moment $\hat{\mu}_z$ (called $\pm m$ above) are $\pm \frac{g_s \mu_B}{2}$, where $g_s = 2.00232$. The eigenvalues of $\hat{S}_z$ are thus $\pm \frac{\hbar}{2}$ to yield the correct value of the magnetic moment. Because $\hat{S}_z$ is proportional to $\hat{\mu}_z$, $| \uparrow z\rangle = \begin{bmatrix} 1 \\ 0 \end{bmatrix}$ and $| \downarrow z\rangle = \begin{bmatrix} 0 \\ 1 \end{bmatrix}$ should also be eigenfunctions of the spin angular momentum operator $\hat{S}_z$.

The existence of electron spin and its associated magnetic dipole moment (μ_s) was proposed by Uhlenbeck and Goudsmith in 1925 to answer questions regarding the splitting of spectral lines for atoms like sodium, which have zero orbital angular momentum. They proposed the idea of spin, i.e., electron rotating around itself. We now know that there is no known classical analog for spin. In 1928, Dirac showed that the spin of the electron is a direct and necessary consequence of solving the Schrödinger equation — by including special relativity, and he found that $g_s = 2$. Further, experimental measurements revealed that $g_s = 2.002319$. This distinction from $g_s = 2$ is due to the interaction of an electron with its own radiation field, which is only calculable with quantum electrodynamics (QED). QED is one of the most successful theories combining electromagnetism and quantum mechanics and explains many phenomena at the atomic level. See (Greiner, 2008) for a detailed derivation.

Following our discussion of operators in bra-ket and matrix forms in Chapter 4, the spin angular momentum operator in bra-ket notation is written as

$$\hat{S}_z = \frac{\hbar}{2}| \uparrow z\rangle\langle \uparrow z| - \frac{\hbar}{2}| \downarrow z\rangle\langle \downarrow z|$$

$$= \frac{\hbar}{2}(| \uparrow z\rangle\langle \uparrow z| - | \downarrow z\rangle\langle \downarrow z|) \tag{15.10}$$

The above operator models the SG_z magnet in Figure 15.5. You can show that $\hat{S}_z|\uparrow z\rangle = \frac{\hbar}{2}|\uparrow z\rangle$ and $\hat{S}_z|\downarrow z\rangle = \frac{\hbar}{2}|\downarrow z\rangle$, which correspond to Figure 15.5(b) and (c).

Now you might ask what happens if we rotate the nonuniform magnetic field to be aligned along the x direction in Figure 15.5 (such an apparatus is referred to as SG_x). The answer is that the beam again splits into two beams along the x direction. It is as if half of the atoms have a magnetic moment along the $+x$ direction, while the other half of the atoms have a magnetic moment along the $-x$ direction. Note that there was no pre-biasing or alignment involved before the beams enter the magnetic field! The measured magnetic moments are again the same as the original experiment with SG_z before, which are $+m$ and $-m$. The magnetic moment operator in the x direction is given by equation (15.9), where the spin angular momentum in the x direction in bra-ket notation is

$$\hat{S}_x = \frac{\hbar}{2}(|\uparrow x\rangle\langle\uparrow x| - |\downarrow x\rangle\langle\downarrow x|) \qquad (15.11)$$

where $\pm\frac{\hbar}{2}$ and $\{|\uparrow x\rangle, |\downarrow x\rangle\}$ are its eigenvalues and eigenvectors.

Finally, assume a new experiment where the beam propagates along the x direction and enters a nonuniform magnetic field pointing along the y direction. Again, the beam splits into two, as if half of the atoms have a magnetic moment along the $+y$ direction, while the other half of the atoms have a magnetic moment along the $-y$ direction. The magnetic moment operator in the y direction is also given by equation (15.9), where the spin angular momentum in the y direction in bra-ket notation is

$$\hat{S}_y = \frac{\hbar}{2}(|\uparrow y\rangle\langle\uparrow y| - |\downarrow y\rangle\langle\downarrow y|) \qquad (15.12)$$

where $\pm\frac{\hbar}{2}$ and $\{|\uparrow y\rangle, |\downarrow y\rangle\}$ are the eigenvalues and eigenvectors.

The basis states $\{|\uparrow x\rangle, |\downarrow x\rangle\}$, $\{|\uparrow y\rangle, |\downarrow y\rangle\}$, and $\{|\uparrow z\rangle, |\downarrow z\rangle\}$ are orthonormal and obey

$$\langle\uparrow x|\uparrow x\rangle = \langle\downarrow x|\downarrow x\rangle = 1 \quad \langle\downarrow x|\uparrow x\rangle = \langle\uparrow x|\downarrow x\rangle = 0$$

$$\langle\uparrow y|\uparrow y\rangle = \langle\downarrow y|\downarrow y\rangle = 1 \quad \langle\downarrow y|\uparrow y\rangle = \langle\uparrow y|\downarrow y\rangle = 0 \qquad (15.13)$$

$$\langle\uparrow z|\uparrow z\rangle = \langle\downarrow z|\downarrow z\rangle = 1 \quad \langle\downarrow z|\uparrow z\rangle = \langle\uparrow z|\downarrow z\rangle = 0$$

We motivated this set of equations based on the experimental observation above.

15.1.3 *Braket and matrix forms of the spin-half operators in the* $\{|\uparrow z\rangle, |\downarrow z\rangle$ *basis*

The spin angular momentum operator in the z direction can be written in matrix form by finding its matrix elements as discussed in Chapter 4,

$$\hat{S}_z = \begin{pmatrix} \langle\uparrow z|\hat{S}_z|\uparrow z\rangle & \langle\uparrow z|\hat{S}_z|\downarrow z\rangle \\ \langle\downarrow z|\hat{S}_z|\uparrow z\rangle & \langle\downarrow z|\hat{S}_z|\downarrow z\rangle \end{pmatrix} \tag{15.14}$$

Using equation (15.10), we can show

$$\hat{S}_z = \frac{\hbar}{2} \begin{pmatrix} 1 & 0 \\ 0 & -1 \end{pmatrix} \tag{15.15}$$

To find the matrix form of the $\hat{S}_x$ operator in the $\{|\uparrow z\rangle, |\downarrow z\rangle$ basis, we would need to find the matrix elements $\langle\uparrow z|\hat{S}_x|\uparrow z\rangle$, $\langle\uparrow z|\hat{S}_x|\downarrow z\rangle$, and so on. We follow the elegant discussion by (Sakurai, 1993). For this, we are again guided by an SG_z experiment but now with a beam of spins directed along $+x$ $(|\uparrow x\rangle)$ incident on the SG apparatus with the nonuniform magnetic field in the z direction. The result of such an experiment is that the beam splits equally into spins oriented along the $+z$ $(|\uparrow z\rangle)$ and $-z$ $(|\downarrow z\rangle)$ directions (see Figure 15.6(a)). Noting that the result of this experiment is the same if a beam directed towards $-x$ $(|\downarrow x\rangle)$ is incident, we can intuitively write the x-directed spins in terms of the z-directed spins,

$$|\uparrow x\rangle = \frac{1}{\sqrt{2}}|\uparrow z\rangle + \frac{e^{i\phi}}{\sqrt{2}}|\downarrow z\rangle \tag{15.16}$$

$$|\downarrow x\rangle = \frac{1}{\sqrt{2}}|\uparrow z\rangle - \frac{e^{i\phi}}{\sqrt{2}}|\downarrow z\rangle \tag{15.17}$$

Note that the $|\uparrow x\rangle$ and $|\downarrow x\rangle$ eigenfunctions are orthogonal by construction, and the magnitudes of the coefficients of the $|\uparrow z\rangle$ and

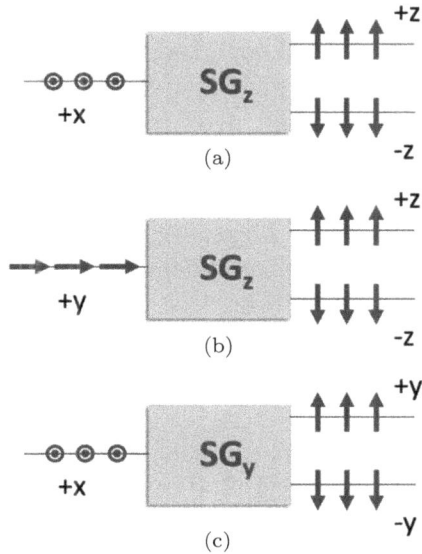

Figure 15.6. (a) Atoms with a $+x$-directed magnetic moment passing through a SG apparatus with the nonuniform field in the z direction split into a $+z$- and $-z$-directed beams. The result is unchanged if the magnetic moment of the beam points in the $-x$ direction. (b) The same statement is true for a beam of atoms with the magnetic moment pointing along the $+y$ or $-y$ direction. (c) Atoms with a $+x$-directed magnetic moment passing through an SG apparatus with the nonuniform field in the y direction split into $+y$- and $-y$-directed beams. The result is unchanged if the magnetic moment of the beam points in the $-x$ direction.

$|\downarrow z\rangle$ terms are equal,

$$|\langle \uparrow x| \uparrow z\rangle| = \frac{1}{\sqrt{2}} \quad \text{and} \quad |\langle \uparrow x| \downarrow z\rangle| = \frac{1}{\sqrt{2}} \qquad (15.18)$$

$$|\langle \downarrow x| \uparrow z\rangle| = \frac{1}{\sqrt{2}} \quad \text{and} \quad |\langle \downarrow x| \downarrow z\rangle| = \frac{1}{\sqrt{2}} \qquad (15.19)$$

The magnitudes of the coefficients of the $|\uparrow z\rangle$ and $|\downarrow z\rangle$ terms are equal because the experiment told us that x-directed spins have 50% chance of getting deflected along the $+z$ ($|\uparrow z\rangle$) and 50% chance of getting deflected along the $-z$ ($|\downarrow z\rangle$) direction. Note that there is a need to have a phase factor of $e^{i\phi}$ in the general case, but it is merely a phase factor and does not impact the 50%–50% values for the probability of measuring along the $|\uparrow z\rangle$ and $|\downarrow z\rangle$ when the spins are incident along $|\uparrow x\rangle$ (or $|\downarrow x\rangle$).

Similarly, to find the matrix form of the $\hat{S}_y$ operator in the $\{|\uparrow z\rangle$, $|\downarrow z\rangle\}$ basis, we would again need to know the matrix elements $\langle\uparrow z|\hat{S}_y|\uparrow z\rangle$, $\langle\uparrow z|\hat{S}_y|\downarrow z\rangle$, and so on. For this, we are inspired by an SG_z experiment with a beam of spins directed along $+y$ ($|\uparrow y\rangle$) incident on an SG apparatus. The result of such an experiment is that the beam splits equally into magnetic moments oriented along the $+z$ ($|\uparrow z\rangle$) and $-z$ ($|\downarrow z\rangle$) directions (see Figure 15.6(b)). Noting that the result of this experiment is unchanged if a $-y$ ($|\downarrow y\rangle$)-directed beam is incident, we can intuitively write the y-directed spins in terms of the z-directed spins,

$$|\uparrow y\rangle = \frac{1}{\sqrt{2}}|\uparrow z\rangle + \frac{e^{i\beta}}{\sqrt{2}}|\downarrow z\rangle \qquad (15.20)$$

$$|\downarrow y\rangle = \frac{1}{\sqrt{2}}|\uparrow z\rangle - \frac{e^{i\beta}}{\sqrt{2}}|\downarrow z\rangle \qquad (15.21)$$

Note again that there is a need to have an arbitrary phase factor of $e^{i\beta}$ in the general case.

Equations (15.16)–(15.21) tell us that $\{|\uparrow x\rangle, |\downarrow x\rangle\}$, $\{|\uparrow y\rangle, |\downarrow y\rangle\}$, and $\{|\uparrow z\rangle, |\downarrow z\rangle\}$ are not orthogonal to each other. That is,

$\langle\uparrow\alpha|\uparrow\alpha\rangle = 1 \quad \langle\uparrow\alpha|\downarrow\alpha\rangle = 0$, where $\alpha = x, y, z$

$\langle\uparrow\alpha|\uparrow\beta\rangle \neq 0 \quad \langle\uparrow\alpha|\downarrow\beta\rangle \neq 0$, where $\alpha \neq \beta$ and $\{\alpha, \beta\} = x, y, z$

To evaluate the value of ϕ and β, we are guided by the results of an experiment where spins oriented along $+x$ ($|\uparrow x\rangle$) go through an SG_y apparatus as shown in Figure 15.6(c). In both experiments, the beams split equally into magnetic moments oriented along the $+y$ ($|\uparrow y\rangle$) and $-y$ ($|\downarrow y\rangle$) directions. So, we reach the conclusion that the magnitude of the overlaps $\langle\uparrow x|\uparrow y\rangle$ and $\langle\uparrow x|\downarrow y\rangle$ should be equal; that is,

$$|\langle\uparrow x|\uparrow y\rangle| = \frac{1}{\sqrt{2}} \qquad (15.22)$$

$$|\langle\uparrow x|\downarrow y\rangle| = \frac{1}{\sqrt{2}} \qquad (15.23)$$

Substituting equations (15.16) and (15.20) into equation (15.22) gives us

$$\langle \uparrow x | \uparrow y \rangle = \left[\frac{1}{\sqrt{2}} \langle \uparrow z | + \frac{e^{-i\phi}}{\sqrt{2}} \langle \downarrow z | \right] \left[\frac{1}{\sqrt{2}} | \uparrow z \rangle + \frac{e^{i\beta}}{\sqrt{2}} | \downarrow z \rangle \right]$$

$$= \frac{1}{2} \langle \uparrow z | \uparrow z \rangle + \frac{e^{i\beta}}{2} \langle \uparrow z | \downarrow z \rangle + \frac{e^{-i\phi}}{2} \langle \downarrow z | \uparrow z \rangle$$

$$+ \frac{e^{i(\beta-\phi)}}{2} \langle \downarrow z | \downarrow z \rangle$$

Using the orthogonality relationships in equation (15.13), we immediately get

$$|\langle \uparrow x | \uparrow y \rangle| = \left| \frac{1}{2} + \frac{e^{i(\beta-\phi)}}{2} \right| = \frac{1}{\sqrt{2}}$$

Similarly, the condition in equation (15.23) gives us

$$|\langle \uparrow x | \downarrow y \rangle| = \left| \frac{1}{2} - \frac{e^{i(\beta-\phi)}}{2} \right| = \frac{1}{\sqrt{2}}$$

The above two conditions can be met only if $\beta - \phi = \pm\frac{\pi}{2}$. We can arbitrarily choose $\phi = 0$, as we can define $| \downarrow z \rangle$ to be $e^{i\phi} | \downarrow z \rangle$ to start with. This leaves us with the condition $\beta = \pm\frac{\pi}{2}$. It turns out that $\beta = +\frac{\pi}{2}$ is the correct solution for a right-handed coordinate system. See below equation (15.45) for further rationale.

Substituting $\beta = +\frac{\pi}{2}$ and $\phi = 0$ in equations (15.16) to (15.21), we have

$$| \uparrow x \rangle = \frac{1}{\sqrt{2}} | \uparrow z \rangle + \frac{1}{\sqrt{2}} | \downarrow z \rangle \qquad (15.24)$$

$$| \downarrow x \rangle = \frac{1}{\sqrt{2}} | \uparrow z \rangle - \frac{1}{\sqrt{2}} | \downarrow z \rangle \qquad (15.25)$$

$$| \uparrow y \rangle = \frac{1}{\sqrt{2}} | \uparrow z \rangle + \frac{i}{\sqrt{2}} | \downarrow z \rangle \qquad (15.26)$$

$$| \downarrow y \rangle = \frac{1}{\sqrt{2}} | \uparrow z \rangle - \frac{i}{\sqrt{2}} | \downarrow z \rangle \qquad (15.27)$$

Now, substituting equations (15.24) and (15.25) in (15.11), we find that the spin angular momentum operator $\hat{S}_x$ is

$$\hat{S}_x = \frac{\hbar}{2}(|\uparrow z\rangle\langle\downarrow z| + |\downarrow z\rangle\langle\uparrow z|) \tag{15.28}$$

We convert this to matrix form using the formalism in Chapter 4,

$$\hat{S}_x = \begin{pmatrix} \langle\uparrow z|\hat{S}_x|\uparrow z\rangle & \langle\uparrow z|\hat{S}_x|\downarrow z\rangle \\ \langle\downarrow z|\hat{S}_x|\uparrow z\rangle & \langle\downarrow z|\hat{S}_x|\downarrow z\rangle \end{pmatrix} \tag{15.29}$$

Using equation (15.28), the matrix elements are found to be

$$\hat{S}_x = \frac{\hbar}{2}\begin{pmatrix} 0 & 1 \\ 1 & 0 \end{pmatrix} \tag{15.30}$$

Similarly substituting equations (15.26) and (15.27) in (15.12), the operator $\hat{S}_y$ is

$$\hat{S}_y = +i\frac{\hbar}{2}(-|\uparrow z\rangle\langle\downarrow z| + |\downarrow z\rangle\langle\uparrow z|) \tag{15.31}$$

We convert this from operator to matrix form using the formalism in Chapter 4 (Section 4.3)

$$\hat{S}_y = \begin{pmatrix} \langle\uparrow z|\hat{S}_y|\uparrow z\rangle & \langle\uparrow z|\hat{S}_y|\downarrow z\rangle \\ \langle\downarrow z|\hat{S}_y|\uparrow z\rangle & \langle\downarrow z|\hat{S}_y|\downarrow z\rangle \end{pmatrix} \tag{15.32}$$

$$\hat{S}_y = \frac{\hbar}{2}\begin{pmatrix} 0 & -i \\ i & 0 \end{pmatrix} \tag{15.33}$$

To summarize, the operators for the magnetic moment discussed above are

$$\hat{S}_x = \frac{\hbar}{2}[|\uparrow z\rangle\langle\downarrow z| + |\downarrow z\rangle\langle\uparrow z|] \tag{15.34}$$

$$\hat{S}_y = \frac{\hbar}{2}[-i|\uparrow z\rangle\langle\downarrow z| + i|\downarrow z\rangle\langle\uparrow z|] \tag{15.35}$$

$$\hat{S}_z = \frac{\hbar}{2}[|\uparrow z\rangle\langle\uparrow z| - |\downarrow z\rangle\langle\downarrow z|] \tag{15.36}$$

Or in the matrix form, they are:

$$\hat{S}_x = \frac{\hbar}{2}\begin{pmatrix} 0 & 1 \\ 1 & 0 \end{pmatrix} \quad \hat{S}_y = \frac{\hbar}{2}\begin{pmatrix} 0 & -i \\ i & 0 \end{pmatrix} \quad \hat{S}_z = \frac{\hbar}{2}\begin{pmatrix} 1 & 0 \\ 0 & -1 \end{pmatrix} \quad (15.37)$$

These operators find extensive use in describing phenomena involving electron spin.

Finally, it is important to note that as the mass of neutrons and protons is 1836 times larger than that of an electron, the magnetic dipole moments of these heavy particles are very small, due to their inherent spin (approximately 1836 times smaller than equation (15.8)). Hence, the name *hyperfine* is assigned to the interactions involving nuclear spins. In this chapter, however, we will only concern ourselves with the *spin angular momentum* of an electron.

15.2 Properties of Spin Angular Momentum Operator

In the previous section, we derived the spin angular momentum operators, which are related to Pauli matrices by

$$\hat{S}_x = \frac{\hbar}{2}\sigma_x, \quad \hat{S}_y = \frac{\hbar}{2}\sigma_y \quad \text{and} \quad \hat{S}_z = \frac{\hbar}{2}\sigma_z \quad (15.38)$$

where the Pauli matrices are defined to be

$$\sigma_x = \begin{bmatrix} 0 & 1 \\ 1 & 0 \end{bmatrix}, \quad \sigma_y = \begin{bmatrix} 0 & -i \\ i & 0 \end{bmatrix}, \quad \text{and} \quad \sigma_z = \begin{bmatrix} 1 & 0 \\ 0 & -1 \end{bmatrix} \quad (15.39)$$

The total Pauli operator is

$$\hat{\sigma} = \sigma_x \hat{j} + \sigma_y \hat{j} + \sigma_z \hat{k} \quad (15.40)$$

where $\hat{i}$ is the unit vector along the x-axis and $\hat{j}$ and $\hat{k}$ are defined similarly. The eigenvalues of all three Pauli matrices are $+1$ and -1, and their eigenvectors are given in Table 15.1.

The following properties of the spin angular momentum follow from the definition of its components in equation (15.38):

(1) The total spin angular momentum $\hat{S}$ is given by

$$\hat{S}^2 = \hat{S}_x^2 + \hat{S}_y^2 + \hat{S}_z^2 \quad (15.41)$$

Table 15.1. Eigenvalues and eigenvectors of the Pauli matrices.

	Eigenvalues	
	1	**−1**
Pauli Matrix	**Eigenvectors**	**Eigenvectors**
$\hat{\sigma}_x = \dfrac{\hbar}{2}\begin{pmatrix} 0 & 1 \\ 1 & 0 \end{pmatrix}$	$\lvert \uparrow x\rangle = \dfrac{1}{\sqrt{2}}\begin{bmatrix} 1 \\ 1 \end{bmatrix}$	$\lvert \downarrow x\rangle = \dfrac{1}{\sqrt{2}}\begin{bmatrix} 1 \\ -1 \end{bmatrix}$
$\hat{\sigma}_y = \begin{pmatrix} 0 & -i \\ i & 0 \end{pmatrix}$	$\lvert \uparrow y\rangle = \dfrac{1}{\sqrt{2}}\begin{bmatrix} 1 \\ -i \end{bmatrix}$	$\lvert \downarrow y\rangle = \dfrac{1}{\sqrt{2}}\begin{bmatrix} i \\ -1 \end{bmatrix}$
$\hat{\sigma}_z = \begin{pmatrix} 1 & 0 \\ 0 & -1 \end{pmatrix}$	$\lvert \uparrow z\rangle = \begin{bmatrix} 1 \\ 0 \end{bmatrix}$	$\lvert \downarrow z\rangle = \begin{bmatrix} 0 \\ 1 \end{bmatrix}$

(2) The **eigenvalues** of $\hat{S}_x, \hat{S}_y$, and $\hat{S}_z$ are $\pm\frac{\hbar}{2}$.   (15.42)

(3) As a result of equations (15.41) and (15.42), the magnitude of total spin angular momentum of an electron is

$$|S|^2 = |S_x|^2 + |S_y|^2 + |S_z|^2 = \frac{\hbar^2}{4} + \frac{\hbar^2}{4} + \frac{\hbar^2}{4} = \frac{3\hbar^2}{4}$$

$$|S| = \frac{\sqrt{3}\hbar}{2} \tag{15.43}$$

This shows that the total spin angular momentum is always larger than its eigenvalues in any one given direction, which as per equation (15.42), can be only $\pm\frac{\hbar}{2}$. Figure 15.7 illustrates that the S_z component of spin is $\frac{\hbar}{2}$, while the S_x and S_y components are nonzero and account for the remaining spin angular momentum.

(4) By multiplying out the Pauli matrices and spin operators defined in equations (15.38) and (15.39), it can be verified that the following commutator relationships hold

$$[\hat{S}_x, \hat{S}_y] = i\hbar\hat{S}_z, \quad [\hat{S}_y, \hat{S}_z] = i\hbar\hat{S}_x, \quad [\hat{S}_z, \hat{S}_x] = i\hbar\hat{S}_y \tag{15.44}$$

$$[\sigma_x, \sigma_y] = i\hbar\sigma_z, \quad [\sigma_y, \sigma_z] = i\hbar\sigma_x, \quad [\sigma_z, \sigma_x] = i\hbar\sigma_y \tag{15.45}$$

If we chose $\beta = -\frac{\pi}{2}$ in the derivation above equation (15.24), the commutation relation in equation (15.45) would have become $[\hat{S}_x, \hat{S}_y] = -i\hbar\hat{S}_z$. But we know from Chapter 16 that the components

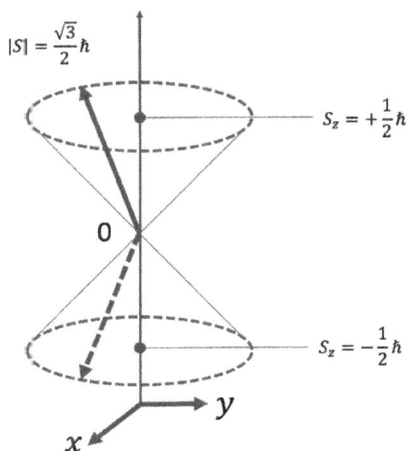

Figure 15.7. The total electron spin has a magnitude of $|S| = \frac{\sqrt{3}\hbar}{2}$ while the maximum magnitude of the spin along any direction can only be $\frac{\hbar}{2}$.

of orbital angular momentum follow the commutation relation with the plus sign on the right-hand side, that is, $[\hat{L}_x, \hat{L}_y] = i\hbar\hat{L}_z$ for a right-handed coordinate system. For a left-handed coordinate system, one would get $[\hat{L}_x, \hat{L}_y] = -i\hbar\hat{L}_z$. In analogy with the case of the orbital angular momentum, we argue that one should choose $\beta = \frac{\pi}{2}$ in the derivation above equation (15.24).

15.2.1 *Simultaneous measurement of spin components*

We know from the previous section that the total spin angular momentum of an electron is $\frac{\sqrt{3}}{2}\hbar$, but its projection along any axis can only have a magnitude of $\frac{1}{2}\hbar$. We now ask if two components of the spin angular momentum can be measured to arbitrary accuracy simultaneously. To answer this question, we rely on the generalized uncertainty relationship discussed in Chapter 5, which helps us determine the extent to which two quantities can be precisely measured. The statement of the generalized uncertainty relationship is

$$\langle(\Delta\hat{A})^2(\Delta\hat{B})^2\rangle \geq \frac{1}{4}|\langle\hat{A}\hat{B} - \hat{B}\hat{A}\rangle|^2 \tag{15.46}$$

Identifying $\hat{A} = \hat{S}_x$ and $\hat{B} = \hat{S}_y$, we have

$$\Delta \hat{S}_x \Delta \hat{S}_y \geq \sqrt{\frac{1}{4}|\langle \hat{S}_x \hat{S}_y - \hat{S}_y \hat{S}_x \rangle|^2} \tag{15.47}$$

Substituting $\hat{S}_x \hat{S}_y - \hat{S}_y \hat{S}_x = i\hbar \hat{S}_z$ from equation (15.44) into equation (15.47), we have

$$\Delta \hat{S}_x \Delta \hat{S}_y \geq \sqrt{\frac{1}{4}|i\hbar \langle \hat{S}_z \rangle|^2}$$

$$\Delta \hat{S}_x \Delta \hat{S}_y \geq \sqrt{\frac{1}{16}\hbar^4}$$

$$\Delta \hat{S}_x \Delta \hat{S}_y \geq \frac{\hbar^2}{4} \tag{15.48}$$

The interpretation of equation (15.48) is that the x- and y-components of spin cannot be determined to arbitrary accuracy in a measurement. This can be verified more generally for any two components of the spin angular momentum,

$$\Delta \hat{S}_i \Delta \hat{S}_j \geq \frac{\hbar^2}{4} \quad (\text{where } i \neq j; \ i, j \in x, y, z)$$

When a magnetic field is applied along the z direction, while the z-component of the spin has a magnitude of $\frac{\hbar}{2}$, the x- and y-components each individually result in an average measurement of zero. The magnitude of the projection of spin on the xy plane is $\sqrt{\frac{3\hbar^2}{4} - \frac{1\hbar^2}{4}} = \frac{\hbar}{\sqrt{2}}$. Below, we show the expectation value of S_x is zero for a spin in the $|\uparrow z\rangle$ eigenfunction:

$$\langle S_x \rangle = \langle \uparrow z | \hat{S}_x | \uparrow z \rangle$$

$$= \langle \uparrow z | \frac{\hbar}{2}[|\uparrow z\rangle\langle \downarrow z| + |\downarrow z\rangle\langle \uparrow z|] | \uparrow z \rangle$$

$$= \frac{\hbar}{2}[\langle \downarrow z | \uparrow z \rangle + \langle \uparrow z | \downarrow z \rangle] = 0 \tag{15.49}$$

We will leave it to the reader to show that the expectation value of S_y is zero.

15.3 Schrödinger Equation with Spin

The time-independent Schrödinger equation for an electron in the absence of an interaction with a magnetic field is

$$-i\hbar\frac{d\Psi}{dt} = \left[-\frac{\hbar^2}{2m_o}\nabla^2 + U(r)\right]\Psi \tag{15.50}$$

In the presence of a magnetic field, the Hamiltonian is affected in two ways as follows:

- inclusion of an additional energy term due to the interaction of the magnetic moment with the magnetic field,
- change of the kinetic energy operator.

From electricity and magnetism, we know that the magnetic moment's interaction with magnetic field results in the energy of (equation (A))

$$E = -\bar{\mu} \cdot \bar{B} \tag{15.51}$$

We now know that the magnetic moment of an electron and its spin operator are related by equation (15.8). Substituting equation (15.8) in (15.51), we have that the energy operator standing for the interaction is equal to

$$\hat{H}_{\text{magnetic moment}} = -\bar{\mu} \cdot \bar{B} = -\frac{g_s\mu_B}{\hbar}\hat{S} \cdot \bar{B} \tag{15.52}$$

Using equation (15.38), the Hamiltonian can also be written as

$$\hat{H}_{\text{magnetic moment}} = -\frac{g_s\mu_B}{2}\hat{\sigma} \cdot \bar{B} \tag{15.53}$$

where

$$\hat{\sigma} = \hat{i}\hat{\sigma}_x + \hat{j}\hat{\sigma}_y + \hat{k}\hat{\sigma}_z \tag{15.54}$$

When all components of the magnetic field are nonzero, equation (15.53) in matrix form becomes

$$\hat{H}_{\text{magnetic moment}} = -\frac{g_s\mu_B}{2}(\sigma_x \cdot B_x + \sigma_y \cdot B_y + \sigma_z \cdot B_z)$$

After substituting the Pauli matrices in the above, we get

$$\hat{H}_{\text{magnetic moment}} = -\frac{g_s\mu_B}{2}\begin{bmatrix} B_z & B_x - iB_y \\ B_x + iB_y & -B_z \end{bmatrix} \tag{15.55}$$

In the above matrix, the off-diagonal terms are called the *transversal field* or *pulse*, and the diagonal terms (B_z) are called *longitudinal field* or *pulse*. This terminology is common in the study of quantum dots, superconducting qubits, and magnetic resonance imaging (MRI). In the special case when $B_x = B_y = 0$, the above Hamiltonian takes the simpler form,

$$\hat{H}_{\text{magnetic moment}} = -\frac{g_s \mu_B}{2} \sigma_z \cdot B_z \quad \text{which has energy eigenvalues}$$

$$E = \pm \frac{g_s \mu_B}{2} B_z \tag{15.56}$$

The second way the Hamiltonian of an electron changes due to interaction with a magnetic field arises from the change in the momentum operator. In a magnetic field, the momentum operator changes to

$$\hat{p} = \begin{cases} \hat{p} - q\bar{A} & \text{(for a positive charge } q) \\ \hat{p} + q\bar{A} & \text{(for a negative charge } - q) \end{cases} \tag{15.57}$$

where $\bar{A}$ is the magnetic vector potential satisfying $\bar{B} = \nabla \times \bar{A}$. We remind the reader that $q = +1.6 \times 10^{-19} \, \text{C}$ throughout this book. The origin of equation (15.57) lies in classical mechanics. Readers can consult with Appendix G of (Singleton, 2001) for a good discussion of this topic. Note that the magnetic vector potential $(\bar{A})$ is indeed momentum per unit charge, hence $q\bar{A}$ is added to the particle momentum $m\bar{v}$, to conserve the total momentum. Regarding the importance of magnetic vector potential in engineering, see Section 15.8.

Adding the new energy term due to the magnetic moment discussed in equation (15.53) and substituting for the modified momentum operator in the presence of a magnetic field (equation (15.57)), the modified Schrödinger equation for an electron becomes a 2×2 matrix equation,

$$-i\hbar \frac{\partial}{\partial t} \Psi = \left\{ \left[\frac{(\hat{p} + q\bar{A})^2}{2m_o} + U(\bar{r}, t) \right] \otimes \hat{I}_{2 \times 2} - \frac{g_s \mu_B}{2} \hat{\sigma} \cdot \bar{B} \right\} \tag{15.58}$$

Ψ is now a 2×1 vector,

$$\Psi = \begin{bmatrix} \psi_1 \\ \psi_2 \end{bmatrix} \tag{15.59}$$

When neither the magnetic field nor the electrostatic potential are time-dependent, the time-independent form of Schrödinger equation is derived from equation (15.58) to yield the eigenvalue problem,

$$\left\{ \left[\frac{(\hat{p} + q\bar{A})^2}{2m_o} + U(\bar{r}) \right] \otimes \hat{I}_{2\times2} - \frac{g_s \mu_B}{2} \hat{\sigma} \cdot \bar{B} \right\} \Psi = E\Psi \tag{15.60}$$

We can rewrite equation (15.60) in matrix form as

$$\begin{bmatrix} \dfrac{(\hat{p} + q\bar{A})^2}{2m_o} + U(\bar{r}) - \dfrac{g_s \mu_B}{2} B_z & -\dfrac{g_s \mu_B}{2}(B_x - iB_y) \\ -\dfrac{g_s \mu_B}{2}(B_x + iB_y) & \dfrac{(\hat{p} + q\bar{A})^2}{2m_o} + U(\bar{r}) + \dfrac{g_s \mu_B}{2} B_z \end{bmatrix} \begin{bmatrix} \psi_1 \\ \psi_2 \end{bmatrix}$$

$$= E \begin{bmatrix} \psi_1 \\ \psi_2 \end{bmatrix} \tag{15.61}$$

The above matrix consists of two equations that couple the wave function components ψ_1 and ψ_2:

$$\left[\frac{(\hat{p} + q\bar{A})^2}{2m_o} + U(\bar{r}) - \frac{g_s \mu_B}{2} B_z \right] \psi_1 - \frac{g_s \mu_B}{2}(B_x - iB_y)\psi_2 = E\psi_1$$

$$-\frac{g_s \mu_B}{2}(B_x + iB_y)\psi_1 + \left[\frac{(\hat{p} + q\bar{A})^2}{2m_o} + U(\bar{r}) + \frac{g_s \mu_B}{2} B_z \right] \psi_2 = E\psi_2$$

If the magnetic field is applied only in the z direction ($B_x = B_y = 0$), the above equations become decoupled,

$$\begin{bmatrix} \dfrac{(\hat{p} + q\bar{A})^2}{2m_o} + U(\bar{r}) - \dfrac{g_s \mu_B}{2} B_z & 0 \\ 0 & \dfrac{(\hat{p} + q\bar{A})^2}{2m_o} + U(\bar{r}) + \dfrac{g_s \mu_B}{2} B_z \end{bmatrix} \begin{bmatrix} \psi_1 \\ \psi_2 \end{bmatrix} = E \begin{bmatrix} \psi_1 \\ \psi_2 \end{bmatrix}$$

$$\tag{15.62}$$

which can be rewritten as the following separable equations for ψ_1 and ψ_2:

$$\left[\frac{(\hat{p} + q\bar{A})^2}{2m_o} + U(\bar{r}) - \frac{g_s \mu_B}{2} B_z\right] \psi_1 = E\psi_1 \qquad (15.63)$$

$$\left[\frac{(\hat{p} + q\bar{A})^2}{2m_o} + U(\bar{r}) + \frac{g_s \mu_B}{2} B_z\right] \psi_2 = E\psi_2 \qquad (15.64)$$

Special case: There are problems where the influence of the vector potential is small. We can neglect the vector potential term, and, the time-independent Schrödinger equation can be written as

$$\begin{bmatrix} \dfrac{\hat{p}^2}{2m_o} + U(\bar{r}) - \dfrac{g_s \mu_B}{2} B_z & -\dfrac{g_s \mu_B}{2}(B_x - iB_y) \\[2ex] -\dfrac{g_s \mu_B}{2}(B_x + iB_y) & \dfrac{\hat{p}^2}{2m_o} + U(\bar{r}) + \dfrac{g_s \mu_B}{2} B_z \end{bmatrix} \begin{bmatrix} \psi_1 \\ \psi_2 \end{bmatrix} = E \begin{bmatrix} \psi_1 \\ \psi_2 \end{bmatrix}$$

$$(15.65)$$

If the magnetic field is uniform in space and oriented only along the z direction, the solution to the above eigenvalue equation is obtained by solving

$$\left[\frac{\hat{p}^2}{2m_o} + U(\bar{r})\right] \psi_1 = \left(E + \frac{g_s \mu_B}{2} B_z\right) \psi_1 = E'\psi_1 \qquad (15.66)$$

$$\left[\frac{\hat{p}^2}{2m_o} + U(\bar{r})\right] \psi_2 = \left(E - \frac{g_s \mu_B}{2} B_z\right) \psi_2 = E''\psi_2 \qquad (15.67)$$

for up- and down-spin electrons respectively.

The spatial part of the wave functions of these equations is identical to the case without any magnetic field, but the energy eigenvalues depend on the electron's spin. By inspection, we can say that if the wave functions and eigenvalues of $H = \frac{1}{2m_o}\hat{p}^2 - U(\bar{r})$ are (ϕ_n, E_n), then the wave functions and eigenvalues for up-spin and down-spin electrons are

$$\left(\phi_n(\bar{r}), E_n - \frac{g_s \mu_B}{2} B_z\right) \quad \text{for \textbf{up-spin} ($\uparrow$) electrons and}$$

$$\left(\phi_n(\bar{r}), E_n + \frac{g_s \mu_B}{2} B_z\right) \quad \text{for \textbf{down-spin} ($\downarrow$) electrons}$$

Recall that $\begin{bmatrix} 1 \\ 0 \end{bmatrix}$ and $\begin{bmatrix} 0 \\ 1 \end{bmatrix}$ are the eigenvectors of $\hat{\sigma}_z$ summarized in Table 15.1. We can then rewrite the above eigenfunctions in the notation of equation (15.59) as

$$\psi_{n\uparrow} = \phi_{\bar{r}}(\bar{r}) \begin{bmatrix} 1 \\ 0 \end{bmatrix} \quad \text{and} \quad E_{n\uparrow} = E_n - \frac{g_s \mu_B}{2} B_z \qquad (15.68)$$

$$\psi_{n\downarrow} = \phi_{\bar{r}}(\bar{r}) \begin{bmatrix} 0 \\ 1 \end{bmatrix} \quad \text{and} \quad E_{n\downarrow} = E_n + \frac{g_s \mu_B}{2} B_z \qquad (15.69)$$

The energy eigenvalues of the up-spin (down-spin) electron linearly decrease (increase) with an increase in the magnitude of a magnetic field.

Another common way to write the wave functions in equations (15.68) and (15.69) is to use the tensor or Kronecker product notation, where the first part is the spatial part of the wave function and the second part is the 2×1 spin eigenvector,

$$\phi_n \otimes | \uparrow \rangle \quad \text{for } \textbf{up-spin } (\uparrow) \text{ electrons} \qquad (15.70)$$

$$\phi_n \otimes | \downarrow \rangle \quad \text{for } \textbf{down-spin } (\downarrow) \text{ electrons} \qquad (15.71)$$

Example 1. Consider a system with quantum numbers $n = 1$ and $n = 2$ with energy eigenvalues E_1 and E_2 in the absence of a magnetic field. What are the energy levels, and how many electrons are there in each of the energy levels, when a uniform magnetic field $B\hat{d}$ is turned on everywhere? Note that $\hat{d}$ is a unit vector along the magnetic field direction. Neglect the role of the vector potential ($q\bar{A}$ term in equation (15.60)).

Solution: In the absence of a magnetic field, each energy level is two-fold degenerate; they are filled by one up-spin and one down-spin electron (see Figure 15.8). In an externally applied uniform magnetic field, one would have to account for both the energy term due to the magnetic moment, and the change in the kinetic energy operator. Without loss of generality, we can take the z direction to be along the externally applied magnetic field $\hat{d}$. As the problem explicitly says that the vector potential can be neglected, Schrödinger equation reduces to the set given by equations (15.66) and (15.67),

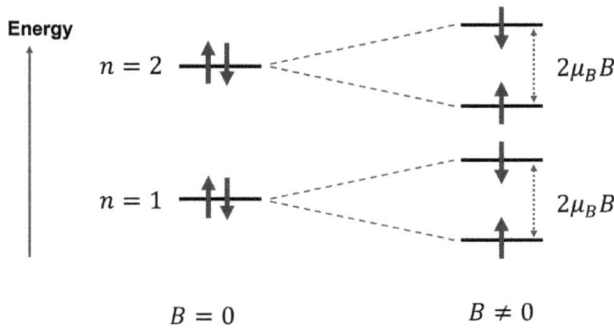

Figure 15.8. Energy levels of an electron assuming that the magnetic field's primary role is to cause interaction with the magnetic moment. (left) Degenerate energy levels without a magnetic field. There are two electrons per energy level: one electron has up-spin and the other has down-spin. (right) Energy levels split when a magnetic field is applied. Now each energy level can have only one electron.

with the eigenvalues and eigenfunctions given by equations (15.68) and (15.69). The lowest energy level (see Figure 15.8) is

$$E_{1\uparrow} = E_1 - \frac{g_S \mu_B}{2} B$$

This corresponds to an up-spin electron with the spin angular momentum pointing along the magnetic field. For a magnetic field with strength $B < \frac{E_2 - E_1}{g_S \mu_B}$, the next highest energy level corresponds to a down-spin electron in the $n = 1$ state, with energy

$$E_{1\downarrow} = E_1 + \frac{g_S \mu_B}{2} B$$

They are followed by the up- and down-spin states in the $n = 2$ quantum level, which are

$$E_{2\uparrow} = E_2 - \frac{g_S \mu_B}{2} B$$

$$E_{2\downarrow} = E_2 + \frac{g_S \mu_B}{2} B$$

The above energy levels show that the magnetic field lifts the degeneracy of the energy levels. If the applied magnetic field $B > \frac{E_2 - E_1}{g_S \mu_B}$, then $E_{2\uparrow}$ has a lower energy than $E_{1\downarrow}$.

In the absence of a magnetic field, the above system has two degenerate (equal) energy levels corresponding to the up- and down-spin electrons. When a magnetic field is applied, these degenerate energy levels become nondegenerate (unequal). This splitting of degenerate energy levels by a magnetic field is referred to as *Zeeman splitting*. One example where this splitting is applied is to measure the magnetic field strength in stars by observing the emission of light from gaseous atoms in the stars.

Note: A unique set of quantum numbers is defined by n and the spin state. In Figure 15.8, the quantum numbers are $(n = 1, \uparrow)$, $(n = 1, \downarrow)$, $(n = 2, \uparrow)$, and $(n = 2, \downarrow)$. It just turns out that when the magnetic field is zero, the energy levels of the quantum number set $(n = 1, \uparrow)$ and $(n = 1, \downarrow)$ are degenerate (have the same energy), and the energy levels of the quantum number set $(n = 2, \uparrow)$ and $(n = 2, \downarrow)$ are also degenerate. For reasons that we cannot delve into here, a fundamental property of an electron in any system is that each electron has a unique set of quantum numbers. Therefore, when the $(n = 1, \uparrow)$ and $(n = 1, \downarrow)$ states have different energies, they can only be occupied by one electron.

15.3.1 *Time-dependent Schrödinger equation only with a magnetic moment*

Suppose the Hamiltonian for a single spin fixed in space consists of the magnetic moment energy, $\hat{H}' = -\hat{\mu} \cdot \bar{B}$ and nothing else (no KE and PE terms). If the magnetic potential points in the z direction, then the time-dependent Schrödinger equation (equation (15.58)) can be written as

$$-i\hbar\frac{d\Psi}{dt} = -\frac{g_s\mu_B}{2}B_z\begin{bmatrix}1 & 0 \\ 0 & -1\end{bmatrix}\Psi \qquad (15.72)$$

or as two separate equations given by

$$-i\hbar\frac{d\psi_1}{dt} = -\frac{g_s\mu_B}{2}B_z\psi_1$$

$$-i\hbar\frac{d\psi_2}{dt} = \frac{g_s\mu_B}{2}B_z\psi_2$$

Recall that in equation (15.61), $\Psi = \begin{bmatrix} \psi_1 \\ \psi_2 \end{bmatrix}$. It is easy to solve the above equations to find that the eigenvectors are equal to

$$\psi_1 = \alpha\, e^{-i\omega_0 t/2}$$

$$\psi_2 = \beta\, e^{i\omega_0 t/2}$$

$$\text{where} \quad \omega_0 = \frac{qB_z}{m_0} \tag{15.73}$$

is called the *Larmor frequency*. The general solution can be written as a linear combination of the eigenvectors,

$$|\Psi\rangle = \begin{bmatrix} \alpha \\ \beta \end{bmatrix} = \alpha e^{-i\omega_0 t/2} \begin{bmatrix} 1 \\ 0 \end{bmatrix} + \beta e^{+i\omega_0 t/2} \begin{bmatrix} 0 \\ 1 \end{bmatrix} \tag{15.74}$$

The significance of the above equation will become apparent below when we calculate the expectation values of spin components along x, y and z directions using equations (15.37),

$$\langle S_x \rangle = \langle \Psi | \hat{S}_x | \Psi \rangle = \begin{bmatrix} \alpha^* e^{+i\omega_0 t/2} & \beta^* e^{-i\omega_0 t/2} \end{bmatrix} \frac{\hbar}{2} \begin{pmatrix} 0 & 1 \\ 1 & 0 \end{pmatrix} \begin{bmatrix} \alpha\, e^{-i\omega_0 t/2} \\ \beta\, e^{+i\omega_0 t/2} \end{bmatrix} \hat{S}_x$$

$$= \frac{\hbar}{2} \begin{bmatrix} \alpha^* e^{+i\omega_0 t/2} & \beta^* e^{-i\omega_0 t/2} \end{bmatrix} \begin{bmatrix} \beta\, e^{+i\omega_0 t/2} \\ \alpha\, e^{-i\omega_0 t/2} \end{bmatrix}$$

$$= \frac{\hbar}{2} (\alpha^* \beta e^{+i\omega_0 t} + \alpha \beta^* e^{-i\omega_0 t})$$

$$\langle S_x \rangle = \hbar \left[\text{Re}(\alpha^* \beta) \cos(\omega_0 t) - \text{Im}(\alpha^* \beta) \sin(\omega_0 t) \right] \tag{15.75}$$

$$\langle S_y \rangle = \langle \Psi | \hat{S}_y | \Psi \rangle = \begin{bmatrix} \alpha^* e^{+i\omega_0 t/2} & \beta^* e^{-i\omega_0 t/2} \end{bmatrix} \frac{\hbar}{2} \begin{pmatrix} 0 & -i \\ i & 0 \end{pmatrix} \begin{bmatrix} \alpha\, e^{-i\omega_0 t/2} \\ \beta\, e^{+i\omega_0 t/2} \end{bmatrix}$$

$$\langle S_y \rangle = \hbar [\text{Re}(\alpha^* \beta) \sin(\omega_0 t) + \text{Im}(\alpha^* \beta) \cos(\omega_0 t)] \tag{15.76}$$

$$\langle S_z \rangle = \langle \Psi | \hat{S}_z | \Psi \rangle = \begin{bmatrix} \alpha^* e^{+i\omega_0 t/2} & \beta^* e^{-i\omega_0 t/2} \end{bmatrix} \frac{\hbar}{2} \begin{pmatrix} 1 & 0 \\ 0 & -1 \end{pmatrix} \begin{bmatrix} \alpha\, e^{-i\omega_0 t/2} \\ \beta\, e^{+i\omega_0 t/2} \end{bmatrix}$$

$$\langle S_z \rangle = \frac{\hbar}{2} (|\alpha|^2 - |\beta|^2) \tag{15.77}$$

If the coefficients α and β in the initial superposition are real, then the expectation values of S_x, S_y, and S_z become

$$\langle S_x \rangle = \hbar \alpha \beta \, \cos(\omega_0 t)$$

$$\langle S_y \rangle = \hbar \alpha \beta \sin(\omega_0 t)$$

$$\langle S_z \rangle = \frac{\hbar}{2}(\alpha^2 - \beta^2)$$

It shows that the average x and y components of spin rotates (precesses) around the z-axis with the Larmor frequency ω_0 while the z-component is time-independent and constant.

Example 2. Show that the square of the total spin angular momentum and its z-component can be measured simultaneously, i.e., prove that

$$[\hat{S}_z, \hat{S}^2] = 0$$

Solution: Since $\hat{S}^2 = \hat{S}_x^2 + \hat{S}_y^2 + \hat{S}_z^2$, the above commutator can be split into three terms as follows:

$$[\hat{S}_z, \hat{S}^2] = [\hat{S}_z, \hat{S}_x^2 + \hat{S}_y^2 + \hat{S}_z^2] = [\hat{S}_z, \hat{S}_x^2] + [\hat{S}_z, \hat{S}_y^2] + [\hat{S}_z, \hat{S}_z^2]$$

It is easy to show that each term individually is zero because the squared terms are proportional to the identity matrix (see equation (15.38)),

$$\hat{S}_x^2 = \hat{S}_y^2 = \hat{S}_z^2 = \frac{\hbar^2}{4} I_{2 \times 2}$$

$$\implies [\hat{S}_z, \hat{S}_i^2] = \frac{\hbar^2}{4}[\hat{S}_z, I] = 0$$

Therefore $[\hat{S}_z, \hat{S}^2] = 0$ and the proof is complete.

15.4 Spin–Orbit Interaction

In Section 15.3, we saw that by applying an external magnetic field, spin-up and spin-down electrons obtain different potential energies determined by the Zeeman interaction term $(-\vec{\mu} \cdot \vec{B})$. As a result, a degenerate energy level splits into two, where the energy splitting is proportional to the magnetic field strength. The final wave function could be decomposed into an outer Kronecker or tensor product of spatial and spin parts, $\phi_n \otimes |\uparrow\rangle$ or $\phi_n \otimes |\downarrow\rangle$.

We discuss the spin–orbit interaction (SOI), which mixes spin and spatial parts (orbit) in a nontrivial manner. It is noteworthy that in contrast to Zeeman interaction, the source of the magnetic field in the SOI interaction is not an external magnetic field but rather internal or applied electric fields! In the following, we show how the electric field emanating from a nucleus acts as an effective magnetic field, using non relativistic physics.

Consider an electron moving around the nucleus with velocity $\bar{v}$ at a radius r (akin to the Bohr model). In the reference frame of the electron, the nucleus is moving around the electron at radius r with a velocity of $-\bar{v}$ (see Figure 15.9). A nuclear charge of Zq moving in a circle with radius r corresponds to a current loop carrying an average current I,

$$I = \frac{\text{Charge} \times \text{speed}}{\text{Cirumference}} = \frac{Zq|\bar{v}|}{2\pi r} \qquad (15.78)$$

The magnitude of the magnetic field at the center of the circular loop is (use the Biot–Savart law (Knight, 2016))

$$B = \frac{\mu_o I}{2r} \qquad (15.79)$$

and the direction is perpendicular to the plane of the loop pointing to the reader.

Substituting equation (15.78) in equation (15.79), the magnitude of the field is

$$B = \frac{\mu_o Zq|\bar{v}|}{4\pi r^2} \qquad (15.80)$$

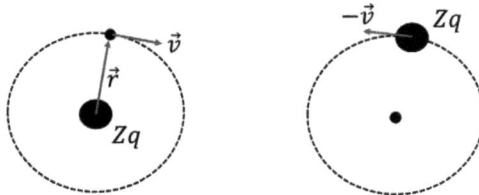

Figure 15.9. (Left) An electron traveling around a nucleus (represented by charge Zq) in a circular orbit with radius r and velocity $\vec{v}$. (Right) In the reference frame of the electron, the nucleus is moving around the electron with a velocity of $-\vec{v}$.

Using the "right-hand rule" (vector cross product rule) associated with the Biot–Savart law, convince yourself that equation (15.80) can be written as a vector,

$$\bar{B} = \frac{\mu_o Z q \, \bar{r} \times \bar{v}}{4\pi r^3} \tag{15.81}$$

Multiplying and dividing equation (15.81) by the vacuum dielectric constant ϵ_o, and noting that the speed of light in a vacuum is $c = \frac{1}{\sqrt{\epsilon_o \mu_o}}$, we can rewrite equation (15.81) as

$$\bar{B} = -\frac{1}{c^2} \bar{v} \times \left(\frac{Z q \bar{r}}{4\pi \epsilon_o r^3} \right) \tag{15.82}$$

The electric field at the location of an electron due to the nucleus is given by $\bar{\mathcal{E}}(\bar{r}) = \frac{Z q \bar{r}}{4\pi \epsilon_o r^3}$. So, we can express the magnetic field felt by the electron as

$$\bar{B}(\bar{r}) = -\frac{1}{c^2} \bar{v} \times \left(\frac{Z q \bar{r}}{4\pi \epsilon_o r^3} \right) = -\frac{1}{c^2} \bar{v} \times \bar{\mathcal{E}}(\bar{r}) \tag{15.83}$$

$\bar{B}(\text{at the location of electron})$

$$= -\frac{1}{c^2} \bar{v} \times \bar{\mathcal{E}}(\text{due to nucleus}) \tag{15.84}$$

A detailed treatment shows that equation (15.84) is incorrect for an electron moving in a circular orbit around a nucleus as it is accelerating. In this case, the correct equation relating the velocity, E-field, and B-field is

$\bar{B}(\text{at location of electron})$

$$= -\frac{1}{2c^2} \bar{v} \times \bar{\mathcal{E}}(\text{due to nucleus}) \tag{15.85}$$

The difference from equation (15.84) is the extra factor of 2 in the denominator, which was worked out by L. H. Thomas (Thomas, 1926) using the theory of relativity.

From now on, we will assume that equation (15.85) is also valid for any source of the electric field which an electron might experience. But this must be taken cautiously as the inclusion of other factors might be necessary, i.e., factors that depend on the band structure of electrons in solids. In this book, we skip these details and refer the reader to more advanced texts (Bandyopadhyay, 2008) (Schäpers 2016).

Now that we have another source of the effective magnetic field, we can substitute equation (15.85) in equation (15.52) to get the energy of interaction due to the new effective magnetic field (generated by an electric field),

Potential Energy $(U_E) = -\bar{\mu} \cdot \bar{B}$

$$= \frac{1}{2c^2} \bar{\mu} \cdot (\bar{v} \times \bar{\mathcal{E}}) = -\frac{1}{2c^2} \bar{\mu} \cdot (\bar{\mathcal{E}} \times \bar{v}) \qquad (15.86)$$

Equation (15.86) is called the **spin–orbit interaction (SOI) Hamiltonian**. The spin–orbit interaction in materials can arise due to an interaction of a moving electron's magnetic moment with an electric field. If the electric field arises due to stress or due to an applied voltage at the surface of a material, it is called the *Rashba interaction*. On the other hand, if the electric field arises due to an intrinsic inversion asymmetry in the material (examples are InAs, InSb, and GaAs), it is called the *Dresselhaus interaction*. We have used and refer the reader to (Bandyopadhyay, 2008), which is a good reference to learn more about SOI Hamiltonians and their device applications.

The Rashba interaction in a solid is written in many ways in literature, but we will consider one form here, which follows easily from equation (15.86),

$$\text{Energy } (U) = -\bar{\mu} \cdot \bar{B} = -\frac{1}{2m^*c^2} \bar{\mu} \cdot (\bar{\mathcal{E}} \times (m^*\bar{v})) \qquad (15.87)$$

where m^* is the effective mass of the electron in the solid. Equation (15.8) is rewritten as

$$\bar{\mu} = \frac{g_s \mu_B \bar{S}}{\hbar} \qquad (15.88)$$

where g_s, the gyromagnetic ratio, is now a material constant. Using the previous equation and noting that the momentum $\bar{p} = m^*\bar{v}$, equation (15.87) can be written as

$$\text{Energy} = -\bar{\mu} \cdot \bar{B} = -\frac{g_s \mu_B}{2\hbar m^* c^2} \bar{S} \cdot (\bar{\mathcal{E}} \times \bar{p}) \qquad (15.89)$$

Converting the spin and momentum in the previous equation to operators, the Hamiltonian of an electron in zero magnetic fields and a non-negligible Rashba term is (similar to equation (15.58))

$$\left\{ \left[\frac{\hat{p}^2}{2m^*} + U(\bar{r}) \right] \otimes \hat{I}_{2\times 2} - \frac{g_s \mu_B}{2\hbar m^* c^2} \hat{S} \cdot (\bar{\mathcal{E}} \times \hat{p}) \right\} \Psi = E\Psi \qquad (15.90)$$

where $U(\bar{r})$ is the electrostatic potential energy.

Example 3. Consider a narrow-diameter nanowire where only the lowest conduction sub-band matters. An electric field is applied along the y direction. See Figure 15.10(a). Assume that the: (a) spin–orbit coupling constant is $\alpha_{so} = \frac{g_s \mu_B}{2\hbar m^* c^2}$, (b) change in electrostatic potential is negligible because the diameter of the nanowire is exceedingly small, and (c) spin–orbit coupling is the dominant factor affecting the energy levels. Find the energy eigenvalues of an electron in the nanowire.

Solution: We start with the Hamiltonian given by equation (15.90) to calculate the band structure. Substituting that the electric field is pointing along the y direction, $\bar{\mathcal{E}} = \mathcal{E}_0 \hat{j}$, the Hamiltonian can be rewritten as

$$\left\{ \frac{1}{2m^*} \hat{p}^2 \otimes \hat{I}_{2\times 2} - \frac{g_s \mu_B}{2\hbar m^* c^2} \vec{S} \cdot (\mathcal{E}_0 \hat{j} \times \hbar k_z \hat{k}) \right\} \Psi = E\Psi \qquad (15.91)$$

$$\left\{ \frac{1}{2m^*} \hat{p}^2 \otimes \hat{I}_{2\times 2} - \alpha_{so} \mathcal{E}_0 \hbar k_z \vec{S} \cdot (\hat{j} \times \hat{k}) \right\} \Psi = E\Psi \qquad (15.92)$$

$$\left\{ \frac{1}{2m^*} \hat{p}^2 \otimes \hat{I}_{2\times 2} - \alpha_{so} \mathcal{E}_0 \hbar k_z S_x \right\} \Psi = E\Psi \qquad (15.93)$$

It is useful to recall that the effective magnetic field created by the electric field ($\bar{B} = \frac{1}{2m^* c^2}(\bar{E} \times \bar{p}) = \frac{\hbar}{2m^* c^2} \mathcal{E}_0 k_j \hat{i}$) points in the x direction.

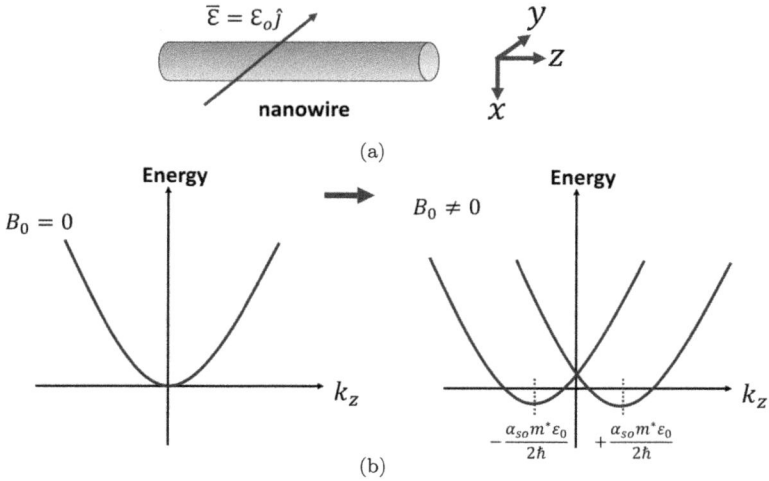

Figure 15.10. (a) Nanowire with an electric field applied along the y direction. (b) Energy bands without and with the electric field. Due to spin–orbit coupling, the up and down spin electrons' energy bands shift in opposite directions. Further, the uniform energy shift in up- and down-spin bands along the energy axis $\left(\frac{\alpha_{so}^2 m^* \mathcal{E}_0^2}{8}\right)$ depends on both the spin–orbit coupling constant and the electric field.

The energy levels are

$$
\begin{aligned}
E_\uparrow &= \frac{\hbar^2}{2m^*}\left(k_z^2 + \alpha_{so}m^*\mathcal{E}_0 k_z\right) \\
&= \frac{\hbar^2}{2m^*}\left(k_z + \frac{\alpha_{so}m^*\mathcal{E}_0}{2}\right)^2 - \frac{\alpha_{so}^2 m^*\mathcal{E}_0^2 \hbar}{8}
\end{aligned}
\tag{15.94}
$$

$$
\begin{aligned}
E_\downarrow &= \frac{\hbar^2}{2m^*}\left(k_z^2 - \alpha_{so}m^*\mathcal{E}_0 k_z\right) \\
&= \frac{\hbar^2}{2m^*}\left(k_z - \frac{\alpha_{so}m^*\mathcal{E}_0}{2}\right)^2 - \frac{\alpha_{so}^2 m^*\mathcal{E}_0^2}{8}
\end{aligned}
\tag{15.95}
$$

The above equations show that the energy band without SOI splits into two sub-bands that are shifted symmetrically around $k_z = 0$ as shown in Figure 15.10(b). For each sub-band, the spin is either up or down. Since the effective magnetic field is along the x direction, the spins are eigenvectors of the σ_x operator, i.e., $\frac{1}{\sqrt{2}} \begin{bmatrix} 1 \\ 1 \end{bmatrix}$ and $\frac{1}{\sqrt{2}} \begin{bmatrix} 1 \\ -1 \end{bmatrix}$.

15.5   Spin-FET (Datta–Das Transistor)

In the previous example, we showed that due to the Rashba interaction, the applied electric field normal to the nanowire axis (y direction) acts as an effective magnetic field along the x direction. Now, if we inject an electron with spin-up along the z direction into this nanowire, the electron's spin precesses around the effective magnetic field (x-direction) with Larmor frequency (equation (15.73)). As the electron travels along the nanowire, the spin rotates in yz-plane. At a later time, the electron's spin would point along the $-z$ direction, i.e., it completes a 180-degree rotation.

Now let's see how the above mechanism can be used to turn on or off the electric current in a device. Figure 15.11(a) shows a simple nanowire terminated by two ferromagnetic contacts. The contacts resemble source and drain in traditional charge-based transistors and an extra electrode on top of the nanowire works as the gate. When the applied gate voltage V_{gate} is zero, meaning the electric field along the x direction is zero, the electrons injected from the left ferromagnetic contact travel through without sensing any effective magnetic field. As a result of this, no rotation of spin occurs, and the electrons reach the right ferromagnetic contacts without precession, thus resulting in a nonzero current. Only the electrons that find their spin parallel to the magnetization direction of the contact will tunnel into the contact because it is energetically favorable for them. In summary, the right contact works as a spin detector, i.e., it determines whether the incoming spin is parallel or antiparallel with the electrical contact's magnetization direction.

Now, as shown in Figure 15.11(b), the electric field in the x direction is provided by a gate electrode, like in a conventional transistor. The electron injected from the source has a spin directed along the $-x$ direction. This electron sees an effective magnetic field $\bar{B}_{\text{eff}}$ along the $+y$ direction. This magnetic field will introduce nonzero spin–orbit interaction and will cause the electron spin to

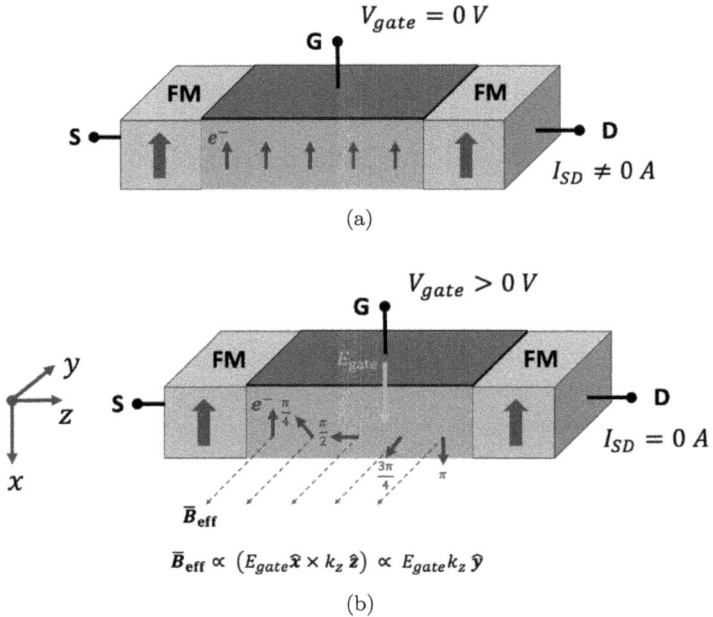

Figure 15.11. Datta–Das or spin-FET transistor. (a) With zero electric field, the electrons do not feel any effective magnetic field. As a result, no rotation occurs, and electrons enter the right contact leading to nonzero current. (b) Application of gate voltage and proper length of the semiconductor under the gate leads to 180 rotations of spin due to Rashba's effective magnetic field. Hence, no current can flow because of the resulting antiparallel directions between the spin of the electron and magnetization of the FM drain contact (spin detector).

precess around the $\bar{B}_{\text{eff}}$ axis, which is shown by the red arrows through the channel. This precession of the spin occurs due to the torque exerted by the magnetic field. The rotation of the spin follows $\frac{d\bar{S}}{dt} = \frac{g_s \mu_B}{\hbar} \bar{S} \times \bar{B}$ (see equation (15.114)).

If the electron spin points along the $-x$ direction when it reaches the drain contact, it is reflected and does not contribute to current because it finds itself anti-parallel to the FM contact and entering into the contact is energetically less favorable. As a result of this, the current is zero and the transistor (switch) is in its off state, in contrast to the previous case when the applied gate voltage was zero and we had nonzero current (the switch was on).

The principle mentioned above is the basis of the operation of a transistor proposed by S. Datta and B. Das in 1990 (Datta, 1990), which is the spin counterpart of a traditional charge-based FET transistor. Datta and Das's transistor was named by *Nature* magazine to be among the top 23 milestones in the Spintronics field in the year 2008. The application of spin-based electronic devices has been an active area of research. Examples include spin-based memories, data storage hard disks, read heads based on giant magnetoresistance, and quantum computing based on the spin of electrons in silicon and germanium quantum dots and nanowires. Nowadays, major silicon chip fabrication companies like IBM, Ever-Spin, and Global Foundries offer CMOS-compatible processes for spin-based devices too.

15.6 Fermions and Bosons

After introducing spin in this chapter, it is time to talk about the difference between two categories of particles, fermions, and bosons.

Imagine you are given two identical and indistinguishable particles, and each one stays in a one-dimensional space, e.g., x direction. How do we write the two-particle wave function $\psi(x_1, x_2)$? Note that x_1 is the one-dimensional space in which the first particle lives, and x_2 is the one for the second particle occupying the same one-dimensional space. The probability density (P) of finding two particles in the intervals dx_1 and dx_2 is found by

$$P(x_1, x_2)dx_1 dx_2 = |\psi(x_1, x_2)|^2 dx_1 dx_2 \qquad (15.96)$$

Now, if we exchange the position of particles, the probability density should remain the same as if nothing has changed because the particles are indistinguisable. For the probability $|\psi(x_1, x_2)|^2$ to remain unchanged under interchange of the particles, $\psi(x_1, x_2)$ and $\psi(x_2, x_1)$ can differ only by a phase factor. That is,

$$\psi(x_1, x_2) = e^{i\theta}\psi(x_2, x_1) \qquad (15.97)$$

To make progress, we assume a specific form of the wave function,

$$\psi(x_1, x_2) = A_1\phi_a(x_1)\phi_b(x_2) + A_2\phi_b(x_1)\phi_a(x_2) \qquad (15.98)$$

where the particles are in quantum states ϕ_a and ϕ_b, respectively. A_1 and A_2 are normalization factors.

Interchanging particles 1 and 2, we have

$$\psi(x_2, x_1) = A_2\phi_a(x_1)\phi_b(x_2) + A_1\phi_b(x_1)\phi_a(x_2) \qquad (15.99)$$

Substituting equations (15.98) and (15.99) in equation (15.97), we observe that the following conditions have to be valid:

$$A_1 = e^{i\theta}A_2 \quad \text{and} \quad A_2 = e^{i\theta}A_1$$

Substituting for A_1 from the first into the second equation, we see that $e^{i2\theta} = 1$ must be satisfied. So, the constraint on the phase is $\theta = 0, \pi, 2\pi, 3\pi, 4\pi, \ldots$ (*any integer times* π). This gives rise to two distinct classes of wave functions:

(1) **Symmetric wave functions:** $\psi(x_1, x_2) = \psi(x_2, x_1)$
    when $\theta = 0$ *or even integer multiple of* π. Particles with a symmetric wave function are called **bosons**.
(2) **Antisymmetric wave functions:** $\psi(x_1, x_2) = -\psi(x_2, x_1)$
    when $\theta = $ *odd integer multiple of* π. Particles with an antisymmetric wave function are called **fermions**.

Now let's assume that both particles want to occupy the same position, e.g., $x_1 = x_2$. This is allowed for bosons because their wave function becomes

$$\psi(x_1, x_1) = \psi(x_1, x_1) \qquad (15.100)$$

However, for fermions with antisymmetric wave function, we have

$$\psi(x_1, x_1) = -\psi(x_1, x_1) \qquad (15.101)$$

This is possible only if $\psi(x_1, x_1) = 0$, which means the probability density becomes zero. As a result of this, two fermions are not allowed to occupy the same state (which is the position in this case). This is called *Pauli's exclusion principle* which forbids two fermions from occupying the same quantum state or the same set of quantum numbers. Electrons are fermions, and they obey Pauli's exclusion principle.

Pauli's exclusion principle is a guide to building the wave function of multi-fermion (read electron) systems, i.e., it mandates that the multi-particle wave function must be antisymmetric, and whenever you put two particles (electrons) in the same states, the wave function must be zero.

Now let's give a specific example for two electrons and build the two-electron wave function. Let's assume we have two electrons, both of which are in the s-orbital, and they have spins $|\alpha\rangle$ and $|\beta\rangle$ respectively. Note that we have not yet decided in which coordinate the spin is measured. For example, if we measure spin in the z direction, then spins $|\alpha\rangle$ and $|\beta\rangle$ are eigenvectors of S_z, i.e., spin up and down. The two electrons are indistinguishable, but we assume they are labeled 1 and 2. The wave functions of electrons 1 and 2 are $\phi_s(x_1)|\alpha_1\rangle$ and $\phi_s(x_2)|\beta_2\rangle$, respectively. As you see in the following, by merely multiplying[1] the wave functions of two electrons, we cannot build the total two-particle wave function because it is not antisymmetric, and it is rejected by Pauli's principle. So the following wavefunction is not possible,

$$\psi(x_1, x_2) = \phi_s(x_1)|\alpha_1\rangle\, \phi_s(x_2)|\beta_2\rangle \qquad (15.102)$$

Now, we build another wave function, $\phi_s(x_1)|\beta_1\rangle\, \phi_s(x_2)|\alpha_2\rangle$ and subtract it from the above we have

$$\psi(x_1, x_2) = \phi_s(x_1)|\alpha_1\rangle\, \phi_s(x_2)|\beta_2\rangle - \phi_s(x_1)|\beta_1\rangle\phi_s(x_2)|\alpha_2\rangle \qquad (15.103)$$

By factoring out the common terms, we have

$$\psi(x_1, x_2) = \phi_s(x_1)\phi_s(x_2)(|\alpha_1\rangle|\beta_2\rangle - |\beta_1\rangle|\alpha_2\rangle) \qquad (15.104)$$

Let's check if ψ obeys Pauli's exclusion principle. By exchanging the particles, you can verify that ψ is indeed antisymmetric:

$$\psi(x_2, x_1) = \phi_s(x_2)\phi_s(x_1)(|\alpha_2\rangle|\beta_1\rangle - |\beta_2\rangle|\alpha_1\rangle) = -\psi(x_1, x_2)$$
$$(15.105)$$

---

[1]Note that here by multiplication we mean the Kronecker type of multiplication which expands the size of Hilbert space.

We see, the spatial part of the wave function $\phi_s(x_1)\phi_s(x_2)$ is symmetric and the spin part $(|\alpha_1\rangle|\beta_2\rangle - |\beta_1\rangle|\alpha_2\rangle)$ is antisymmetric, and as a result, the total wave function ends up being antisymmetric. Now let's assume both electrons want to be in the same position state (state of s-orbital), i.e., $x_2 = x_1$. Putting $x_2 = x_1$, we have

$$\psi(x_1, x_1) = \phi_s(x_1)\phi_s(x_1)(|\alpha_2\rangle|\beta_1\rangle - |\beta_2\rangle|\alpha_1\rangle) \tag{15.106}$$

It is noteworthy that the above wave function is nonzero only if two electrons have different spins, i.e., $\alpha \neq \beta$. Otherwise, if they have the same spin $(\alpha = \beta)$, they cannot occupy the same state because

$$\psi(x_1, x_1) = \phi_s(x_1)\phi_s(x_1)(|\alpha_2\rangle|\alpha_1\rangle - |\alpha_2\rangle|\alpha_1\rangle) = 0 \tag{15.107}$$

For example, Figure 15.8 shows that only if two electrons have opposite spins, then they can occupy the same energy state.

In summary, Pauli's exclusion principle says two electrons cannot occupy the same state or quantum number (position, momentum) unless they have different spins. The two-particle wave function which satisfies this condition is the one that is antisymmetric. The recipe for building a two-particle antisymmetric wave function using the wave function of each electron is to use the following determinant. Readers can confirm that the following determinant is indeed the same equation (15.104).

$$\psi_{antisym}(x_2, x_1) = \frac{1}{\sqrt{2}} \begin{vmatrix} \phi_s(x_1)|\alpha_1\rangle & \phi_s(x_1)|\beta_1\rangle \\ \phi_s(x_2)|\alpha_2\rangle & \phi_s(x_2)|\beta_2\rangle \end{vmatrix} \tag{15.108}$$

The factor of $\frac{1}{\sqrt{2}}$ is for normalizing the two-particle wave function. The above determinant is called the *Slater determinant*, named after J. C. Slater.

For bosons, you can use the plus $(+)$ sign in equation (15.104). You can show that if particles 1 and 2 were bosons, this symmetric two-particle wave function would allow them to have the same spin and occupy the same state as opposed to fermions (electrons)

$$\psi_{sym}(x_1, x_2) = \frac{1}{\sqrt{2}}\phi_s(x_1)\phi_s(x_2) (|\alpha_1\rangle|\beta_2\rangle + |\beta_1\rangle|\alpha_2\rangle) \tag{15.109}$$

You can check that by exchanging the particles, the above wave function remains symmetric.

It suffices here to say that particles with **half-integer spin** values $(\pm\frac{\hbar}{2}, \pm\frac{3\hbar}{2}, \pm\frac{5\hbar}{2}, \ldots)$ are called **fermions**. Particles like electrons, neutrons, protons, and He-3 nuclei are fermions. Half-integer spin particles follow the Fermi–Dirac statistics. Also, the combined wave function of N number of fermions must be antisymmetric. If any two particles are exchanged, the sign of the wave function flips (positive to negative or vice versa).

Particles with **integer spin** values $(\pm\hbar, \pm2\hbar, \pm3\hbar, \ldots)$ are called **bosons**. The fundamental property of a boson is that there can be many particles present in a state with a unique set of quantum numbers. That is, there can be many particles in each of the states (levels) shown in Figure 15.8. Photons, helium-4 nuclei, and phonons are bosons. Bosons follow Bose–Einstein statistics.

The reader may ask now that if we have a particle with integer spin, what happens if a beam of bosons with a total spin of 1 enters the nonuniform magnetic field along the z-direction in the Stern–Gerlach experiment? We will have three split beams on the screen in Figure 15.3, which correspond to $m_s = \{-1, 0, 1\}$ (one beam goes up, one goes down, and the center one for $m_s = 0$ is not deflected.

15.7 Appendix A: Relationship between Spin Angular Momentum and Torque

Classically torque $(\bar{\tau})$ and angular momentum $(\bar{L} = \bar{r} \times \bar{p})$ are related by

$$\bar{\tau} = \frac{d\bar{L}}{dt}$$

For a magnetic moment $(\bar{\mu})$, the torque is related to the applied magnetic field by

$$\bar{\tau} = \bar{\mu} \times \bar{B} \tag{15.110}$$

Quantum mechanically, the torque is related to a quantity called the generalized angular momentum $(\bar{J})$ by

$$\bar{\tau} = \frac{d\bar{J}}{dt} \tag{15.111}$$

The generalized angular momentum (Chapter 16) is the sum of orbital angular momentum and spin angular momentum by:

$$\bar{J} = \bar{L} + \bar{S} \quad \text{(in operator notation, } \hat{J} = \hat{L} + \hat{S}) \qquad (15.112)$$

In cases where the orbital angular momentum is not significant, the relationship between torque and spin angular momentum is

$$\bar{\tau} = \frac{d\bar{S}}{dt} \qquad (15.113)$$

Noting that the magnetic moment due to spin of an electron is given by $\bar{\mu} = \frac{g_s \mu_B \bar{S}}{\hbar}$, using equation (15.110), equation (15.113) can be written as

$$\frac{d\hat{S}}{dt} = -\frac{g_s \mu_B}{\hbar} \hat{S} \times \bar{B} \qquad (15.114)$$

15.8   Appendix B: Applications and Importance of Magnetic Vector Potential

It is often told to electrical engineering students that magnetic vector potential $\bar{A}$ is nothing but a mathematical object whose curl or rotation is the magnetic field $\bar{B}$, i.e., $\bar{B} = \text{curl}\,(\bar{A})$ or $\nabla \times \bar{A}$, as this is needed to make the magnetic field divergence free ($\nabla \cdot \bar{B} = 0$). However, the magnetic vector potential is essential in understanding the quantum mechanical effects like the Aharonov–Bohm effect and the working principle of superconducting devices like the Josephson Junction or the SQUID (superconducting quantum interreference device).

According to James Clerk Maxwell, $\bar{A}$ is as important as electric potential (ϕ or V). The electric potential is energy per unit charge and $\bar{A}$ is momentum per unit charge (Maxwell, 1954) (Semon, 1996). As much as adding qV to other energy terms in a Hamiltonian is important to keep the energy conserved, adding $q\bar{A}$ to the momentum is necessary to keep the total momentum of a charged particle in an electromagnetic field conserved. That's why in quantum mechanics the momentum is updated by adding the extra term ($q\bar{A}$ for an electron) as

$$\bar{p} = \bar{p}_o + q\bar{A} \qquad (15.115)$$

where the first term $\bar{p}_o$ corresponds to the classical definition of momentum and can be written as $m\bar{v}$, where $\bar{v}$ is the velocity.

Also, when it is said the electric potential is ϕ, we assume it is measured with respect to an arbitrary reference, i.e., we are free to choose the reference from where the potential is measured. The electric field is independent of the potential and remains the same as

$$\bar{\mathcal{E}} = -\nabla(V + V_o) = -\nabla(V) + 0 = -\nabla(V) \qquad (15.116)$$

Then this should not be of surprise that we have freedom to choose which $\bar{A}$ gives the magnetic field $\bar{B} = \nabla \times \bar{A}$. This freedom in choosing $\bar{A}$ is called gauge. For example, if we add the gradient of a scalar field χ to the vector potential so that $\bar{A}_{\text{new}} = \bar{A}_{\text{old}} + \nabla\chi$, it does not change the magnetic potential because we have

$$\bar{B} = \nabla \times (\bar{A}_{\text{old}} + \nabla\chi)$$
$$= \nabla \times \bar{A}_{\text{old}} + \nabla \times (\nabla\chi) = \nabla \times \bar{A}_{\text{old}} \qquad (15.117)$$

It is noteworthy that $\bar{A}$ has many important applications in engineering and physics. For example, it is used in modern electromagnetic simulators to simulate electric motors and calculate the mutual inductance in RFID tags, among other applications.

Imagining the $\bar{A}$ is not difficult if we use the following analogy as suggested in (Semon, 1996). From Maxwell equations, we can write the relationship between the current density $(\bar{J})$ and the magnetic field $(\bar{B})$ which it produces

$$\nabla \times \bar{B} = \mu\bar{J}, \nabla \cdot \bar{B} = 0, \quad \mu_0 I = \oint \bar{B} \cdot d\bar{l} \qquad (15.118)$$

If we assume a form of the vectoral potential such that $\nabla \cdot \bar{A} = 0$, then we have the following equations relating the magnetic field and the vector potential:

$$\nabla \times \bar{A} = \bar{B}, \nabla \cdot \bar{A} = 0, \quad \Phi = \oint \bar{A} \cdot d\bar{l} \qquad (15.119)$$

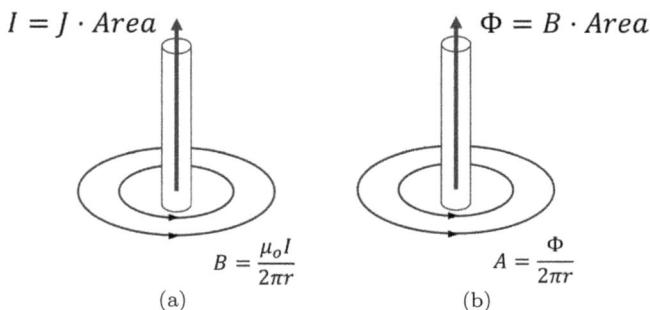

Figure 15.12. Relationship between current density in a wire and the magnetic field. By analogy, the same pattern can be visualized for a magnetic field inside a solenoid and the magnetic vector potential creating the field. Adapted from (Semon, 1996).

That is, by replacing $\bar{J}$ with $\bar{B}$ and $\bar{B}$ with $\bar{A}$, we can imagine how $\bar{A}$ looks like in an electromagnetic problem. Figure 15.12(a) shows how the magnetic field $\bar{B}$ around a long wire with current density $\bar{J}$ is created. The current I is the area of the wire multiplied by the current density. The magnetic flux Φ in the solenoid is $\bar{B}$ times the area of the solenoid (called Area in the figure). Using the above method, we can see how a magnetic vector potential looks if we are given a magnetic field inside a long solenoid.

15.9 Problems

Section 15.1

(1) Show that the expectation value of S_x is zero for a particle in the $|\uparrow z\rangle$ eigenfunction.

Section 15.3

(2) Consider an electron in a quantum system with eigenenergies of 1 and $2\,\text{eV}$ when the magnetic field is zero. In the presence of a magnetic field of $2.5\,\text{Tesla}$,

   (i) What is the lowest energy possible for the electron?
   (ii) What is the energy of the second highest energy level?

Assume that (a) the energy is modified by the magnetic field only due to the $E = -\bar{\mu} \cdot \bar{B}$ term and (b) gyromagnetic ratio is 2.

(3) Show that when coefficients α and β are complex in equation (15.74), the expectation value of $\langle S_x \rangle$ and $\langle S_y \rangle$ rotate around z-axis with Larmor frequency.

(4) We have an electron that is fixed at the origin and is in a magnetic field that is applied along the z direction. Show that the eigenstates corresponding to the simplified Hamiltonian $-\bar{\mu} \cdot \bar{B}$ has an expectation value for spin angular moment along the y direction $\langle S_y \rangle$ to be zero. Neglect Hamiltonian terms other than the magnetic dipole moment's energy.

Section 15.4

(5) Consider a narrow diameter nanowire where only the lowest conduction subband is important. That is, it is a single-mode nanowire. The axis of the nanowire is along the z-axis. You are given that the $E(k_z)$ relationship is parabolic with a minimum at $k_z = 0$.

  (a) What are the energy levels (energy band structure) when a magnetic field B_o points only along the z-direction?
  (b) What are the energy levels (energy band structure) when a magnetic field B_o points only along the y direction?

  Hint: Approximate the Hamiltonian as a one-dimensional free particle Hamiltonian but assume that the magnetic field and vector potential are in three-dimensional space.

(6) Consider a narrow-diameter nanowire such that the energy levels of only the lowest conduction subband are important. For all practical reasons, you can assume that the kinetic energy operator only depends on the momentum operator along the z direction. The axis of the nanowire is along the z-axis. An electric field is applied pointing only in the y-direction.

  Hint: The Hamiltonian is $H = \frac{\hat{p}^2}{2m^*} - \alpha_{so} \hat{S} \cdot (\bar{\mathcal{E}} \times \bar{p})$. Calculate and plot the energy band structure. m^* is the effective mass.

References

Bandyopadhyay, S. and Cahay, M. (2008). *Introduction to Spintronics*, 1st edn. CRC Press.

Datta, S. and Das, B. (1990). Electronic analog of the electro-optic modulator. *Applied Physics Letter* **56**, 666–667.

Greiner, W. (1998). *Quantum Electrodynamics*. Springer Verlag, Heidelberg.

Knight, R. D. (2016). *Physics for Scientists and Engineers with Modern Physics*, Chapter 32, 4th ed. Pearson.

Maxwell, J. C. (1954). *A Treatise on Electricity and Magnetism*. Dover Publications, New York. Article 405.

Sakurai, J. J. (1993). *Modern Quantum Mechanics*. Pearson.

Schäpers, T. (2016). Semiconductor Spintronics. Walter de Gruyter GmbH, Berlin.

Semon, M. D. and Taylor, J. R. (1996). Thoughts on the magnetic vector potential. *American Journal of Physics* **64**(11).

Singleton, J. (2001). *Band Theory and Electronic Properties of Solids*. Oxford University Press, New York.

Thomas, L. H. (1926). The motion of spinning electron. *Nature* **117**, 514.

Chapter 16

ANGULAR MOMENTUM

Contents

16.1 Introduction

In the previous chapters, we were concerned mainly with a particle's linear motion and problems with Hamiltonians separable into the three Cartesian coordinates. Examples included the free particle, nanowire, quantum well, and quantum dot. In these problems, we expressed the Hamiltonian in terms of Cartesian coordinates and momentum components $\hat{p}_x$, $\hat{p}_y$, and $\hat{p}_z$. In the case of a free particle, we saw that an eigenfunction of the Hamiltonian, e^{ikx}, is also an eigenfunction of the momentum operator (see Chapter 3). Mathematically, we can have the same set of eigenfunctions for momentum and Hamiltonian, if they commute.

In this chapter, we discuss the angular momentum operator. Starting from the operators for position and momentum, we define the operator for angular momentum, determine its commutation relationships, and find its eigenvalues and eigenfunctions using the differential form of the operators. We will then discuss the energy and angular momentum eigenvalues of a simple two-dimensional rotor, where the Hamiltonian and angular momentum operator commute. We will end this chapter by discussing raising and lowering operators for the angular momentum, which provides us with an elegant method to calculate the eigenvalues of the total angular momentum operator. Using the raising and lowering operators, we will discuss the concept of generalized angular momentum.

16.2   Orbital Angular Momentum

This section starts with the classical expression for the angular momentum of an object rotating around an arbitrary axis. We then generalize this to the quantum mechanical operators and call it *orbital angular momentum*. Figure 16.1 shows an object with mass M, which moves with velocity $\bar{v}$ and rotates on a circular path in a plane which makes an arbitrary angle with the z-axis. We could use the x, y or any other axis, but to keep things simple, we use the z-axis. The orbital angular momentum of this object is

$$\bar{L} = \bar{r} \times \bar{p} \tag{16.1}$$

which is directed normal to the plane of rotation (see Figure 16.1). In quantum mechanics, the angular momentum operators are defined by converting $\bar{r}$ and $\bar{p}$ in the above classical expression to operators $\hat{r}$ and $\hat{p}$. Recalling that the momentum operator $\hat{p}$ is

$$\hat{p} = -i\hbar\nabla \tag{16.2}$$

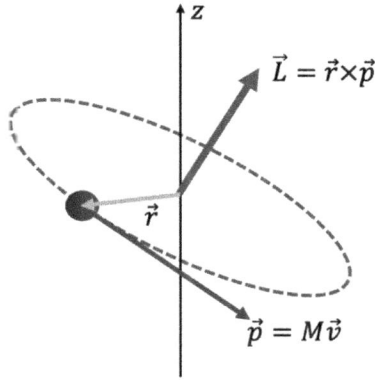

Figure 16.1. Definition of classical orbital angular momentum, $\vec{L} = \vec{r} \times \vec{p}$.

we can rewrite the orbital angular momentum operator as

$$\hat{L} = \hat{r} \times \hat{p} = \hat{r} \times -i\hbar\nabla \tag{16.3}$$

The gradient operator is a vector, and in Cartesian coordinates, it is given by

$$\nabla = \hat{i}\frac{\partial}{\partial x} + \hat{j}\frac{\partial}{\partial y} + \hat{k}\frac{\partial}{\partial z} \tag{16.4}$$

where $\hat{i}$, $\hat{j}$, and $\hat{k}$ are unit vectors along the x, y, and z-axis. Taking the cross product, the angular momentum components are

$$\hat{L}_x = y\hat{p}_z - z\hat{p}_y = -i\hbar\left(y\frac{\partial}{\partial z} - z\frac{\partial}{\partial y}\right) \tag{16.5}$$

$$\hat{L}_y = z\hat{p}_x - x\hat{p}_z = -i\hbar\left(z\frac{\partial}{\partial x} - x\frac{\partial}{\partial z}\right) \tag{16.6}$$

$$\hat{L}_z = x\hat{p}_y - y\hat{p}_x = -i\hbar\left(x\frac{\partial}{\partial y} - y\frac{\partial}{\partial x}\right) \tag{16.7}$$

Note that there is a cyclic relation between x, y, and z. As an example, by memorizing equation (16.5), you can write down equations (16.6) and (16.7) simply by converting $x \to y$, $y \to z$, and $z \to x$.

Let us find the commutation relationship between the components $\hat{L}_x$, $\hat{L}_y$, and $\hat{L}_z$. These relations determine if $\hat{L}_x$, $\hat{L}_y$, and $\hat{L}_z$ can be measured simultaneously and help in finding their eigenvalues and matrix representation. Before continuing, we prove the following helpful relation between three operators, $\hat{A}$, $\hat{B}$, and $\hat{C}$ which comes in handy in our derivations. We claim that

$$[\hat{A}, \hat{B}\hat{C}] = [\hat{A}, \hat{B}]\hat{C} + \hat{B}[\hat{A}\hat{C}] \tag{16.8}$$

To prove this, we start from the definition of commutation, and add subtract a $\hat{B}\hat{A}\hat{C}$ to get

$$[\hat{A}, \hat{B}\hat{C}] = \hat{A}\hat{B}\hat{C} - \hat{B}\hat{C}\hat{A} \tag{16.9}$$

$$= [\hat{A}\hat{B}\hat{C} - \hat{B}\hat{A}\hat{C}] + [\hat{B}\hat{A}\hat{C} - \hat{B}\hat{C}\hat{A}] \tag{16.10}$$

$$= [\hat{A}, \hat{B}]\hat{C} + \hat{B}[\hat{A}\hat{C}] \tag{16.11}$$

Using equation (16.8), we verify the commutation relation between angular momentum operators can be simplified as follows:

$$[\hat{L}_x, \hat{L}_y] = [y\hat{p}_z - z\hat{p}_y, z\hat{p}_x - x\hat{p}_z]$$

$$= [y\hat{p}_z, z\hat{p}_x] - [y\hat{p}_z, x\hat{p}_z] - [z\hat{p}_y, z\hat{p}_x] + [z\hat{p}_y, x\hat{p}_z]$$

$$= [y\hat{p}_z, z\hat{p}_x] - 0 - 0 + [z\hat{p}_y, x\hat{p}_z]$$

$$= z[y\hat{p}_z, \hat{p}_x] + [y\hat{p}_z, z]\hat{p}_x + x[z\hat{p}_y, \hat{p}_z] + [z\hat{p}_y, x]\hat{p}_z$$

$$= 0 + [y\hat{p}_z, z]\hat{p}_x + x[z\hat{p}_y, \hat{p}_z] + 0$$

$$= y[\hat{p}_z, z]\hat{p}_x + [y, z]\hat{p}_z\hat{p}_x + xz[\hat{p}_y, \hat{p}_z] + x[z, \hat{p}_z]\hat{p}_y \tag{16.12}$$

In reaching equation (16.12), we used that the components of linear momentum operator commute with each other, and that two different components of spatial coordinate and linear momentum operators commute with each other,

$$[\hat{p}_a, \hat{p}_b] = 0 \quad \text{and} \tag{16.13}$$

$$[a, \hat{p}_{b \neq a}] = 0 \quad \text{for } a, b \in \{x, y, z\} \tag{16.14}$$

For the z-component, we can use the above to write

$$[\hat{p}_z, z] = -i\hbar \tag{16.15}$$

Substituting equation (16.15) in equation (16.12) and using equations (16.13) and (16.14), we have

$$[\hat{L}_x, \hat{L}_y] = i\hbar(x\hat{p}_y - y\hat{p}_x) = i\hbar\hat{L}_z \tag{16.16}$$

$$[\hat{L}_x, \hat{L}_y] = i\hbar\hat{L}_z \tag{16.17}$$

Similarly,

$$[\hat{L}_y, \hat{L}_z] = i\hbar\hat{L}_x \tag{16.18}$$

$$[\hat{L}_z, \hat{L}_x] = i\hbar\hat{L}_y \tag{16.19}$$

All three equations ((16.17), (16.18), (16.19)) can be found from cyclic rotation of x, y, and z subscripts which are represented by p, q, and r in the following equation:

$$[\hat{L}_p, \hat{L}_q] = i\hbar\hat{L}_r \quad p \to q, \; q \to r, \quad \text{and} \quad r \to p \tag{16.20}$$

In Appendix A, the commutation relations are expressed using the Levi-Civita symbol, which helps memorize them.

Now, we find the commutation relation between the total angular momentum – squared $(\hat{L}^2)$ and the components of $\hat{L}$,

$$[\hat{L}_z, \hat{L}^2] = [\hat{L}_z, \hat{L}_x^2 + \hat{L}_y^2 + \hat{L}_z^2] = [\hat{L}_z, \hat{L}_x^2] + [\hat{L}_z, \hat{L}_y^2] + [\hat{L}_z, \hat{L}_z^2]$$

$$= \hat{L}_x[\hat{L}_z, \hat{L}_x] + [\hat{L}_z, \hat{L}_x]\hat{L}_x + \hat{L}_y[\hat{L}_z, \hat{L}_y] + [\hat{L}_z, \hat{L}_y]\hat{L}_y$$

$$= i\hbar\hat{L}_x\hat{L}_y + i\hbar\hat{L}_y\hat{L}_x - i\hbar\hat{L}_y\hat{L}_x - i\hbar\hat{L}_x\hat{L}_y \tag{16.21}$$

$$[\hat{L}_z, \hat{L}^2] = 0 \tag{16.22}$$

Similarly,

$$[\hat{L}_x, \hat{L}^2] = 0 \tag{16.23}$$

$$[\hat{L}_y, \hat{L}^2] = 0 \tag{16.24}$$

Equations (16.22), (16.23), and (16.24) mean that the angular momentum squared commutes with any of its components. On the other hand, equations (16.17), (16.18), and (16.19) show that any two components of angular momentum do not commute with each other. The previous statements imply that infinite precision in simultaneous measurement of (i) two orbital angular momentum components is impossible while (ii) total orbital angular momentum and one of its components are possible. On the other hand, infinite precision in simultaneous measurement of the components of linear momentum p_x, p_y, and p_z is possible because their operator's commute (equation (16.13)).

16.2.1 *Eigenvalues of orbital angular momentum*

The first step in finding the eigenvalues of $\hat{L}^2$ and $\hat{L}_z$ is noting that they commute (equation (16.22)). So, we can find a set of eigenfunctions $\{|u\rangle\}$ that simultaneously are solutions to the eigenvalue problems of both $\hat{L}_z$ and $\hat{L}^2$,

$$\hat{L}_z|u\rangle = a|u\rangle \tag{16.25}$$

$$\hat{L}^2|u\rangle = b|u\rangle \tag{16.26}$$

a and b are the eigenvalues of $\hat{L}_z$ and $\hat{L}^2$, respectively. It is convenient to transform the angular momentum components written in Cartesian coordinates in equations (16.5), (16.6), and (16.7) to spherical coordinates. Figure 16.2 shows the relationship between the Cartesian coordinates (x, y, z) and their corresponding spherical coordinates (r, θ, ϕ).

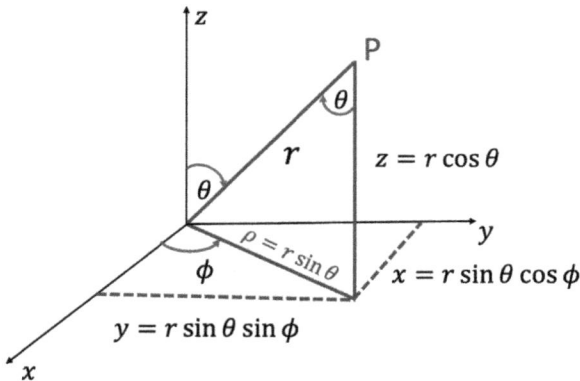

Figure 16.2. Conversion of Cartesian coordinates to spherical coordinates.

The coordinates and the gradient operator in the Cartesian and spherical coordinate systems are related by

$$x = r \sin \theta \cos \phi \tag{16.27}$$

$$y = r \sin \theta \sin \phi \tag{16.28}$$

$$z = r \cos \theta \tag{16.29}$$

$$\frac{\partial}{\partial x} = \sin \theta \cos \phi \frac{\partial}{\partial r} + \cos \theta \cos \phi \frac{1}{r} \frac{\partial}{\partial \theta} - \frac{\sin \phi}{r \sin \theta} \frac{\partial}{\partial \phi} \tag{16.30}$$

$$\frac{\partial}{\partial y} = \sin \theta \sin \phi \frac{\partial}{\partial r} + \cos \theta \sin \phi \frac{1}{r} \frac{\partial}{\partial \theta} - \frac{\cos \phi}{r \sin \theta} \frac{\partial}{\partial \phi} \tag{16.31}$$

$$\frac{\partial}{\partial z} = \cos \theta \frac{\partial}{\partial r} - \sin \theta \frac{1}{r} \frac{\partial}{\partial \theta} \tag{16.32}$$

Using equations (16.30), (16.31), and (16.32) in equations (16.5), (16.6), and (16.7), the differential operators for the angular momentum components in spherical coordinates are (see Appendix B for details)

$$\hat{L}_x = -\frac{\hbar}{i} \left[\sin \phi \frac{\partial}{\partial \theta} + \cot \theta \cos \phi \frac{\partial}{\partial \phi} \right] \tag{16.33}$$

$$\hat{L}_y = \frac{\hbar}{i} \left[\cos \phi \frac{\partial}{\partial \theta} - \cot \theta \sin \phi \frac{\partial}{\partial \phi} \right] \tag{16.34}$$

$$\hat{L}_z = \frac{\hbar}{i} \frac{\partial}{\partial \phi} \tag{16.35}$$

The $\hat{L}_z$ operator is mathematically simple to handle as it depends only on the azimuthal angle (ϕ).

From the equation for $\hat{L}_x$, $\hat{L}_y$, and $\hat{L}_z$, we have (see Appendix B)

$$\hat{L}^2 = [\hat{L}_x^2 + \hat{L}_y^2 + \hat{L}_z^2] = -\hbar^2 \left[\frac{\partial^2}{\partial \theta^2} + \cot \theta \frac{\partial}{\partial \theta} + \frac{1}{\sin^2 \theta} \frac{\partial^2}{\partial \phi^2} \right] \tag{16.36}$$

We can rewrite the eigenvalue equations (16.25) and (16.26) as partial differential equations using $|u\rangle = u(r, \theta, \phi)$.

$$\frac{\hbar}{i} \frac{\partial}{\partial \phi} u(r, \theta, \phi) = au(r, \theta, \phi) \tag{16.37}$$

$$-\hbar^2 \left[\frac{\partial^2}{\partial\theta^2} + \cot\theta \frac{\partial}{\partial\theta} + \frac{1}{\sin^2\theta} \frac{\partial^2}{\partial\phi^2} \right] u(r,\theta,\phi) = bu(r,\theta,\phi) \quad (16.38)$$

We are going to use the separation of variables to simplify the above equations. We first note that the differential operators do not depend on the radial coordinate r. So, we can write $u(r,\theta,\phi) = R(r)v(\theta,\phi)$, and the above two differential equations become

$$\frac{\hbar}{i} \frac{\partial}{\partial\phi} v(\theta,\phi) = av(\theta,\phi) \quad (16.39)$$

$$-\hbar^2 \left[\frac{\partial^2}{\partial\theta^2} + \cot\theta \frac{\partial}{\partial\theta} + \frac{1}{\sin^2\theta} \frac{\partial^2}{\partial\phi^2} \right] v(\theta,\phi) = bv(\theta,\phi) \quad (16.40)$$

In equation (16.39), the operator on the left-hand side depends only on the angle ϕ, so it is easy to verify that its solution is

$$v(\theta,\phi) = A(\theta)e^{\frac{i}{\hbar}a\phi} \quad (16.41)$$

where the function A depends on the angle θ. If we impose the boundary condition that for a given r and θ, the wave function, $v(\theta,\phi)$, is continuous for all values of ϕ between 0 and 2π, then the following condition holds:

$$v(\theta,\phi) = v(\theta,\phi + 2\pi) \quad (16.42)$$

This implies that

$$e^{\frac{i}{\hbar}a\phi} = e^{\frac{i}{\hbar}a(\phi+2\pi)} \Rightarrow e^{\frac{2\pi i a}{\hbar}} = 1 \quad (16.43)$$

For the above condition to hold, a can only be $\hbar$ times an integer,

$$a = m\hbar, \ m \in \mathbb{Z} \quad (16.44)$$

So, the eigenfunction in equation (16.40) becomes

$$v(\theta,\phi) = A(\theta)e^{im\phi} \quad (16.45)$$

Substituting equation (16.45) in equation (16.40) gives

$$-\hbar^2 \left[\frac{\partial^2}{\partial\theta^2} + \cot\theta \frac{\partial}{\partial\theta} - \frac{m^2}{\sin^2\theta} \right] A(\theta) = bA(\theta) \quad (16.46)$$

Equation (16.46) is a well-known equation in mathematical physics. It will suffice for us to know that $A(\theta)$ needs to be square integrable, meaning that (1) the integral of $|A(\theta)|^2$ on its domain $0 \le \theta \le \pi$, $\int_0^\pi |A(\theta)|^2 d\theta$ must be finite, and (2) the eigenvalues m and b should be of the form

$$b = l(l+1)\hbar^2 \quad \text{where } l = 0, 1, 2, 3, \ldots (l \in \mathbb{Z}^+, 0) \quad (16.47)$$

$$\text{and} \quad -l < m < +l \text{ (for a given } l) \quad (16.48)$$

In mathematical physics, equation (16.46) appears as

$$-\left[\frac{\partial^2}{\partial\theta^2} + \cot\theta \frac{\partial}{\partial\theta} - \frac{m^2}{\sin^2\theta}\right] P_l^m(\cos\theta) = l(l+1)P_l^m(\cos\theta) \quad (16.49)$$

where $P_l^m(\cos\theta)$ are called Legendre polynomials. The solution $v(\theta, \phi)$ in terms of l and m is then given by

$$v(\theta, \phi) = Y_l^m(\theta, \phi) = \frac{1}{\sqrt{4\pi}} e^{im\phi} P_l^m(\cos\theta) \quad (16.50)$$

The solutions $Y_l^m(\theta, \phi)$ are called spherical harmonics and listed in Table 16.1. They satisfy the following normalization condition:

$$\int_{\theta=0}^{\pi} \int_{\phi=0}^{2\pi} |Y_l^m(\theta, \phi)|^2 \sin\theta\, d\theta\, d\phi = 1 \quad (16.51)$$

Note that for each value of l, there are $2l + 1$ different values for m. The absolute value of spherical harmonics, $Y_l^m(\theta, \phi)$, for $l = 0, 1$,

Table 16.1. Spherical harmonics $Y_l^m(\theta, \phi)$ listed for $l = 0, 1$, and 2.

l	m	$Y_l^m(\theta\phi)$
0	0	$\frac{1}{2\sqrt{\pi}}$
1	0	$\frac{1}{2}\sqrt{\frac{3}{\pi}}\cos\theta$
	± 1	$\pm\frac{1}{2}\sqrt{\frac{3}{2\pi}}\sin\theta e^{\pm i\phi}$
2	0	$\frac{1}{4}\sqrt{\frac{5}{\pi}}(2\cos^2\theta - \sin^2\theta)$
	± 1	$\pm\frac{1}{2}\sqrt{\frac{15}{2\pi}}\cos\theta\sin\theta e^{\pm i\phi}$
	± 2	$\frac{1}{4}\sqrt{\frac{15}{2\pi}}\cos^2\theta e^{\pm i2\phi}$

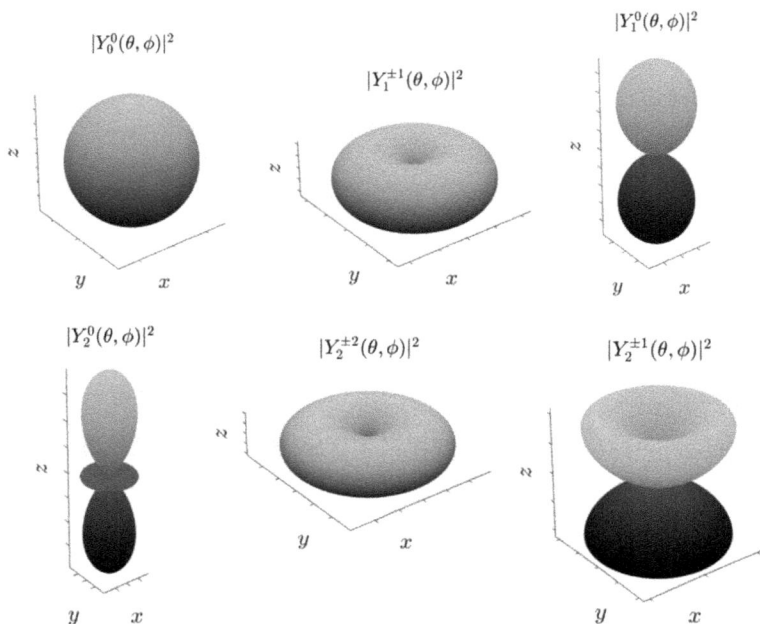

Figure 16.3. The spherical harmonics, $Y_l^m(\theta, \phi)$, for $l = 0, 1, 2, 3$.

and 2 in terms of spherical coordinates are plotted in Figure 16.3. Note that the shapes of the absolute value of Y_l^m corresponding to $+|m|$ and $-|m|$ are identical to one another because they differ only by the phase factors $e^{+i|m|\phi}$ and $e^{-i|m|\phi}$.

Summary of findings:

(1) The eigenvalues of $\hat{L}^2$ are $b = l(l+1)\hbar^2$, where

$$l = 0, 1, 2, 3, \ldots \qquad (16.52)$$

(2) Given an eigenvalue of $\hat{L}^2$, the eigenvalue of $\hat{L}_z$ is $a = m\hbar$, and

$$m \text{ ranges from } -l, -(l-1), \ldots, -1, 0, 1, \ldots, l-1, l \quad (16.53)$$

Table 16.2 summarizes the above results.

(3) The common eigenstates of $\hat{L}^2$ and $\hat{L}_z$ are usually indexed by l and m, i.e., $|u\rangle = |l, m\rangle$.

(4) The eigenstates are called spherical harmonics and are shown as $|l, m\rangle = Y_l^m(\theta, \phi)$.

Table 16.2. Eigenvalues of $\hat{L}^2$ and $\hat{L}_z$ listed for $l = 0, 1, 2, 3$.

l	Eigenvalues of $\hat{L}^2$, $[l(l+1)\hbar^2]$	Eigenvalue of $\hat{L}_z$, $(m\hbar)$
0	0	0
1	$1(1+1)\hbar^2 = 2\hbar^2$	$-\hbar, 0, +\hbar$
2	$2(2+1)\hbar^2 = 6\hbar^2$	$-2\hbar, -\hbar, 0, +\hbar, +2\hbar$
3	$3(3+1)\hbar^2 = 12\hbar^2$	$-3\hbar, -2\hbar, -\hbar, 0, +\hbar, +2\hbar, +3\hbar$

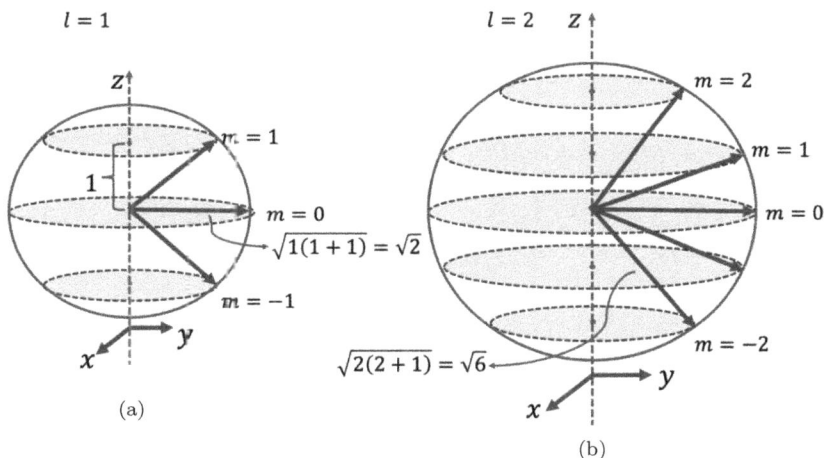

Figure 16.4. Orbital angular momentum of (a) $l = 1$ and (b) $l = 2$ and their projections on the z-axis. The tip of $\bar{L}$ is located on the periphery of the yellow-colored circles around the z-axis.

(5) The z-component of the angular momentum $(m\hbar)$ is smaller than the total angular momentum $\sqrt{l(l+1)}\hbar$, i.e., $m < \sqrt{l(l+1)}$. Figure 16.4 shows the eigenvalues of $\hat{L}_z$ for two different values of the angular momentum quantum number l. By increasing the total angular momentum value (l), the number of possible projections on the z-axis increases, but it never becomes parallel to the z-axis. The smallest angle between the total angular momentum and its projection on z-axis is

$$\cos(\theta_{min}) = \frac{|L_z|_{\max}}{\sqrt{|L|^2}} = \frac{m_{\max}\hbar}{\sqrt{l(l+1)\hbar^2}} = \frac{l}{\sqrt{l(l+1)}} \qquad (16.54)$$

The above equation implies that as the angular momentum l increases to large numbers, θ_{min} tends to 0 ($\lim_{l\to\infty} \theta_{min} = 0$), and we get the classical result that total angular momentum can point (almost) along any axis (here, the z-axis). Again, note that the z-axis is not special. We could have started calling the vertical axis in Figure 16.1 the x-axis, and we would have reached the same conclusions as above, for $\hat{L}^2$ and $\hat{L}_x$.

(6) Simultaneous measurement of the total angular momentum and only a single component of it, with infinite precision, is possible. This fact is a consequence of $\hat{L}^2$ and any component of $\hat{L}$ commuting. That is, $[\hat{L}_\alpha, \hat{L}^2] = 0$, for $\alpha \in x, y, z$.

The eigenvalues of the orbital angular momentum are

$$\hat{L}^2|l, m\rangle = l(l+1)\hbar^2|l, m\rangle$$

$$\hat{L}_z|l, m\rangle = m\hbar|l, m\rangle$$

$$l = 0, 1, 2, 3, \ldots$$

$$m = -l, -l+1, \ldots, l-1, l$$

Common eigenstates are spherical harmonics $|l, m\rangle = Y_l^m(\theta, \phi)$

Example 1. Find the total angular momentum and its component along the z-axis for the state $\psi(\theta, \phi) = \frac{\cos\theta}{\sqrt{2}} + \frac{1}{\sqrt{2}}$.

Solution: We can rewrite the given state as a superposition of spherical harmonics with the help of Table 16.1. This gives

$$\psi(\theta, \phi) = \frac{\cos\theta}{\sqrt{2}} + \frac{1}{\sqrt{2}} = \sqrt{\frac{2\pi}{3}}Y_1^0 + \sqrt{2\pi}Y_0^0 = \sqrt{\frac{2\pi}{3}}|1, 0\rangle + \sqrt{2\pi}|0, 0\rangle$$

By applying the $\hat{L}^2$ and $\hat{L}_z$ operators to the above superposition, we have

$$\hat{L}_z\left\{\sqrt{\frac{2\pi}{3}}|1, 0\rangle + \sqrt{2\pi}|0, 0\rangle\right\} = \sqrt{\frac{2\pi}{3}} \times 0\hbar + \sqrt{2\pi} \times 0\hbar = 0$$

$$\hat{L}^2 \left\{ \sqrt{\frac{2\pi}{3}}|1,0\rangle + \sqrt{2\pi}|0,0\rangle \right\} = \sqrt{\frac{2\pi}{3}}1(1+1)\hbar^2 + \sqrt{2\pi}0(0+1)\hbar^2$$

$$= 2\hbar^2 \sqrt{\frac{2\pi}{3}}$$

As a result, the total angular momentum is $\sqrt{2}\left(\frac{2\pi}{3}\right)^{1/4}\hbar$, and it has zero projection on the z-axis.

Example 2 (2D rotor). Derive the Hamiltonian of an object with mass M which is rotating at a constant speed (magnitude of the velocity $\bar{v}$) at a fixed distance from an axis in a 2D plane (see Figure 16.5).

Solution: For simplicity, we assume the axis of rotation is the z-axis. In this case the orbital angular momentum has only the z-component; this is because the cross products of the position ($\bar{r}$) and momentum ($\bar{p} = M\bar{v}$) vectors are parallel to the z-axis. Noting that the radius of rotation is a constant, R_0, we can write

$$\bar{L} = \bar{r} \times \bar{p} = R_0 \hat{r} \times M|\bar{v}|\hat{\phi} = R_0 M|\bar{v}|\hat{k} \rightarrow L_z = R_0 M|\bar{v}| \quad (16.55)$$

where we have used $\hat{r} \times \hat{\phi} = \hat{k}$ ($\hat{r}$ and $\hat{\phi}$ are unit vectors along the radial and azimuthal directions. $\hat{k}$ is a unit vector along the

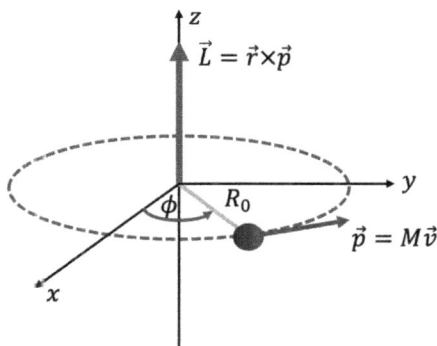

Figure 16.5. Two-dimensional rotor. A particle of mass M is rotating in the xy plane around the z-axis with a constant radius R_o.

z-direction.). The Hamiltonian of the particle can be written in terms of L_z as

$$\hat{H} = \frac{|\hat{p}|^2}{2M} = \frac{M^2 v^2}{2M} = \frac{L_z^2}{2MR_0^2} \quad \text{or} \quad \frac{1}{2}\mathcal{J}\omega^2 = \frac{L_z^2}{2\mathcal{J}} \tag{16.56}$$

where $\mathcal{J}$ is the rotational moment of inertia,

$$\mathcal{J} = MR_0^2 \tag{16.57}$$

Using equation (16.35), the Hamiltonian corresponding to equation (16.56) is

$$\hat{H} = \frac{\hat{p}^2}{2M} = \frac{\hat{L}_z^2}{2MR_0^2} = -\frac{\hbar^2}{2MR_0^2}\frac{\partial^2}{\partial^2\phi} \tag{16.58}$$

Therefore, the Schrödinger equation takes the following form

$$\hat{H}X = EX \Rightarrow -\frac{\hbar^2}{2MR_0^2}\frac{\partial^2}{\partial^2\phi}X = EX \tag{16.59}$$

The solution of the above equation is an exponential function of angle, ϕ,

$$X = |\phi\rangle = Ae^{im\phi} \tag{16.60}$$

Let us compare the wave function at locations (R, ϕ) and $(R, \phi+2\pi)$, which we reach after completing a full rotation by 2π radians around the z-axis. As (R, ϕ) and $(R, \phi+2\pi)$ represent the same point in space and the wave function is single valued at any location, we write

$$X(\phi) = X(\phi + 2\pi) \tag{16.61}$$

$$Ae^{im\phi} = Ae^{im(\phi+2\pi)} \tag{16.62}$$

$$\Rightarrow e^{i2\pi m} = 1 \tag{16.63}$$

The above condition is satisfied if m is an integer, i.e.,

$$m = \ldots, -2, -1, 0, +1, +2, \ldots \quad \text{or} \quad m \in \mathbb{Z} \tag{16.64}$$

To find the eigenvalues of the z-component of the orbital angular momentum operator, we first note that the Hamiltonian and $\hat{L}_z$ commute because the Hamiltonian is proportional to L_z^2,

$$[\hat{H}, \hat{L}_z] = \left[\frac{\hat{L}_z^2}{2MR_0^2}, \hat{L}_z\right] = 0 \tag{16.65}$$

As they commute, both $\hat{H}$ and $\hat{L}_z$ can have the same set of eigenfunctions. After applying $\hat{L}_z$ to the eigenfunction of the Hamiltonian $(Ae^{im\phi})$, one can at once see that it is also an eigenvalue of $\hat{L}_z$ with

$$\text{Eigenvalue of } \hat{L}_z : a = m\hbar \quad \text{and} \quad m \in \mathbb{Z} \text{ (integers)} \qquad (16.66)$$

This example showed that for a 2D rotor, eigenvalues of the angular momentum in z direction are equal to integer times $\hbar$. In the case of circular rotation in the $x - y$ plane, the total orbital angular momentum does not have x and y components and is directed only along the z-axis. Therefore, the length of the total angular momentum vector is equal to its projection on the z-axis, i e., $|L|^2 = |L_z|^2$ in 2D.

Example 3. Calculate the eigenvalues of the Hamiltonian for the 2D rotor in equation (16.58).

Solution: Substituting $X = Ae^{im\phi}$ in $HX = EX$, we find the eigenvalues to be

$$\hat{H}X = -\frac{\hbar^2}{2MR_0^2}\frac{\partial^2}{\partial^2\phi}Ae^{im\phi} = EAe^{im\phi} \rightarrow E = \frac{\hbar^2 m^2}{2MR_0^2}$$

$$= \frac{|L_z|^2}{2MR_0^2} \quad \text{for } m = 0, \pm 1, \pm 2, \ldots$$

Both the energy levels and angular momentum values are discrete. Except for $m = 0$, the energy eigenvalues are two-fold degenerate except for $m = 0$ because the clockwise $(+|m|)$ and anticlockwise $(-|m|)$ rotations have the same eigenenergy.

Example 4. A small electric motor is rotating a 10-gram mass around its rotation axis (see Figure 16.6). The mass is 10 cm off center. The rotational speed of the motor is 100 rpm. What is the quantum number m corresponding to this rotation? By how much should the rotational speed be increased so that m is increased by only 1?

This example illustrates that for a particle with a large mass, the classical value of angular velocity corresponds to a huge value of the quantum number m. The change in angular momentum when

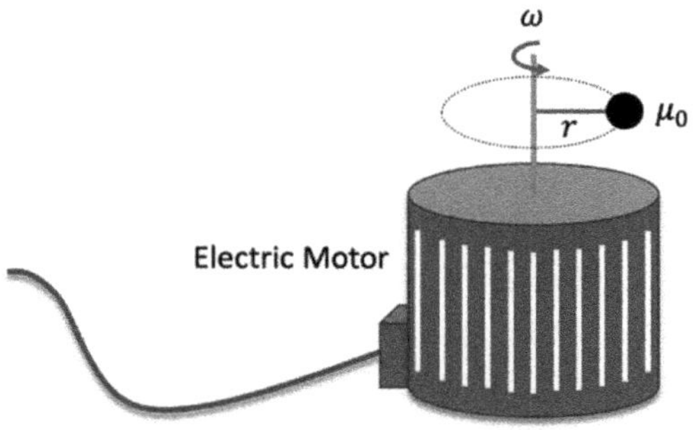

Figure 16.6. Example of a classical rotor.

the quantum number m increases by one unit is so small that it seems for all practical purposes that the angular momentum is a continuous variable.

Solution: First, we calculate the moment of inertia using $J = \mu_o r^2$.

$$J = \mu_o r^2 = 10 \times 10^{-3}\,\text{kg} \times (0.1\,\text{m})^2 = 10^{-4}\,\text{kg.m}^2$$

The rotational speed in terms of radian per second is $\omega = 100\,\text{rpm} = 100 \times \frac{2\pi}{60\,s} = 10.47\,\frac{\text{rad}}{\text{s}}$.

The rotational energy of the 2D rotor is given as follows from which the quantum number m can be extracted:

$$E = \frac{\hbar^2 m^2}{2J} = \frac{1}{2}J\omega^2$$

$$\rightarrow m = \frac{J\omega}{\hbar} = \frac{10^{-4} \times 10.47\,\text{kg.m}^2.\text{s}^{-1}}{6.62 \times 10^{-34}\,\text{J.s}} = 1.58 \times 10^{30}!$$

Thus, the quantum number is 1580000000000000000000000000000 !!

To increase this quantum number by 1, the motor must rotate faster, but how much?

$$\Delta m = 1 = \frac{J\Delta\omega}{\hbar} \rightarrow \Delta\omega = \frac{\hbar}{J} = \frac{6.62 \times 10^{-34}\,\text{J.s}}{10.47\,\text{kg.m}^2} = 63.22 \times 10^{-36}\,\frac{\text{rad}}{\text{s}}!$$

But controlling the speed of a motor with such a fine resolution is impossible! We don't see the quantum or discrete nature of angular momentum for such a big classical rotor. The angular momentum and energy look like continuous quantities as we cannot sense the change of energy or angular momentum or rotational speed by increasing or decreasing m by one.

What did we learn in this section? A classical object rotating around an axis has three well-defined components of orbital angular momentum. The components of angular momentum can take any continuous value such that $\hat{L}_x^2 + \hat{L}_y^2 + \hat{L}_z^2 = \hat{L}^2$. However, in quantum mechanics, the projections of angular momentum along a given axis (let us call this the z-axis) can take only discrete values that depend on the magnitude of the value of $|L|^2$. Increasing the magnitude of $|L|^2$ or $l(l+1)$ only increases the number of possible quantized projections along the z-axis. This fact is called *quantization of space or direction* and does not have a classical interpretation.

We also note that while $\hat{L}^2$ and $\hat{L}_z$ operators commute, they can both be determined simultaneously. But because $\hat{L}_z$ does not commute with $\hat{L}_x$ and $\hat{L}_y$, the components of angular momentum on the x- and y-directions cannot be simultaneously determined. Also, as a reminder, nothing is special about the z-axis. You can start with the x-axis and find the same arguments as above, i.e., we can measure both L^2 and $\hat{L}_x$ simultaneously and the other two components are left undetermined. It is not possible to measure all three components of the angular momentum with infinite precision at the same time.

16.3   Raising and Lowering Operators

This section introduces the concept of raising and lowering operators, which is a powerful technique to find the eigenvalues of angular momentum. We found the eigenvalues of the orbital angular momentum in the previous section by considering the differential form of the operators. The reader will find that the alternate derivation in this section leads to the idea of generalized angular momentum which includes the spin angular momentum as well as the orbital angular momentum.

Because $\hat{L}^2$ and $\hat{L}_z$ commute (equation (16.22)), we can find a set of eigenfunction $|\chi\rangle$ that simultaneously are solutions to the eigenvalue problems of both $\hat{L}_z$ and $\hat{L}^2$. To keep this derivation distinct from Section 16.2.1, we use a different symbol for the eigenfunctions here.

From equations (16.25) and (16.26), we at once obtain

$$\langle \chi | \hat{L}^2 - \hat{L}_z^2 | \chi \rangle = (b - a^2)\langle \chi | \chi \rangle \tag{16.67}$$

Since the expectation values of squared operators $\langle \hat{L}^2 \rangle$ and $\langle \hat{L}_z^2 \rangle$ are both positive, and the length of a vector is always greater than or equal to one of its components ($\langle \hat{L}^2 \rangle \geq \langle \hat{L}_z^2 \rangle$), we have

$$\langle \hat{L}_x^2 + \hat{L}_y^2 \rangle = \langle \hat{L}^2 - \hat{L}_z^2 \rangle \geq 0 \tag{16.68}$$

$$\Rightarrow b - a^2 \geq 0 \tag{16.69}$$

$$\Rightarrow a \leq \sqrt{b} \tag{16.70}$$

Let us assume that

$$a = \gamma \hbar \tag{16.71}$$

where we are not constraining γ to be an integer. Our next step is to find the eigenvalue of $\hat{L}^2$, which is b. We are going to prove that $b = l(l+1)\hbar^2$ and the allowable values for γ are $\{-l, -l+1, \ldots, l-1, l\}$, where l is an integer $(0, 1, 2, 3, \ldots)$. To do this, we define two operators called "raising" and "lowering" operators for reasons that will be clear shortly,

$$\hat{L}_+ = \hat{L}_x + i\hat{L}_y \quad \text{(raising operator)} \tag{16.72}$$

$$\hat{L}_- = \hat{L}_x - i\hat{L}_y \quad \text{(lowering operator)} \tag{16.73}$$

$\hat{L}_+$ and $\hat{L}_-$ obey the following properties:

$$\hat{L}_+ \hat{L}_- = (\hat{L}_x + i\hat{L}_y)(\hat{L}_x - i\hat{L}_y)$$

$$= \hat{L}_x^2 + \hat{L}_y^2 + i[\hat{L}_y, \hat{L}_x] = \hat{L}_x^2 + \hat{L}_y^2 + \hbar \hat{L}_z \tag{16.74}$$

$$[\hat{L}_z, \hat{L}_+] = [\hat{L}_z, \hat{L}_x] + i[\hat{L}_z, \hat{L}_y] \tag{16.75}$$

$$= i\hbar \hat{L}_y - i \times i\hbar \hat{L}_x \tag{16.76}$$

$$= \hbar(\hat{L}_x + i\hat{L}_y) \tag{16.77}$$

$$[\hat{L}_z, \hat{L}_+] = \hbar \hat{L}_+ \tag{16.78}$$

Using the above commutation relation, we will show that $\hat{L}_+ | \chi \rangle$ is also an eigenstate of $\hat{L}_z$. We operate by $\hat{L}_z \hat{L}_+$ from the left-hand side of $|\chi\rangle$ and use the relation $\hat{L}_z\hat{L}_+ = \hbar\hat{L}_+ + \hat{L}_+\hat{L}_z$ (which follows from equation (16.78)) to get

$$\hat{L}_z\hat{L}_+|\chi\rangle = (\hbar\hat{L}_+ + \hat{L}_+\hat{L}_z)|\chi\rangle = \hbar\hat{L}_+|\chi\rangle + \hat{L}_+\hat{L}_z|\chi\rangle \quad (16.79)$$

$$\hat{L}_z(\hat{L}_+|\chi\rangle) = (a + \hbar)(\hat{L}_+|\chi\rangle) = (\gamma + 1)\hbar(\hat{L}_+|\chi\rangle) \quad (16.80)$$

where $\hat{L}_z|\chi\rangle = a|\chi\rangle$. Equation (16.80) tells us that $\hat{L}_+|\chi\rangle$ is an eigenvector of $\hat{L}_z$ with eigenvalue $(a + \hbar) = (\gamma + 1)\hbar$, while $|\chi\rangle$ is an eigenvector of $\hat{L}_z$ with eigenvalue a. $\hat{L}_+$ is called the *raising* operator as it increases eigenvalue a by $\hbar$ or increases γ by 1.

Similarly, you can show that

$$[\hat{L}_z, \hat{L}_-] = -\hbar\hat{L}_- \quad \text{and} \quad (16.81)$$

$$\hat{L}_z(\hat{L}_-|\chi\rangle) = (a - \hbar)(\hat{L}_-|\chi\rangle) = (\gamma - 1)\hbar(\hat{L}_-|\chi\rangle) \quad (16.82)$$

$\hat{L}_-|\chi\rangle$ is an eigenvector of $\hat{L}_z$ with eigenvalue $(a - \hbar) = (\gamma - 1)\hbar$, while $|\chi\rangle$ is an eigenvector of $\hat{L}_z$ with eigenvalue a. As $\hat{L}_-$ lowers the eigenvalue by "$\hbar$" (reduces γ by 1), it is called the *lowering* operator.

We are now going to show that $\hat{L}_+|\chi\rangle$ is the eigenvector of $\hat{L}^2$ with the same eigenvalue b as $|\chi\rangle$. To show this, we first note that $[\hat{L}^2, \hat{L}_x] = [\hat{L}^2, \hat{L}_y] = 0$ (see equations (16.23) and (16.24)). Using this, we can at once see that

$$[\hat{L}^2, \hat{L}_+] = 0 \quad (16.83)$$

$$[\hat{L}^2, \hat{L}_-] = 0 \quad (16.84)$$

$$\hat{L}^2(\hat{L}_+|\chi\rangle) = \hat{L}_+\hat{L}^2|\chi\rangle = \hat{L}_+b|\chi\rangle \quad (16.85)$$

$$= b(\hat{L}_+|\chi\rangle) \quad (16.86)$$

Equations (16.26) and (16.86) say that both $|\chi\rangle$ and $\hat{L}_+|\chi\rangle$ are eigenvectors of $\hat{L}^2$ with the same eigenvalue b.

By mathematical induction, you can prove that $\hat{L}_+^n|\chi\rangle$ is an eigenvector of the $\hat{L}_z$ operator with eigenvalue $a + n\hbar$,

$$\hat{L}_z(\hat{L}_+^n|\chi\rangle) = (a + n\hbar)(\hat{L}_+^n|\chi\rangle) \quad (16.87)$$

and $\hat{L}_+^n|\chi\rangle$ is an eigenfunction of $\hat{L}^2$ with eigenvalue b,

$$\hat{L}^2(\hat{L}_+^n|\chi\rangle) = b(\hat{L}_+^n|\chi\rangle) \quad (16.88)$$

But from equation (16.70), we know for eigenfunction $\hat{L}_+^n|\chi\rangle$, the eigenvalue b of $\hat{L}^2$ must be larger than the square of the eigenvalue of $\hat{L}_z$ which is $(a + n\hbar)^2$. So, for some integer n_{max}, raising $\hat{L}_+^n|\chi\rangle$ further should yield zero. That is, after some maximum value of n, $\hat{L}_+^{n_{max}+1}|\chi\rangle = 0$, which can be written as

$$\hat{L}_+[\hat{L}_+^{n_{max}}|\chi\rangle] = 0 \quad \text{if } [a + (n_{max} + 1)\hbar]^2 > b \qquad (16.89)$$

From the above, we find n_{max} as follows:

$$\hat{L}_-\hat{L}_+ = \hat{L}_x^2 + \hat{L}_y^2 + i\hbar(\hat{L}_x\hat{L}_y - \hat{L}_y\hat{L}_x) \qquad (16.90)$$

$$\hat{L}_-\hat{L}_+ = \hat{L}^2 - \hat{L}_z^2 - \hbar\hat{L}_z \qquad (16.91)$$

$$\hat{L}_-\hat{L}_+[\hat{L}_+^{n_{max}}|\chi\rangle] = 0 \implies (\hat{L}^2 - \hat{L}_z^2 - \hbar\hat{L}_z)[\hat{L}_+^{n_{max}}|\chi\rangle] = 0 \qquad (16.92)$$

$$[b - (a + n_{max}\hbar)^2 - \hbar(a + n_{max}\hbar)] = 0 \qquad (16.93)$$

$$b - a_{max}^2 - \hbar a_{max} = 0 \implies b = a_{max}(a_{max} + \hbar) \qquad (16.94)$$

$$\text{where } a_{max} = a + n_{max}\hbar \qquad (16.95)$$

Now, we will see what happens when $\hat{L}^2$ acts on $\hat{L}_-|\chi\rangle$,

$$\hat{L}^2(\hat{L}_-|\chi\rangle) = \hat{L}_-\hat{L}^2|\chi\rangle \qquad (16.96)$$

$$= b(\hat{L}_-|\chi\rangle) \qquad (16.97)$$

From equation (16.82), we know that $\hat{L}_-|\chi\rangle$ is an eigenfunction of $\hat{L}_z$ with eigenvalues $(a - \hbar)$. Acting further by $\hat{L}_-$ a number of times (n times), we have

$$\hat{L}_z(\hat{L}_-^n|\chi\rangle) = (a - n\hbar)(\hat{L}_-^n|\chi\rangle) \qquad (16.98)$$

Because $\hat{L}_z\hat{L}^2 = \hat{L}^2\hat{L}_z$, $\hat{L}_-^n|\chi\rangle$ is also an eigenfunction of $\hat{L}^2$ with eigenvalue b,

$$\hat{L}^2(\hat{L}_-^n|\chi\rangle) = b(\hat{L}_-^n|\chi\rangle) \qquad (16.99)$$

Equations (16.87) and (16.98) show that consecutive eigenvalues of $\hat{L}_z$ can change only by an integer times $\hbar$. Equations (16.88) and (16.99) show that the corresponding eigenfunctions of $\hat{L}_z$ ($\hat{L}_-^n|\chi\rangle$ and $\hat{L}_+^n|\chi\rangle$) are also eigenfunctions of $\hat{L}^2$ with the same eigenvalue b.

But we know that for eigenfunction $\hat{L}_-^n|\chi\rangle$, the eigenvalue b of $\hat{L}^2$ must be larger than $(a - n\hbar)^2$, the squared eigenvalue of $\hat{L}_z$ (see equation (16.70)). So, for some integer $n_{\min}$, lowering $\hat{L}_-^n|\chi\rangle$ further should yield zero. This means there is an $n_{\min}$ for which $\hat{L}_-^{n_{\min}-1}|\chi\rangle = 0$.

$$\hat{L}_-[\hat{L}_-^{n_{\min}}|\chi\rangle] = 0 \quad \text{if } (a - (n_{\min} - 1)\hbar)^2 > b \tag{16.100}$$

Noting that

$$\hat{L}_+\hat{L}_- = \hat{L}_x^2 + \hat{L}_y^2 - i\hbar(\hat{L}_x\hat{L}_y - \hat{L}_y\hat{L}_x) \tag{16.101}$$

$$\hat{L}_+\hat{L}_- = (\hat{L}^2 - \hat{L}_z^2) + \hbar\hat{L}_z \tag{16.102}$$

we have

$$\hat{L}_+\hat{L}_-[\hat{L}_-^{n_{\min}}|\chi\rangle] = 0$$

$$\Rightarrow (\hat{L}^2 - \hat{L}_z^2 + \hbar\hat{L}_z)[\hat{L}_-^{n_{\min}}|\chi\rangle] = 0 \tag{16.103}$$

$$[b - (a - n_{\min}\hbar)^2 + \hbar(a - n_{\min}\hbar)] = 0 \tag{16.104}$$

$$b - a_{\min}^2 + \hbar a_{\min} = 0 \Rightarrow b = a_{\min}(a_{\min} - \hbar) \tag{16.105}$$

$$\text{where} \quad a_{\min} = a - n_{\min}\hbar \tag{16.106}$$

The LHS of equations (16.94) and (16.105) are both b, the eigenvalue of $\hat{L}^2$ corresponding to the eigenfunction $|\chi\rangle$, which is also the eigenfunction of $\hat{L}_z$ with eigenvalue a. For equations (16.94) and (16.105) to hold good, the following condition is necessary:

$$b = a_{\min}(a_{\min} - \hbar) = a_{\max}(a_{\max} + \hbar) \tag{16.107}$$

$$a_{\max}^2 - a_{\min}^2 + \hbar(a_{\max} + a_{\min}) = (a_{\max} + a_{\min})(a_{\max} - a_{\min} + \hbar)$$

$$= 0 \tag{16.108}$$

$$a_{\max} = -a_{\min} \tag{16.109}$$

This means that the minimum of a is negative of maximum of a or vice versa. So, let's represent the maximum and minimum values by

$$a_{\max} = +\beta\hbar \quad \text{and} \tag{16.110}$$

$$a_{\min} = -\beta\hbar \tag{16.111}$$

Equation (16.109) shows that the minimum eigenvalue of $\hat{L}_z(a_{\min} = -\beta\hbar)$ is equal to negative of its maximum eigenvalue $(a_{\max} = \beta\hbar)$.

To find the value of β, we are guided by equations (16.94) and (16.105), which informed us that the eigenvalues of $\hat{L}^2$ are

$$b = a_{max}(a_{max} + \hbar) = \beta(\beta + 1)\hbar^2 \quad \text{(eigenvalues of } \hat{L}^2) \quad (16.112)$$

$$b = a_{min}(a_{min} - \hbar) = -\beta(-\beta - 1)\hbar^2 = \beta(\beta + 1)\hbar^2 \quad (16.113)$$

Equation (16.112) shows that the eigenvalues of $\hat{L}^2$ (namely, b) are related to the largest eigenvalue of $\hat{L}_z(\beta)$ by $b = \beta(\beta + 1)\hbar^2$.

From equations (16.95) and (16.106), we saw that $a_{max} = a + n_{max}\hbar$ and $a_{min} = a - n_{min}\hbar$, where both n_{max} and n_{min} are integers. Using equation (16.109) which says $a_{max} = -a_{min}$, we infer that

$$2a = (n_{min} - n_{max})\hbar \quad (16.114)$$

$$a = \frac{k}{2}\hbar \quad (16.115)$$

where k is an integer.

Equation (16.115) tells us that the eigenvalue of $\hat{L}_z$ can be both integer and half-integer multiples of $\hbar$. However, in discussing equation (16.44), we found that constraining the orbital angular momentum operator's eigenfunction $v(\theta, \phi)$ to be single valued when the azimuthal angle changes from ϕ to $\phi + 2\pi$,

$$v(\theta, \phi) = v(\theta, \phi + 2\pi)$$

resulted in the eigenvalues of $\hat{L}_z$ being $m\hbar$, where m is an integer. So, we conclude that for orbital angular momentum, the only allowed values of k in equation (16.115) are *even numbers*. We will keep the even values of k and discard the odd values for the time being. **So, $\gamma = m$, is an integer in equation (16.71).**

Our main observations based on the statements noted above in bold are as follows:

(1) The orbital angular momentum $\hat{L}^2$ and its z-component $\hat{L}_z$ commute (equation (16.22)). This means we can find a set of eigenfunctions $|\chi\rangle = |lm\rangle$ that simultaneously are solutions to the eigenvalue problems of both $\hat{L}_z$ and $\hat{L}^2$. Note that lm means l and m, not l multiplied by m, hence we may use $|\chi\rangle = |lm\rangle$ or $|l, m\rangle$.

(2) The eigenvalues of $\hat{L}_z$ are equal to $m\hbar$, where m is an integer.

(3) If $l\hbar$ is the maximum eigenvalue of $\hat{L}_z$ corresponding to a given total angular momentum, then we have the following:

    a. The minimum eigenvalue of $\hat{L}_z$ is $-l\hbar$ (Note $\beta = l$ in equations (16.110) and (16.111).)

    b. Consecutive eigenvalues of $\hat{L}_z$ can only change by an integer times $\hbar$ (equations (16.87) and (16.98)). So, the eigenvalues of $\hat{L}_z$ for a given l are

$$m\hbar = -l, -(l-1), \ldots, -1, 0, +1, \ldots, +(l-1), +l \text{ (times } \hbar)$$

    There are $(2l+1)$ eigenvalues of $\hat{L}_z$ for a given l.

    c. Corresponding to every value of m listed above, the eigenvalue of the total orbital angular momentum $\hat{L}^2$ is $l(l+1)\hbar^2$ (see equation (16.112)).

(4) The reader ought to be disturbed that the **odd values** of k in equation (16.115) are ignored; this will be the topic of the following section.

$|l, m\rangle$ is an eigenfunction of both $\hat{L}^2$ and $\hat{L}_z$ with eigenvalues,

$$\hat{L}^2|l, m\rangle = l(l+1)\hbar|l, m\rangle \quad \text{and} \qquad (16.116)$$

$$\hat{L}_z|l, m\rangle = m\hbar|l, m\rangle \qquad (16.117)$$

where $l = 0, 1, 2, 3, \ldots$.

For a given l, there are $(2l+1)$ value of quantum number m.

The range of $m = -l, -(l-1), \ldots, -1, 0, +1, \ldots, +(l-1), +l$.

The raising and lowering operators of the orbital angular momentum obey the following:

$\hat{L}_+|l, m\rangle$ is an eigenfunction of $\hat{L}_z$ with eigenvalue $(m+1)\hbar$

$\hat{L}_-|l, m\rangle$ is an eigenfunction of $\hat{L}_z$ with eigenvalue $(m-1)\hbar$

$$\hat{L}_+|l, m = l\rangle = 0 \quad \text{(i.e., } \hat{L}_+ \text{ does not raise anymore)}$$

$$\hat{L}_-|l, m = -l\rangle = 0 \quad \text{(i.e., } \hat{L}_- \text{ does not lower anymore)}$$

Example 5. Show that $\hat{L}_+$ and $\hat{L}_-$ are not Hermitian.

Solution: Recall from equations (16.72) and (16.73) that $\hat{L}_+ = \hat{L}_x + i\hat{L}_y$ and $\hat{L}_- = \hat{L}_x - i\hat{L}_y$.

The Hermitian conjugate of $\hat{L}_+$ is $\hat{L}_+^\dagger = (\hat{L}_x + i\hat{L}_y)^\dagger = \hat{L}_x^\dagger - i\hat{L}_y^\dagger = \hat{L}_x - i\hat{L}_y = \hat{L}_-$.

Since $\hat{L}_+^\dagger \neq \hat{L}_+$, the *raising* operator is not Hermitian. Similarly, the *lowering* operator $\hat{L}_-$ is not Hermitian since $\hat{L}_-^\dagger = \hat{L}_+$.

In summary, the raising and lowering operators are related by the following relations:

$$\hat{L}_+^\dagger = \hat{L}_- \quad \text{and} \tag{16.118}$$

$$\hat{L}_-^\dagger = \hat{L}_+ \tag{16.119}$$

Finally, note that $\hat{L}_+$ and $\hat{L}_-$ do not correspond to physical observables as they are not Hermitian.

16.4 Generalized Angular Momentum ($\hat{J}$)

In Section 16.2, we defined the orbital angular momentum based on the rotation of an object around an axis. The classical position and momentum vectors were replaced by their quantum mechanical operator analogs to form the orbital angular momentum operator. We found that constraining the orbital angular momentum operator's eigenfunction $v(\theta, \phi)$ to be single valued when the azimuthal angle changes from ϕ to $\phi + 2\pi$,

$$v(\theta, \phi) = v(\theta, \phi + 2\pi)$$

resulted in the eigenvalues of $\hat{L}_z$ to take only values equal to $m\hbar$, where m is an integer (that is, k can only be even in equation (16.115)).

This section will assume that the operators obey the commutation relations discussed in equations (16.17), (16.22) and (16.24) and (16.24). But we will not constrain the eigenfunction to be single valued when $\phi \to \phi + 2\pi$. That is, $v(\theta, \phi)$ does not have to be equal to $v(\theta, \phi + 2\pi)$. To differentiate this situation from the *orbital angular momentum*, the operators in this section are called the *generalized angular momentum* and are referred to by the symbol $\hat{J}$. They are thought of as abstract operators that do not have a classical analog. The three components of $\hat{J}$ satisfy

$$[\hat{J}_x, \hat{J}_y] = i\hbar \hat{J}_z \tag{16.120}$$

$$[\hat{J}_y, \hat{J}_z] = i\hbar \hat{J}_x \tag{16.121}$$

$$[\hat{J}_z, \hat{J}_x] = i\hbar \hat{J}_y \tag{16.122}$$

$\hat{J}^2$ also commutes with its components as follows:

$$[\hat{J}_x, \hat{J}^2] = 0 \tag{16.123}$$

$$[\hat{J}_y, \hat{J}^2] = 0 \tag{16.124}$$

$$[\hat{J}_z, \hat{J}^2] = 0 \tag{16.125}$$

Before we proceed, the reader should note that equations (16.120)–(16.125) for $\hat{J}_x$, $\hat{J}_y$, $\hat{J}_z$, and $\hat{J}$ are identical to equations (16.17)–(16.19), and (16.22)–(16.24) for the orbital angular momentum components $\hat{L}_x$, $\hat{L}_y$, $\hat{L}_z$, and $\hat{L}$. However, there are no differential operators for the components of $\hat{J}$. Table 16.3 compares the orbital and the generalized angular momentum.

Table 16.3. Comparison of the orbital and the generalized angular momentum.

	Orbital angular momentum	Generalized angular momentum
Total	$\hat{L}$	$\hat{J}$
Components	$\hat{L}_x$, $\hat{L}_y$, $\hat{L}_z$	$\hat{J}_x$, $\hat{J}_y$, $\hat{J}_z$
Is this wave function single-valued?	$v(\theta, \phi) = v(\theta, \phi + 2\pi)$ Wave function is single-valued	$v(\theta, \phi)$ is not always equal to $v(\theta, \phi + 2\pi)$ Single valued nature is not enforced.
$\hat{L}^2$ and $\hat{J}^2$ eigenvalues	$l(l + 1)\hbar^2$ where $l = 0, 1, 2, 3, \ldots$ (integers)	$j(j + 1)\hbar^2$ where $j = 0, \frac{1}{2}, 1, \frac{3}{2}, 2, \ldots$ (half integers)
$\hat{L}_z$ and $\hat{J}_z$ eigenvalues	$m\hbar$ a total of $(2l + 1)$ values $m = -l, -(l - 1), \ldots, -1, 0, 1, \ldots (l - 1), l$	$m\hbar$ a total of $(2j + 1)$ values $m = -j, -(j - 1), \ldots, -1, 0, 1, \ldots (j - 1), j$

16.4.1 *Eigenvalues of generalized angular momentum*

As $\hat{J}^2$ and $\hat{J}_z$ commute, we would be able to find an eigenfunction $|u\rangle$ that simultaneously is a solution to the eigenvalue problems of both operators:

$$\hat{J}_z|u\rangle = a|u\rangle \tag{16.126}$$

$$\hat{J}^2|u\rangle = b|u\rangle \tag{16.127}$$

We note that equations (16.67)–(16.115) also hold good for the components of the generalized angular momentum ($\hat{J}_z$ and $\hat{J}^2$) as their commutation relations are identical to the orbital angular momentum operators. The eigenvalues of $\hat{J}_z$ are given in equation (16.115), *but k can now be both even and odd integers*. This means that m can be both integers and half integers,

$$a = \frac{k}{2}\hbar = m\hbar, \tag{16.128}$$

where $m = 0, \pm\frac{1}{2}, \pm 1, \pm\frac{3}{2}, 2, \ldots$ is the set of all half integers

Recall that in contrast, for the orbital angular momentum component k could only be even integers because the wave function was single valued, $v(\theta, \phi) = v(\theta, \phi + 2\pi)$.

The equations for the derivation of b (eigenvalues of the total angular momentum), i.e., equations (16.93), (16.94), (16.104), (16.105), (16.114), and (16.115), are all intact. But we do not know the exact range of m in equation (16.128). Equation (16.109) tells us that the magnitudes of the maximum and minimum values of $\hat{J}_z$ are equal. Let us represent the maximum value by j. Then,

$$a_{\max} = +j\hbar \quad \text{and} \tag{16.129}$$

$$a_{\min} = -j\hbar \tag{16.130}$$

To find the value of j, we are guided by equations (16.94) and (16.105), which inform us that the eigenvalues of $\hat{J}^2$ are

$$b = a_{\max}(a_{\max} + \hbar) = j(j+1)\hbar^2 \tag{16.131}$$

$$b = a_{\min}(a_{\min} - \hbar) = -j(-j-1)\hbar^2 = j(j+1)\hbar^2 \tag{16.132}$$

The eigenvalues and common eigenstates of $\hat{J}^2$ and $\hat{J}_z$ are summarized as

$$\hat{J}^2|j,m\rangle = j(j+1)\hbar^2|j,m\rangle; \quad j = 0, \frac{1}{2}, 1, \frac{3}{2}, 2, \frac{5}{2}, 3, \ldots \quad (16.133)$$

$$\hat{J}_z|j,m\rangle = m\hbar|j,m\rangle; \quad m = -j, -j+1, \ldots, j-1, j \quad (16.134)$$

$|j,m\rangle$ is an eigenfunction of both $\hat{J}^2$ and $\hat{J}_z$ with eigenvalues,

$$\hat{J}^2|j,m\rangle = j(j+1)\hbar^2|j,m\rangle$$

$$\hat{J}_z|j,m\rangle = m\hbar|j,m\rangle$$

where $j = 0, \frac{1}{2}, 1, \frac{3}{2}, 2, \frac{5}{2}, \ldots$.

For a given j, there are $(2j+1)$ different values for quantum number m.

The range of $m = -j, -(j-1), \ldots + (j-1), +j$.

The difference between the eigenvalues of the orbital and generalized angular momentum is that the former can have only integer values for the eigenvalues of $\hat{L}^2$, while the later can have half integer values for the eigenvalues of $\hat{J}^2$.

16.5 Matrix Representation of Orbital Angular Momentum

We will show that the matrix forms of the raising and lowering operators are upper and lower triangulars, respectively. From equation (16.80), we inferred that $\hat{L}_+|\chi\rangle = \hat{L}_+|l,m\rangle$ is the eigenstate of $\hat{L}_z$ with eigenvalue $(m+1)\hbar$. By comparing equation (16.80) with $\hat{L}_z|l,m\rangle = m\hbar|l,m\rangle$ we can write $\hat{L}_+|l,m\rangle = \alpha|l,m+1\rangle$. That is,

$$\hat{L}_z(\hat{L}_+|l,m\rangle) = (m+1)\hbar\hat{L}_+|l,m\rangle \rightarrow \hat{L}_+|l,m\rangle = \alpha|l,m+1\rangle \quad (16.135)$$

where α is an unknown factor, determined by calculating the norm of $\hat{L}_+|l,m\rangle$. Using equations (16.91) and (16.118), we have

$$|\alpha|^2 = \langle l,m|\hat{L}_+^\dagger\hat{L}_+|l,m\rangle = \langle l,m|\hat{L}_-\hat{L}_+|l,m\rangle \quad (16.136)$$

$$= \langle l,m|\hat{L}^2 - \hat{L}_z^2 - \hbar\hat{L}_z|l,m\rangle \quad (16.137)$$

Using the eigenvalues of $\hat{L}^2$ and $\hat{L}_z$, we can write

$$|\alpha|^2 = \{l(l+1)\hbar^2 - m^2\hbar^2 - m\hbar^2\}\langle l,m|l,m\rangle \qquad (16.138)$$

$$|\alpha|^2 = \{l(l+1) - m(m+1)\}\hbar^2 \qquad (16.139)$$

Substituting equation (16.139) in equation (16.135), upto an ordinary phase factor,

$$\hat{L}_+|l,m\rangle = \sqrt{l(l+1) - m(m+1)}\hbar|l,m+1\rangle \qquad (16.140)$$

A similar calculation for the lowering operator yields (see Problem 10)

$$\hat{L}_-|l,m\rangle = \sqrt{l(l+1) - m(m-1)}\hbar|l,m-1\rangle \qquad (16.141)$$

$$\hat{L}_+|l,m\rangle = \sqrt{l(l+1) - m(m+1)}\hbar|l,m+1\rangle \quad \text{where } |m| \leq l$$

$$\hat{L}_-|l,m\rangle = \sqrt{l(l+1) - m(m-1)}\hbar|l,m-1\rangle \quad \text{where } |m| \leq l$$

Using equations (16.140) and (16.141), it quickly follows that raising and lowering operators possess the following properties:

$$\hat{L}_+|l,m_{\max} = l\rangle = 0 \qquad (16.142)$$

$$\hat{L}_-|l,m_{\min} = -l\rangle = 0 \qquad (16.143)$$

Example 6. Find the matrix representation of orbital angular momentum $(\hat{L})$ for a particle in the $l = 1$ eigenstate.

Solution: For $l = 1$, there are $2l + 1 = 3$ different values of m, i.e., $m = -1, 0, 1$. This means that the orbital angular momentum component $\hat{L}_z$ has three different projections on the z-axis. Therefore, the Hilbert space to represent the simultaneous eigenfunctions of $\hat{L}^2$ and $\hat{L}_z$, i.e., $|l,m\rangle$, must be composed of $2l + 1$ vectors and the matrices representing the operators are of size $(2l + 1) \times (2l + 1)$. Since there are three allowed values for m, three vectors are required to represent the state $|l, m\rangle$.

$$|l, m\rangle \in \{|1, 1\rangle, |1, 0\rangle, |1, -1\rangle\} = \left\{ \begin{bmatrix} 1 \\ 0 \\ 0 \end{bmatrix}, \begin{bmatrix} 0 \\ 1 \\ 0 \end{bmatrix}, \begin{bmatrix} 0 \\ 0 \\ 1 \end{bmatrix} \right\} \quad (16.144)$$

The orthonormality relation for the basis states is

$$\langle l, m|l', m'\rangle = \delta_{l,l'}\delta_{m,m'} \quad (16.145)$$

To find the matrix elements of $\hat{L}_x$ and $\hat{L}_y$, it is easier to find the matrix elements of $\hat{L}_+$ and $\hat{L}_-$ using equations (16.140) and (16.141). For the raising operator, we write

$$\hat{L}_+|1, 1\rangle = \sqrt{1(1 + 1) - 1(1 + 1)}\hbar|1, 1 + 1\rangle = 0 \quad (16.146)$$
$$\hat{L}_+|1, 0\rangle = \sqrt{1(1 + 1) - 0(0 + 1)}\hbar|1, 0 + 1\rangle$$
$$= \sqrt{2}\hbar|1, 1\rangle \quad (16.147)$$
$$\hat{L}_+|1, -1\rangle = \sqrt{1(1 + 1) + 1(-1 + 1)}\hbar|1, -1 + 1\rangle$$
$$= \sqrt{2}\hbar|1, 0\rangle \quad (16.148)$$

From the above three equations, we can deduce that all diagonal elements of the raising operator are zero because

$$\langle 1, -1|\hat{L}_+|1, -1\rangle = \sqrt{2}\hbar\langle 1, -1|1, 0\rangle = 0 \quad (16.149)$$
$$\langle 1, 0|\hat{L}_+|1, 0\rangle = \sqrt{2}\hbar\langle 1, 0|1, 1\rangle = 0 \quad (16.150)$$
$$\langle 1, 1|\hat{L}_+|1, 1\rangle = 0\langle 1, -1|1, 2\rangle = 0 \quad (16.151)$$

Therefore, the only nonzero matrix elements of the raising operator are

$$\langle 1, 0|\hat{L}_+|1, -1\rangle = \sqrt{2}\hbar\langle 1, 0|1, 0\rangle = \sqrt{2}\hbar \quad (16.152)$$
$$\langle 1, 1|\hat{L}_+|1, 0\rangle = \sqrt{2}\hbar\langle 1, 1|1, 1\rangle = \sqrt{2}\hbar \quad (16.153)$$

The raising operator $\hat{L}_+$ can be written as

$$\hat{L}_+ = \hbar \begin{bmatrix} 0 & \sqrt{2} & 0 \\ 0 & 0 & \sqrt{2} \\ 0 & 0 & 0 \end{bmatrix} \tag{16.154}$$

This matrix is upper triangular, and it is not Hermitian. Equations (16.118) and (16.119) can be used to find raising and lowering operators if one of them is known. For example, the lowering operator is found from the raising operator by taking its Hermitian conjugate:

$$\hat{L}_- = (\hat{L}_+)^\dagger = \hbar \begin{bmatrix} 0 & 0 & 0 \\ \sqrt{2} & 0 & 0 \\ 0 & \sqrt{2} & 0 \end{bmatrix} \tag{16.155}$$

The matrix form of the x and y components of the orbital angular momentum operator for $l = 1$ is found using the raising and lowering operators. Using equations (16.72) and (16.73), the x- and y-component operators are as follows:

$$\hat{L}_x = \frac{\hat{L}_+ + \hat{L}_-}{2} = \frac{\hbar}{\sqrt{2}} \begin{bmatrix} 0 & 1 & 0 \\ 1 & 0 & 1 \\ 0 & 1 & 0 \end{bmatrix} \tag{16.156}$$

$$\hat{L}_y = \frac{\hat{L}_+ - \hat{L}_-}{2i} = \frac{\hbar}{\sqrt{2}} \begin{bmatrix} 0 & -i & 0 \\ i & 0 & -i \\ 0 & i & 0 \end{bmatrix} \tag{16.157}$$

For $\hat{L}^2$ and $\hat{L}_z$, we use $\hat{L}^2|l,m\rangle = l(l+1)\hbar^2|l,m\rangle$ and $\hat{L}_z|l,m\rangle = m\hbar|l,m\rangle$, respectively. It is then easy to show that $\hat{L}^2$ and $\hat{L}_z$ are diagonal matrices. For $\hat{L}^2$, all three eigenvalues are equal to $l(l+1)\hbar^2 = 2\hbar^2$, which is consistent with the properties discussed in Sections 16.2.1 and 16.3,

$$\hat{L}^2 = \begin{bmatrix} 2\hbar^2 & 0 & 0 \\ 0 & 2\hbar^2 & 0 \\ 0 & 0 & 2\hbar^2 \end{bmatrix} = 2\hbar^2 \begin{bmatrix} 1 & 0 & 0 \\ 0 & 1 & 0 \\ 0 & 0 & 1 \end{bmatrix} \tag{16.158}$$

The diagonal elements of $\hat{L}_z$ are $m\hbar = -1\hbar, 0, +1\hbar$.

$$\hat{L}_z = \hbar \begin{bmatrix} -1 & 0 & 0 \\ 0 & 0 & 0 \\ 0 & 0 & 1 \end{bmatrix} \qquad (16.159)$$

16.6 Matrix Representation of Generalized Angular Momentum

The raising and lowering operators of the generalized angular momentum are constructed in the same way as we did for $\hat{L}$. The properties extracted before hold, except that l must be replaced by j and noting that j and m can be half integers too.

$$\hat{J}_+ = \hat{J}_x + i\hat{J}_y \qquad (16.160)$$
$$\hat{J}_- = \hat{J}_x - i\hat{J}_y \qquad (16.161)$$

The matrix form of the raising and lowering operators can be found using the following equations which are similar to (16.140) and (16.141):

$$\hat{J}_+|j, m\rangle = \sqrt{j(j+1) - m(m+1)}\hbar|j, m+1\rangle \qquad (16.162)$$
$$\hat{J}_-|j, m\rangle = \sqrt{j(j+1) - m(m-1)}\hbar|j, m-1\rangle \qquad (16.163)$$

where $|m| \leq j$.

Particles with generalized angular momentum $j = \frac{1}{2}$ and $j = \frac{3}{2}$

Electrons, protons, and neutrons are called fermions. They have a generalized angular momentum of $j = \frac{1}{2}$ which often is written as $s = \frac{1}{2}$ (s for spin). The nucleus of some atoms like $_3^7Li$ has spin of $3/2$.

Electron spin $(j = \frac{1}{2})$: The generalized angular momentum squared in this case is $(j+1)\hbar^2 = \frac{3}{4}\hbar^2$. There are two different values for m (representing the projection of spin or angular momentum on z-axis) $m = -\frac{1}{2}, \frac{1}{2}$ because $2j + 1 = 2$. The Hilbert space is simply composed of two basis vectors corresponding to two different $|j, m\rangle$:

$$|j, m\rangle \in \left\{ \left| \frac{1}{2}, +\frac{1}{2} \right\rangle, \left| \frac{1}{2}, -\frac{1}{2} \right\rangle \right\} = \left\{ \begin{bmatrix} 1 \\ 0 \end{bmatrix}, \begin{bmatrix} 0 \\ 1 \end{bmatrix} \right\} \qquad (16.164)$$

The above basis vectors are also represented using a more compact notation, $\{|up\rangle, |down\rangle\}$ or $\{| \uparrow\rangle, | \downarrow\rangle,\}$ or $\{|+\frac{1}{2}\rangle, |-\frac{1}{2}\rangle\}$ among others. It is easy to show that $\hat{J}^2$ and $\hat{J}_z$ in this basis are

$$\hat{J}^2 = \frac{3\hbar^2}{4} \begin{bmatrix} 1 & 0 \\ 0 & 1 \end{bmatrix} \quad \text{and} \quad \hat{J}_z = \frac{\hbar}{2} \begin{bmatrix} 1 & 0 \\ 0 & -1 \end{bmatrix} \qquad (16.165)$$

The x and y components are found from the raising and lowering operators, which are

$$\hat{J}_+ \left| \frac{1}{2}, \frac{1}{2} \right\rangle = 0 \qquad (16.166)$$

$$\hat{J}_+ \left| \frac{1}{2}, -\frac{1}{2} \right\rangle = \hbar \sqrt{\frac{1}{2} \left(\frac{1}{2} + 1 \right) + \frac{1}{2} \left(-\frac{1}{2} + 1 \right)} \, \hbar \left| \frac{1}{2}, -\frac{1}{2} + 1 \right\rangle$$

$$= \hbar \left| \frac{1}{2}, \frac{1}{2} \right\rangle \qquad (16.167)$$

For the lowering operator, we have

$$\hat{J}_- \left| \frac{1}{2}, \frac{1}{2} \right\rangle = \hbar \sqrt{\frac{1}{2} \left(\frac{1}{2} + 1 \right) - \frac{1}{2} \left(\frac{1}{2} - 1 \right)} \, \hbar \left| \frac{1}{2}, \frac{1}{2} - 1 \right\rangle$$

$$= \hbar \left| \frac{1}{2}, -\frac{1}{2} \right\rangle \qquad (16.168)$$

$$\hat{J}_- \left| \frac{1}{2}, -\frac{1}{2} \right\rangle = 0 \qquad (16.169)$$

The raising operator cannot raise the topmost eigenstate, and returns zero. And, the lowering operator cannot lower the bottom-most eigenstate anymore, and returns zero.

The only nonzero element of $\hat{J}_+$ is $\langle \frac{1}{2}, \frac{1}{2} | \hat{J}_+ | \frac{1}{2}, -\frac{1}{2} \rangle = \hbar$ and the only nonzero element of $\hat{J}_-$ is $\langle \frac{1}{2}, -\frac{1}{2} | \hat{J}_+ | \frac{1}{2}, \frac{1}{2} \rangle = \hbar$. Therefore, we can write

$$\hat{J}_+ = \hbar \begin{bmatrix} 0 & 1 \\ 0 & 0 \end{bmatrix} \quad \text{and} \quad \hat{J}_- = \hbar \begin{bmatrix} 0 & 0 \\ 1 & 0 \end{bmatrix} \tag{16.170}$$

From the above, the x and y components of the generalized angular momentum are as follows:

$$\hat{J}_x = \frac{\hbar}{2} \begin{bmatrix} 0 & 1 \\ 1 & 0 \end{bmatrix} \quad \text{and} \quad \hat{J}_y = \frac{\hbar}{2} \begin{bmatrix} 0 & -i \\ i & 0 \end{bmatrix} \tag{16.171}$$

The reader should note that these are the same spin operators $\hat{S}_x$ and $\hat{S}_y$ derived in the spin angular momentum chapter (Chapter 15).

Figure 16.7 shows the projections of spin (angular momentum) of $j = \frac{1}{2}$ on z-axis.

Particle with angular momentum ($j = \frac{3}{2}$): The process of deriving the matrix form of the angular momentum operators is like the previous cases. The total angular momentum squared in this

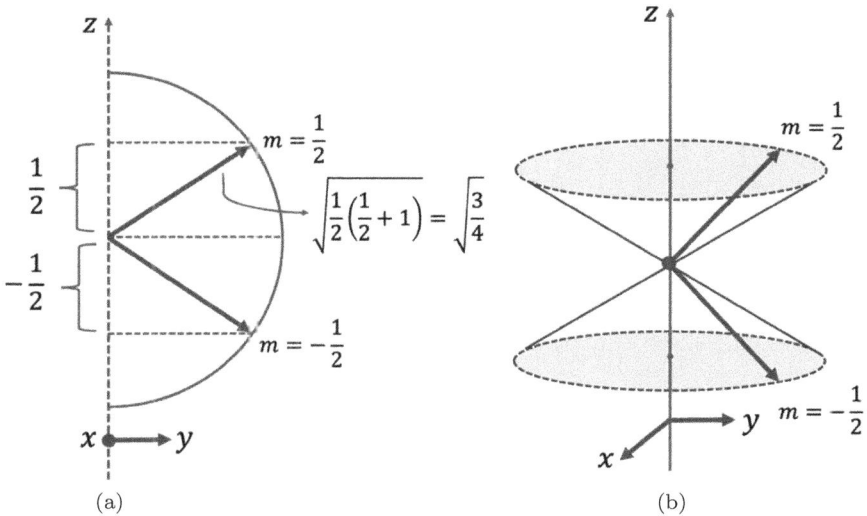

Figure 16.7. (a) Projections of generalized angular momentum on the z-axis for a particle with $j = \frac{1}{2}$. (b) The tip of $\bar{J}$ is located somewhere on the periphery of the yellow-colored circles around the z-axis.

case is

$$j(j+1)\hbar^2 = \frac{3}{2}\left(\frac{3}{2}+1\right)\hbar^2 = \frac{15}{4}\hbar^2$$

There are four different values for m (the projection of angular momentum on z-axis) which are: $m = -\frac{3}{2}, -\frac{1}{2}, +\frac{1}{2}, +\frac{3}{2}$, because $2j+1 = 4$. The Hilbert space is then made of four vectors to represent four different $|j, m\rangle$. We arrange them from the highest to the lowest value of m. Note that choosing the order is arbitrary (but it must be kept fixed while working out a problem),

$$|j, m\rangle \in \left\{\left|\frac{3}{2}, +\frac{3}{2}\right\rangle, \left|\frac{3}{2}, +\frac{1}{2}\right\rangle, \left|\frac{3}{2}, -\frac{1}{2}\right\rangle, \left|\frac{3}{2}, -\frac{3}{2}\right\rangle\right\}$$

$$= \left\{\begin{bmatrix} 1 \\ 0 \\ 0 \\ 0 \end{bmatrix}, \begin{bmatrix} 0 \\ 1 \\ 0 \\ 0 \end{bmatrix}, \begin{bmatrix} 0 \\ 0 \\ 1 \\ 0 \end{bmatrix}, \begin{bmatrix} 0 \\ 0 \\ 0 \\ 1 \end{bmatrix}\right\} \tag{16.172}$$

It is easy to show that $\hat{J}^2$ and $\hat{J}_z$ in this basis are:

$$\hat{J}^2 = \frac{15\hbar^2}{4}\begin{bmatrix} 1 & 0 & 0 & 0 \\ 0 & 1 & 0 & 0 \\ 0 & 0 & 1 & 0 \\ 0 & 0 & 0 & 1 \end{bmatrix} \quad \text{and} \quad \hat{J}_z = \frac{\hbar}{2}\begin{bmatrix} +3 & 0 & 0 & 0 \\ 0 & +1 & 0 & 0 \\ 0 & 0 & -1 & 0 \\ 0 & 0 & 0 & -3 \end{bmatrix} \tag{16.173}$$

Readers can show that the x and y components of the $\hat{J}$ are

$$\hat{J}_x = \frac{\hbar}{2}\begin{bmatrix} 0 & \sqrt{3} & 0 & 0 \\ \sqrt{3} & 0 & 2 & 0 \\ 0 & 2 & 0 & \sqrt{3} \\ 0 & 0 & \sqrt{3} & 0 \end{bmatrix} \quad \text{and}$$

$$\hat{J}_y = \frac{\hbar}{2} \begin{bmatrix} 0 & -i\sqrt{3} & 0 & 0 \\ i\sqrt{3} & 0 & 2i & 0 \\ 0 & 2i & 0 & -i\sqrt{3} \\ 0 & 0 & i\sqrt{3} & 0 \end{bmatrix} \tag{16.174}$$

Figure 16.8 shows the projections of angular momentum $\left(j = \frac{3}{2}\right)$ on the z-axis.

16.7 Appendix A: Levi-Civita Symbol

Equation (16.20) can be written in a general form using the Levi-Civita symbol, ε_{pqr},

$$[\hat{L}_p, \hat{L}_q] = i\hbar\varepsilon_{pqr}\hat{L}_r$$

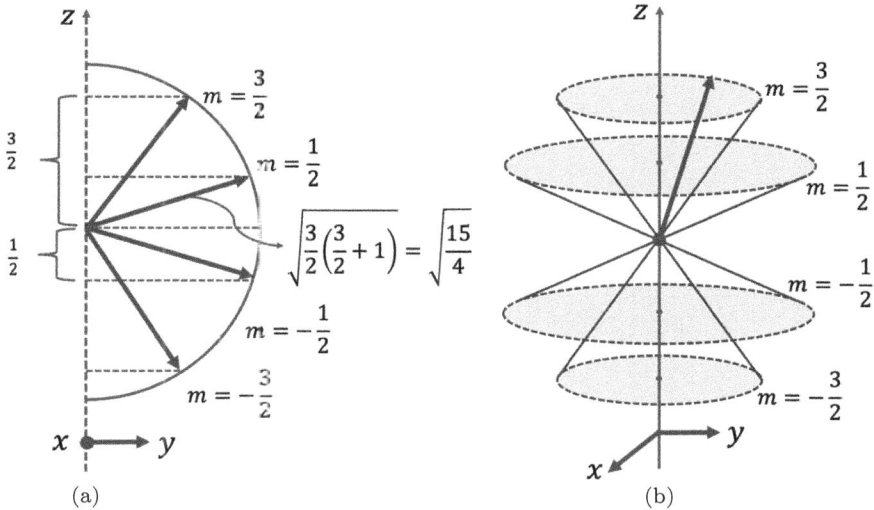

Figure 16.8. Projections of generalized angular momentum on z-axis for a particle with $j = 3/2$. (b) Shows the 3D rendering of the tip of $\bar{J}$ which is located on the periphery of the yellow-colored circles around the z-axis.

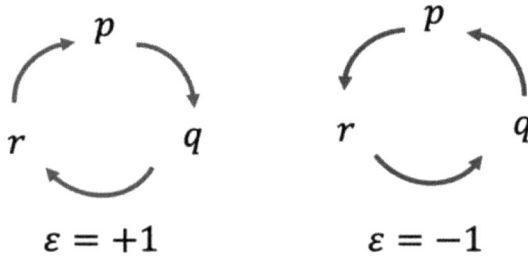

$$\varepsilon = +1 \qquad\qquad \varepsilon = -1$$

Figure 16.9. Clockwise rotation (left) and counterclockwise (right) rotation of indices.

The value of ε_{pqr} depends on how the indices p, q and r are arranged. Note that p, q, and r can stand for any index from the xyz triplet.

Permutation of the indices so that they rotate and convert to each other on a clockwise rotating cycle makes $\varepsilon_{pqr} = +1$ (see Figure 16.9). Arranging the indices on a counterclockwise rotating circle makes $\varepsilon_{pqr} = -1$. Repetition of the indices makes $\varepsilon_{pqr} = 0$.

Then, the symbol value is summarized as follows:

$$\varepsilon_{pqr} = \begin{cases} 1 & \text{if } p \to q \to r \to p \\ -1 & \text{if } r \to q \to p \to r \\ 0 & \text{if } p = q, \text{ or } p = r, \text{or } q = r \end{cases}$$

Here are a few examples:

$[\hat{L}_x, \hat{L}_y] = i\hbar\varepsilon_{xyz}\hat{L}_z = i\hbar \cdot 1 \cdot \hat{L}_z$ because the xyz triplet follows the first pattern of the above equation.

$[\hat{L}_x, \hat{L}_z] = i\hbar\varepsilon_{xzy}\hat{L}_y = i \cdot (-1) \cdot \hat{L}_y = -i\hbar\hat{L}_y$ because the order of x to z requires counterclockwise rotation of the xyz indices. Hence, there is a -1 for the symbol.

$[\hat{L}_y, \hat{L}_y] = 0$, because it contains two similar indices, and the symbol is 0 according to its definitions.

Note that the Levi-Civita symbol is only a *memory aid*. It helps us to memorize the commutation relations between any two components of orbital ($\hat{L}$) as well as generalized angular momentum ($\hat{J}$).

16.8 Appendix B: Angular Momentum Operators in Spherical Coordinate

Substituting equations (16.27) to (16.32) in equations (16.5), (16.6), and (16.7), we can derive $\hat{L}_x$, $\hat{L}_y$, and $\hat{L}_z$ in terms of polar coordinates

$$\frac{i}{\hbar}\hat{L}_x = r\sin\theta \sin\phi \left[\cos\theta\frac{\partial}{\partial r} - \sin\theta\frac{1}{r}\frac{\partial}{\partial\theta}\right]$$

$$- r\cos\theta\left[\sin\theta\sin\phi\frac{\partial}{\partial r} + \cos\theta\sin\phi\frac{1}{r}\frac{\partial}{\partial\theta} - \frac{\cos\phi}{r\sin\theta}\frac{\partial}{\partial\phi}\right]$$

$$= -\sin^2\theta\sin\phi\frac{\partial}{\partial\theta} - \cos^2\theta\sin\phi\frac{\partial}{\partial\theta} - \frac{\cos\theta\cos\phi}{\sin\theta}\frac{\partial}{\partial\phi}$$

$$\hat{L}_x = -\frac{\hbar}{i}\left[\sin\phi\frac{\partial}{\partial\theta} + \cot\theta\cos\phi\frac{\partial}{\partial\phi}\right]$$

$$\frac{i}{\hbar}\hat{L}_y = r\cos\theta\left[\sin\theta\cos\phi\frac{\partial}{\partial r} + \cos\theta\cos\phi\frac{1}{r}\frac{\partial}{\partial\theta} - \frac{\sin\phi}{r\sin\theta}\frac{\partial}{\partial\phi}\right]$$

$$- r\sin\theta\cos\phi\left[\cos\theta\frac{\partial}{\partial r} - \sin\theta\frac{1}{r}\frac{\partial}{\partial\theta}\right]$$

$$= \cos^2\theta\cos\phi\frac{\partial}{\partial\theta} - \cot\theta\sin\phi\frac{\partial}{\partial\phi} + \sin^2\theta\cos\phi\frac{\partial}{\partial\theta}$$

$$\hat{L}_y = \frac{\hbar}{i}\left[\cos\phi\frac{\partial}{\partial\theta} - \cot\theta\sin\phi\frac{\partial}{\partial\phi}\right]$$

$$\hat{L}_z = \frac{\hbar}{i}\left[r\sin\theta\cos\phi\left[\sin\theta\sin\phi\frac{\partial}{\partial r} + \cos\theta\sin\phi\frac{1}{r}\frac{\partial}{\partial\theta} - \frac{\cos\phi}{r\sin\theta}\frac{\partial}{\partial\phi}\right]\right.$$

$$\left. - r\sin\theta\sin\phi\left[\sin\theta\cos\phi\frac{\partial}{\partial r} + \cos\theta\cos\phi\frac{1}{r}\frac{\partial}{\partial\theta} - \frac{\sin\phi}{r\sin\theta}\frac{\partial}{\partial\phi}\right]\right]$$

$$\hat{L}_z = \frac{\hbar}{i}\frac{\partial}{\partial\phi}$$

From the equations for $\hat{L}_x$ and $\hat{L}_y$, we have the equations for $\hat{L}_x^2$ and $\hat{L}_y^2$ as follows:

$$\hat{L}_x^2 = \left(-\frac{\hbar}{i}\right)^2\left[\sin\phi\frac{\partial}{\partial\theta} + \cot\theta\cos\phi\frac{\partial}{\partial\phi}\right]\left[\sin\phi\frac{\partial}{\partial\theta} + \cot\theta\cos\phi\frac{\partial}{\partial\phi}\right]$$

$$\hat{L}_x^2 f = -\hbar^2 \left[\sin^2 \phi \frac{\partial^2}{\partial\theta^2} + \sin\phi\cos\phi \frac{\partial}{\partial\phi}\frac{\partial}{\partial\theta}\cot\theta \right.$$

$$\left. + \cot\theta \frac{\partial}{\partial\theta}\cos\phi \frac{\partial}{\partial\phi}\sin\phi + \cot^2\theta\cos\phi \frac{\partial}{\partial\phi}\cos\phi \frac{\partial}{\partial\phi} \right] f$$

$$= -\hbar^2 \left[\sin^2\phi \frac{\partial^2}{\partial\theta^2} + \sin\phi\cos\phi\cot\theta \frac{\partial}{\partial\phi}\frac{\partial}{\partial\theta} - \frac{\sin\phi\cos\phi}{\sin^2\theta}\frac{\partial}{\partial\phi} \right.$$

$$+ \cot\theta\cos\phi\sin\phi \frac{\partial}{\partial\theta}\frac{\partial}{\partial\phi} + \cot\theta\cos^2\phi \frac{\partial}{\partial\theta}$$

$$\left. - \cot^2\theta\cos\phi\sin\phi \frac{\partial}{\partial\phi} + \cot^2\theta\cos^2\phi \frac{\partial^2}{\partial\phi^2} \right] f$$

$$\hat{L}_y^2 = \left(\frac{\hbar}{i}\right)^2 \left[\cos\phi \frac{\partial}{\partial\theta} - \cot\theta\sin\phi \frac{\partial}{\partial\phi} \right]\left[\cos\phi \frac{\partial}{\partial\theta} - \cot\theta\sin\phi \frac{\partial}{\partial\phi} \right]$$

$$\hat{L}_y^2 f = -\hbar^2 \left[\cos^2\phi \frac{\partial^2}{\partial\theta^2} - \sin\phi\cos\phi \frac{\partial}{\partial\phi}\frac{\partial}{\partial\theta}\cot\theta \right.$$

$$\left. - \cot\theta \frac{\partial}{\partial\theta}\sin\phi \frac{\partial}{\partial\phi}\cos\phi + \cot^2\theta\sin\phi \frac{\partial}{\partial\phi}\sin\phi \frac{\partial}{\partial\phi} \right] f$$

$$= -\hbar^2 \left[\cos^2\phi \frac{\partial^2}{\partial\theta^2} - \sin\phi\cos\phi\cot\theta \frac{\partial}{\partial\phi}\frac{\partial}{\partial\theta} + \frac{\sin\phi\cos\phi}{\sin^2\theta}\frac{\partial}{\partial\phi} \right.$$

$$- \cot\theta\cos\phi\sin\phi \frac{\partial}{\partial\theta}\frac{\partial}{\partial\phi} + \cot\theta\sin^2\phi \frac{\partial}{\partial\theta}$$

$$\left. + \cot^2\theta\cos\phi\sin\phi \frac{\partial}{\partial\phi} + \cot^2\theta\sin^2\phi \frac{\partial^2}{\partial\phi^2} \right] f$$

The sum of x and y components squared is then

$$[\hat{L}_x^2 + \hat{L}_y^2]f = -\hbar^2 \left[\sin^2\phi \frac{\partial^2}{\partial\theta^2} + \cos^2\phi \frac{\partial^2}{\partial\theta^2} + 0 - 0 + 0 \right.$$

$$+ \cot\theta\cos^2\phi \frac{\partial}{\partial\theta} + \cot\theta\sin^2\phi \frac{\partial}{\partial\theta} - 0$$

$$\left. + \cot^2\theta\cos^2\phi \frac{\partial^2}{\partial\phi^2} + \cot^2\theta\sin^2\phi \frac{\partial^2}{\partial\phi^2} \right] f$$

$$= -\hbar^2 \left[\frac{\partial^2}{\partial\theta^2} + \cot\theta \frac{\partial}{\partial\theta} + \cot^2\theta \frac{\partial^2}{\partial\phi^2} \right] f$$

Using

$$\hat{L}_z^2 f = -\hbar^2 \frac{\partial^2 f}{\partial \phi^2}$$

we can immediately write down

$$\hat{L}^2 = [\hat{L}_x^2 + \hat{L}_y^2 + \hat{L}_z^2] f = -\hbar^2 \left[\frac{\partial^2}{\partial \theta^2} + \cot \theta \frac{\partial}{\partial \theta} + \cot^2 \theta \frac{\partial^2}{\partial \phi^2} + \frac{\partial^2}{\partial \phi^2} \right] f$$

$$= -\hbar^2 \left[\frac{\partial^2}{\partial \theta^2} + \cot \theta \frac{\partial}{\partial \theta} + \frac{\cos^2 \theta}{\sin^2 \theta} \frac{\partial^2}{\partial \phi^2} + \frac{\sin^2 \theta}{\sin^2 \theta} \frac{\partial^2}{\partial \phi^2} \right] f$$

$$= -\hbar^2 \left[\frac{\partial^2}{\partial \theta^2} + \cot \theta \frac{\partial}{\partial \theta} + \frac{1}{\sin^2 \theta} \frac{\partial^2}{\partial \phi^2} \right] f$$

which completes the proof of equation (16.36).

16.9 Problems

Section 16.2

(1) Figure 16.10 shows how the unit vectors in spherical and Cartesian coordinates are related. Use these relations and the ones given for the derivative of the unit vectors to derive the expression for L^2 from its definition $L = r \times P$ (Gupta, 1976).

Hint: Use the following relations, i.e., decompose each spherical coordinate unit vector with the help of its projections along each Cartesian coordinate (x, y, z) as follows:

$$\mathbf{1}_r = \mathbf{1}_x \sin \theta \cos \varphi + \mathbf{1}_y \sin \theta \sin \varphi + \mathbf{1}_z \cos \theta$$

$$\mathbf{1}_\theta = \mathbf{1}_x \cos \theta \cos \varphi + \mathbf{1}_y \cos \theta \sin \varphi - \mathbf{1}_z \sin \theta$$

$$\mathbf{1}_\varphi = -\mathbf{1}_x \sin \varphi + \mathbf{1}_y \cos \varphi$$

The following relations are also required, which can be proven from the above:

$$\frac{\partial \mathbf{1}_\theta}{\partial \varphi} = \mathbf{1}_\theta \cos \theta$$

$$\frac{\partial \mathbf{1}_\varphi}{\partial \theta} = 0$$

$$\frac{\partial \mathbf{1}_\varphi}{\partial \varphi} = -\mathbf{1}_r \sin \theta - \mathbf{1}_\theta \cos \theta$$

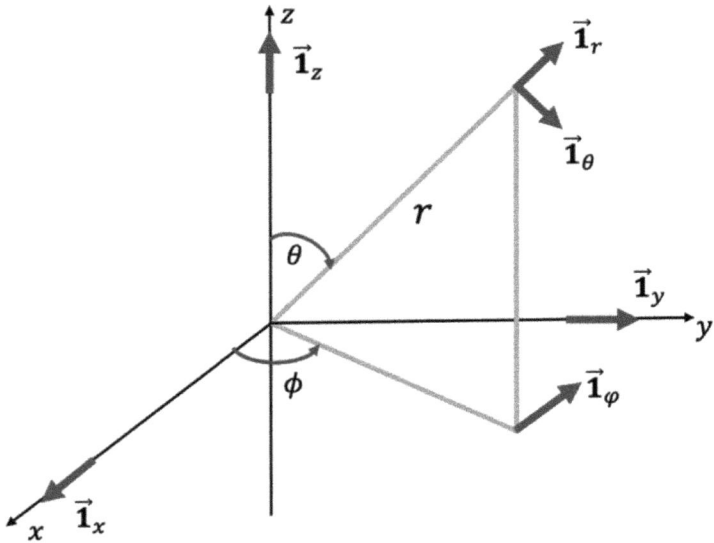

Figure 16.10. The unit vectors in the Cartesian coordinate are indexed by x, y and z. The unit vectors in spherical coordinates are indexed by r, θ and φ. These vectors point towards the direction of growth of their respective coordinate.

1 represents a unit vector in the direction defined by the subscript.

(2) Show that normalization factor, A, in the wave function $Ae^{im\phi}$ (equation (16.60)) is equal to

$$A = \frac{1}{\sqrt{2\pi}}$$

(3) For the 2D rotor discussed in section 16.2.1: (a) Find the eigenvalues and eigenfunction of L_z corresponding to the quantum number $m = \pm1$. (b) If the wave function is $X = \frac{2}{\sqrt{2}}\cos(2\phi)$, find the expectation value of L_z and energy. (c) For the wave function in (b), what will be the results of a measurement of the total angular momentum? (**Hint:** Apply operator $\hat{L}^2$ to X.)

(4) For a rotating mass in a 2D plane, show that the probability of finding the particle in an angle from to β radians is

$$P_\beta = \frac{\beta}{2\pi}$$

Convince yourself that the probability density is uniform and equal to $1/2\pi$.

(5) Using $[\hat{L}_x, \hat{L}^2] = 0$ and $[\hat{L}_y, \hat{L}^2] = 0$, prove that $[\hat{L}^2, \hat{L}_x^2] = [\hat{L}^2, \hat{L}_y^2] = 0$.

Section 16.3

(6) Show that $[\hat{L}_+, \hat{L}_-] = 2\hbar \hat{L}_z$. The same can be said for $\hat{J}$, i.e., $[\hat{J}_+, \hat{J}_-] = 2\hbar \hat{J}_z$.

(7) Prove $\hat{L}_z(\hat{L}_+^n |\chi\rangle) = (a + n\hbar)(\hat{L}_+^n |\chi\rangle)$ and $\hat{L}^2(\hat{L}_+^n |\chi\rangle) = b(\hat{L}_+^n |\chi\rangle)$ using mathematical induction. This involves showing that they are true for $n = 1$, $n = 2$ and $n + 1$.

(8) Prove $\hat{L}_z(\hat{L}_-^n |\chi\rangle) = (a - n\hbar)(\hat{L}_-^n |\chi\rangle)$ and $\hat{L}^2(\hat{L}_-^n |\chi\rangle) = b(\hat{L}_-^n |\chi\rangle)$ using mathematical induction. This involves showing that they are true for $n = 1$, $n = 2$ and $n + 1$.

(9) Find the minimum angle between the angular momentum and its projection on the z-axis for the classical rotor of Example 3 in the text.

Section 16.5

(10) Show that $\hat{L}_- |l, m\rangle = \sqrt{l(l+1) - m(m-1)}\hbar |l, m-1\rangle$ (equation (16.141)).

(11) Redrive equation (16.155) using equation (16.141).

(12) Find the matrix representation of orbital angular momentum $(\hat{L}^2)$ of a particle in the $l = 2$ state. Discuss a physical scenario where you can find a particle in such an angular momentum state.

Section 16.6

(13) Using the properties of the raising operator, show that the matrix element $\langle j, m | \hat{J}_+ | j, m' \rangle$ (row m and column m') is nonzero only if m is one unit larger than m i.e., when $m = m' + 1$.

(14) Using the properties of the lowering operator, show that the matrix element $\langle j, m | \hat{J}_- | j, m' \rangle$ (row m and column m') is

nonzero only if m is one unit smaller than m i.e., when $m = m' - 1$.

(15) Find the matrix representation of $\hat{J}_x$, $\hat{J}_y$, $\hat{J}_z$, $\hat{J}^2$, $\hat{J}_+$, and $\hat{J}_-$ of a particle with generalized angular momentum $j = 1$.

(16) Show that the raising and lowering operators for $j = 3/2$ in matrix form are given by:

$$\hat{J}_+ = \hbar \begin{bmatrix} 0 & \sqrt{3} & 0 & 0 \\ 0 & 0 & 2 & 0 \\ 0 & 0 & 0 & \sqrt{3} \\ 0 & 0 & 0 & 0 \end{bmatrix} \quad \text{and} \quad \hat{J}_- = \hbar \begin{bmatrix} 0 & 0 & 0 & 0 \\ \sqrt{3} & 0 & 0 & 0 \\ 0 & 2 & 0 & 0 \\ 0 & 0 & \sqrt{3} & 0 \end{bmatrix}$$

(17) From Problem 16, show that the matrix forms of $\hat{J}_x$ and $\hat{J}_y$ for $j = 3/2$ are given by:

$$\hat{J}_x = \frac{\hbar}{2} \begin{bmatrix} 0 & \sqrt{3} & 0 & 0 \\ \sqrt{3} & 0 & 2 & 0 \\ 0 & 2 & 0 & \sqrt{3} \\ 0 & 0 & \sqrt{3} & 0 \end{bmatrix} \quad \text{and}$$

$$\hat{J}_y = \frac{\hbar}{2} \begin{bmatrix} 0 & -i\sqrt{3} & 0 & 0 \\ i\sqrt{3} & 0 & 2i & 0 \\ 0 & 2i & 0 & -i\sqrt{3} \\ 0 & 0 & i\sqrt{3} & 0 \end{bmatrix}$$

Reference

Gupta, P. D. (1976). A new derivation of quantum-mechanical angular momentum operator L^2. *American Journal of Physics* **44**, 888.

Chapter 17

TIME-INDEPENDENT PERTURBATION THEORY

Contents

17.1 Introduction

In Chapter 9, we learned that a physical system with two eigenfunctions (which can be called an artificial atom or two-level system) could represent a qubit, but leaving the qubit in an eigenstate has no practical application. To make a logic gate and finally a quantum computer, we must write or change the qubit's value and control its characteristics by some means. We need to know how to *perturb* our quantum system at hand and predict its response to this *perturbation*.

   Perturbation theory, as discussed in this chapter, is a method to solve the eigenvalue problem of a system *approximately* from the knowledge of a closely related system whose eigenvalues and eigenfunctions are known. This chapter deals with a perturbation that does not depend on time; the following chapter considers a time-dependent perturbation.

We will assume that somehow, we have managed to solve the original system with Hamiltonian H_o exactly, and we know its exact eigenfunctions and eigenenergies (φ_{no}, E_{no}). We will look for a mathematical method to find the new eigenvalues and eigenstates (wave functions) of the perturbed system due to this newly added **time-independent** perturbation U to the Hamiltonian.

A few decades ago, when powerful computer simulations were not as abundant as they are today, microwave engineers used time-independent perturbation theory to find eigenfrequencies and eigenmodes of electromagnetic waves in waveguides with small structural imperfections. Refer to (Collin, 2001) for further reading.

While many time-independent perturbation problems can be solved by computer simulation, learning time-independent perturbation theory gives additional insight into predicting and gaining a qualitative understanding of a physical system.

17.2   Perturbation Theory

Consider a system with Hamiltonian H_o whose solution (φ_{no}, E_{no}) is precisely known. Now, consider that the Hamiltonian changes by a small perturbation so that the *perturbed* Hamiltonian is $H = H_o + U$. Is there a simple way to find the eigenfunctions and energy levels of the perturbed Hamiltonian H?

The eigenvalues (E_{ko}) and eigenfunctions $(|\varphi_{ko}\rangle)$ of the unperturbed Hamiltonian are given by,

$$H_o|\varphi_{ko}\rangle = E_{ko}|\varphi_{ko}\rangle \quad \text{(unperturbed)} \qquad (17.1)$$

and the eigenfunctions are orthonormal,

$$\langle\varphi_{mo}|\varphi_{no}\rangle = \delta_{nm} \quad \text{(unperturbed)} \qquad (17.2)$$

When the perturbation U is small, one anticipates that the eigenvalues and eigenfunctions of the perturbed Hamiltonian are a power series of the perturbation strength U. First, we will assume the strength of the weak perturbation is controllable by a parameter λ, which controls the degree to which the perturbing potential U is applied

$$H_\lambda = H_o + \lambda U \tag{17.3}$$

The eigenvalues and eigenfunctions of the above Hamiltonian follow from

$$H_\lambda|\varphi_k\rangle = (H_o + \lambda U)|\varphi_k\rangle = E_k|\varphi_k\rangle \quad \text{(perturbed)} \tag{17.4}$$

When $\lambda = 0$ and 1, we get the eigenvalues and eigenfunctions of the unperturbed (H_o) and perturbed $(H = H_o + U)$ Hamiltonians, respectively. Because the perturbation is small, we assume that $|\varphi_k\rangle$ can be expanded as a linear combination of the unperturbed eigenfunctions $\{|\varphi_{n}\rangle\}$,

$$|\varphi_k\rangle = |\varphi_{ko}\rangle + \sum_{m \neq k} a_m|\varphi_{mo}\rangle \quad \text{(not normalized)} \tag{17.5}$$

When $\lambda = 0$, as expected $|\varphi_k\rangle = |\varphi_{ko}\rangle$. The additional terms corresponding to quantum numbers $m \neq k$ are corrections due to the perturbing potential U (we will see this later), meaning that perturbation causes unperturbed states with different quantum numbers to affect each other. The eigenfunction $|\varphi_k\rangle$ is not normalized when $\lambda \neq 0$. We write a_m and E_k as polynomials of the perturbation parameter λ,

$$a_m = \lambda a_m^{(1)} + \lambda^2 a_m^{(2)} + \cdots \tag{17.6}$$

$$E_k = E_{ko} + \lambda E_k^{(1)} + \lambda^2 E_k^{(2)} + \cdots \tag{17.7}$$

The quantities with (1) and (2) as superscripts are unknown in the above equations, and are not exponents. The index k refers to the quantum numbers of the unperturbed Hamiltonian. Let us plugin equations (17.5), (17.6), and (17.7) in equation (17.4),

$$(H_o + \lambda U)\left[|\varphi_{ko}\rangle + \sum_{m \neq k}(\lambda a_m^{(1)} + \lambda^2 a_m^{(2)} + \cdots)|\varphi_{mo}\rangle\right]$$

$$= (E_{ko} + \lambda E_k^{(1)} + \lambda^2 E_k^{(2)} + \cdots) \left[|\varphi_{ko}\rangle \right.$$

$$\left. + \sum_{m \neq k} (\lambda a_m^{(1)} + \lambda^2 a_m^{(2)} + \cdots)|\varphi_{mo}\rangle \right] \qquad (17.8)$$

After expanding the terms and factoring out different powers of λ we have,

$$\left. \begin{array}{l} H_o|\varphi_{ko}\rangle \\[4pt] +\lambda \left[U|\varphi_{ko}\rangle + \sum_{m \neq k} a_m^{(1)} H_o|\varphi_{mo}\rangle \right] \\[4pt] +\lambda^2 \left[\sum_{m \neq k} a_m^{(1)} U|\varphi_{mo}\rangle + a_m^{(2)} H_o|\varphi_{mo}\rangle \right] \\[4pt] + \cdots \end{array} \right\}$$

$$= \left\{ \begin{array}{l} E_{ko}|\varphi_{ko}\rangle \\[4pt] +\lambda \left[E_k^{(1)}|\varphi_{ko}\rangle + E_{ko} \sum_{m \neq k} a_m^{(1)}|\varphi_{mo}\rangle \right] \\[4pt] +\lambda^2 \left[E_k^{(2)}|\varphi_{ko}\rangle + E_{ko} \sum_{m \neq k} a_m^{(2)}|\varphi_{mo}\rangle + E_k^{(1)} \sum_{m \neq k} a_m^{(1)}|\varphi_{mo}\rangle \right] \\[4pt] + \cdots \end{array} \right.$$

$$(17.9)$$

The above equation must be valid for all values of λ between 0 and 1, which controls the strength of the perturbation. For this to be true, the coefficients of any λ^n on the LHS should be equal to the ones on the RHS. Equating the coefficients of λ^0 (i.e., constant terms with respect to λ), we immediately get

$$H_o|\varphi_{ko}\rangle = E_{ko}|\varphi_{ko}\rangle \qquad (17.10)$$

which is the Schrödinger equation for the unperturbed system. Equating the coefficients of λ, we get

$$U|\varphi_{ko}\rangle + \sum_{m \neq k} a_m^{(1)} H_o|\varphi_{mo}\rangle = E_k^{(1)}|\varphi_{ko}\rangle + E_{ko} \sum_{m \neq k} a_m^{(1)}|\varphi_{mo}\rangle \quad (17.11)$$

Operating from the left-hand side by $\langle \varphi_{ko} |$ and using equations (17.1) and (17.2), we get the first-order correction to energy which is

$$E_k^{(1)} = \langle \varphi_{ko} | U | \varphi_{ko} \rangle \tag{17.12}$$

Substituting this in equation (17.7) (setting $\lambda = 1$), the first-order corrected eigenenergy is found to be

$$E_k = E_{ko} + \langle \varphi_{ko} | U | \varphi_{ko} \rangle \quad \text{(the 1}^{\text{st}} \text{ order corrected energy)} \tag{17.13}$$

Now to find $a_m^{(1)}$, we operate equation (17.11) from the left-hand side by $\langle \varphi_{no} |$ $(n \neq k)$ and use equations (17.1) and (17.2) to get

$$U_{nk} + \sum_{m \neq k} a_m^{(1)} E_{mo} \delta_{nm} = \sum_{m \neq k} E_{ko} a_m^{(1)} \delta_{nm} \tag{17.14}$$

from which $a_n^{(1)}$ is found to be

$$a_n^{(1)} = \frac{U_{nk}}{E_{ko} - E_{no}} \quad (n \neq k) \tag{17.15}$$

Using this in equations (17.5) and (17.6) (by setting $\lambda = 1$), the first-order corrected eigenfunction is

$$|\varphi_k\rangle = |\varphi_{ko}\rangle + \sum_{m \neq k} \frac{U_{mk}}{E_{ko} - E_{mo}} |\varphi_{mo}\rangle$$

$$\text{(the 1}^{\text{st}} \text{ order corrected eigenfunction)} \tag{17.16}$$

Now we continue to find the second-order corrections to the eigenenergy and eigenfunction. This is done by equating the coefficients of λ^2 in equation (17.9) to get,

$$\sum_{m \neq k} (a_m^{(1)} U | \varphi_{mo}\rangle + a_m^{(2)} H_o | \varphi_{mo}\rangle)$$

$$= E_k^{(2)} |\varphi_{ko}\rangle + E_{ko} \sum_{m \neq k} a_m^{(2)} |\varphi_{mo}\rangle + E_k^{(1)} \sum_{m \neq k} a_m^{(1)} |\varphi_{mo}\rangle \tag{17.17}$$

Operating from the left-hand side by $\langle \varphi_{ko} |$ and using equations (17.1) and (17.2), the second-order correction to energy is,

$$E_k^{(2)} = \sum_{m \neq k} a_m^{(1)} U_{km} \tag{17.18}$$

The coefficient $a_m^{(1)}$ is replaced by equation (17.15), from which we obtain

$$E_k^{(2)} = \sum_{m \neq k} \frac{U_{km} U_{mk}}{E_{ko} - E_{mo}} = \sum_{m \neq k} \frac{|U_{mk}|^2}{E_{ko} - E_{mo}} \quad \text{(since } U_{mk} = U_{km}^* \text{)}$$

$$(17.19)$$

Substituting $E_k^{(1)}$ and $E_k^{(2)}$ in equation (17.7) and setting $\lambda = \lambda^2 = 1$, the second-order corrected eigenenergy is

$$E_k = E_{ko} + \langle \varphi_{ko} | U | \varphi_{ko} \rangle$$

$$+ \sum_{m \neq k} \frac{|U_{mk}|^2}{E_{ko} - E_{mo}} \quad \text{(the 2}^{\text{nd}} \text{ order corrected energy)} \quad (17.20)$$

To find $a_m^{(2)}$, we operate equation (17.17) from the left-hand side by $\langle \varphi_{no} | \ (n \neq k)$ and use equations (17.1) and (17.2), to get

$$\sum_{m \neq k} a_m^{(1)} U_{nm} + \sum_{m \neq k} a_m^{(2)} E_{mo} \delta_{nm}$$

$$= E_{ko} \sum_{m \neq k} a_m^{(2)} \delta_{nm} + E_k^{(1)} \sum_{m \neq k} a_m^{(1)} \delta_{nm} \quad (17.21)$$

After rearranging, we get

$$a_n^{(2)} (E_{ko} - E_{no}) = \sum_{m \neq k} a_m^{(1)} U_{nm} - E_k^{(1)} a_n^{(1)} \quad (17.22)$$

Substituting for $E_k^{(1)}$ and $a_m^{(1)}$ from equations (17.12) and (17.15), we get

$$a_n^{(2)} = \sum_{m \neq k} \frac{U_{nm} U_{mk}}{(E_{ko} - E_{mo})(E_{ko} - E_{no})} - \frac{U_{kk} U_{nk}}{(E_{ko} - E_{no})^2} \quad (17.23)$$

Using the expressions for $a_n^{(1)}$ and $a_n^{(2)}$ in equations (17.5) and (17.6) (setting $\lambda = \lambda^2 = 1$), the second-order corrected eigenfunction is

$$|\varphi_k\rangle = |\varphi_{ko}\rangle + \sum_{n \neq k} \frac{U_{nk}}{E_{ko} - E_{no}} |\varphi_{no}\rangle$$

$$+ \sum_{n \neq k} \sum_{m \neq k} \frac{U_{nm} U_{mk}}{(E_{ko} - E_{mo})(E_{ko} - E_{no})} |\varphi_{no}\rangle$$

$$- \sum_{n \neq k} \frac{U_{kk} U_{nk}}{(E_{ko} - E_{no})^2} |\varphi_{no}\rangle \tag{17.24}$$

One can continue calculating higher-order expansions for the eigenenergy and eigenfunction by equating the coefficients of λ^3, λ^4 and so on in equations (17.9).

Let's summarize the first-order and second-order correction to eigenenergies and eigenfunctions, before we move to the examples showing how to use them.

	The first-order perturbation
Eigenenergy	$E_k = E_{ko} + \langle \varphi_{ko} | U | \varphi_{ko} \rangle$
Eigenfunction	$|\varphi_k\rangle = |\varphi_{ko}\rangle + \sum_{m \neq k} \frac{U_{mk}}{E_{ko} - E_{mo}} |\varphi_{mo}\rangle$
	The second-order perturbation
Eigenenergy	$E_k = E_{ko} + \langle \varphi_{ko} | U | \varphi_{ko} \rangle + \sum_{m \neq k} \frac{U_{km} U_{mk}}{E_{ko} - E_{mo}}$
Eigenfunction	$|\varphi_k\rangle = |\varphi_{ko}\rangle + \sum_{n \neq k} \frac{U_{nk}}{E_{ko} - E_{no}} |\varphi_{no}\rangle$
	$+ \sum_{n \neq k} \sum_{m \neq k} \frac{U_{nm} U_{mk}}{(E_{ko} - E_{mo})(E_{ko} - E_{no})} |\varphi_{no}\rangle$
	$- \sum_{n \neq k} \frac{U_{kk} U_{nk}}{(E_{ko} - E_{no})^2} |\varphi_{no}\rangle$

Some general remarks are:

(1) Perturbation theory works well when $|\Delta E| > |U_{mk}|$, where $\Delta E = E_{ko} - E_{mo}$. That is, the energy level spacing between consecutive energy levels must be much larger than the perturbation. We can observe from equation (17.24) that the first- and second-order corrections depend on $\left|\frac{U}{\Delta E}\right|$ and $\left|\frac{U}{\Delta E}\right|^2$ in the perturbation series expansion of the wave function. The series will usually diverge when $|\Delta E| < |U_{mk}|$.

(2) The perturbation theory discussed above fails in the case of degenerate energy levels. Degenerate energy levels refer to the case where more than one quantum number has the same numerical value of the energy. As an example, recall that in a cubic quantum dot in which every side is L, the first three eigenenergies are the same (degenerate). This is because these combinations of quantum numbers $(m, n, l) = (2, 1, 1), (1, 2, 1), (1, 1, 2)$ all lead to $m^2 + n^2 + l^2 = 6$. Therefore, there are three sets of quantum numbers with the same energy value, so we may say that the energy has a degeneracy of 3.

(3) Perturbation theory also fails when the perturbing potential U introduces new energy levels not contained in the Hamiltonian H_o. That is, there should be a one-to-one mapping between the energy eigenvalues of the unperturbed (E_{no}) and perturbed (E_n) Hamiltonians.

A motivation for perturbation theory is to perform analytical calculations, understand trends, and gain intuition about the problem. In the following, we will provide examples to help understand time-independent perturbation theory.

Example 1. For the PiB, we found the wave functions and eigenenergies to be (see Chapter1)

$$\varphi_{no}(x) = \sqrt{\frac{2}{L}} \sin\left(\frac{n\pi}{L}x\right), \quad E_{no} = \frac{\hbar^2}{2m}\left(\frac{n\pi}{L}\right)^2 \qquad (17.25)$$

If the perturbation inside the well is an added potential energy term $U(x) = q\mathcal{E}x$, find (φ_n, E_n) to the first order in the perturbing potential. Note that this perturbation in potential energy arises from applying an electric field of strength $\mathcal{E}$ along x direction.

Solution: Using equation (17.12), the first-order correction to the eigenenergy is

$$E_n^{(1)} = \langle\varphi_{no}|U|\varphi_{no}\rangle = \frac{2}{L}\int_{x'=0}^{x'=L} U(x')\sin\left(\frac{n\pi}{L}x'\right)\sin\left(\frac{n\pi}{L}x'\right)dx'$$

$$\qquad (17.26)$$

The first-order corrected eigenenergy $E_n = E_{no} + E_n^{(1)}$ is,

$$E_n = \frac{\hbar^2}{2m} \left(\frac{n\pi}{L}\right)^2 + \frac{2}{L} \int_{x'=0}^{x'=L} U(x') \sin\left(\frac{n\pi}{L}x'\right) \sin\left(\frac{n\pi}{L}x'\right) dx'$$

(17.27)

Using equation (17.16), we find that the wave function corrected to the first order in the perturbation $U(x)$ is

$$|\varphi_n\rangle = |\varphi_{no}\rangle + \sum_{p\neq n} \frac{U_{pn}}{E_{no} - E_{po}} |\varphi_{po}\rangle$$

$$= \sqrt{\frac{2}{L}} \sin\left(\frac{n\pi}{L}x\right) + \sum_{p\neq n} \frac{\frac{2}{L}\int dx' U(x') \sin\left(\frac{p\pi}{L}x'\right) \sin\left(\frac{n\pi}{L}x'\right)}{\frac{\hbar^2}{2m}\left(\frac{\pi}{L}\right)^2 (n^2 - p^2)}$$

$$\times \left[\sqrt{\frac{2}{L}} \sin\left(\frac{p\pi}{L}x\right)\right] \qquad (17.28)$$

The above expressions for the first-order corrected eigenfunction and eigenvalue are valid for an arbitrary perturbing potential $U(x)$. If we use $U(x) = q\mathcal{E}x$ in the above integral, the perturbed eigenenergies can be simplified to

$$E_n = \frac{\hbar^2}{2m}\left(\frac{n\pi}{L}\right)^2 + \frac{2q\mathcal{E}L}{8\pi^2 n^2}\left(2\pi^2 n^2 - 2\pi n \sin(2\pi n) - \cos(2\pi n) + 1\right)$$

(17.29)

For $n = 1$, the first-order corrected eigenenergy is

$$E_1 = \frac{\hbar^2}{2m}\left(\frac{\pi}{L}\right)^2 + \frac{q\mathcal{E}L}{2} \qquad (17.30)$$

Example 2. In Chapter 1, we considered the numerical solution of the Schrödinger equation. Here, we will compare that to the eigenvalues obtained from solution with perturbation theory. In this example, we will restrict ourselves to the first-order perturbation theory. Consider a PiB in a constant (spatially independent) electric field. Compare the first eigenenergy numerically obtained with the first-order perturbation theory result (equation (17.30)). The box length is 10 nm, the applied electric field of $5\,\text{V}/\mu\text{m}$ along the length of the box, and the effective mass of the electron is the same as the rest mass, m_o.

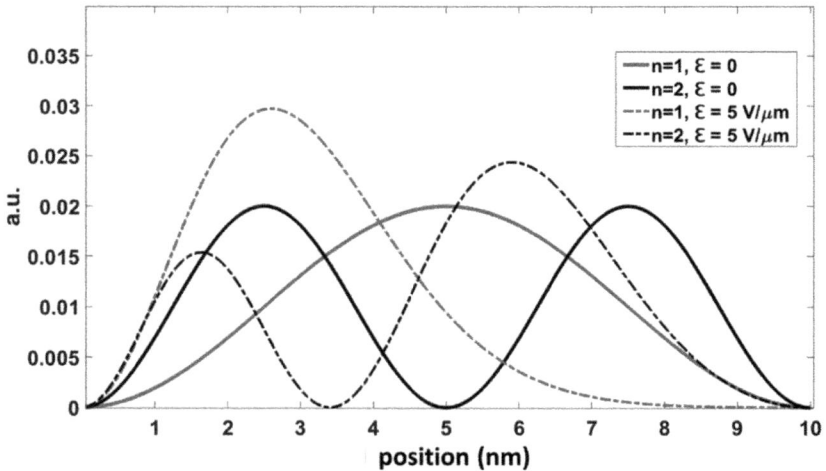

Figure 17.1. The unperturbed (solid) and perturbed (dashed) probability densities for a particle in a box. The perturbing potential is due to the applied constant electric field along the length of the box, which is PE(x) = q $\mathcal{E}$ x. The red and black lines correspond to the $n = 1$ and 2 quantum numbers, respectively.

Solution: We solve Schrödinger's equation numerically with 10,000 grid points evenly spaced over the 10 nm length of the box (Figure 17.1). The energy eigenvalues obtained numerically and using the first-order corrected perturbation theory result in equation (17.30) are shown in Table 17.1. We first remark that the numerically calculated eigenenergies at zero electric field (see Table 17.1) match the values obtained from the analytical formula for a PiB in Chapter 1 to within 0.04%. For the $n = 1$, 2, and 3, the analytical formula $E_{no} = \frac{\hbar^2}{2m} \left(\frac{n\pi}{L}\right)^2$ gives $E_1 = 3.76\,\text{meV}$, $E_2 = 15.0398\,\text{meV}$, and $E_3 = 33.8395\,\text{meV}$.

When compared to the numerical solution, Table 17.1 shows that the eigenenergies calculated using the first-order perturbation theory (equation (17.30)) have an error of 25%, 2%, and 1.95%, for the $n = 1, 2$, and 3 states, respectively. The reason for the significant error in the case of the $n = 1$ state is the large value of the electric

Table 17.1. Comparison of the lowest three eigenenergies of the PiB calculated by the numerical method and first-order perturbation theory.

	Energy levels(meV)		
	Electric field $= 0$	Electric field $= 5\frac{V}{\mu m}$	
Quantum number	Numerical solution	Numerical solution	First-order corrected energy
$n = 1$	3.7584	23.0062	28.7599
$n = 2$	15.0337	40.8290	40.0040
$n = 3$	33.8260	60.0088	58.8395

field. In this case, it is necessary to include higher-order terms to find a more precise result.

Example 3. Repeat Example 2 by solving the perturbed PiB with a weaker perturbation that is one-hundredth of the electric field in Example 2, i.e., $\mathcal{E}_0 = 0.05\,\text{V}/\mu\text{m}$. Compare the results obtained by solving Schrödinger's equation numerically with the perturbation theory result.

Solution: We use the same grid spacing over the 10 nm box length as in Example 2. Table 17.2 summarizes the results. When the electric field is weaker than in the previous example, the error of using the first-order perturbation theory is smaller than in Example 2. The eigenenergies are closer to the ones calculated by the brute force numerical method.

When compared to the numerical solution, Table 17.2 shows that the eigenenergies calculated using the first-order perturbation theory (equation (17.30)) match well at this weaker electric field.

Example 4. In Example 2, the eigenenergy calculated using the first-order perturbation theory had a rather large error of 25% for the $n = 1$ state. In this example, calculate the energy eigenvalues using

Table 17.2. Comparison of the lowest three eigenenergies of a PiB calculated by the numerical method and first-order perturbation theory.

	Energy levels(meV)		
	Electric field $= 0$	Electric field $= 0.05\frac{V}{\mu m}$	
Quantum number	Numerical solution	Numerical solution	First-order corrected energy
$n = 1$	3.7584	4.0077	4.0099
$n = 2$	15.0337	15.284	15.2898
$n = 3$	33.8260	34.0761	34.0895

Note: The perturbing electric field is $0.05 \frac{V}{\mu m}$ along the box length. Compare these results to those from Example 2.

second-order perturbation theory (equation (17.20)). Show that the energy eigenvalues approach the values obtained from the numerical solution.

Solution: We use equation (17.20) and write the second-order corrected energy to be

$$E_n = E_{no} + U_{nn} + \sum_{n \neq p} \frac{|U_{np}|^2}{E_{no} - E_{po}}$$

$$= \frac{\hbar^2}{2m}\left(\frac{n\pi}{L}\right)^2 + \frac{q\mathcal{E}L}{2} - \sum_{n \neq p} \frac{|\langle \varphi_{po}|U|\varphi_{no}\rangle|^2}{E_{po} - E_{no}} \qquad (17.31)$$

Now, we need the matrix elements of the electric potential energy $U = q\mathcal{E}x$, between the energy states with $n = 1, 2$ and 3, and every other state with quantum number p. Note that for higher quantum numbers, the eigenenergy, $E_{po} = p^2 E_{1o}$. This means that the denominator in the last term of equation (17.31) get bigger by going to larger quantum numbers. So, in the following second-order perturbation calculation, we will only include a few states p for every energy that we will calculate.

The matrix element of the electric potential energy $U = q\mathcal{E}x$ between states $n = 1$ and p is

$$\int_{x=0}^{x=L} \sin\left(\frac{\pi x}{L}\right) q\mathcal{E}x \sin\left(\frac{p\pi x}{L}\right) dx$$

$$= \frac{q\mathcal{E}L^2[-\pi(p^2-1)\sin(p\pi)-2p(\cos(p\pi)+1)]}{\pi^2(p^2-1)^2} \qquad (17.32)$$

When p is an odd integer, the above matrix element is zero (see Problem 3). Therefore, the odd-numbered eigenstates do not contribute to the second-order corrected result for E_1. So, let's include only the $p = 2$ and $p = 4$ terms and ignore higher values of p. We find

$$|\langle \varphi_{2o}|U|\varphi_{1o}\rangle|^2 = \left(16\frac{q\mathcal{E}L}{9\pi^2}\right)^2 \quad \text{and} \qquad (17.33)$$

$$|\langle \varphi_{4o}|U|\varphi_{1o}\rangle|^2 = \left(32\frac{q\mathcal{E}L}{225\pi^2}\right)^2 \qquad (17.34)$$

Substituting the above expressions in equation (17.31), the second-order corrected energy for the $n = 1$ state is

$$E_1 = \frac{\hbar^2}{2m}\left(\frac{\pi}{L}\right)^2 + \frac{q\mathcal{E}L}{2} + \frac{\left(16\frac{q\mathcal{E}L}{9\pi^2}\right)^2}{\frac{\hbar^2}{2m}\left(\frac{\pi}{L}\right)^2 - \frac{\hbar^2}{2m}\left(\frac{2\pi}{L}\right)^2}$$

$$+ \frac{\left(32\frac{q\mathcal{E}L}{225\pi^2}\right)^2}{\frac{\hbar^2}{2m}\left(\frac{\pi}{L}\right)^2 - \frac{\hbar^2}{2m}\left(\frac{4\pi}{L}\right)^2} \qquad (17.35)$$

$E_1 = (28.7599 - 7.1910 - 0.0092)\,\text{meV} = 21.5597\,\text{meV}$

As can be seen, using the second-order corrected expression for the eigenenergy and including a few nonzero matrix elements of the electric potential energy leads to a value that is closer to the numerical solution ($E_1 = 23.00616692\,\text{meV}$). The error has decreased from 25% (Example 2) to 6.28% when using second order perturbation theory

The matrix element of the electric potential energy $U = q\mathcal{E}x$, between states $n = 2$ and p is

$$\int_{x=0}^{x=L} \sin\left(\frac{2\pi x}{L}\right) U(x) \sin\left(\frac{p\pi x}{L}\right) dx$$

$$= \frac{2q\mathcal{E}L^2[\pi(p^2-4)\sin(p\pi) + 2p(\cos(p\pi) - 1)]}{\pi^2(p^2-4)^2} \qquad (17.36)$$

The nonzero matrix elements for the above are those with odd p-values.

The matrix element of the electric potential energy $U = q\mathcal{E}x$ between states $n = 3$ and p is

$$\int_{x=0}^{x=L} \sin\left(\frac{3\pi x}{L}\right) U(x) \sin\left(\frac{p\pi x}{L}\right) dx$$

$$= \frac{-3q\mathcal{E}L^2[\pi(p^2-9)\sin(p\pi)+2p+(2p)\cos(p\pi)]}{\pi^2(p^2-9)^2} \qquad (17.37)$$

The nonzero matrix elements for the above are those with even p-values. We limit ourselves to the following terms:

$$\text{for } n = 2, \quad |\langle \varphi_{2o}|U|\varphi_{1o}\rangle|^2 = \left(16\frac{q\mathcal{E}L}{9\pi^2}\right)^2$$

$$\text{and} \quad |\langle \varphi_{2o}|U|\varphi_{3o}\rangle|^2 = \left(-48\frac{q\mathcal{E}L}{25\pi^2}\right)^2 \qquad (17.38)$$

$$\text{for } n = 3, \quad |\langle \varphi_{3o}|U|\varphi_{2o}\rangle|^2 = \left(-48\frac{q\mathcal{E}L}{25\pi^2}\right)^2$$

$$\text{and} \quad |\langle \varphi_{3o}|U|\varphi_{4o}\rangle|^2 = \left(-96\frac{q\mathcal{E}L}{49\pi^2}\right)^2 \qquad (17.39)$$

Using the above, we have that the second-order corrected energy for the $n = 2$ state is

$$E_2 = E_{2o} + U_{22} - \sum_{p \neq 2} \frac{|U_{2p}|^2}{E_{po} - E_{2o}}$$

$$= \frac{4\hbar^2}{2m}\left(\frac{\pi}{L}\right)^2 + \frac{q\mathcal{E}L}{2} - \sum_{p \neq 2} \frac{|\langle \varphi_{po}|U|\varphi_{2o}\rangle|^2}{E_{po} - E_{2o}} \qquad (17.40)$$

$$E_2 = \frac{4\hbar^2}{2m}\left(\frac{\pi}{L}\right)^2 + \frac{q\varepsilon L}{2} + \frac{\left(16\frac{q\varepsilon L}{9\pi^2}\right)^2}{\frac{\hbar^2}{2m}\left(\frac{2\pi}{L}\right)^2 - \frac{\hbar^2}{2m}\left(\frac{\pi}{L}\right)^2}$$

$$+ \frac{\left(48\frac{q\varepsilon L}{25\pi^2}\right)^2}{\frac{\hbar^2}{2m}\left(\frac{2\pi}{L}\right)^2 - \frac{\hbar^2}{2m}\left(\frac{3\pi}{L}\right)^2} \tag{17.41}$$

$$E_2 = (40.00 + 7.1910 - 5.0326)\,\text{meV} = 42.1624\,\text{meV}$$

When compared to the numerical solution, the second-order perturbation calculation gives a slightly larger error than the first-order calculation for the $n = 2$ state (see Table 17.1).

The second-order corrected energy for the $n = 3$ state is

$$E_3 = E_{3o} + U'_{33} - \sum_{p \neq 3} \frac{|U_{3p}|^2}{E_{po} - E_{3o}}$$

$$= \frac{9\hbar^2}{2m}\left(\frac{\pi}{L}\right)^2 + \frac{q\varepsilon L}{2} - \sum_{p \neq 3} \frac{|\langle \varphi_{po}|U|\varphi_{3o}\rangle|^2}{E_{po} - E_{3o}} \tag{17.42}$$

$$E_3 = \frac{9\hbar^2}{2m}\left(\frac{\pi}{L}\right)^2 + \frac{q\varepsilon L}{2} + \frac{\left(48\frac{q\varepsilon L}{25\pi^2}\right)^2}{\frac{\hbar^2}{2m}\left(\frac{3\pi}{L}\right)^2 - \frac{\hbar^2}{2m}\left(\frac{2\pi}{L}\right)^2}$$

$$+ \frac{\left(96\frac{q\varepsilon L}{49\pi^2}\right)^2}{\frac{\hbar^2}{2m}\left(\frac{3\pi}{L}\right)^2 - \frac{\hbar^2}{2m}\left(\frac{4\pi}{L}\right)^2} \tag{17.43}$$

$$E_3 = (58.8395 + 5.0326 - 3.7429)\,\text{meV} = 60.1292\,\text{meV}$$

When compared to the numerical solution, the second-order perturbation calculation gives a smaller error of 0.2% while the first-order calculation for the $n = 3$ state resulted in an error of -1.95% (see Table 17.1). Going to a higher order perturbation improved the results in this case.

17.3 Systems with Degenerate Eigenvalues

Note that the results from first- and second-order perturbation theory in equations (17.16), (17.20), and (17.24) fail if the unperturbed

Hamiltonian has degenerate states (orthogonal states with the same eigenenergy). If for some n and p values, $E_{no} - E_{po} = 0$, then we have a singularity on the RHS of these equations. This section will discuss the computation of eigenvalues and eigenstates of a perturbed system in situations when the unperturbed Hamiltonian has degenerate eigenvalues.

Let us assume the unperturbed Hamiltonian (H_o) of a system has two degenerate eigenenergies $E_1 = E_2 = E_o$, with orthonormal eigenfunctions $|\varphi_1\rangle$ and $|\varphi_2\rangle$. These eigenstates satisfy the following relations:

$$H_o|\varphi_1\rangle = E_o|\varphi_1\rangle \quad \text{and} \quad H_o|\varphi_2\rangle = E_o|\varphi_2\rangle \qquad (17.44)$$

$$\langle\varphi_1|\varphi_2\rangle = 0, \langle\varphi_1|\varphi_1\rangle = 1, \quad \text{and} \quad \langle\varphi_2|\varphi_2\rangle = 1 \qquad (17.45)$$

Note that any superposition of the above states is also an eigenstate of the unperturbed Hamiltonian, i.e., if $|\psi\rangle = a_1|\varphi_1\rangle + a_2|\varphi_2\rangle$, then

$$H_o|\psi\rangle = H_o(a_1|\varphi_1\rangle + a_2|\varphi_2\rangle) = E_o a_1|\varphi_1\rangle + E_o a_1|\varphi_2\rangle = E_o|\psi\rangle \qquad (17.46)$$

After applying the perturbation U, the Hamiltonian is $H = H_o + U$. We look for a new (perturbed) eigenstate (ψ) and eigenenergy $(E = E_o + \Delta E)$. The Schrödinger equation for the perturbed system is

$$H|\psi\rangle = E|\psi\rangle \qquad (17.47)$$

We assume that the new eigenstate ψ with the applied perturbation is still a superposition of the two degenerate eigenstates without perturbation,

$$|\psi\rangle = b_1|\varphi_1\rangle + b_2|\varphi_2\rangle \qquad (17.48)$$

The coefficients b_1 and b_2 are unknown. We plug this trial wave function in equation (17.47) to obtain

$$H|\psi\rangle = H(b_1|\varphi_1\rangle + b_2|\varphi_2\rangle) = E|\psi\rangle$$

We multiply the above equation from the left side by $\langle\varphi_1|$ and $\langle\varphi_2|$ individually to get

$$b_1\langle\varphi_1|H|\varphi_1\rangle + b_2\langle\varphi_1|H|\varphi_2\rangle = E(b_1\langle\varphi_1|\varphi_1\rangle + b_2\langle\varphi_1|\varphi_2\rangle)$$

$$b_1\langle\varphi_2|H|\varphi_1\rangle + b_2\langle\varphi_2|H|\varphi_2\rangle = E(b_1\langle\varphi_2|\varphi_1\rangle + b_2\langle\varphi_2|\varphi_2\rangle)$$

Using the orthonormality of eigenfunctions $\langle \varphi_i | \varphi_j \rangle = \delta_{ij}$, we have the following eigenvalue equations:

$$H_{11}b_1 + H_{12}b_2 = Eb_1 \qquad (17.49)$$

$$H_{21}b_1 + H_{22}b_2 = Eb_2 \qquad (17.50)$$

which in matrix form is

$$\begin{pmatrix} H_{11} & H_{12} \\ H_{21} & H_{22} \end{pmatrix} \begin{pmatrix} b_1 \\ b_2 \end{pmatrix} = E \begin{pmatrix} b_1 \\ b_2 \end{pmatrix} \qquad (17.51)$$

Solving the above equations, we can find both the new eigenvalues E and the new eigenvectors $\begin{pmatrix} b_1 \\ b_2 \end{pmatrix}$. Substituting the values of b_1 and b_2 corresponding to each eigenvalue E in equation (17.48) gives the two eigenfunctions of the Hamiltonian H. Having nontrivial (nonzero) solutions for b_1 and b_2 mandates that the following secular determinant is zero:

$$\begin{vmatrix} H_{11} - E & H_{12} \\ H_{21} & H_{22} - E \end{vmatrix} = 0 \qquad (17.52)$$

We remark that by observing the following identities,

$$H_{11} = \langle \varphi_1 | H | \varphi_1 \rangle = \langle \varphi_1 | H_o + U | \varphi_1 \rangle = E_o + U_{11} \qquad (17.53)$$

$$H_{22} = \langle \varphi_2 | H | \varphi_2 \rangle = \langle \varphi_2 | H_o + U | \varphi_2 \rangle = E_o + U_{22} \qquad (17.54)$$

$$H_{12} = \langle \varphi_1 | H | \varphi_2 \rangle = \langle \varphi_1 | H_o + U | \varphi_2 \rangle$$
$$= \langle \varphi_1 | U | \varphi_2 \rangle = U_{12} \qquad (17.55)$$

$$H_{21} = \langle \varphi_2 | H | \varphi_1 \rangle = \langle \varphi_2 | H_o + U | \varphi_1 \rangle$$
$$= \langle \varphi_2 | U | \varphi_1 \rangle = U_{21} \qquad (17.56)$$

it is easy to also rewrite equation (17.51) in terms of the perturbing potential U, and the change in energy ΔE from the unperturbed energy E_o as

$$\begin{pmatrix} U_{11} & U_{12} \\ U_{21} & U_{22} \end{pmatrix} \begin{pmatrix} b_1 \\ b_2 \end{pmatrix} = \Delta E \begin{pmatrix} b_1 \\ b_2 \end{pmatrix} \qquad (17.57)$$

Solving equation (17.57) for each value of E also gives us the coefficients b_1 and b_2, corresponding to each new energy level.

General formulation: The method presented above is generalizable to the case of N degenerate eigenenergies. That is, when the system has N equal eigenenergies $E_i = E_o$ $(i = 1, 2, \ldots, N)$ and N orthonormal eigenstates corresponding to each of them which obey the following:

$$\langle \varphi_i | \varphi_j \rangle = \delta_{ij} \quad (i, j = 1, 2, \ldots, N) \tag{17.58}$$

In this case, we assume the eigenstate of the perturbed system is a superposition of the N degenerate states,

$$|\psi\rangle = \sum_{i=1}^{N} b_i |\varphi_i\rangle \tag{17.59}$$

Having the matrix elements of the perturbed Hamiltonian, $H = H_o + U$, we set up a system of N equations to find the unknown coefficients (b_i),

$$\begin{pmatrix} H_{11} - E & H_{12} & \cdots & H_{1N} \\ H_{21} & H_{22} - E & \cdots & H_{2N} \\ \vdots & \vdots & \cdots & \vdots \\ H_{N1} & H_{N2} & \cdots & H_{NN} - E \end{pmatrix} \begin{pmatrix} b_1 \\ b_2 \\ \vdots \\ b_N \end{pmatrix} = E \begin{pmatrix} b_1 \\ b_2 \\ \vdots \\ b_N \end{pmatrix} \tag{17.60}$$

To have nonzero solutions, the following secular determinant must be zero,

$$\begin{vmatrix} H_{11} - E & H_{12} & \cdots & H_{1N} \\ H_{21} & H_{22} - E & \cdots & H_{2N} \\ \vdots & \vdots & \cdots & \vdots \\ H_{N1} & H_{N2} & \cdots & H_{NN} - E \end{vmatrix} = 0 \tag{17.61}$$

Solving the above equation gives an Nth degree algebraic equation, which we can solve to find N eigenenergies.

Example 5. Let's assume an unperturbed Hamiltonian has three eigenvalues, two of which are degenerate, i.e., $E_1 = 1\,\text{eV}$, $E_2 = E_3 = 2\,\text{eV}$. Find the new eigenenergies and eigenstates of the Hamiltonian when the perturbation is

$$U = \begin{bmatrix} 0 & 0 & 0 \\ 0 & 0 & 0.1 \\ 0 & 0.1 & 0 \end{bmatrix}$$

when written in terms of the eigenstates of the unperturbed Hamiltonian.

Solution: The perturbed Hamiltonian is written as

$$H = H_o + U = \begin{bmatrix} 1 & 0 & 0 \\ 0 & 2 & 0 \\ 0 & 0 & 2 \end{bmatrix} + \begin{bmatrix} 0 & 0 & 0 \\ 0 & 0 & 0.1 \\ 0 & 0.1 & 0 \end{bmatrix} = \begin{bmatrix} 1 & 0 & 0 \\ 0 & 2 & 0.1 \\ 0 & 0.1 & 2 \end{bmatrix}$$

Now we solve the secular determinant of equation (17.61) for $N = 3$ to find the perturbed eigenenergies,

$$\begin{vmatrix} H_{11} - E & H_{12} & H_{13} \\ H_{21} & H_{22} - E & H_{23} \\ H_{31} & H_{32} & H_{33} - E \end{vmatrix} = \begin{vmatrix} 1 - E & 0 & 0 \\ 0 & 2 - E & 0.1 \\ 0 & 0.1 & 2 - E \end{vmatrix} = 0$$

The third-order equation to find E is

$$(1 - E)[(2 - E)(2 - E) - 0.1 \times 0.1] = 0 \rightarrow \begin{cases} E_1 = 1\,\text{eV} \\ E_2 = 1.9\,\text{eV} \\ E_3 = 2.1\,\text{eV} \end{cases}$$

As you see, the previously degenerate eigenvalues $(E = 2eV)$ are now different because of perturbation. For the first eigenenergy $(E_1 = 1\,\text{eV})$, the eigenstate is found by solving

$$\begin{bmatrix} 1 - 1 & 0 & 0 \\ 0 & 2 - 1 & 0.1 \\ 0 & 0.1 & 2 - 1 \end{bmatrix} \begin{bmatrix} b_1 \\ b_2 \\ b_3 \end{bmatrix} = \begin{bmatrix} 0 \\ 0 \\ 0 \end{bmatrix}$$

which gives $b_2 = -0.1b_3$, $0.1b_2 = -b_3$ implying that $b_2 = b_3 = 0$. We are free to choose $b_1 = 1$ to find the following normalized eigenvector:

$$|\psi_1\rangle = \sum_{i=1}^{N} b_i |\varphi_i\rangle = \begin{bmatrix} 1 \\ 0 \\ 0 \end{bmatrix}$$

For the second eigenenergy ($E_2 = 1.9\,\text{eV}$), the eigenstate is found from,

$$\begin{bmatrix} 1-1.9 & 0 & 0 \\ 0 & 2-1.9 & 0.1 \\ 0 & 0.1 & 2-1.9 \end{bmatrix} \begin{bmatrix} b_1 \\ b_2 \\ b_3 \end{bmatrix} = \begin{bmatrix} 0 \\ 0 \\ 0 \end{bmatrix}$$

which gives $b_1 = 0$ and $b_2 = -b_3$. We can choose $b_2 = \frac{1}{\sqrt{2}}$ and $b_3 = -\frac{1}{\sqrt{2}}$ to have a normalized eigenvector

$$|\psi_2\rangle = \sum_{i=1}^{N} b_i|\varphi_i\rangle = \begin{bmatrix} 0 \\ 1 \\ \frac{1}{\sqrt{2}} \\ -\frac{1}{\sqrt{2}} \end{bmatrix}$$

For the third eigenenergy ($E_3 = 2.1\,\text{eV}$), the eigenstate is found from

$$\begin{bmatrix} 1-2.1 & 0 & 0 \\ 0 & 2-2.1 & 0.1 \\ 0 & 0.1 & 2-2.1 \end{bmatrix} \begin{bmatrix} b_1 \\ b_2 \\ b_3 \end{bmatrix} = \begin{bmatrix} 0 \\ 0 \\ 0 \end{bmatrix}$$

which gives $b_1 = 0$, and $b_2 = b_3$. Choosing $b_2 = b_3 = \frac{1}{\sqrt{2}}$, the eigenstate is

$$|\psi_3\rangle = \sum_{i=1}^{N} b_i|\varphi_i\rangle = \begin{bmatrix} 0 \\ 1 \\ \frac{1}{\sqrt{2}} \\ \frac{1}{\sqrt{2}} \end{bmatrix}$$

As can be seen, the first eigenstate is not affected by perturbation since the matrix elements U_{11}, U_{12} and U_{21} are all zero.

17.3.1 *More on degeneracy*

As we saw in Section 17.2, for the case of two degenerate eigenstates, the new eigenenergies are found by solving equation (17.52). Let's solve this equation and see how each new eigenenergy behaves as the value of the interaction (perturbation) matrix element ($H_{12} = U_{12}$) changes.

The quadratic equation resulting from equation (17.52),

$$(H_{11} - E)(H_{22} - E) - H_{21}H_{12} = 0 \tag{17.62}$$

simplifies to

$$E^2 - E(H_{11} + H_{22}) + H_{11}H_{22} - |U_{12}|^2 = 0 \tag{17.63}$$

This equation has two solutions which are

$$E = \frac{(H_{11} + H_{22}) \pm \sqrt{(H_{11} - H_{22})^2 + 4|U_{12}|^2}}{2} \tag{17.64}$$

For a two-level system, equation (17.64) includes the correction to all orders in perturbation meaning that it is an exact result. If the perturbation is very small, i.e., $U_{12} \ll |U_{11} - U_{22}|$, a Taylor series expansion yields the two eigenvalues

$$E \sim E_o + U_{11} + \frac{|U_{12}|^2}{U_{11} - U_{22}} \tag{17.65}$$

$$E \sim E_o + U_{22} - \frac{|U_{12}|^2}{U_{11} - U_{22}} \tag{17.66}$$

The above two equations are of the same form as those from the second-order perturbation theory in equation (17.20) or (17.91).

Example 6. If $H_{11} = H_{22} = 1\,eV$, plot E versus the interaction matrix element, $|U_{12}|$.

Solution: We assume $|U_{12}|$, the interaction energies between the two electronic states, vary from 0 eV (no interaction) to large values. The plot of the perturbed eigenenergies in equation (17.64) is shown in Figure 17.2. It shows that even a tiny $|U_{12}|$ *lifts the degeneracy* (meaning that the energy levels are no longer degenerate). The energy level splitting increases as $|U_{12}|$ increases.

Example 7. Assume that $H_{11} = 2\,eV$ and H_{22} changes linearly from 1 to 3 eV. Plot E in equation (17.64) versus H_{22} when $|U_{12}| = 0.1\,eV$.

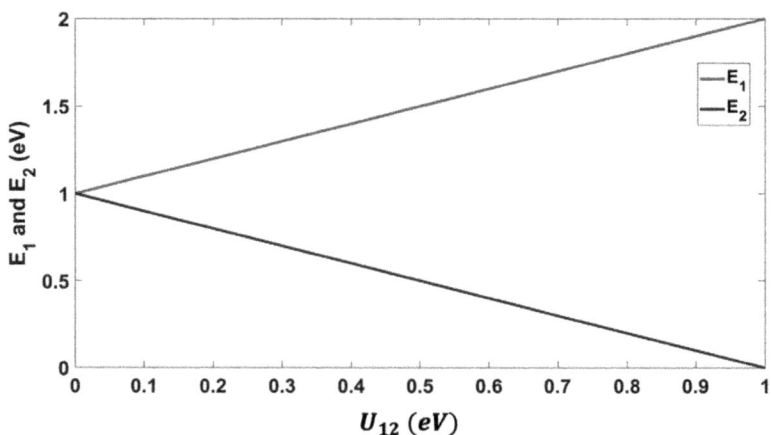

Figure 17.2. Splitting of degenerate eigenenergies as the off-diagonal matrix element of perturbing energy increases from 0 to 1.

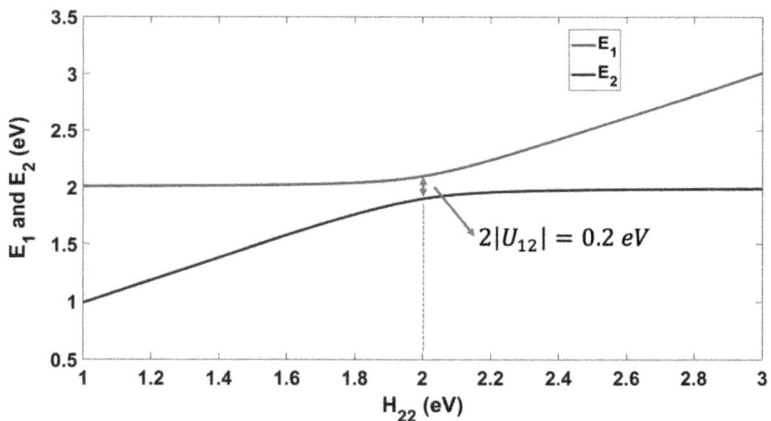

Figure 17.3. The graph of E_1 and E_2 versus H_{22} shows an avoided crossing at $H_{22} = 2\,\text{eV} = H_{11}$, as a result of nonzero interaction between states 1 and 2. For H_{22} much smaller and larger than $2\,\text{eV}$, one of the eigenenergies is always $2\,\text{eV}$.

Solution: As can be seen in Figure 17.3, when the difference between H_{11} and H_{22} is very large, the first term under the square root in equation (17.64) is much larger than the second term; therefore, solutions for E in equation (17.64) are almost $E = H_{11}$ and $E = H_{22}$.

However, when H_{22} value gets closer to H_{11}, i.e., when $(H_{11} - H_{22}) \rightarrow 0$, the dominant term under the square root sign is $4|U_{12}|^2$. In this case, the solutions of equation (17.64) are different and given by

$$E = \frac{(H_{11} + H_{22})}{2} \pm |U_{12}| = \begin{cases} 2 + 0.1 \, \text{eV} \\ 2 - 0.1 \, \text{eV} \end{cases}$$

The above equation says that as a result of the interaction between states 1 and 2 through a nonzero U_{12}, the degeneracy of the unperturbed energy levels $H_{11} = H_{22}$ is lifted. This is called an *avoided crossing* between the two energy levels as they get close to each other. Figure 17.3 shows how both solutions of equation (17.64) behave as a function of H_{22}. As H_{22} approaches H_{11}, both plots avoid crossing each other and their difference is exactly $2|U_{12}| = 2 \times 0.1 \, \text{eV} = 0.2 \, \text{eV}$.

In the chapter on periodic solids (Chapter 14), we saw when two hydrogen atoms are brought together, the nonzero overlapping of their s orbitals, will cause electrons to hop/tunnel between the two atoms. Due to this, the degeneracy of their eigenenergy is lifted, and we saw that the H_2 molecule acquires two eigenenergies, which are split by $2t$, where t is the hopping or tunneling energy.

Example 8. Coupling between a transmon qubit and a resonator: A tunable transmon qubit is coupled to a quarter-wavelength resonator at $f_{res} = 4.5 \, \text{GHz}$ The qubit uses a SQUID which acts as a nonlinear tunable inductor. The magnetic flux is induced by the flux line or the Z control line in Figure 17.4. The direct current in the flux line (or Z-line) changes the external magnetic flux winding the SQUID loop; which changes the qubit frequency as a function of the magnetic flux according to the following equation:

$$\hbar \omega_{\text{qubit}} \approx \sqrt{8 E_J E_C \left| \cos \left(\frac{\pi \Phi_{ext}}{\Phi_0} \right) \right|} - E_C \tag{17.67}$$

Use equation (17.64) and plot the frequencies of the qubit and resonator as a function of the external flux value (Φ_{ext}). Assume the

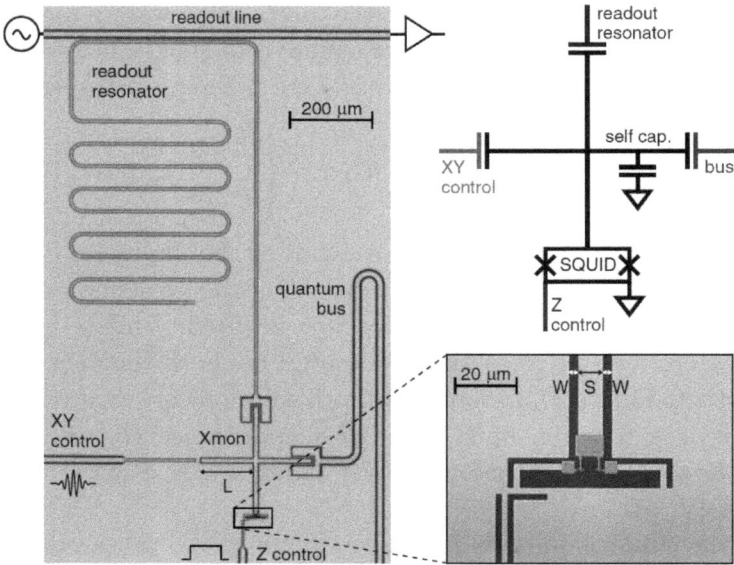

Figure 17.4. A tunable transmon qubit coupled to a meandering quarter-wavelength resonator, which is termed, readout resonator. The qubit is made of a SQUID and controlled by a magnetic flux. The magnetic flux arises from the direct current going through the Z-line at the bottom-left of the inset. Courtesy of John Martinis, UCSB.

charging energy (E_C) and Josephson energy (E_J) in units of $\hbar$ are $2\pi \times 0.2\,\text{GHz}$ and $2\pi \times 20\,\text{GHz}$, respectively. The capacitive coupling between the qubit and the resonator is $U_{12} = 2\pi \times 60\,\text{MHz}$. E_C is the energy stored in the capacitor that is connected in parallel with the Josephson junction. E_J is the energy stored in the nonlinear inductance created by the Josephson junction.

Solution: In this problem, H_{11} and H_{22} of equation (17.64) are energies of the qubit and the resonator, respectively, and U_{12} is the coupling. H_{22} is a fixed number, i.e., $4.5\,\text{GHz}$ and H_{11} is given in equation (17.67). We factor out 2π and express the quantities in terms of GHz. Note that Φ_0 is the quantum of magnetic flux

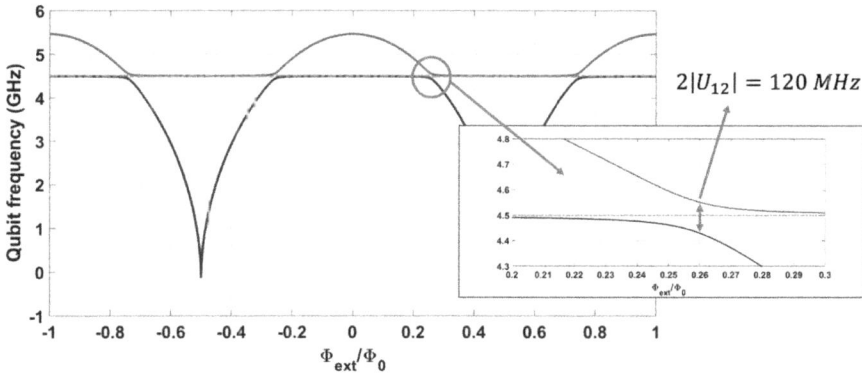

Figure 17.5. Avoided crossing in a qubit: The resonator and qubit frequencies are plotted versus applied external flux Φ_{ext}, which is normalized to the flux quantum Φ_0. The horizontal dashed green line at 4.5 GHz is the resonator frequency when the resonator and qubit do not interact with each other. The inset on the right side shows that when $\frac{\Phi_{ext}}{\Phi_0} = 0.26$, the qubit and resonator frequencies are equal, but this degeneracy *is lifted* due to the capacitive coupling of the resonator and qubit. The frequencies at the avoided crossing are 120 MHz apart.

which is 2.07×10^{-15} Weber or Volt-second. Figure 17.5 shows how the qubit frequency is tuned by the external flux $\frac{\Phi_{ext}}{\Phi_0}$, and once it matches the resonator frequency, the qubit frequency splits by $2|U_{12}| = 120$ MHz. Note that the qubit frequency is maximum when the applied flux is zero. For this example, it is 5.461 GHz. This process is an experimental method to extract the coupling between a qubit and resonator, which is an important parameter in designing quantum gates. That is, U_{12} (and as a result the coupling capacitance) is extracted experimentally by measuring the splitting of the red and blue lines that represent the frequency.

17.4 Secular Determinant Method for Perturbation Theory

This section outlines an alternative derivation of the perturbation method based on the secular determinant of the underlying system of equations (Raimes, 1961) (Ryde, 1976). We consider a system with Hamiltonian H_o whose solution (φ_{no}, E_{no}) is precisely known. Our goal is to find a simple way to determine the energy levels of the perturbed Hamiltonian H.

Schrödinger equations for the unperturbed and perturbed systems are as follows:

$$H_o|\varphi_{no}\rangle = E_{no}|\varphi_{no}\rangle \tag{17.68}$$

$$H|\varphi\rangle = (H_o + U)|\varphi\rangle = E|\varphi\rangle \tag{17.69}$$

The eigenvalues of the unperturbed system form an orthonormal set. That is,

$$\langle\varphi_{mo}|\varphi_{no}\rangle = \delta_{nm} \quad \text{(unperturbed, } \varphi_{no} \text{ and } E_{no} \text{ are known)} \tag{17.70}$$

We assume that the eigenfunctions and eigenvalues of the perturbed Hamiltonian H can be written as a superposition of eigenfunctions of the unperturbed Hamiltonian,

$$|\varphi\rangle = \sum_m a_m|\varphi_{mo}\rangle \tag{17.71}$$

By substituting equation (17.71) in equation (17.69), we get

$$\sum_m a_m E_{mo}|\varphi_{mo}\rangle + \sum_m a_m U|\varphi_{mo}\rangle = E\sum_m a_m|\varphi_{mo}\rangle \tag{17.72}$$

Recall that the index o refers to the unperturbed solutions of the system. By multiplying both sides of the above equation from the left side by $\langle\varphi_{ko}|$, we get

$$\sum_m a_m E_{mo}\langle\varphi_{ko}|\varphi_{mo}\rangle + \sum_m a_m\langle\varphi_{ko}|U|\varphi_{mo}\rangle = E\sum_m a_m\langle\varphi_{ko}|\varphi_{mo}\rangle$$
$$\tag{17.73}$$

The matrix element of the perturbing potential is

$$U_{km} = \langle\varphi_{ko}|U|\varphi_{mo}\rangle \tag{17.74}$$

Using the orthogonality property of the basis functions, we can write equation (17.73) as

$$\sum_m E_{mo}\delta_{km}a_m + \sum_m U_{km}a_m = E\sum_m a_{nm}\delta_{km} \qquad (17.75)$$

$$(E_{ko} - E)a_k + \sum_m U_{km}a_m = 0 \qquad (17.76)$$

Equation (17.76) is a set of N equations, one for each value of k running from $1, 2, \ldots N$. By expanding the values of m and k, the following N equations are formed

$$\begin{pmatrix} E_{1o} + U_{11} - E & U_{12} & \cdots & U_{1N} \\ U_{21} & E_{2o} + U_{22} - E & \cdots & U_{2N} \\ \vdots & \vdots & \cdots & \vdots \\ U_{N1} & U_{N2} & \cdots & E_{No} + U_{NN} - E \end{pmatrix}$$

$$\times \begin{pmatrix} a_1 \\ a_2 \\ \vdots \\ a_N \end{pmatrix} = \begin{pmatrix} 0 \\ 0 \\ \vdots \\ 0 \end{pmatrix} \qquad (17.77)$$

Solving this eigenvalue problem, returns the N eigenenergies and eigenfunctions for the perturbed Hamiltonian. However, instead of solving the eigenvalue problem in equation (17.77) numerically, it is possible to use an approximate method which returns analytic solutions for the eigenenergies if the perturbation U is small. We will show that this method returns the same first and second order corrections we found in previous sections (equations (17.13) and (17.20)).

Without loss of generality, we assume the secular determinant of equation (17.61) or equation (17.77) is to be solved for a perturbed system with four eigenenergies and eigenstates, i.e., $N = 4$. We have

$$\begin{vmatrix} H_{11} - E & U_{12} & U_{13} & U_{14} \\ U_{21} & H_{22} - E & U_{23} & U_{24} \\ U_{31} & U_{32} & H_{33} - E & U_{34} \\ U_{41} & U_{42} & U_{43} & U_{44} - E \end{vmatrix} = 0 \qquad (17.78)$$

It is noteworthy that in deriving the above secular determinant, *everything was exact*, and no approximation has been made so far.

In the first-order approximation, however, we assume the perturbing terms are very small as if the off-diagonal terms are almost zero. In this case, the determinant is reduced to

$$\begin{vmatrix} H_{11} - E & 0 & 0 & 0 \\ 0 & H_{22} - E & 0 & 0 \\ 0 & 0 & H_{33} - E & 0 \\ 0 & 0 & 0 & H_{44} - E \end{vmatrix} = 0 \qquad (17.79)$$

where

$$(H_{11} - E)(H_{22} - E)(H_{33} - E)(H_{44} - E) = 0 \qquad (17.80)$$

The solutions of the above equation are

$$E_i = H_{ii} = E_{io} + U_{ii} \quad i = 1, 2, 3, 4 \quad \text{(to 1}^{\text{st}}\text{ order in perturbation)}$$
$$(17.81)$$

This is the same result that we derived from equation (17.13).

Now, we are going to find the second-order correction to the first nondegenerate eigenenergy. To proceed, we need two more assumptions.

Firstly, we assume all off-diagonal perturbations involving the 1st state (index 1) are nonzero; however, terms like U_{23}, etc. are zero. This is to say the matrix elements of the perturbing energy like U_{23} only indirectly affect the eigenstate number 1, and therefore we ignore them.

Secondly, we replace E in rows 2, 3 and 4 with H_{11}. This makes the secular determinant of equation (17.78) to be the first order in E:

$$\begin{vmatrix} H_{11} - E & U_{12} & U_{13} & U_{14} \\ U_{21} & H_{22} - H_{11} & 0 & 0 \\ U_{31} & 0 & H_{33} - H_{11} & 0 \\ U_{41} & 0 & 0 & H_{44} - H_{11} \end{vmatrix} = 0 \quad (17.82)$$

Note that every matrix element (U_{1i} or U_{i1}) with index 1 is kept and everything else is assumed to be zero.

As we learned in linear algebra, we can multiply a row with a constant and subtract it from another row, and this will not change the value of the determinant. This helps us to change the first row such that all elements except the leftmost one become zero (you will see that this helps calculate the 1st eigenenergy). You may have seen this method in solving a set of N linear equations for a circuit with N nodes (branches) by writing Kirchhoff's laws.

Let's multiply and divide the second row by U_{12} and $H_{22} - H_{11}$, respectively, and subtract it from the first row. After doing this, we have

$$\begin{vmatrix} H_{11} - E - \dfrac{U_{21}U_{12}}{H_{22} - H_{11}} & 0 & U_{13} & U_{14} \\ U_{21} & H_{22} - H_{11} & 0 & 0 \\ U_{31} & 0 & H_{33} - H_{11} & 0 \\ U_{41} & 0 & 0 & H_{44} - H_{11} \end{vmatrix} = 0$$

$$(17.83)$$

Now, multiply and divide the third row by U_{13} and $H_{33} - H_{11}$, respectively, and subtract it from the first row. This helps get rid of another nonzero element in the first row which is U_{13}. After doing this, we have

$$\begin{vmatrix} H_{11} - E - \dfrac{U_{21}U_{12}}{H_{22} - H_{11}} - \dfrac{U_{31}U_{13}}{H_{33} - H_{11}} & 0 & 0 & U_{14} \\ U_{21} & H_{22} - H_{11} & 0 & 0 \\ U_{31} & 0 & H_{33} - H_{11} & 0 \\ U_{41} & 0 & 0 & H_{44} - H_{11} \end{vmatrix} = 0$$

$$(17.84)$$

Repeat the same procedure as above to get rid of U_{14} in the first row and you will get

$$\begin{vmatrix} H_{11} - E - \dfrac{U_{21}U_{12}}{H_{22} - H_{11}} - \dfrac{U_{31}U_{13}}{H_{33} - H_{11}} - \dfrac{U_{41}U_{14}}{H_{44} - H_{11}} & 0 & 0 & 0 \\ U_{21} & H_{22} - H_{11} & 0 & 0 \\ U_{31} & 0 & H_{33} - H_{11} & 0 \\ U_{41} & 0 & 0 & H_{44} - H_{11} \end{vmatrix}$$
$$= 0 \qquad\qquad\qquad (17.85)$$

The above determinant is the multiplication of all diagonal elements. Since the terms like $H_{33} - H_{11}$ are generally nonzero, only the first term makes the determinant equal zero. By equating that with zero we find E to be,

$$E = H_{11} - \frac{U_{21}U_{12}}{H_{22} - H_{11}} - \frac{U_{31}U_{13}}{H_{33} - H_{11}} - \frac{U_{41}U_{14}}{H_{44} - H_{11}} \tag{17.86}$$

Recalling that $H_{11} = E_{1o} + U_{11}$, we obtain

$$E = E_{1o} + U_{11} - \frac{U_{21}U_{12}}{H_{22} - H_{11}} - \frac{U_{31}U_{13}}{H_{33} - H_{11}} - \frac{U_{41}U_{14}}{H_{44} - H_{11}}$$

$$= E_{1o} + U_{11} - \sum_{\substack{i=1 \\ i \neq 1}}^{4} \frac{U_{i1}U_{1i}}{H_{ii} - H_{11}} \tag{17.87}$$

Since $U_{i1} = U_{1i}^{*}$, we may write $U_{i1}U_{1i} = |U_{1i}|^2$ and summarize the second-order perturbation correction to the first eigenenergy as follows. The index i runs over every state except 1, i.e., 2, 3, and 4. The first eigenenergy is

$$E_1 = E_{1o} + U_{11} - \sum_{\substack{i=1 \\ i \neq 1}}^{4} \frac{|U_{1i}|^2}{H_{ii} - H_{11}} \tag{17.88}$$

Before proceeding, we make another approximation to simplify the denominator of the last term in the above equation. As the perturbation energies are small, it is reasonable to assume their difference is smaller and insignificant. Hence, we can simplify $H_{ii} - H_{11}$ to

$$H_{ii} - H_{11} = (E_{io} + U_{ii}) - (E_{1o} + U_{11})$$

$$= E_{io} - E_{1o} + U_{ii} - U_{11} \cong E_{io} - E_{1o} \tag{17.89}$$

As a result of this approximation, we have the second-order corrected E_1,

$$E_1^{(2)} = E_{1o} + U_{11} - \sum_{\substack{i=1 \\ i \neq 1}}^{4} \frac{|U_{1i}|^2}{E_{io} - E_{1o}} \quad \text{(to 2}^{\text{nd}}\text{ order in perturbation)} \tag{17.90}$$

Note that this is what we found in equation (17.20). Observe that the other unperturbed eigenenergies are always larger than the first one, i.e., $E_{io} - E_{1o} > 0$, therefore *the second-order correction to the first perturbed eigenenergy is always negative*.

The same procedure is repeated for the second, third, and fourth row to find the second, third, and fourth eigenstates, respectively. Every time we do this for row m, we apply the two aforementioned assumptions: 1) Those matrix elements of U which do not have the index (m) in them are set to be zero as they only indirectly affect state m and hence E_m. 2) Every other E (expect E_m) is assumed to be equal to H_{nn} in row n. With this, the mth eigenenergy corrected up to the second-order perturbation is given as

$$E_m = E_{mo} + U_{mm} - \sum_{i \neq m}^{N} \frac{|U_{mi}|^2}{E_{io} - E_{mo}}$$

(to 2nd order in perturbation U)   (17.91)

In this case, depending on the sign of $E_{io} - E_{mo}$, the contribution to the second-order correction to the mth perturbed eigenenergy might be negative or positive. This is the same second-order correction to the energy that we found in equation (17.20).

Sometimes due to the special symmetry of the wave functions and the nature of the perturbing potential, the second term in equations (17.20) and (17.91) is zero. Therefore, we must move to higher orders of approximations.

17.5 Problems

Section 17.2

(1) Consider a particle in a box with box length L that lies between $x = 0$ and $x = L$. A small perturbation with a potential energy of $U(x) = U\delta(x - L/2)$ is applied. Find the first-order correction of

    (a) The lowest three energy levels ($n = 1, 2, 3$)
    (b) The lowest energy wave function ($n = 1$). You do not have to normalize the wave function.

(2) Recall that the first-order perturbation lets us write the perturbed eigenenergy of a PiB according to equation (17.13), $E_n = E_{no} + \langle \varphi_{no} | U(x) | \varphi_{no} \rangle$, where $U(x) = q\mathcal{E}x$ is the perturbation due to an electric field $\mathcal{E}$. Find the perturbed eigenenergy for $n = 1$, $n = 100$, and $n = 200$. Does the difference between the perturbed and unperturbed energy depend on n?

(3) Show that for a particle in a box, the matrix element of the potential energy $U(x) = q\mathcal{E}x$ for the 1st and p^{th} state is zero if p is odd, i.e.,

$$\int_{x=0}^{x=L} \sin\left(\frac{\pi x}{L}\right) U(x) \sin\left(\frac{p\pi x}{L}\right) dx = 0$$

(4) Expand equation (17.9) by factoring out the terms with λ^3, and show that the third-order correction of the eigenenergy is given by

$$E_k^{(3)} = \sum_{m \neq k} \sum_{n \neq k} \frac{U_{km} U_{mn} U_{nk}}{(E_{ko} - E_{mo})(E_{ko} - E_{no})} - \sum_{m \neq k} \frac{U_{km} U_{mk} U_{kk}}{(E_{ko} - E_{mo})^2}$$

Hint: In equation (17.9), assume $m = k$ and plug the values found for $E_k^{(1)}$, $E_k^{(2)}$, $a_m^{(1)}$, and $a_m^{(2)}$ to find $E_k^{(3)}$.

Section 17.3

(5) Use the eigensolver in a mathematical package to find the eigenenergies and eigenvalues of the perturbed Hamiltonian in Example 5 and compare your results.

(6) Use the same perturbation matrix U of Example 5 and assume unperturbed energy levels $E_1 = 1\,\text{eV}$, $E_2 = 2\,\text{eV}$, and $E_3 = 5\,\text{eV}$ and calculate the perturbed energy levels using the 2nd order perturbation theory. Compare the results with exact numerical method.

(7) Plot the first-order corrected eigenenergies in Example 1 as a function of electric field $\mathcal{E}$ or voltage drop $(\mathcal{E}L)$ for the lowest five eigenenergies.

(8) Repeat Example 5 and find the new eigenenergies and states for the perturbed Hamiltonian when degenerate states 1 and 2

are now mixed by the following perturbation:

$$U = \begin{bmatrix} 0 & 0.2 & 0 \\ 0.2 & 0 & 0.1 \\ 0 & 0.1 & 0 \end{bmatrix}$$

(9) Find the analytic form of the two solutions of equation (17.63). Plot both solutions for the energy as a function of the inter-action matrix element of the perturbation, i.e., $|U_{12}|^2$. Assume $H_{11} = 1.1\,\text{eV}$ and $H_{22} = 1.11\,\text{eV}$.

(10) Using the two solutions found for the eigenenergies in Problem 9, find the unknown coefficients for the wave function, i.e., b_1 and b_2, using equation (17.51). Assume $H_{11} = 1.1\,\text{eV}$, $H_{22} = 1.11\,\text{eV}$, and $U_{12} = 0.5\,\text{eV}$.

References

Collin, R. E. (2001). *Foundations of Microwave Engineering*. Wiley-IEEE Press.

Raimes, S. (1961). *The Wave Mechanics of Electrons in Metals*. North-Holland Publishing Company, Amsterdam.

Ryde, N. (1976). *Atoms and Molecules in Electric Fields*. Almqvist & Wiksell, Stockholm.

Chapter 18

SCATTERING RATES AND FERMI GOLDEN RULE

Contents

In this chapter, we will learn how to use the time-dependent
perturbation theory to calculate the properties of quantum systems
in the presence of a perturbing potential. We will consider the
transition rate of electrons between energy levels, which leads to
photon emission and absorption.

Consider a time-independent Hamiltonian $(\widehat{H}_o)$ whose eigenfunctions and eigenvalues are known. An application of a time-dependent

perturbing potential $U(\bar{r}, t)$ causes the resulting wave function to be in a superposition of the eigenfunctions of the unperturbed Hamiltonian $(\widehat{H}_o)$. The time evolution of the coefficients of this superposition state is governed by a differential equation, which is derived and discussed in this chapter. The formulae derived will help us understand interesting scenarios such as the scattering of electrons between eigenfunctions of macroscopially large solids, nanowires, and quantum dots due to electron–photon and electron–phonon interactions.

18.1 Rewriting the Time-Dependent Schrödinger Equation

We will discuss how an electron transitions between the eigenstates of an unperturbed Hamiltonian (see $\widehat{H}_o$ in the following), in response to a time-dependent potential. The reference states in our discussion will refer to the eigenstates of the unperturbed or stationary system.

Consider a time-dependent perturbation (such as a new potential energy term) $\widehat{U}(\bar{r}, t)$ added to a time-independent Hamiltonian $\widehat{H}_o$ (also called the unperturbed Hamiltonian),

$$\widehat{H} = \widehat{H}_o + \widehat{U}(\bar{r}, t) \tag{18.1}$$

The eigenenergies and eigenfunctions of $\widehat{H}_o$ are assumed to be known and denoted by (E_{no}, ϕ_{no}). The time-dependent potential $\widehat{U}(\bar{r}, t)$ is also known. Then, the wave function of the time-dependent Schrödinger equation with Hamiltonian $\widehat{H}$ can be obtained by solving (Chapter 1)

$$i\hbar \frac{\partial \psi(\bar{r}, t)}{\partial t} = \widehat{H} \psi(\bar{r}, t) \tag{18.2}$$

We will expand the solution to the above differential equation as

$$\psi = \sum_n a_n(t)\, e^{-\frac{iE_{no}t}{\hbar}} \phi_{no} \tag{18.3}$$

Note that without a perturbation, the coefficients a_n would be constants. In the presence of a time-dependent perturbation, we find that $a_n(t)$ is time-dependent. We will find $a_n(t)$ by forming and solving coupled differential equations that govern these coefficients. Substituting equation (18.3) in equation (18.2) gives us the equations for their time evolution:

$$i\hbar \sum_n e^{-\frac{iE_{no}t}{\hbar}} \phi_{no} \frac{\partial a_n(t)}{\partial t} = \sum_n a_n(t) e^{-\frac{iE_{no}t}{\hbar}} \widehat{U} \phi_{no} \qquad (18.4)$$

Applying the operator $\int dv\, \phi_{po}^*$ to both sides of equation (18.4) from the left side gives us

$$i\hbar \sum_n e^{-\frac{iE_{no}t}{\hbar}} \left[\int dv\, \phi_{po}^* \phi_{no} \right] \frac{\partial a_n t}{\partial t}$$

$$= \sum_n a_n(t) e^{-\frac{iE_{no}t}{\hbar}} \left[\int dv\, \phi_{po}^* \widehat{U} \phi_{no} \right] \qquad (18.5)$$

Using the orthonormality of the wave functions, $\int dv\, \phi_{po}^* \phi_{no} = \delta_{pn}$, we note that the terms on the LHS only survive when $n = p$. Equation (18.5) then becomes

$$e^{-\frac{iE_{po}t}{\hbar}} \frac{\partial a_p t}{\partial t} = \frac{1}{i\hbar} \sum_n a_n(t) e^{-\frac{iE_{no}t}{\hbar}} \left[\int dv\, \phi_{po}^* \widehat{U} \phi_{no} \right] \qquad (18.6)$$

$$\frac{\partial a_p(t)}{\partial t} = \frac{1}{i\hbar} \sum_n a_n(t) e^{\frac{i(E_{po}-E_{no})t}{\hbar}} \left[\int dv\, \phi_{po}^* \widehat{U} \phi_{no} \right] \qquad (18.7)$$

The remaining integral on the RHS is called the matrix element of the perturbing potential $U_{pn}(t)$, and it is represented by

$$U_{pn}(t) = \int dv\, \phi_{po}^*(\bar{r})\, \widehat{U}(\bar{r}, t)\phi_{no}(\bar{r}) = \langle \phi_{po}|\widehat{U}|\phi_{no}\rangle \qquad (18.8)$$

The resonant frequency ω_{pn} is defined as the difference of energy levels n and p divided by Planck's constant,

$$\omega_{pn} = \frac{E_{po} - E_{no}}{\hbar} \qquad (18.9)$$

With these substitutions, equation (18.7) can be rewritten as

$$\frac{\partial a_p(t)}{\partial t} = \frac{1}{i\hbar} \sum_n e^{i\omega_{pn}t} U_{pn}(t) a_n(t) \qquad (18.10)$$

Equation (18.10) represents a set of coupled linear differential equations which can be solved to find the coefficients $a_p(t)$. As a result, the wave function in equation (18.3) can be constructed. U_{pn} and ω_{pn} are known because we already know the unperturbed eigenenergies and eigenstates (those with subscript o), and the form of the perturbation $\widehat{U}(\bar{r}, t)$.

18.2   Scattering rates/Transition Probability

Equation (18.10) can be solved for closed systems such as quantum dots and the two-level system discussed in Chapter 9. However, if the system under study has many quantum numbers, then solving equation (18.10) becomes more difficult. We will discuss one limiting case that leads to an extensively used formula for scattering rates called the *Fermi golden rule*. The terminology "scattering rate" refers to the transition rate between states of the unperturbed Hamiltonian, due to interaction with the time-dependent perturbation. It has units of 1/second. For this, we will make a few assumptions to simplify equation (18.10) further.

Assumption 1. The timescale over which $a_p(t)$ varies is much larger than ω_{pn}^{-1}.

Assumption 2. At time $t = 0$, $a_p(t) = \delta_{ps}$. That is, the system is in state ϕ_{so} at time $t = 0$.

We can then rewrite equation (18.10) as,

$$\frac{\partial a_p(t)}{\partial t} = \frac{1}{i\hbar} e^{i\omega_{ps}t} U_{ps}(t) a_s(t) + \frac{1}{i\hbar} \sum_{n \neq s} e^{i\omega_{pn}t} U_{pn}(t) a_n(t)$$

Assumption 3. We are interested in timescales where the system is still primarily in the state ϕ_{so}, except for a tiny deviation to other states where $p \neq s$. Therefore, $a_s(t) \cong 1 - N\varepsilon$ and $a_{p \neq s}(t) \cong \varepsilon$, where N is the total number of states other than s and ε is a tiny number. Then, the second term on the RHS can be neglected and the above equation can be written approximately as

$$\frac{\partial a_p(t)}{\partial t} \sim \frac{1}{i\hbar} e^{i\omega_{ps}t} U_{ps}(t) a_s(t) \qquad (18.11)$$

The coefficient $a_p(t)$ is found by integrating both sides of equation (18.11) with respect to time,

$$a_p(t) = \frac{1}{i\hbar} \int_0^t dt' \, e^{i\omega_{ps}t'} U_{ps}(t') \, a_s(t') \tag{18.12}$$

We will next consider a particular form for the time-dependent perturbing potential,

$$U(\bar{r}, t) = u(\bar{r}) \cos (\bar{k} \cdot \bar{r} - \omega_o t) = \frac{1}{2} u(r) [e^{i\bar{k}\cdot\bar{r} - i\omega_o t} + e^{-i\bar{k}\cdot\bar{r} + i\omega_o t}] \tag{18.13}$$

where ω_o is the angular frequency of the perturbing potential. This is a scalar plane wave potential with a single wave vector. The matrix element of this potential is found by substituting equation (18.13) in equation (18.8), which gives

$$U_{ps}(t) = \int dv \, \phi_{po}^*(\bar{r}) U(\bar{r}, t) \phi_{so}(\bar{r}) = [e^{-i\omega_o t} u_{ps} + e^{+i\omega_o t} \tilde{u}_{ps}] \tag{18.14}$$

where,

$$u_{ps} = \frac{1}{2} \int dv \, \phi_{po}^*(\bar{r}) \, u(\bar{r}) e^{+i\bar{k}\cdot\bar{r}} \phi_{so}(\bar{r}) \text{ and} \tag{18.15}$$

$$\tilde{u}_{ps} = \frac{1}{2} \int dv \, \phi_{po}^*(\bar{r}) \, u(\bar{r}) e^{-i\bar{k}\cdot\bar{r}} \phi_{so}(\bar{r}) \tag{18.16}$$

It is easy to see that

$$\tilde{u}_{ps} = u_{sp}^* \tag{18.17}$$

Substituting equation (18.14) into equation (18.12) and using Assumption 3 $(a_s \cong 1 - N\varepsilon \cong 1)$ gives

$$a_p(t) = \frac{1}{i\hbar} \int_0^t dt' \left[u_{ps} e^{i(\omega_{ps} - \omega_o)t'} + u_{sp}^* e^{i(\omega_{ps} + \omega_o)t'} \right] \tag{18.18}$$

Upon integration, equation (18.18) becomes

$$a_p(t) = \left[\frac{u_{ps}}{\hbar} \frac{e^{i(\omega_{ps} - \omega_o)t} - 1}{\omega_{ps} - \omega_o} + \frac{u_{sp}^*}{\hbar} \frac{e^{i(\omega_{ps} + \omega_o)t} - 1}{\omega_{ps} + \omega_o} \right] \tag{18.19}$$

The first term of equation (18.19) dominates when $\hbar\omega_{ps} = E_p - E_s = \hbar\omega_o$ (the final state energy E_p is larger than the initial state energy E_s). We will see that this term corresponds to the absorption of a quantum of energy $\hbar\omega_o$. The second term dominates when $\hbar\omega_{ps} = E_p - E_s \cong -\hbar\omega_o$ (the final state energy E_p is smaller than the initial state energy E_s). This term corresponds to the emission of a quantum of energy $\hbar\omega_o$. We will first analyze the absorption term, which dominates when $\omega_{ps} \sim \omega_o$,

$$a_p\left(t\right) \sim \frac{u_{ps}}{\hbar} \frac{e^{i(\omega_{ps}-\omega_o)t} - 1}{\omega_{ps} - \omega_o} \tag{18.20}$$

Using the formula $\sin\left(\theta\right) = \frac{1}{2i}\left[e^{i\theta} - e^{-i\theta}\right]$, equation (18.20) can be written as follows by factoring out a term $e^{\frac{i(\omega_{ps}-\omega_o)t}{2}}$:

$$a_p\left(t\right) \sim \frac{u_{ps}}{\hbar} \frac{e^{\frac{i(\omega_{ps}-\omega_o)t}{2}} - e^{-\frac{i(\omega_{ps}-\omega_o)t}{2}}}{\omega_{ps} - \omega_o} e^{\frac{i(\omega_{ps}-\omega_o)t}{2}}$$

$$= \frac{2iu_{ps}}{\hbar} \frac{\sin\left[\frac{(\omega_{ps}-\omega_o)t}{2}\right]}{\omega_{ps} - \omega_o} e^{\frac{i(\omega_{ps}-\omega_o)t}{2}} \tag{18.21}$$

The probability $P_{ps}\left(t\right)$ of finding the electron in state ϕ_{po} when the electron started out in the state ϕ_{so} at time $t = 0$ is

$$P_{ps}\left(t\right) = |a_p\left(t\right)|^2 = \frac{4\left|u_{ps}\right|^2}{\hbar^2} \frac{\sin^2\left[(\omega_{ps} - \omega_o)\,t/2\right]}{(\omega_{ps} - \omega_o)^2} \tag{18.22}$$

It is instructive to plot the second part of the above equation as a function of detuning $\Delta = \omega_{ps} - \omega_o$ for various interaction times. The probability $P_{ps}\left(t\right)$ is zero at the instances when the detuning $\Delta = \frac{2\pi n}{t}$, where n is any integer. For example, with interaction time $t = 1\,\mathrm{ns}$, the function is zero when the detuning is equal to integer multiples of 2π GHz. Note that in Figure 18.1(a), the y-value is large even for nonzero detuning; this means that the probability of making a transition from state s to state p is nonzero even though the driving time-dependent potential does not have the same frequency as ω_o.

On the other hand, as the duration of the perturbation (t) increases, the y-value in Figure 18.1(a) becomes sharper, and the number of zeros increases. Additionally, the value of the function distant from resonant frequencies drops significantly. This means that

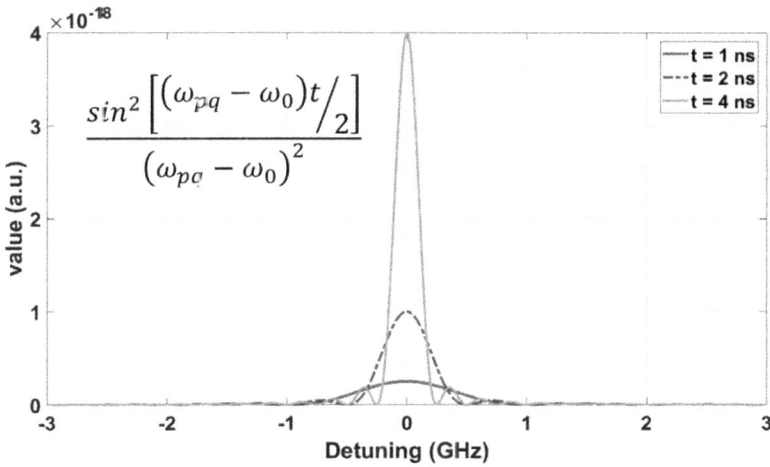

(a)

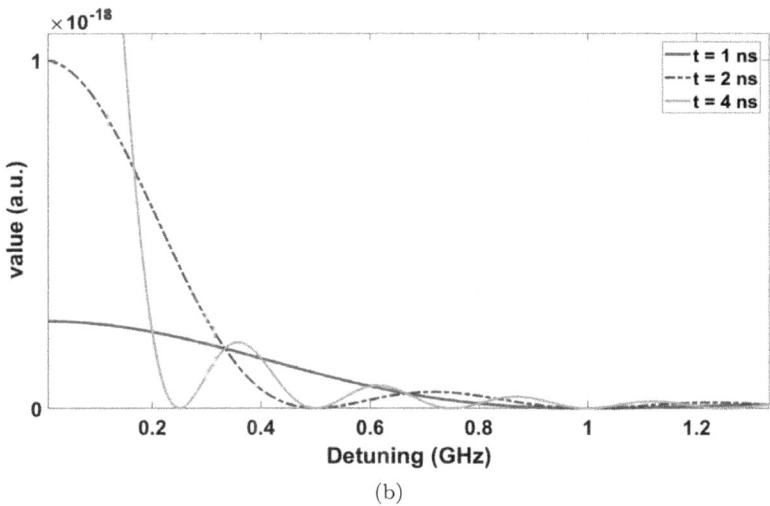

(b)

Figure 18.1. (a) The behavior of equation (18.22) as a function of detuning $(f_{ps} - f_o) = \frac{\omega_{ps} - \omega_o}{2\pi}$ at different interaction times. As the interaction time increases, the peak becomes sharper at resonance (zero detuning). (b) An enlarged version of (a) shows how the probability drops off-resonance (nonzero detuning), and the peaks further from zero-detuning decay faster as the duration of the perturbing field increases.

Note: a.u. stands for arbitrary units; the coefficient $\frac{4|u_{ps}|^2}{\hbar^2}$ in equation (18.22) has been set to unity. The probability of the transition from state s to p has an identical shape to the above graph.

when t is large, the probability of transition (y-axis) is significant if the driving time-dependent potential is in resonance with the two-level atom or qubit. Figure 18.1(b) shows the enlarged version of Figure 18.1(a), zooming in where the detuning Δ is 0 to $2\pi \times 1.2\,\text{GHz}$.

We rewrite equation (18.22) by multiplying and dividing the last term by $\frac{1}{4}t^2$ to obtain

$$P_{ps}\left(t\right) = \left|a_p\left(t\right)\right|^2 = \frac{4\left|u_{ps}\right|^2}{\hbar^2}\frac{t^2}{4}\frac{\sin^2\left[\left(\omega_{ps} - \omega_o\right)t/2\right]}{\left[\left(\omega_{ps} - \omega_o\right)t/2\right]^2} \tag{18.23}$$

It can be proven that the last term on the right side, called the Sinc function, is proportional to a Dirac delta function as time t tends to infinity (see Chapter 20) and can be written as

$$\frac{\sin^2\left[\left(\omega_{ps} - \omega_o\right)t/2\right]}{\left[\left(\omega_{ps} - \omega_o\right)t/2\right]^2} = \frac{2\pi}{t}\delta(\omega_{ps} - \omega_o) \tag{18.24}$$

Figure 18.2 shows the LHS of the above equation in terms of detuning $(f_{ps} - f_o)$ for four different interaction times. The function

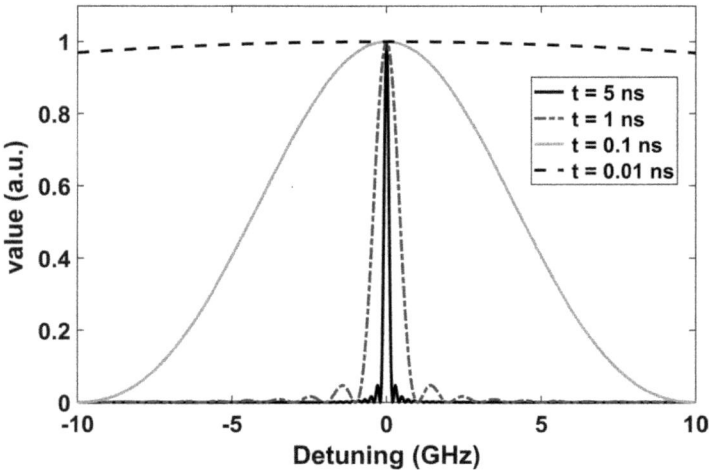

Figure 18.2. The Sinc function versus the detuning frequency $(f_{ps} - f_o)$. The energy spacing corresponds to $f_{ps} = 4\,\text{GHz}$. At larger values of time, the Sinc function approaches a Dirac delta function. The curves are all normalized to have a maximum value of 1.

is very sharp and narrow at large time intervals, meaning the transition probability almost vanishes for the detuned time-dependent potential.

Substituting equation (18.24) in equation (18.23), we get

$$P_{ps}(t) = |a_p(t)|^2$$

$$= \frac{4|u_{ps}|^2 t^2}{\hbar^2} \frac{2\pi}{4} \frac{}{t} \delta(\omega_{ps} - \omega_o) = \frac{2\pi|u_{ps}|^2}{\hbar^2} t\, \delta(\omega_{ps} - \omega_o) \quad (18.25)$$

We will now define the **transition probability** (R), which is the probability per unit time that an electron in state ϕ_{so} will transition to state ϕ_{po},

$$R_{ps} = \frac{P_{ps}(t)}{t} = \frac{|a_p(t)|^2}{t} = \frac{2\pi|u_{ps}|^2}{\hbar^2} \delta(\omega_{ps} - \omega_o) \quad (18.26)$$

Recall that P_{ps} (R_{ps}) is the probability (transition probability) to go from initial state s to final state p. Noting that $E_p - E_s = \hbar\omega_{ps}$ and $\delta(ax) = \frac{1}{a}\delta(x)$ as discussed in Chapter 20, we can write

$$\delta(\omega_{ps} - \omega_o) = \delta\left(\frac{E_p - E_s - \hbar\omega_o}{\hbar}\right) = \hbar\delta(E_p - E_s - \hbar\omega_o) \quad (18.27)$$

As a result, the transition probability given by equation (18.26) is rewritten as,

$$R_{ps} = \frac{2\pi|u_{ps}|^2}{\hbar} \delta(E_p - E_s - \hbar\omega_o) \qquad \text{(absorption)} \qquad (18.28)$$

The delta function in equation (18.28) means that the electron in state ϕ_{so} with energy E_s can transition to state ϕ_{po} with energy E_p only if energy is conserved by absorbing a quantum of the perturbing field having energy $\hbar\omega_o$ $(E_p = E_s + \hbar\omega_o)$.

Through a similar process to our above discussion, we can also show that the second term of equation (18.19) is

$$R_{ps} = \frac{2\pi|u_{sp}^*|^2}{\hbar} \delta(E_p - E_s + \hbar\omega_o) \qquad \text{(emission)} \qquad (18.29)$$

The delta function in equation (18.29) means that the electron in state ϕ_{so} with energy E_s can transition to state ϕ_{po} with energy E_p

only if energy is conserved by emitting a quantum of the perturbing field having energy $\hbar\omega_o$ ($E_p = E_s - \hbar\omega_o$).

18.2.1   *Fermi golden rule*

Now, in the case that the final states ϕ_{po} are continuous rather than discrete, we can sum over the final states to find the total transition probability of an electron scattering from its initial state ϕ_{so} to any of the final states $\{\phi_{po}\}$. Then equations (18.28) and (18.29) become,

$$R_{\{p\}s} = \sum_p \frac{2\pi \left|u_{ps}\right|^2}{\hbar} \delta(E_p - E_s - \hbar\omega_o) \text{ and } \quad \text{(absorption)} \quad (18.30)$$

$$R_{\{p\}s} = \sum_p \frac{2\pi \left|u_{sp}^*\right|^2}{\hbar} \delta(E_p - E_s + \hbar\omega_o) \quad \text{(emission)} \quad (18.31)$$

where the curly bracket $\{p\}$ denotes that $R_{\{p\}s}$ is the transition probability to any one of the many final states $\{\phi_{po}\}$.

The above equations are called **the Fermi golden rule**. It gives the rate at which electrons in state s are scattered into one of the states p by the potential $U(\bar{r}, t)$. Solving a scattering problem starts with determining the potential $U(\bar{r}, t)$ that is added to the unperturbed Hamiltonian (H_o). As a result, we can find the matrix element of U between any two states ϕ_{po} and ϕ_{so} (i.e., u_{ps}). The last step is to use the matrix element to numerically or analytically calculate the transition probability R_{ps} in equations (18.30) and (18.31).

The above **Fermi golden rule** is for the cases when the perturbing potential is weak such that Assumption 3 is valid. When this assumption is invalid, other methods should be used. See Chapter 9 of this book for the specific case of a two-level system and (Rosencher, 2002) for the Fermi golden rule applied to higher orders of perturbation.

In the following section, we assume the scatterer is the electromagnetic field of light. Materials can absorb light/photons and cause electrons in the material to get excited from a lower to a higher energy level. The absorption rate, the number of photons a material can absorb per unit time, will vary from material to material. For example, at certain wavelengths, a thinner slab of InSb would absorb light more efficiently than a thicker slab of Si. Absorption coefficients of elemental and compound semiconductors can be found in (Madelung, 2004).

The same concept also applies to photon emission. Electrons with sufficient energy within a material can drop from a higher to a lower energy level and cause photons to be emitted. The emission rate of photons is an important concept in technology. For example, to design a TV screen based on PN junction diodes (LED-TV), one would choose a material where an electron injected at a higher energy level (conduction band) in a diode will fall to a lower energy level (valence band) while emitting a photon in the process.

In the next section, we will calculate the emission and absorption rate of photons due to the transition of electrons between energy levels. But before that, let's wrap up this section by making two more simplifications. First, it is assumed that the transition matrix elements, u_{ps} and $\tilde{u}_{ps}$ are independent of the final indices, p. As a result, the matrix elements are factored out of the summation over p to give,

$$R_{\{p\}s} = \frac{2\pi \left|u_{ps}\right|^2}{\hbar} \sum_p \delta(E_p - E_s - \hbar\omega_o) \quad \text{(absorption)} \qquad (18.32)$$

$$R_{\{p\}s} = \frac{2\pi \left|u_{sp}^*\right|^2}{\hbar} \sum_p \delta(E_p - E_s + \hbar\omega_o) \quad \text{(emission)} \qquad (18.33)$$

Second, assuming a free electron wave function $\phi_{po}(\bar{r}) = \frac{1}{\sqrt{V}}e^{i\bar{k}\cdot\bar{r}}$, the delta function term inside the summation signs in equations (18.32) and (18.33) can be replaced by the density of states at energy $E_p = E_s + \hbar\omega_o$ and $E_p = E_s - \hbar\omega_o$, multiplied by the volume of the material V. Note that in derivations involving summation over final states, the end result is volume independent. Equations (18.32) and

(18.33) can then be rewritten as

$$R_{\{p\}s} = \frac{2\pi \left|u_{ps}\right|^2}{\hbar} V \cdot \mathrm{DOS}\left(E_p = E_s + \hbar\omega_o\right) \quad \text{(absorption)} \quad (18.34)$$

$$R_{\{p\}s} = \frac{2\pi \left|u_{sp}^*\right|^2}{\hbar} V \cdot \mathrm{DOS}\left(E_p = E_s - \hbar\omega_o\right) \quad \text{(emission)} \quad (18.35)$$

The Fermi golden rule: The rate of transition of an electron from an initial (or starting) state s to one of the different possible final states indexed by p is given by

$$R_{\{p\}s} = \sum_p \frac{2\pi \left|u_{ps}\right|^2}{\hbar} \delta(E_p - E_s - \hbar\omega_o) \quad \text{(absorption)}$$

$$R_{\{p\}s} = \sum_p \frac{2\pi \left|u_{sp}^*\right|^2}{\hbar} \delta(E_p - E_s + \hbar\omega_o) \quad \text{(emission)}$$

The reader should also remember that in many cases, the scattering rate is directly proportional to the square of the matrix element of the perturbing potential u_{ps} and the density of final states (DOS):

$$R_{\{p\}s} \propto \left|u_{ps}\right|^2 \mathrm{DOS}\left(E_p = E_s + \hbar\omega_o\right) \quad \text{(absorption)}$$

$$R_{\{p\}s} \propto \left|u_{ps}\right|^2 \mathrm{DOS}\left(E_p = E_s - \hbar\omega_o\right) \quad \text{(emission)}$$

18.3   Interaction of Atoms and Nanostructures with Light

The momentum operator in the Hamiltonian of an electron changes due to interaction with an electromagnetic (EM) field. In a field, the momentum operator changes to

$$\hat{p} = \begin{cases} \hat{p} - q\bar{A} & \text{(for a positive charge } q) \\ \hat{p} + q\bar{A} & \text{(for a negative charge } - q) \end{cases}$$

where $\bar{A}$ is the magnetic vector potential satisfying $\bar{B} = \nabla \times \bar{A}$. We remind the reader that $q = +1.6 \times 10^{-19}$ C throughout this book. The origin of the above equation lies in classical mechanics. Readers can consult Appendix G of (Singleton, 2001) for a discussion of this topic. Note that the magnetic vector potential $(\bar{A})$, is indeed momentum per unit charge, hence $q\bar{A}$ is added to the particle momentum $m\bar{v}$ to conserve the total momentum of a charged particle. The Hamiltonian for the interaction of an electron in vacuum with an electromagnetic field is,

$$\hat{H} = \frac{1}{2m_o} \left(\hat{p} + q\bar{A} \right)^2 + U_{\text{static}}(\bar{r}) \tag{18.36}$$

U_{static} is part of the original time-independent Hamiltonian and does not depend on the EM field. In previous chapters, we simply called this term $U(\bar{r})$, but in this chapter, we use $U(\bar{r}, t)$ to represent the time-dependent perturbation. The reader should recall that equation (18.36) was also discussed in Chapter 15. The first term can be expanded out to give

$$\hat{H} = \frac{\hat{p}^2}{2m_o} + U_{\text{static}}(\bar{r}) + \frac{q}{2m_o} \left(\hat{p} \cdot \bar{A} + \bar{A} \cdot \hat{p} \right) + \frac{1}{2m_o} q^2 A^2 \tag{18.37}$$

The size of the last term $q^2 A^2$ is negligible unless the electromagnetic field is intense, e.g., if we use high-power electromagnetic sources. Including this term in the physics of light–atom interaction is the subject of nonlinear optics and often requires higher order perturbation or nonperturbative methods. We will not discuss perturbation by intense fields, and so we will neglect the last term.

The Hamiltonian of an electron in a weak EM field is approximated as

$$\hat{H} = \frac{\hat{p}^2}{2m_o} - U_{\text{static}}(\bar{r}) + \frac{q}{2m_o} \left(\hat{p} \cdot \bar{A} + \bar{A} \cdot \hat{p} \right) \tag{18.38}$$

To simplify the above equation, we first show that,

$$\hat{p} \cdot \bar{A} = \bar{A} \cdot \hat{p} \tag{18.39}$$

Operating $\hat{p} \cdot \bar{A}$ on a function f, we have

$$\begin{aligned}
\left[\hat{p} \cdot \bar{A} \right] f &= \left[-i\hbar \nabla \cdot \bar{A} \right] f = f \left[-i\hbar \nabla \cdot \bar{A} \right] + \bar{A} \cdot \left[-i\hbar \nabla \right] f \\
&= f \left[-i\hbar \nabla \cdot \bar{A} \right] + \left[\bar{A} \cdot \hat{p} \right] f
\end{aligned} \tag{18.40}$$

In the Coulomb gauge for an EM field (see an introductory electromagnetics textbook),

$$\nabla \cdot \bar{A} = 0 \tag{18.41}$$

Substituting equation (18.41) in equation (18.40), we get equation (18.39). Using equation (18.39) in equation (18.38), the Hamiltonian becomes

$$\hat{H} = \frac{\hat{p}^2}{2m_o} + U_{\text{static}}(\bar{r}) + \frac{q}{m_o}\bar{A} \cdot \hat{p} \tag{18.42}$$

In equation (18.42), the third term is the perturbation created by the electromagnetic field. We will consider a plane wave EM field with a magnetic vector potential given by

$$\bar{A}(\bar{r},t) = \vec{\eta}A_o \cos\left(\bar{k} \cdot \bar{r} - \omega_o t\right) = \frac{A_o}{2}\vec{\eta}\left[e^{i\left(\bar{k}\cdot\bar{r} - \omega_o t\right)} + e^{-i\left(\bar{k}\cdot\bar{r} - \omega_o t\right)}\right] \tag{18.43}$$

where $\bar{k}$ is the wave vector, $\vec{\eta}$ is the unit polarization vector, and ω_o is the frequency.

Then, the perturbation $U(\bar{r},t)$ is

$$U(\bar{r},t) = -\frac{q}{m_o}\bar{A} \cdot \hat{p} = -\frac{qA_o}{2m_o}\left[e^{i\bar{k}\cdot\bar{r} - i\omega_o t} + e^{-i\bar{k}\cdot\bar{r} + i\omega_o t}\right]\vec{\eta} \cdot \hat{p} \tag{18.44}$$

Note that this perturbation has a similar form as equation (18.13). Using the above perturbation due to the EM field and equations (18.28) and (18.29), the transition probability or scattering rate for electrons from the state ϕ_{so} to the state ϕ_{po} is

$$R_{ps} = \frac{2\pi \left|u_{ps}\right|^2}{\hbar}\delta(E_p - E_s - \hbar\omega_o) \quad \text{(absorption)} \tag{18.45}$$

$$R_{ps} = \frac{2\pi \left|u_{sp}^*\right|^2}{\hbar}\delta(E_p - E_s + \hbar\omega_o) \quad \text{(emission)} \tag{18.46}$$

where the matrix element is the one written for the absorption part as in equation (18.15),

$$u_{ps} = \frac{qA_o}{2m_o}\int \phi_{po}^*(\bar{r})\,e^{+i\bar{k}\cdot\bar{r}}\vec{\eta} \cdot \hat{p}\,\phi_{so}(\bar{r})dv \quad \text{(absorption)} \tag{18.47}$$

And, similarly for the emission part, we use equation (18.16),

$$u_{sp}^* = \tilde{u}_{ps} = \frac{qA_0}{2m_c} \int \phi_{po}^*(\bar{r}) e^{-i\bar{k}\cdot\bar{r}} \vec{\eta} \cdot \hat{p}\, \phi_{so}(\bar{r}) dv \quad \text{(emission)} \quad (18.48)$$

Using the shorthand notation $\langle e^{\pm i\bar{k}\cdot\bar{r}} \vec{\eta} \cdot \hat{p}\rangle_{ps} = \int d^3r\, \phi_{po}^*(\bar{r}) e^{\pm i\bar{k}\cdot\bar{r}} \vec{\eta} \cdot \hat{p}\, \phi_{so}(\bar{r})$, we write the scattering rates as

$$R_{ps} = \frac{2\pi q^2 A_o^2}{4\, m_o^2\, \hbar} \left| \langle e^{i\bar{k}\cdot\bar{r}} \vec{\eta} \cdot \hat{p}\rangle_{ps} \right|^2 \delta(E_p - E_s - \hbar\omega_o) \quad \text{(absorption)}$$

$$(18.49)$$

$$R_{ps} = \frac{2\pi q^2 A_o^2}{4\, m_o^2 \hbar} \left| \langle e^{-i\bar{k}\cdot\bar{r}} \vec{\eta} \cdot \hat{p}\rangle_{ps} \right|^2 \delta(E_p - E_s + \hbar\omega_o) \quad \text{(emission)}$$

$$(18.50)$$

The delta functions in the above equations enforce energy conservation.

18.3.1 *Optical Transition Rates in Quantum Dots*

For the case where the spatial extent of the quantum dot is much smaller than the wavelength of light, the vector potential (and the corresponding electric and magnetic fields) does not change significantly with position at the location of the quantum dot. Such an assumption is valid only for a small-closed system such as a quantum dot or atom but not a nanowire, quantum well, or bulk semiconductor. This is because the spatial extent of these systems is large and comparable to the wavelength of the photon.

In this special case ($r \ll \lambda$), the $e^{\pm i\bar{k}\cdot\bar{r}}$ term in equations (18.49) and (18.50) can be expanded as a Taylor series and approximated to 1 because $\bar{k} \cdot \bar{r} = \frac{2\pi}{\lambda} r \ll 1$,

$$e^{\pm i\bar{k}\cdot\bar{r}} = 1 \pm \bar{k} \cdot \bar{r} \pm \frac{1}{2}\left(\bar{k} \cdot \bar{r}\right)^2 \pm \ldots \approx 1 \quad (18.51)$$

As a result, $|\langle e^{\pm i\bar{k}\cdot\bar{r}} \vec{\eta} \cdot \hat{p}\rangle_{ps}|$ can be approximated as $|\langle \vec{\eta} \cdot \hat{p}\rangle_{ps}|$ where,

$$\langle \vec{\eta} \cdot \hat{p}\rangle_{ps} = \int \phi_{po}^*(\bar{r})\, \vec{\eta} \cdot \hat{p}\, \phi_{so}(\bar{r})\, dv$$

Assuming that the polarization $\vec{\eta}$ of light is fixed in space, we get,

$$\langle \vec{\eta} \cdot \hat{p}\rangle_{ps} = \vec{\eta} \cdot \int \phi_{po}^*(\bar{r})\, \hat{p}\, \phi_{so}(\bar{r}) dv = \vec{\eta} \cdot \bar{p}_{ps}$$

where we have defined $\bar{p}_{ps}$ as

$$\bar{p}_{ps} = \int \phi_{po}^* (\bar{r}) \, \hat{p} \, \phi_{so}(\bar{r}) dv$$

For a planar electromagnetic wave with wave vector $\bar{k}$, the two possible polarizations are normal to the wave vector as shown in Figure 18.3. Note that because $e^{\pm i \vec{k} \cdot \vec{r}} \sim 1$, we can see from equations (18.47) and (18.48) that $u_{ps} = u_{sp}^*$.

Recall that the scattering rate in equation (18.49) is

$$R_{ps} = \frac{2\pi q^2 A_o^2}{4\, m_o^2 \hbar} \left| \langle e^{i\bar{k}\cdot\bar{r}} \vec{\eta} \cdot \hat{p} \rangle_{ps} \right|^2 \delta(E_p - E_s - \hbar\omega_o) \qquad (18.52)$$

The transition between states ϕ_{po} and ϕ_{so} depends on the momentum matrix elements between these states. Substituting (see equation (18.103))

$$|A_o| = \frac{\mathcal{E}_o}{\omega_o} \qquad (18.53)$$

in the above expression, we get,

$$R_{ps} = \frac{2\pi q^2 \mathcal{E}_o^2}{4\, m_o^2 \hbar \omega_o^2} \left| \langle e^{i\bar{k}\cdot\bar{r}} \vec{\eta} \cdot \hat{p} \rangle_{ps} \right|^2 \delta(E_p - E_s - \hbar\omega_o) \qquad (18.54)$$

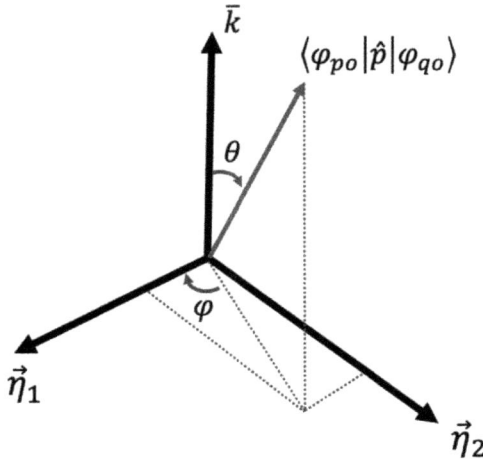

Figure 18.3. Two possible polarizations for a photon, which are normal to the wave vector $\bar{k}$.

We assume that the incident field is different from equation (18.43) in that waves are propagating in all possible directions (no longer a single $\bar{k}$ but light is propagating in all 4π steradians) and that the vector potential A_z does not vary when the magnitude of $\bar{k}$ changes (for the $\bar{k}$ values of importance to the calculation below). You will see that we have a Dirac delta function that comes out in the calculation that makes only a narrow range of $|\bar{k}|$ important. This value of $|\bar{k}|$ corresponds to energy level separation of the QD which is $\omega_{ps} = c|\bar{k}|$. The total scattering rate is found by summing equation (18.54) over all possible photon wave vectors $(\bar{k})$ and two possible polarizations (see Figure 18.3),

$$R_{ps}^{\text{total}} = \frac{2\pi q^2 \mathcal{E}_o^2}{4\,m_o^2 \hbar} \sum_{\eta_1,\eta_2} \sum_{k} \left| \langle e^{i\bar{k}\cdot\bar{r}} \vec{\eta} \cdot \hat{p} \rangle_{ps} \right|^2 \frac{1}{\omega_o^2} \delta(E_p - E_s - \hbar\omega_o)$$

(18.55)

The summation over the polarizations $(\vec{\eta}_1, \vec{\eta}_2)$ can be written as

$$\sum_{\eta_1,\eta_2} \left| \langle e^{i\bar{k}\cdot\bar{r}} \vec{\eta} \cdot \hat{p} \rangle_{es} \right|^2 = \sum_{\eta_1,\eta_2} \left| \langle \vec{\eta} \cdot \hat{p} \rangle_{ps} \right|^2 = |\vec{\eta}_1 \cdot \bar{p}_{ps}|^2 + |\vec{\eta}_2 \cdot \bar{p}_{ps}|^2$$

(18.56)

where we have assumed $e^{i\bar{k}\cdot\bar{r}} \sim 1$. Using the angles defined in the spherical coordinate system, we write (see Figure 18.3),

$$|\vec{\eta}_1 \cdot \bar{p}_{ps}|^2 + |\vec{\eta}_2 \cdot \bar{p}_{ps}|^2 = |\bar{p}_{ps}|^2 \sin^2\theta \cos^2\varphi + |\bar{p}_{ps}|^2 \sin^2\theta \sin^2\varphi$$

$$= |\bar{p}_{ps}|^2 \sin^2\theta$$

(18.57)

The summation over photon wave vectors can be converted to integration according to the following equation using the element of volume in a spherical coordinate system,

$$\sum_{k} \ldots = \frac{V_{em}}{8\pi^3} \int \ldots d^3k = \frac{V_{em}}{8\pi^3} \int_{\varphi=0}^{\varphi=2\pi} \int_{\theta=0}^{\theta=\pi} \int \ldots d\varphi \sin\theta \; d\theta \; k^2 dk$$

(18.58)

where V_{em} is the volume of the environment in which the EM field exists. Note that here k means $|\bar{k}|$. Putting back equations (18.56),

(18.57), and (18.58) in (18.55), we have,

$$R_{ps}^{\text{total}} = \frac{2\pi q^2 \mathcal{E}_o^2}{4 m_o^2 \hbar} \frac{V_{em}}{8\pi^3} \int_{\varphi=0}^{\varphi=2\pi} \int_{\theta=0}^{\theta=\pi} \int \frac{|\bar{p}_{ps}|^2}{\omega_o^2} \sin^2\theta$$

$$\times \ \delta(E_p - E_s - \hbar\omega_o) d\varphi \sin\theta \ d\theta \ k^2 dk \qquad (18.59)$$

The integration over angular coordinates returns $\frac{8\pi}{3}$ and the integration over $\bar{k}$ can be simplified to,

$$R_{ps}^{\text{total}} = \frac{2\pi q^2 \mathcal{E}_o^2}{4 m_o^2 \hbar} \frac{V_{em}}{8\pi^3} \frac{8\pi}{3} \int \frac{|\bar{p}_{ps}|^2}{\omega_o^2} \delta(E_p - E_s - \hbar\omega_o) k^2 dk \qquad (18.60)$$

For a photon, $\omega_o = c|\bar{k}|$, so we can write the argument of delta function as a function of k. From Chapter 20 which deals with delta functions, we observe $\delta(E_p - E_s - \hbar\omega_o) = \frac{1}{\hbar c} \delta\left(\frac{E_p - E_s}{\hbar c} - k\right)$. Using this and the sifting property of delta functions, equation (18.60) becomes

$$R_{ps}^{\text{total}} = \frac{q^2 \mathcal{E}_o^2 V_{em}}{6 \, m_o^2 \, \hbar^2 \pi c^3} |\bar{p}_{ps}|^2 \quad \text{(absorption)} \qquad (18.61)$$

Note that the delta function required the value of photon energy to be $E_p - E_s = \hbar\omega_{ps} = \hbar\omega_o$ (conservation of energy). The transition rate is also commonly expressed in terms of the dipole matrix element, which is defined as,

$$\bar{r}_{ps} = \int dv \ \phi_{po}^*(\bar{r}) \ \bar{r} \ \phi_{so}(\bar{r}) \qquad (18.62)$$

The relationship between the momentum and dipole matrix element is,

$$\bar{p}_{ps} = i m_o \omega_{ps} \bar{r}_{ps} \qquad (18.63)$$

as shown in Section 18.9. Using equation (18.63), the transition rate in equation (18.61) can be rewritten as

$$R_{ps}^{\text{total}} = \frac{q^2 \mathcal{E}_o^2 V_{em} \omega_{ps}^2}{6\hbar^2 \pi c^3} |\bar{r}_{ps}|^2 \quad \text{(absorption)} \qquad (18.64)$$

We can exploit the relationship between classical and quantum energies for photons to simplify the expressions in equations (18.61) and (18.64) further. If there are n_{ph} photons per unit energy in the volume V_{em}, then equating the classical expression for energy per unit volume $\left(\frac{1}{2}\epsilon_o\mathcal{E}_o^2\right)$ in an EM field and the quantum expression $(n_{ph}\hbar\omega_o/V_{em})$, we have

$$\frac{1}{2}\epsilon_o\mathcal{E}_o^2 = \frac{n_{ph}\hbar\omega_o}{V_{em}} \rightarrow \mathcal{E}_o^2 = \frac{2n_{ph}\hbar\omega_o}{\epsilon_o V_{em}} \qquad (18.65)$$

Substituting equation (18.65) into equations (18.61) and (18.64), we have two commonly used forms for the transition rate:

$$R_{ps}^{\text{total}} = \frac{q^2\omega_o n_{ph}}{3\pi\epsilon_o m_o^2 \hbar c^3}\,|\bar{p}_{ps}|^2 \qquad (18.66)$$

$$R_{ps}^{\text{total}} = \frac{q^2\omega_o^3 n_{ph}}{3\pi\epsilon_o \hbar c^3}\,|\bar{r}_{ps}|^2 \qquad (18.67)$$

Equations (18.66) and (18.67) are applicable for both the absorption and emission rates of photons. However, they only include the **stimulated components of absorption and emission** (see Figure 18.4). That is, the transition rate depends on the number of photons, which means that if the number of photons in the cavity is zero, then the transition rate is zero. Experimentally, we know that an atom in the excited state can still emit photons without external EM fields or photons by spontaneously transitioning to a lower energy level (see Figure 18.4). This process of spontaneous emission is not accounted for. The spontaneous emission rate is obtained by setting $n_{ph} = \frac{1}{2}$ in equation (18.66),

$$\tau_{spon} = \tau_{ps} = \frac{6\pi\epsilon_o \hbar c^3}{q^2\omega_o^3}\frac{1}{|\bar{r}_{ps}|^2} \qquad (18.68)$$

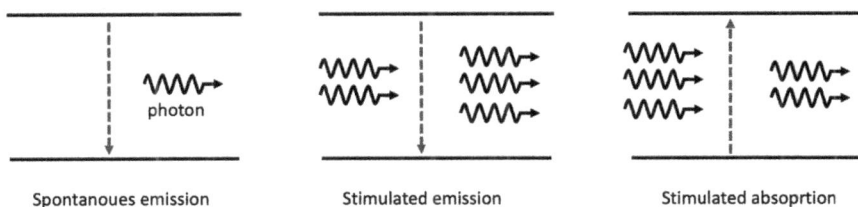

Figure 18.4. Stimulated and spontaneous emission parts of the transition rate. Black waves from the left show the number of incoming photons. In spontaneous emission, a photon is emitted even when $n_{ph} = 0$ (when no photons are incident in the system). In the case of stimulated absorption or emission, a photon is absorbed or emitted only if n_{ph} is nonzero, which means that a few photons must already exist in the system.

In equations (18.62) and (18.68), if the quantum dot or atom is embedded in a medium rather than a vacuum, replace the velocity of light in vacuum (c) by its value in the medium ($\tilde{c}$), the dielectric constant of vacuum (ϵ_o) by its value in the medium ($\epsilon_r \epsilon_o$). Additionally, in equation (18.61), replace m_o with the effective mass of the electron in the quantum dot m^*.

Optical absorption and emission: From equations (18.66), (18.67), and (18.68), we see that the transition rate of electrons between any two energy levels increases as the dipole matrix element increases. This means that photons are absorbed more efficiently in materials with a large dipole matrix element. If we want to build a light-emitting device based on quantum dots, designing the dots to have a large dipole matrix element would be helpful.

Example 1. Consider a particle in a box with width $L = 10\,\text{Å}$. Find the radiative lifetime for an electron in states $n = 2$, 3, and 4 to spontaneously emit a photon and transition to the state $n = 1$. Assume that the PiB is embedded in a medium with a refractive index of $n_r = 3.42$. The effective mass of electrons in the PiB is $m^* = 0.36\,m_0$.

Solution: We use equation (18.68) to obtain the values of τ_{21}, τ_{31}, and τ_{41}. But because the medium is no longer vacuum, we should first convert equation (18.68) to the appropriate formula in the medium. For this, we observe that the speed of light in a medium, $\tilde{c}$, is related to the vacuum speed of light c by $\tilde{c} = c/n_r$, where n_r is the refractive index of the medium (given by $n_r = \sqrt{\epsilon_r \mu_r}$). The relative electric permittivity and relative magnetic permeability of the medium are ϵ_r and μ_r, respectively. We will assume $\mu_r = 1$. Then equation (18.68) becomes

$$\tau_{spon} = \tau_{ps} = \frac{6\pi\epsilon_o \hbar c^3}{n_r q^2 \omega_o^3} \frac{1}{|\bar{r}_{ps}|^2}$$

For a radiative transition from state s to p, we must find the position matrix elements. The pth eigenenergy of a PiB is $E_p = \frac{p^2 \pi^2 \hbar^2}{2m^* L^2}$ and the eigenfunction is $\psi_p = \sqrt{\frac{2}{L}}\sin\left(\frac{p\pi x}{L}\right)$. The position matrix element is

$$x_{ps} = \langle \psi_p | x | \psi_s \rangle = \frac{2}{L} \int_0^L x \sin\left(\frac{p\pi x}{L}\right) \sin\left(\frac{s\pi x}{L}\right) dx$$

$$= \begin{cases} 0 & \text{if } p - s \text{ is even} \\ \dfrac{2^3 L}{\pi^2} \dfrac{ps}{(p^2 - s^2)^2} & \text{if } p - s \text{ is odd} \end{cases}$$

The frequency of the photon is found from $\omega_{ps} = \frac{E_p - E_s}{\hbar} = \frac{\pi^2 \hbar (p^2 - s^2)}{2m^* L^2}$.

After combining the above relations, the radiative lifetime from state s to p is written as

$$\tau_{ps} = \frac{6\pi\epsilon_o \hbar c^3}{n_r q^2 \omega_{ps}^3 \, |x_{ps}|^2} = \frac{6\,\epsilon_o c^3 m^{*3} L^4}{8\pi \hbar^2 n_r q^2} \times \frac{p^2 - s^2}{(ps)^2} = (2.057\,\text{ns}) \times \frac{p^2 - s^2}{(ps)^2}$$

Using $p = 2$ and $s = 1$,

$$\tau_{21} = (2.057\,\text{ns}) \times \frac{3}{4} = 1.543\,\text{ns}$$

Similarly, the other lifetimes are

$\tau_{31} = $ infinity (rate $= 0$ because the position matrix element is zero if $p - s$ is even)

$$\tau_{41} = (2.057\,\text{ns}) \times \frac{15}{16} = 1.93\,\text{ns}$$

Infinite time or zero rate means that the event (photon absorption or emission) will not happen between $n = 3$ and $n = 1$ states.

Example 2. Radiative lifetime of a state in a harmonic oscillator.

(a) Calculate the matrix element of position operator for a transition between two different states of a harmonic oscillator.
(b) Calculate the spontaneous emission lifetime of a state to the ground state (state 0) if it is already excited to states 1 and 2.

Assume that the harmonic oscillator is in vacuum and the electron mass is m^*.

Solution: Recall that the eigenenergy of the state n in a harmonic oscillator is $E_n = \left(n + \frac{1}{2}\right)\hbar\omega$ and the eigenfunction written in terms of Hermite polynomials (H_n) is,

$$\psi_n = \frac{1}{\sqrt{2^n n!}} \left(\frac{m^*\omega}{\pi\hbar}\right)^{\frac{1}{4}} e^{-\frac{m\omega}{2\hbar}x^2} H_n\left(\sqrt{\frac{m^*\omega}{\hbar}}x\right) \qquad n = 0, 1, 2, \ldots$$

m^* is the effective mass. The wave functions for ground state $(n = 0)$ and the first and second excited states $(n = 1, 2)$ are

$$\psi_0 = \left(\frac{m^*\omega}{\pi\hbar}\right)^{\frac{1}{4}} e^{-\frac{m^*\omega}{2\hbar}x^2}, \quad \psi_1 = \left(\frac{m^*\omega}{\pi\hbar}\right)^{\frac{1}{4}} \sqrt{\frac{2m^*\omega}{\hbar}} x\, e^{-\frac{m^*\omega}{2\hbar}x^2} \quad \text{and}$$

$$\psi_2 = \left(\frac{m^*\omega}{\pi\hbar}\right)^{\frac{1}{4}} \frac{1}{\sqrt{2}} \left(\frac{2m^*\omega}{\hbar}x^2 - 1\right) e^{-\frac{m^*\omega}{2\hbar}x^2}$$

(a) The matrix elements of position (x) which correspond to the optical transitions of interest are,

$$\langle\psi_1|x|\psi_0\rangle = \sqrt{2}\left(\frac{m^*\omega}{\pi\hbar}\right)^{\frac{1}{2}} \sqrt{\frac{2m^*\omega}{\hbar}} \int_{-\infty}^{+\infty} dx\, x^2\, e^{-\frac{m^*\omega}{\hbar}x^2} = \sqrt{\frac{\hbar}{2m^*\omega}}$$

$$\langle\psi_2|x|\psi_0\rangle = \frac{1}{\sqrt{2}}\left(\frac{m^*\omega}{\pi\hbar}\right)^{\frac{1}{2}} \int_{-\infty}^{+\infty} dx\, x\left(\frac{2m^*\omega}{\hbar}x^2 - 1\right) e^{-\frac{m^*\omega}{\hbar}x^2} = 0$$

$$\langle\psi_2|x|\psi_1\rangle = 1\left(\frac{m^*\omega}{\pi\hbar}\right)^{\frac{1}{2}} \sqrt{\frac{2m^*\omega}{\hbar}} \int_{-\infty}^{+\infty} dx\, x^2\left(\frac{2m^*\omega}{\hbar}x^2 - 1\right) e^{-\frac{m^*\omega}{\hbar}x^2}$$

$$= \sqrt{\frac{\hbar}{m^*\omega}} = \sqrt{2}\,\langle\psi_1|x|\psi_0\rangle$$

As can be seen, the matrix element is nonzero only if the transition occurs between states with even and odd symmetry. For example, from 1 to 0 or vice versa, from 1 to 2 or vice versa, but not from 2 to 0.

(b) Since the allowed transitions occur between neighboring states, we can write

$$\omega_{ps} = \frac{E_{ps}}{\hbar} = \left(p + \frac{1}{2}\right)\omega - \left(s + \frac{1}{2}\right)\omega = (p - s)\,\omega = \omega$$

$$\tau_{ps} = \frac{6\pi\epsilon_o\hbar c^3}{q^2\omega_{ps}^3}\frac{1}{|x_{ps}|^2} = \frac{6\pi\epsilon_o\hbar c^3}{q^2\omega_{ps}^3}\frac{2m^*\omega}{\hbar}$$

With the above, τ_{10}, τ_{21}, and τ_{20} are as follows:

$$\tau_{10} = \frac{12\pi\epsilon_o m^{*2}c^3}{q^2\omega^2}$$

$$\tau_{21} = \frac{6\pi\epsilon_o m^{*2}c^3}{q^2\omega^2} = \frac{1}{2}\tau_{10}$$

$\tau_{20} = \infty$ (Since this transition is not possible.)

18.4 Optical Absorption in 3D Materials

An application of the time-dependent perturbation theory is to extract the optical properties of materials from their electronic structure. Consider photons from a vacuum incident on a slab as shown in Figure 18.5. The absorption coefficient in the slab $\alpha(\omega)$ is defined as the optical power absorbed per unit volume divided by the incident power per unit surface area (I_o),

$$\alpha(\omega) = \frac{\text{Power absorbed/volume}}{\text{Incident power/surface area}} = \frac{(\hbar\omega\,W_{cv})/V_{\text{slab}}}{n_{ph}\,\hbar\omega\,\tilde{c}/V_{\text{slab}}} = \frac{W_{cv}\,n_r}{n_{ph}\,c} \tag{18.69}$$

where $\tilde{c}$ is the velocity of light in the slab, V_{slab} is the volume of the slab and n_{ph} is the total number of photons in the slab (Figure 18.5). In the definition above, we have assumed that the absorption rate W_{cv} is the rate with which an electron in the full valence band (v) of the solid absorbs a photon and transits to an empty conduction band (c). If the population of the electrons in

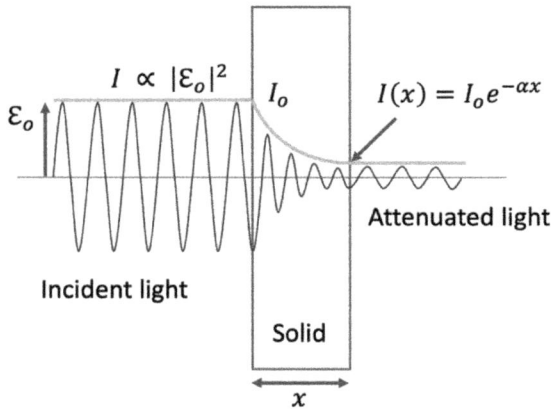

Figure 18.5. Absorption of light wave intensity in a slab of thickness x characterized by the absorption coefficient.

the conduction band is appreciable, then equation (18.69) should be modified. The dimension of absorption is $1/m$ and it tells how rapidly the intensity of an incoming light wave drops as it enters the material and travels through the slab of thickness x according to

$$I(x) = I_o e^{-\alpha x} \tag{18.70}$$

According to the Fermi golden rule, the absorption rate, W_{cv}, is proportional to the momentum or position matrix element squared and density of states (see equation (18.34)). We leave the detailed derivation of $\alpha(\omega)$ to Appendix C and we only state the final formula which is

$$\alpha(\omega) \propto \sum_{c,v} \int_{BZ} |\bar{p}_{cv}|^2 \, \delta(E_{cv}(\bar{k}') - \hbar\omega) \, d^3 k'$$

$$\propto |\bar{p}_{cv}|^2 \, \text{JDOS}(E_{cv}(\bar{k}')) \tag{18.71}$$

where the difference between conduction and valence energy at each k' is $E_{cv}(\bar{k}') = E_c(k') - E_v(k')$. The summation over c and v includes different combinations of transitions from valence to conduction sub-bands. The integration is along k' within the BZ. As opposed

Scattering Rates and Fermi Golden Rule

to usual density of states for each band which depends only on $E_c(k')$ or $E_v(k')$, the new quantity, JDOS (*joint density of states*), depends on the energy difference between two bands involving photon absorption. The formula for JDOS is identical to the derivation for DOS in Chapter 12 except that the expression for energy is $E_{cv}(\bar{k}')$, and JDOS $= \frac{k}{2\pi^2\bar{n}^3}\sqrt{2\mu(E-E_g)}u(E-E_g)$. Here, $\mu = \frac{m_c m_v}{m_c + m_v}$ depends on the effective mass of the conduction and valence band, m_c and m_v and E_g is the bandgap. μ is called the reduced effective mass. A more advanced treatment of density of states (which is not required here) will show that

$$\text{JDOS}\left(E_{cv}\left(k'\right)\right)$$

$$= \left|\frac{\partial k'}{\partial E_{cv}(k')}\right|_{\text{where } E_{cv}(k')=\hbar\omega} \qquad \text{(joint density of states)} \quad (18.72)$$

This is the inverse of slope of energy difference between the two bands. The exact mathematical form of the JDOS depends on how both conduction and valence bands change as a function of k'. As an example, in the band structure of bulk silicon there are two bands whose energy difference is almost constant, (for example, $E_{cv}(k') \approx 4\,\text{eV}$). As a result of this, the inverse of slope or JDOS is very large. Due to this, silicon has a very sharp peak in its absorption spectrum when photon energy is at $\hbar\omega = 4\,\text{eV}$. For example, see (Madelung, 2004).

In equation (18.71), it is assumed that the final and initial momentum of electron is the same, i.e., the transitions are direct. Further simplification is possible by making the following assumptions: (a) only one valence and conduction band is included and (b) the momentum matrix element is constant around the bandgap, which is usually the case. Then, the absorption coefficient can be written as

$$\alpha_{cv} = \frac{q^2 n_r \mu}{\pi \, \epsilon_r \epsilon_o \, c \, m_o^2 \, \hbar^3 \omega} \, |\bar{p}_{cv}|^2 \, \sqrt{2\mu(\hbar\omega - E_g)}$$

$$= \frac{2q^2 \mu}{\pi \, n_r \epsilon_o \, c \, m_o \, \hbar^3 \omega} E_P \, \sqrt{2\mu(\hbar\omega - E_g)} \qquad (18.73)$$

where E_P is called the Kane energy and it is found from the momentum matrix element squared as

$$E_P = \frac{|\bar{p}_{cv}|^2}{2m_o} \tag{18.74}$$

Figure 18.6(a) shows the absorption spectra of GaAs using equation (18.73) in which we have used the numerical values of GaAs bandgap ($E_g = 1.45\,\text{eV}$), Kane energy ($E_P = 22.71\,\text{eV}$), reduced effective mass ($\mu = 0.058$), and refractive index ($n = 3.6$). These values are taken from (Rosencher, 2002). The experimental measurement of GaAs's light absorption is plotted in Figure 18.6(b). The figure included only the highest and lowest valence and conduction bands, respectively. Further, it assumed a full valence band and an empty conduction band, which is why the model only qualitatively agrees with the experimental data between the bandgap and 2 eV. To calculate other features of the absorption spectrum at higher photon energies, the summation in equation (18.71) must include other combinations of c (conduction bands) and v (valence bands).

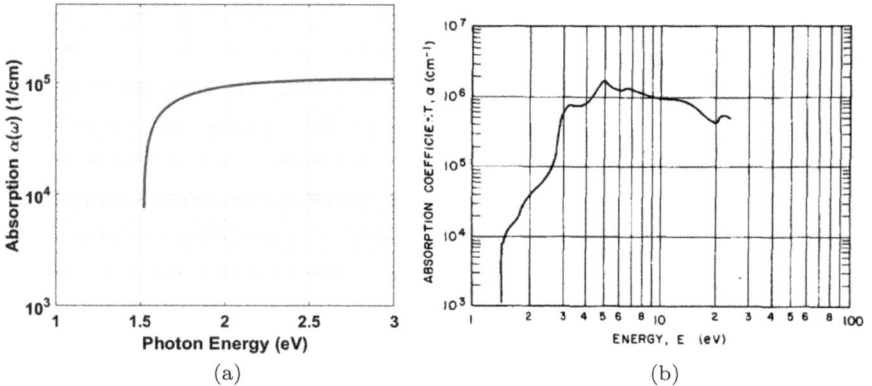

Figure 18.6. (a) Absorption coefficient of GaAs in terms of 1/cm calculated from equation (18.73) in semilogarithmic scale. (b) Data obtained experimentally in Casey *et al.* (1975). Note that plots (a) and (b) agree in the energies close to the bandgap up to 2 eV. To have a better match with the experiment, higher energy bands must be included in the calculation.

Looking at equation (18.73) reveals some interesting physics of photon absorption. First, the absorption is proportional to the reduced effective mass (μ), meaning that if both bands are heavy (high effective mass), the absorption is stronger. This is not completely surprising as a higher effective mass for a band implies a higher density of states (see equation (12.40)), which makes transitions more probable. Second, the absorption is proportional to the momentum matrix element between conduction and valence state wave functions (equation (18.15)). The value of this integral depends on the symmetry of wave functions, which can in turn be adjusted by external stimuli. For example, by applying electric or magnetic field or by applying mechanical strain to the solid. This allows for engineering/modulating the value of photon absorption by external stimuli. Third and importantly, within the effective mass approximation, the absorption shape resembles the usual DOS. For example, in case of a 3D solid, it is proportional to $(\hbar\omega - E_g)^{1/2}$ which resembles $(E - E_{cmin})^{1/2}$ or $(E_{vmax} - E)^{1/2}$ (see Chapter 12). As you may surmise, the absorption for a nanowire (1D solid) is proportional to DOS of nanowire (where $\hbar\omega > E_g$),

$$\alpha_{cv} \propto \frac{1}{\sqrt{(\hbar\omega - E_g)}} \quad \text{(nanowire)}$$

and for 2D solids (e.g., quantum well) it is proportional to the step function,

$$\alpha_{cv} \propto u(\hbar\omega - E_g) \quad \text{(quantum well)}$$

In 3D solids, absorption value is zero when photon energy is equal to the bandgap energy, $\hbar\omega = E_g$. In contrast, in 1D solids, the absorption has a sharp peak when the photon energy is equal to the bandgap due to the singularity in DOS. For 2D solids, there is a constant step like jump at the band edge when the photon energy is E_g.

18.5 Electron–Phonon Scattering Rate

When the time-dependent scattering potential is created by a crystal lattice vibration instead of electromagnetic waves, the perturbing potential $U(x, t)$ is expressed in terms of a mechanical strain wave propagating in the crystal. Here, we only consider the one-dimensional case for simplicity. The time-dependent perturbing potential energy for a strain wave is

$$U\left(x,t\right) = E_D\, s\left(x,t\right) = E_D\frac{\partial d(x,t)}{\partial x} \qquad (18.75)$$

The first term, E_D is called the *deformation potential energy*, and the second term is a unitless quantity called *strain* $s\left(t\right)$. E_D is a measure of the change in potential energy of an electron in the solid per unit strain. For example, when silicon is compressed or stretched, the energy of an electron in the conduction band increases or decreases proportional to the amount of strain. Deformation potential for silicon and germanium is 13.24 eV and 8.65 eV, respectively (Fischetti, 1996). Strain is a measure of the displacement of an atom from its stationary (equilibrium) position due to vibration. In other words, it is the percentage of contraction or dilatation and is expressed as $\frac{\partial d(x,t)}{\partial x}$ in equation (18.75). It is the derivative of displacement of each atom, $d(x,t)$, with respect to its stationary position at x (see Figure 18.7).

As shown in Figure 18.7, and similar to equation (18.13), the displacement $d(x,t)$ can be written as the sum of two counterpropagating plane waves with wave vector k and frequency Ω in the solid. This wave is a periodic elastic contraction and dilatation

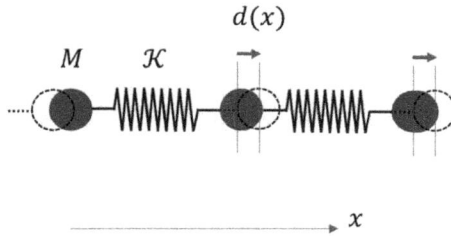

Figure 18.7. A one-dimensional crystal made of atoms with mass M connected by interatomic forces modeled as springs with spring constant $\mathcal{K}$. When there is no vibration, each atom is at its equilibrium position. In case of vibration, each atom is displaced from its equilibrium position by amount d which is a function of x. The rate of change of $d(x)$ with x is strain or simply put, percentage of contraction or dilatation by the inter atomic force.

of the solid along the x-axis. In this case, $\hbar k$ is called the momentum of the *phonon* and it is parallel to the arrows. Phonons are quanta of elastic waves, and their displacement is given by

$$d\left(x,t\right) = \frac{1}{2}\left(u_k e^{ikx-i\Omega t} + u_k^* e^{-ikx+i\Omega t}\right) \tag{18.76}$$

where u_k is the amplitude of an oscillating strain wave with wave vector k and frequency Ω. The strain is then found by taking the derivative of the above equation. Substituting the strain $\frac{\partial d(x,t)}{\partial x}$ in equation (18.75) gives

$$U(x,t) = \frac{iE_D k}{2}\left(u_k e^{ik\cdot x-i\Omega t} - u_k^* e^{-ik\cdot x+i\Omega t}\right) \tag{18.77}$$

Recalling equations (18.28) and (18.29), the first and second terms above correspond to absorption and emission of a quantum of the oscillating phonon field, respectively. Figure 18.8 shows these processes pictorially. The time axis is from left to right, and the solid and broken arrows represent the path of the electron and phonon, respectively. In these processes, an electron with a wave vector k_s either absorbs or emits a phonon with momentum k, and ends up being in the final state with wave vector k_p. Diagrams like Figure 18.8 are called Feynman diagrams.

For simplicity, we will consider the calculation of the scattering rate when an electron transitions from a higher to lower energy due to phonon emission. Further, we will assume a single phonon mode

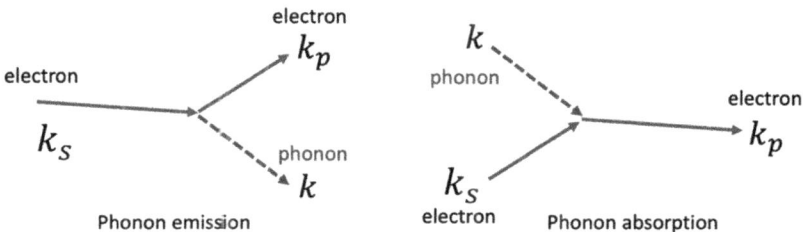

Figure 18.8. Single phonon emission (left) and single phonon absorption processes (right). Time increases from left to right in both figures.

with energy $\hbar\Omega$. The scattering rate when an electron in state ϕ_{so} with energy E_s transitions to state ϕ_{po} with energy E_p by emitting a phonon with energy $\hbar\Omega$ is given by (equation 18.29),

$$R_{ps} = \frac{2\pi \left| u_{sp}^* \right|^2}{\hbar} \delta(E_p - E_s + \hbar\Omega) \tag{18.78}$$

The matrix element u_{sp}^* was given in equation (18.16) for the general perturbation case. Since our solid is 1D, we replaced the volume element dv with dx, and electron wave functions only depend on x. So u_{ps}^* is

$$u_{sp}^* = \int dx \, \phi_{po}^*(x) \left\{ \frac{iE_D k}{2} u_k e^{-ik \cdot x} \right\} \phi_{so}(x) \tag{18.79}$$

Note that we used $u_k = u_k^*$. To calculate the matrix element for spontaneous emission, we need the amplitude of the strain wave, which is (Gurevich, 1986)

$$|u_k| = \sqrt{\frac{\hbar}{2MN\Omega}} = \sqrt{\frac{\hbar}{2\rho_l L\Omega}} = \sqrt{\frac{\hbar}{2\rho_l L v_s |k|}} \tag{18.80}$$

where M is the mass of the atom composing the crystal and N is the number of atoms in the solid. The total mass, MN, can be replaced by the solid's linear mass density times the total length of the solid, i.e., $MN = \rho_l L$. The phonon frequency can be replaced by $\Omega = v_s |k|$, where v_s is the group velocity of the elastic wave propagating in the crystal. Like the group velocity for electrons which is found from the slope of the energy band, v_s is found from the first derivative of phonon dispersion, i.e., the energy of phonons ($\hbar\Omega$) versus their wave vectors (k). The assumption of a linear relationship between phonon frequency and wave vector is a good approximation in most cases. However in general, the relationship $\Omega(k)$ should be calculated.

The matrix element of the perturbation due to the strain wave is then equal to

$$u_{sp}^* = \frac{iE_D k}{2} \sqrt{\frac{\hbar}{2\rho_l L v_s |k|}} \int dx \, \phi_{po}^*(x) \, e^{-ik \cdot x} \, \phi_{so}(x) \tag{18.81}$$

In the above process, an electron with a wave vector k_s emits a phonon with momentum k, and ends up being in the final state with wave vector k_p. By substituting equation (18.81) in equation (18.78), we obtain the scattering rate,

$$R_{ps} = \frac{2\pi}{\hbar}\frac{E_D^2 k^2}{2}\frac{\hbar}{2\rho_l L v_s |k|}|\langle\phi_{po}(x)|e^{-ik\cdot x}|\phi_{so}(x)\rangle|^2\delta(E_p - E_s + \hbar\Omega)$$

$$R_{ps} = \frac{\pi E_D^2 |k|}{2\rho_l L v_s}|\langle\phi_{po}(x)|e^{-ik\cdot x}|\phi_{so}(x)\rangle|^2\delta(E_p - E_s + \hbar\Omega) \quad \text{(emission)}$$

$$(18.82)$$

Note that the conservation of momentum mandates $k_p = k_s \pm k$. Further, the difference between initial and final energies of the electron must be equal to the phonon energy $\hbar\Omega$ (energy conservation).

We will now calculate the *scattering rate due to the emission of a phonon* in a 1D solid. The readers can see how the technique works and generalize it to nanowires, quantum wells, and 3D solids by consulting (Hamaguchi, 2017).

The wave functions for free electrons in the two states s (before phonon emission) and p (after phonon emission) are:

$$\phi_{so}(x) = \frac{1}{\sqrt{L}}e^{ik_s x} \tag{18.83}$$

$$\phi_{po}(x) = \frac{1}{\sqrt{L}}e^{ik_p\cdot x} \tag{18.84}$$

The matrix element of these wave functions is calculated as follows:

$$\langle\phi_{po}(x)|e^{-ik\cdot x}|\phi_{so}(x)\rangle = \int dx\, \phi_{po}^*(x)\, e^{-ik\cdot x}\phi_{so}(x)$$

$$= \frac{1}{L}\int dx\, e^{i(-k_p+k_s-k)\cdot x} = \delta_{k_p, k_s-k} \tag{18.85}$$

using the normalization of plane waves discussed in Chapter 1.

The right side is a Kronecker delta (see Problem 8) which imposes the conservation of momentum. This means that the new momentum of scattered electron changes by $\hbar k$ which is the momentum of the emitted phonon, i.e., $\hbar k_p = \hbar k_s - \hbar k$.

The scattering rate because of the electron losing energy $\hbar\Omega$ is found by plugging the matrix element of equation (18.85) in (18.82),

$$R_{ps} = \frac{\pi E_D^2 |k|}{2\rho_l L v_s} \delta(E_p - E_s + \hbar\Omega)\, \delta_{k_p, k_s - k} \quad \text{(phonon emission)} \quad (18.86)$$

In the above, energy conservation is imposed by the Dirac delta function and conservation of momentum is guaranteed by the Kronecker delta. Energy conservation also mandates that the difference between initial and final energy of the electron is equal to the phonon energy $\hbar\Omega$.

When phonons have different momenta (or wave vectors k) the total scattering rate of electrons is found by summing up equation (18.86) over all possible values and directions of phonon momentum k. In our 1D solid example, summation is only on values of k, and the total scattering rate is

$$R_{\text{total}} = \sum_k R_{ps} = \frac{\pi E_D^2}{2\rho_l L v_s} \sum_k |k| \delta(E_p - E_s + \hbar\Omega) \cdot \delta_{k_p, k_s - k} \quad (18.87)$$

The summation over phonon wave vectors can be converted to integration using $\sum_k \ldots = \frac{L}{2\pi} \int \ldots dk$. With this, $\delta_{k_p, k_s - k} = 1$, and using the assumption of linear dependency of phonon frequency and wave vector (i.e., linear dispersion or $\Omega = v_s k$), we have

$$R_{\text{total}} = \frac{\pi E_D^2}{2\rho_l L v_s} \frac{L}{2\pi} \int dk |k| \frac{1}{\hbar v_s} \delta\left(\frac{E_p - E_s}{\hbar v_s} + k\right) \quad (18.88)$$

The sifting property of the Dirac delta function imposes $k = \frac{E_s - E_p}{\hbar v_s}$ and leads to further simplification

$$R_{\text{total}} = \frac{E_D^2}{4\rho_l} \frac{E_s - E_p}{\hbar^2 v_s^3} \quad \text{(single phonon emission)} \quad (18.89)$$

As an exercise, check that the unit of the above quantity is $1/s$.

The above equation was derived for the emission of one phonon. However, for the case of emission of many phonons, the equations should be updated by an extra factor of $(N_k + 1)$ where $N_k = 1/(e^{\frac{\hbar\Omega}{k_B T}} - 1)$ is the Bose–Einstein distribution at temperature T, and the Boltzmann constant $k_B = 1.38 \times 10^{-23}$ J/K. Without proof, we write down the expressions in the case of nonzero N_k,

$$R_{ps} = \frac{\pi \bar{E}_D^2 |k|}{2\rho_l L v_s} \delta(E_p - E_s + \hbar\Omega)\, \delta_{k_p, k_s - k}(N_k + 1) \quad \text{(emission)}$$
$$(18.90)$$

$$R_{ps} = \frac{\pi E_D^2 |k|}{2\rho_l L v_s} \delta(E_p - E_s - \hbar\Omega)\, \delta_{k_p, k_s + k}\, N_k \quad \text{(absorption)} \quad (18.91)$$

18.6 Appendix A: Review of Some Equations Governing EM Field

Consider an electromagnetic (EM) field,

$$\bar{\mathcal{E}} = \bar{\mathcal{E}}_o e^{i(\bar{k}\cdot\bar{r} - \omega t)} \tag{18.92}$$
$$\bar{B} = \bar{B}_o e^{i(\bar{k}\cdot\bar{r} - \omega t)} \tag{18.93}$$

The relationship between frequency and wave vector for an electromagnetic field is

$$\omega = ck \tag{18.94}$$

The relationship between the electric field and the vector potential $\bar{A}$ is

$$\bar{\mathcal{E}} = -\frac{\partial \bar{A}}{\partial t} \tag{18.95}$$

The classical expression for energy density (energy per unit volume) in an EM wave is

$$\text{Energy per unit volume} = \frac{1}{2}\epsilon_r \epsilon_o \mathcal{E}_o^2 \tag{18.96}$$

where ϵ_o is the vacuum dielectric constant and ϵ_r is the relative permittivity of the material/medium. The energy per unit volume of n_{ph} number of photons of energy $\hbar\omega_o$ in a box of volume V_{em} is

$$\text{Energy per unit volume} = \frac{n_{ph}\hbar\omega}{V_{em}} \tag{18.97}$$

where V_{em} denotes the volume of the box in which the EM wave is present. For an electromagnetic field where the vector potential is given by

$$\bar{A}(\bar{r}, t) = A_o \vec{\eta} \cos (\bar{k} \cdot \bar{r} - \omega t) \tag{18.98}$$

we use the following Maxwell's equations in free space

$$\bar{\mathcal{E}} = -\frac{\partial \bar{A}}{\partial t} \tag{18.99}$$

and

$$\bar{B} = \nabla \times \bar{A} \tag{18.100}$$

to find the electric and magnetic fields which are

$$\bar{\mathcal{E}}(\bar{r}, t) = |\mathcal{E}_o|\vec{\eta} \sin (\bar{k} \cdot \bar{r} - \omega t) \tag{18.101}$$

and,

$$\bar{B}(\bar{r}, t) = |B_o|\vec{b} \sin (\bar{k} \cdot \bar{r} - \omega t) \tag{18.102}$$

where the amplitudes and polarizations of the fields are related by

$$\mathcal{E}_o = \omega A_o, \quad B_o = |\bar{k}|A_o \quad \text{and} \quad \vec{b} = \bar{k} \times \vec{\eta}/|\bar{k}| \tag{18.103}$$

In these equations, $\bar{k}$ is the direction of wave propagation, $\vec{\eta}$ is the polarization, and ω is the angular frequency of the electromagnetic wave.

Using equations (18.96) and (18.97), we obtain

$$\mathcal{E}_o^2 = \frac{2n_{ph}\hbar\omega}{\epsilon_r \epsilon_o V_{em}}$$

Now using (18.103) and the previous equation, we have

$$A_o = \sqrt{\frac{2n_{ph}\hbar}{\epsilon_r \epsilon_o \omega V_{em}}}$$

18.7   Appendix B: Number of EM Modes per Unit Volume per Unit Energy

The number of modes per unit volume per unit frequency $[n(\omega)]$ of the electromagnetic field is

$$n(\omega) = 2 \times \frac{\text{Volume occupied in wave vector by a small window } \Delta\omega}{\text{Volume occupied by a single wave vector point}}$$

$$\times \frac{1}{V_{em}\Delta\omega} \tag{18.104}$$

where the multiplicative factor of 2 accounts for the two polarization modes perpendicular to the direction of the EM wave propagation and V_{em} is the volume of space in which the electromagnetic modes exist. We will assume the volume V_{em} to consist of a large cuboid box with each side length equal to L. We will finally set L equal to infinity to mimic free space.

To evaluate $n(\omega)$ in equation (18.104), we will first consider an EM wave directed in the x direction ($k_x \neq 0$ and $k_y = k_z = 0$). The EM wave is in a large 3D box with all side lengths equal to L. To find the denominator of equation (18.104), which is the volume occupied by a single wave vector point, periodic boundary conditions are applied to equation (18.92) which gives us

$$\bar{\mathcal{E}}(x = 0) = \bar{\mathcal{E}}(x = L) \tag{18.105}$$

Equation (18.105) implies that

$$\bar{\mathcal{E}}_o e^{i(k_x x - \omega t)} = \bar{\mathcal{E}}_o e^{i(k_x (x+L) - \omega t)} \tag{18.106}$$

which gives

$$e^{ik_x L} = 1 \tag{18.107}$$

Equation (18.107) can be satisfied only if

$$k_x = \frac{2n\pi}{L} \quad (\text{where } n = 0, 1, 2, 3, \ldots) \tag{18.108}$$

Equation (18.108) means that a single wave vector point occupies a length $(2\pi/L)$ along the x direction. Identically, it can be argued that the single wave vector point occupies a length $(2\pi/L)$ in both the y

and z directions. As a result, the denominator of equation (18.104) is,

$$\text{The volume occupied by a single wave}$$
$$\text{vector point in } k\text{-space} = (2\pi/L)^3 \qquad (18.109)$$

To evaluate the numerator of equation (18.104), we note that the frequency versus wave vector (k) relationship in equation (18.94) is isotropic in k, meaning that it depends only on the size of k. So, all points on a sphere with radius k have the same frequency $\omega = ck$. Then, it is easy to see that the

$$\text{Volume in wave vector space occupied at } \omega$$
$$\text{by a small frequency window } \Delta\omega = 4\pi k^2 \, \Delta k \qquad (18.110)$$

Substituting equations (18.109) and (18.110) in equation (18.104) gives

$$n(\omega) = \frac{8\pi k^2 \Delta k}{(2\pi/L)^3} \cdot \frac{1}{V_{em}\Delta\omega} \qquad (18.111)$$

Noting that $V_{em} = L^3$, the above equation can be rewritten as

$$n(\omega) = \frac{k^2}{\pi^2} \frac{\Delta k}{\Delta\omega} \qquad (18.112)$$

Substituting equation (18.94) in equation (18.112) and noting that equation (18.94) implies that $\Delta\omega = c\Delta k$, we have that the number of modes per unit volume per unit frequency $[n(\omega)]$

$$n(\omega) = \frac{\omega^2}{c^3\pi^2} \qquad (18.113)$$

18.8   Appendix C: Optical Absorption Coefficient of a Bulk 3D Crystal

As we saw in the Fermi golden rule, it is possible to find the transition rates between different states as given in equations (18.30) and (18.31) using the matrix element of the perturbation Hamiltonian. In this appendix, we will calculate the formula for optical absorption of bulk (3D) semiconductors such as GaAs, which have a direct bandgap. We see how the energy bandgap, matrix elements, and effective masses determine the strength of absorption. See (Anselm, 1981) for optical absorption in indirect bandgap materials.

In a periodic solid, the Hamiltonian of an electron in the presence of a perturbing EM field is given by (neglecting the A^2 term as in equation (18.42))

$$\hat{H} = \frac{\hat{p}^2}{2m_o} + U_{\text{static}}(\bar{r}) + \frac{q}{m_o}\bar{A} \cdot \hat{p}$$

$U_{\text{static}}(\bar{r})$ is the periodic potential created by all the atoms comprising the solid. The first two terms of the Hamiltonian $\hat{H}_o = \frac{\hat{p}^2}{2m_o} + U_{\text{static}}(\bar{r})$ form the unperturbed Hamiltonian that gives rise to the conduction and valence bands in the solid. The time-dependent perturbation is the third term $\frac{q}{m_o}\bar{A} \cdot \hat{p}$, which after using the momentum operator is

$$U(\bar{r}, t) = -\frac{iq\hbar}{m_o}\bar{A}(\bar{r}, t) \cdot \nabla \qquad (18.114)$$

Note that the denominator of the perturbing Hamiltonian is the free electron mass m_o as per the above argument. After inserting the electromagnetic vector potential $A_o \sin(\bar{k}_p \cdot \bar{r} - \omega t)\bar{\eta}$ in the above equation, we have

$$U(\bar{r}, t) = -\frac{iq\hbar A_o}{2m_o}\left[e^{i(\bar{k}_p \cdot \bar{r} - \omega t)} + e^{-i(\bar{k}_p \cdot \bar{r} - \omega t)}\right]\bar{\eta} \cdot \nabla \qquad (18.115)$$

$\bar{\eta}$ is the polarization of the EM wave. We use $\bar{k}_p$ for the photon wave vector to avoid confusing it with the electron wave vector k, k', and k''. To find the matrix elements of the perturbing Hamiltonian between two states p and s (see equations (18.30) and (18.45)), we first write the wave functions of those states. According to Figure 18.9, we assume one state belongs to the valence band with wave vector $\bar{k}'$, and the second one is within the conduction band with wave vector $\bar{k}''$.

$$\psi_v = \frac{1}{\sqrt{V}}e^{i\bar{k}' \cdot \bar{r}}\phi_{vk'}(\bar{r}) \quad \text{and} \quad \psi_c = \frac{1}{\sqrt{V}}e^{i\bar{k}'' \cdot \bar{r}}\phi_{ck''}(\bar{r}) \qquad (18.116)$$

where V, the volume of the crystal, appears in the normalization factor $\frac{1}{\sqrt{V}}$. The periodic part of the wave functions of the conduction and valence states are represented by $\phi_{ck''}$ and $\phi_{vk'}$, respectively. The

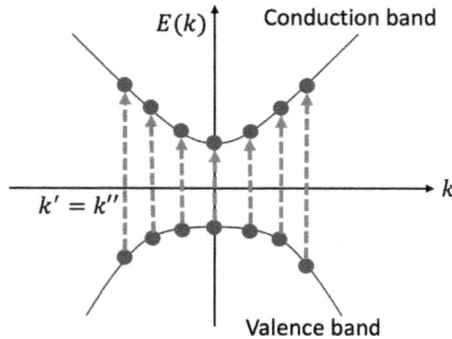

Figure 18.9. Transition of electron from states k' in the valence band to states k'' in conduction band due to absorption of a photon. Direct transition mandates that $k' = k''$. It is assumed that the valence band is full and the conduction band is completely empty.

matrix element of the perturbing Hamiltonian u_{cv} corresponding to the absorption process is then found by

$$u_{cv} = \langle \psi_c | U(\bar{r}, t) | \psi_v \rangle = \int \psi_c^* \left[\frac{-iq\hbar A_o}{2m_o} e^{i\bar{k}_p \cdot \bar{r}} \vec{\eta} \cdot \nabla \right] \psi_v \, dv$$

$$= \frac{-iq\hbar A_o}{2m_o V} \int \phi_{ck''}^* e^{-i\bar{k}'' \cdot \bar{r}} e^{i\bar{k}_p \cdot \bar{r}} \vec{\eta} \cdot \nabla \left[e^{i\bar{k}' \cdot \bar{r}} \phi_{vk'} \right] dv \qquad (18.117)$$

We let the gradient operator act on the valence wave function, and the result has two terms:

$$u_{cv} = \frac{-iq\hbar A_o}{2m_o V} \int \phi_{ck''}^* e^{-i\bar{k}'' \cdot \bar{r}} e^{i\bar{k}_p \cdot \bar{r}} [\vec{\eta} \cdot \nabla \phi_{vk'} + \vec{\eta} \cdot \bar{k}'(i\phi_{vk'})] e^{i\bar{k}' \cdot \bar{r}} dv$$

$$= \frac{-iq\hbar A_o}{2m_o V} \int \phi_{ck''}^* e^{i(\bar{k}' + \bar{k}_p - \bar{k}'') \cdot \bar{r}} [\vec{\eta} \cdot \nabla \phi_{vk'} + \vec{\eta} \cdot \bar{k}'(i\phi_{vk'})] dv$$

$$(18.118)$$

We remind the reader that $\bar{k}_p$ is the wave vector of the photon, and k' and k'' are the wave vectors of the electron in the valence and conduction bands, respectively.

Before we continue, let's stop here and see how the exponential factor is simplified. Since the photon has a tiny momentum (see equation (18.153)), it cannot induce a significant momentum change of the electron. So, we will assume that $\bar{k}' - \bar{k}'' = 0$. In dealing with quantum dots in Section 18.3.1, we assumed that the wavelength of

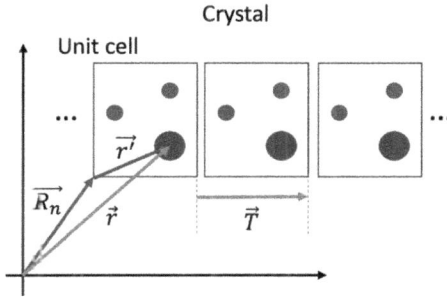

Figure 18.10. A simple crystal model built by repetition of a unit cell with three atoms.

the light (λ) is much larger than the geometric size of the structures involved (nanowires and quantum dots). As a result, we assumed that $\bar{k}_p \cdot r = 2\pi r/\lambda$ is a very small number. Therefore, $e^{i(\bar{k}' \cdot \bar{r} + \bar{k}_p \cdot \bar{r} - \bar{k}'' \cdot \bar{r})} \approx 1$. While this assumption is valid for nanostructures like nanowires and quantum dots, it is not valid for any 3D bulk-like solids with macroscopic dimensions. Thus, to simplify $e^{i(\bar{k}' \cdot \bar{r} + \bar{k}_p \cdot \bar{r} - \bar{k}'' \cdot \bar{r})}$, we must use the properties of periodicity of the solid. As shown in Figure 18.10, the integration over space can be decomposed into integration over one unit cell multiplied by the number of unit cells in the solid since the crystalline solid is formed by repeating a unit cell using translation vectors ($\vec{T}$). We can then write $\bar{r} = \bar{R}_n + \bar{r}'$, where $\bar{R}_n$ is the position of a fixed atom in unit cell number n, with respect to the whole crystal. This means that $\bar{r}'$ is the local coordinate of each atom within a unit cell. As a result,

$$e^{i(\bar{k}' \cdot \bar{r} + \bar{k}_p \cdot \bar{r} - \bar{k}'' \cdot \bar{r})} = e^{i(\bar{k}' + \bar{k}_p - \bar{k}'') \cdot (\bar{R}_n + \bar{r}')} \tag{18.119}$$

With this, the integration of the matrix element in equation (18.118) is decomposed into a summation over each unit cell (ucell) indexed by n and an integration over a unit cell based on the local coordinate (r').

$$u_{cv} = \frac{-iq\hbar A_o}{2m_o V} \sum_n e^{i(\bar{k}' + \bar{k}_p - \bar{k}'') \cdot \bar{R}_n} \int_{\text{ucell}} \phi^*_{ck''} e^{i(\bar{k}' + \bar{k}_p - \bar{k}'') \cdot r'}$$

$$\times \left[\vec{\eta} \cdot \nabla \varphi_{vk'} + \vec{\eta} \cdot \bar{k}'(i\phi_{vk'}) \right] dv' \tag{18.120}$$

The integral does not depend on n, so it can be factored out and put before the summation. To simplify the summation S, we have

$$S = \sum_n e^{i(\bar{k}' + \bar{k}_p - \bar{k}'') \cdot \bar{R}_n} \tag{18.121}$$

Since the momentum of the photon is negligible (wave vector $\bar{k}_p \sim 0$), only direct transitions are possible, meaning those with $\bar{k}' \sim \bar{k}''$. See Example 3 at the end of this appendix for a clear proof of this in 1D. As a result,

$$S = \sum_n e^{i(\bar{k}' + \bar{k}_p - \bar{k}'') \cdot \bar{R}_n} \sim \sum_n e^{i(0) \cdot \bar{R}_n} = \sum_n 1$$

$$= N \text{ (number of unit cells in the crystal)} \tag{18.122}$$

Note that we are not assuming anything about the size of the solid with respect to the wavelength.

If the volume of a unit cell is Δ, then the total volume of the crystal is $V = N\Delta$. Using equation (18.122) and the fact that $e^{i(\bar{k}' + \bar{k}_p - \bar{k}'') \cdot \bar{r}'} \sim 1$ because $\bar{k}_p \sim 0$ and $\bar{k}' \sim \bar{k}''$ for optical transitions, equation (18.120) becomes

$$u_{cv} = \frac{-iq\hbar A_o}{2\,m_o\,(N\Delta)} N \int_{\text{ucell}} \phi^*_{ck'} \left[\vec{\eta} \cdot \nabla \phi_{vk'} + \vec{\eta} \cdot \bar{k}'\,(i\phi_{vk'}) \right] dv' \tag{18.123}$$

The integral is decomposed into two terms:

$$u_{cv} = \frac{-iq\hbar A_o}{2\,m_o\,\Delta} \left\{ \int_{\text{ucell}} \phi^*_{ck'}\vec{\eta} \cdot \nabla \phi_{vk'} dv' + i\vec{\eta} \cdot \bar{k}' \int_{\text{ucell}} \phi^*_{ck'}\,\phi_{vk'} dv' \right\} \tag{18.124}$$

The second term becomes zero due to orthonormality of valence and conduction band states, meaning that $\int_{\text{ucell}} \phi^*_{ck'}\phi_{vk'}\,dv' = 0$. The matrix element is then equal to the following:

$$u_{cv} = \frac{-iq\hbar A_o}{2\,m_o\,\Delta} \int_{\text{ucell}} \phi^*_{ck'}\vec{\eta} \cdot \nabla \phi_{vk'}\,dv' \tag{18.125}$$

Noting that the momentum operator is $\hat{p} = -i\hbar\nabla$, the above matrix element can instead be written in terms of the momentum operator matrix element,

$$u_{cv} = \frac{+qA_o}{2\,m_o\,\Delta}\,\vec{\eta} \cdot \int_{\text{ucell}} \phi_{ck'}^* \,\hat{p}\,\phi_{vk'}\,dv' \tag{18.126}$$

Using the Fermi golden rule (equation (18.45)), the rate of photon absorption which causes an electron's transition from the initial state $(\phi_{vk'})$ in the valence band to the final state $(\phi_{ck'})$ in the conduction band is

$$R_{cv} = \frac{2\pi}{\hbar}\,|u_{cv}|^2\,\delta(E_{cv} - \hbar\omega), \quad E_{cv} = E_c - E_v \tag{18.127}$$

$$R_{cv} = \frac{2\pi}{\hbar}\frac{q^2 A_0^2}{4\,m_o^2}\,|\vec{\eta}\cdot\bar{p}_{cv}|^2\,\delta(E_{cv} - \hbar\omega) \tag{18.128}$$

where we have defined the momentum matrix element as

$$\bar{p}_{cv} = \frac{1}{\Delta}\int_{\text{ucell}} \phi_{ck'}^*\,\hat{p}\,\phi_{vk'}dv' \tag{18.129}$$

We add that E_{cv} is a function of $\bar{k}'$ and its precise functional form is determined by the shape of the band structure. As Figure 18.9 shows, the transition between the valence and conduction band can take place at many different initial wave vectors. Hence, the *total transition rate* (W_{cv}) between these sub-bands is the sum of all individual direct transition rates. After adding an extra multiplicative factor of 2 to include the spin degeneracy, the total transition rate is

$$W_{cv} = 2\sum_{\bar{k}'}\sum_{\bar{k}''} R_{cv} = 2\sum_{\bar{k}'=\bar{k}''} R_{cv} \tag{18.130}$$

where $\bar{k}'$ and $\bar{k}''$ represent the wave vectors of electrons in the conduction and valence bands as used in equation (18.114). $\bar{k}' = \bar{k}''$ for reasons discussed above equation (18.122). In the case of a 3D solid, the summation over wave vectors $\bar{k}'$ can be converted to an integral as follows:

$$2\sum_{\bar{k}'}\ldots = \frac{V}{4\pi^3}\int_{\text{BZ}}\ldots d^3k \tag{18.131}$$

From the value of R_{cv} in equation (18.128) and using equation (18.131),

$$W_{cv} = \frac{V}{4\pi^3} \frac{2\pi}{\hbar} \frac{q^2 A_o^2}{4m_o^2} \int_{BZ} |\vec{\eta} \cdot \bar{p}_{cv}|^2 \, \delta(E_{cv}(\bar{k}') - \hbar\omega) \, d^3 k' \quad (18.132)$$

where $E_{cv}(\bar{k}') = E_c(\bar{k}') - E_v(\bar{k}')$ is the value of the energy gap at each $\bar{k}'$ point in the first BZ.

Before finding the absorption coefficient, $\alpha(\omega)$, let's replace the amplitude of the oscillating vector potential A_o by the following quantity using equations (18.96), (18.97), and (18.113) and electric field $\mathcal{E}_o = \omega A_o$ from equation (18.103) (see Problem 9):

$$A_o^2 = \frac{2n_{ph}\hbar}{\epsilon_r\epsilon_o\omega V_{em}} \quad (18.133)$$

where we used $V = V_{em} = V_{slab}$, meaning that the electromagnetic radiation is assumed to fill the entire 3D solid but no more.

The absorption coefficient, $\alpha(\omega)$, is defined as the power absorbed per unit volume divided by the incident power per unit surface (see equation (18.69)),

$$\alpha(\omega) = \frac{W_{cv} \, n_r}{n_{ph} c} \quad (18.134)$$

Inserting equations (18.132) and (18.133) in equation (18.134), the absorption coefficient in the above equation can be written as

$$\alpha(\omega) = \frac{n_r}{c} \frac{1}{4\pi^2} \frac{q^2}{m_o^2} \frac{1}{\epsilon_r\epsilon_o\omega} \int_{BZ} d^3 k' \, |\vec{\eta} \cdot \bar{p}_{cv}|^2 \delta(E_{cv}(\bar{k}') - \hbar\omega) \quad (18.135)$$

The absorption coefficient in equation (18.135) is found by assuming that all states in the valence band are occupied by electrons and that all the states in the conduction band are empty. This is called *intrinsic absorption*. The concept of *absorption* is important in discussing devices like solar cells and lasers but equation (18.135) should be modified if the conduction band has nonnegligible occupancy. It is interesting to note the dependency of absorption value on the polarization of incoming light ($\vec{\eta}$). To show this better, we use equation (18.63) to convert the momentum matrix element to position matrix element, and as a result,

$$\alpha(\omega) = \frac{n_r}{c} \frac{q^2}{4\pi^2} \frac{1}{\epsilon_r \epsilon_o \omega} \int_{BZ} d^3k' \, \omega_{cv}^2 |\bar{\eta} \cdot \bar{r}_{cv}|^2 \delta(E_{cv}(\bar{k}') - \hbar\omega) \quad (18.136)$$

In the above equation, we see that the absorption coefficient depends on the angle between the polarization of the EM field $\bar{\eta}$ and the dipole matrix element $\bar{r}_{cv}$ though their dot product. This means that the absorption of photons will depend on the orientation of polarization of incident light with respect to $\bar{r}_{cv}$, which is an intrinsic quantity for a given material. For example, if $\bar{\eta} = \hat{x}$, we have,

$$\hat{x} \cdot \bar{r}_{cv} = x_{cv} = \frac{1}{\Delta} \int_{ucell} \phi_{ck'}^* x \phi_{vk'} \, dv' \quad (18.137)$$

This means the absorption of a x-polarized photon depends on the precise value of the above integral which depends on the symmetry of the wave functions $\phi_{ck'}$ and $\phi_{vk'}$ with respect to x. The above integral can be different for light with polarization along the y direction. As a result of this, materials can have *anisotropic (direction-dependent) absorption*.

Effective mass approximation: We will now assume that both valence and conduction bands can be approximated to be isotropic and parabolic, with effective masses of m_c^* and m_v^*, respectively,

$$E_c(\bar{k}') = \frac{\hbar^2 k'^2}{2m_c^*} + E_c \quad \text{and} \quad E_v(\bar{k}') = \frac{-\hbar^2 k'^2}{2m_v^*} + E_v \quad (18.138)$$

where E_c and E_v are the minimum and maximum of conduction and valence bands, respectively. Because the transition of electrons from valence to conduction bands occurs at the same value of $\bar{k}'$, E_{cv} can be written as

$$E_{cv}(\bar{k}') = \frac{\hbar^2 k'^2}{2\mu} + E_g, \quad \frac{1}{\mu} = \frac{1}{m_c^*} + \frac{1}{m_v^*},$$

$$E_g = E_{cv}(k' = 0) = E_c - E_v \quad (18.139)$$

where μ is the reduced effective mass and E_g is the bandgap at the BZ center ($k' = 0$). Taking the differential of equation (18.139), we have

$$dE_{cv} = \frac{\hbar^2 k'}{\mu} dk' = \hbar \sqrt{\frac{2(E_{cv} - E_g)}{\mu}} dk'$$

$$\rightarrow dk' = \frac{1}{\hbar} \sqrt{\frac{\mu}{2(E_{cv} - E_g)}} dE_{cv} \qquad (18.140)$$

In the isotropic case, $d^3k = 4\pi k^2 dk$, and equation (18.135) for the absorption coefficient becomes

$$\alpha(\omega) = \frac{n_r}{c} \frac{1}{4\pi^2} \frac{q^2}{m_o^2} \frac{1}{\epsilon_o \omega} \int_{BZ} 4\pi k'^2 \, |\vec{\eta} \cdot \bar{p}_{cv}|^2 \, \delta(E_{cv}(\bar{k}') - \hbar\omega) \, dk' \qquad (18.141)$$

We will assume that we are interested in bandgap energies close to $\bar{k}' = 0$ and that we can neglect the wave vectors dependence of $\bar{p}_{cv}$ on $\bar{k}'$. Then, we can use

$$|\vec{\eta} \cdot \bar{p}_{cv}(\bar{k}')|^2 = |\vec{\eta} \cdot \bar{p}_{cv}(\bar{k}' = 0)|^2 = |\vec{\eta} \cdot \bar{p}_{cv}|^2 \qquad (18.142)$$

Substituting for k' from equation (18.139), dk' from equation (18.140) and (18.142) in equation (18.141), we get

$$\alpha(\omega) = \frac{n_r}{c} \frac{1}{4\pi^2} \frac{q^2}{m_o^2} \frac{1}{\epsilon_r \epsilon_o \omega} |\vec{\eta} \cdot \bar{p}_{cv}|^2 \int_{BZ} 4\pi \frac{2\mu(E_{cv} - E_g)}{\hbar^2} \delta(E_{cv} - \hbar\omega)$$

$$\times \frac{1}{\hbar} \sqrt{\frac{\mu}{2(E_{cv} - E_g)}} dE_{cv}$$

$$\alpha(\omega) = \frac{|\vec{\eta} \cdot \bar{p}_{cv}|^2 q^2 n_r (2\mu)^{1.5}}{2\pi \, c \, m_o^2 \, \epsilon_r \epsilon_o \omega \hbar^3} \sqrt{\hbar\omega - E_g}$$

$$\alpha(\omega) = \frac{|\vec{\eta} \cdot \bar{p}_{cv}|^2 q^2 n_r \mu}{\pi \, \epsilon_r \epsilon_o \, c \, m_o^2 \, \hbar^3 \omega} \sqrt{2\mu(\hbar\omega - E_g)} \qquad (18.143)$$

If we assume that the relative permeability of the medium $\mu_r = 1$, then the refractive index $n_r = \sqrt{\epsilon_r}$, and the above equation can be

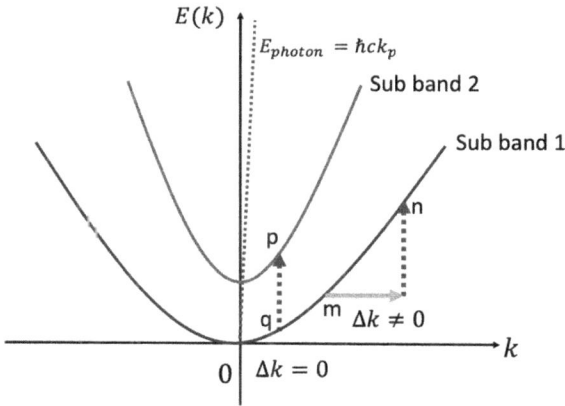

Figure 18.11. Allowed and forbidden photon-assisted transitions between two electronic states.

written as

$$\alpha(\omega) = \frac{|\vec{\eta} \cdot \bar{p}_{cv}|^2 q^2 \mu}{\pi\, n_r \epsilon_o\, c\, m_o^2\, \hbar^3 \omega} \sqrt{2\mu(\hbar\omega - E_G)} \tag{18.144}$$

Example 3 (Photon has a tiny momentum in comparison to electrons). Using the principles of momentum and energy conservation, show that the absorption of an electron involves photons with a tiny wave vector or momentum. For simplicity, show this only for the case of a 1D nanowire.

Solution: Consider a nanowire with the band structure $(E(k))$ as shown in Figure 18.11. Let the initial and final wave vectors of the electron before and after photon absorption be s and p in sub-bands 1 and 2, respectively. Figure 18.11 also shows the dispersion relation for a photon. $E_{photon} = \hbar c k_p$, where c and k_p are the speed and wave vector of the photon in the material. When an electron is excited by a photon, both energy and momentum should be conserved. The underlying equations for these conditions to be satisfied are as follows:

$$E_1(k_i) + \hbar\omega = E_2(k_f) \quad \text{and} \tag{18.145}$$

$$\hbar k_i + \hbar k_p = \hbar k_f \tag{18.146}$$

The sub-band energies in equation (18.145) are given by

$$E_1(k_i) = E_{1o} + \frac{\hbar^2 k_i^2}{2m} \tag{18.147}$$

$$E_2(k_f) = E_{2o} + \frac{\hbar^2 k_f^2}{2m} \quad \text{and} \tag{18.148}$$

$$\hbar\omega = \hbar c |k_p| \tag{18.149}$$

where we have assumed the same effective mass m for both sub-bands. Substituting equations (18.146), (18.147), (18.148), and (18.149) in equation (18.145) gives us

$$E_{1o} + \frac{\hbar^2 k_i^2}{2m} + \hbar c |k_p| = E_{2o} + \frac{\hbar^2 (k_i + k_p)^2}{2m} \tag{18.150}$$

Solving the above equation, we can find the magnitude of the photon wave vector k_p that satisfies the conservation of energy. Assuming that the initial wave vector of the electron is $k_i = 0$, the previous equation gives us

$$E_{1o} + \hbar c |k_p| = E_{2o} + \frac{\hbar^2 k_p^2}{2m} \rightarrow k_p^2 - \frac{2mc}{\hbar}|k_p| + \frac{2m}{\hbar^2}(E_{2o} - E_{1o}) = 0$$

$$|k_p| = \frac{mc}{\hbar} \pm \sqrt{\left(\frac{mc}{\hbar}\right)^2 - \frac{2m}{\hbar^2}(E_{2o} - E_{1o})} \tag{18.151}$$

This equation has two solutions. The smaller value of $|k_p|$ that satisfies this solution is

$$|k_p| = \frac{mc}{\hbar} - \sqrt{\left(\frac{mc}{\hbar}\right)^2 - \frac{2m}{\hbar^2}(E_{2o} - E_{1o})}$$

$$= \frac{mc}{\hbar}\left[1 - \sqrt{1 - \frac{2(E_{2o} - E_{1o})}{mc^2}}\right] \tag{18.152}$$

In most materials, $E_{2o} - E_{1o}$ is at most 5 eV and $mc^2 \approx 5.12 \times 10^4$ eV; we assume that $m = 9.1 \times 10^{-31}$ kg. So, using the Taylor series expansion, $|k_p|$ in the above equation can be approximated to be

$$|k_p| \approx \frac{E_{2o} - E_{1o}}{\hbar c} \tag{18.153}$$

Assuming that $E_{2o} - E_{1o} \approx 5\,\mathrm{eV}$, we immediately see that $|k_p| \approx 2.6 \times 10^7\,\mathrm{m}^{-1}$. This is small considering the maximum Brillouin zone wave vector is typically $3 \times 10^9\,\mathrm{m}^{-1}$. So, we can say that the final state in sub-band 2 with wave vector k_f is nearly the same as the initial state in sub-band 1 with wave vector k_i. That is, the transition from the initial to the final state is almost vertical as shown in Figure 18.11.

18.9 Appendix D: Momentum and Dipole Matrix Elements

To find the relationship between the momentum and dipole matrix elements, we can evaluate $\langle \phi_p | \widehat{H}_o \bar{r} - \bar{r} \widehat{H}_o | \phi_s \rangle$ using two different methods. The first method gives us

$$\langle \phi_p | \widehat{H}_o \bar{r} - \bar{r} \widehat{H}_o | \phi_s \rangle = (E_p - E_s)\langle \phi_p | \bar{r} | \phi_s \rangle \tag{18.154}$$

by simply noting that $\widehat{H}_o | \phi_s \rangle = E_s | \phi_s \rangle$ and $\widehat{H}_o | \phi_p \rangle = E_p | \phi_p \rangle$. The second method gives us

$$\langle \phi_p | \widehat{H}_o \bar{r} - \bar{r} \widehat{H}_o | \phi_s \rangle = \left\langle \phi_p \left| \left[\frac{\hat{p}^2}{2m_o} + U(\bar{r}) \right] \bar{r} - \bar{r} \left[\frac{\hat{p}^2}{2m} + U(\bar{r}) \right] \right| \phi_s \right\rangle \tag{18.155}$$

$$\langle \phi_p | \widehat{H}_o \bar{r} - \bar{r} \widehat{H}_o | \phi_s \rangle = -\frac{\hbar^2}{2m_o} \langle \phi_p | \nabla^2 \bar{r} - \bar{r} \nabla^2 | \phi_s \rangle \tag{18.156}$$

To simply the above equation further, let us evaluate $\nabla^2 \bar{r}\, \phi_s(\bar{r})$ first. We note that

$$\frac{\partial^2}{\partial x^2} [\bar{r}\, \phi_s(\bar{r})] = \frac{\partial^2}{\partial x^2} \left[(x\hat{i} + y\hat{j} + z\hat{k})\, \phi_s(\bar{r}) \right]$$

$$= \frac{\partial}{\partial x} \left[\hat{i}\, \phi_s(\bar{r}) + \bar{r} \frac{\partial \phi_s}{\partial x} \right] = \hat{i} \frac{\partial \phi_s}{\partial x} + \bar{r} \frac{\partial^2 \phi_s}{\partial x^2} + \hat{i} \frac{\partial \phi_s}{\partial x}$$

$$= 2\hat{i} \frac{\partial \phi_s}{\partial x} + \bar{r} \frac{\partial^2 \phi_s}{\partial x^2}$$

Similarly,

$$\frac{\partial^2}{\partial y^2}\left[\bar{r}\,\phi_s(\bar{r})\right] = 2\hat{j}\frac{\partial\phi_s}{\partial y} + \bar{r}\frac{\partial^2\phi_s}{\partial y^2}$$

$$\frac{\partial^2}{\partial z^2}\left[\bar{r}\,\phi_s(\bar{r})\right] = 2\hat{k}\frac{\partial\phi_s}{\partial z} + \bar{r}\frac{\partial^2\phi_s}{\partial z^2}$$

Using the above three equations, only the gradient term remains in equation (18.156) and we see that

$$\nabla^2\,\bar{r}\,\phi_s(\bar{r}) = \bar{r}\,\nabla^2\phi_s(\bar{r}) + 2\nabla\phi_s \qquad (18.157)$$

Using the above equation in equation (18.156), we get

$$\langle\phi_p|\widehat{H}_o\bar{r} - \bar{r}\widehat{H}_o|\phi_s\rangle = -\frac{\hbar^2}{m_o}\langle\phi_p|\nabla|\phi_s\rangle = -\frac{i\hbar}{m_o}\langle\phi_p|-i\hbar\nabla|\phi_s\rangle$$

$$= -\frac{i\hbar}{m_o}\langle\phi_p|\hat{p}|\phi_s\rangle \qquad (18.158)$$

Using $E_p - E_s = \hbar\omega_{ps}$, equation (18.154) and equation (18.158), we get

$$\langle\phi_p|\hat{p}|\phi_s\rangle = im_o\omega_{ps}\langle\phi_p|\bar{r}|\phi_s\rangle$$

$$\bar{p}_{ps} = im_o\omega_{ps}\bar{r}_{ps} \qquad (18.159)$$

While the above derivation was for a free particle with mass m_o, equation (18.159) is also used in a material when m_o is replaced by the effective mass m^*.

18.10   Problems

Section 18.2

(1) In equation (18.13), assume that $u(\bar{r},t) = V_o\sin(kx - \omega t)$ and H_o is the free particle Hamiltonian. Derive the equivalent of equations (18.28) and (18.29), simplifying u_{ps} as much as possible. What conservation laws do you see coming out of this?

(2) The properties of $\frac{\sin^2[(\omega-\omega_o)t]}{[(\omega-\omega_o)t]^2}$ are essential in the derivation of transition rates. In this problem, you are going to convince

yourself that $\frac{\sin^2[(\omega-\omega_o)t]}{[(\omega-\omega_o)t]^2} = \frac{2\pi}{t}\delta(\omega - \omega_o)$. Choose $t = 100\,ns$ for all parts.

(a) Plot $\frac{\sin^2[(\omega-\omega_o)t]}{[(\omega-\omega_o)t]^2}$ as a function of $(\omega - \omega_o)$.

(b) Tabulate the "area under the curve divided by $2\pi/t$" from

   (i) $(\omega - \omega_o) = 0$ to ∞

   (ii) $(\omega - \omega_o) = \frac{1}{100\,ns}$ to ∞

   (iii) $(\omega - \omega_o) = \frac{10}{100\,ns}$ to ∞

   (iv) $(\omega - \omega_o) = \frac{100}{100\,ns}$ to ∞

(c) What is the area under the curve in part (a) equal to? How close is it to $\frac{2\pi}{t}$?

Section 18.3

(3) If the harmonic oscillator frequency is $\omega = 2\pi \times 6\,\text{GHz}$, calculate the radiative lifetimes τ_{10}, τ_{21}, and τ_{23}. Note that τ in equation (18.68) is the inverse of the transition rate R.

(4) Prove that the spontaneous emission time from state $n + 1$ to state n is, $\tau_{n+1\,n} = \frac{1}{n+1}\tau_{10}$ and the absorption time from state $n - 1$ to n is, $\tau_{n-1\,n} = \frac{1}{n}\tau_{10}$, where $\tau_{10} = \frac{12\pi\epsilon_o m^{*2}c^3}{q^2\omega^2}$.

(5) Calculate the lifetime of an electron in a hydrogen atom that is excited from $1s$ state to $2p$ state. That is, how long does it take for an electron to *spontaneously* emit a photon and return to $1s$ state from the excited $2p$ state?

(6) From Chapter 16, recall the spherical harmonic part of the wave functions, $Y_m^l(\theta, \varphi)$. Show that the matrix elements for an optical dipole transition between two states $|l, m\rangle$ and $|l', m'\rangle$ is forbidden unless $l - l' = \pm 1$ and $m - m' = 0$.

(Hint: Write x, y, and z in terms of spherical harmonics $\left(z = \sqrt{\frac{4\pi}{3}} r Y_{m=0}^{l=1} e^{i0\varphi}\right)$ and find $\langle Y_{m'}^{l'} | \hat{z} | Y_m^l \rangle$.)

(7) Use the representation of the position operator using annihilation and creation operators to prove that the only possible transitions between state n and m of harmonic oscillators are the ones

with $n - m = \pm 1$. In Example 7 of Chapter 4, we defined how annihilation and creation operators act on a given state $|n\rangle$. Hint: Write the position operator as

$$x = \sqrt{\frac{\hbar}{2m\omega}}(\hat{B} + \hat{B}^+)$$

We represent the states ψ_n by their corresponding ket in number state, i.e., $|n\rangle$.

$$\langle \psi_n | x | \psi_m \rangle = \sqrt{\frac{\hbar}{2m\omega}}\,\langle n | \hat{B} + \hat{B}^+ | m \rangle$$

$$= \sqrt{\frac{\hbar}{2m\omega}}\{\langle n | \hat{B} | m \rangle + \langle n | \hat{B}^+ | m \rangle\}$$

Using the properties of annihilation and creation operators, we have

$$\langle \psi_n | \hat{x} | \psi_m \rangle = \sqrt{\frac{\hbar}{2m\omega}}\,\sqrt{n}\,\delta_{m,(n-1)} + \sqrt{\frac{\hbar}{2m\omega}}\,\sqrt{n+1}\,\delta_{m,(n+1)}$$

This means that the optical transition between state n and m is only possible when $m = n - 1$ or $m = n + 1$.

Section 18.5

(8) Prove equation (18.85).
(Hint: Calculate the integral and then assume the length L approaches infinity.)

Section 18.8

(9) Show that the amplitude of the electromagnetic wave's vector potential wave can be written in vacuum as $A_o = \sqrt{\frac{2n_{ph}\hbar}{\epsilon_r \epsilon_o \omega V_{em}}}$, where V_{em} is the volume of the cavity over which the EM field exists.

References

Anselm, A. (1981). *An Introduction to Semiconductor Theory*. Mir Publishers, Moscow.

Casey Jr., H. C., Sell, D. D., and Wecht, K. W. (1975). Concentration dependence of the absorption coefficient for n- and p-type GaAs between 1.3 and 1.6 eV. *Journal of Applied Physics* **46**, 250–257. https://doi.org/10.1063/1.321330

Fischetti, M. V. and Laux, S. E. (1996). Band structure, deformation potentials, and carrier mobility in strained Si, Ge, and SiGe alloys. *Journal of Applied Physics* **80**, 2234.

Gurevich, V. L. (1986). *Transport in Phonon Systems, Modern Problems in Condensed Matter*, Vol. 18. North Holland, Amsterdam.

Hamaguchi, C. (2017). *Basic Semiconductor Physics*, 3rd edn. Springer-Verlag, Berlin.

Madelung, O. (2004). *Semiconductors: Data Handbook*, 3rd edn. Springer-Verlag, Berlin, Heidelberg.

Rosencher, E. and Vinter, B. (2002). *Optoelectronics*. Cambridge University Press, Cambridge.

Singleton, J. (2001). *Band Theory and Electronic Properties of Solids*. Oxford University Press, New York.

Chapter 19

HYDROGEN ATOM AND SPHERICAL HARMONICS

Contents

In Chapter 11, we discussed and learned how to solve the Schrödinger equation for quantum wires, wells, and dots using the separation of variables. The study of materials begins by solving the Schrödinger equation for the simplest of atoms, hydrogen. Important concepts like atomic orbitals and quantization of angular momentum also arise from the study of hydrogen. This is because hydrogen atoms and other simple ions like He^+ and Li^{2+} are systems for which the Schrödinger equation can be solved analytically. Multi-electron atoms and molecules, such as F or H_2O, require numerical methods and approximations to solve the Schrödinger equation.

Note that the analytic solution of the Schrödinger equation for the hydrogen atom explains only a part of the experimental observations about its spectra. Explanation of other quantum effects such as fine and hyperfine structures in the spectra and the anomalous Zeeman effect requires including relativistic effects and the spin of both the electron and the nucleus in the solution. Demtröder (2010) has a good discussion of these.

In a hydrogen atom, the electron and proton rotate around the center of mass located at $\bar{r}_{cm} = \frac{m_p \bar{r}_p + m_e \bar{r}_e}{m_p + m_e}$, where $\bar{r}_p$ and $\bar{r}_e$ are coordinates of the proton and electron, respectively. In order to simplify this model further, we approximate the center of mass to be at the proton, with the electron at radial distance $\bar{r}$ from the proton. We assume that the proton with charge $+q$ is stationary and the electron is the only moving part. This is a good approximation as the proton's mass is 1836 times larger than that of the electron.

The Coulomb force is attractive and points along the line connecting the proton and electron. It is calculated by Coulomb's law

$$\bar{F} = \frac{(+q)(-q)}{4\pi\epsilon_0 r^3}\bar{r} \tag{19.1}$$

The electrical potential energy is found by calculating how much work is required to bring the electron from a very far distance (think infinity) to a point $\bar{r}$ away from the proton

$$U(\bar{r}) = -\int_\infty^r \bar{F} \cdot d\bar{r} = \int_\infty^r \frac{q^2}{4\pi\epsilon_0 r^2}dr = -\frac{q^2}{4\pi\epsilon_0 r} \tag{19.2}$$

The above potential energy is only a function of electron–proton radial distance (r) and has no dependence on the azimuthal and elevation angles (θ, ϕ). Since $U(\bar{r})$ has spherical symmetry, it is a natural choice to use spherical coordinates (r, θ, ϕ) in solving the Schrödinger equation for the hydrogen atom. The potential energy is plotted in Figure 19.1. There is a deep attractive well at the location of the proton.

Further, ions like He^+, Li^{2+}, and Be^{3+} have only one electron on their outer shells and can be analyzed similarly. The only modification is to replace $(+q)$ with $+Zq$ as the charge of the nucleus, where Z, the atomic number, is the number of protons in the

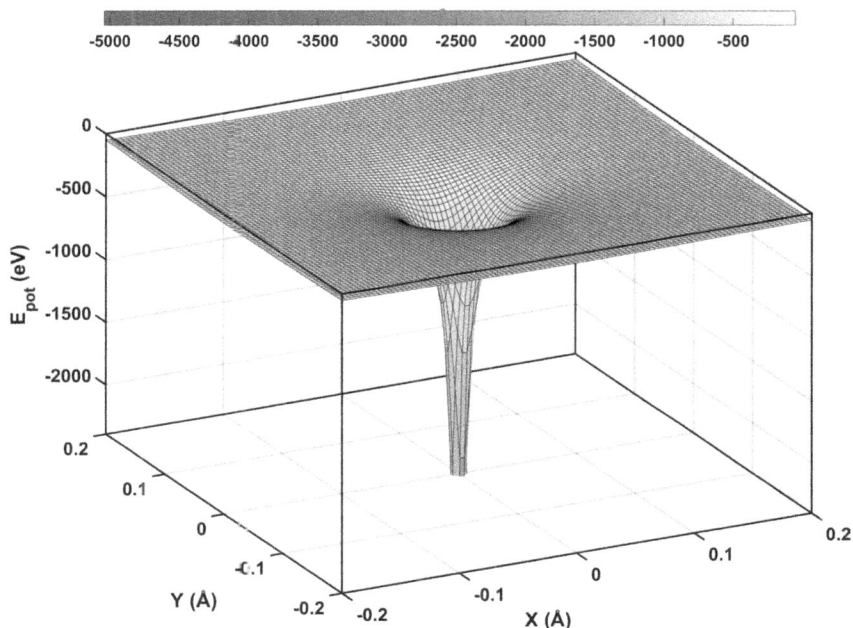

Figure 19.1. The shape of Coulomb potential centered on the hydrogen atom nucleus located at $(0, 0\ 0)$. As $r \to \infty$, the potential flattens out and approaches zero. Note that the potential is shown only at the $z = 0$ plane.

nucleus (recall that neutrons have zero charges and do not affect the electrical potential energy). In atoms other than hydrogen, there are many electrons, and the electrical potential energy at any location is determined by the position of the nucleus and other electrons.

19.1 Schrödinger Equation in Spherical Coordinate

The Laplacian operator in spherical coordinates is,

$$
\begin{aligned}
\nabla^2 &= \frac{1}{r^2} \frac{\partial}{\partial r} \left(r^2 \frac{\partial}{\partial r} \right) + \frac{1}{r^2} \left(\frac{1}{\sin \theta} \frac{\partial}{\partial \theta} \left(\sin \theta \frac{\partial}{\partial \theta} \right) + \frac{1}{\sin^2 \theta} \frac{\partial^2}{\partial \phi^2} \right) \\
&= \frac{1}{r^2} \frac{\partial}{\partial r} \left(r^2 \frac{\partial}{\partial r} \right) + \frac{1}{r^2} \left(\nabla_{\theta, \phi} \right)
\end{aligned}
\tag{19.3}
$$

Substituting the Laplacian operator in Schrödinger equation from Chapter 1, we get

$$\left[-\frac{\hbar^2}{2m} \left(\frac{1}{r^2} \frac{\partial}{\partial r} \left(r^2 \frac{\partial}{\partial r} \right) + \frac{1}{r^2 \sin\theta} \frac{\partial}{\partial\theta} \left(\sin\theta \frac{\partial}{\partial\theta} \right) \right. \right.$$
$$\left. \left. + \frac{1}{r^2 \sin^2\theta} \frac{\partial^2}{\partial\phi^2} \right) + U(\bar{r}) \right] \psi = E\psi \tag{19.4}$$

The above equation is solved using the separation of variables method that was used in Chapter 11. We now need to look for solutions of the wave function that are separable in the $r, \theta,$ and ϕ components of the spherical coordinate system. We try the decomposition

$$\psi(r, \theta, \phi) = R(r) \, f(\theta) \, g(\phi) \tag{19.5}$$

where R depends only on the radius r, f depends only on the elevation angle θ, and g depends only on the azimuthal angle ϕ. Putting this back in equation (19.4), we get three separable equations in r, θ, and ϕ,

$$\frac{d}{dr} \left(r^2 \frac{dR}{dr} \right) - r^2 \left[\frac{2m}{\hbar^2} (U(r) - E) \right] R = \beta R$$

$$\left[\sin\theta \frac{d}{d\theta} \left(\sin\theta \frac{df}{d\theta} \right) \right] + \beta f \sin^2\theta = m^2 f$$

$$\frac{\partial^2 g}{\partial\phi^2} = -m^2 g$$

See Appendix for more detailed steps outlining this derivation. Solving the above three equations, we find eigenfunctions defined by three quantum numbers (n, l, m), given by

$$\psi_{n,l,m} = \psi(r, \theta, \phi) = A \, R_{n,l}(r) Y_l^m(\theta, \phi) \tag{19.6}$$

The functions (i) $R_{n,l}(r)$ represent the radial part of the solution $R(r)$ and (ii) $Y_l^m(\theta, \phi)$ represents the $f(\theta) \, g(\phi)$ parts of the solution. The first quantum number n is the *principal quantum number* and it only takes positive integer values $\{1, 2, 3, \ldots\}$. Y_l^m is the angular part of the solution and they are referred to as the *spherical harmonics*; they depend only on $\cos(\theta), \sin(\theta), \cos(\phi),$ and $\sin(\phi)$. Note that spherical harmonics also appear in other engineering problems like acoustics and electromagnetism. See Table 19.2 for the mathematical form of Y_m^l and note that they always appear as a product $f(\theta)\,g(\phi)$. The prefactor A is the normalization factor of the wave function and is found from the following equation:

$$\int |\psi_{n,l,m}|^2 dv$$

$$= |A|^2 \int_{r=0}^{\infty} \int_{\theta=0}^{\theta=\pi} \int_{\phi=0}^{\phi=2\pi} r^2 \sin\theta d\theta d\phi dr (R_{n,l})^* R_{n,l} (Y_l^m)^* Y_l^m = 1 \tag{19.7}$$

The electron eigenenergy E_n is inversely proportional to the principal quantum number squared,

$$E_n = -\frac{E_1}{n^2} = -\frac{13.6\,\text{eV}}{n^2} \tag{19.8}$$

For $n = 1$, the electron has the lowest energy within the potential trap of the proton, which is $E_1 = -13.6\,\text{eV}$.

The integer numbers l and m are called the *angular* (*or azimuthal*) and *magnetic* quantum numbers, respectively; they are constrained to have the following values for a given principal quantum number n,

$$\text{for a given } n \to l = 0, 1, 2, \ldots, n-1 \tag{19.9}$$

$$\text{for any } l, \quad |m| \le l \; (m = -l, -l+1, \ldots - 1, 0, 1, \ldots, l-1, l) \tag{19.10}$$

From the above, we see that for each l, there are $2l + 1$ allowed values of m.

Also, it is noteworthy that for each quantum number n, the allowed set of l and m values all correspond to the same eigenenergy E_n, as given in equation (19.8). The wave functions $\psi_{n,l,m}$ are called atomic orbitals, and their magnitude squared is the probability density distribution of electrons in orbitals at different radii and angles around the nucleus. The number of distinct sets of quantum numbers l and m with the same energy E_n is given by

$$\sum_{l=0}^{n-1} (2l + 1) = n^2 \tag{19.11}$$

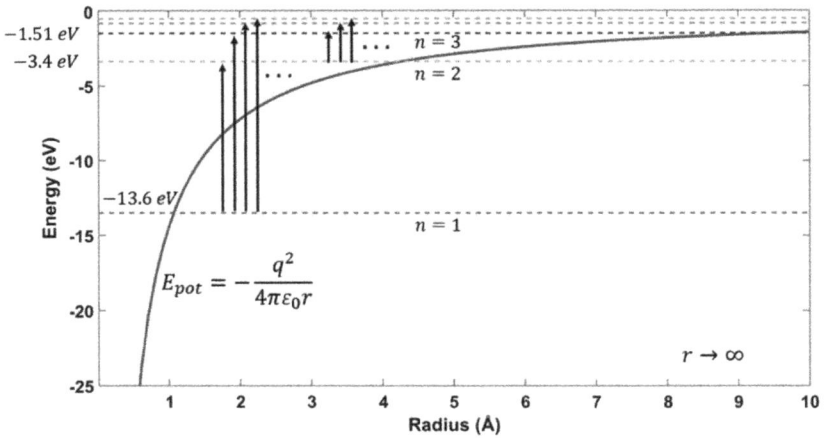

Figure 19.2. The energy levels in a hydrogen atom scale as $1/n^2$. The first and the second group of optical transitions (from left to right) are called Lyman and Balmer series. The curved line is the potential energy plotted versus r.

For this reason, n^2, which is the total number of quantum numbers with the same energy, is called the *degeneracy*. The energy eigenvalues are represented by horizontal levels in the electrical potential energy trap of the hydrogen nucleus, as shown in Figure 19.2. The vertical arrows represent some of the possible optical transitions from the energy levels corresponding to $n = 1$ and 2, etc. The transitions are grouped and named depending on which energy level is the first starting level. These are the absorption lines observed in the spectroscopy of hydrogen gas. For example, in Figure 19.2, the first group of lines are called Lyman series and the second group from left is called Balmer series. Note that as $n \to \infty$, the eigenenergy becomes zero. The eigenenergies above zero correspond to electrons that are free and no longer bound to the nucleus.

Tables 19.1 and 19.2 show the radial and angular parts of the electron wave functions in the hydrogen atom for different quantum numbers. In Table 19.1, the normalization factor $A = (1/na_o)^{3/2}$. The substitution x used in the table is the radius scaled to be unitless and is defined to be $x = r/na_o$, where,

$$a_o = \frac{4\pi\epsilon_0\hbar^2}{mq^2} = 0.529 \text{ Å} \tag{19.12}$$

Table 19.1. The radial parts of wave function R_n for $n = 1, 2$, and 3, for allowable values for l.

n	l	$R_{n,l}(r)$
1	0	$2e^{-x}$
2	0	$2e^{-x}(1 - x)$
	1	$\dfrac{2}{\sqrt{3}} e^{-x} x$
3	0	$2e^{-x}\left(1 - 2x + \dfrac{2x^2}{3}\right)$
	1	$\dfrac{2}{3}\sqrt{2} e^{-x} x(2 - x)$
	2	$\dfrac{4}{3\sqrt{10}} e^{-x} x^2$

Note: $x = r/na_o$.

Table 19.2. The angular parts of the wave function for $l = 0, 1$, and 2.

l	m	$Y_l^m(\theta, \phi)$
0	0	$\dfrac{1}{2\sqrt{\pi}}$
1	0	$\dfrac{1}{2}\sqrt{\dfrac{3}{\pi}} \cos\theta$
	± 1	$\pm\dfrac{1}{2}\sqrt{\dfrac{3}{2\pi}} \sin\theta e^{\pm i\phi}$
2	0	$\dfrac{1}{4}\sqrt{\dfrac{5}{\pi}}(2\cos^2\theta - \sin^2\theta)$
	± 1	$\pm\dfrac{1}{2}\sqrt{\dfrac{15}{2\pi}} \cos\theta \sin\theta e^{\pm i\phi}$
	± 2	$\dfrac{1}{4}\sqrt{\dfrac{15}{2\pi}} \cos^2\theta e^{\pm i2\phi}$

Note: For each value of l, the allowable values of m are listed in the second column.

is the Bohr radius of an electron that was derived in Chapter 1 using semiclassical methods for the hydrogen atom (as opposed to the fully quantum treatment in this chapter).

Example 1 (1s orbital). For quantum number $n = 1$, (a) plot the radial and angular parts of the electron wave function in hydrogen atom, (b) plot the probability density to find the electron at radius r from the nucleus (integrated over all angles θ and ϕ), and (c) calculate the expectation value or average value of the radial location of the electron around the nucleus.

Solution: (a) For $n = 1$, the allowed values of l and m are 0, i.e., $n = 1$, $l = 0$, $m = 0$. In this case, the eigenenergy is $E_1 = -13.6\,\text{eV}$. The wave function is found using the first rows of Tables 19.1 and 19.2.

$$\psi_{1,0,0} = AR_{1,0}(r) \cdot Y_0^0(\theta, \phi) = \frac{1}{\sqrt{\pi}} \left(\frac{1}{a_o}\right)^{3/2} e^{-\frac{r}{a_o}} \tag{19.13}$$

As an exercise, check that the wave function is normalized. The angular part of the wave function $Y_0^0(\theta, \phi)$ is $1/2\sqrt{\pi}$ according to Table 19.2. It has no dependence on the angles, meaning that it has spherical symmetry as shown in Figure 19.3(a). The radial part of the wave function has a maximum at the center $(r = 0)$ and decays exponentially as we go farther away from the nucleus as shown in Figure 19.3(b).

(b) The volume element dv in spherical coordinates is $dv = r^2 \sin\theta d\phi d\theta dr$. The total probability to find an electron in a spherical shell of thickness dr at radius r is $|\psi_{1,0,0}|^2 dv$ integrated over all angles and is given by,

$$\int_0^\pi \int_0^{2\pi} |\psi_{1,0,0}|^2 r^2 \sin\theta d\phi d\theta dr = 4\pi r^2 |\psi_{1,0,0}|^2 dr = \wp(r)dr \tag{19.14}$$

The symbol $\wp(r)$ is called the radial probability density, and it represents the probability per unit length to find an electron at radius r. From the above, we infer that radial probability density $\wp(r)$ is $4\pi r^2$ times the probability density $|\psi_{1,0,0}|^2$.

$$\wp(r) = 4\pi r^2 |\psi_{1,0,0}|^2 \tag{19.15}$$

(a)

(b)

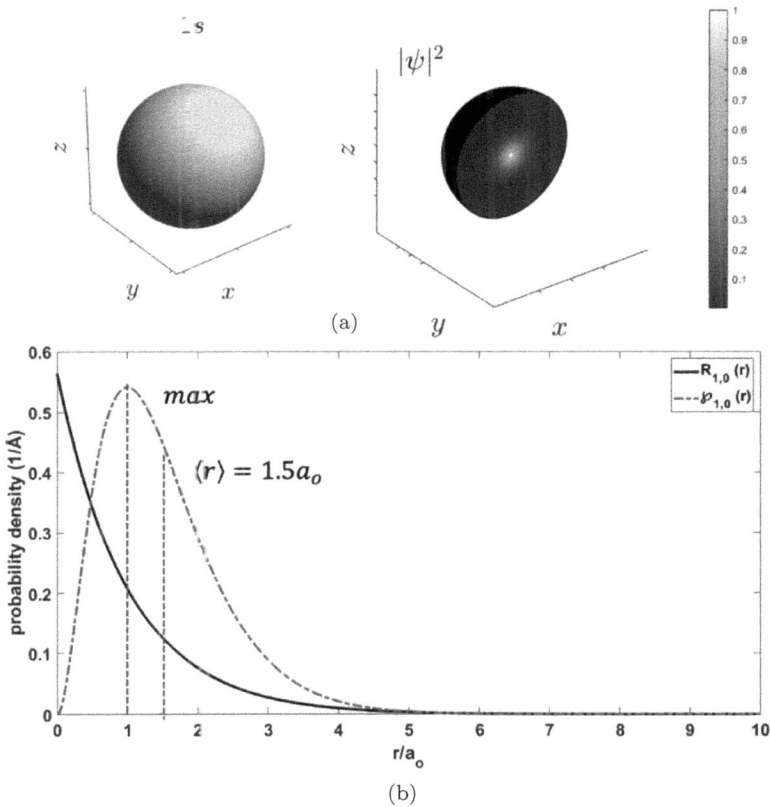

Figure 19.3. (a) (Left) The angular part of $\psi_{1,0,0}$ from Table 19.2, Y_0^0, is spherically symmetric and thus does not have an angular dependence. (Right) A cross-section of the probability density $|\psi_{1,0,0}|^2$. See the color bar for values of the probability density. (b) The solid black line is the radial part of the wave function which is $R_{1,0}$ from Table 19.1. The dashed blue line is the radial probability density shown in equation (19.15). The maxima in the probability density occurs at $r = a_o$, while the average or expectation value of radius is $1.5a_o$ (shown by vertical lines).

While the magnitude of the wave function has a maximum at $r = 0$, the radial probability density $\wp(r)$ shown in Figure 19.3(b) has a maximum at the Bohr radius (a_o). It is found by taking the maxima of $\wp(r)$,

$$\frac{d}{dr}\wp(r) = 0 \rightarrow r_{max} = a_o \tag{19.16}$$

Table 19.3. Common names of the orbitals based on the value of the angular quantum number, l.

Angular quantum number l	Name of the state
0	s (sharp)
1	p (principal)
2	d (diffuse)
3	f (fundamental)

(c) The average or expectation value of r is found from the following equation:

$$\langle\psi_{1,0,0}|r|\psi_{1,0,0}\rangle = \int_0^\infty r \times 4\pi r^2 |\psi_{1,0,0}|^2 dr$$

$$= \frac{4}{a_o^3}\int_0^\infty dr\, r^3 e^{-2r/a_o} = \frac{3a_o}{2} \quad (19.17)$$

This shows that the average value of the radial location of an electron is 1.5 times the Bohr radius, i.e., $\langle r\rangle = 3a_o/2$. The above integral can be calculated using tables of integrals or integration by parts.

Table 19.3 shows the numbers and symbols assigned to the different angular quantum numbers l. These symbols originated from the names given to lines observed in the light absorption or emission spectrum of atoms. The $\psi_{1,0,0}$ calculated in the previous example is called 1s orbital.

Example 2 (2s and 2p orbitals). For all orbitals corresponding to quantum number $n = 2$, (a) plot the radial and angular parts of the electron wave function, (b) plot the radial probability density, and (c) calculate the expectation value (average) of the radius of electron orbit around the nucleus.

Solution: (a) For $n = 2$, the allowed values of l and m are listed as follows:

$$(n, l, m) = (2,0,0), (2,1,-1), (2,1,0), (2,1,1)$$

In this case, the eigenenergy is $E_2 = \frac{-13.6\,\text{eV}}{2^2} = -3.4\,\text{eV}$. This energy is four-fold degenerate meaning that the electron can have four different orthogonal wave functions with energy E_2. The radial and angular components of each wave function are shown in Tables 19.1 and 19.2. For $n = 2$, $l = 0$, $m = 0$, we have

$$\psi_{2,0,0} = AR_{2,0}(r) \cdot Y_0^0(\theta, \phi) = \frac{1}{4\sqrt{2\pi}} \left(\frac{1}{a_o}\right)^{\frac{3}{2}} \left(2 - \frac{r}{a_o}\right) e^{-\frac{r}{2a_o}}$$

(19.18)

The above wave function is called the $2s$ orbital. As an exercise, check that the wave function is normalized. The angular part of the wave function according to Table 19.2 is the same as $1s$, which is $Y_0^0 = 1/2\sqrt{\pi}$. It is spherically symmetric as shown in Figure 19.4(a). The radial part has a maximum at the center ($r = 0$) and has a zero at $r = 2a_o$ after which it decays exponentially as we go farther away from the nucleus as shown in Figure 19.4(b).

(b) To find the radial probability density of the $2s$ orbital at radius r, we calculate the probability of finding an electron in a spherical shell of thickness dr around the nucleus. This is found by integrating the probability density $|\psi_{2,0,0}|^2$ over all angles,

$$\int_r^{r+dr} \int_0^\pi \int_0^{2\pi} r^2 \sin\theta \, d\theta \, d\phi \, dr |\psi_{2,0,0}|^2 = \wp(r)dr$$

(19.19)

From the above, the radial probability density is given by

$$\wp(r) = 4\pi r^2 |\psi_{2,0,0}|^2 = \frac{r^2}{8a_o^3} \left(2 - \frac{r}{a_o}\right)^2 e^{-\frac{r}{a_o}}$$

(19.20)

The radial probability density shown in Figure 19.4(b) is zero at $r = 2a_o$ and it has two maxima. Furthermore, the number of maxima is equal to the principal quantum number n. The radii of maximum radial probability bands are found from

$$\frac{d}{dr}\wp(r) = 0 \rightarrow r_{max} = \frac{6 \pm \sqrt{20}}{2} a_o$$

$$= 0.7639a_o \text{ and } 5.2361a_o$$

(19.21)

(a)

(b)

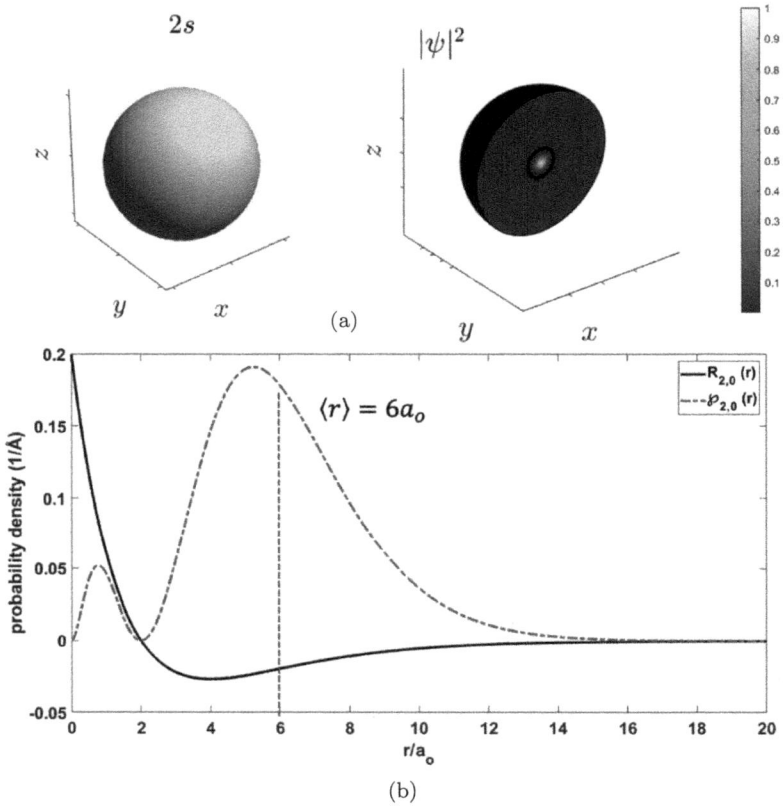

Figure 19.4. (a) (Left) The angular part of $\psi_{2,0,0}$ from Table 19.2, Y_0^0, is spherically symmetric. (Right) A cross-section of the probability density $|\psi_{2,0,0}|^2$. See the color bar for values of the probability density. (b) The solid black line is the radial part of the wave function, which is $R_{2,0}$ from Table 19.1. The dashed blue line is the radial probability density $\wp(r)$ shown in equation (19.20). The maxima in the probability density occurs at $r = 0.7639a_o$ and $r = 5.2361a_o$, while the average or expectation value of radius is $6a_o$ (shown by a vertical line).

(c) The average or expectation value of r is found from the following integral:

$$\langle\psi_{2,0,0}|r|\psi_{2,0,0}\rangle = \int_0^\infty r \times 4\pi r^2 |\psi_{2,0,0}|^2 dr$$

$$= \frac{1}{8a^3}\int_0^\infty dr\, r^3 \left(2 - \frac{r}{a_o}\right)^2 e^{-\frac{r}{a_o}} = 6a_o \quad (19.22)$$

This shows the average radial location of an electron in the $2s$ state is $\langle r \rangle = 6a_o$ from the nucleus.

For $n = 2, l = 1$, there are three values for m, i.e., $-1, 0, 1$. The corresponding wave functions are called $2p$ orbitals. Note that p_x and p_y orbitals are written as a superposition of $\psi_{2,1,-1}$ and $\psi_{2,1,1}$, which results in orbitals aligned along the x and y axes respectively. Since the radial part of the wave function, $R_{2,1}(r)$, is common, it can be factored out. This leaves the angular parts as a superposition of $Y_1^{+1}(\theta, \phi)$ and $Y_1^{-1}(\theta, \phi)$ as follows (using Tables 19.1 and 19.2):

$$2p_z = \psi_{2,1,0} = AR_{2,1}(r) \cdot Y_1^0(\theta, \phi)$$

$$= \frac{1}{4\sqrt{2\pi}} \left(\frac{1}{a_o}\right)^{\frac{3}{2}} \left(\frac{r}{a_o}\right) e^{-\frac{r}{2a_o}} \cos\theta \qquad (19.23)$$

$$2p_x = AR_{2,1}(r) \cdot \frac{1}{\sqrt{2}}[Y_1^{+1} + Y_1^{-1}]$$

$$= \frac{1}{8\sqrt{2\pi}} \left(\frac{1}{a_o}\right)^{\frac{3}{2}} \left(\frac{r}{a_o}\right) e^{-\frac{r}{2a_o}} \sin\theta \cos\phi \qquad (19.24)$$

$$2p_y = AR_{2,1}(r) \cdot \frac{1}{i\sqrt{2}}[Y_1^{+1} - Y_1^{-1}]$$

$$= \frac{1}{8\sqrt{2\pi}} \left(\frac{1}{a_o}\right)^{\frac{3}{2}} \left(\frac{r}{a_o}\right) e^{-\frac{r}{2a_o}} \sin\theta \sin\phi \qquad (19.25)$$

The orbitals are plotted in Figure 19.5(a) and (b). It is interesting that the sum of the radial probability densities in the $2p_x$, $2p_y$, and $2p_z$ orbitals is

$$\wp(r) = 4\pi r^2(|2p_z|^2 + |2p_x|^2 + |2p_y|^2) = \frac{r^4}{8a_o^5}e^{-\frac{r}{a_o}} \qquad (19.26)$$

Even though the $2p_x$, $2p_y$, and $2p_z$ orbitals have both angular and radial dependence, the sum of the probability densities of the three orbitals depends only on the radius from the nucleus. The radial probability density $\wp(r)$ shown in the previous equation has a maximum at $r = 4a_o$, which can be found from $\frac{d\wp(r)}{dr} = 0$. The expectation value is $\langle r \rangle = 5.6604a_o$ using the methods explained in the previous examples (see Figure 19.5(c)).

(a)

(c)

(b)

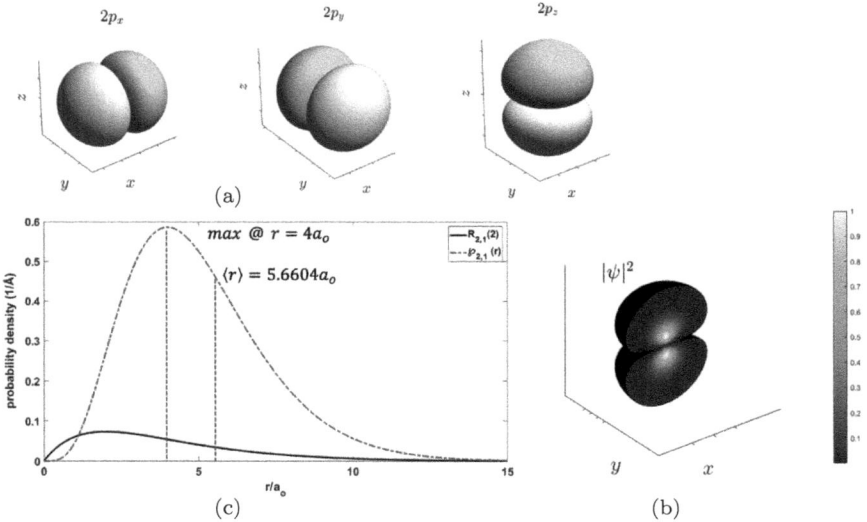

Figure 19.5. (a) The angular part of $2p_x$, $2p_y$, and $2p_z$. (b) A cross-section of the probability density for $2p_z$ orbital or $|\psi_{2,1,0}|^2$. See the color bar for values of the probability density. (c) The solid black line is the radial part of the wave function, which is $R_{2,1}$ from Table 19.1. The dashed blue line is the radial probability density $\wp(r)$ as a function of r/a_o. The maximum of the density occurs at $r = 4a_o$ while the average or expectation value of radius is $5.6604a_o$ (as shown by vertical lines).

Example 3 ($3s, 3p$ and $3d$ orbitals). For all orbitals corresponding to quantum number $n = 3$, plot the (a) radial and angular parts of the electron wave function and (b) radial probability density. Do this for $3s$, $3p$, and $3d$ orbitals.

Solution: (a) For $n = 3$, the allowed values of l and m are given as follows:

$$(n, l, m) = (3, 0, 0), (3, 1, 0), (3, 1, \pm 1), (3, 2, 0), (3, 2, \pm 1), (3, 2, \pm 2)$$

That means there are $n^2 = 9$ orbitals (wave functions) associated with the single energy level E_3, meaning that it is nine-fold degenerate. In this case, the eigenenergy is $E_3 = \frac{-13.6\,\text{eV}}{3^2} = -1.51\,\text{eV}$.

3s orbital: For $n = 3$, $l = 0$, $m = 0$, from Tables 19.1 and 19.2, we have

$$\psi_{3,0,0} = AR_{3,0}(r) \cdot Y_0^0(\theta, \phi)$$

$$= \frac{1}{81\sqrt{3\pi}} \left(\frac{1}{a_o}\right)^{\frac{3}{2}} \left(27 - \frac{18r}{a_o} + 2\frac{r^2}{a_o^2}\right) e^{-\frac{r}{3a_o}} \quad (19.27)$$

The above wave function is called the 3s orbital and the angular part $Y_0^0 = 1/2\sqrt{\pi}$ is the same spherically symmetric function we saw in the cases of the 1s and 2s orbitals. The radial part depends on the principal quantum number $n = 3$ and has three maxima. The radial dependence of the 3s orbital is shown in Figure 19.6(a). It has a maximum at the center ($r = 0$) and has two zeros at $r = 1.5(3-\sqrt{3})a_o$ and $r = 1.5(3 + \sqrt{3})a_o$ after which it decays exponentially as we go farther away from the nucleus.

(b) To find the radial probability density of the 3s orbital, we proceed in a manner like the previous examples to find

$$\wp(r) = 4\pi r^2 |\psi_{3,0,0}|^2 \quad (19.28)$$

The above radial probability density is spherically symmetric and plotted in Figure 19.6(b). It has three maxima at $0.74a_o$, $4.18593a_o$, and $13.074a_o$. These are found from $\frac{d}{dr}\wp(r) = 0$.

3p orbitals: For $n = 3$, $l = 1$, $m = -1, 0, +1$, from Tables 19.1 and 19.2, we have the following three wave functions whose radial part, $R_{3,1}$, is the same. However, they have three different angular parts, determined by quantum number m. Similar to the equations (19.24) and (19.25), the angular parts of p_x and p_y orbitals are written as superpositions of $Y_1^{-1}(\theta, \phi)$ and $Y_1^{-1}(\theta, \phi)$.

$$3p_z = \psi_{3,\,0} = AR_{3,1}(r) \cdot Y_1^0(\theta, \phi)$$

$$= \frac{\sqrt{2}}{81\sqrt{\pi}} \left(\frac{1}{a_o}\right)^{\frac{3}{2}} \left(6 - \frac{r}{a_o}\right) \frac{r}{a_o} e^{\frac{-r}{3a_o}} \cos\theta \quad (19.29)$$

(a)

(b)

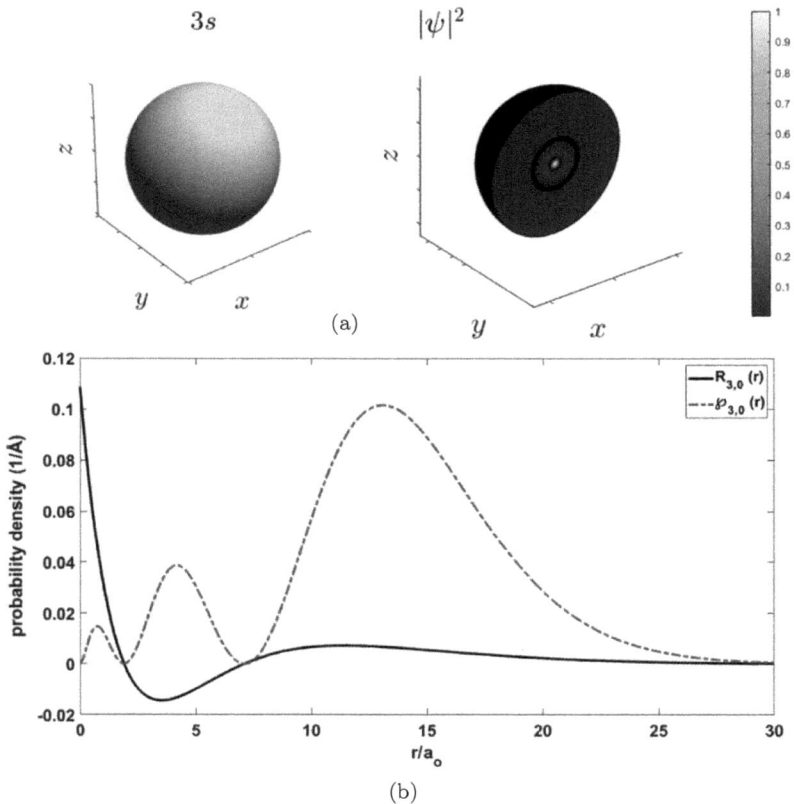

Figure 19.6. (a) (Left) The angular part of $\psi_{3,0,0}$ from Table 19.2. (Right) A cross-section of the probability density $|\psi_{3,0,0}|^2$. See the color bar for values of the probability density. (b) The solid black line is the radial part of the wave function $R_{3,0}$ from Table 19.1. The dashed blue line is the radial probability density $\wp(r)$.

$$3p_x = AR_{3,1}(r) \cdot \frac{1}{\sqrt{2}}[Y_1^{+1} + Y_1^{-1}]$$

$$= \frac{1}{81\sqrt{\pi}} \left(\frac{1}{a_o}\right)^{\frac{3}{2}} \left(6 - \frac{r}{a_o}\right) \frac{r}{a_o} e^{\frac{-r}{3a_o}} \sin\theta \cos\phi \qquad (19.30)$$

$$3p_y = AR_{3,1}(r) \cdot \frac{1}{i\sqrt{2}}[Y_1^{+1} - Y_1^{-1}]$$

$$= \frac{1}{81\sqrt{\pi}} \left(\frac{1}{a_o}\right)^{\frac{3}{2}} \left(6 - \frac{r}{a_o}\right) \frac{r}{a_o} e^{\frac{-r}{3a_o}} \sin\theta \sin\phi \qquad (19.31)$$

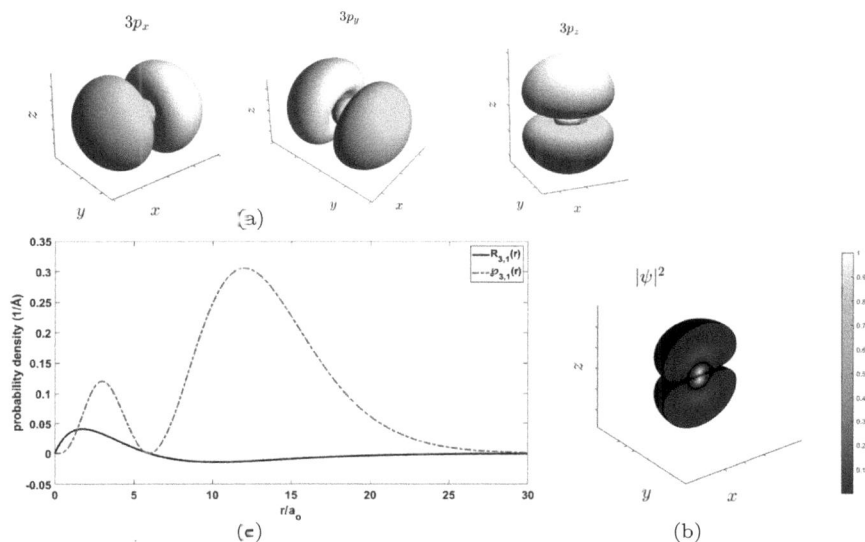

Figure 19.7. (a) The angular part of $3p_x$, $3p_y$, and $3p_z$. (b) A cross-section of the probability density for $3p_z$ orbital or $|\psi_{3,1,0}|^2$. See the color bar for values of the probability density. (c) The radial part (solid line) of the wave function $R_{3,1}$ from Table 19.1 has zeros at $r = 0$ and $r = 6a_o$. (c) The radial probability density $\wp(r)$ (dashed line) as a function of r/a_o (equation (19.32)).

These wave functions are called $3p$ orbitals and their angular parts are shown in Figure 19.7(a). The radial part has one zero at the center ($r = 0$) and another one at $r = 6a_o$ and it decays exponentially as the radius increases in Figure 19.7(c).

(b) To find the radial probability density of $3p$ orbitals, we write

$$\wp(r) = 4\pi r^2 |R_{3,1}|^2 \tag{19.32}$$

which is plotted in Figure 19.7(c). It has two zeros and two maxima found from $\frac{d}{dr}\wp(r) = 0$.

3d orbitals: For $n = 3$, $l = 2$, $m = -2, -1, 0, +1, +2$, from Tables 19.1 and 19.2, we have the following five wave functions whose radial part, $R_{3,2}$, is the same, but they have different angular parts determined by the value of m.

$$\psi_{3,2,0} = AR_{3,2}(r) \cdot Y_2^0(\theta,\phi)$$

$$= \frac{1}{81\sqrt{6\pi}} \left(\frac{1}{a_o}\right)^{\frac{3}{2}} \left(\frac{r}{a_o}\right)^2 e^{-\frac{r}{3a_o}} (3\cos^2\theta - 1) \qquad (19.33)$$

$$\psi_{3,2,\pm1} = AR_{3,2}(r) \cdot Y_2^{\pm1}(\theta,\phi)$$

$$= \frac{1}{81\sqrt{\pi}} \left(\frac{1}{a_o}\right)^{\frac{3}{2}} \left(\frac{r}{a_o}\right)^2 e^{-\frac{r}{3a_o}} \sin\theta\cos\theta e^{\pm i\phi} \qquad (19.34)$$

$$\psi_{3,2,\pm2} = AR_{3,2}(r) \cdot Y_2^{\pm2}(\theta,\phi)$$

$$= \frac{1}{162\sqrt{\pi}} \left(\frac{1}{a_o}\right)^{\frac{3}{2}} \left(\frac{r}{a_o}\right)^2 e^{-\frac{r}{3a_o}}, \sin^2\theta e^{\pm i2\phi} \qquad (19.35)$$

The $3d$ orbitals for $m \neq 0$ are written as a different superposition of the above wave functions. The radial part is the same for all, and the angular part will be the superposition of spherical harmonics $Y_2^{\pm1}(\theta,\phi)$ and $Y_2^{\pm2}(\theta,\phi)$ as follows. The subscripts like xy, yz in the naming of $3d$ orbitals reflect their symmetry and the plane on which they are formed.

$$3d_{z^2} = \psi_{3,2,0} = AR_{3,2}(r) \cdot Y_2^0(\theta,\phi)$$

$$= \frac{1}{81\sqrt{6\pi}} \left(\frac{1}{a_o}\right)^{\frac{3}{2}} \left(\frac{r}{a_o}\right)^2 e^{-\frac{r}{3a_o}} (3\cos^2\theta - 1) \qquad (19.36)$$

$$3d_{xz} = AR_{3,2}(r) \cdot \frac{1}{\sqrt{2}}[Y_2^1 + Y_2^{-1}]$$

$$= \frac{1}{81\sqrt{2\pi}} \left(\frac{1}{a_o}\right)^{\frac{3}{2}} \left(\frac{r}{a_o}\right)^2 e^{-\frac{r}{3a_o}} \sin\theta\cos\theta\cos\phi \qquad (19.37)$$

$$3d_{yz} = AR_{3,2}(r) \cdot \frac{1}{i\sqrt{2}}[Y_2^1 - Y_2^{-1}]$$

$$= \frac{1}{81\sqrt{2\pi}} \left(\frac{1}{a_o}\right)^{\frac{3}{2}} \left(\frac{r}{a_o}\right)^2 e^{-\frac{r}{3a_o}} \sin\theta\cos\theta\sin\phi \qquad (19.38)$$

$$3d_{x^2-y^2} = AR_{3,2}(r) \cdot \frac{1}{\sqrt{2}}[Y_2^2 + Y_2^{-2}]$$

$$= \frac{1}{162\sqrt{2\pi}} \left(\frac{1}{a_o}\right)^{\frac{3}{2}} \left(\frac{r}{a_o}\right)^2 e^{-\frac{r}{3a_o}} \sin^2\theta \cos 2\phi \qquad (19.39)$$

$$3d_{xy} = AR_{3,2}(r) \cdot \frac{1}{i\sqrt{2}}[Y_2^2 - Y_2^{-2}]$$

$$= \frac{1}{162\sqrt{2\pi}} \left(\frac{1}{a_o}\right)^{\frac{3}{2}} \left(\frac{r}{a_o}\right)^2 e^{-\frac{r}{3a_o}} \sin^2\theta \sin 2\phi \qquad (19.40)$$

The $3d$ orbitals are plotted in Figure 19.8(a). The radial part has a zero at the center ($r = 0$) and it decays exponentially as we go farther away from the nucleus as shown in Figure 19.8(c).

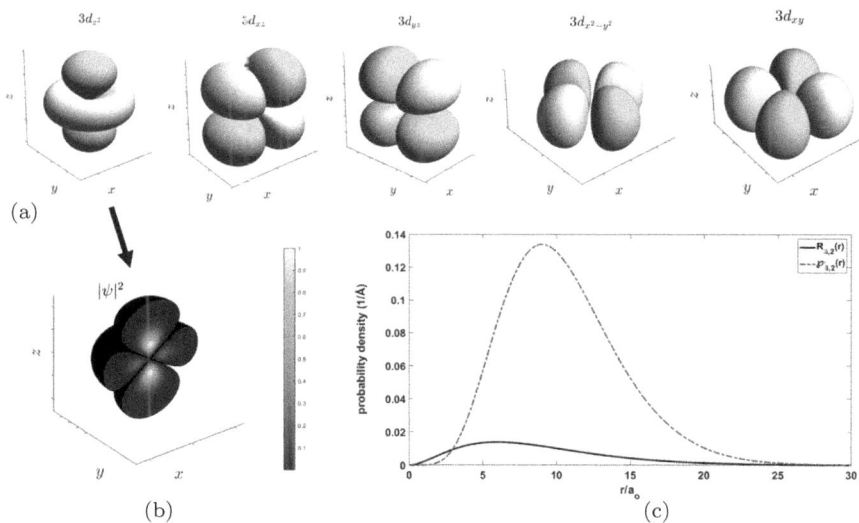

Figure 19.8. (a) The angular part of $3d$ orbitals. (b) A cross-section of the probability density for $|\psi_{3,2,0}|^2$ or orbital $3d_{z^2}$ which has an angular part Y_2^0. See the color bar for values of the probability density. (c) The radial part of the wave function $R_{3,2}$ (solid line) and radial probability density $\wp(r)$ (dashed line) as a function of r/a_o. The maximum density occurs at $r = 9a_o$.

(b) The sum of the radial probability densities of the five $3d$ orbitals is

$$\wp(r) = 4\pi r^2 |R_{3,2}|^2 \tag{19.41}$$

Figure 19.8(c) shows that it has a maximum at $r = 9a_o$.

19.2 Quantum Numbers (n, l, m)

To recap, the solution of Schrödinger equation for a Hydrogen atom yields three quantum numbers:

1. n is the principal quantum number and it determines the energy of the system and the size of the orbital shells; $n = 1$ is the ground state or the lowest energy state.
2. l is the azimuthal (angular momentum) quantum number and describes the shape of the wave functions (orbitals) that an electron can occupy.
3. m is the magnetic quantum number and it specifies the spatial orientation of an orbital for a given n and l. For a given l, there are $2l + 1$ values for m.

(n, l, m) must be integer values and are determined by the following relations:

$$n = 1, 2, 3, \ldots \tag{19.42}$$

$$l = 0, 1, 2, \ldots, n - 1 \tag{19.43}$$

$$|m| \leq l \tag{19.44}$$

The ground state energy $(n = 1)$ for the H atom is $-13.6\,\text{eV}$. This means that to ionize a hydrogen atom, we need to supply $13.6\,\text{eV}$ of energy to the electron in the ground state to move it infinitely far away. The radius of the $n = 1$ state is called the **Bohr radius,** $a_o = 0.529\text{Å}$, which is the mean orbital radius of an electron around the nucleus in the ground state of the H atom, i.e., the $1s$ state. Note that by increasing the value of n, the mean radius of the electron orbit increases.

19.3 Appendix: Overview of Solving Schrödinger Equation

This appendix provides an overview of solving Schrödinger equation for a hydrogen atom for mathematically curious students who want to understand some of the nuances involved in the solution. It is based on using separation of variables in spherical coordinates. The detailed derivation is discussed in many texts

We substitute equation (19.5) into equation (19.4) to get

$$-\frac{\hbar^2}{2m}\left(\frac{fg}{r^2}\frac{d}{dr}\left(r^2\frac{dR}{dr}\right) + \frac{Rg}{r^2\sin\theta}\frac{\partial}{\partial\theta}\left(\sin\theta\frac{\partial f}{\partial\theta}\right) + \frac{Rf}{r^2\sin^2\theta}\frac{\partial^2 g}{\partial\phi^2}\right)$$
$$+URfg = ERfg \tag{19.45}$$

We divide both sides of the above equation by Rfg and multiply both sides by $-r^2\frac{2m}{\hbar^2}$. After rearranging radial variables to one side and angular ones to the other, we get

$$\left\{\frac{1}{R}\frac{d}{dr}\left(r^2\frac{dR}{dr}\right) - r^2\frac{2m}{\hbar^2}[U(r) - E]\right\}$$
$$= -\frac{1}{fg}\left\{\frac{g}{\sin\theta}\frac{\partial}{\partial\theta}\left(\sin\theta\frac{\partial f}{\partial\theta}\right) + \frac{f}{\sin^2\theta}\frac{\partial^2 g}{\partial\phi^2}\right\} \tag{19.46}$$

Note the LHS of equation (19.46) is only a function of r and the RHS is dependent only on θ and ϕ. The equality of both sides only holds if both sides are equal to a constant β. As a result, we now arrive at two separate second-order differential equations.

$$\frac{1}{R}\frac{d}{dr}\left(r^2\frac{dR}{dr}\right) - r^2\left[\frac{2m}{\hbar^2}[U(r) - E]\right] = \beta \tag{19.47}$$

$$\frac{1}{fg}\left\{\frac{g}{\sin\theta}\frac{\partial}{\partial\theta}\left(\sin\theta\frac{\partial f}{\partial\theta}\right) + \frac{f}{\sin^2\theta}\frac{\partial^2 g}{\partial\phi^2}\right\} = \frac{1}{gf}(\nabla_{\theta,\phi}gf) = -\beta \tag{19.48}$$

19.3.1 *The angular part of the wave function*

Before solving equation (19.47), which is general for any spherically symmetric potential $U(r)$, we will focus on equation (19.48) and notice that it is a second-order differential equation in two variables. First, we are going to rearrange it by multiplying both sides by $\sin^2 \theta$,

$$\frac{1}{f} \sin \theta \frac{\partial}{\partial \theta} \left(\sin \theta \frac{\partial f}{\partial \theta} \right) + \frac{1}{g} \frac{\partial^2 g}{\partial \phi^2} = -\beta \sin^2 \theta \tag{19.49}$$

Moving the terms that depend on θ only and ϕ only to the LHS and RHS, respectively, we get

$$\frac{1}{f} \left[\sin \theta \frac{d}{d\theta} \left(\sin \theta \frac{df}{d\theta} \right) \right] + \beta \sin^2 \theta = -\frac{1}{g} \frac{\partial^2 g}{\partial \phi^2} \tag{19.50}$$

The equality only holds if both sides are equal to some constant that we will call m^2 (there is no assumption on if m^2 is real or integer or complex). That is,

$$\left[\sin \theta \frac{d}{d\theta} \left(\sin \theta \frac{df}{d\theta} \right) \right] + \beta f \sin^2 \theta = m^2 f \tag{19.51}$$

$$\frac{\partial^2 g}{\partial \phi^2} = -m^2 g \tag{19.52}$$

The latter differential equation (19.52) is simple to solve and its general solution is

$$g(\phi) = A e^{im\phi} + B e^{-im\phi} \tag{19.53}$$

However, if we let m run over both positive and negative values, the solution can be written as

$$g(\phi) = e^{im\phi} \tag{19.54}$$

By doing this, we remove factors like A or B by letting them be absorbed into normalization factors of other parts of the wave function. Note also that advancing around the sphere by 2π radians returns us to the same point in space. Now to solve for the constant m, we must invoke the single-valued property of the wave function,

$$g(\phi) = g(\phi + 2\pi) \rightarrow \exp(im2\pi) = 1$$

This can only be true if m takes on integer values,

$$m = 0, \pm 1, \pm 2, \ldots \tag{19.55}$$

The constant m is called the **magnetic quantum number** and is the third quantum number.

Now we go back to the differential equation (19.51), set $\beta = l(l+1)$, and rewrite the equation as

$$\left[\sin\theta \frac{d}{d\theta}\left(\sin\theta \frac{df}{d\theta}\right)\right] + [l(l+1)\sin^2\theta - m^2]f = 0 \qquad (19.56)$$

This equation has well-behaved (does not blowup) and nontrivial (non-zero) solutions only when (i) l is an integer and (ii) the magnitude of m are constrained to be either equal to or smaller than the value of l. We will not prove this here and instead just state this mathematical result. The integer l is the azimuthal quantum number and is the second quantum number. The solutions to equation (19.56) that meets the conditions just discussed were known at the time of the hydrogen atom derivation and called the **associated Legendre polynomials**. They are represented by the symbol P_l^m. So, $f(\theta)$ is

$$f(\theta) = AP_l^m(\cos\theta) \qquad (19.57)$$

where A is an arbitrary constant. The solutions P_l^m are given by

$$P_l^m(x) = (-1)^m(1-x^2)^{m/2}\frac{d^m P_l(x)}{dx^m} \qquad (19.58)$$

where

$$P_l(x) = \frac{1}{2^l l!}\left(\frac{d}{dx}\right)^l (x^2-1)^l \qquad (19.59)$$

Combining equations (19.54) and (19.57), we get the angular part of the solution $f(\theta)g(\phi)$,

$$f(\theta)g(\phi) = AP_l^m(\cos\theta) \cdot e^{im\phi} \qquad (19.60)$$

To find the constant A, we must normalize the angular part of the wave function $f(\theta)g(\phi)$. The normalization condition for the angular variables in spherical coordinates is found by integrating the wave function over the angular coordinates,

$$\int_0^{2\pi} \int_0^{\pi} |f(\theta)g(\phi)|^2 \sin\theta d\theta d\phi = 1 \qquad (19.61)$$

Working out the above integral, the normalized angular wave functions are

$$f(\theta)g(\phi) = Y_l^m(\theta, \phi) = (-1)^m \sqrt{\frac{(2l+1)}{4\pi} \frac{(l-|m|)!}{(l+|m|)!}} P_l^m(\cos\theta) e^{im\phi}$$

$$(19.62)$$

The above wave functions are called the **spherical harmonics** and are usually represented by the symbol $Y_l^m(\theta, \phi)$. As a reminder, the allowed value of l and m for well-behaved wave functions correspond only to

$$l = 0, 1, 2, 3, \ldots \qquad (19.63)$$

$$|m| \le l \ (m = -l, -l+1, \ldots -1, 0, 1, 2, \ldots, l-1, l) \quad (19.64)$$

The latter means that m is allowed to have $2l+1$ different values.

The first few angular wave functions for the values of $\{l = 0, m = 0\}$ and $\{l = 1, m = -1, 0, 1\}$ are

$$Y_0^0(\theta, \phi) = \frac{1}{2}\sqrt{\frac{1}{\pi}} \qquad (19.65)$$

$$Y_1^0(\theta, \phi) = \frac{1}{2}\sqrt{\frac{3}{\pi}} \cos\theta \qquad (19.66)$$

$$Y_0^{\pm 1}(\theta, \phi) = \frac{1}{2}\sqrt{\frac{3}{2\pi}} e^{\pm i\phi} \sin\theta \qquad (19.67)$$

Before moving to the radial part of the wave function, let's pause a bit and look at equation (19.48), the angular part of the Laplacian,

$$\frac{1}{fg}(\nabla_{\theta,\phi})fg = -\beta \qquad (19.68)$$

Table 19.4. List of azimuthal quantum number l values from 0 to 4, their corresponding values of m, names given to the wave functions corresponding to each value of l.

l	m	Name	Degeneracy $2l + 1$
0	0	s	1
1	$-1, 0, 1$	p	3
2	$-2, -1, 0, 1, 2$	d	5
3	$-3, -2, -1, 0, 1, 2, 3$	f	7
4	$-4, -3, -2, -1, 0, 1, 2, 3, 4$	g	9

Note: The last column is the total number of m values called degeneracy, and it is $2l + 1$.

If we multiply both sides by $-\hbar^2$ and use $\beta = l(l+1)$, we have

$$-\hbar^2 \nabla_{\theta,\phi} f(\theta)g(\phi) = \hbar^2 l(l+1) f(\theta)g(\phi) \qquad (19.69)$$

This is an eigenvalue equation showing that the spherical wave function $f(\theta)g(\phi)$ or Y is the eigenfunction of the operator $-\hbar^2 \nabla_{\theta,\phi}$ and the eigenvalues are $\hbar^2 l(l+1)$. As we saw in the chapter on angular momentum (Chapter 16), the operator $\hat{L}^2 = -\hbar^2 \nabla_{\theta,\phi}$ is called the total angular momentum operator, and the eigenvalues $\hbar^2 l(l+1)$ are quantized values of the orbital angular momentum of the electron orbiting around the nucleus. For this reason, l is also called the *quantum number of angular momentum*.

Table 19.4 shows the l values from 0 to 4 and the allowable values of m. The third column lists the name given to the wave functions corresponding to the quantum numbers. The number of available m values is called degeneracy and is listed in the last column.

The angular parts of wave functions are listed in Table 19.2 and plotted in Figure 19.9 for $l = 0, 1, 2$. Note that the superposition of the following spherical harmonics (multiplied by the radial part of the wave function) determines the shape of the $2p$, $3p$, and $3d$ orbitals that we studied in Section 19.1.

19.3.2 *The radial part of the wave function*

We can now turn our attention to the differential equation in (19.47) for the radial wave equation $R(r)$. Recall that the potential energy

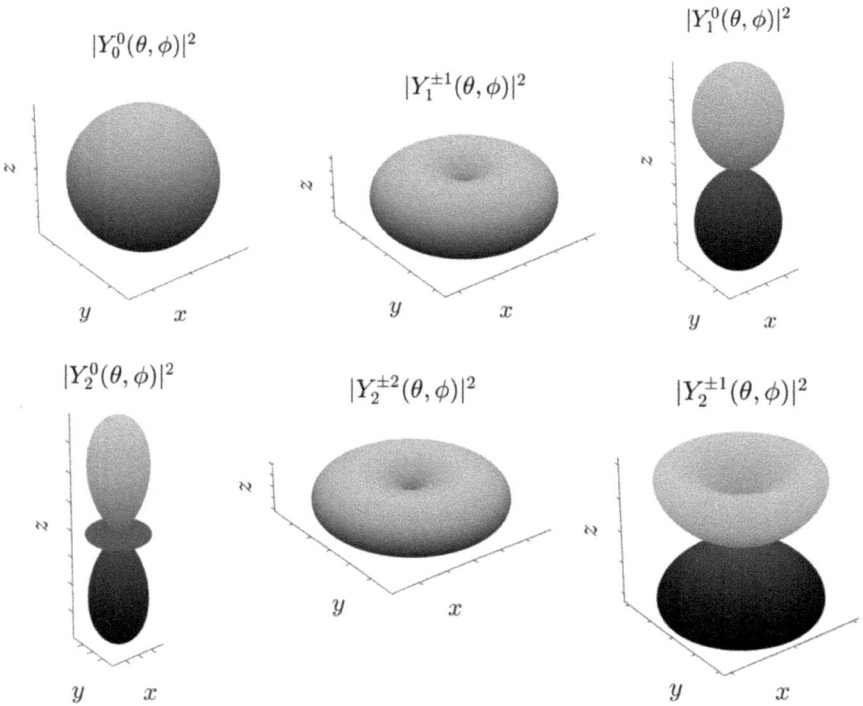

Figure 19.9. The angular parts of wave functions for the quantum numbers l, m where $l = 0, 1, 2$.

for the electron in the hydrogen atom is

$$U(\bar{r}) = -\frac{C}{r}, \quad \text{where } C = \frac{q^2}{4\pi\epsilon_o} \tag{19.70}$$

Now the radial equation (19.47), after plugging in for $U(\bar{r})$ and β, gives

$$\frac{d}{dr}\left(r^2 \frac{dR}{dr}\right) + r^2 \left[\frac{2m}{\hbar^2}\left(\frac{C}{r} + E\right)\right] R = l(l+1)R \tag{19.71}$$

$$\frac{d^2 R}{dr^2} + \frac{2}{r}\frac{dR}{dr} + \left[\frac{2m}{\hbar^2}\left(\frac{C}{r}\right) - \frac{l(l+1)}{r^2}\right] R = \frac{2m}{\hbar^2}ER \tag{19.72}$$

The eigenvalues E_n and eigenfunctions $R(r)$ of the above equation are given by

$$R_n^l(r) = \sqrt{\left(\frac{2}{na_o}\right)^3 \frac{(n-l-1)!}{2n[(n+l)!]^3}} e^{-\frac{r}{na_o}} \left(\frac{2r}{na_o}\right)^l \left[L_{n-l-1}^{2l+1}\left(\frac{2r}{na_o}\right)\right]$$

(19.73)

$$E_n = -\frac{mq^4}{32\pi^2 \epsilon_0^2 \hbar^2}\frac{1}{n^2} = -\frac{13.6}{n^2}$$

(19.74)

where physically valid solutions exist for a given value of l only for integers $n > l$. Note that the eigenvalues do not depend on the quantum numbers n, l, and m.

Here, L_{n-l-1}^{2l+1} is known as a **Laguerre polynomial** which is defined by

$$L_p^q(\rho) = \frac{d^q}{d\rho^q} L_p(\rho)$$

where,

$$L_p(\rho) = e^\rho \frac{d^p}{d\rho^p}(\rho^p e^{-\rho}).$$

The normalized eigenfunctions (equation (19.6)) and eigenvalues based on equations (19.62), (19.73), and (19.74) are

$$\psi_{nlm} = \sqrt{\left(\frac{2}{na_o}\right)^3 \frac{(n-l-1)!}{2n[(n+l)!]^3}} e^{-\frac{r}{na_o}} \left(\frac{2r}{na_o}\right)^l \left[L_{n-l-1}^{2l+1}\left(\frac{2r}{na_o}\right)\right]$$

$$\times (-1)^m \sqrt{\frac{(2l+1)}{4\pi}\frac{(l-|m|)!}{(l+|m|)!}} P_l^m(\cos\theta)e^{im\phi}$$

In summary, by solving the Schrödinger equation using the separation of variables, we obtain three differential equations (19.52), (19.56), and (19.72) in terms of coordinates ϕ, θ, and r respectively. The differential equation in terms of ϕ, gives the magnetic quantum number m which can assume both positive and negative integer values and eigenfunction $e^{im\phi}$. The differential equation in variable θ depends on the quantum number m; this differential equation yields the azimuthal/angular quantum number l which can assume $l = 0, 1, \ldots, n - 1$. The eigenfunctions P_l^m are known as associated Legendre polynomials (equation (19.57)). These solutions in the ϕ and θ coordinates also play a role in other problems with radially symmetric potential energy functions. The differential equation in terms of the radial coordinate r depends on the azimuthal quantum number l (equation (19.72)) and has solutions that depend on how the spherically symmetric potential varies along the radial direction. In the case of the hydrogen atom, whose potential energy has $\frac{1}{r}$ dependence, by solving equation (19.72), we get the quantum number n and the eigenfunctions (Laguerre polynomials) given in equation (19.73). In the case of the hydrogen atom, physically valid solutions (those which do not diverge when $r \to 0$ or $r \to \infty$ and are single valued) for a given value of integer n determine that the values of l range from $0, 1, 2, \ldots, n{-}1$, and the values of m corresponding to each l are integers ranging from $-l, -(l-1), \ldots, -1, 0, +1, \ldots, +(l-1), +l$.

19.4   Problems

(1) Show that the Laplacian operator in spherical coordinates is written as

$$\nabla^2 = \frac{1}{r^2}\frac{\partial}{\partial r}\left(r^2\frac{\partial}{\partial r}\right) + \frac{1}{r^2 \sin\theta}\frac{\partial}{\partial \theta}\left(\sin\theta\frac{\partial}{\partial \theta}\right) + \frac{1}{r^2 \sin^2\theta}\frac{\partial^2}{\partial \phi^2}$$

(2) Use the Rodrigues formula in equation (19.59) and find the spherical harmonics for $l = 2, m = -2, -1, 0, 1, 2$.

(3) Find the average or expectation value of the electron radius in a $3s$ orbital. Hint: Calculate the integral resulting from the expectation value $\langle \psi_{3,0,0}|r|\psi_{3,0,0}\rangle$.

(4) Show that the maxima of probability densities in a $3p$ orbital are at $r = 3a_o$ and $r = 12a_o$.

(5) Find the average or expectation value of the electron radius in a $3p$ orbital. Hint: Calculate the integral resulting from the expectation value $\langle \psi_{3,1,0}|r|\psi_{3,1,0}\rangle$.

(6) Using the Rodrigues formula and recursive relation (equations (19.58) and (19.59)), show that the properties of m and

l claimed in (19.63) and (19.64), holds. Consult with Griffiths and Schroeter (2018).

(7) Prove that the spherical harmonics are orthogonal, i.e.,

$$\int_0^{2\pi} \int_0^{\tau} [Y_l^m(\theta, \phi)]^* \cdot Y_{l'}^{m'}(\theta, \phi) \sin \theta d\theta d\phi = \delta_{m,m'} \delta_{l,l'}$$

(8) Use the properties of the Legendre polynomials to prove that the normalization factor in equation (19.60) is indeed,

$$A = (-1)^m \sqrt{\frac{(2l+1)}{4\pi} \frac{(l-|m|)!}{(l+|m|)!}}$$

(9) Calculate the following (you will see the probability becoming smaller as n increases):

a. Probability of finding an electron in $r \le a_o$ for $n = 1$, $l = 0$.
b. Probability of finding an electron in $r \le a_o$ for $n = 2$, $l = 0$.
c. Probability of finding an electron in $r \le a_o$ for $n = 2$, $l = 1$.
d. Probability of finding an electron in $r \le a_o$ for $n = 3$, $l = 0$.
e. Probability of finding an electron in $r \le a_o$ for $n = 3$, $l = 1$.
f. Probability of finding an electron in $r \le a_o$ for $n = 3$, $l = 2$.

References

Demtröder, W. (2010). *Atoms, Molecules and Photons: An Introduction to Atomic-, Molecular- and Quantum Physics, Graduate Texts in Physics.* Springer Berlin, Heidelberg, eBook ISBN 978-3-642-10298-1. https://doi.org/10.1007/978-3-642-10298-1.

Griffiths, D J. and Schroeter, D. F. (2018). *Introduction to Quantum Mechanics*, 3rd edn. Cambridge University Press, Cambridge.

Chapter 20

PROPERTIES OF DIRAC AND KRONECKER DELTA FUNCTIONS

Contents

20.1 Definition of Dirac Delta Function

The delta function is defined by the following properties, where E is an arbitrary one-dimensional real variable and E_n is a constant number. The function has an infinitely large peak wherever its argument is zero, and it is zero otherwise.

$$\delta(E - E_n) = \begin{cases} \infty & at\ E = E_n \\ 0 & \text{otherwise} \end{cases} \qquad (20.1)$$

The area under the delta function is one. The only contribution to the area comes from immediately before and immediately after E_n,

$$\int_{-\infty}^{+\infty} \delta(E - E_n)dE = \int_{E_n-\varepsilon}^{E_n+\varepsilon} \delta(E - E_n)dE = 1 \qquad (20.2)$$

where ε is a very small number. This means that even though the delta function's maximum value is infinity, the area under the delta function is one. Note that the dimension of the Dirac delta function according to equation (20.2) is the inverse of the dimension of its argument (here E or energy). This means if the unit of x is meters, the dimension of $\delta(x)$ is m^{-1}.

In this chapter, we will review a few useful properties of the Dirac delta function and its approximations without proving them. Afterwards, we will discuss the Kronecker delta function.

20.2   Approximations of Dirac Delta Function

It is often useful to approximate the delta function $\delta(x)$ with another parametrized function $f(x)$, which is zero when x is far from zero, and has a sharp peak at $x = 0$. The value of the parameter then controls the function's sharpness. This is useful when numerically computing quantities that involve Fermi golden rule, the density of states, etc.

Rectangular pulse: The simplest approximation is a rectangular pulse of duration a centered at $x = 0$ with height $1/a$, when a approaches zero. As shown in Figure 20.1, the pulse resembles a Dirac delta function if $a \to 0$ and the area under the pulse is always 1.

$$f(x) = \begin{cases} \dfrac{1}{a} & \text{when } \dfrac{-a}{2} < x < \dfrac{a}{2} \\ 0 & \text{everywhere else} \end{cases} \qquad (20.3)$$

Lorentzian function: The second example is a Lorentzian function given as

$$\delta(x) = \lim_{a \to 0} \frac{1}{\pi} \frac{a}{x^2 + a^2} \qquad (20.4)$$

It is easy to prove that the area under the Lorentzian function is unity. As the parameter a approaches zero, the function becomes sharper and resembles the delta function (see Figure 20.1).

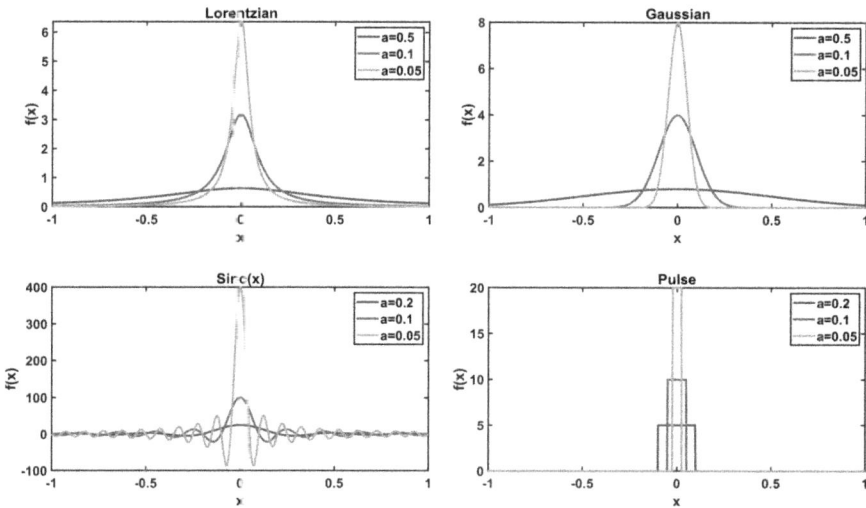

Figure 20.1. Approximations of $\delta(x)$ using a Lorentzian, Gaussian, Sinc, and rectangular pulse functions. As $a \to 0$, the functions become better approximations of the delta function.

Sinc function: The third example is called the Sinc(x) function as the parameter a approaches zero. See Figure 20.1, which shows that as the parameter a becomes smaller, the central peak grows very quickly, and the magnitude away from $x = 0$ becomes smaller.

$$\lim_{a \to 0} \frac{\sin\left(\frac{\pi x}{a}\right)}{\pi x} = \lim_{a \to 0} \frac{1}{a}\text{Sinc}\left(\frac{\pi x}{a}\right) = \delta(x) \qquad (20.5)$$

Gaussian function: A Gaussian function with the variance parameter a can also approximate the delta function when $a \to 0$. The identity in equation (20.6) is proven by integrating the function and showing that the area under it is always one regardless of the value of the parameter a and showing that the function sharpens and becomes narrower as $a \to 0$. See Figure 20.1, which plots the Gaussian function for a few different variance parameter a:

$$\lim_{a \to 0} \frac{1}{2\sqrt{\pi a}}e^{-\frac{x^2}{4a}} = \delta(x) \qquad (20.6)$$

20.3   Properties of Dirac Delta Function

Sifting property: If, in equation (20.2), there is an extra function in the integrand that multiplies the delta function as in equation 20.7, the integral returns the value of the function at the points where the delta function's argument is zero. This is best visualized with,

$$\int_{-\infty}^{+\infty} f(E)\delta(E - E_n)dE = f(E_n) \tag{20.7}$$

$$\int_{-\infty}^{+\infty} g(x)\delta(x)dx = g(0) \tag{20.8}$$

Substitution rule: If the argument of the delta function is itself a function, e.g., $f(x)$, the delta function can be expanded using the roots of the $f(x)$. If $f(x) = 0$ has N roots, i.e., $f(x_i) = 0$ for $i = 1, 2, \ldots, N.$, then

$$\delta(f(x)) = \sum_{i=1}^{N} \frac{\delta(x - x_i)}{\left|\frac{df}{dx}\right|_{x=x_i}} \tag{20.9}$$

This means that the delta function of $f(x)$ is expanded as a series of delta functions with peaks on each root of $f(x)$. See Figure 20.2. The peak (or area under each new delta function) is scaled by the inverse of the absolute value of the slope of $f(x)$ evaluated at points where it crosses zero. This is of importance when you integrate over-scattering rates given by Fermi golden rule. Here are some examples of this property as well as the proof of (20.9).

Example 1. Expand the function $\delta(x^2 - 9)$.

Solution: The roots of $x^2 - 9 = 0$ are $x = +3$ and $x = -3$. The values of $|df/dx|$ at the roots, are both 6. Using equation (20.9), we obtain

$$\delta(x^2 - 9) = \sum_{i=1}^{2} \frac{\delta(x - x_i)}{\left|\frac{df}{dx}\right|_{x=x_i}} = \frac{\delta(x - 3)}{|2x|_{x=3}} + \frac{\delta(x + 3)}{|2x|_{x=-3}}$$

$$= \frac{1}{6}\delta(x - 3) + \frac{1}{6}\delta(x + 3) \tag{20.10}$$

Example 2. Expand the function $\delta(\cos(x))$.

Solution: The roots of $\cos(x) = 0$ are $x = \frac{k\pi}{2}$, where k is an odd integer. The value of $|d\cos(x)/dx| = |-\sin(x)|$ is always 1 for the roots of $\cos(x)$. Therefore,

$$\delta(\cos(x)) = \sum_{k \text{ odd}} \frac{\delta\left(x - \frac{k\pi}{2}\right)}{|-\sin(x)|_{x=\frac{k\pi}{2}}} = \sum_{k \text{ odd}} \delta\left(x - \frac{k\pi}{2}\right) \qquad (20.11)$$

Example 3. Prove that the delta function has even parity, i.e., it is an even function or mathematically put $\delta(-x) = \delta(x)$.

Solution: To prove the above, we first prove that $\delta(ax) = \frac{1}{|a|}\delta(x)$, $a \neq 0$. As the argument, $ax = 0$ has only one root, i.e., $x_i = 0$, we can write

$$\delta(ax) = \sum_i \frac{\delta(x - x_i)}{\left|\frac{d(ax)}{dx}\right|_{x=x_i}} = \frac{1}{|a|}\delta(x) \qquad (20.12)$$

If we use $a = -1$, it is easy to show that $\delta(-x) = \delta(x)$.

Example 4. Show that when $\epsilon \to 0$, the following function (called Lorentzian distribution) approaches a delta function:

$$\delta(x) = \lim_{\epsilon \to 0} \frac{1}{\pi} \frac{\epsilon}{x^2 + \epsilon^2} \qquad (20.13)$$

Solution: First, it is easy to show that the area under the function is 1,

$$\int_{-\infty}^{+\infty} \frac{1}{\pi} \frac{\epsilon}{x^2 + \epsilon^2} dx = \frac{1}{\pi} \tan^{-1}\left(\frac{x}{\epsilon}\right)\Big|_{-\infty}^{+\infty} = \frac{1}{\pi}\left(\frac{2\pi}{2}\right) = 1 \qquad (20.14)$$

Now we want to show that as $\epsilon \to 0$, the sifting property holds, i.e., if $\varphi(x)$ is a test function, we have

$$\lim_{\epsilon \to 0} \int_{-\infty}^{+\infty} \frac{1}{\pi} \frac{\epsilon}{x^2 + \epsilon^2} \varphi(x) dx = \varphi(0)$$

We will follow the method given in [Debnath 2005]. Without loss of generality, we can assume that $\epsilon = \frac{1}{n}$, where n is a large integer and approaches infinity in stepwise manner. Therefore, the Lorentzian function can be rewritten as

$$\frac{1}{\pi}\frac{\epsilon}{x^2 + \epsilon^2} = \frac{n}{\pi(n^2x^2 + 1)} = f_n(x) \quad n = 1, 2, 3, \ldots$$

Now, it must be shown that as $n \to \infty$, the sequence of functions, $f_n(x)$, converges towards the delta function (i.e., the bigger the n, the sharper the function $f_n(x)$). In the proof, it is assumed that the test function $\varphi(x)$ is nonzero only on a limited interval around $x = 0$, e.g., $[-a, a]$. Then we must show,

$$\lim_{n \to \infty} \int_{-\infty}^{+\infty} f_n(x)\varphi(x)dx = \varphi(0)$$

Since we showed that $\int_{-\infty}^{+\infty} f_n(x) = 1$, we can write the above as,

$$\lim_{n \to \infty} \int_{-\infty}^{+\infty} f_n(x)(\varphi(x) - \varphi(0))dx = 0$$

Note that in the above step, we can move the $\varphi(0)$ term inside the integral because the $\int_{-\infty}^{+\infty} f_n(x)dx = 1$ (this follows from equation 20.14 by setting $\epsilon = \frac{1}{n}$). The above integral is split into three intervals: $[-\infty, -a]$, $[-a, a]$, and $[a, +\infty]$,

$$\left| \int_{-\infty}^{+\infty} f_n(x)(\varphi(x) - \varphi(0))dx \right| \leq \left| \int_{-\infty}^{-a} f_n(x)(\varphi(x) - \varphi(0))dx \right|$$

$$+ \left| \int_{-a}^{a} f_n(x)(\varphi(x) - \varphi(0))dx \right|$$

$$+ \left| \int_{a}^{+\infty} f_n(x)(\varphi(x) - \varphi(0))dx \right|$$

$$(20.15)$$

Since the test function, $\varphi(x)$, is zero outside the interval $[-a, a]$, the first and the third integrals in the above equation can be simplified to yield,

$$\left| \int_{-\infty}^{+\infty} f_n(x)(\varphi(x) - \varphi(0))dx \right| \leq \left| \varphi(0) \int_{-\infty}^{-a} f_n(x)dx \right|$$

$$+ \left| \int_{-a}^{a} f_n(x)(\varphi(x) - \varphi(0))dx \right|$$

$$+ \left| \varphi(0) \int_{a}^{+\infty} f_n(x)dx \right|$$

The first and third integrals are zero when $n \to \infty$. To show this, we integrate the first term,

$$\lim_{n \to \infty} \left| \varphi(0) \int_{-\infty}^{-a} f_n(x)dx \right|$$

$$= \lim_{n \to \infty} |\varphi(0)| \left| \frac{1}{\pi}(tan^{-1}(-na) - tan^{-1}(-\infty)) \right| = 0$$

Similarly, the third integral is also zero. As a result, we must show the second term is zero. Since $\left| \int u \, dx \right| \leq \int |u \, dx|$, we write,

$$\left| \int_{-a}^{+a} f_n(x)(\varphi(x) - \varphi(0))dx \right| \leq \int_{-a}^{+a} |f_n(x)(\varphi(x) - \varphi(0))|dx$$

Using the mean value theorem, it can be said $|\varphi(x) - \varphi(0)| \leq max|\varphi'(x)||x|$. Hence, the RHS of the previous inequality can be written as,

$$\int_{-a}^{+a} |f_n(x)(\varphi(x) - \varphi(0))|dx \leq max|\varphi'(x)| \int_{-a}^{+a} |xf_n(x)|dx$$

The last term is zero as n approaches infinity (or equivalently as $\epsilon \to 0$),

$$\lim_{n \to \infty} \int_{-a}^{+a} |xf_n(x)|dx = \lim_{n \to \infty} \frac{\ln(1 + n^2a^2)}{\pi n} = 0 \qquad (20.16)$$

Thus, we showed that as $n \to \infty$ (or $\epsilon \to 0$),

$$\lim_{n \to \infty} \int_{-\infty}^{+\infty} f_n(x)(\varphi(x) - \varphi(0))dx \to 0,$$

and the proof is complete.

Example 5. Show that in a spherical coordinate system,

$$\delta(x - x_0, y - y_0, z - z_0) = \frac{\delta(r - r_0)\delta(\theta - \theta_0)\delta(\varphi - \varphi_0)}{r^2 \sin \theta} \quad (20.17)$$

Solution: One example of the application of the 3D delta function is the density of a point charge, e.g., $\rho(\bar{r}) = Q\delta(\bar{r} - \bar{r}_0)$. So, if we integrate the density over a small element of volume around the point charge, we obtain

$$\int_{\Delta V} \rho(\bar{r})dv = \int_{\Delta V} Q\delta(\bar{r} - \bar{r}_0)dV = \begin{cases} Q & \text{Inside } \Delta V \\ 0 & \text{Outside } \Delta V \end{cases} \quad (20.18)$$

We will now integrate the delta function on the right side of equation (20.17) over the volume and convert the volume element from Cartesian to spherical one using $dxdydz = r^2 \sin \theta dr d\theta d\varphi$, to get,

$$\int_{3D \text{ space around } (x_0, y_0, z_0)} \delta(x - x_0, y - y_0, z - z_0)dxdydz$$

$$= \int_{3D \text{ space around } (x_0, y_0, z_0)} \delta(x - x_0, y - y_0, z - z_0)r^2 \sin \theta dr d\theta d\varphi$$

$$= 1 \quad (20.19)$$

The right side is an integral over the three spherical variables, which returns 1. So, we may conclude that the integrand itself is a delta function around the point $(r_0, \theta_0, \varphi_0)$ and the proof of equation (20.17) is complete,

$$\delta(x - x_0, y - y_0, z - z_0)r^2 \sin \theta = \delta(r - r_0)\delta(\theta - \theta_0)\delta(\varphi - \varphi_0) \quad (20.20)$$

Example 6. As we saw in equation (20.9), the substitution property claims that if the argument of the delta function is itself a function, e.g., $f(x)$, then it can be expanded as a series of delta functions located on each root of $f(x) = 0$ and weighted by the inverse of the slope of $f(x)$ on each root. So,

$$\delta(f(x)) = \sum_{i=1}^{N} \frac{\delta(x - x_i)}{\left| \frac{df}{dx} \right|_{x=x_i}}$$

Prove the above equation.

Solution: We will show a simple proof by assuming that $f(x)$ is a smooth function and it has only two roots, e.g., x_1 and x_2. The general case is proved in problem 2.19 of Chapter 2 of (Zeidler, 1995). When the argument of the delta function is nonzero, the value of the function is zero, and it has a considerable value only around the roots of the $f(x)$, i.e., x_1 and x_2. This means that we can split the integral into two terms, around each root,

$$\int_{-\infty}^{+\infty} \delta(f(x)) \cdot \varphi(x)dx = \sum_i \int_{x_i-\epsilon}^{x_i+\epsilon} \delta(f(x)) \cdot \varphi(x)dx \qquad (20.21)$$

Since $f(x)$ is a smooth function, there exists an inverse function for it around each root such that $y = f(x) \Rightarrow g(y) = x$. We replace x with y and note that the element of length dx can be written as dy divided by the absolute value of slope around each x_i (see Figure 20.2).

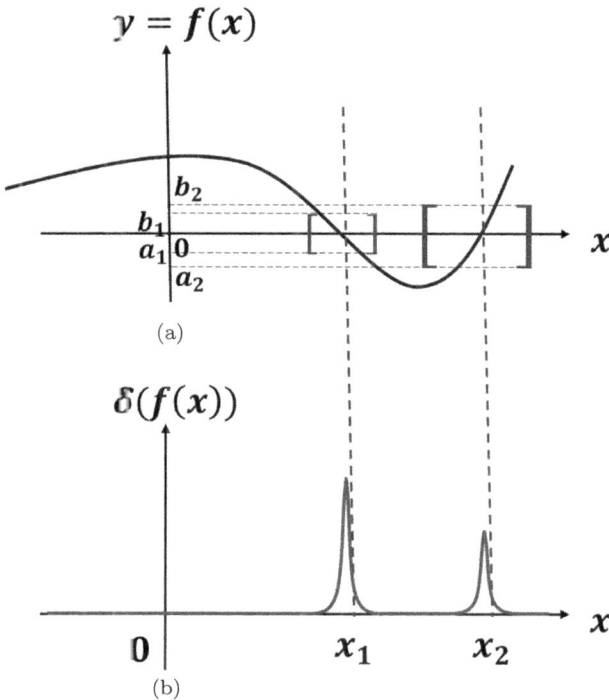

Figure 20.2. A function with (a) two roots (zero crossings) and (b) the plot of $\delta(f(x))$.

Therefore,

$$dx = |g_i'(y)| dy \qquad (20.22)$$

By replacing the above in RHS of equation (20.21), we have

$$\text{LHS of equation}(20.21) = \sum_i \int_{a_i}^{b_i} \delta(y) \cdot \varphi(g_i(y)) |g_i'(y)| dy$$

$$= \sum_i \varphi(g_i(y = 0)) |g_i'(y = 0)| \qquad (20.23)$$

In which we used the sifting property of equation (20.8). Using the property of inverse function, i.e., $g' = \frac{1}{f'}$, we can write

$$\text{LHS of equation } (20.21) = \sum_i \varphi(x_i) \frac{1}{|f'(x_i)|}$$

$$= \int_{-\infty}^{+\infty} \sum_i \delta(x - x_i) \frac{1}{|f'(x)|} \varphi(x) dx$$

$$(20.24)$$

where we have written the RHS as the result of integrating with a delta function over x. We equate LHS of (20.21) and RHS of equation (20.24) as follows:

$$\text{RHS of equation}(20.24) = \int_{-\infty}^{+\infty} \sum_i \delta(x - x_i) \frac{1}{|f'(x)|} \varphi(x) dx$$

$$= \sum_i \frac{1}{|f'(x_i)|} \varphi(x_i) \qquad (20.25)$$

The proof of equation (20.9) is then complete.

20.4   The Kronecker Delta

This function has a discrete (integer) argument instead of a continuous argument as in the Dirac delta function. The Kronecker delta function's value is one when the argument is zero, and it is zero everywhere else. Figure 20.3 shows two Kronecker delta functions centered on $n = -2$ and $n = 4$ with different amplitudes:

$$\delta[n] = \begin{cases} 1 & n = 0 \\ 0 & n \neq 0 \end{cases} \qquad (20.26)$$

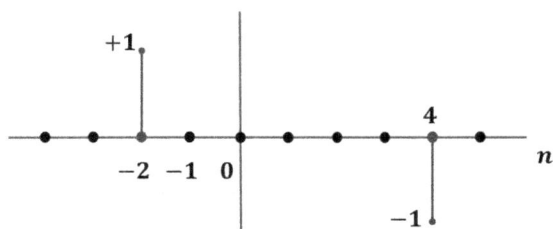

Figure 20.3. The diagram of $\delta[n+2] - \delta[n-4]$.

In case two different integer arguments are covering a given range, the two-variable Kronecker delta is one if both arguments are equal, and zero if the arguments are different:

$$\delta_{mn} = \begin{cases} 1 & n = m \\ 0 & n \neq m \end{cases} \tag{20.27}$$

For example, the nonzero elements of a 3×3 identity matrix can be written as

$$I_{3\times3} = \delta_{mn} \quad m, n = 1, 2, 3 \tag{20.28}$$

Example 7. Show that

$$\sum_{k=0}^{k=N-1} e^{i2\pi k(n-m)/N} = N\delta_{mn} \tag{20.29}$$

Solution: We assume that $n - m = Q$, and find the summation for $Q = 0$ and $Q \neq 0$.

If $Q = 0$, we have

$$\sum_{k=0}^{k=N-1} 1 = N \tag{20.30}$$

If $Q \neq 0$, we use the geometrical series summation formula:

$$\sum_{k=0}^{k=N-1} e^{i2\pi k(Q)/N} = \frac{1 - u^N}{1 - u} = \frac{1 - e^{i2\pi(Q)N/N}}{1 - e^{i2\pi(Q)/N}}$$

$$= \frac{1 - 1}{1 - e^{i2\pi(Q)/N}} = 0 \tag{20.31}$$

Since the summation is N for $n = m$ and zero for $n \neq m$, we have just proved (20.29).

The sifting property for the Kronecker delta can be used in the following way. As you see, it reduces (contracts) one index,

$$\sum_m A_m \delta_{mj} = A_j \qquad (20.32)$$

By simply expanding the above summation, you can show that the only term that survives is the one with $m = j$, and the rest are zero because of the Kronecker delta.

Interested readers can find more examples and read on the mathematical foundations in (Zeidler, 1995). It has a special chapter dedicated to Dirac algebra and the Dirac delta function applications in quantum mechanics.

20.5   Problems

(1) Show that $\delta(\sin(x)) = \sum_{n=-\infty}^{+\infty} \delta(x - n\pi)$, where n is an integer,
(2) If $a \neq 0$, show that

$$\delta(ax + b) = \frac{1}{|a|} \delta \left(x + \frac{b}{a} \right)$$

Reference

Debnath, L. and Mikusinki, P. (2005). *Introduction to Hilbert spaces with Applications*. 3rd Edition, Elsevier Academic Press.
Zeidler, E. (1995). *Applied Functional Analysis: Applications to Mathematical Physics*. Springer Verlag, New York.

INDEX

Milton Keynes UK
Ingram Content Group UK Ltd.
UKHW021306110624
444062UK00014B/110

9 789811 275326